C0-AKT-558

AMERICAN MATHEMATICAL SOCIETY
COLLOQUIUM PUBLICATIONS
VOLUME XXVII

ALGEBRAIC TOPOLOGY

BY
SOLOMON LEFSCHETZ
HENRY B. FINE PROFESSOR EMERITUS OF MATHEMATICS
PRINCETON UNIVERSITY

Published with aid from the Marion Reilly Fund,
American Mathematical Society

PUBLISHED BY THE
AMERICAN MATHEMATICAL SOCIETY
531 WEST 116TH STREET, NEW YORK CITY
1942

Reprinted 1963

PHOTOLITHOPRINTED BY CUSHING - MALLOY, INC.
ANN ARBOR, MICHIGAN, UNITED STATES OF AMERICA

PREFACE

When the present volume was first contemplated some five years ago it was primarily meant to be a second edition of the author's *Topology* (1930, Volume XII of the American Mathematical Society Colloquium Series). It soon became evident however that the subject had moved too rapidly for a mere revised edition, and that a completely new book would have to be written. With the consent of the Colloquium Committee the task was undertaken by the author and resulted in the present work. Its basic topic, often referred to as "Combinatorial Topology," is in substance the theory of complexes and its applications. Many factors have contributed to a great increase in the role of algebra in this subject. For this reason it is more appropriately described as "Algebraic Topology," and this explains the title of the volume.

The purely topological (non-algebraic) part has been concentrated in the first chapter, and all the necessary group-theoretic material in the second, thus resulting in a great economy and simplification in the treatment of many questions, notably duality and intersections. The next three chapters deal with the theory of complexes proper. The basic type selected is A. W. Tucker's modified in that the elements may also take negative dimensions. As is well known one of the important recent advances has been the extension to complexes of the duality and intersection properties of manifolds. This may be accomplished by means of special "dual" cycles (the "pseudocycles" of *Topology*, Chapter VI), or by a special dual complex as done by Tucker (companion algebraic development by W. Mayer), or else again with Alexander and Whitney without new elements but with a new boundary operator for the chains. By utilizing negative dimensions it has been possible to associate with each complex X a dual complex X^* such that the relation between the two is wholly symmetrical. As a consequence the "co-theory" of X (Whitney's terminology) appears as the ordinary theory of X^*, and all the duality and intersection properties are obtained by combining the X, X^* relationship with group-duality and group-multiplication in the sense of Pontrjagin. There emerges thus a theory of complexes of purely algebraic nature, with manifolds relegated to the second plane.

The homology theory of topological spaces is taken up in Chapter VII, the necessary limiting processes constituting the theory of nets and webs being dealt with in Chapter VI. We have chosen as our basic theory the Čech homology theory and in substance reduced to it the other known theories thus unifying a domain which has definitely stood in need of it for some time.

The relative concepts which played such an important role in the previous volume have not been neglected in the present. They appear chiefly in the guise of certain binary dissections which run right through complexes, nets and topological spaces, and are at the root of the mechanism of webs.

The last chapter contains the applications to polyhedra and certain related questions, notably a very concise and very general treatment of fixed points. The book concludes with an appendix by Eilenberg and MacLane on the homology groups of infinite complexes and another by Paul Smith on his theory of fixed points of periodic transformations.

Owing to limitations of time and space it has not been possible to take up the applications of algebraic topology. However with Marston Morse's *Calculus of Variations in the Large* (1934, Volume XVIII of the Colloquium Series), W. V. D. Hodge's *The Theory and Applications of Harmonic Integrals* (1941, Cambridge University Press), and a forthcoming volume by Hassler Whitney on sphere spaces, the reader interested in the applications will readily satisfy his curiosity.

Certain deviations from standard usage have been adopted in the text and should be kept in mind. Thus "compact" replaces "bicompact," and "complex" replaces "abstract complex." (A nomenclature of complexes and manifolds is given at the end of Chapter VIII.) All groups are topological (the topology may be discrete); unless otherwise stated homomorphisms are supposed to be continuous and group-isomorphisms topological, exceptions being indicated by the mention "in the algebraic sense." For vector spaces over a field there is a special set of conventions indicated in Chapter II (22.2).

The literature in topology has grown to such proportions that it has been impossible to provide more than a scanty bibliography. References are given by the author's name followed by an appropriate letter in square brackets. Those to the present volume are of the form (IV, 16.3), where IV stands for Chapter IV and 16.3 for the numbering in the chapter.

It has been my good fortune to have obtained sympathetic cooperation and advice from many sources. In preparation of the manuscript invaluable assistance was received from Samuel Eilenberg, W. W. Flexner, N. E. Steenrod, John Tukey, and as regards the second chapter, Claude Chevalley practically acted as a collaborator. Parts of the manuscript in more or less final form or important parts of the proofs were carefully read by Hubert Arnold, E. G. Begle, Paco Lagerstrom, Saunders MacLane, Moses Richardson, Seymour Sherman, J. D. Tamarkin, A. D. Wallace and Hassler Whitney. To one and all it is a great pleasure to express here my appreciation and thanks.

S. Lefschetz

Princeton, N. J.
October, 1941

CONTENTS

CHAPTER I

INTRODUCTION TO GENERAL TOPOLOGY

The scope of the chapter is sufficiently clear from its title. Particular attention has been paid to compactness and there is also a thoroughgoing treatment of inverse mapping systems which come strongly to the fore in (II), and also in (VI, VII) in connection with the homology theory of topological spaces.

General references: The standard treatises and in addition: Alexandroff-Urysohn [a], Čech [g], Steenrod [a], Tukey [T], Wallace [a], Wallman [a].

§1. PRIMITIVE CONCEPTS

1. We introduce a few formal abbreviations:

$A \rightarrow B$ means "A implies B";

$A \leftrightarrow B$ means "A is equivalent to B";

$A \cong B$ means "A is isomorphic with B."

We shall assume that the reader is familiar with the basic concepts of point sets. The null-set is designated by $\emptyset$ and if X is a set, $X = \emptyset$ signifies that X is empty. If X, Y are sets we write $X \subset Y$ or $Y \supset X$ for: "every element of X is an element of Y", or: "X is a subset of Y". We shall also say of three sets X, Y, Z that "*Y is between X and Z*" whenever $X \subset Y \subset Z$ or else $X \supset Y \supset Z$.

The statement "x is an element of the set X" is written symbolically $x \in X$ or $X \ni x$. Frequently the different elements of a set X are denoted by the same letter x with additional affixes as: x_i^p, x_{an}, $\cdots$, or say by x_a with complementary affixes as: x_{ai}, $\cdots$. In that case the set will sometimes be designated by $\{x\}$, $\{x_a\}$, $\cdots$. We shall also write $X = \{x\}$, $X = \{x_a\}$, $\cdots$ whenever it is the intention to designate the different elements in the manner just stated. However, the symbol $\{\ \ \}$ is too convenient to be reserved strictly for the preceding usage; deviations will be allowed but their meaning will generally be clear from the context.

Let $\{X_a\}$ be a collection of sets which may or may not be distinct. Let particularly $X_a = \{x_a\}$. Then the set of all the x_a for all a is called the *union* of the X_a and designated by $\bigcup_a X_a$. In this and similar symbols the subscript a will often be omitted, and we shall write $\bigcup$ in place of $\bigcup_a$, wherever the "a" is clear from the context. Similarly the set of all the elements which are in every X_a (i.e., common to all the X_a) is called the *intersection* of the X_a and denoted by $\bigcap_a X_a$. If the number of X_a is finite, say consisting of the collection $X_1, \cdots, X_r$ (r an integer), we also designate the union and intersection, respectively, by $X_1 \cup X_2 \cup \cdots \cup X_r$ and $X_1 \cap X_2 \cap \cdots \cap X_r$.

Given two sets X, Y, the set of all the elements of X which are not in Y is called the *complement* of Y in X, also the *difference* of X and Y, and is denoted by $X - Y$.

If P is a property and $X = \{x\}$, the totality of all the elements x which satisfy P is denoted by $\{x \mid x \text{ has the property } P\}$. As an example of this notation, if $\{x\}$ is the set of all real numbers, the set of all those between 0, 1 is denoted by $\{x \mid 0 < x < 1\}$.

Negation of any relation shall be indicated by a bar drawn through its symbol as in $Y \not\subset X$ (Y not contained in X), $Y \neq X$ (Y different from X), etc.

The sets X_a are said to be *disjoint* whenever any two are disjoint ($X_a \cap X_b = \emptyset$, for $a \neq b$).

2. Transformations or functions. Let $X = \{x\}$, $Y = \{y\}$ be two sets and let G be a subset of the set whose elements are the ordered pairs (x, y). We suppose that G has the following property: every element x is found in precisely one pair $(x, y_x) \in G$. There results then an assignment to each $x \in X$ of a definite element $y_x \in Y$ and this assignment is known as a *transformation of X into Y* or *function on X to Y*. The statement "T is a transformation of X into Y" will be generally written in one of the symbolic forms: "$T : X \to Y$," "$T : x \to y_x$," "$x \to y_x$ defines T." The set X is the *range* of T and y_x is the *value* of T at x. The element y_x is frequently designated by Tx, and called the *transform* or *image of x under T*. The set Y' of all the values Tx for all $x \in X$ is a subset of Y called the *transform* or *image* of X under T, and we write $Y' = TX$. It may happen that $Y' = Y$, i.e., that every element y occurs in some pair $(x, y) \in G$ (every y is a Tx), in which case T is said to transform X *onto* Y.

The transformation T is said to be *univalent* whenever $x \neq x' \to Tx \neq Tx'$. It is said to be *one-one* when it is both univalent and a transformation "onto." That is to say, every y occurs in one and only one pair (x, y).

The set G of the pairs (x, y) serving to define T is known as the *graph* of T.

Example. x is a real variable and $X = Y = \{x\}$, while $T: x \to x^2$. Then $X' = TX \neq X$, so T is a transformation of X *into* X but not *onto* X. Suppose now the same situation except that x is a complex variable. This time T is a transformation of X *onto* X.

Let X, Y, T be as before and let $Z \subset X$. Then if $x \in Z$ the assignment of Tx to x defines a transformation T_1: $Z \to Y$ denoted by $T \mid Z$. We also say that T is an *extension* of $T \mid Z$ to X.

(2.1) *Multi-valued transformations.* Let the sets X, Y, G be as before, except that this time G is not subjected to any restrictions. The elements y in any pair $(x, y) \in G$ in which x occurs make up a set $Y_x \subset Y$, which may be $\emptyset$ (an automatic subset of every set). The assignment T to any x of the set Y_x is called a *multi-valued transformation* of X into Y. The terms "value, image, transform" and designations "Tx, TX," are carried over to multi-valued transformations. If it is known that every Tx consists of n elements, T is sometimes said to be *n-valued*. The earlier transformations correspond to $n = 1$, and are sometimes designated as *single-valued*.

A multi-valued transformation $X \to Y$ may be considered as a (single-valued) transformation of X into the set Y' of all the subsets of Y.

Example. If X is the set of all complex numbers then $x \to x^{1/n}$ is a multi-valued transformation $X \to X$, the values Y_x being sets of n complex numbers.

(2.2) *Indexed system of sets.* A multi-valued transformation $T : X \to Y$, with $Tx = Y_x$, is also called a *system of sets indexed by* X, or more simply an *indexed system*, and denoted by $\{Y_x\}$. It may be said that this designation will be used chiefly whenever X plays a minor rôle. As an example of an indexed system we may mention *set sequences*. We have then $X = \{1, 2, \cdots, n, \cdots\}$, and the set sequence is here a system $\{Y_n\}$ indexed by $\{n\}$. Whenever every Y_n is a single point the set sequence becomes a point sequence or more simply a *sequence*.

(2.3) *Inverse transformations, one-one transformations.* Let $T : X \to Y$ be single-valued or multi-valued, and set $X_y = \{x \mid Tx \ni y\}$. Then $y \to X_y$ defines a multi-valued transformation known as the *inverse* of T and denoted by T^{-1}. Thus if x is a complex variable then $T : x \to x^n$ is a transformation $X \to X$ whose inverse is $T^{-1} : x \to x^{1/n}$, already considered above.

If both T and T^{-1} are single-valued T is one-one. In terms of the set G the transformation T is one-one whenever every x and every y each occur in a single pair $(x, y) \in G$.

(2.4) *Identification.* Let R be a relation of equivalence between the elements of a set $X = \{x\}$ and let the resulting equivalence classes be taken as elements of a new set $Y = \{y\}$. The set Y is said to be obtained from X by identification of the elements in each class y.

There is an obvious connection between "identification" and "transformation." Indeed if we define T by $Tx =$ the class $y \ni x$ then T is a transformation $X \to Y$. Conversely, if T is a transformation $X \to Y$ and we define the relation R by "x and x' are in the relation R whenever x and x' are elements of the same set $T^{-1}y$," then R is a relation of equivalence and Y is derived from X by identification of the elements in each class.

EXAMPLES. (2.5) A "book" may be obtained from a collection of rectangles $\{R_a\}$ by identification of points on a set of edges $\{E_a\}$, one in each rectangle. Each equivalence class consists of the points at a specified distance from one vertex in each E_a or of a single point not on an E_a.

(2.6) Let X consist of a circular region with its boundary circumference Z and let the relation R be defined as follows: each interior point is in the relation R with itself and itself alone; two end points x, x' of the same diameter are in the relation R with one another and with no other points. The resulting identification yields the *projective plane*. Similarly the Euclidean set $x_1^2 + \cdots + x_n^2 \leqq 1$ gives rise to *projective n-space*.

(2.7) *Imbedding.* Let T be a univalent transformation $X \to Y$ and let $X' = TX$. Then the process of replacing Y by $(Y - X') \cup X$ is known as *imbedding* X in Y.

3. **Cartesian products.**

(3.1) DEFINITION. *Let $\{X_a\}$ be a system of sets indexed by $A = \{a\}$, with $X_a = \{x_a\}$. The cartesian product, or merely product of the X_a is the set of all the single-valued functions $\xi(a)$ on A to $\bigcup X_a$ such that $\xi(a) \in X_a$ for every a. The product is denoted by $\mathbf{P}X_a$ or also by $X_1 \times \cdots \times X_n$ when $A = \{1, 2, \cdots, n\}$.*

If the sets X_a are merely the same set X repeated we also write the product as a power: X^A.

EXAMPLES. (3.2) Take two disjoint sets $X_1 = \{x_1\}$, $X_2 = \{x_2\}$. Then $X_1 \times X_2$ is in one-one correspondence with the collection of all the pairs (x_1, x_2), $x_i \in X_i$. Similarly if $X_i = \{x_i\}$, $i = 1, 2, \cdots, n$, are disjoint sets then $X_1 \times \cdots \times X_n$ is the collection of all the sets $(x_1, \cdots, x_n)$, $x_i \in X_i$.

(3.3) Let $X_1 = X_2 = X = \{x\}$. Then the product $X \times X$, also written X^2, is in one-one correspondence with the set of all *ordered* pairs (x', x''), x' and $x'' \in X$. Here then $(x', x'') = (x'', x')$ when and only when $x' = x''$. Similarly $X \times \cdots \times X$ (r factors), written also X^r, is in one-one correspondence with the set of all *ordered* r-uples $(x^{(1)}, \cdots, x^{(r)})$ of elements of X.

(3.4) *Application to functions.* By a function f on the sets X_a, or of the variables x_a, to a set Y, is meant a function on $\mathbf{P}X_a$ to Y. If there are only a finite number of x_a, say $x_1, \cdots, x_r$, f is often designated by $f(x_1, \cdots, x_r)$.

(3.5) *Graphs.* Let X, Y designate the sets of points x, y on two cartesian axes $x' O x$, $y' O y$ in an Euclidean plane π. Then the points of π are in one-one correspondence with the pairs of coordinates x, y, i.e., with the elements of $X \times Y$. With a function f on X to Y there is associated the set G of all points $(x, f(x))$, in which we recognize the graph of f in the sense of (2).

By interchanging x, y and X, Y throughout, G may be viewed as the graph of f^{-1}. If f is single-valued every vertical meets G in a single point.

This simple configuration is so effective that its terminology has been increasingly borrowed. Explicitly, given the product $\mathbf{P}X_a$ and an element x in the product, we call *projection* of x on X_a, or ath *coordinate* of x, its value $x(a)$ at a. In the case of a product of two factors $X \times Y$, we call x and y the *horizontal* and *vertical* projections of the point (x, y) of the product. The horizontal, vertical, ath projection of a set in $X \times Y$ or $\mathbf{P}X_a$ is the aggregate of those of its elements.

(3.6) *Other products.* Interesting generalizations of the cartesian product may be obtained. For example, we may take as elements all the unordered pairs (x', x'') of elements of X, thus obtaining the *symmetric* product of X by itself. Similarly the unordered sets of r elements give the symmetric product of X by itself r times.

4. **Partially ordered and directed systems.** A set A is said to be *partially ordered* or merely *ordered*, if certain pairs of elements (a, b) of A satisfy an ordering relation denoted by $a \prec b$ and subjected to the sole condition of transitivity: $a \prec b$ and $b \prec c \rightarrow a \prec c$. Instead of $a \prec b$ we also write $b \succ a$. The ordering is said to be: *reflexive* if $a \prec a$ for every $a \in A$, *proper* if $a \prec a'$ and $a' \prec a \rightarrow a = a'$. The set A is said to be *simply ordered* whenever every pair of elements a, b are ordered: one of $a \prec b$ or $b \prec a$ or both must hold.

Let A be ordered by $\prec$. Then A is said to be *directed* by $\succ$ [by $\prec$] whenever given any two elements a, b of A there exists a third c such that $c \succ a$ and $c \succ b$ [$c \prec a$ and $c \prec b$]. We also write accordingly $A = \{a; \succ\}$ [$A = \{a; \prec\}$].

EXAMPLES. (4.1) A is the set of all real numbers and $a \prec b \leftrightarrow a \leqq b$. This set is simply ordered and directed both by $\prec$ and $\succ$.

(4.2) A is the Euclidean plane referred to the coordinates (x, y) and $(x, y) \prec (x', y')$ means that $x = x'$, $y \leqq y'$. This set is ordered but not simply ordered and not directed.

(4.3) A consists of all the subsets of a given set E and $a \prec b \leftrightarrow a \subset b$. This system is directed by $\succ$ and not simply ordered. Its ordering will occur frequently and is sometimes called *ordering by inclusion*.

A subset A' of a directed set $A = \{a; >\}$ [$= \{a; <\}$] is said to be *cofinal* [*coinitial*] in A whenever for every $a \in A$ there is an $a' \in A'$ such that $a < a'$ [$a > a'$]. Thus if $A = \{a_n\}$ is a monotone numerical sequence, then any subsequence A' is cofinal in A. In point of fact, cofinal systems play in many respects a role analogous to that of subsequences.

In a partially ordered system A the subset A' is said to have a_0 for an *upper* [*lower*] *bound* whenever $a < a_0$ [$a > a_0$] for every $a \in A'$. The element a_0 is said to be *maximal* for A if $a > a_0 \rightarrow a_0 > a$. If the ordering is proper then this definition specializes to the usual one: a_0 is maximal if no $a \neq a_0$ is such that $a > a_0$.

(4.4) *If $\{\lambda; >\}$ is a countable directed system, either it contains a maximal element or it contains a cofinal simply ordered sequence.*

Let $\{\lambda\} = \{\lambda_1, \lambda_2, \cdots\}$. Choose $\lambda_1' = \lambda_1$, and choose λ_n' so that $\lambda_n' > \lambda_n$, λ_{n-1}' and $\lambda_{n-1}' \not> \lambda_n'$. If such a choice is impossible at the nth step, then λ_{n-1}' is a maximal element. If the choice can always be carried out, then $\{\lambda_n'\}$ is a sequence cofinal in $\{\lambda\}$.

5. **Zorn's theorem.** We now introduce a theorem which will be used in a number of proofs. It is logically equivalent to the well-ordering postulate, but in a form which can be used to replace arguments based on well-ordering, particularly transfinite induction, by a simpler procedure.

We give three statements of the theorem, which are easily proved equivalent.

(5.1) THEOREM OF ZORN. *If in a partially ordered system A each simply ordered subset has an upper bound in the system, then there exists at least one maximal element $a \in A$, with $a > a_0$ for a preassigned a_0.*

(a) A property P of sets is said to have *finite character* if whenever it holds for every finite subset of a set X it also holds for X itself, and conversely.

(b) *If a property P of some subsets of a set X has finite character then there exists at least one subset Y of X with property P such that any subset containing Y which has property P is equal to Y.*

(c) *Every partially ordered system contains at least one maximal simply ordered subset; that is, a subset B which cannot be extended in simple order by an element greater than or less than all elements of B.*

This last form of the theorem is perhaps most intuitive; but (b) brings out more clearly the basis for the proofs making use of Zorn's theorem, since the properties involved usually are first defined for finite subsets of some set and then extended.

In the formulation (a) and for the subsets of a given set ordered by inclusion the theorem was given by R. L. Moore [M, 84] but the first general formulation, and particularly its usage as a substitute for transfinite induction are due to Zorn.

§2. TOPOLOGICAL SPACES

6. We shall understand by *topological space* $\mathfrak{R}$ an aggregate of elements, the *points* of $\mathfrak{R}$, and an aggregate $\mathfrak{U}$ of subsets, the *open sets* of $\mathfrak{R}$, which satisfy the following axioms:

OS1. *The null-set and $\mathfrak{R}$ itself are open.*

OS2. *The union of any number of open sets is open.*

OS3. *The intersection of two (and hence of any finite number of) open sets is open.*

Although a space, as we have defined the concept, is made up of points, the points themselves will not be used in a major way until (§6) is reached.

(6.1) *Open base.* An *open base*, or merely a *base*, for $\mathfrak{R}$ is an aggregate $\{W_a\}$ of open sets ($\neq\emptyset$) of $\mathfrak{R}$ such that every open set of $\mathfrak{R}$ is a union of these basic open sets. The empty set is understood to be a union of an empty aggregate of sets.

If we wish to make a set R into a space by choosing an aggregate $\{W_a\}$ of its subsets as a base, we will need

(6.2) *$\{W_a\}$ is a base for a topological space if and only if*: (a) *R is a union of W_a's*; (b) *the intersection of every two W_a's is a union of W_a's.*

The necessity of these conditions is clear from OS123. If R is a union of W_a's, then OS1 will be satisfied. OS2 is automatically satisfied. The intersection of two unions of W_a's is the union of intersections of pairs of W_a's; hence if the intersections of pairs of W_a's are unions of W_a's, then the intersection of two unions of W_a's is a union of W_a's and OS3 holds.

In the applications it is more convenient to replace (6.2) by the equivalent condition:

(6.3) *$\{W_a\}$ is a base whenever*: (a) *every point x is in some W_a*; (b) *if $x \in W_a \cap W_b$ there is a W_c such that $x \in W_c \subset W_a \cap W_b$.*

Two bases $\{W_a\}$, $\{W'_b\}$ are said to be *equivalent* if they are bases for the same topological space. The condition for this is that every W_a is a union of sets W'_b, and conversely. Or more conveniently in terms of points: if $x \in W_a$ there is a W'_b such that $x \in W'_b \subset W_a$ and likewise with W_a, W'_b interchanged.

(6.4) *Subbase.* An aggregate $\{W_a\}$ of open sets of $\mathfrak{R}$ such that their finite intersections constitute a base is known as a subbase for $\mathfrak{R}$. If a space has a subbase it is necessarily topological.

(6.5) *Base and subbase at a point.* Let x be a point of $\mathfrak{R}$. An aggregate $\{W_a\}$ of open sets, all containing x, is a *base at x*, whenever if U is any open set containing x there is a set W_a such that $x \in W_a \subset U$. An aggregate $\{W_a\}$ is a *subbase* at x whenever the finite intersections of its sets constitute a base at x.

(6.6) *Countable bases.* The presence of countable bases is often an important property of a space. In this connection we must mention the two well known axioms due to Hausdorff:

First countability axiom. *There is a countable base at each point of $\mathfrak{R}$.*

Second countability axiom. *The space $\mathfrak{R}$ has a countable base.*

Clearly the second implies the first.

In connection with countable bases we have also the classical:

(6.7) Theorem of Lindelöf. *If $\mathfrak{R}$ has a countable base and $V = \bigcup V_a$, where the V_a are open sets, then there is a countable subcollection $\{V_{a_n}\}$ of $\{V_a\}$ such that $V = \bigcup V_{a_n}$.*

Let $\{W_n\}$ be a countable base. Every V_a is a union of sets W_n. The totality of the sets W_n which are contained in some V_a is thus a subcollection $\{W'_n\}$ of $\{W_n\}$ and so necessarily countable. We have then $V = \bigcup W'_n$. Now for each W'_n there is a

$$V_{a_n} \supset W'_n$$

and clearly $\{V_{a_n}\}$ behaves as required.

From the theorem we deduce

(6.8) *If $\mathfrak{R}$ has a countable base $\{W_n\}$ then every base $\{V_a\}$ contains a countable subaggregate $\{V_{a_n}\}$ which is already a base.*

By the theorem just proved out of the sets V_a whose union is W_p there may be selected a countable subcollection $\{V_{a_{pq}}\}$ whose union is again W_p. Therefore $\{V_{a_{pq}}\}$ (all p, q) is a countable base.

7. **Closed sets.** The complement $F = \mathfrak{R} - U$ of an open set is known as a *closed set.* The properties of closed sets are the duals of those of open sets; explicitly, the duals of OS123 are:

CS1. *$\mathfrak{R}$ and $\emptyset$ are closed.*

CS2. *Any interesection of closed sets is closed.*

CS3. *The union of two (and hence of a finite number of) closed sets is closed.*

Conversely, if we had the closed sets satisfying CS123 and defined the open sets as the complements of closed sets, then the open sets would satisfy OS123.

(7.1) *Closed base.* An aggregate $\{F_a\}$ of closed sets of $\mathfrak{R}$ is a closed base whenever every closed set is an intersection of basic closed sets. Clearly:

(7.2) *$\{F_a\}$ is a closed base for $\mathfrak{R}$ if and only if $\{\mathfrak{R} - F_a\}$ is an open base for the space.*

8. **Transformations between spaces.** A single- or multi-valued transformation T is called *open* [*closed*] if it takes open sets of $\mathfrak{R}$ onto open sets of $T\mathfrak{R}$ [closed sets of $\mathfrak{R}$ onto closed sets of $T\mathfrak{R}$]. Since the image of a union is the union of the images we have

(8.1) *If $\{U_a\}$ is a base for $\mathfrak{R}$, then a transformation T of $\mathfrak{R}$ onto $\mathfrak{S}$ is open if and only if each TU_a is open in $\mathfrak{S}$.*

A *continuous transformation* or *mapping* is a transformation T, whose inverse T^{-1} is open. An argument similar to that for (8.1) yields

(8.2) *If $\{V_a\}$ is a subbase for $\mathfrak{S}$, then a transformation T of $\mathfrak{R}$ into $\mathfrak{S}$ is continuous if and only if each $T^{-1}V_a$ is open in $\mathfrak{R}$.*

If both T and T^{-1} are single-valued and continuous, then T is called a *topological transformation* or a *homeomorphism.* Clearly:

(8.3) *A one-one transformation T is topological if and only if both T and T^{-1} are open.*

(8.4) *A transformation of $\mathfrak{R}$ into $\mathfrak{S}$ is continuous if and only if its inverse is closed.*

A formal description of topology may now be given:

(8.5) DEFINITIONS. *A topological property of a topological space $\mathfrak{R}$ is a property of $\mathfrak{R}$ which remains invariant under topological transformations. Topology is the study of the topological properties of topological spaces.*

EXAMPLES. (8.6) A rigid plane motion is a topological transformation. A folding over of the plane is a continuous transformation but it is not topological.

(8.7) *Topological equivalence.* The relation expressing that one space is the topological transform of another is evidently an equivalence, and is called *topological equivalence.*

9. **Some examples of topological spaces.**

(9.1) *Euclidean spaces.* Consider the set X^n of all the ordered sets of n real numbers $\{x_1, \cdots, x_n\}$. The subsets of X defined by inequalities

$$a_i < x_i < b_i, \qquad i = 1, 2, \cdots, n,$$

are known as *n-intervals*, written I^n, or merely *intervals* when $n = 1$. Since $\{I^n\}$ is immediately seen to verify (6.2) it may be chosen as a base in a topology for X^n. The resulting topological space, or any other topologically equivalent, is known as an *Euclidean n-space*, written $\mathfrak{E}^n$, also as a *real line* for $n = 1$. The open sets of $\mathfrak{E}^n$ are sometimes called *regions.*

Strictly speaking "Euclidean n-space" should be applied only to certain metric spaces described more accurately in (44.1). However, in this and other similar instances it will be generally more convenient to enlarge the meaning of a well known term in the above manner, rather than to have recourse to a more involved terminology.

Let $\mathfrak{U}$ be the base just defined for $\mathfrak{E}^n$ and let $\mathfrak{B}$ be the set of the *rational n-intervals*, i.e., corresponding to the a_i, b_i all rational. If $x \in I^n$, there is an element of $\mathfrak{B}$ between x and I^n; hence every I^n is a union of sets of $\mathfrak{B}$. Moreover every element of $\mathfrak{B}$ is an I^n. Therefore $\mathfrak{B}$ may serve as a base for $\mathfrak{E}^n$. In other words $\mathfrak{E}^n$ *has a countable base* namely $\mathfrak{B}$.

(9.2) Let R be any point set and let the open sets be defined as all the subsets of R so that the points themselves are open. The verification of the axioms OS123 is now trivial. The topology thus affixed to R is known as *the discrete topology.* Its chief function is to make statements for topological spaces valid for arbitrary point sets, it being always understood when this is done that the discrete topology is assigned to the set.

(9.3) Let $\mathfrak{R}$ be a set ordered by $\prec$. Define as an open set any subset U such that $x \in U$ and $x \prec x' \rightarrow x' \in U$. Then the sets U verify OS123. In fact OS3 is fulfilled in the stronger form:

OS3′. *Any intersection of open sets is open.*

$\mathfrak{R}$ is known as an *ordered* space. Let a set F have the property that $x \in F$ and $x' \prec x \rightarrow x' \in F$. Then $\mathfrak{R} - F = U$ is open, and so F is closed. Conversely, if F is closed it has the property just considered. From this follows that the *closed sets of* $\mathfrak{R}$ *satisfy the same axioms* OS123 *as the open sets of* $\mathfrak{R}$. Noteworthy examples of ordered spaces are the *complexes* (III, 1). In their theory however the fact that they are topological spaces is not important.

(9.4) An interesting example of ordered space is the real line L: $-\infty < x < +\infty$, considered as a set ordered by $\leqq$. The open sets are then the "rays" $a \leqq x < \infty$, and the closed sets the rays $-\infty < x \leqq a$. This topology is manifestly different from the customary topology of L as an $\mathfrak{E}^1$.

10. **Additional topological concepts.** The new concepts to be introduced must of course be expressed directly or indirectly in terms of the primitive elements, the open sets.

(10.1) The *interior* of a set A, written Int A, is the open set which is the union of all the open sets $\subset A$ (greatest open set contained in A). If $x \in \text{Int } A$, x is said to be an *interior* point of A.

(10.2) The *closure* $\bar{A}$ of A is the closed set which is the intersection of all the closed sets $\supset A$ (least closed set containing A).

(10.3) A is *dense* in $\mathfrak{R}$ if $\bar{A} = \mathfrak{R}$.

(10.4) The *boundary* $\mathfrak{B}A$ of A is the intersection of the closures of A and its complement: $\mathfrak{B}A = \bar{A} \cap \overline{\mathfrak{R} - A}$.

(10.5) A *neighborhood* of A is any open set containing A.

Many formal properties may be derived directly from the definitions. Thus:

(10.6) *Interiors are open sets, and the interior of an open set U is U itself*: $\text{Int } U = U$. *Closures and boundaries are closed sets, and the closure of a closed set F is F itself*: $\bar{F} = F$.

(10.7) *A is dense in $\mathfrak{R} \leftrightarrow A \cap U_a \neq \emptyset$ for every set U_a of a base $\{U_a\}$.*

Noteworthy and readily proved properties of the closure are:

$$A \subset \bar{A}, \tag{10.8a}$$

$$\bar{\emptyset} = \emptyset, \tag{10.8b}$$

$$\bar{\bar{A}} = \bar{A}, \tag{10.8c}$$

$$\overline{A \cup B} = \bar{A} \cup \bar{B}. \tag{10.8d}$$

It may be shown that if we take (10.8 abcd) as axioms for a closure operator and define a set F as closed by: $\bar{F} = F$ (10.6), then we obtain a collection $\{F\}$ satisfying CS123. Thus following F. Riesz and Kuratowski, one may describe topological spaces in terms of a suitably restricted closure operator.

Additional properties of the closure needed later will now be considered.

$$\overline{\bigcap A_a} \subset \bigcap \bar{A}_a \quad \text{Int} \bigcup A_a \supset \bigcup \text{Int } A_a \tag{10.9}$$

$$\mathfrak{R} - \bar{A} = \text{Int } (\mathfrak{R} - A). \tag{10.10}$$

This last property may also be expressed as: the complement of $\bar{A}$ is the union of all the open sets which do not meet A. It leads to the following important property (the only one of the present set where "points" are mentioned):

(10.11) *The closure $\bar{A}$ is the set of all the points x such that every neighborhood of x meets A.*

For if $x \in \bar{A}$, no neighborhood U of x can be in $\mathfrak{R} - A$ and so every such neighborhood meets A. On the other hand if this last property holds then $x \notin \mathfrak{R} - \bar{A}$, since otherwise $\mathfrak{R} - \bar{A}$ would be a neighborhood of x disjoint from A. Therefore $x \in \bar{A}$.

(10.12) *Let $\mathfrak{R}$, $\mathfrak{S}$ be topological spaces and T a mapping $\mathfrak{R} \rightarrow \mathfrak{S}$. Then if A is any subset of $\mathfrak{R}$ we have $T(\bar{A}) \subset \overline{TA}$.*

For $T^{-1}(\overline{TA})$ is closed and $\supset A$, hence $T^{-1}(\overline{TA}) \supset \bar{A}$ which yields at once (10.12).

11. **Topologization of subsets.** Let A be any subset of the space $\mathfrak{R}$ and let $\{U_a\}$, $\{F_a\}$ be the aggregates of open sets and closed sets of $\mathfrak{R}$. We see at

once that any of the properties OSi which hold for $\{U_a\}$ also hold for $\{A \cap U_a\}$, provided only that we replace $\mathfrak{R}$ by A in their statement. The same is true for $\{F_a\}$, CSi, and $\{A \cap F_a\}$. This leads to adopting throughout the present work the rule

(11.1) Principle of relativization. *Any subset A of a topological space $\mathfrak{R}$ is turned into a topological space by choosing as its open sets the intersections with A of the open sets of $\mathfrak{R}$. In this statement "open sets" may be replaced by "closed sets."*

(11.2) Example. Under the principle of relativization the subsets of any Euclidean space are topological spaces.

Observe that B might well be closed in A but not closed in $\mathfrak{R}$. For example, let L be the real line. If A is the interval $0 < x < 1$, and B the set $0 < x \leqq 1/2$, then B is closed in A but not in L.

(11.3) *The closure in A of a subset of A is the intersection with A of its closure in $\mathfrak{R}$.*

(11.4) *Application.* Let T be a mapping $\mathfrak{R} \to \mathfrak{S}$ and let $A \subset \mathfrak{R}$. Then $T \mid A$ is a mapping of A onto a subset B of $\mathfrak{S}$, for $T \mid A$ is continuous in the relative topologies. In particular T is a mapping of $\mathfrak{R}$ onto its image $T\mathfrak{R}$.

12. **Topological products.** Let $\{\mathfrak{R}_a\}$ be a collection of topological spaces, and let $\{U_{a,h}\}$ be the aggregate of open sets of $\mathfrak{R}_a$. We have already defined the set-product $\mathbf{P}\mathfrak{R}_a$. We now agree to topologize it by choosing as a base the sets $V = \mathbf{P}U_{a,h(a)}$, where $U_{a,h(a)} = \mathfrak{R}_a$ except for a finite set of a's depending on V. These sets might well be called "basic prisms." It is easily seen that (6.2) is fulfilled and so the product is a topological space $\mathfrak{R}$.

We notice that $\{U_a \times \mathbf{P}_{b \neq a}\mathfrak{R}_b\}$ is a subbase for $\mathfrak{R}$.

(12.1) *The projection $\pi_a : \mathfrak{R} \to \mathfrak{R}_a$ is an open mapping. More generally if $\{a\} = \{b\} \cup \{c\}$, $\mathfrak{R}' = \mathbf{P}\mathfrak{R}_b$, $\mathfrak{R}'' = \mathbf{P}\mathfrak{R}_c$, so that $\mathfrak{R} = \mathfrak{R}' \times \mathfrak{R}''$, then the projection $\pi : \mathfrak{R} \to \mathfrak{R}'$ is an open mapping.*

It is only necessary to consider the projection π_a. If U_a is open in $\mathfrak{R}_a$ then $V_a = \{x \mid x_a \in U_a\} = \pi_a^{-1}U_a$ is open in $\mathfrak{R}$, so π_a is a mapping. Since $\pi_a V_a = U_a$, π_a is open by (8.1).

The following proposition is expressed in the form in which it usually occurs:

(12.2) *Let $\mathfrak{S} = \{y\}$ be a topological space and let $f_a(y)$ be a continuous function on $\mathfrak{S}$ to $\mathfrak{R}_a$. If we set $x = \{f_a(y)\} = \varphi(y)$, then $\varphi(y)$ is likewise continuous on $\mathfrak{S}$. We may also say that φ is a mapping: $\mathfrak{S} \to \mathfrak{R}$.*

Since $x_a = f_a(y)$ is continuous $f_a^{-1}U_a = \varphi^{-1}V_a$ is open and since $\{V_a\}$ is a subbase for $\mathfrak{R}$, φ is continuous (8.2).

(12.3) *Application to the continuity of functions of several variables.* To simplify matters consider a function of two variables $f(x, x')$ with ranges $\mathfrak{R}$, $\mathfrak{R}'$ and values in a space $\mathfrak{S}$. By definition f is merely a function of the point (x, x') of $\mathfrak{R} \times \mathfrak{R}'$ i.e., with range $\mathfrak{R} \times \mathfrak{R}'$, with values in $\mathfrak{S}$; and f is said to be *continuous in both* x, x', when it is a continuous mapping $\mathfrak{R} \times \mathfrak{R}' \to \mathfrak{S}$. Let $\{U\}$, $\{U'\}$ be the open sets of $\mathfrak{R}$, $\mathfrak{R}'$. Since $\{U \times U'\}$ is a base for $\mathfrak{R} \times \mathfrak{R}'$ a n. a. s. c. for the function f to be continuous is that if $f(x_0, x_0') = y_0$ and V is any neighborhood of y_0 then there exists a neighborhood $U \times U'$ of $(x_0, x_0') \in \mathfrak{R} \times \mathfrak{R}'$ such that the values of f on $U \times U'$ are in V. This is the well known condition: There exist neighborhoods U, U' of x_0, x_0' such that $x \in U$, $x' \in U' \rightarrow f(x, x') \in V$. The extension to any number of variables is obvious.

(12.4) *The graph* G *of a mapping* $T: \mathfrak{R} \rightarrow \mathfrak{R}'$ *is topologically equivalent to* $\mathfrak{R}$. *More precisely if* π *is the projection* $\mathfrak{R} \times \mathfrak{R}' \rightarrow \mathfrak{R}$ *then* $\pi \mid G$ *is a topological mapping* $G \rightarrow \mathfrak{R}$.

It is already known that $\pi \mid G$ is one-one, and it is continuous since π is continuous. Therefore we only need to prove $\pi \mid G$ open. If $\{U\}$, $\{U'\}$ are the open sets of $\mathfrak{R}$, $\mathfrak{R}'$ then $\{U \times \mathfrak{R}'\}$ and $\{\mathfrak{R} \times U'\}$ together form a subbase for $\mathfrak{R} \times \mathfrak{R}'$. Hence $\{G \cap U \times \mathfrak{R}'\}$ and $\{G \cap \mathfrak{R} \times U'\}$ together form a subbase for the subset G. Now $\pi(G \cap U \times \mathfrak{R}') = U$, $\pi(G \cap \mathfrak{R} \times U') = T^{-1}U' =$ an open set of $\mathfrak{R}$, since T is continuous. Since $\pi \mid G$ maps the elements of a subbase of G onto open sets it is open, and (12.4) follows.

(12.5) *If* $\mathfrak{R} = \mathbf{P}\mathfrak{R}_a$, $A_a \subset \mathfrak{R}_a$, $A = \mathbf{P}A_a$, *then* $\bar{A} = \mathbf{P}\bar{A}_a$. *Or, explicitly: the closure of a product is the product of the closures.*

Since A is a product in order that $x = \{x_a\} \in \bar{A}$ a n. a. s. c. is that every neighborhood of x meet A, and hence that every set of the subbase $\{U_a \times \mathbf{P}_{b \neq a}\mathfrak{R}_b\}$ (U_a are the open sets of $\mathfrak{R}_a$) containing x meet A. Hence the n.a.s.c. is: for every a every neighborhood of x_a must meet A_a, or $x_a \in \bar{A}_a$ for every a, and this is (12.5).

If the A_a are closed then $\bar{A}_a = A_a$, and hence $\bar{A} = A$ or:

(12.6) *A product of closed sets is a closed set.*

This may also be proved directly as follows. If the A_a are closed then $A = \bigcap G_a$, $G_a = A_a \times \mathbf{P}_{b \neq a}\mathfrak{R}_b$. Since $\mathfrak{R} - G_a = (\mathfrak{R}_a - A_a) \times \mathbf{P}\mathfrak{R}_b$ is open, G_a is closed and so is A.

(12.7) *An Euclidean n-space* $\mathfrak{E}^n$ is the product of n real lines: $L_1 \times \cdots \times L_n$. If we replace in this product one factor say L_i by an interval λ of L_i, there is obtained a strip "perpendicular" to L_i, and the totality of these strips forms a subbase. Thus in the Euclidean plane the horizontal and vertical strips form a subbase.

(12.8) *Parallelotopes, cells, spheres.* We have already defined the interval as a subset $a < x < b$ of the real line. Its closure $a \leqq x \leqq b$ $(a \neq b)$ is known as a *segment*. Let $\lambda_1, \cdots, \lambda_n$ be intervals, and $l_i = \bar{\lambda}_i$ the corresponding segments. The product $\mathbf{P}\lambda_i$ is an n-interval I^n. The product $P^n = \mathbf{P}l_i$ is known as an *n-parallelotope*. The set $S^{n-1} = P^n - I^n = \mathfrak{B}P^n$ is called a *topological* $(n-1)$*-sphere*. The topological zero-sphere consists of two points, the topological one-sphere of the perimeter of a rectangle.

The terms "parallelotope," "sphere" are also applied to any sets topologically equivalent to P^n, S^{n-1}. However, a set topologically equivalent to I^n is generally called an *n-cell*.

The number n for the n-parallelotope, n-cell or n-sphere is called its *dimension*. For the present this designation is merely to be understood in a formal way. Later (VIII, 15), we shall identify n with the topological dimension.

Owing to their importance it is advisable to recognize parallelotopes, cells and spheres even when they occur in a form unrelated to the products. Most models may be deduced from:

(12.9) *A bounded convex region* Ω *(region with bounded coordinates) in an Euclidean n-space* $\mathfrak{E}^n$ *is an n-cell; its closure* $\bar{\Omega}$ *is an n-parallelotope and its boundary* $\mathfrak{B}\Omega$ *is a topological* $(n - 1)$*-sphere.*

Let x_i be running coordinates for the space and choose for n-cell the set: $\Omega_0 : 0 < x_i < 1$. Take a point a on Ω and a point a_0 on Ω_0. Any ray λ from a meets the boundary $\mathfrak{B}\Omega$ in a single point p. Draw from a_0 a ray λ_0 parallel to λ and in the same direction, and let it meet $\mathfrak{B}\Omega_0$ in p_0. Let T be the transformation whereby a point x_0 dividing a_0p_0 in a given ratio between 0 and 1 goes into the point x dividing ap in the same ratio, while $a_0 \to a$, $p_0 \to p$. T is manifestly a topological transformation, and since $T\Omega_0 = \Omega$, $T\mathfrak{B}\Omega_0 = \mathfrak{B}\Omega$, (12.9) follows.

Application. Let $\mathfrak{E}^n$ be referred to the coordinates x_i. Then the Euclidean spherical region

$$\sum x_i^2 < 1$$

is an n-cell. Its closure, the set

$$\sum x_i^2 \leqq 1,$$

is an n-parallelotope. The boundary, the Euclidean $(n - 1)$-sphere, is a topological $(n - 1)$-sphere. This is, of course, the justification for the term "sphere."

(12.10) Let $\{l_n\}$ be a countable collection of segments. The product $P^\omega = \mathbf{P}l_n$ is known as the *Hilbert parallelotope*. If P^ω is parametrized by $0 \leqq x_n \leqq 1$, then the "strips" defined by one condition of the form $a_n < x_n < b_n$, $0 \leqq x_n < b_n$, $a_n < x_n \leqq 1$, make up a subbase for P^ω.

Here again strictly speaking the Hilbert parallelotope as we have defined it, is only a topological image of the set commonly designated by that name.

(12.11) Let X, Y denote, respectively, the segments $0 \leqq x \leqq 1$, $0 \leqq y \leqq 1$ and let $\{Y_x\}$ be a system indexed by X, where $Y_x = Y$. Then $\mathbf{P}Y_x = Y^X$ is the set of all functions f on X to Y. Let U_x be an interval of Y_x. Then the set $V_x = \{f \mid f(x) \epsilon U_x\}$ is open in Y^X and $\{V_x\}$ is a subbase for this space. The space thus obtained has many important properties of great interest in analysis. For instance, the subset Q of Y^X which represents the continuous functions on $[0 - 1]$ to $[0 - 1]$ is "very thinly spread" in Y^X.

The space Y^X is also interesting as a special case of what we shall call later (25.2) a "compact parallelotope."

13. **Topological identification.** We have seen (2.4) that a relation of equivalence R between the elements of a set X yields a new set Y by identification of the elements in each equivalence class y. We also have an associated transformation T of X onto Y whereby $Tx = y$. Suppose that X is a topological space and let Y be topologized by specifying $V \subset Y$ to be open whenever $T^{-1}V$ is open in X. Then OS123 are readily verified, and so Y is a topological space. This space is said to be obtained from X by topological identification.

EXAMPLES. (13.1) The set described as "projective plane" in (2.6) receives by topological identification a definite topology and it is the set thus topologized which is referred to henceforth as *projective plane*. Similarly for projective n-space.

(13.2) Let $X = \{x\}$ be the real line and let the relation R be defined by the condition: x and x' are in the relation R whenever $x \equiv x' \bmod 1$ ($x - x'$ is an integer). This is manifestly a relation of equivalence and topological identification yields the space Y referred to as *the real line* mod 1, or also the *circumference*.

(13.3) *Topological imbedding.* Let $\mathfrak{R}$ contain a set S such that there is a topological mapping $t: \mathfrak{S} \to S$. If we replace every point $x \in S$ by $t^{-1}x$ both in $\mathfrak{R}$ and in its open sets, we obtain what is known as a *topological imbedding* or *immersing* of $\mathfrak{S}$ into $\mathfrak{R}$. It is an imbedding in the sense of (2.7) since it replaces $\mathfrak{R}$ by $(\mathfrak{R} - S) \cup \mathfrak{S}$.

§3. AGGREGATES OF SETS. COVERINGS. DIMENSION

14. In view of the fundamental role of aggregates of sets and coverings it is important to settle the nomenclature as rapidly as possible.

We shall be dealing with aggregates of subsets of a given space $\mathfrak{R}$. Let $\mathfrak{A} = \{A_\alpha\}$ be such an aggregate. The set $\{\bar{A}_\alpha\}$ of the closures of the A_α is denoted by $\bar{\mathfrak{A}}$. Given a second collection $\mathfrak{B} = \{B_\beta\}$ we shall write:

$\mathfrak{A} \cup \mathfrak{B}$ or $\mathfrak{A} \vee \mathfrak{B}$ = the union of $\mathfrak{A}$ and $\mathfrak{B}$;

$\mathfrak{A} \wedge \mathfrak{B} = \{A_\alpha \cap B_\beta\}$;

$\mathfrak{A} \times \mathfrak{B} = \{A_\alpha \times B_\beta\}$;

$\mathfrak{A} > \mathfrak{B}$ = every A_α is in some B_β ; we say also that $\mathfrak{A}$ is a *refinement* of $\mathfrak{B}$ or *refines* $\mathfrak{B}$.

As a special case of $\mathfrak{A} \wedge \mathfrak{B}$ one of the collections, say $\mathfrak{A}$, may consist of a single set A so that $\mathfrak{A} \wedge \mathfrak{B}$ is now $\{A \cap B_\beta\}$.

The *order* of $\mathfrak{A}$ is the largest number p if one exists such that some $p + 1$ sets of $\mathfrak{A}$ intersect; if p does not exist the order is said to be infinite.

The finiteness properties of the aggregates are important. We say that $\mathfrak{A}$ is:

point-finite whenever every point of $\mathfrak{R}$ belongs to at most a finite number of A_α ;

neighborhood-finite whenever every point of $\mathfrak{R}$ has a neighborhood N which meets at most a finite number of A_α ;

locally finite whenever every A_α meets at most a finite number of A_α ;

finitely covered by $\mathfrak{B}$ whenever $\mathfrak{A} > \mathfrak{B}$ and every B_β contains at most a finite number of A_α .

Notice that when $\mathfrak{A}$, $\mathfrak{B}$ are point-finite or neighborhood-finite so is $\mathfrak{A} \vee \mathfrak{B}$

Two aggregates $\mathfrak{A} = \{A_\alpha\}$, $\mathfrak{B} = \{B_\beta\}$ are said to be *similar* whenever there may be established a one-one transformation τ: $\{\alpha\} \to \{\beta\}$ such that $A_\alpha \cap \cdots \cap A_{\alpha'} \neq \emptyset \leftrightarrow B_{\tau\alpha} \cap \cdots \cap B_{\tau\alpha'} \neq \emptyset$. The transformation $\mathfrak{A} \to \mathfrak{B}$ defined by $A_\alpha \to B_{\tau\alpha}$ is known also as a *similitude*.

By a *covering* of $\mathfrak{R}$ is meant an aggregate $\mathfrak{A}$ whose sets A_α have $\mathfrak{R}$ for unicn: $\bigcup A_\alpha = \mathfrak{R}$ (every point x belongs to an A_α). An *open* [*closed*] covering is a covering by open [closed] sets.

Let $\{\mathfrak{A}_\alpha\}$ be a collection of coverings. We say that the subcollection $\{\mathfrak{A}_\beta\}$ is *cofinal* in $\{\mathfrak{A}_\alpha\}$ whenever every $\mathfrak{A}_\alpha$ has an $\mathfrak{A}_\beta$ refinement.

Notice that a neighborhood-finite covering $\mathfrak{A}$ may be characterized thus: There exists an open covering each of whose sets meet at most a finite number of sets of $\mathfrak{A}$.

15. **Dimension.** The general theory of dimension, of Menger and Urysohn, fully developed for separable metric spaces, has not reached very far beyond these spaces. An important reason is that several equivalent definitions, all natural and which agree for separable metric spaces, seem to part company for other, less simple, spaces. A full treatment of these questions is wholly outside the scope of the present treatise, and they will be touched upon here and there only in those phases of interest in algebraic topology. Let us say at all events, that while in the early definition of Menger-Urysohn the "local" point of view predominates, we shall adopt the definition, inspired by Lebesgue, in terms of the order of coverings, as it is most closely related to our general purpose.

(15.1) DEFINITION. *Let* $K = \{\mathfrak{A}_\lambda ; >\}$ *be a class of coverings of a topological space* $\mathfrak{R}$, *which is directed by refinement*: $\mathfrak{A}_\lambda > \mathfrak{A}_\mu \leftrightarrow \mathfrak{A}_\lambda$ *refines* $\mathfrak{A}_\mu$. *For a given* $\mathfrak{A}_\mu$ *consider all the* $\mathfrak{A}_\lambda > \mathfrak{A}_\mu$, *and let* n_μ *be the least order of all such* $\mathfrak{A}_\lambda$. *The K-dimension of* $\mathfrak{R}$ *is* sup n_μ.

Among the noteworthy classes K are: all the finite open or all the finite closed coverings, all the point-finite, or neighborhood-finite open or closed coverings. If $\mathfrak{R}$ is topological then each of these has the property that $\mathfrak{A}_\mu \wedge \mathfrak{A}_\nu$ refines both $\mathfrak{A}_\mu$ and $\mathfrak{A}_\nu$ and is in the class. Hence each may serve to define a dimension. We have thus the dimensions by finite open or closed coverings, $\cdots$. The most generally utilized is the first, and it is to this dimension by finite open coverings that the term *dimension*, written dim $\mathfrak{R}$, is applied in the sequel.

Little is known regarding the mutual relations between the various dimensions and there are few, if any, very general properties. A simple property is:

(15.2) *If F is a closed set in $\mathfrak{R}$ then* $\dim F \leq \dim \mathfrak{R}$.

Any finite open covering of F is of the form $\{F \cap U_i\}$, where $\mathfrak{U} = \{U_i\}$ is a finite collection of open sets of $\mathfrak{R}$. Since $\{U_i, \mathfrak{R} - F\}$ is a finite covering of $\mathfrak{R}$, if $\dim \mathfrak{R} = n$, the covering has a refinement $\mathfrak{B} = \{V_j\}$ whose order does not exceed n. The sets $\{F \cap V_j\}$ are then a refinement of the covering $\{F \cap U_i\}$ of F, whose order does not exceed n. Therefore $\dim F \leq n = \dim \mathfrak{R}$.

§4. CONNECTEDNESS

16. There is perhaps no simpler intuitive property of a space than connectedness.

(16.1) DEFINITION. *A topological space* $\mathfrak{R}$ *is said to be connected when it is not the union of two non-void disjoint open sets.*

If $\mathfrak{R} = U \cup V$, where U and V are open and disjoint, then U and V are also closed, so that $\mathfrak{R}$ is the union of two disjoint closed sets, and conversely. Therefore in the definition "open sets" may be replaced by "closed sets." Moreover the property of the definition is seen to be equivalent to the following: The null-set and $\mathfrak{R}$ itself are the only subsets of $\mathfrak{R}$ which are both open and closed.

Let A, B be two subsets of $\mathfrak{R}$ which satisfy the so-called *Hausdorff-Lennes separation condition*:

$$(\bar{A} \cap B) \cup (A \cap \bar{B}) = \emptyset. \tag{16.2}$$

Explicitly neither set meets the closure of the other. We prove:

(16.3) THEOREM. *A n. a. s. c. for the connectedness of a subset C of $\mathfrak{R}$ is that it admits of no decomposition $C = A \cup B$ wherein A, B are not empty and satisfy the Hausdorff-Lennes condition.*

This characteristic property is frequently taken as the definition of connectedness.

If $C = A \cup B$ and (16.2) holds, then $C \cap \bar{A} \subset C - B \subset A$, and hence A is closed in C. Similarly B is closed in C, and hence B and A are open in C and C is not connected.

If C is not connected, then $C = A \cup B$, where A and B are disjoint, nonempty, open and closed in C; hence $(\bar{A} \cap B) \cup (A \cap \bar{B}) = (C \cap \bar{A} \cap B) \cup (A \cap C \cap \bar{B}) = (A \cap B) \cup (A \cap B) = A \cap B = \emptyset$. Thus the theorem is proved.

17. **Connected aggregates.** The definition of connectedness for aggregates of sets rests upon the simple properties of certain finite collections, the *chains*.

Let us call *topological chain* or merely *chain* a finite collection $A = A_1$, $A_2, \cdots, A_r = A'$ such that consecutive sets of the collection intersect. The A_i are the *links* of the chain. The chain is said to *join* A to A', and if every A_i is member of a collection $\mathfrak{A} = \{A_\alpha\}$, the chain is said to join A to A' in $\mathfrak{A}$, and it is called an $\mathfrak{A}$-*chain*. In particular when $r = 1$ then $A = A'$ and the chain consists of one link A.

Let us take a particular set A in $\mathfrak{A}$ and let $\mathfrak{A}_1$ be the subaggregate of $\mathfrak{A}$ consisting of all the sets which may be joined to A by a chain in $\mathfrak{A}$. The aggregate $\mathfrak{A}_1$ is called a *component* of $\mathfrak{A}$. The following properties are immediate:

(17.1) *The set A belongs to the component $\mathfrak{A}_1$ which it serves to determine.*

(17.2) *The component determined by any set of $\mathfrak{A}_1$ is $\mathfrak{A}_1$ itself.*

In other words the components are independent of the individual sets which serve to determine them and they depend on $\mathfrak{A}$ alone.

(17.3) *Two components of $\mathfrak{A}$ with a common set A coincide, or equivalently: distinct components are disjoint.*

(17.4) *Each component of a locally finite aggregate is composed of a countable number of sets.*

For we may then obtain, say $\mathfrak{A}_1$ as follows: Take the sets A_i (finite in number) which meet $A_1 \in \mathfrak{A}_1$, then the sets A_{ij} (finite in number) which meet the sets A_i, etc. The totality of the sets thus obtained is $\mathfrak{A}_1$ and it is countable.

An aggregate $\mathfrak{A}$ consisting of a single component is said to be *connected.* It is characterized by the property that any two of its sets may be joined by an $\mathfrak{A}$-chain. This corresponds in every way to the intuitive concept of connectedness.

(17.5) *Let $\mathfrak{A}$, $\mathfrak{B}$ be two aggregates such that $\mathfrak{A} > \mathfrak{B}$ and that every set of $\mathfrak{B}$ contains a set of $\mathfrak{A}$. Then if $\mathfrak{A}$ is connected so is $\mathfrak{B}$.*

For let $B, B' \in \mathfrak{B}$. By assumption we have in $\mathfrak{A}$ two sets $A \subset B$ and $A' \subset B'$. Since $\mathfrak{A}$ is connected there is a chain $A, A_1, \cdots, A_r, A'$ joining A to A' in $\mathfrak{A}$. Choose for each A_i a set $B_i \supset A_i$. Then $B, B_1, \cdots, B_r, B'$ is a chain joining B to B' in $\mathfrak{B}$. Therefore $\mathfrak{B}$ is connected.

18. We now link up connectedness in sets and in aggregates by:

(18.1) *A n.a.s.c. for connectedness of a space $\mathfrak{R}$ is that all the coverings of any one of the following families be connected*: (a) *all the open coverings*; (b) *all the locally finite open coverings*; (c) *all the finite open coverings*; (d) *a family cofinal in any one of* (a)–(c) (refinements as in 17.5).

The proof of (a, b, c) is the same, while combined with (17.5) they yield (d), so we merely consider (a). If the condition holds all the open coverings are connected. Therefore in every decomposition $\mathfrak{R} = U \cup V$, U and V open, necessarily $U \cap V \neq \emptyset$, or $\mathfrak{R}$ is connected. Thus the condition is sufficient. To prove necessity let $\mathfrak{R}$ be connected and let the open covering $\mathfrak{U}$ be disconnected. $\mathfrak{U}$ has then at least two components. Let U be the open set which is the union of all the elements of one of the components, and V the open set which is the union of the remaining elements. We then have $\mathfrak{R} = U \cup V$, U and V are open and $U \cap V = \emptyset$. But this is ruled out since $\mathfrak{R}$ is connected. This proves necessity and hence (18.1).

19. We shall utilize the result just proved to derive a certain number of simple properties of connected sets. Unless otherwise stated they are supposed to be subsets of a given topological space $\mathfrak{R}$.

(19.1) *A union of connected sets of which every pair intersect is itself connected* (18.1, 17.1).

(19.2) *Whenever in a sequence of connected sets $A_1, A_2, \cdots$ each meets the next one, their union is connected.*

For $A_1, A_1 \cup A_2, A_1 \cup A_2 \cup A_3, \cdots$ are all connected and contain A_1. Hence their union which is $\bigcup A_i$ is connected.

(19.3) *If A is connected so is $\bar{A}$.*

Let $\mathfrak{U}$ be a covering of $\bar{A}$ without sets not meeting $\bar{A}$. Then $A \wedge \mathfrak{U}$ is a covering of A. Since $A \wedge \mathfrak{U} > \mathfrak{U}$ these two aggregates are related as in (17.5). It follows that in the sequence of sets and aggregates of sets: A, $A \wedge \mathfrak{U}$, $\mathfrak{U}$, $\bar{A}$ the connectedness of each implies that of the following one. This proves (19.3).

(19.4) *If A, B are connected so is $A \times B$.*

If $(x, y), (x_1, y_1) \in A \times B$, both are in the union of the connected sets $A \times y_1$, $x \times B$ with the common point (x, y_1) and (19.4) follows.

(19.5) *The image of a connected set under a mapping is connected.*

For let T map A continuously onto B and let $\mathfrak{U} = \{U_a\}$ be an open covering of B. Then $\{T^{-1}U_a\}$ is an open covering of A and hence connected. Therefore any two sets $T^{-1}U_a$ and $T^{-1}U_b$ may be joined by a chain of such sets. Since T is single-valued, $TT^{-1}U_a = U_a$, and so the images of the links of this chain make up a $\mathfrak{U}$-chain joining U_a to U_b, hence $\mathfrak{U}$ is connected and so is B.

20. From (19.1) follows that the union $C(x)$ of all the connected sets containing a given point x is connected. If $y \neq x$ and $y \in C(x)$ then $C(y) = C(x)$. For otherwise their union would be a connected set containing x and $\not\subset C(x)$ which contradicts the definition of $C(x)$. Thus $C(x)$ is uniquely defined by any one of its points. The set $C(x)$ is called a *component* of $\mathfrak{R}$. We have similarly of course components of any subset of $\mathfrak{R}$.

Since $C(x)$ contains all the connected sets containing x and is itself connected, it is the maximal connected set containing the point.

(20.1) *The n-cell, n-parallelotope, Hilbert parallelotope, n-sphere* $(n > 0)$, *are all connected.*

We first show that a segment $l: 0 \leqq t \leqq 1$ is connected. For if it is not we have $l = A \cup B$, where A, B are closed and disjoint. Let $0 \in A$ and set $a = \sup \{t \mid t \in A; t < B\}$. Since A is closed $a \in A$. Furthermore whatever $\eta > 0$ there is a point of B in the interval $a, a + \eta$. Hence $a \in \bar{B} = B$, and so A, B are not disjoint, contrary to assumption.

Since any two points of one of the sets in (20.1) can be joined by a "closed arc" (one-parallelotope, topological image of a segment), by the result just proved the sets are connected.

§5. COMPACT SPACES

21. It is not too much to say that all the spaces of chief interest in general topology, and even more so in algebraic topology, are compact spaces or their subsets. This is largely due to the fact that in dealing with compact spaces one may frequently replace infinite collections by finite collections.

We emphasize at the outset the following important departures from hitherto accepted terminology: (a) with Bourbaki we shall replace the term *bicompact* of Alexandroff-Urysohn by the term *compact*; (b) what has been known hitherto as *compact* (following Fréchet who introduced the concept) shall be called *countably compact*; (c) following Alexandroff-Hopf, a compact metric space shall be called a *compactum*. It is important that these modifications be kept in mind. The chief justification for adopting them, aside from convenience, are first that "bicompact metric" = "compact metric" and that "compact" (non-metric) spaces in the earlier sense occur but rarely.

22. (22.1) Definition. *A collection of sets is said to have the finite intersection property whenever every finite subcollection has a non-empty intersection.*

(22.2) *The following properties of a topological space are equivalent.*

P1. *If* $\{U_a\}$ *is any open covering of* $\mathfrak{R}$, *then some finite subcollection of* $\{U_a\}$ *is already a covering.*

P2. *If* $\{F_a\}$ *is a collection of closed sets with the finite intersection property, then the intersection of the whole collection is non-empty.*

For P2 is equivalent to: if the sets of $\{F_a\}$ have a void intersection the same holds for some finite subcollection, and this is the dual of P1, hence equivalent to P1.

On the strength of (22.2) we lay down the

(22.3) DEFINITION. *A topological space satisying any one of* P12 (*and hence both*) *is said to be compact.*

Notice that the concept of compactness is primitive in the sense that it may be expressed without reference to the other properties of open or closed sets.

In the applications it is convenient to have:

(22.4) *If the* U_a *in* P1 *are restricted to a particular open base, or the* F_a *in* P2 *to a particular closed base, then we still have equivalent conditions.*

Let in fact $\mathfrak{B} = \{V_b\}$ be any covering of $\mathfrak{R}$ and suppose P1 to hold in the restricted manner. Let $\mathfrak{U}' = \{U'_a\}$ be the set of the elements of the base which are contained in any V_b. Since every V_b is a union of elements of $\mathfrak{U}'$, $\mathfrak{U}'$ is also a covering. By assumption it has a finite subcovering $\{U'_{a_1}, \cdots, U'_{a_p}\}$. Each U'_{a_i} is in some set V_{b_i} of $\mathfrak{B}$. Hence $\{V_{b_i}\}$ is a finite subcovering of $\mathfrak{B}$ and so P1 holds. The treatment of P2 is wholly similar and is omitted.

(22.5) COMPACTNESS OF SUBSETS. *A subset* A *of a topological space* $\mathfrak{R}$ *is compact when and only when one of the following two equivalent properties holds*: (a) *if* $\{U_a\}$ *is any open covering of* A *by open sets of* $\mathfrak{R}$ *then some finite subcollection of* $\{U_a\}$ *is already a covering*; (b) *if* $\{F_a\}$ *is a collection of closed sets of* $\mathfrak{R}$ *such that* $\{A \cap F_a\}$ *has the finite intersection property, then the sets* F_a *have a non-empty intersection which meets* A.

This is an immediate consequence of the principle of relativization (11.1).

23. (23.1) *A closed subset* F *of a compact space* $\mathfrak{R}$ *is also compact.*

For a collection $\{F_a\}$ of closed subsets of F with the finite intersection property is also a similar collection for $\mathfrak{R}$ itself, and so $\bigcap F_a \neq \emptyset$, proving F compact.

(23.2) *If a compact space* $\mathfrak{R}$ *is mapped onto a subset* A *of a topological space* $\mathfrak{S}$ *then* A *is compact.*

Let τ be the mapping and $\{F_a\}$ a collection of closed sets of A with the finite intersection property. Then $\{\tau^{-1}F_a\}$ is a similar collection for $\mathfrak{R}$. Hence

$$\bigcap \tau^{-1}F_a \neq \emptyset$$

and therefore $\bigcap F_a \neq \emptyset$, proving A compact.

(23.3) *The union of a finite number of closed subsets of a space is compact if and only if each subset is compact.*

Let $F = \bigcup F_a$. If F is compact, then, since each F_a is a closed subset of F, each F_a is compact. If each F_a is compact, consider any open covering of F; each F_a is covered by a finite number of its sets (22.5a) and hence F is covered by a finite number of its sets. Thus P1 holds and F is compact.

24. Products of compact spaces.

(24.1) THEOREM. *An arbitrary product of compact spaces is compact* (Tychonoff; proof after Bourbaki).

Let $\mathfrak{R} = \mathbf{P}\mathfrak{R}_\lambda$ where the $\mathfrak{R}_\lambda$ are compact. Given in $\mathfrak{R}$ a family $\mathfrak{F} = \{F_a\}$ of closed sets with the finite intersection property we must show that there is a point common to all the F_a.

By Zorn's theorem $\mathfrak{F}$ is contained in a family $\mathfrak{G} = \{G_b\}$ of sets (not necessarily closed) with the finite intersection property and maximal relatively to this property. As a consequence: (a) any finite intersection of sets of $\mathfrak{G}$ is in $\mathfrak{G}$; (b) a set meeting every set of $\mathfrak{G}$ is in $\mathfrak{G}$.

Let π_λ be the projection $\mathfrak{R} \to \mathfrak{R}_\lambda$ and set $\mathfrak{G}_\lambda = \{\overline{\pi_\lambda G_b}\}$. Since $\mathfrak{G}_\lambda$ is a family of closed sets in $\mathfrak{R}_\lambda$ with the finite intersection property there is an x_λ common to all its sets. Let $x = \{x_\lambda\}$ and let N be any neighborhood of x. If $\{N_\lambda\}$ are the neighborhoods of x_λ then $\{\pi_\lambda^{-1} N_\lambda\}$ is a subbase at x, and so for some finite set $\lambda_1, \cdots, \lambda_k$ we have: $x \in \bigcap(\pi_{\lambda_i}^{-1} N_{\lambda_i}) \subset N$. Since $x_{\lambda_i} \in \overline{\pi_{\lambda_i} G_b}$, N_{λ_i} meets $\pi_{\lambda_i} G_b$. Hence $\pi_{\lambda_i}^{-1} N_{\lambda_i}$ meets G_b, and so it is in $\mathfrak{G}$. It follows that $\bigcap(\pi_{\lambda_i}^{-1} N_{\lambda_i}) \in \mathfrak{G}$, hence $N \in \mathfrak{G}$. Therefore N meets F_a and consequently $x \in \bar{F}_a = F_a$. This proves the theorem.

(24.2) *If $\mathfrak{R}$, $\mathfrak{S}$ are compact then every finite open covering $\mathfrak{W} = \{W_i\}$ of $\mathfrak{R} \times \mathfrak{S}$ has a refinement $\mathfrak{U} \times \mathfrak{V}$ where $\mathfrak{U}$ and $\mathfrak{V}$ are finite open coverings of $\mathfrak{R}$ and $\mathfrak{S}$.*

Let $\{U\}$, $\{V\}$ be the open sets of $\mathfrak{R}$, $\mathfrak{S}$, so that $\{U \times V\}$ is a base for $\mathfrak{R} \times \mathfrak{S}$. For each $(x, y) \in \mathfrak{R} \times \mathfrak{S}$ select a $W_i \ni (x, y)$, then a $U_x' \times V_y'$ between (x, y) and W_i. Thus $\{U_x' \times V_y'\}$, y fixed, is a covering of the compact set $\mathfrak{R} \times y$, and so there is a finite subcovering $\{U_y^i \times V_y^i\}$. If $V_y'' = \bigcap V_y^i$ then $\{U_y^i \times V_y''\}$ is a covering of $\mathfrak{R} \times V_y''$. Since $\{V_y''\}$ is a covering of the compact set $\mathfrak{S}$ it has a finite subcovering $\mathfrak{V} = \{V_{y_i}''\}$. If $\mathfrak{U}_j = \{U_{x_j}^i\}$ and $\mathfrak{U} = \mathfrak{U}_1 \wedge \cdots \wedge \mathfrak{U}_r$, then $\mathfrak{U} \times \mathfrak{V}$ is readily shown to behave as asserted.

25. Applications.

(25.1) *Segments are compact.*

Let l: $0 \leqq x \leqq 1$, be a segment. It has for base all the sets V of the following types: $0 \leqq x < a$, $a < x < b$, $b < x \leqq 1$, where a, b are rational and $0 < a$, $b < 1$. Therefore by (22.4) we merely need to show that a covering $\mathfrak{V}$ by such sets has a finite subcovering. Since the set of all the V's is countable so is $\mathfrak{V}$. Let then $\mathfrak{V} = \{V_n\}$ and suppose that it has no finite subcovering. The sets $W_n = l - (V_1 \cup \cdots \cup V_n)$ are never empty. Since $W_{n+1} \subset W_n$, and each W_n is a finite set of disjoint segments, there may be selected among these segments one, say l_n, such that $l_{n+1} \subset l_n$ throughout. It follows then from elementary properties of the Dedekind cut that $\bigcap l_n \neq \emptyset$, and so it contains a point x. Since $x \in W_n$ for every n, it is contained in no V_n, which contradicts the fact that $\mathfrak{V}$ is a covering and (25.1) follows.

(25.2) *Let α be any cardinal number and l a segment. Then l^α is a compact set known as "compact parallelotope"* (24.1, 25.1).

(25.3) *The n-parallelotope* P^n *and the Hilbert parallelotope* P^ω *are compact.*

(25.4) *All spheres* S^n *are compact.*

For S^n is closed in P^{n+1}, and so (25.4) follows from (23.1) and (25.2).

(25.5) *Any power* C^α *of a circumference* C *is compact and is known as a "toroid" and* α *is called the "dimension" of the toroid* (24.1).

(25.6) *Every closed and bounded subset of an Euclidean space* $\mathfrak{E}^n$ *is compact.*

By "bounded subset of $\mathfrak{E}^n$" we mean here a set A such that the coordinates of its points are bounded. Since A is in some P^n, and closed in P^n, (25.6) follows from (25.3) and (23.1).

(25.7) *Real projective spaces are compact.*

For the set A of all points of $\mathfrak{E}^n$ which satisfy

$$\sum x_i^2 \leqq 1$$

is closed and bounded, and hence compact. Since a real projective n-space is the image of A under a mapping it is likewise compact (23.2).

26. **Compacting.** This refers to the operation of imbedding topologically a space in a compact space. The basic result is the

(26.1) THEOREM. *Every topological space* $\mathfrak{R}$ *may be mapped topologically onto a dense subset of a compact space* $\mathfrak{S}$ *such that* $\dim \mathfrak{S} \leqq \dim \mathfrak{R}$ (Wallman [a]).

We will define the points, the closed sets and the open sets of $\mathfrak{S}$, and for later purposes develop their properties somewhat beyond the immediate requirements of the theorem.

Notations. The points, open sets and closed sets of $\mathfrak{R}$ are denoted by x, u, f, and the same for $\mathfrak{S}$ by X, U, F.

The points of $\mathfrak{S}$. Let φ denote the union of a closed set of $\mathfrak{R}$ and of a finite point set of $\mathfrak{R}$. By a *basic set* is meant a collection $\xi = \{\varphi_a\}$ with the finite intersection property. By Zorn's theorem ξ is contained in a similar collection which is maximal with respect to this property. Such a collection will be called a *maximal basic set* (= m.b.s.). The points X of $\mathfrak{S}$ are the m.b.s. and if $X = \{\varphi_a\}$ then the φ_a are called the *coordinates* of X. As in (24) the maximality of X implies that: (a) every finite intersection of coordinates of X is a coordinate of X; (b) every set φ, and in particular, every set f meeting every coordinate of X is also a coordinate.

Every $x \in \mathfrak{R}$ is a set φ, and so it is a coordinate of at least one maximal basic set $X = \{\varphi_a\}$. Since X is a basic set $x \cap \varphi_a \neq \emptyset$, hence $x \in \varphi_a$ and $\bigcap \varphi_a = x$. Suppose that x is a coordinate of $X' = \{\varphi_b'\} \neq X$. Then some $\varphi_b' \notin X$. Since $x \in \varphi_b'$, X may be augmented by φ_b' without ceasing to be a basic set, which contradicts the assumption that X is maximal. Therefore $X' = X$. Thus:

(26.2) *Every point* x *of* $\mathfrak{R}$ *is a coordinate of a unique m.b.s. which is written* $X(x)$.

If $x \neq x'$ then $x \cap x' = \emptyset$, hence $x' \notin X$ and $X(x') \neq X(x)$. Therefore

(26.3) *The transformation* $T\colon x \to X(x)$ *is univalent.*

Let X be a point which is not an $X(x)$: $X \epsilon \mathfrak{S} - T\mathfrak{R}$. If $\varphi_a \epsilon X$ then $\varphi_a = f_a \cup x_1 \cup \cdots \cup x_r$. Since $X \neq Tx_i$, none of the x_i may belong to all the coordinates of X. Therefore for each i there is a φ_{a_i} which does not contain x_i. Hence $\varphi_b = \bigcap \varphi_{a_i}$ does not contain any x_i. Now $(\varphi_a \cap \varphi_b \cap \varphi_c \neq \emptyset) \rightarrow (f_a \cap \varphi_b \cap \varphi_c \neq \emptyset)$ whatever the coordinate φ_c of X. Therefore $f_a \cap \varphi_c \neq \emptyset$ and hence f_a is a coordinate of X also. Thus:

(26.4) *If $X \notin T\mathfrak{R}$ has the coordinate $\varphi_a = f_a \cup x_1 \cup \cdots \cup x_r$, then the closed set f_a is also a coordinate of X.*

27. *The closed sets of $\mathfrak{S}$.* Let f be a closed set of $\mathfrak{R}$ and $\Phi(f)$ the set of all the points X with the coordinate f. We verify at once:

$$(27.1) \qquad \Phi(\emptyset) = \emptyset, \qquad \Phi(\mathfrak{R}) = \mathfrak{S}, \qquad f \neq f' \leftrightarrow \Phi(f) \neq \Phi(f');$$

$$(27.2) \qquad \Phi(\bigcap f_a) = \bigcap \Phi(f_a)$$

whenever $\{f_a\}$ is finite;

$$(27.3) \qquad f \subset f' \rightarrow \Phi(f) \subset \Phi(f').$$

Less obvious is the relation

$$(27.4) \qquad \Phi(f \cup f') = \Phi(f) \cup \Phi(f').$$

Let $f_1 = f \cup f'$. Since $f_1 \supset f$, by (27.3): $\Phi(f) \subset \Phi(f_1)$, and similarly $\Phi(f') \subset \Phi(f_1)$. Therefore

$$(27.5) \qquad \Phi(f) \cup \Phi(f') \subset \Phi(f \cup f').$$

Suppose now $X \epsilon \Phi(f_1) - \Phi(f')$. Since f_1 is a coordinate of X and f' is not, there is a finite intersection φ of coordinates of X which include f_1, and such that φ does not meet f'. Since X is a maximal basic set, φ itself must be a coordinate of X and so $\varphi \subset f_1 - f' \subset f$, hence $X \epsilon \Phi(f)$. This proves (27.5) with the inclusion reversed and (27.4) follows.

Referring to (6.2, 7.2) and by (27.1, 27.4) the collection $\{\Phi(f)\}$ may be chosen as a closed base for $\mathfrak{S}$, turning it into a topological space. We shall show that *$\mathfrak{S}$ is compact.* Since $\{\Phi(f)\}$ is a base, we merely need to show that if $\mathfrak{F} = \{\Phi(f_a)\}$ has the finite intersection property then $\bigcap \Phi(f_a) \neq \emptyset$. Now by (27.2) when $\mathfrak{F}$ has the finite intersection property so has $\{f_a\}$. That is to say, $\{f_a\}$ is a basic set. It follows that there is a m.b.s. X with the f_a as coordinates. Thus $X \epsilon \Phi(f_a)$, $\bigcap \Phi(f_a) \neq \emptyset$, and $\mathfrak{S}$ is compact.

The open sets of $\mathfrak{S}$. Let $u = \mathfrak{R} - f$ and set $\Omega(u) = \mathfrak{S} - \Phi(f) =$ an open set of $\mathfrak{S}$. A point X is in $\Omega(u)$ when and only when it does not have f as a coordinate, or when and only when it has a coordinate $\varphi \subset u$. Since $\{\Phi(f)\}$ is a closed base for $\mathfrak{S}$, $\{\Omega(u)\}$ is an open base. From (27.1), $\cdots$, (27.4) follows then by dualization:

$$(27.6) \qquad \Omega(\emptyset) = \emptyset, \qquad \Omega(\mathfrak{R}) = \mathfrak{S}, \qquad u \neq u' \leftrightarrow \Omega(u) \neq \Omega(u');$$

$$(27.7) \qquad \Omega(\bigcup u_a) = \bigcup \Omega(u_a),$$

whenever $\{u_a\}$ is finite;

(27.8) $$u \subset u' \rightarrow \Omega(u) \subset \Omega(u');$$

(27.9) $$\Omega(u \cap u') = \Omega(u) \cap \Omega(u').$$

Further properties of the $\Omega(u)$ follow.

(27.10) *If* $\mathfrak{U} = \{u_1, \cdots, u_r\}$ *is a finite open covering of* $\mathfrak{R}$ *then* $\Omega(\mathfrak{U}) = \{\Omega(u_i)\}$ *is one for* $\mathfrak{S}$ *and in addition*: (a) $\mathfrak{U} < \mathfrak{U}' \rightarrow \Omega(\mathfrak{U}) < \Omega(\mathfrak{U}')$; (b) order $\Omega(\mathfrak{U}) =$ order $\mathfrak{U}$; (c) $u_i \rightarrow \Omega(u_i)$ *defines a similitude* $\mathfrak{U} \rightarrow \Omega(\mathfrak{U})$.

The covering property is a consequence of (27.6) and (27.7), while (a), (b), (c) follow from (27.6, $\cdots$, 27.9).

(27.11) *Any finite open covering* $\mathfrak{B}$ *of* $\mathfrak{S}$ *has a finite refinement* $\Omega(\mathfrak{U})$.

Since $\{\Omega(u)\}$ is a base for $\mathfrak{S}$ there is a refinement $\{\Omega(u_{1\alpha})\} = \Omega(\mathfrak{U}_1)$ of $\mathfrak{B}$. Since $\mathfrak{S}$ is compact $\Omega(\mathfrak{U}_1)$ contains a finite subcovering $\Omega(\mathfrak{U})$ which refines $\mathfrak{B}$.

28. All the elements for the proof of the compacting theorem are now at hand. We have already shown that $\mathfrak{S}$ is compact and we have a univalent transformation $T: \mathfrak{R} \rightarrow \mathfrak{S}$. Let $T\mathfrak{R} = R$. If $x \in u$ then $X(x) = Tx$ has the coordinate x in u, and so $X(x) \in \Omega(u)$. Conversely, $X(x) \in \Omega(u)$ implies that $X(x)$ has a coordinate φ in u, and since $x \in \varphi$, likewise $x \in u$. Therefore $x \in u \leftrightarrow X(x) \in \Omega(u) \cap R$. It follows that T induces a one-one transformation of the elements of the base $\{u\}$ for $\mathfrak{R}$ into those of the base $\{\Omega(u) \cap R\}$ for R. Consequently T imbeds $\mathfrak{R}$ topologically as a subset R of $\mathfrak{S}$.

Every $u \neq \emptyset$ contains at least one point x of $\mathfrak{R}$ and so $\Omega(u)$ contains $X(x) = Tx \in R$. Therefore $\Omega(u)$ meets R and $\mathfrak{S} - R$ contains no $\Omega(u)$, hence no open set since $\{\Omega(u)\}$ is a base. It follows that $\mathfrak{S} = \bar{R}$. Thus the imbedding is dense.

Let finally dim $\mathfrak{R} = n$ and let $\mathfrak{B}$ be a finite open covering of $\mathfrak{S}$. By (27.11) it has a finite refinement $\Omega(\mathfrak{U})$, where $\mathfrak{U}$ is a finite open covering of $\mathfrak{R}$. Since dim $\mathfrak{R} = n$, $\mathfrak{U}$ has a refinement $\mathfrak{U}'$ of order not exceeding n, and $\Omega(\mathfrak{U}')$ is a refinement of $\mathfrak{B}$ of order likewise not exceeding n. Therefore dim $\mathfrak{S} \leqq n$. This completes the proof of the theorem.

29. **Locally compact spaces.** The compacting process just given, while very general, usually provides a far more involved space than one would wish to have. Consider, for example, the interval λ: $0 < x < 1$, and let f, f' be two infinite convergent sequences tending towards 0 or 1, but having no common terms. Each is the coordinate of a m.b.s., and the two m.b.s., say X, X' thus obtained must be distinct since $f \cap f' = \emptyset$. Thus in the case under consideration the space $\mathfrak{S}$ of (26.1) is such that $\mathfrak{S} - R$ contains at least as many points as there are disjoint sequences $\rightarrow 0, 1$. On the other hand if C is a circumference and $y \in C$, λ is topologically equivalent to $R = C - y$, and so C is a compacting space such that $C - R$ is a point. This is a special case of a theorem which we shall now prove. First a

(29.1) Definition. A *topological space* $\mathfrak{R}$ *is said to be locally compact whenever every point* x *of* $\mathfrak{R}$ *has a neighborhood* N *whose closure* $\bar{N}$ *is compact. Thus the interval, the real line, indeed any Euclidean space, are locally compact but not compact. They show that the locally compact class is very extensive.*

(29.2) *A n.a.s.c. for $\mathfrak{R}$ to be locally compact is the existence of an open base whose elements have compact closures.*

Sufficiency is obvious. Suppose $\mathfrak{R}$ locally compact and let $\{U\}$ be a base and $\{V\}$ the open sets with compact closures. Since $U \cap V \subset V$, we have $\overline{U \cap V} \subset \bar{V}$, and hence $\overline{U \cap V}$ is compact. If W is any open set and $x \in W$, there is a U between x and W and a $V \ni x$. Hence $U \cap V$ is an open set between x and W whose closure is compact, and so $\{U \cap V\}$ is a base whose sets have compact closures. This proves necessity and hence also (29.2).

(29.3) THEOREM. *A locally compact space $\mathfrak{R}$ may be compacted by the addition of a single point y.*

Let F denote the closed sets of $\mathfrak{R}$. Define the closed sets of $\mathfrak{R}' = \mathfrak{R} \cup y$ as all the sets $F \cup y$ and all the sets F which are compact. The verification for $\mathfrak{R}'$ of the conditions CSi of (7) is a consequence of the same for $\mathfrak{R}$, and so $\mathfrak{R}'$ is a topological space. Since the closed sets of $\mathfrak{R}$ are the intersections with $\mathfrak{R}$ of those of $\mathfrak{R}'$, $\mathfrak{R}$ is topologically imbedded in $\mathfrak{R}'$. Let $\{f_a\}$ be a collection of closed sets of $\mathfrak{R}'$ possessing the finite intersection property. Separate the f_a into two groups. The first made up of sets f_b which are compact, and also closed sets of $\mathfrak{R}$ itself. The second group consists of sets f_c such that $f_c = F_c \cup y$, where F_c is closed in $\mathfrak{R}$. Suppose that there exist sets f_b, and let f_{b_0} be one of them. We have $\bigcap f_a = \bigcap (f_a \cap f_{b_0})$. The sets $f_a \cap f_{b_0}$ are closed in the compact set f_{b_0} and their collection has the finite intersection property. Hence their intersection is non-empty and the same holds for $\{f_a\}$.

Suppose now that there are no sets f_b. We have then

$$\bigcap f_a = \bigcap f_c = \bigcap (F_c \cup y) \ni y \neq \emptyset.$$

Since $\bigcap f_a \neq \emptyset$, $\mathfrak{R}'$ is compact. The theorem is therefore proved.

(29.4) REMARK. Since local compactness has not been utilized in the proof, the theorem is valid for any topological space $\mathfrak{R}$. However, if $x \in \mathfrak{R}$ has the neighborhood N and $\bar{N}$ is not compact then $\bar{N} \ni y$. Hence when $\mathfrak{R}$ is not locally compact the open sets of $\mathfrak{R}'$ do not behave very well and so the theorem is of value only in the locally compact case.

§6. SEPARATION AXIOMS

30. The theory developed so far rests exclusively upon the axioms OSi in which the points are nowhere mentioned. Thus the points have merely been the primitive elements of which the sets considered are composed. To express it in another way the properties with which open or closed sets have been implemented do not as yet enable us to distinguish between the individual points by means of these sets: It may well happen that there exist pairs of distinct points x, y such that every open or closed set containing one of the two also contains the other. This is certainly remote from the situation in the familiar spaces, where usually the points are closed, and where in fact no two are on the same total aggregate of open sets.

We require then a suitable "separation" axiom for the points. The most frequently utilized are the following which we describe in the "T_i-nomenclature" of Alexandroff-Hopf [A–H, 58]:

AXIOM T_0. *Of each pair of distinct points at least one has a neighborhood which does not contain the other.*

AXIOM T_1. *Each point of every pair of distinct points has a neighborhood which does not contain the other.*

AXIOM T_2. *(Hausdorff's separation axiom.) Every pair of distinct points have disjoint neighborhoods.*

A topological space which verifies Axiom T_i is known as a *T_i-space*, although T_2-spaces are commonly called *Hausdorff spaces*. They include all the spaces of classical geometry and analysis.

The following proposition is an immediate consequence of the T_i-axioms together with the principle of relativization:

(30.1) *The subsets of a T_i-space are T_i-spaces.*

Again from the T_i-axioms together with the definition of topological products we deduce:

(30.2) *A product of T_i-spaces is a T_i-space.*

In our ascending scale of axioms the following property shows that with the T_1-class points begin to assume their customary properties:

(30.3) *A n.a.s.c. in order that the points of a topological space $\mathfrak{R}$ be closed sets is that $\mathfrak{R}$ be a T_1-space.*

For let x, y be any two distinct points of $\mathfrak{R}$. A n.a.s.c. for x to be closed is that $\mathfrak{R} - x$ be open. Since y is merely any point of $\mathfrak{R} - x$ a n.a.s.c. is that given any $y \neq x$ there exist an open set U between y and $\mathfrak{R} - x$, i.e., such that $y \in U$, $x \in \mathfrak{R} - U$. In other words the required condition is that $\mathfrak{R}$ satisfies Axiom T_1.

We prove also for an ulterior purpose:

(30.4) *If $\mathfrak{S}$ is a Hausdorff space and T is a mapping $\mathfrak{R} \to \mathfrak{S}$, then the graph G of T in $\mathfrak{R} \times \mathfrak{S}$ is closed.*

Let $(x_0, y_0) \in G$, or $y_0 = Tx_0$ and let $(x_0, y_1) \notin G$, or $y_1 \neq y_0$. Since $\mathfrak{S}$ is a Hausdorff space y_0, y_1 have disjoint neighborhoods V_0, V_1. Since T is continuous x_0 has a neighborhood U such that $x \in U \to Tx \in V_0$ hence $Tx \notin V_1$. Hence $(U \times V_1) \cap G = \emptyset$ or (x_0, y_1) has the neighborhood $U \times V_1$ free from points of G. Therefore $\mathfrak{R} \times \mathfrak{S} - G$ is open and G is closed.

EXAMPLES. (30.5) All the examples considered hitherto in the chapter except ordered spaces (9.3) are T_2-spaces. We state explicitly that *cells, spheres, parallelotopes, Euclidean and projective spaces as well as all their subsets, finally discrete spaces, are all T_2-spaces.*

(30.6) The following example essentially due to Alexandroff-Urysohn [a], describes a space which is T_1 but not T_2. The space is the real line L: $-\infty < x < +\infty$. For $x_0 \leqq 0$ a base at x_0 consists of the intervals with the center x_0. For $x_0 > 0$ a base $\{U\}$ at x_0 is made up as follows: U is an interval $0 < a < x < b$ of center x_0 together with the interval $-b < x < -a$ with the center $-x_0$ removed. It is readily seen that as between x_0 and $-x_0$ Axiom T_1 holds but T_2 fails to hold.

(30.7) Any ordered space which contains at least one ordered pair: $x \prec x'$, is a T_0-space but not a T_1-space. Thus the real line L with the points ordered as in (9.4) by $\leqq$ is a T_0-space but not even a T_1-space. Under its customary topology however L is a T_2-space.

31. **Limits.** A few words about this important concept will not be out of place. Let $\mathfrak{R}$ be a topological space. A sequence $\{x_n\}$ of points of $\mathfrak{R}$ is said to have for *limit* the point x of $\mathfrak{R}$ or to *tend* or *converge* to x, written $\{x_n\} \to x$, or $x_n \to x$, whenever corresponding to any neighborhood U of x there is an integer p such that $n > p \rightarrow x_n \in U$. When $\{x_n\}$ has a limit it is said to be *convergent.*

(31.1) THEOREM. *Let $\mathfrak{R}$ have countable bases at each point and let T be a transformation $\mathfrak{R} \to \mathfrak{R}'$. Then a n.a.s.c. for T to be continuous, i.e., that it be a mapping, is that if $\{x_n\} \to x$ then $\{Tx_n\} \to Tx$. This is the situation in particular when $\mathfrak{R}$ has a countable open base.*

The proof of necessity is elementary and requires no restriction on the bases. Conversely, suppose the condition fulfilled and yet T fail to be continuous. There exist then x and $x' = Tx$, with a neighborhood U' of x' such that x is not an interior point of $T^{-1}U'$. We may construct a countable base $\{U_n\}$ at x such that $U_{n+1} \subset U_n$. Then U_n contains a point $x_n \notin T^{-1}U'$. Therefore $x_n \to x$ and yet $Tx_n \nrightarrow x'$, contrary to assumption. This proves (31.1).

(31.2) THEOREM. *In a Hausdorff space limits are unique.*

Suppose that a sequence $\{x_n\}$ converges to two distinct limits x, x'. There exist disjoint neighborhoods U, U' of x, x' such that for n sufficiently high $x_n \in U \cap U' = \emptyset$ which is absurd. This proves (31.2).

32. **Compact subsets of Hausdorff spaces.** Many of the important and better known characteristic properties of compact sets appear first in the Hausdorff class.

(32.1) *A compact subset A of a Hausdorff space $\mathfrak{R}$ is closed.*

Let $x \in \mathfrak{R} - A$ and $y \in A$. There exist disjoint open neighborhoods $U_y(x)$ of x and $U_x(y)$ of y. Since A is compact and has the open covering $\{A \cap U_x(y)\}$ there is a finite subcovering. Hence there is a finite set $\{y_i\}$ such that $A \subset V = \bigcup U_x(y_i)$. If $W = \bigcap U_{y_i}(x)$, we have then $A \subset V$, $x \in W$, $V \cap W = \emptyset$, and since W is open, so is $\mathfrak{R} - A$ which implies (32.1).

(32.2) *A continuous image of a compact space into a Hausdorff space is closed* (23.2, 32.1).

(32.3) *A continuous transformation of a compact space into a Hausdorff space is a closed transformation* (32.2).

(32.4) *A one-one mapping of a compact space onto a Hausdorff space is topological; hence both spaces must be compact Hausdorff spaces.*

For T, the mapping, is continuous and closed hence both T and T^{-1} are continuous.

(32.5) *If A and B are disjoint compact subsets of a Hausdorff space, then they have disjoint neighborhoods.*

Let $y \in A$, $x \in B$. As shown in the proof of (32.1) there exist disjoint neighborhoods V_x of A, and $W(x)$ of x (the V, W there considered). Since $\{B \cap W(x)\}$ covers the compact set B, it has a finite subcovering $\{B \cap W(x_i)\}$. Hence this time $V = \bigcap V_{x_i}$ and $W = \bigcup W(x_i)$ are disjoint neighborhoods of A, B.

33. **Normality.** The separation axioms alone are in general not powerful enough to reach down to the usual spaces. For example, they do not suffice to characterize metric spaces. For this reason further restrictions are required and one of the most important, given presently, is a separation axiom for closed sets analogous to Hausdorff's axiom. The basic definition is:

(33.1) DEFINITION. *A topological space $\mathfrak{R}$ is said to be normal whenever every two disjoint closed sets F, F' have disjoint neighborhoods: $F \subset U$, $F' \subset U'$, $U \cap U' = \emptyset$.*

In point of fact normality, like compactness, is a *primitive* concept, in the sense that it may likewise be expressed without reference to the other properties of open or closed sets. One must also bear in mind that normality does not imply, nor is implied by any one of the separation axioms T_i. Of course mutual relations do exist. Thus if $\mathfrak{R}$ is T_1 and normal it is necessarily a Hausdorff space.

The dual form of (33.1) is

(33.2) *If $\{U, U'\}$ is an open covering of $\mathfrak{R}$ then there exists a closed covering $\{F, F'\}$ such that $F \subset U$ and $F' \subset U'$.*

(33.3) DEFINITION. *Given an open covering $\mathfrak{U} = \{U_\alpha\}$ of $\mathfrak{R}$, if there exists for each α an open set V_α such that $\bar{V}_\alpha \subset U_\alpha$, and that $\mathfrak{V} = \{V_\alpha\}$ is a covering, we shall say that $\mathfrak{U}$ has been shrunk to $\mathfrak{V}$, also that $\mathfrak{U}$ is shrinkable.*

A stronger result than (33.2) is:

(33.4) *Every point-finite (in particular every finite or locally finite) open covering $\mathfrak{U}$ of a normal space $\mathfrak{R}$ is shrinkable.*

(a) *$\mathfrak{U}$ is finite.* Although the proof for this case is covered by the general proof, it is so simple, that we give it first. Let $\mathfrak{U} = \{U_1, \cdots, U_n\}$. Since the closed sets $F = \mathfrak{R} - U_1$, $F' = \mathfrak{R} - (\bigcup \{U_i \mid i \neq 1\})$ are disjoint, they have disjoint open neighborhoods U, V_1. From $V_1 \subset \mathfrak{R} - U$ follows $\bar{V}_1 \subset \mathfrak{R} - U \subset U_1$. Since $\mathfrak{R} - (\bigcup \{U_i \mid i \neq 1\}) \subset V_1$ we know that $\{V_1, U_2, \cdots, U_n\}$ is an open covering. We proceed to shrink the U_i, $i \neq 1$, in the same way, proving the theorem.

(b) *General case.* Let $\mathfrak{U} = \{U_\alpha\}$, $A = \{\alpha\}$ and let $\varphi(\alpha)$ be a function on A such that: (a) $\varphi(\alpha) = U_\alpha$ or else $\varphi(\alpha) = V_\alpha$, $\bar{V}_\alpha \subset U_\alpha$; (b) $\{\varphi(\alpha)\}$ is a covering.

Order $\Phi = \{\varphi\}$ by the relation: $\varphi < \varphi'$ whenever $\varphi'(\alpha) = \varphi(\alpha)$ if $\varphi(\alpha) = V_\alpha$. It is readily shown that if $\Phi' = \{\varphi'\} \subset \Phi$ is simply ordered, and $\varphi''(\alpha) = \bigcap \varphi'(\alpha)$ then $\varphi'' \in \Phi$ and $\varphi'' = \sup \Phi'$. Therefore by Zorn's theorem Φ contains an element φ_1 such that $\varphi > \varphi_1 \rightarrow \varphi = \varphi_1$. It remains to be shown that $\varphi_1(\alpha) = V_\alpha$ for every α. Suppose indeed that $\varphi_1(\beta) = U_\beta$ and set $F = \mathfrak{R} - \bigcup\{\varphi_1(\alpha) \mid \alpha \neq \beta\}$. We show as above that there is an open set V_β such that $F \subset V_\beta$, $\bar{V}_\beta \subset U_\beta$. Hence φ_2 such that $\varphi_2(\alpha) = \varphi_1(\alpha)$, $\alpha \neq \beta$, and $\varphi_2(\beta) = V_\beta \neq \varphi_1$ and $> \varphi_1$. This contradiction proves that $\varphi_1(\beta) = V_\beta$ and (33.4) follows.

Notice that point-finiteness is required to prove that φ'' is a covering: Let $U_{\alpha_1}, \cdots, U_{\alpha_r}$ be the sets U_α containing x. From some φ_0' on, none of them will be modified so that if $\varphi' > \varphi_0'$ then $\varphi'(\alpha_i) = \varphi_0'(\alpha_i) = \varphi''(\alpha_i) = U_{\alpha_i}$ or V_{α_i}. Since φ_0' is a covering, for some $\alpha_i : x \in \varphi_0'(\alpha_i) = \varphi''(\alpha_i)$.

A direct generalization of the property of (33.1) is:

(33.5) *If $\{F_i\}$ is a finite collection of closed sets in the normal space $\mathfrak{R}$, there can be found for each F_i an open set $U_i \supset F_i$ such that $F_i \cap \cdots \cap F_j = \emptyset \leftrightarrow U_i \cap \cdots \cap U_j = \emptyset$.*

Suppose first that $\{G_i\}$ is a finite collection of nonintersecting closed sets. Then $\{(\mathfrak{R} - G_i)\}$ is a finite open covering and so by (33.4) there exist open sets W_i such that $\bar{W}_i \subset \mathfrak{R} - G_i$ and that $\{W_i\}$ is a covering. Therefore $U_i = \mathfrak{R} - \bar{W}_i$ is an open set such that $G_i \subset U_i$ and that $\bigcap U_i = \emptyset$.

In the general case consider all the combinations $\alpha = \{\alpha_1, \cdots, \alpha_j\}$ of indices such that $\bigcap F_{\alpha_i} = \emptyset$. By the result just obtained there exist corresponding neighborhoods $U^\alpha_{\alpha_i}$ of the F_{α_i} such that $\bigcap U^\alpha_{\alpha_i} = \emptyset$. For a given i let $U_i = \bigcap\{U^\alpha_{\alpha_i} \mid \alpha_i \in \alpha\}$. Clearly $\{U_i\}$ is such that $F_i \subset U_i$ and that $F_i \cap \cdots \cap F_j = \emptyset \rightarrow U_i \cap \cdots \cap U_j = \emptyset$. The implication in the other direction is obvious.

(33.6) *A compact Hausdorff space is normal* (23.1, 32.5).

34. **Urysohn's characteristic function.** Normality is intimately related to the existence of nonconstant real continuous functions. Let A, B be disjoint sets and f a real continuous function on $\mathfrak{R}$ whose values on A, B are constant and distinct. Then for suitable constants a, b the function $a + b \arctan f$ has the same properties and takes its values in the segment $[0 - 1]$. Urysohn has considered more particularly continuous functions f on $\mathfrak{R}$ to $[0-1]$ such that $f(A) = 0$, $f(B) = 1$. If such a function exists it is called a *characteristic function* of the pair (A, B). Of particular importance is:

(34.1) Urysohn's lemma. *Normality is equivalent to the existence of a characteristic function for every pair of disjoint closed sets (F, F').*

Suppose that F, F' have the characteristic function f and set $U = \{x \mid f(x) < 1/4\}$, $U' = \{x \mid f(x) > 3/4\}$. Since f is continuous U, U' are open and as they are thus disjoint neighborhoods of F, F', $\mathfrak{R}$ is normal.

Conversely, let $\mathfrak{R}$ be normal and F, F' disjoint closed sets. There exists an open set $U(1/2)$ such that $F \subset U(1/2)$, $\bar{U}(1/2) \subset \mathfrak{R} - F'$. We treat similarly the pairs of disjoint closed sets $(F, \mathfrak{R} - U(1/2))$, $(\bar{U}(1/2), F')$, and thus obtain

$U(1/4)$, $U(3/4)$, etc. This produces an open set $U(t)$ for every dyadic proper fraction $t = m/2^n$ with the property: $t < t' \rightarrow \overline{U(t)} \subset U(t')$. Given any $x \in \mathfrak{R}$ we define $f(x) = y$, where $y = \sup \{t \mid x \notin U(t)\}$. The function f is single-valued and has the proper range. Furthermore $f = 0$, at F, $f = 1$ at F'. If $\lambda = ab$ is a subinterval of $0 - 1$, or else closed at an end point a, b which is then 0 or 1, we have $f^{-1}(\lambda) = \bigcup\{U(t) \mid t < b\} - \bigcap\{\overline{U(t)} \mid t > a\} =$ an open set. Therefore f is continuous, and the lemma is proved.

A noteworthy application of Urysohn's lemma is the proof (after Alexandroff-Hopf [A–H, 75]) of:

(34.2) TIETZE'S EXTENSION THEOREM. *Any mapping f of a closed subset F of a normal space $\mathfrak{R}$ into an n-parallelotope P^n or into the Hilbert parallelotope P^ω has an extension $\varphi: \mathfrak{R} \rightarrow P^n$ or P^ω as the case may be.*

If P^n or P^ω are referred to coordinates $x_1, x_2, \cdots$, each of these will be a real continuous single-valued function on F and (34.2) will follow from:

(34.3) *Any real continuous single-valued function $f(x)$ on F has an extension $\varphi(x)$ to $\mathfrak{R}$: $\varphi \mid F = f$, which is also real continuous and single-valued.*

Evidently f may be replaced by $1/2 + (1/\pi) \arctan f$ whose values are in the segment $[0 - 1]$. Therefore we may assume that this is already the case for f. Given any two disjoint closed sets A, B in $\mathfrak{R}$, we denote their characteristic function in $\mathfrak{R}$ by $\Phi(A, B; x)$. Functions $\{f_n, \varphi_n\}$, where the range of the f_n is F, and the range of the φ_n is $\mathfrak{R}$, are now introduced as follows:

$$f_0 = f, f_{n+1} = f_n - \varphi_n;$$

setting now $F_n = \{x \mid f_n(x) \leqq (1/3)(2/3)^n\}$, $F'_n = \{x \mid f_n(x) \geqq (2/3)(2/3)^n\}$, we have two disjoint closed sets, and take

$$\varphi_n = \tfrac{1}{3}(\tfrac{2}{3})^n \Phi(F_n, F'_n; x).$$

These relations yield a determination of the functions in the order f_0, φ_0, f_1, $\varphi_1, \cdots$, and an elementary recurrence leads to the inequalities:

$$0 \leqq \varphi_n \leqq \tfrac{1}{3}(\tfrac{2}{3})^n, \qquad 0 \leqq f_n \leqq (\tfrac{2}{3})^n. \tag{34.4}$$

Introduce now

$$s_n(x) = \varphi_0 + \cdots + \varphi_n.$$

From (34.4) follows that the series $\sum \varphi_n(x)$ is uniformly convergent on $\mathfrak{R}$, and since the φ_n are continuous and single-valued $\lim s_n(x)$ exists and is a continuous and single-valued function $\varphi(x)$ on $\mathfrak{R}$. From the relation

$$s_n(x) = f(x) - f_{n+1}(x), \qquad x \in F,$$

follows then that $\varphi \mid F = f$. This proves (34.3) and hence (34.2).

35. **Tychonoff spaces.** These spaces originally called *completely regular* by their discoverer Tychonoff are given by the

(35.1) DEFINITION. *A Tychonoff space* $\mathfrak{R}$ *is a Hausdorff space such that for every point* x *and neighborhood* U *of* x *there is a characteristic function of the pair* $x, \mathfrak{R} - U$.

We verify immediately:

(35.2) *Every subset of a Tychonoff space is a Tychonoff space.*

(35.3) *Every normal Hausdorff space is a Tychonoff space.*

The following two definitions are designed to introduce two important concepts needed immediately:

(35.4) DEFINITION. *Let* $\mathfrak{R}$ *be a topological space,* $\{\mathfrak{R}_a\}$ *an indexed system of topological spaces and for each* a *let* f_a *be a function on* $\mathfrak{R}$ *to* $\mathfrak{R}_a$. *Then* $\kappa = \{f_a\}$ *is said to be a separating class for* $\mathfrak{R}$ *whenever for any two distinct points* x, y *of* $\mathfrak{R}$ *there is an* a *such that* $f_a(x) \neq f_a(y)$.

(35.5) DEFINITION. *Under the same conditions let* $\{V_{a\lambda}\}$ *be the open sets of* $\mathfrak{R}_a$. *Then* κ *is said to be a basic class for* $\mathfrak{R}$ *whenever* $\{f_a^{-1}V_{a\lambda}\}$ *is a subbase for* $\mathfrak{R}$.

(35.6) *When* $\mathfrak{R}$ *is a* T_0*-space, a basic class* κ *is necessarily separating.*

Since $\mathfrak{R}$ is a T_0-space if $x \neq y$ there is an open set U such that say $x \in U$, $y \notin U$, and hence an $f_a^{-1}V_{a\lambda}$ such that $x \in f_a^{-1}V_{a\lambda}$, $y \notin f_a^{-1}V_{a\lambda}$ and so clearly $f_a(x) \neq f_a(y)$.

Returning to Tychonoff spaces we prove:

(35.7) LEMMA. *Let* $\mathfrak{R}$ *be a Tychonoff space. Then the class* $\kappa = \{f_a\}$ *of all continuous functions on* $\mathfrak{R}$ *to the segment* $[0 - 1]$ *is a basic class for* $\mathfrak{R}$.

Consider the f_a as mapping $\mathfrak{R}$ on the segment $0 \leqq y \leqq 1$. Since f_a is continuous the set $V_a = f_a^{-1}\{y \mid 0 \leqq y < 1\}$ is open. Take now any open set U of $\mathfrak{R}$, and $x \in U$ and let f_a be the characteristic function of $x, \mathfrak{R} - U$. Evidently $x \in V_a \subset U$. Therefore $\{V_a\}$ is a base and κ is basic.

The fundamental theorem for the spaces under consideration is:

(35.8) THEOREM. *Every Tychonoff space* $\mathfrak{R}$ *may be mapped topologically into a compact parallelotope and every subset of such a parallelotope and indeed of any compact Hausdorff space is a Tychonoff space* (Tychonoff).

Let $\kappa = \{f_a\}$ be as in (35.7) and set $A = \{a\}$. For each a introduce a segment $l_a : 0 \leqq y_a \leqq 1$ and set $P^A = \mathbf{P}l_a$. If $y = \{y_a\}$, $y_a = f_a(x)$, then $x \to y$ defines a transformation $T: \mathfrak{R} \to P^A$.

(a) T *is univalent* (35.6, 35.7).

(b) T *is open.* For $\{V_a\}$ being as before, $TV_a = \{y \mid y_a < 1\} \cap T\mathfrak{R}$ is open in $T\mathfrak{R}$ and since $\{V_a\}$ is a base, T is open.

(c) T *is continuous* (12.2).

Properties (a), (b), (c) prove that T imbeds $\mathfrak{R}$ topologically in P^A.

Since every compact Hausdorff space $\mathfrak{R}$ is normal (33.6), by (35.3) and (35.2) $\mathfrak{R}$ and its subsets are Tychonoff spaces.

(35.9) *Remark.* The proof of (35.8) goes through step by step if κ is replaced by any subclass $\kappa_0 = \{f_b\}$ such that $\{V_b\}$ is a base.

36. **Separation properties and compacting.** The influence of separation on the two compacting processes that have been given is described in the two propositions to follow.

(36.1) Theorem. *The space* $\mathfrak{R}$ *and the compact space* $\mathfrak{S}$ *of the general compacting theorem* (26.1) *are related in their separation properties as follows: When* $\mathfrak{R}$ *is* T_0, T_1, *or Hausdorff normal, so is* $\mathfrak{S}$.

Let the notations be those of (26, 27, 28). Take two distinct points X, X' of $\mathfrak{S}$ both in $T\mathfrak{R} = R$. That is to say, $X = X(x)$, $X' = X(x')$, $x \neq x'$. Suppose first $\mathfrak{R}$ to be T_0. There exists then an open set u of $\mathfrak{R}$ containing say x but not x'. Since x, x' are coordinates of X, X' we have $X \in \Omega(u)$, $X' \notin \Omega(u)$, and so the T_0-axiom holds for the pair (X, X'). Suppose now $X \notin R$. There exist disjoint coordinates φ_a, φ_b' of X, X'. If $\varphi_a = f_a \cup x_1 \cup \cdots \cup x_r$, by (26.4) f_a is also a coordinate of X and $f_a \cap \varphi_b' = \emptyset$. Hence if $u = \mathfrak{R} - f_a$ then $X \notin \Omega(u)$, $X' \in \Omega(u)$ and the situation is as before. Therefore $\mathfrak{S}$ is T_0.

Suppose now $\mathfrak{R}$ to be T_1. Since the points of $\mathfrak{R}$ are closed sets all the coordinates φ_a are closed sets. If $X \neq X'$ they have disjoint coordinates f_a, f_b' and so $\Omega(\mathfrak{R} - f_b')$, $\Omega(\mathfrak{R} - f_a)$ are neighborhoods of X, X' which are, respectively, free from X', X. Therefore $\mathfrak{S}$ is T_1.

Suppose finally $\mathfrak{R}$ to be normal Hausdorff. We have again disjoint coordinates f_a, f_b' of X, X'. Since $\mathfrak{R}$ is normal there exist disjoint open neighborhoods u, u' of f_a, f_b'. Therefore $\Omega(u)$, $\Omega(u')$ are disjoint neighborhoods of X, X' in $\mathfrak{S}$ and so $\mathfrak{S}$ is Hausdorff. Since it is compact it is also normal, and the proof of (36.1) is completed.

37. For locally compact spaces we have the stronger

(37.1) Theorem. *When the locally compact space* $\mathfrak{R}$ *of* (29.3) *is a* T_i*-space so is the associated compact space*

$$\mathfrak{R}' = \mathfrak{R} \cup y.$$

We may as well exclude at the outset the trivial case of $\mathfrak{R}$ compact. We denote by $\mathfrak{U} = \{U\}$ the collection of all the open sets U of $\mathfrak{R}$ such that $\overline{U}$ is compact.

(37.2) *A set* U *is also open in* $\mathfrak{R}'$.

For $F = \mathfrak{R} - U$ is not compact, else $\mathfrak{R}$ would be the union of the two compact sets F, $\overline{U}$, and hence compact (23.3). It follows that $F' = F \cup y$ is closed in $\mathfrak{R}'$ and so $U = \mathfrak{R}' - (F' \cup y)$ is open in $\mathfrak{R}'$.

(37.3) $\mathfrak{U}$ *is a base for* $\mathfrak{R}$.

Let V be an open set of $\mathfrak{R}$ and $x \in V$. Then some U contains x and $\overline{V \cap U} \subset \overline{U} \rightarrow V \cap U \in \mathfrak{U}$. Since $x \subset V \cap U \subset V$, $\mathfrak{U}$ is a base.

The proof of our theorem is now a simple matter. Any point $x \in \mathfrak{R}$ has a neighborhood U. The set $\overline{U}$ is thus closed in $\mathfrak{R}'$ and so $V = \mathfrak{R}' - \overline{U}$ is a neighborhood of y. By (37.2) U is also a neighborhood of x in $\mathfrak{R}'$. Since $U \cap V = \emptyset$, Axiom T_2 holds for the pair (x, y).

Suppose now $\mathfrak{R}$ to be T_i and let x, x' be distinct points of $\mathfrak{R}$. The T_i condition may be fulfilled with neighborhoods out of any base for $\mathfrak{R}$, and in particular out of the base $\mathfrak{U}$ of (37.3). Since the elements of $\mathfrak{U}$ are open in $\mathfrak{R}'$ also, the T_i condition is fulfilled for the pair (x, x'). Since T_2 is fulfilled for the pairs (x, y) so is T_i. Therefore $\mathfrak{R}'$ is a T_i-space.

§7. INVERSE MAPPING SYSTEMS

38. The spaces which are to be introduced here are especially important in the applications to homology. For our purposes it will be quite sufficient to restrict the treatment to Hausdorff spaces.

Let then $\{\mathfrak{R}_\lambda\}$ be a system of Hausdorff spaces indexed by a directed set $\Lambda = \{\lambda; >\}$ and suppose that whenever $\lambda > \mu$ there is given a mapping, also known as a *projection*, $\pi_\mu^\lambda : \mathfrak{R}_\lambda \rightarrow \mathfrak{R}_\mu$ such that $\lambda > \mu > \nu \rightarrow \pi_\nu^\mu \pi_\mu^\lambda = \pi_\nu^\lambda$. The system $\Sigma = \{\mathfrak{R}_\lambda ; \pi_\mu^\lambda\}$ of the $\mathfrak{R}_\lambda$ and the π_μ^λ is called an *inverse mapping system.*

Let $\mathfrak{R}^* = \mathbf{P}\mathfrak{R}_\lambda$ and in $\mathfrak{R}^*$ let $\mathfrak{R}$ be the set of all the points $x = \{x_\lambda\}$ such that $\lambda > \mu \rightarrow \pi_\mu^\lambda x_\lambda = x_\mu$. We call $\mathfrak{R}$ the *limit-space* of the inverse mapping system Σ.

Notice incidentally that for $\lambda < \lambda$ we have $\pi_\lambda^\lambda x_\lambda = x_\lambda$ or $\pi_\lambda^\lambda = 1$.

EXAMPLE. $\{\mathfrak{R}_\lambda\}$ is a sequence $\{C_n\}$ of circumferences, where C_n is the image of a real variable x_n reduced mod 1. Choose $\pi_n^{n+1} = k_n x$, k_n an integer, and define $\pi_n^{n+p} = \pi_n^{n+1} \cdots \pi_{n+p-1}^{n+p}$. Then $\Sigma = \{C_n ; \pi_n^{n+p}\}$ is an inverse mapping system. Its limit-space $\mathfrak{R}$, introduced by Vietoris, is known as a *solenoid.*

As a subset of $\mathfrak{R}^*$ the limit-space $\mathfrak{R}$ receives the relative topology and by (30.1) and (30.2)

(38.1) *The limit-space $\mathfrak{R}$ is a Hausdorff space.*

(38.2) The topology of the limit-space $\mathfrak{R}$ is frequently described as follows: For each open set $U_{\lambda a}$ of $\mathfrak{R}_\lambda$ introduce the set $V_{\lambda a} = \{x \mid x \in \mathfrak{R}, x_\lambda \in U_{\lambda a}\}$ and choose $\{V_{\lambda a}\}$ as a base for the topology. This topology is readily identified with the relative topology. Define in fact $V_{\lambda a}^* = \{x \mid x \in \mathfrak{R}^*, x_\lambda \in U_{\lambda a}\}$. Then $V_{\lambda a} = V_{\lambda a}^* \cap \mathfrak{R}$, and since $\{V_{\lambda a}^*\}$ is a subbase for $\mathfrak{R}^*$, $\{V_{\lambda a}\}$ is one for $\mathfrak{R}$ in the relative topology. Furthermore given $V_{\mu b}$, $V_{\nu c}$ let $\lambda > \mu, \nu$. Owing to the continuity of the projections $(\pi_\mu^\lambda)^{-1} U_{\mu b}$, $(\pi_\nu^\lambda)^{-1} U_{\nu c}$ are open, and so is their intersection $U_{\lambda a}$. From this follows that $V_{\mu b} \cap V_{\nu c} = V_{\lambda a}$, so that $\{V_{\lambda a}\}$ is actually a base for the relative topology, making the identity of the two topologies obvious.

The set $\mathfrak{R}$ as a subset of $\mathfrak{R}^*$ may be viewed as the graph of the set of relations $\pi_\mu^\lambda x_\lambda = x_\mu$. We find therefore the natural analogue of the property (30.4) for graphs:

(38.3) *The limit-space $\mathfrak{R}$ is closed in the product space $\mathfrak{R}^*$.*

For $\lambda > \mu$ introduce $S_\mu^\lambda = \{x \mid \pi_\mu^\lambda x_\lambda = x_\mu\}$, and let G_μ^λ be the graph of π_μ^λ. We have: $S_\mu^\lambda = G_\mu^\lambda \times \mathbf{P}_{\nu \neq \lambda, \mu} \mathfrak{R}_\nu$. Since G_μ^λ is closed in $\mathfrak{R}_\lambda \times \mathfrak{R}_\mu$ (30.4), S_μ^λ is closed in $\mathfrak{R}^*$ (12.6). Therefore $\mathfrak{R} = \bigcap S_\mu^\lambda$ is also closed in $\mathfrak{R}^*$.

From (23.1, 38.3) and (24.1) we deduce:

(38.4) *When the $\mathfrak{R}_\lambda$ are compact so is the limit-space $\mathfrak{R}^*$.*

39. We now come to the important:

(39.1) THEOREM. *If the $\mathfrak{R}_\lambda$ are compact and not empty then the limit-space is likewise not empty* (Steenrod [a]).

Let the notations be those of (38.3). Since $\mathfrak{R}^*$ is compact we only need to prove

(39.2) $\{S_\mu^\lambda\}$ *has the finite intersection property.*

Given a finite set $\{S_{\mu_i}^{\lambda_i}\}$, $i = 1, 2, \cdots, r$, choose $\lambda_0 > \lambda_1, \cdots, \lambda_r$. Since $\lambda_i > \mu_i$ we also have $\lambda_0 > \mu_i$. Take any $x_{\lambda_0} \in \mathfrak{R}_{\lambda_0}$, define $x_{\lambda_i} = \pi_{\lambda_i}^{\lambda_0} x_{\lambda_0}$, $x_{\mu_i} = \pi_{\mu_i}^{\lambda_0} x_{\lambda_0}$, and let x be any point of $\mathfrak{R}^*$ with the coordinates x_{λ_i}, x_{μ_i}. From $\pi_{\mu_i}^{\lambda_0} = \pi_{\mu_i}^{\lambda_i} \pi_{\lambda_i}^{\lambda_0}$ there follows $\pi_{\mu_i}^{\lambda_i} x_{\lambda_i} = x_{\mu_i}$, and so $x \in S_{\mu_i}^{\lambda_i}$. This proves (39.2) and hence also (39.1).

A complementary property is:

(39.3) *If the $\mathfrak{R}_\lambda$ are compact and x_μ is such that for every $\lambda > \mu$ the set $(\pi_\mu^\lambda)^{-1} x_\mu \neq \emptyset$ then $\mathfrak{R}$ contains a point x with the coordinate x_μ.*

If π_μ is the natural projection $\mathfrak{R}^* \to \mathfrak{R}_\mu$ then $F = \pi_\mu^{-1} x_\mu$ is closed in $\mathfrak{R}^*$ and (39.3) reduces to: $F \cap \mathfrak{R} \neq \emptyset$, and hence to:

(39.4) $\{F, S_\mu^\lambda\}$ *has the finite interection property.*

Choose this time $\lambda_0 > \lambda_1, \cdots, \lambda_r, \mu$. By hypothesis $(\pi_\mu^{\lambda_0})^{-1} x_\mu \neq \emptyset$, and so we may take x_{λ_0} in that set. We now take x as before save that in addition its μ coordinate is to be x_μ, a condition which may manifestly be fulfilled. Thus $x \in F$ and still $x \in S_{\mu_i}^{\lambda_i}$. Hence (39.4) holds and (39.3) follows.

40. Let $\mathrm{M} = \{\mu; >\}$ be a directed subsystem of $\Lambda = \{\lambda; >\}$. Then the spaces $\mathfrak{R}_\mu$ and projections $\pi_{\mu'}^\mu$ give rise to a new inverse mapping system $\Sigma_1 = \{\mathfrak{R}_\mu ; \pi_{\mu'}^\mu\}$ known as a *partial system* of Σ. If M is cofinal in Λ we say that Σ_1 is *cofinal* in Σ.

Let Σ_1 be a partial system of Σ and $\mathfrak{S}$ its limit-space. If $x = \{x_\lambda\} \in \mathfrak{R}$ the coordinates x_μ of x determine a unique point x' of $\mathfrak{S}$. The transformation $\tau: \mathfrak{R} \to \mathfrak{S}$ whereby $\tau x = x'$ is known as the *projection of $\mathfrak{R}$ into $\mathfrak{S}$.*

(40.1) *Let Σ_1 with limit-space $\mathfrak{S}$ be a partial system of Σ. Then*: (a) *the projection $\tau: \mathfrak{R} \to \mathfrak{S}$ is a mapping*; (b) *when Σ_1 is cofinal in Σ then τ is topological, so that $\mathfrak{R}$, $\mathfrak{S}$ are then topologically equivalent.*

Let $\mathfrak{S}^* = \mathbf{P}\mathfrak{R}_\mu$. We have then $\mathfrak{R}^* = \mathfrak{S}^* \times \mathbf{P}_{\lambda \neq \mu} \mathfrak{R}_\lambda$ and the projection $\tau^*: \mathfrak{R}^* \to \mathfrak{S}^*$ is continuous. Now if $x = \{x_\lambda\} \in \mathfrak{R}$ the point $\tau^* x$ is the point of $\mathfrak{S}^*$ whose coordinates are the μ coordinates of x, i.e., $\tau^* x = \tau x$. Hence $\tau = \tau^* \mid \mathfrak{R}$, and so τ is also continuous.

Suppose now that Σ_1 is cofinal in Σ. We shall show that τ is a one-one mapping of $\mathfrak{R}$ onto $\mathfrak{S}$. Given $x' = \{x_\mu\} \in \mathfrak{S}$ if $x = \{x_\lambda\} \in \mathfrak{R}$ is to be such that

$\tau x = x'$ then we must have $x_\lambda = \pi_\lambda^\mu x_\mu$, $\mu > \lambda$. Choose some $\mu > \lambda$ and let x_λ be defined by this relation. If $\mu_1 > \mu > \lambda$ then $\pi_\lambda^{\mu_1} x_{\mu_1} = \pi_\lambda^\mu(\pi_\mu^{\mu_1} x_{\mu_1}) = \pi_\lambda^\mu x_\mu$ so that μ_1 yields the same value of x_λ as μ. Take any two indices μ_1, μ_2. There is a $\mu > \mu_1$, μ_2 and since μ, μ_1 and μ, μ_2 yield the same value for x_λ, so do μ_1, μ_2. Therefore x_λ is unique. If $\lambda > \lambda'$ choose $\mu > \lambda$. From $x_{\lambda'} = \pi_{\lambda'}^\mu x_\mu = \pi_{\lambda'}^\lambda \pi_\lambda^\mu x_\mu = \pi_{\lambda'}^\lambda x_\lambda$ follows that $\{x_\lambda\}$ are in fact the coordinates of a point $x \in \mathfrak{R}$. Since this point has the coordinates x_μ of x' we have $\tau x = x'$. Thus every point x' is the image under τ of one and only one x and so τ is a one-one mapping. To prove τ topological there remains to show that it is open. Let $U_{\lambda a}$, $V_{\lambda a}$ be as in (38.2). Then $x \in V_{\lambda a} \rightarrow x_\lambda \in U_{\lambda a}$. Choose any $\mu > \lambda$. Since π_λ^μ is continuous $(\pi_\lambda^\mu)^{-1} U_{\lambda a} = U_{\mu b}$ is open in $\mathfrak{R}_\mu$ and $V'_{\mu b} = \{x' \mid x_\mu \in U_{\mu b}\}$ is open in $\mathfrak{S}$. Since $\tau V_{\lambda a} = V'_{\mu b}$, τ is open and hence topological. This proves (40.1).

Two inverse mapping systems are said to be *equivalent* if there exists a third in which both are cofinal. From (40.1) we deduce:

(40.2) *Equivalent inverse mapping systems have topologically equivalent limit-spaces.*

41. When $\{\lambda\}$ is a sequence it is more convenient to use 1, 2, $\cdots$ as the indices. We write therefore $\mathfrak{R}_n$ for $\mathfrak{R}_{\lambda_n}$ and π_n^{n+1} for $\pi_{\lambda_n}^{\lambda_{n+1}}$, and require that for $n < p < q$ we have $\pi_n^p \pi_p^q = \pi_n^q$. The system $\{\mathfrak{R}_n ; \pi_n^p\}$ is then called an *inverse mapping sequence.* From (4.4) and (40.2) we see that

(41.1) *When* $\{\lambda\}$ *is countable,* $\mathfrak{R} = \lim \{\mathfrak{R}_\lambda ; \pi_\mu^\lambda\}$ *is topologically equivalent either to some* $\mathfrak{R}_\lambda$ *or to the limit-space of an inverse mapping sequence cofinal in* $\{\mathfrak{R}_\lambda ; \pi_\mu^\lambda\}$.

It is clear that if $\{U_{\lambda a}\}$ is a base for each $\mathfrak{R}_\lambda$, then $\{V_{\lambda a}\}$ (in the notation of 38) is a base for $\mathfrak{R}$. Thus we have

(41.2) *When* $\{\lambda\}$ *is countable and the* $\mathfrak{R}_\lambda$ *all have countable bases,* $\mathfrak{R} = \lim \{\mathfrak{R}_\lambda ; \pi_\mu^\lambda\}$ *has likewise a countable base.*

§8. METRIZATION

42. Guided by the Euclidean situation, given a point set R we call *distance-function* or merely *distance* a real function $d(x, y)$ defined for all $x, y \in R$ and possessing the following properties:

D1. $d(x, y) = 0$ *when and only when* $x = y$.

D2. (*Triangle axiom*): $d(y, z) \leqq d(x, y) + d(x, z)$.

From D12 we derive, with Lindenbaum, the other two noted properties of the distance:

D3. $d(x, y) \geqq 0$.

D4. $d(x, y) = d(y, x)$.

We prove D3 by making $z = y$ in D2. Regarding D4 making $x = z$ in D2 and taking account of D1 we have $d(y, x) \leqq d(x, y)$. Since the inequality may be proved also in reverse order D4 follows.

The set R with an associated distance-function $d(x, y)$ is called a *metric space.* We also say that $d(x, y)$ defines a *metric* for R.

We shall now discuss the first properties of metric spaces.

(42.1) *Subsets of a metric space.* Let $\mathfrak{R}$ be a metric space with the distance $d(x, y)$. If $A \subset \mathfrak{R}$ and we take $x, y \in A$, $d(x, y)$ becomes a suitable distance-function for A, so that A is metric also.

(42.2) *Distance between two sets. Diameter of a set. Spheres.* The *distance* $d(A, B)$ between two subsets A and B of $\mathfrak{R}$ is inf $\{d(x, y) \mid x \in A, y \in B\}$. The *diameter* of A (diam A) is sup $\{d(x, y) \mid x, y \in A\}$. The set of all points x such that $d(x, A) < \epsilon$ is called an ϵ *neighborhood* of A and denoted by $\mathfrak{S}(A, \epsilon)$. When $A = x_0$, a single point, $\mathfrak{S}(x_0, \epsilon)$ is commonly called a *spheroid* or *sphere*; the point x_0 is the *center* of the spheroid and ϵ its *radius*. The analogy with Euclidean spherical regions is obvious.

(42.3) ϵ *aggregates*, ϵ *coverings*, ϵ *transformations.* This type of designation with appropriate variations is frequently convenient. The *mesh* of an aggregate is the supremum of the diameters of its sets. An ϵ aggregate or covering is one whose mesh is less than ϵ. If a space has a finite ϵ covering for every $\epsilon > 0$ it is said to be *totally bounded.* An ϵ transformation of a metric space $\mathfrak{R}$ into a set $Q = \{y\}$ is a transformation such that mesh $\{T^{-1}y\} < \epsilon$.

(42.4) In terms of the spheres we may define *regions* as in Euclidean geometry: U is a region whenever $x \in U$ implies $\mathfrak{S}(x, \epsilon) \subset U$ for some $\epsilon > 0$.

43. The chief justification for considering metric spaces at this juncture lies in the

(43.1) THEOREM. *If a metric space is topologized by choosing regions as open sets it becomes a normal Hausdorff space with a countable base at each point.*

The verification of OS123 (6) is immediate. If $x \neq y$, then $d(x, y) = \epsilon > 0$. Hence $\mathfrak{S}(x, \epsilon/3)$ and $\mathfrak{S}(y, \epsilon/3)$ are disjoint neighborhoods of x and y. If A and B are disjoint closed sets, then $\{x \mid 3d(x, A) < d(x, B)\}$, $\{x \mid 3d(x, B) < d(x, A)\}$, are disjoint open sets containing A and B. Thus the space is a normal Hausdorff space. Clearly $\{\mathfrak{S}(x_0, 1/n)\}$ is a countable base at the point x_0.

(43.2) *The spheroids form a base, and those of center x form a base at x.*

On the strength of (43.2) we shall say that two distinct metrics are *equivalent* whenever if $\mathfrak{S}(x, \epsilon)$, $\mathfrak{S}'(x, \epsilon)$ are the corresponding spheroids then given ϵ and x there is an η such that $\mathfrak{S}'(x, \eta) \subset \mathfrak{S}(x, \epsilon)$, and *vice versa.* That is to say the two metrics are equivalent if they induce the same topology in $\mathfrak{R}$.

It is easy to see that

(43.3) *The distance-function in $\mathfrak{R}$ defines a topology in the subsets which is in accordance with the principle of relativization.*

A topological space $\mathfrak{R}$ is said to be *metrizable* whenever it is possible to assign it a *metric*, i.e., a distance function $d(x, y)$ inducing the topology of the space. *Metrization* is the process of assigning a metric to a metrizable space. Frequently for shortness a space is described as metric when it is merely metrizable. In each case the context shows clearly what is meant.

(43.3a) *The closure $\bar{A}$ is the set of all points x such that $d(x, A) = 0$.*

(43.4) *Limits, continuity.* Since metric spaces are Hausdorff spaces with a countable base in each point, limits in such spaces are unique (31.2) and continuity of mappings of metric spaces into one another may be expressed in terms of limits as in (31.1). Moreover in so far as such spaces alone are involved all questions of continuity and convergence may be dealt with by the "ϵ, δ" method of classical analysis. Many, if not most of the well known concepts of the latter may be introduced here also. We merely recall the useful concept of *uniform continuity*: $\mathfrak{R}$, $\mathfrak{R}'$ being metric the mapping $T: \mathfrak{R} \to \mathfrak{R}'$ is said to be uniformly continuous whenever if $x \in \mathfrak{R}$ and $x' = Tx$ then given any $\epsilon > 0$, there is an $\eta > 0$ independent of x such that $T\mathfrak{S}(x, \eta) \subset \mathfrak{S}(x', \epsilon)$.

(43.5) *Completeness.* Since $\{\mathfrak{S}(x, \epsilon\}$ is a base at x, $\{x_n\} \to x$ whenever $\{d(x, x_n)\} \to 0$. As is well known a necessary condition in order that $\{x_n\}$ converge (to some point) is that Cauchy's condition hold: given any $\epsilon > 0$ there is an n such that $p, q > n \to d(x_p, x_q) < \epsilon$. Whenever Cauchy's condition implies the convergence of any sequence for which it holds the space $\mathfrak{R}$ is said to be *complete*.

(43.6) *Separability.* A space is said to be *separable* whenever it has a countable dense subset. This property is only of interest in connection with metrization, and largely owing to:

(43.7) *For a metrizable space $\mathfrak{R}$ separability is equivalent to the existence of a countable base.*

For this reason we shall call such spaces "separable metric."

At all events whether $\mathfrak{R}$ is metric or otherwise, when it has a countable base $\{U_n\}$ we may choose a point x_n on U_n, and since $\{x_n\}$ is a countable dense set, $\mathfrak{R}$ is separable. Conversely, let $\mathfrak{R}$ be metric with the countable dense set $\{x_n\}$. To show that $\mathfrak{R}$ has a countable base it will be sufficient to show that $\mathfrak{B} = \{\mathfrak{S}(x_n, \rho_p)\}$, ρ_p rational, which is composed of a countable number of spheroids, is a base. Let U be any neighborhood of x. There is a spheroid $\mathfrak{S}(x, \rho) \subset U$. Since the x_n are dense in $\mathfrak{R}$ we may find an x_n such that $d(x, x_n) < \rho/4$, then choose ρ_p between $\rho/4$ and $\rho/2$. As a consequence $x \in \mathfrak{S}(x_n, \rho_p) \subset \mathfrak{S}(x, \rho) \subset U$. Therefore $\mathfrak{B}$ is a base.

(43.8) *Metric product.* Let $\{\mathfrak{R}_n\}$ be a countable collection of metrizable spaces. Choose a distance $d_n(x, y)$ for $\mathfrak{R}_n$ and metrize the product $\mathfrak{R} = \mathbf{P}\mathfrak{R}_n$ (for the present merely set-product) as follows. If $x = (x_1, \cdots)$, $y = (y_1, \cdots)$ are two points of $\mathfrak{R}$ we choose a distance-function

$$d(x, y) = \sum k_n \frac{d_n(x_n, y_n)}{1 + d_n(x_n, y_n)} \tag{43.9}$$

where $\sum k_n$ is any convergent series of strictly positive terms. For instance we may take $k_n = 2^{-n}$ but any other choice will do. In particular if the number of factors $\mathfrak{R}_n$ is finite we may choose

$$d(x, y) = \sum d_n(x_n, y_n).$$

The verification of D12 is elementary, so that $d(x, y)$ defines a metric for $\mathfrak{R}$

and $\mathfrak{R}$ is metrizable. We call $\mathfrak{R}$ with the metric (43.9) a *metric product* of the $\mathfrak{R}_n$. We shall now prove

(43.10) THEOREM. *The metric (43.9) determines the same topology as previously assigned to the topological product. In particular it is independent of the special choice of $\{k_n\}$ and of the metrics d_n.*

Since the spheroids form a base on each $\mathfrak{R}_n$, we may choose as a base for $\mathfrak{R}$ the sets $U = \mathbf{P}U_n$, where $U_n = \mathfrak{R}_n$ for $n > m$, and U_n, $n \leqq m$, is a spheroid of $\mathfrak{R}_n$. Given $x \in U$ there is a similar $U'_n \ni x$, $U' \subset U$, with spheroidal factors $U'_n = \mathfrak{S}(x_n, r_n)$, $n \leqq m$, where the x_n are the projections of x. Let $r = \inf r_n$, $k = \inf k_n$ for $n \leqq m$ and let $R = kr/(1 + r)$. Then if $y \in \mathfrak{S}(x, R)$, we have from (43.9)

$$\delta_n = k_n \frac{d_n(x_n, y_n)}{1 + d_n(x_n, y_n)} < \frac{kr}{1 + r}, \qquad n \leqq m,$$

and hence $d(x_n, y_n) < r$. Therefore $y_n \in U_n$, $y \in U$, and finally $\mathfrak{S}(x, R) \subset U' \subset U$. Thus there is a spheroid between x and U.

Conversely, given $\mathfrak{S}(x, R)$ we may choose m so large that the remainder of $\sum k_n$ after m terms is less than $R/2$, then take $U_n = \mathfrak{S}(x_n, r_n)$, $r_n < R/2mk_n$ when $n \leqq m$. As a consequence $y \in U \rightarrow d(x, y) < R$ and hence $x \in U \subset \mathfrak{S}(x, R)$. Therefore $\{U\}$ and $\{\mathfrak{S}(x, R)\}$ are equivalent bases for $\mathfrak{R}$.

44. **Examples.**

(44.1) The n-dimensional number space $\mathfrak{E}^n$ referred to the coordinates $x_1, \cdots, x_n$ has the distance-function

$$d(x, y) = [\sum (x_i - y_i)^2]^{1/2}. \tag{44.2}$$

The space with this metric is an Euclidean n-space. It is an elementary matter to identify the resulting topology with that of (9.1).

If we consider $\mathfrak{E}^n$ as the product of the n lines of the variables x_i, metrized by $d_i = | x_i - y_i |$, the method of (43.10) leads for $\mathfrak{E}^n$ to the distance-function

$$d'(x, y) = \sum | x_n - y_n | \tag{44.3}$$

which yields again the same topology as (9.1), and hence as (44.2). Therefore for topological purposes the two metrics are interchangeable.

(44.4) Since $\mathfrak{E}^n$ is metrizable so are all its subsets.

(44.5) Consider the Hilbert parallelotope $P^\omega = \mathbf{P}l_n$, and let l_n be parametrized as: $0 \leqq x_n \leqq 1/n$. Thus P^ω is now the set of all sets: $(x_1, x_2, \cdots)$, $0 \leqq x_n \leqq 1/n$. By (43.10) P^ω is metrizable and admits the distance-function

$$d(x, y) = \sum \frac{| x_n - y_n |}{n^2}. \tag{44.6}$$

Consider on the other hand the Euclidean metric for P^ω defined by

$$d'(x, y) = [\sum (x_n - y_n)^2]^{1/2}. \tag{44.7}$$

If $\mathfrak{S}(x, \epsilon)$ and $\mathfrak{S}'(x, \epsilon)$ are the spheroids corresponding to the two metrics, it is readily shown that for fixed x each $\mathfrak{S}(x, \epsilon)$ contains an $\mathfrak{S}'(x, \eta)$, and conversely. Hence $\{\mathfrak{S}'(x, \epsilon)\}$ is a base for P^ω, and so $d'(x, y)$ defines an admissible metric for the Hilbert parallelotope. It is in fact the metric customarily assigned to it.

45. **Compacta.** A compactum is a compact metric space. The category of compacta partakes therefore of the combined advantages of compactness and metrizability. Its importance is sufficiently indicated if we observe that closed and bounded subsets of Euclidean spaces are compacta.

(45.1) *Every compactum $\mathfrak{R}$ is separable and hence possesses a countable base.*

Define a set A as ϵ *dense* in $\mathfrak{R}$ whenever every point of $\mathfrak{R}$ is nearer than ϵ to A. I say that we merely have to prove that there exists in $\mathfrak{R}$ a countable set A which is ϵ dense whatever ϵ. For in that case every sphere $\mathfrak{S}(x, \rho)$ about a given point x will contain points of A. As these spheres form a base for x, we will have $x \epsilon \bar{A}$, hence $\bar{A} = \mathfrak{R}$.

Now $\{\mathfrak{S}(x, \epsilon)\}$, $x \epsilon \mathfrak{R}$, is an open covering of the compact space $\mathfrak{R}$. Hence there is a finite subcovering $\{\mathfrak{S}(x_i, \epsilon)\}$. Clearly the set $A(\epsilon) = \{x_i\}$ is ϵ dense. The set $A = \bigcup A(1/n)$ is countable and ϵ dense for each $\epsilon > 0$.

(45.2) *If F is closed in $\mathfrak{R}$ and $x \epsilon \mathfrak{R} - F$ then $d(x, F) > 0$* (43.3a).

(45.3) *If $\mathfrak{F} = \{F_1, \cdots, F_r\}$ is a finite aggregate of nonintersecting closed sets in the compactum $\mathfrak{R}$ there is a constant $c(\mathfrak{F}) > 0$ such that every $x \epsilon \mathfrak{R}$ is at a distance not less than $c(\mathfrak{F})$ from at least one F_i.*

For otherwise whatever n there are points x whose distance from every F_i is not more than $1/n$. The set G_n of all such points is closed and $G_n \supset G_{n+1}$. Since $\mathfrak{R}$ is compact and $\bigcap G_n \neq \emptyset$ it contains a point x. Clearly $d(F_i, x) = 0$ and since F_i is closed $x \epsilon F_i$. Therefore $x \epsilon \bigcap F_i \neq \emptyset$ contrary to assumption. This proves (45.3).

(45.4) *If F, F' are closed in the compactum $\mathfrak{R}$ and $d(F, F') = 0$ then F and F' intersect.*

For otherwise if $x \epsilon F$ we have $d(x, F') \geqq c(F, F') > 0$, and hence $d(F, F') \geqq c(F, F') > 0$.

(45.5) *For every finite aggregate of closed sets $\mathfrak{F}$ in the compactum $\mathfrak{R}$ there exists a positive constant $d(\mathfrak{F})$ called the Lebesgue number of $\mathfrak{F}$, such that if $A \subset \mathfrak{R}$,* diam $A < d(\mathfrak{F})$, *and A meets a collection of sets of $\mathfrak{F}$, then these sets have a nonvacuous intersection.*

Let $\mathfrak{F}_i$, $i = 1, 2, \cdots, s$, be the subaggregates of $\mathfrak{F}$ whose sets do not meet, and let $d(\mathfrak{F}) = \inf c(\mathfrak{F}_i)$. If A behaves as stated it cannot meet the sets of $\mathfrak{F}_i$, since otherwise $x \epsilon A$ would imply that x is nearer than $c(\mathfrak{F}_i)$ to every set of $\mathfrak{F}_i$, which contradicts the definition of $c(\mathfrak{F})$. This proves (45.5).

(45.6) *For every finite open covering $\mathfrak{U} = \{U_i\}$ of the compactum $\mathfrak{R}$ there exists a positive constant $d_1(\mathfrak{U})$ called the Lebesgue number of $\mathfrak{U}$, such that:* (a) *every point x of $\mathfrak{R}$ is on some set U_i and at a distance at least $d_1(\mathfrak{U})$ from $\mathfrak{R} - U_i$;* (b) *if $A \subset \mathfrak{R}$ and* diam $A < d_1(\mathfrak{U})$ *then A is in some set U_i.*

Since $\bigcup U_i = \mathfrak{R}$ we have $\bigcap(\mathfrak{R} - U_i) = \emptyset$. Therefore $d_1(\mathfrak{U}) = d(\mathfrak{R} - U_1, \cdots)$ has property (a). If A is chosen in accordance with (b), and $x \epsilon A$ then for some i property (a) holds and hence $A \subset U_i$.

(45.7) *A continuous mapping f of a compactum $\mathfrak{R}$ into a metric space $\mathfrak{S}$ is uniformly continuous.*

By (23.2) the values of f make up a compactum R in $\mathfrak{S}$. Given then any $\epsilon > 0$ there is a finite open ϵ covering $\mathfrak{U} = \{U_n\}$ of R. It follows that $\mathfrak{B} = \{f^{-1}U_n\}$ is a finite open covering of $\mathfrak{R}$. Let $\eta = d_1(\mathfrak{B})$. If $x' = f(x)$, $y' = f(y)$ and $d(x, y) < \eta$, some $f^{-1}U_n$ contains both x and y and hence some U_n contains both x' and y', which implies $d(x', y') < \epsilon$. Therefore f is uniformly continuous.

The following two properties of compacta are obvious but often useful:

(45.8) *A compactum is totally bounded* (42.3).

(45.9) *A decreasing sequence of closed sets $\{F_n\}$: $F_{n+1} \subset F_n$, has a non-void intersection and if* diam $F_n \to 0$ *the intersection is a point.*

Sequential compactness. A familiar and very important fact in analysis is the close connection between *compactness* and *convergence* (see notably J. Tukey [T]). The specialization to separable metric spaces brings to the fore the

(45.10) DEFINITION. *The space $\mathfrak{R}$ is said to be sequentially compact whenever every sequence $\{x_n\}$ has a subsequence $\{x_{n'}\}$ which converges to a point of $\mathfrak{R}$.*

(45.11) *A compactum is sequentially compact.*

By (45.8) $\mathfrak{R}$ possesses a finite open ϵ covering. The closures of its sets make up a finite ϵ closed covering $\mathfrak{F} = \{F_{11}, \cdots, F_{1r}\}$. Let $\{x_n\} \subset \mathfrak{R}$. One of the F_{1i}, say F_{1i_1}, contains an infinite subsequence $\{x_{1m}\}$. Since F_{1i_1} is a compactum it has an $\epsilon/2$ finite closed covering $\{F_{2i}\}$, one of whose sets F_{2i_2} contains an infinite subsequence $\{x_{2m}\}$ of $\{x_{1m}\}$, etc. By (45.9): $\bigcap F_{ni_n} = x_0$ is a point and clearly

$$\{x_{nn}\} \to x_0 \text{ (diagonal process).}$$

(45.12) *A sequentially compact metric space $\mathfrak{R}$ is a compactum.*

We first prove $\mathfrak{R}$ *separable.* For any ϵ the space has a finite ϵ dense set $A(\epsilon)$. For if this were false we could find a sequence $\{x_n\}$ such that $d(x_m, x_n) \geqq \epsilon$ whatever $m, n, m \neq n$, and no subsequence could converge. It follows that $\bigcup A(1/n)$ is a countable dense set, and so $\mathfrak{R}$ is separable.

Since $\mathfrak{R}$ is separable it has a countable base (43.7). Hence (6.7) an open covering $\{U_a\}$ of $\mathfrak{R}$ has a countable subcovering $\{U'_n\}$. Suppose that the latter has no finite subcovering. Then we may choose an $x_n \in \mathfrak{R} - (U'_1 \cup \cdots \cup U'_n)$. By hypothesis a subsequence $\{x_{n'}\}$ of $\{x_n\}$ has a limit x_0. Since $\{U'_m\}$ is a covering we have $x_0 \in U'_m$ for some m, hence $x_{n'} \in U'_m$ for n' above a certain value. Since this is ruled out (45.12) is proved.

(45.13) *For metric spaces compactness and sequential compactness are equivalent* (45.11, 45.12).

An interesting consequence of (45.11) is:

(45.14) *A compactum is complete.*

46. **Urysohn's metrization theorems.** We have now all the elements necessary for dealing with these classical theorems.

(46.1) THEOREM. *Every Tychonoff space with a countable base can be imbedded topologically in the Hilbert parallelotope P^ω and hence it is metrizable, and for that matter also normal.*

(46.2) THEOREM. *A n. a. s. c. for a Hausdorff space with a countable base to be metrizable is normality.*

(46.3) THEOREM. *Separable metric spaces are those and only those which may be imbedded topologically in P^ω.*

(46.4) THEOREM. *A n. a. s. c. for a compact Hausdorff space to be a compactum is that it possess a countable base.*

PROOF OF (46.1). Referring to (35.8) and (35.9), the mapping considered in (35.8) exists when the base $\{V_a\}$ there considered is replaced by any subcollection forming a base. Now under the hypothesis of (46.1), and by (6.8), there is a countable subcollection $\{V_n\}$ which is a base and the mapping of (35.8) is then into P^ω. This proves (46.1).

PROOF OF (46.2). Since normal Hausdorff spaces are also Tychonoff spaces (35.3) sufficiency is a consequence of (46.1); and necessity follows from (43.1).

PROOF OF (46.3). Since P^ω is a compactum it has a countable base. Hence the subsets of P^ω are metric with a countable base, and therefore also separable. Conversely, if $\mathfrak{R}$ is separable metric it is normal with a countable base and hence by (46.1) it may be imbedded topologically in P^ω.

PROOF OF (46.4). Necessity is a consequence of (45.1). Since a compact Hausdorff space is normal (33.6), sufficiency follows from (46.2).

§9. HOMOTOPY. DEFORMATION. RETRACTION

47. These concepts are important not only in their strict form, but also in view of certain noteworthy algebraic analogues which occur in the theory of complexes.

Homotopy, deformation. The intuitive concept of a deformation or displacement is clear enough. Duly generalized and made fully rigorous it gives rise to the:

(47.1) DEFINITIONS. *Let A, B be topological spaces, and l the segment $0 \leqq u \leqq 1$. Two mappings $t_1, t_2 : A \rightarrow B$ are said to be homotopic whenever there is a mapping T of the product $l \times A \rightarrow B$ such that $T(0 \times x) = t_1x$, $T(1 \times x) = t_2x$, $x \in A$. If $t_1 = 1$, which implies $A \subset B$, then t_2 is a deformation. The set $T(l \times x)$ is the path of x. Whenever the space is metric and the paths are all of diameter less than ϵ we have an ϵ homotopy, or ϵ deformation as the case may be.*

In a more geometric form the images of t_1A and t_2A are homotopic whenever the "cylinder" $l \times A$ may be so mapped in B that its bases agree with the images t_1A, t_2A.

(47.2) *Homotopy is an equivalence relation.*

Homotopy is:

symmetric, for if T is as above then T_1 such that $T_1(u \times x) = T((1 - u) \times x)$ bears the same relation to t_1, t_2 as T but in reverse order;

reflexive, for $T(u \times x) = t_1 x$ is a mapping $l \times A \to A$ making t_1 homotopic to itself;

transitive, for let (t_1, t_2) and (t_2, t_3) be homotopic pairs of mappings $A \to B$ with T', T'' as the analogues of T. Define

$$T(u \times x) = T'((2u) \times x), \qquad 0 \leqq u \leqq 1/2;$$

$$T(u \times x) = T''((2u - 1) \times x), \qquad 1/2 \leqq u \leqq 1.$$

It is clear that $T(u \times x)$ is continuous in $u \times x$. We have at once $T(0 \times x) = t_1 x$, $T(1 \times x) = t_3 x$, and so t_1, t_3 are homotopic which proves transitivity, hence also (47.2).

Since for fixed A, B homotopy is an equivalence there are corresponding classes, which are known as *homotopy-classes.*

(47.3) *Let t_1, t_2 be homotopic mappings $A \to B$ and let t be a mapping $B \to C$. Then tt_1, tt_2 are homotopic mappings $A \to C$.*

The notations being as before tT is a mapping $l \times A \to C$ such that $tT(0 \times x) = tt_1 x$, $tT(1 \times x) = tt_2 x$, proving our assertion.

For mappings into subsets of an Euclidean space $\mathfrak{E}^n$ or parallelotope P a convenient and intuitive sufficiency condition for homotopy is:

(47.4) *The notations being the same suppose that B is a subset of $\mathfrak{R} = \mathfrak{E}^n$ or P. If for every x the points $t_1 x$ and $t_2 x$ coincide or else may be joined by a segment of $\mathfrak{R}$ which is in B then t_1 and t_2 are homotopic.*

For let $\lambda(x) = t_1 x$ when $t_1 x = t_2 x$, and $\lambda(x) =$ the segment joining $t_1 x$, $t_2 x$ when they are distinct. Then (in vector notation)

$$T(u \times x) = (1 - u)(t_1 x) + u(t_2 x)$$

defines a mapping $l \times A \to \mathfrak{R}$ making t_1, t_2 homotopic mappings $A \to \mathfrak{R}$. Since $T(u \times x) \in \lambda(x)$ we have $T(l \times A) \subset B$, so that t_1, t_2 are in fact homotopic as mappings $A \to B$ also.

(47.5) *Retraction.* This convenient concept, formulated by Borsuk, is closely related to homotopy.

(47.6) Definitions. *Let A, B be topological spaces, with $A \subset B$. A retraction of B onto A is a mapping $t: B \to A$ such that $t \mid A = 1$. When t exists A is called a retract of B. If t is a deformation keeping every point x of A fixed (i.e., x is its own path) then t is also called a deformation retraction and A is then said to be a deformation retract of B.*

The notations being the same A is called a *neighborhood retract* of B when it has a neighborhood in B for which it is a retract.

CHAPTER II

ADDITIVE GROUPS

The content of this chapter consists of the group-theoretic material required later. All the groups are assumed topological, and everywhere except in a few places the topology is significant. Particular attention has been paid to the Pontrjagin duality theory and related group multiplication, which we will find most useful in connection with intersections. This need cause no surprise since group multiplication may be considered as obtained by abstraction from the Kronecker index or topological intersections. We have also devoted considerable space to inverse and direct systems and their limit-groups which will come very much to the fore in (VI, VII). In addition a full extension of the Pontrjagin theory is made to vector spaces over a field.

General references: Alexander-Zippin [a], Chevalley [a], Freudenthal [a], van Kampen [a], Pontrjagin [P, b], Steenrod [a], Weil [W].

§1. GENERAL PROPERTIES

1. We are dealing exclusively with abelian groups: they will always be written additively. The groups are denoted by $G, H, \cdots$, their elements by $g, h, \cdots$ with eventual supplementary indices. The zero element (unit element, neutral element) will be written 0.

If A, A' are two subsets of a group G, we denote, respectively, by $A + A'$ and $A - A'$ the sets of elements $\{g + g'\}$ and $\{g - g'\}$, $g \in A$, $g' \in A'$. The set $-A$ is the set of elements $\{-g \mid g \in A\}$. Observe that in this notation $A - A$ is not the element 0 but represents the set of all the differences $g - g'$ where $g, g' \in A$.

If $\{A_\lambda\}$ is an indexed system of subsets of G then $\sum A_\lambda$ is the set $\bigcup(A_{\lambda_1} + \cdots + A_{\lambda_n})$ where $\{\lambda_1, \cdots, \lambda_n\}$ is a finite subset of $\{\lambda\}$. If the A_λ are subgroups, $\sum A_\lambda$ is also a subgroup, namely the smallest subgroup containing all the A_λ.

(1.1) DEFINITION. *Let the group $G = \{g\}$ as a set of elements be assigned a topology turning it into a topological space. Then G thus topologized is called a topological group whenever it is a T_0-space and in addition $g - g'$ is a continuous function of (g, g') in this topology.*

Since it will occur repeatedly it is important to bear in mind that the continuity condition imposed upon $g - g'$ means that it must be a mapping $G \times G \to G$. That is to say, if V is any neighborhood of $g_0'' = g_0 - g_0'$ then there must exist neighborhoods U, U' of g_0, g_0' such that $g \in U, g' \in U' \to g - g' \in V$, or equivalently such that $U - U' \subset V$.

Hereafter all groups are assumed topological. Since the discrete topology always makes any group G a topological group, G may always be assigned a topology turning it into a topological group.

(1.2) DEFINITIONS. *A neighborhood of zero in a topological group G is known as a nucleus of G. A base or subbase at zero will be called a nuclear base or subbase as the case may be.*

EXAMPLES. (1.3) The additive group of the real numbers (real line) with its usual topology is a topological group; likewise the additive group of the real vectors with the usual topology is a topological group. However, in (25) and for certain special purposes an altogether different topology will be assigned to vector spaces, under which they will still be topological groups.

(1.4) Consider the additive group $\mathfrak{P}$ of the reals mod 1. Take a circumference C referred to an angular variable $\theta: 0 \leqq \theta < 2\pi$, and identify $\mathfrak{P}$ with C so that $p \in \mathfrak{P}$ goes into the angle $\theta = 2p\pi \bmod 2\pi$. If the usual topology of C is assigned to $\mathfrak{P}$ it becomes a compact topological group and it is this group that is referred to hereafter as "the group of the reals mod 1," with the notation $\mathfrak{P}$ attached. Its fundamental importance will become clear when we deal with Pontrjagin's duality theory.

2. From the definitions we deduce at once:

(2.1) *The mapping $g \to -g$ is a topological mapping of G onto itself. Likewise for the mapping $g \to g_0 + g$ (g_0 fixed).*

(2.2) *If N is a nucleus so is $-N$. Furthermore if $g_0 \in G$ then $g_0 + N$ is a neighborhood of g_0 and every neighborhood of g_0 is of this form.*

It is a consequence of (2.2) that the nuclei of G determine its topology.

Does a given assignment of open sets or nuclei turn a group G into a topological group? This question is answered by the following two propositions:

(2.3) *Let a group G be topologized by the assignment of a non-empty family $\{U\}$ as its family of open sets. N. a. s. c. in order that G be a topological group under this assignment of open sets are*: (a) *if U is open and $g_0 \in G$ then $g_0 + U$ is also open*; (b) *if $g_0 \neq 0$ there is an open set U such that $0 \in U$, $g_0 \notin U$*; (c) *if $0 \in U$ there is a U' containing* 0 *and such that $U' - U' \subset U$* (Pontrjagin, [P, 54]).

Regarding necessity, (a) is a consequence of (2.1) and (c) is implied by the continuity of $g - g'$. As for (b), since G is a T_0-space there exists an open set U containing one of 0, g_0 but not the other. If $0 \in U$ then (b) holds. If $g_0 \in U$, $0 \notin U$, we may set $U = g_0 + N$, where N is a nucleus $\not\ni -g_0$. Hence $-N$ is a nucleus $\not\ni g_0$ and this is (b). Passing to sufficiency let g_0, g_0' be distinct elements. Then $g_0 - g_0' \neq 0$ and so by (b) there is a U such that $0 \in U$, $g_0 - g_0' \notin U$, or $g_0' \in g_0' + U = U'$, $g_0 \notin U'$. Therefore G is a T_0-space. Whether g_0, g_0' are distinct or not let $U \ni g_0 - g_0'$. Then $(g_0' - g_0) + U$ is open and contains 0. By (c) there is a U' such that $0 \in U'$, $U' - U' \subset (g_0' - g_0) + U$, or $(g_0 + U') - (g_0' + U') \subset U$. Since $g_0 + U'$ and $g_0' + U'$ are open sets containing g_0, g_0', the mapping $(g, g') \to g - g'$ is continuous and (2.3) follows.

(2.4) *Let G be a group. N. a. s. c. in order that a family $\{N\}$ of subsets may serve as a nuclear base for a topology turning G into a topological group are*: (a) *the family is non-empty*; (b) *every* N *contains* 0; (c) *if* $g \in N$ *there is an* N' *such that* $g + N' \subset N$; (d) *the intersection of two sets* N *contains an* N; (e) *if* $g \neq 0$ *there is an* $N \not\ni g$; (f) *for every* N *there is an* N' *such that* $N' - N' \subset N$.

Necessity is immediate. To prove sufficiency we will first show that $\{N + g\}$ may serve as a base for a topology. At all events the union of its sets is G. Let $g_1 \in (N + g_0) \cap (N' + g_0')$. Then $g_1 - g_0 \in N$, $g_1 - g_0' \in N'$. Therefore by (c) there exist N_1, N_1' such that $g_1 - g_0 + N_1 \subset N$, $g_1 - g_0' + N_1' \subset N'$. The intersection $N_1 \cap N_1'$ contains N_2 such that $g_1 + N_2$ is between g_1 and $(g_0 + N) \cap (g_0' + N')$. Hence by (I, 6.3) $\{N + g\}$ may be chosen as a base. We must now verify the conditions of (2.3). Take $g_0 + U$, U open, and $g_1 \in g_0 + U$. Since $g_1 - g_0 \in U$ there is a $g_2 + N$ between $g_1 - g_0$ and U and so $g_1 \in (g_0 + g_2) + N \subset g_0 + U$. Therefore $g_0 + U$ is open or (2.3a) holds. Property (2.3b) is a consequence of (e). Let now $0 \in U$. Again some $g + N$ is between 0 and U. Hence $-g \in N$ and so by (c) we have some $-g + N_1 \subset N$, then by (f) an N' such that $N' - N' \subset N_1$. From this follows $N' - N' \subset U$ which is (2.3c). Thus G has been made a topological group. There remains to show that in its topology $\{N\}$ is a nuclear base. We have just seen that if $0 \in U$ then some $N' - N' \subset U$. This implies that if $g \in N'$ then $g - 0 = g \in U$ or that $N' \subset U$ which proves the asserted base property and also (2.4).

A nucleus N is said to be *symmetrical* whenever $N = -N$.

(2.5) *Every nucleus* N *contains a symmetrical nucleus* (*for example* $N \cap (-N)$).

(2.6) *A topological group is a Hausdorff space.*

That is to say, for a topological group G the separation axiom T_0 implies T_2. Let indeed g, g' be distinct elements of G. Since $g - g' \neq 0$ there is a nucleus $N \not\ni g - g'$, then another N_1 such that $N_1 - N_1 \subset N$. Now $g - g' \notin N \rightarrow g - g' \notin N_1 - N_1 \rightarrow (g + N_1) \cap (g' + N_1) = \emptyset$. Thus g, g' have the disjoint neighborhoods $g + N_1$, $g' + N_1$ which proves (2.6).

A stronger result is:

(2.7) *A topological group is a Tychonoff space* (Pontrjagin).

In view of (2.2) it is only necessary to establish the Tychonoff property for 0. That is to say, we are to show that if N is a nucleus there exists a characteristic function for 0 and the complement M of N in G. We first construct symmetrical nuclei $N_0, N_1, \cdots$ such that $N_0 \subset N$, $N_{k+1} + N_{k+1} \subset N_k$. That this may be done is a consequence of (2.4f) and (2.5). Any rational dyadic number α of the interval $0 - 1$ may be written as a finite dyadic fraction $0.a_1a_2 \cdots a_p$, $a_i = 0, 1$. Corresponding to α let $N(\alpha)$ denote the set of all the elements $\sum a_ig_i$, $g_i \in N_i$. Since $N(\alpha)$ is a union of open sets it is open. Moreover $\alpha < \beta \rightarrow \overline{N(\alpha)} \subset N(\beta)$. To prove it we merely need to show it for $\beta = 0.a_1 \cdots a_p1$. Now if $g \in \overline{N(\alpha)}$ the set $g + N_{p+1}$ meets $N(\alpha)$ or $g \in N(\alpha) - N_{p+1} = N(\alpha) + N_{p+1} = N(\beta)$ which proves our assertion. Referring now to (I, 34.1) we find that $\{N(\alpha)\}$ has precisely the properties required in the construction for a characteristic function of $(0, M)$.

(2.8) *A n. a. s. c. for* G *to be locally compact is that it possess a nucleus* N *whose closure* $\bar{N}$ *is compact.*

For a neighborhood of g is of the form $g + N$. Its closure $g + \bar{N}$ is topologically equivalent to $\bar{N}$ and hence compact when and only when $\bar{N}$ is compact.

3. **Subgroups.** A subgroup of G is a subset H of G such that $H - H \subset H$. Under the principle of relativization H is assigned a definite topology.

(3.1) *A subgroup is a topological group.*

For as a subset of G it is already a T_0-space by (2.6). Moreover since the transformation $\tau: G \times G \to G$ sending (g, g') into $g - g'$ is continuous so is $\tau \mid H \times H: H \times H \to H$ which sends (g, g'), where $g, g' \in H$, into $g - g'$. This proves (3.1).

(3.2) *If H is a subgroup so is $\bar{H}$.*

We must show that $\bar{H} - \bar{H} \subset \bar{H}$ or that $g_1, g_2 \in \bar{H} \to g = g_1 - g_2 \in \bar{H}$, or again that if N is a nucleus then $g + N$ meets H. There is a nucleus N' such that $N' - N' \subset N$. Since $g_i \in \bar{H}$ the set $g_i + N'$ contains an element $g_i' \in H$ and since H is a group $g' = g_1' - g_2' \in H$. Therefore $g' \in g_1 - g_2 + N' - N' \subset g + N$, and since $g' \in H$, (3.2) follows.

As an interesting special result needed later we have:

(3.3) *Every closed proper subgroup G of the group $\mathfrak{P}$ of the reals* mod 1 *is finite and cyclic (i.e., consists of the multiples of a single element).*

Identify $\mathfrak{P}$ with a circumference. About every point $p \in \mathfrak{P}$ there is an arc λ such that λ contains no element $g \in G$ other than p. For otherwise G must contain elements on the arc $0 - 1/n$ whatever n, and so G is dense in $\mathfrak{P}$, hence $\bar{G} = G = \mathfrak{P}$, contrary to assumption. Since G is compact it may be covered by a finite number of arcs each containing at most one g. Thus G is finite. Assuming, as we may, $G \neq 0$ let $g_1 \neq 0$ be an element of G as near to 0 as possible. Then G consists of the multiples of g_1. For if $g_2 \in G$ is not an mg_1 then some $mg_1 - g_2$ is nearer 0 than g_1. Thus G is cyclic.

4. **Homomorphisms.** Let G, H be two groups. The purely algebraic concept of a homomorphism τ of G into H is that of a transformation $\tau: G \to H$ such that $\tau(g - g') = \tau g - \tau g'$. Since we are dealing with topological groups we must ascribe suitable continuity properties to τ. There arise thus a certain number of mixed concepts which we must now examine.

(4.1) Definitions. *A homomorphism $\tau: G \to H$ is a mapping $G \to H$ such that $\tau(g - h) = \tau g - \tau h$. If τ has merely the algebraic property just stated (i.e., is perhaps not continuous) we shall call it a homomorphism in the algebraic sense.*

(4.2) Definitions. *A homomorphism $\tau: G \to H$ which is a topological transformation is called an isomorphism. If τ is merely a one-one homomorphism in the algebraic sense τ is called an isomorphism in the algebraic sense.*

A homomorphism τ is said to be open if it is an open mapping. A univalent open mapping $\tau: G \to H$ is thus an isomorphism of G with a subgroup of H.

(4.3) *Let τ be a homomorphism $G \to H$ in the algebraic sense. If τ is continuous at 0 it is a homomorphism. If moreover τ maps every nucleus of G onto a nucleus of τG, it is an open homomorphism.*

Suppose τ continuous at 0 and let $g_0 \in G$. The mapping $g \to \tau(g - g_0)$ is continuous at g_0 and hence this is also the case for $\tau: g \to \tau g = \tau g_0 + \tau(g - g_0)$. Thus τ is continuous everywhere. A similar proof applies to the second assertion of (4.3).

(4.4) *Let τ be a homomorphism $G \to H$. The set of the elements g such that $\tau g = 0$ is a closed subgroup of G called the kernel of τ. A n. a. s. c. for τ to be univalent is that the kernel reduces to the element* 0.

(4.5) *Let τ and θ be homomorphisms $G \to H$ and $H \to G$. If $\tau\theta = 1_H$, $\theta\tau = 1_G$ (the identity mappings of H, G into themselves) then both τ and θ are isomorphisms.*

It is readily seen that both kernels are zero and then that $\theta = \overset{-1}{\tau}$. Hence τ is an isomorphism and similarly for θ.

5. Factor-groups. Let G be a group and G' a closed subgroup of G. If $g \in G$ the set $g + G'$ is known as the *coset of* g mod G'. If g, g_1 are in the same coset mod G' then they are said to be *congruent* mod G' and this relation is manifestly an equivalence. If h, h_1 are the cosets of g, g_1 mod G' then $h \pm h_1$ are those of $g \pm g_1$ mod G'. Therefore under this addition the cosets form a group H, which, duly topologized as shown in a moment, is known as *the factor-group of G by G' or* mod G', and denoted by G/G'.

The assignment to g of the coset h containing it defines a homomorphism in the algebraic sense π of G onto G/G' known as the *natural projection* or merely the *projection* of G onto the factor-group.

We now define a set $V \in G/G'$ as open whenever $\overset{-1}{\pi} V$ is open in G. If the resulting topology turns G/G' into a topological group it will also automatically turn π into a homomorphism.

The topology assigned to G/G' is immediately seen to make it a topological group. The proof merely requires that we verify the conditions of (2.3).

Let g^* be any coset mod G' and V an open set of G/G'. We have: $\overset{-1}{\pi}(g^* + V) = \bigcup\{g + \overset{-1}{\pi} V \mid g \in g^*\}$ = a union of open sets of G. Therefore $g^* + V$ is open in G/G', or (2.3a) holds.

We notice now that if U is open in G then πU is open in G/G'. For $\overset{-1}{\pi}(\pi U) = G' + U = \bigcup\{g + U \mid g \in G'\}$ = an open set of G.

Let $g^* \in G/G'$ and $\neq 0$ and let g be an element of the coset $\overset{-1}{\pi} g^*$ of G mod G'. Since G' is closed so is $g + G'$. Hence its complement U in G is open and it contains G'. Therefore $\pi U = V$ is a nucleus not containing g^*. This proves (2.3b) for G/G'.

Let V be any nucleus of G/G'. Then $\overset{-1}{\pi} V$ is a nucleus of G. Hence it contains another U_1 such that $U_1 - U_1 \subset \overset{-1}{\pi} V$. Therefore πU_1 is a nucleus of G/G' such that $\pi U_1 - \pi U_1 \subset V$ and this is (2.3c) for G/G'. We have thus proved:

(5.1) *The topology assigned to G/G' turns it into a topological group. It is this topological group which is henceforth denoted by G/G' and called the factor-group of G by G' or* mod G'. *Furthermore since congruence is an equivalence we may identify elements congruent* mod G' (I, 13) *and the resulting group is precisely G/G'.*

We have also obtained, incidentally:

(5.2) *The natural projection* $\pi: G \to G/G'$ *is an open homomorphism of* G *onto* G/G'.

We now prove:

(5.3) *Let* φ *be a transformation of* G *into a topological space* X *such that* $\varphi(g)$ *depends solely upon the coset* g^* *of* g *modulo a closed subgroup* G'. *If we set* $\varphi^*(g^*) = \varphi(g)$ *then* φ^* *is a transformation* $G/G' \to X$ *and we have*: (a) *if* φ *is a mapping* (*i.e., continuous*) *so is* φ^*; (b) *if* φ *is open so is* φ^*; (c) *if* X *is a group and* φ *a homomorphism* $G \to X$ *then* φ^* *is a homomorphism* $G/G' \to X$.

Let U be open in X. If φ is continuous then $\varphi^{-1}(U)$ is open in G and hence $\varphi^{*-1}(U) = \pi\varphi^{-1}(U)$, where π is the natural projection $G \to G/G'$, is open in G/G'. Therefore φ^* is continuous, which proves (a). Assume now φ open and let V be open in G/G'. Then $\pi^{-1}V$ is open in G and so $\varphi^*(V) = \varphi(\pi^{-1}V)$ is open, which proves (b). Finally if $g_1 \in g_1^*$, $g_2 \in g_2^*$ and φ is a homomorphism then φ^* is continuous by (a) and in addition:

$$\varphi^*(g_1^* + g_2^*) = \varphi(g_1 + g_2) = \varphi(g_1) + \varphi(g_2) = \varphi^*(g_1^*) + \varphi^*(g_2^*),$$

so φ^* is a homomorphism, and this is (c).

(5.4) *Let* τ *be a homomorphism* $G \to H$ *and* G', H' *closed subgroups of* G, H. *If* $\tau G' \subset H'$ *there corresponds to* τ *a homomorphism* $\tau': G/G' \to H/H'$ *given as follows: if* π, ω *are the natural projections of* G, H, *onto* G/G', H/H' *then* $\tau'\pi = \omega\tau$. *If* τ *is open so is* τ'.

Since $\omega\tau$ depends solely upon the coset of g mod G', (5.4) is a consequence of (5.3).

(5.5) *Let* G' *be a closed subgroup of* G. *If any two of the groups* G, G', G/G' *are compact so is the third.*

It is already known that when G is compact so is G' (I, 23.1) and hence likewise $G/G' = \pi G$ (I, 23.2) where π is the natural projection $G \to G/G'$. Thus there remains to show that when G' and G/G' are compact so is G. Let $\mathfrak{F} = \{F_a\}$ be an indexed system of closed sets of G with the finite intersection property. We are to prove that $\bigcap F_a \neq \emptyset$. We can obviously augment $\mathfrak{F}$ by the finite intersections of the F_a without disturbing the situation, and so we assume them already in $\mathfrak{F}$.

Now the indexed system $\{\overline{\pi F_a}\}$ of closed sets of G/G' has also the finite intersection property. Since G/G' is compact $\bigcap(\overline{\pi F_a}) \neq \emptyset$ and so it contains an element g^*. The set $\pi^{-1}g^*$ is a coset $g_0 + G'$ of G mod G'. Since addition is a topological operation and G' is closed and compact so is $g_0 + G'$.

Let U be any neighborhood of $g_0 + G'$. We propose to show that F_a *meets* U. If $g \in g_0 + G'$, there is a nucleus N such that $g + N \subset U$ then a symmetrical one N' such that $N' - N' = N' + N' \subset N$. It follows that $g + N'$ is such that if $g' \in g + N'$ then $g' + N' \subset U$. Since $\{g + N'\}$ is an open covering of the compact set $g_0 + G'$ there is a finite subcovering $\{g_i + N'_i\}$, and $g \in g_i + N'_i \to g + N'_i \subset U$. Therefore if $\bigcap N'_i = N_0$ then whatever $g \in g_0 + G'$ we have $g + N_0 \subset U$. Now the union of all such sets $g + N_0$ is a set $\pi^{-1}V$ where V

is a neighborhood of g^* in G/G', and all sets $\pi^{-1}V$ are of the type in question. Therefore to prove that U meets F_a we only need to show that V meets πF_a, and this is a consequence of the fact that $g^* \in \overline{\pi F_a}$.

Since F_a meets every neighborhood U of $g_0 + G'$ it must meet this set itself, otherwise the complement of F_a in G would be a U which F_a would fail to meet.

We have found then that $\{F_a \cap (g_0 + G')\}$ is an indexed system of closed sets in the compact set $g_0 + G'$ with the finite intersection property. Therefore $\bigcap(F_a \cap (g_0 + G')) \neq \emptyset$, hence $\bigcap F_a \neq \emptyset$, and G is compact.

For locally compact groups we may prove the weaker property:

(5.6) *If G is locally compact and G' is a closed subgroup of G then G/G' is likewise locally compact.*

Let π still denote the natural projection $G \to G/G'$. Under the assumption G has a nucleus N with $\bar{N}$ compact. Hence G/G' has the nucleus $N^* = \pi N$. Now: $N^* \subset \pi(\bar{N}) \subset \overline{\pi N}$. Since $\pi(\bar{N})$ is compact in a Hausdorff space, it is closed and since it is between N^* and $\bar{N}^*$ we must have $\pi(\bar{N}) = \bar{N}^*$. Thus N^* is a nucleus of G/G' with compact closure, proving (5.6).

6. Product.

(6.1) Definition. *Let $\{G_\lambda\}$ be a system of groups indexed by $\Lambda = \{\lambda\}$. The topological product $G = \mathbf{P}G_\lambda$ is a Hausdorff space* (I, 30.2). *If $g = \{g_\lambda\}$, $g' = \{g'_\lambda\}$ are elements of G we define $g + g' = \{g_\lambda + g'_\lambda\}$ and $0 = \{0_\lambda\}$ (0_λ is the zero of G_λ). Under these conditions G becomes an additive group. Since g_λ, g'_λ are continuous in (g, g') so is $g_\lambda - g'_\lambda$ and hence also $g - g'$* (I, 12.2). *Therefore under its topology G is a topological group.*

(6.2) *Let $M = \{\mu\} \subset \Lambda$ and $G' = \mathbf{P}G_\mu$. Then the projection $\pi: G \to G'$ is an open homomorphism.*

For π is a homomorphism in the algebraic sense and an open mapping as well (I, 12.1).

(6.3) Example. If Λ is any set, and $\mathfrak{P}$ as in (1.4) then $G = \mathfrak{P}^\Lambda$ is a compact topological group known as *toroidal*. The cardinal number $|\Lambda|$ is the *dimension* of G. (See I, 25.5.)

7. Weak product.

(7.1) Definition. *Consider now a system $\{G_\lambda\}$ of discrete groups indexed by $\Lambda = \{\lambda\}$. The elements of $\mathbf{P}G_\lambda$ which have at most a finite number of coordinates different from 0 form a subgroup G which taken discrete is called the weak product of the G_λ, written $\mathbf{P}^w G_\lambda$.*

(7.2) Remark. If Λ is infinite, the discrete topology is not the topology that G is to receive as a subgroup of the product, and so it is not to be considered as a subgroup of $\mathbf{P}G_\lambda$ in the topological sense.

(7.3) Let $M = \{\mu\} \subset \Lambda$ and $G' = \mathbf{P}^w G_\mu$. To every $g' \in G'$ there may be assigned $g = \{g_\lambda\} = \eta(g') \in G$ defined thus: $g_\mu = g'_\mu$, $g_\lambda = 0$ for $\lambda \notin M$. Clearly

η is an isomorphism of G' with a subgroup of G. It is called the *injection* of G' into G. In particular we may choose M as consisting of the single element λ and we then obtain the injection η_λ of the single group G_λ into G. For simplicity and wherever no confusion arises, we continue to denote by g_λ the element $\eta_\lambda(g_\lambda)$ of G. Under this convention any element $g \in G$ may be written as a finite sum $\sum g_\lambda$ (at most a finite number of terms different from 0).

8. **Chains.** The chains may be considered as a convenient symbolism for dealing with certain products. This symbolism will be used extensively in the theory of complexes, where owing to the similarity with linear forms, it has become traditional.

Let again $\Lambda = \{\lambda\}$ be any set of indices and let $'G = G^\Lambda$. Introduce now a set of symbols $X = \{x_\lambda\}$ called *elementary chains* such that $\lambda \to x_\lambda$ is one-one. The elements of $'G$ may now be represented by the symbolic sums, called *infinite chains*: $C = \sum g_\lambda x_\lambda$, $g_\lambda \in G$, with the convention that

$$\sum g_\lambda x_\lambda \pm \sum g'_\lambda x_\lambda = \sum (g_\lambda \pm g'_\lambda) x_\lambda . \tag{8.1}$$

$'G$ is also referred to as the *group of the infinite chains over G based on X.*

Suppose now the group G *discrete.* We could introduce a "weak Λ power" of G, but it is more convenient to take an indexed collection $\{G_\lambda\}$ in which $G_\lambda = G$ and then the weak power $G^* = \mathbf{P}^w G_\lambda$. We introduce again $X = \{x_\lambda\}$ with $\lambda \to x_\lambda$ one-one. The elements of G^* may now be represented by the finite sums $C = \sum g_\lambda x_\lambda$, $g_\lambda \in G$, called *finite chains*, with the same addition convention (8.1). G^* is then referred to as the *group of the finite chains over G based on X.*

In both types of chains and groups the base X will often be evident from the context and mention of it omitted.

If Λ is finite the two groups $'G$, G^* are in algebraic isomorphism, their chains being finite in both cases. They are then described more adequately as groups *with topology* (for $'G$) *or discrete* (for G^*).

Chain-homomorphisms. Let H, K be two groups of chains over G based, respectively, on $X = \{x_\lambda\}$, $Y = \{y_\mu\}$ where $\Lambda = \{\lambda\}$, $\mathrm{M} = \{\mu\}$. The elements of the two groups are then represented as

$$h = \sum h_\lambda x_\lambda , \; k = \sum k_\mu y_\mu , \qquad h_\lambda , k_\mu \in G. \tag{8.2}$$

Let now $A = \| a_{\lambda\mu} \|$ be a matrix of integers with *finite columns* (for given μ at most a finite number of $a_{\lambda\mu}$ are different from 0). We associate with A the transformation $H \to K$ given by

$$\tau \sum h_\lambda x_\lambda = \sum_\mu \left(\sum_\lambda h_\lambda a_{\lambda\mu}\right) y_\mu . \tag{8.3}$$

(8.4) *τ is a homomorphism $H \to K$.*

It is clearly so in the algebraic sense. The continuity proof is best carried through by means of the powers. We have then $H = G^\Lambda$, $K = G^\mathrm{M}$ and (8.3) states that τ sends the element h with coordinates h_λ into the element k with the coordinates k_μ given by

(8.5) $$k_\mu = \sum_\lambda h_\lambda a_{\lambda\mu}.$$

Now h_λ is a continuous function of h (I, 12.1) and k_μ is a continuous function of the h_λ in (8.5) since their number is finite and group addition is continuous. It follows that k is also a continuous function of h (I, 12.2), and so (8.4) holds.

Suppose now G discrete with H, K as the groups of finite chains over G based on X, Y. This time it will be necessary to impose *finiteness of the rows* upon A (for a given λ at most a finite number of $a_{\lambda\mu}$ are $\neq 0$) and then τ given by (8.3) will still represent a homomorphism $H \to K$. Since the groups H, K are discrete the continuity considerations do not arise.

Homomorphisms of one of the two types just considered will be referred to as *chain-homomorphisms.*

(8.6) We will now consider certain chain-groups required in connection with the so-called "dissections" of a complex (III, 23).

Let Y, X be as before, except that now $Y \subset X$ and let $X' = \{x'_\nu\}$ be the complement of Y in X. The chains $h' = \sum h'_\nu x'_\nu$, $h'_\nu \in G$, make up a group H'. Corresponding to any open set U of G introduce the sets

$$V_\nu = \{h \mid h'_\nu \in U\}, \qquad V'_\nu = \{h' \mid h'_\nu \in U\},$$

$$W_\mu = \{h \mid k_\mu \in U\}, \qquad W'_\mu = \{k \mid k_\mu \in U\}.$$

Let also $'K$ denote the subgroup of H consisting of the elements whose coordinates $h'_\nu = 0$. By (I, 12.6) $'K$ is closed in H. Furthermore $k = \sum k_\mu y_\mu \to \sum 0x'_\nu + \sum k_\mu y_\mu$ defines a homomorphism: $K \to H$ called the *injection* of K into H.

(8.7) *The injection η is an isomorphism of K with $'K$.*

It is clearly an isomorphism in the algebraic sense. Furthermore since $\{V_\nu ; W_\mu\}$ is a subbase for H, $\{'K \cap W_\mu\}$ is one for $'K$. We have then in $\{W'_\mu\}$ a subbase for K such that $\eta W'_\mu = {'K} \cap W_\mu$. Hence η maps into one another the elements of subbases for K, $'K$ and so it is topological, proving (8.7).

(8.8) Taking advantage of (8.7) the identification $k \to \eta k$ induced by η will cause the topological imbedding of K as the closed subgroup $'K$ of H. This identification is assumed henceforth, so that K is now a closed subgroup of H.

Let then $H^* = H/K$. If $h = \sum h'_\nu x'_\nu + \sum k_\mu y_\mu$ then $h' = \sum h'_\nu x'_\nu$ is a representative of the coset h^* of h mod K. Clearly $h^* \to h'$ defines an isomorphism θ of H^* with H' in the algebraic sense.

(8.9) *θ is an isomorphism of H^* with H'.*

If π is the natural projection $H \to H^*$ then $\{\pi V_\nu\}$ is a subbase for H^*. Since $\theta\pi V_\nu = V'_\nu$, θ maps the elements of a subbase for H^* into those of a subbase for H'. Hence θ is topological and (8.9) follows.

§2. GENERATORS OF A GROUP

9. The present section is a digression from topological groups necessitated by the requirements of the theory of complexes. For the present the topology of the groups will not be utilized.

Let $G = \{g\}$ be a group and $B = \{g'\}$ a (finite or infinite) subset of G. The elements g' of B are said to be *linearly independent* whenever there exists no relation between a finite number of elements of B of the form:

$$(9.1) \qquad a_1 g_1' + \cdots + a_r g_r' = 0,$$

where the a_i are integers not all zero. The *rank* of G is the maximum number of linearly independent elements of G. Evidently:

(9.2) *The rank of a finite product is the sum of the ranks of the factors.*

The elements of the set B are said to be *generators* of G if it is possible to express every $g \in G$ in the form

$$(9.3) \qquad g = a_1 g_1' + \cdots + a_r g_r', \qquad g_i' \in B,\ a_i \text{ an integer}.$$

If in (9.3) the coefficients a_i are always unique for a given g, or which is the same, if the generators are independent, then B is said to be a *base* of G. A group possessing a base is said to be *free.*

10. (10.1) *A subgroup $H \neq 0$ of a free group G is a free group.*

Let $B = \{b_\alpha\}$ be a base for G and let $A = \{\alpha\}$ be well-ordered in a definite way. Any $h \in H$ may be written uniquely in the normal form $h = m_1 b_{\alpha_1} + \cdots + m_r b_{\alpha_r}$, $\alpha_1 < \cdots < \alpha_r$. We call α_r the *index* of h. Given α if there is an h of index α choose one, b_α', such that in the normal form the coefficient of b_α has its least absolute value different from 0. If no such h exists choose $b_\alpha' = 0$. Then $B' = \{b_\alpha'\}$, with the elements zero omitted, is a base for H. Consider in fact h as above and let $b_{\alpha_r}' = \cdots + n_r b_{\alpha_r}$ (normal form). If ν is the h.c.f. of m_r, n_r we have $mm_r + nn_r = \nu$ for some integers m, n and hence $mh + nb_{\alpha_r}' = \cdots + \nu b_{\alpha_r}$ (normal form). Since $|\nu| \geqq |n_r|$, we have $n_r = \pm\nu$, $m_r = \mu_r n_r$. It follows that $h_r = h - \mu_r b_{\alpha_r}' \in H$ is of index $\beta_{r-1} < \alpha_r$. Repeating the reasoning we obtain $h_s = h - (\mu_s b_{\beta_s}' + \cdots + \mu_r b_{\alpha_r}')$ of index $\beta_{s+1} < \beta_s < \cdots < \alpha_r$. Since A is well-ordered the process must stop for some s. We will then have $h_s = 0$. Therefore B' is a set of generators for H. Since its elements are obviously linearly independent, B' is a base for H and (10.1) is proved.

Remark. The preceding proof based on well-ordering is shorter and more intuitive than alternates resting upon Zorn's theorem. Incidentally this is the only instance where well-ordering will be utilized in the present work.

(10.2) *A (discrete) group G is isomorphic with a factor-group of a free group.*

Let $G = \{g\}$ and let

$$h = \sum a_i g_i = 0 \text{ (finite sums)}$$

be the relations between the generators. For each g_i introduce a new symbol g_i' and let G' be the free group based on the g_i'. Then the symbols $h' = \sum a_i g_i'$ generate a subgroup H' of G' and $G \cong G'/H'$.

11. **Digression on integral matrices.** The reduction properties of integral matrices are very useful in the treatment of groups with a finite number of generators, and hence in the theory of complexes.

Consider then the integral matrix:

$$A = \| a_{ij} \|, \quad i = 1, 2, \cdots, p; j = 1, 2, \cdots, q.$$

We say that A is *unimodular* when it is square ($p = q$) and its determinant $|A| = \pm 1$.

(11.1) *The product of two unimodular matrices is a unimodular matrix.*

(11.2) *The inverse A^{-1} of a unimodular matrix A exists and is unimodular.*

Let again A be any integral matrix whatever. The following three types of operations on A are known as *elementary*: (a) transposition (interchange of two rows or columns); (b) change in sign of a row or column; (c) adding to the elements of a row or column the corresponding elements of another row or column multiplied by an integer. Each corresponds to the multiplication of A on one or the other side by a unimodular matrix. As a consequence, the effect of a finite number of elementary operations on A is to replace it by a matrix BAC, where B, C are unimodular.

(11.3) Definition. *Two matrices A, A_1 such that $A_1 = BAC$, B and C unimodular, are said to be equivalent. It is clear in fact that their relation to one another is a true relation of equivalence.*

Let ρ be the rank of A and $\delta_p(A)$ a number which is zero for $p > \rho$, and equal to the h.c.f. of the minors of order p for $p \leqq \rho$. If A_1 is equivalent to A the minors of order p of A_1 are linear integral combinations of those of A and so $\delta_p(A)$ divides $\delta_p(A_1)$. Since the reverse is also true $\delta_p(A_1) = \delta_p(A)$, or equivalent matrices have the same numbers $\delta_p(A)$ and hence the same rank ρ. The numbers $d_p = \delta_p(A)/\delta_{p-1}(A)$, ($\delta_0(A) = 1$), are the *invariant factors* of A.

(11.4) Theorem. *An integral matrix A is equivalent to a "diagonal" matrix*

$$\left\| \begin{matrix} d_1 & & & & & & \\ & \cdot & & & & & \\ & & \cdot & & & & \\ & & & \cdot & & & \\ & & & & d_\rho & & \\ & & & & & \cdot & \\ & & & & & & \cdot \\ & & & & & & & \cdot \end{matrix} \right\| \tag{11.5}$$

such that d_p divides d_{p+1}.

As a consequence of (11.4) we shall have:

(11.6) *$d_1, \cdots, d_\rho$ is the sequence of invariant factors of A, and so each invariant factor is divisible by its predecessor.*

(11.7) *A n.a.s.c. for two integral matrices with the same numbers of rows and columns to be equivalent is that they have the same sequence of invariant factors $d_1, \cdots, d_\rho$.*

It is clear that (11.4) and (11.6) yield (11.7). As for (11.6), in the matrix (11.5) with the d_p as stated, $d_1 d_2 \cdots d_p$ divides every minor of order p and is itself such a minor. Therefore $\delta_p(A) = d_1 \cdots d_p$, hence $\delta_{p+1}(A)/\delta_p(A) = d_{p+1}$, and so $d_1, \cdots, d_\rho$ is the sequence of invariant factors of A. Furthermore d_p divides d_{p+1} and so (11.6) will follow from (11.4).

Turning then our attention to (11.4) we dismiss the trivial case where the terms of A are all zero. Then in each matrix equivalent to A there are terms different from 0 and in one of these $B = \| b_{ij} \|$ there will occur a term d_1 (we make as yet no assertion as to d_1) which is the least positive term occurring in all the matrices equivalent to A. By transpositions we may dispose of B so that $b_{11} = d_1$. Let $b_{1i} = d_1 q + r$, $i > 1$, $0 \leqq r < d_1$. By multiplying the first column by q and subtracting it from the ith we replace b_{1i} by $r < d_1$, hence $r = 0$. Thus we may replace B by a matrix in which the b_{1i}, and similarly the b_{i1}, $i > 1$, are all zero, or assume:

$$(11.8) \qquad B = \begin{Vmatrix} d_1 & 0 \\ 0 & B_1 \end{Vmatrix}$$

Since by adding the ith row, $i > 1$, to the first we may bring b_{hi}, $h > 1$, to the position of b_{1i}, d_1 must divide b_{hi} and so all the terms of B_1. Now if A has a single row, the reduction to (11.8) already proves (11.4). Therefore we may use induction on the number of columns, and so assuming (11.4) for B_1 prove it for B. Under the circumstances B_1 is reducible by operations on B to the form (11.5) with diagonal terms $d_1 d_2'$, $\cdots$, $d_1 d_\rho'$, where $d_1 d_p'$ are the invariant factors of B_1 and $d_1 d_p'$ divides $d_1 d_{p+1}'$. Hence B is reducible to the form (11.5) with diagonal terms d_1, $d_1 d_2'$, $\cdots$, $d_1 d_\rho'$ which proves (11.4).

12. **Groups with a finite number of generators.** We shall discuss certain properties of these groups culminating in the basic product decomposition (12.8).

(12.1) Definition. *Let $B = \{g_1, \cdots, g_n\}$, $B' = \{g_1', \cdots, g_n'\}$ be two sets of elements of G containing the same number n of elements. By a unimodular transformation $\tau: B \to B'$ is meant a system of relations,*

$$(12.2) \qquad g_i' = \sum a_{ij} g_j, \ \| a_{ij} \| \ \textit{unimodular}.$$

(12.3) *If a group G has a finite base $B = \{g_1, \cdots, g_n\}$ its rank is (obviously) n. A n.a.s.c. for a second set $B' = \{g_1', \cdots, g_n'\}$ to be a base is that it may be obtained from B by a unimodular transformation.*

Whatever the set B' of n elements there exist relations

$$(12.4) \qquad g_i' = \sum c_{ij} g_j, \qquad C = \| c_{ij} \|.$$

A n.a.s.c. in order that $\{g_i'\}$ be a base is that the g_j be expressible as linear combinations of the g_i', or that there exist relations:

$$(12.5) \qquad g_i = \sum d_{ij} g_j', \qquad D = \| d_{ij} \|.$$

From this follows

$$g_i \doteq \sum d_{ij} c_{jk} g_k ,$$

and so since B is a base we must have $DC = 1$. This matrix relation yields $|D| \cdot |C| = 1$, and since the determinants are integers we must have $|C| = \pm 1$. Thus in order that B' be a base C must be unimodular, or the condition of (12.3) must be fulfilled. Conversely, if it is fulfilled, C is unimodular and (12.5) holds with $D = C^{-1}$, from which follows readily that B' is a base.

(12.6) *Let G be a free group of rank n, and H a subgroup of G. Then bases $\{g_1, \cdots, g_n\}$, $\{h_1, \cdots, h_m\}$ may be chosen for G, H such that*

$$h_i = d_i g_i , \qquad i = 1, 2, \cdots, m; \; d_i \textit{ divides } d_{i+1} . \tag{12.7}$$

Furthermore the set $d_1, \cdots, d_m$ is uniquely determined by H.

Let $\{g_i\}$ be a base for G. Since H is a free group (10.1) it has also a base $\{h_i\}$ and there subsist relations

$$h_i = \sum a_{ij} g_j , \qquad A = \| a_{ij} \|.$$

By a change in the bases one may replace A by any other equivalent matrix and so (12.6) is a consequence of (11.4).

(12.8) THEOREM. *Let G be a group with a finite number n of generators and of rank ρ. Then*

$$G \cong G_0 \times G_1 \times \cdots \times G_m , \tag{12.9}$$

where: (a) *G_0 is a free group of rank ρ when $\rho > 0$ and $G_0 = 0$ when $\rho = 0$*; (b) *G_p is cyclic of finite order e_p where e_p divides e_{p+1}*; (c) $m + \rho \leqq n$; (d) *the sequence $e_1, \cdots, e_m$ is uniquely determined by G.*

By (10.2) $G \cong G'/H'$, where G' is a free group. We may therefore assume $G = G'/H'$. Referring to (12.6) bases $\{g_1', \cdots, g_n'\}$, $\{h_1', \cdots, h_r'\}$ may be chosen for G', H' such that

$$h_i' = d_i g_i' , \qquad i = 1, 2, \cdots, r; \; d_i \text{ divides } d_{i+1} .$$

If π is the natural projection $G' \rightarrow G$ and we set $\pi g_i' = g_i$, $\pi h_i' = h_i$, the generators $g_1, \cdots, g_n$ are derived from the initial set by a unimodular transformation and we shall have

$$h_i = d_i g_i = 0, \qquad i = 1, 2, \cdots, r.$$

There may be a certain number of the first d_i equal to unity. We cast away the corresponding generators, denote the remaining d_i by e_p and so have a new system of generators which we denote by $g_1, \cdots, g_{m+\rho}$ such that

$$h_p = e_p g_p = 0, \qquad p = 1, 2, \cdots, m \leqq r; \; e_p \text{ divides } e_{p+1} .$$

In any case ρ is the rank of G. Since the g_{m+i} are linearly independent if $\rho > 0$ they generate a free group G_0 of rank ρ, while if $\rho = 0$ we set $G_0 = 0$. As for

g_p it generates a cyclic group of order e_p and we have manifestly the asserted decomposition (12.9).

(12.10) There remains to prove the unicity of the sequence $e_1, \cdots, e_m$. The decomposition (12.9) depends only on the choice of G' and the relation between G' and H' and hence upon the set of generators $g_1, \cdots, g_n$ and so $\{e_p\}$ depends solely upon these. If we choose $\{h'_1, \cdots, h'_r\}$ as a base for the group H' and write down the relations expressing its elements in terms of the g'_i :

$$h'_i = \sum a_{ij}g'_j, \qquad A = \| a_{ij} \|, \tag{12.11}$$

then $\{e_p\}$ is simply the set of the invariant factors of A which are greater than 1. Suppose now that we add a new generator g_{n+1} to the set $g_1, \cdots, g_n$. We have then a relation

$$-g_{n+1} = a_{n+1,1}g_1 + \cdots + a_{n+1,n}g_n .$$

There correspond new groups G'_1, H'_1 based on $\{g'_1, \cdots, g'_{n+1}\}$, $\{h'_1, \cdots, h'_{r+1}\}$ with the relations (12.11) and

$$h'_{r+1} = a_{n+1,1}g'_1 + \cdots + a_{n+1,n}g'_n + g'_{n+1} .$$

Thus A is now replaced by

$$A_1 = \left\| \begin{array}{c|c} A & \begin{array}{c} 0 \\ \cdot \\ \cdot \\ \cdot \\ 0 \end{array} \\ \hline a_{n+1,1}, \cdots, & 1 \end{array} \right\|$$

whose invariant factors greater than 1 are the same as those of A. Thus by adding new generators we do not modify $\{e_p\}$. Suppose then that we have two systems of generators $\{g_i\}$ and $\{\bar{g}_i\}$. Together they form again a system of generators which is obtained from each by adding new generators. Therefore the three systems yield the same set $\{e_p\}$ and in particular this is true for $\{g_i\}$ and $\{\bar{g}_i\}$ thus proving (12.8).

§3. LIMIT-GROUPS

13. **Inverse systems.** We now return to topological groups and investigate certain systems of groups $\{G_\lambda\}$ indexed by a directed set $\Lambda = \{\lambda; >\}$.

(13.1) DEFINITION. *Let* $\{G_\lambda\}$ *be an inverse mapping system* (I, 38) *whose projections* π^λ_μ *are homomorphisms. Then* $S = \{G_\lambda ; \pi^\lambda_\mu\}$ *is said to be an inverse system of groups, or merely an inverse system.*

(13.2) *The limit* $'G$ *of* S *is a closed subgroup of* $G = \mathbf{P}G_\lambda$ *and is known as the "limit-group" of* S.

By (I, 38.3) $'G$ is closed in G. Then if $g, g' \in 'G$ we have: $\pi_\mu^\lambda g_\lambda = g_\mu$, $\pi_\mu^\lambda g_\lambda' = g_\mu'$, hence $\pi_\mu^\lambda(g_\lambda - g_\lambda') = g_\mu - g_\mu'$, or $g - g' \in 'G$. Therefore $'G$ is a subgroup of G.

(13.3) *Let* $\mathrm{M} = \{\mu; >\}$ *be a directed subsystem of* Λ, S' *the corresponding partial system of* S (I, 40), $'G'$ *the limit-group of* S'. *Then the projection* τ (I, 40.1) : $'G \to 'G'$ *is a homomorphism, and when* S' *is cofinal in* S *then* τ *is an isomorphism.*

If $G' = \mathbf{P}G_\mu$ then $\tau = \pi \mid 'G$ where π is the projection $G \to G'$, and since π is here a homomorphism so is τ. When S' is cofinal in S the mapping τ is topological and hence it is an isomorphism.

(13.4) *If* $g = \{g_\lambda\} \in 'G$ *then*: (a) $g \to g_\lambda$ *defines a homomorphism* π_λ: $'G \to G_\lambda$; (b) *if the* π_μ^λ *(all* λ, μ, $\lambda > \mu$*) are isomorphisms then* π_λ *is likewise an isomorphism, so that* $'G \cong G_\lambda$.

Property (a) is a special case of (13.3) obtained when M consists of the single element λ.

Regarding (b), let λ be kept fixed and take any g_λ. Whatever μ choose $\nu > \lambda, \mu$ and set $g_\mu = \pi_\mu^\nu(\pi_\lambda^\nu)^{-1}g_\lambda$. We see immediately that: $(\alpha) g_\mu$ is independent of ν; (β) $\{g_\mu\} \in 'G$; (γ) any element $g \in 'G$ with the coordinate g_λ will have its μ coordinate determined by the same relation as g_μ and so it must be g_μ. Therefore π_λ is a univalent homomorphism onto.

To complete the proof there remains to show that π_λ is open. Let U_μ be any open set of G_μ and set $V_\mu = \{g \mid g_\mu \in U_\mu\}$. By (I, 38.2), $\{V_\mu\}$ is a base for $'G$. Choose again $\nu > \lambda, \mu$. Since the π_μ^λ are isomorphisms $\pi_\lambda V_\mu = \pi_\lambda^\nu(\pi_\mu^\nu)^{-1}U_\mu$ is open in G_λ. Hence π_λ is open and (13.4) follows.

(13.5) *Let* $S = \{G_\lambda; \pi_\mu^\lambda\}$, $\Sigma = \{H_\lambda; \omega_\mu^\lambda\}$ *be inverse systems both indexed by* $\Lambda = \{\lambda; >\}$ *and with limit-groups* $'G$, $'H$. *Suppose that for each* λ *there is a homomorphism* $\tau_\lambda : G_\lambda \to H_\lambda$ *such that:* $\omega_\mu^\lambda\tau_\lambda = \tau_\mu\pi_\mu^\lambda$, $\lambda > \mu$. *There exists a homomorphism* τ: $'G \to 'H$ *such that if* $g = \{g_\lambda\} \in 'G$ *then* $\tau g = \{\tau_\lambda g_\lambda\}$.

Let $H = \mathbf{P}H_\lambda$. If $g \in 'G$ then $g \to \{\tau_\lambda g_\lambda\}$ defines a homomorphism τ: $'G \to H$ in the algebraic sense. Since every coordinate $\tau_\lambda g_\lambda$ of τg is a continuous function of g so is τg. Hence τ is a homomorphism. From $\omega_\mu^\lambda(\tau_\lambda g_\lambda) = \tau_\mu\pi_\mu^\lambda g_\lambda = (\tau_\mu g_\mu)$ follows that $\tau g \in 'H$, so τ is actually a homomorphism $'G \to 'H$.

(13.6) *Under the same assumptions as in* (13.5) *let the* G_λ *be compact. If* $H_\lambda' = \tau_\lambda G_\lambda$ *then* $\Sigma' = \{H_\lambda'; \omega_\mu^\lambda\}$ *is an inverse system with limit-group say* $'H'$ *and* τ *is an open homomorphism of* $'G$ *onto* $'H'$.

We have found that τ is a homomorphism of $'G$ *into* $'H'$. To prove that τ is *onto* take $h' = \{h_\lambda'\} \in 'H'$ and let $F_\lambda = \tau_\lambda^{-1}h_\lambda'$. Since H_λ' is a Hausdorff space h_λ' is closed, and since τ_λ is continuous F_λ is closed in G_λ and so like G_λ it is a compact Hausdorff space. If $g_\lambda \in F_\lambda$ then $\tau_\lambda g_\lambda = h_\lambda'$, hence for $\lambda > \mu$: $\omega_\mu^\lambda\tau_\lambda g_\lambda = \omega_\mu^\lambda h_\lambda' = h_\mu' = \tau_\mu(\pi_\mu^\lambda g_\lambda)$. Therefore $\pi_\mu^\lambda g_\lambda \in F_\mu$ or $\pi_\mu^\lambda F_\lambda \subset F_\mu$. It follows that $\{F_\lambda; \pi_\mu^\lambda\}$ is an inverse mapping system of compact Hausdorff spaces. By (I, 39.1) its limit-space contains an element $g = \{g_\lambda\}$ which is also in $'G$. Since $g_\lambda \in F_\lambda$ we have $\tau_\lambda g_\lambda = h_\lambda'$ or $h' = \tau g$. Therefore τ is a mapping onto.

Let $'G_1$ be the kernel of τ. Then τ defines a univalent homomorphism τ_1:

$'G/'G_1 \to 'H'$. Since $'G$ is compact so is $'G_1$ and hence also $'G/'G_1$ (5.5). Since τ_1 is a continuous one-one mapping of a compact space it is topological.

Now the natural projection π: $'G \to 'G/'G_1$ is an open homomorphism. Since $\tau = \tau_1\pi$, and τ_1 is topological, τ is also open, thus proving (13.6).

(13.7) *Generalization.* An important type of group arises in the homology theory of nets (Čech theory, VI), which is analogous and closely related to the limit-group of inverse systems. As before we have a system of groups $\{G_\lambda\}$ indexed by $\{\lambda; >\}$ and for each λ there exists a closed subgroup G'_λ of G_λ. Corresponding to every ordered pair $\lambda > \mu$ there are given one or more homomorphisms or *projections* $\pi^\lambda_\mu : G_\lambda \to G_\mu$ such that:

I. If $\lambda > \mu > \nu$ and π^λ_μ, π^μ_ν are projections so is $\pi^\mu_\nu \pi^\lambda_\mu$.

II. $\pi^\lambda_\mu G'_\lambda \subset G'_\mu$.

III. If π^λ_μ, $\bar{\pi}^\lambda_\mu$, $\lambda > \mu$, are projections then $\pi^\lambda_\mu g_\lambda - \bar{\pi}^\lambda_\mu g_\lambda \subset G'_\mu$, $g_\lambda \in G_\lambda$.

Let $H_\lambda = G_\lambda/G'_\lambda$ and let τ_λ be the natural projection: $G_\lambda \to H_\lambda$. By (5.4) there is a homomorphism $\omega^\lambda_\mu : H_\lambda \to H_\mu$, $\lambda > \mu$, such that $\omega^\lambda_\mu \tau_\lambda = \tau_\mu \pi^\lambda_\mu$ and III has for consequence that ω^λ_μ is independent of the particular π^λ_μ in its definition: ω^λ_μ is *unique.* We may also say that all the π^λ_μ induce the same projection of the cosets of G_λ mod G'_λ into those of G_μ mod G'_μ. It follows then readily that $\{H_\lambda ; \omega^\lambda_\mu\}$ is an inverse system and its limit-group is denoted by H.

A decidedly different type of group, which generalizes limits of inverse systems, arises now as follows. Let $G = \mathbf{P}G_\lambda$ and let $g = \{g_\lambda\} \in G$ be such that $\lambda > \mu \to \pi^\lambda_\mu g_\lambda - g_\mu \in G'_\mu$. By III this remains true if π^λ_μ is replaced by any other projection $\bar{\pi}^\lambda_\mu$. It is clear that the set $^\circ G$ of all such elements is a subgroup of G. It is the group which we had in view.

(a) *$^\circ G$ is closed in G.*

The function $f_{\lambda\mu}(g) = \pi^\lambda_\mu g_\lambda - g_\mu$ is a continuous function on G to G_μ. Hence $F_{\lambda\mu} = f^{-1}_{\lambda\mu} G'_\mu$ is closed in G and so is $^\circ G = \bigcap F_{\lambda\mu}$.

(b) *$G' = \mathbf{P}G'_\lambda$ is a closed subgroup of G* (I, 12.6) *and hence also of $^\circ G$.*

(c) *$^\circ G/G' \cong H$.*

Let $G^* = {}^\circ G/G'$. If $g = \{g_\lambda\} \in {}^\circ G$ then $\{\tau_\lambda g_\lambda\} = h \in H$ and h depends solely upon the coset g^* of g mod G'. From this we infer that $g^* \to h$ defines a homomorphism θ: $G^* \to H$. If $\theta g^* = 0$ then $g_\lambda \in G'_\lambda$, and so $g \in G'$, hence $g^* = 0$, or θ is univalent. Given $h = \{h_\lambda\}$ take in the coset h_λ of G_λ mod G'_λ a definite element g_λ. Evidently $\lambda > \mu \to \pi^\lambda_\mu g_\lambda - g_\mu \in G'_\mu$ and so $g = \{g_\lambda\} \in {}^\circ G$ is such that $\theta g^* = h$. Thus θ is onto and so it is an isomorphism in the algebraic sense.

Let U_λ be open in G_λ. Since the sets $\{g \mid g_\lambda \in U_\lambda\}$ make up a subbase for G, if $V_\lambda = \{g \mid g \in {}^\circ G, g_\lambda \in U_\lambda\}$ then by the principle of relativization $\{V_\lambda\}$ is a subbase for $^\circ G$. Since τ_λ is open the set $\tau_\lambda U_\lambda$ is open in H_λ and $\theta V_\lambda = \{h \mid h_\lambda \in \tau_\lambda U_\lambda\}$ is open in H. Therefore θ is open and so it is an isomorphism. This proves (c).

(d) *When the G_λ are compact so are the groups $^\circ G$, G', H.*

For G is then compact, hence $^\circ G$, G' are compact as closed subgroups of G, and likewise H by (5.5).

14. Direct systems.

(14.1) DEFINITION. *Let this time* $\{G_\lambda\}$ *be a system of discrete groups still indexed by* $\Lambda = \{\lambda; >\}$ *and with projections running the other way: for* $\lambda > \mu$ *there is a homomorphism* $\pi_\lambda^{*\mu}: G_\mu \to G_\lambda$ *such that* $\lambda > \mu > \nu \to \pi_\lambda^{*\mu} = \pi_\lambda^{*\mu}\pi_\mu^{*\nu}$. *Then the system* S^* *of the* G_λ *and the* $\pi_\lambda^{*\mu}$ *is said to form a direct system of groups or merely a direct system, written* $S^* = \{G_\lambda ; \pi_\lambda^{*\mu}\}$.

If $G = \mathbf{P}^w G_\lambda$ (weak product), the elements $g_\mu - \pi_\lambda^{*\mu} g_\mu$ $(\lambda > \mu)$ generate a subgroup H of G, and the factor-group $G^* = G/H$ is called the *limit-group* of S^*. Since G is discrete so are H and G^*.

Let $g_\lambda \in G_\lambda$. Under the convention in (7.3) g_λ represents also an element of G and g^* its coset mod H is an element of G^*. We call g_λ a *representative* of g^* in G_λ. The mapping $g_\lambda \to g^*$, which is clearly a homomorphism $G_\lambda \to G^*$ is again called the *injection of* G_λ *into* G^*.

(14.2) *Every* $g^* \in G^*$ *has a representative in some* G_μ, *and in fact for every element of a set* $\mathrm{M} = \{\mu\}$ *cofinal in* Λ.

In g^* (now a coset of G mod H) take any element g. We may write g as a finite sum $g = g_{\lambda_1} + \cdots + g_{\lambda_k}$. Choose any $\mu > \lambda_1, \cdots, \lambda_k$. We have then $g_{\lambda_i} - \pi_\mu^{*\lambda_i} g_{\lambda_i} \in H$, and hence $g - \Sigma\pi_\mu^{*\lambda_i} g_{\lambda_i} \in H$. Therefore g^* has the representative $\Sigma\pi_\mu^{*\lambda_i} g_{\lambda_i} \in G_\mu$. It is clear that $\mathrm{M} = \{\mu\}$ is cofinal in Λ.

(14.3) *H consists of those and only those elements* $g = g_{\lambda_1} + \cdots + g_{\lambda_k}$ *of G such that there exists a* $\lambda_0 > \lambda_1, \cdots, \lambda_k$ *such that* $\Sigma\pi_{\lambda_0}^{*\lambda_i} g_{\lambda_i} = 0$.

Let (α) denote the property of the statement.

(a) *If g has the property (α) then $g \in H$.*

For $g = \Sigma(g_{\lambda_i} - \pi_{\lambda_0}^{*\lambda_i} g_{\lambda_i}) \in H$.

(b) *A generator* $h = g_\mu - \pi_\lambda^{*\mu} g_\mu$ *of H has property (α).*

We may choose $\lambda_0 > \lambda > \mu$ and then $\pi_{\lambda_0}^{*\mu} g_\mu - \pi_{\lambda_0}^{*\lambda}(\pi_\lambda^{*\mu} g_\mu) = 0$. Since $\pi_\lambda^{*\mu} g_\mu \in G_\lambda$ this means that h has property (α).

(c) *The elements g with property (α) form a subgroup H_1 of H.*

Let $g = \Sigma g_\lambda$, $g' = \Sigma g'_\lambda$ be such that there are corresponding λ_0, λ'_0 such that $g_\lambda = 0$ for $\lambda > \lambda_0$, $g'_\lambda = 0$ for $\lambda > \lambda'_0$ and that $\Sigma\pi_{\lambda_0}^{*\lambda} g_\lambda = 0$, $\Sigma\pi_{\lambda'_0}^{*\lambda'} g_{\lambda'} = 0$. If we choose $\mu > \lambda_0, \lambda'_0$ we find at once $\Sigma_{\lambda \prec \mu}\pi_\mu^{*\lambda}(g_\lambda - g'_\lambda) = 0$ proving that $g - g'$ has property (α), and hence the asserted group property.

By (a) we have $H_1 \subset H$ and by (b) and (c): $H \subset H_1$ hence $H = H_1$, which proves (14.3).

(14.4) *The representatives of the zero of G^* in G_λ make up a set H_λ which consists of those and only those elements g_λ such that* $\pi_{\lambda_0}^{*\lambda} g_\lambda = 0$ *for some* $\lambda_0 > \lambda$.

If η_λ is the injection $G_\lambda \to G^*$ then by (14.3): $\eta_\lambda H_\lambda = \eta_\lambda(G_\lambda \cap H)$ which proves (14.4).

(14.5) *Let* $\mathrm{M} = \{\mu; >\}$ *be a directed subsystem of* $\Lambda = \{\lambda; >\}$ *and let* $S'^* = \{G_\mu ; \pi_\mu^{*\mu'}\}$ *be the direct system attached to* M *and* G'^* *its limit-group. There is a homomorphism* $\tau: G'^* \to G^*$ *such that if* g_μ *is a representative of* $g'^* \in G'^*$ *then it is also a representative of* $\tau g'^*$. *If* M *is cofinal in* Λ (*we then say:* S'^* *is cofinal in* S^*) *then* τ *is an isomorphism.*

Let G', H' be the analogues of G, H for S'^*, and η the injection $G' \to G$. We have $G^* = G/H$, $G'^* = G'/H'$, also immediately $\eta H' \subset H$. Let φ be the mapping $G' \to G^*$ whereby $\varphi(g')$ is the coset g^* of $\eta g'$ mod H. Since $\varphi(g')$ depends solely upon the coset of g' mod H', by (5.3) it induces a homomorphism τ: $G'^* \to G^*$ which obviously behaves as required.

Suppose now M cofinal in Λ. Then: (a) τ is univalent. For suppose $\tau g'^* = 0$ and let g'_μ be a representative of g'^*. Since it is also a representative of $\tau g'^*$, by (14.4) for some $\lambda > \mu$ we have $\pi_\lambda^{*\mu} g'_\mu = 0$. Since M is cofinal in Λ there is a $\mu' > \lambda$ and so $\pi_{\mu'}^{*\lambda} \pi_\lambda^{*\mu} g'_\mu = 0 = \pi_{\mu'}^{*\mu} g'_\mu$. Therefore $g'^* = 0$. (b) $\tau G'^* = G^*$. For let $g^* \in G^*$ have the representative g_λ. There exists a $\mu > \lambda$ and hence $\pi_\mu^{*\lambda} g_\lambda = g'_\mu$ is a representative of g^* and likewise of a $g'^* \in G'^*$ such that $\tau g'^* = g^*$. From (a), (b) follows now that τ is the asserted isomorphism.

(14.6) *There is a homomorphism $\eta_\mu : G_\mu \to G^*$ such that $g^* = \eta_\mu g_\mu$ has g_μ for representative. If the $\pi_\lambda^{*\mu}$ are all isomorphisms then η_μ is an isomorphism so that $G^* \cong G_\mu$.*

That η_μ is a homomorphism is a consequence of (14.5) obtained when M consists of the single element μ.

Suppose now that the $\pi_\lambda^{*\mu}$ are isomorphisms. If $\eta_\mu g_\mu = 0$ then some $\pi_\lambda^{*\mu} g_\mu = 0$, and since $\pi_\lambda^{*\mu}$ is an isomorphism $g_\mu = 0$. Therefore η_μ is univalent. Take now any $g^* \in G^*$ and let it have the representative g_ν. Choose $\lambda > \mu, \nu$. Since the $\pi_\lambda^{*\mu}$ are all isomorphisms $(\pi_\lambda^{*\mu})^{-1} \pi_\lambda^{*\nu} g_\nu = g_\mu$ is likewise a representative of g^* and $\eta_\mu g_\mu = g^*$. Hence η_μ maps G_μ onto G^*. Since η_μ is univalent "onto" and G^*, G_μ are discrete they are isomorphic under η_μ.

The indirect definition of the limit of a direct system in terms of the weak product is most suitable from the point of view of group theory. In the applications (VI, VII) we shall find convenient to have the direct:

(14.7) *Alternate definition.* Let $S^* = \{G_\lambda ; \pi_\lambda^{*\mu}\}$ be a direct system. Let g_μ, g_ν be identified whenever for some $\lambda > \mu, \nu$ we have $\pi_\lambda^{*\nu} g_\nu = \pi_\lambda^{*\mu} g_\mu$. A collection of identified elements $g_\lambda, g_\mu, \cdots$ is now denoted by $'g$, and $g_\lambda, g_\mu, \cdots$ are called the *representatives* of $'g$. The set $'G = \{'g\}$ of the elements thus obtained is turned into a group as follows. *First*, the zeros of the G_λ are representatives of a single element which is taken as the zero of $'G$. *Second*, if g, g' have representatives g_μ, g_ν we choose any $\lambda > \mu, \nu$ and find that $(\pi_\lambda^{*\mu} g_\mu \pm \pi_\lambda^{*\nu} g_\nu) \in G_\lambda$ is the representative of a unique element which is by definition $'g \pm 'g'$. The verification of the group axioms is elementary and $'G$, now denoting the new group taken discrete, is by definition the limit-group of S^*.

We shall now show that $'G = G^*$. It follows from the definition and (14.4) that the representatives of a given $'g \in 'G$ are in a coset g^* of G mod H. That is to say, the representatives of $'g$ are all representatives of the element g^* of G^*. Furthermore it is a simple matter to show that $'g \to g^*$ is a homomorphism θ: $'G \to G^*$. Suppose now that g_μ, g_ν are representatives of g^*. Then $g_\mu - g_\nu \in H$ and by (14.3) for some λ: $\pi_\lambda^{*\mu} g_\mu = \pi_\lambda^{*\nu} g_\nu = g_\lambda \in G_\lambda$. Therefore g_μ, g_ν represent the same element $'g$ of $'G$ and $\theta 'g = g^*$. It follows that $\theta 'G = G^*$. Let finally g_μ represent the 0 of G^*. Then $g_\mu \in H$ and so for some $\lambda > \mu$: $\pi_\lambda^{*\mu} g_\mu = 0$, and hence $\theta^{-1}0 = 0$. Therefore θ is an isomorphism. If we identify the elements $'g$ and $g^* = \theta 'g$ the two groups $'G$, G^* become identical.

(14.8) *Generalization.* It is essentially parallel to (13.7) and likewise required later. This time we have a system of discrete groups $\{G_\lambda\}$ still indexed by $\{\lambda; >\}$ and for each λ a subgroup G'_λ of G_λ. For each pair $\lambda > \mu$ there exist one or more homomorphisms or *projections* $\pi_\lambda^{*\mu}: G_\mu \to G_\lambda$ such that:

I. If $\lambda > \mu > \nu$ and $\pi_\mu^{*\nu}$, $\pi_\lambda^{*\mu}$ are projections so is $\pi_\lambda^{*\mu}\pi_\mu^{*\nu}$.

II. $\pi_\lambda^{*\mu}G'_\mu \subset G'_\lambda$, $\lambda > \mu$.

III. If $\pi_\lambda^{*\mu}$, $\bar{\pi}_\lambda^{*\mu}$, $\lambda > \mu$, are projections then $\pi_\lambda^{*\mu}g_\mu - \bar{\pi}_\lambda^{*\mu}g_\mu \subset G'_\lambda$.

Let $H_\lambda = G_\lambda/G'_\lambda$ and let τ_λ be the natural projection $G_\lambda \to H_\lambda$. By (5.4) there is a homomorphism $\omega_\lambda^{*\mu}: H_\mu \to H_\lambda$, $\lambda > \mu$, such that $\omega_\lambda^{*\mu}\tau_\mu = \tau_\lambda\pi_\lambda^{*\mu}$ and it is unique, i.e., independent of the particular $\pi_\lambda^{*\mu}$ in its definition. From this follows readily that $\{H_\lambda; \omega_\lambda^{*\mu}\}$ is a direct system and its limit-group is denoted by H.

A new type of group is now introduced in the following manner. The elements g_μ, g_ν are identified whenever for some $\lambda > \mu, \nu$ there exist projections $\pi_\lambda^{*\mu}$, $\pi_\lambda^{*\nu}$ such that $\pi_\lambda^{*\mu}g_\mu - \pi_\lambda^{*\nu}g_\nu \in G'_\lambda$. We will now turn the set $°G$ of the identified elements into a group. The zeros of the G_λ are manifestly representatives of a single element which is by definition the zero of $°G$. If $g, g' \in °G$ have representatives g_μ, g'_ν we choose a $\lambda > \mu, \nu$ and find that $\pi_\lambda^{*\mu}g_\mu \pm \pi_\lambda^{*\nu}g'_\lambda$ is the representative of a unique element written $g \pm g'$ and $°G$ is the group arising under these rules.

The elements g of $°G$ which have a representative g_μ such that for some $\lambda > \mu$: $\pi_\lambda^{*\mu}g_\mu \in G'_\lambda$ form a subgroup $°G'$ of $°G$.

(a) $°G/°G' \cong H$.

Let $G^* = °G/°G'$. If $g \in °G$ has the representative g_μ then $\tau_\mu g_\mu$ is the representative of an $h \in H$ which depends solely upon the coset g^* of g mod $°G'$ and $g^* \to h$ defines a homomorphism $\theta: G^* \to H$. If $\theta g^* = 0$ then for some $\lambda > \mu$: $\omega_\lambda^{*\mu}\tau_\mu g_\mu = \tau_\lambda(\pi_\lambda^{*\mu}g_\mu) = 0$. Therefore $\pi_\lambda^{*\mu}g_\mu \in G'_\lambda$ and so $g \in °G'$, $g^* = 0$. Thus θ is univalent. If h_μ is a representative of $h \in H$ and g_μ is in the coset h_μ of G_μ mod G'_μ then g_μ is the representative of a $g \in °G$ whose coset g^* mod $°G'$ is such that $\theta g^* = h$. Hence θ is onto and so it is an isomorphism. This proves (a).

§4. GROUP MULTIPLICATION

15. This section initiates the study of the group properties which lie at the root of the duality theorems of topology. It has been shown by Pontrjagin that these theorems consist of two parts, a *group duality* and what may be termed a *geometric duality*. The former is based essentially upon Pontrjagin's concept of *group multiplication* which may be considered as obtained by abstraction from the Kronecker index of topology.

(15.1) Definitions. *Two groups G, H are said to be paired to a third group K, whenever there is given a function $\varphi(g, h)$, continuous and distributive in both* *variables, and whose values are in K. The operation φ is generally written as a product gh and called a group multiplication or merely multiplication.*

(15.2) DEFINITIONS. *Let G, H be paired to K under a multiplication gh and let H' be a subgroup of H. If $gh' = 0$ for every $h' \in H'$, g is said to annul H'. If all the elements of a subset A of G annul H', A is said to annul H'. The totality of all the elements of G which annul H' is a subgroup of G known as the annihilator of H' in G. These terms may also be applied when H' is not a subgroup, but merely a subset of H. Furthermore the same terms may be applied with the roles of G, H interchanged throughout.*

(15.3) DEFINITIONS. *Under the same conditions G, H are said to be orthogonal to K or K-orthogonal or merely orthogonal, whenever the annihilators of G in H and of H in G are both zero. This means that if $gh = 0$ for every h then $g = 0$, and conversely.*

(15.4) *Let G, H be paired to K under a multiplication gh, and let G', H' be subgroups of G, H. Then if G' annuls H' it also annuls $\bar{H}'$ and the annihilators of G', H' are closed subgroups.*

Let $B(g')$ denote the annihilator of $g' \in G'$ in H. Since K is a Hausdorff space and $B(g')$ is the inverse image of a point of K under the mapping $h \to g'h$, $B(g')$ is closed. Since the annihilator of G' is the subgroup $B(G') = \bigcap\{B(g') \mid g' \in G'\}$ it is a closed subgroup. If G' annuls H' then $H' \subset B(G')$ and hence $\bar{H}' \subset B(G')$, or $\bar{H}'$ annuls G' also, and hence G' annuls $\bar{H}'$. Since G, H and their subsets may be interchanged throughout, (15.4) is proved.

(15.5) THEOREM. *Let G, H be paired to K and let H' be a closed subgroup of H and G' a subgroup of G which annuls H'. Then:*

(a) *G' and H/H' are paired to K under a multiplication defined as follows: if $g' \in G'$ and $h^* \in H/H'$ then $g'h^*$ is the common value of all the products $g'h$ for $h \in h^*$.*

(b) *If G, H are orthogonal and H' is the annihilator of G' then G' and H/H' are orthogonal under the same multiplication.*

(c) *Similarly with G, H interchanged.*

The pairing described in (a) shall be referred to as *induced by the pairing of G, H.*

A generalization of (15.5) needed in the homology theory of complexes is:

(15.6) *Let G, H be paired to K and let $G \supset G_1 \supset G_2$, $H \supset H_1 \supset H_2$, where G_2, H_2 are closed subgroups of G and H, and H_2, G_2 annul G_1, H_1. Then:*

(a) *G_1/G_2 and H_1/H_2 are paired to K under a multiplication defined as follows: if g^* and h^* are elements of the two factor-groups, then g^*h^* is the common value of all the products gh for $g \in g^*$, $h \in h^*$.*

(b) *If H_2, G_2 are the annihilators of G_1, H_1 then G_1/G_2 and H_1/H_2 are orthogonal under the same multiplication.*

If we make $G_1 = G'$, $G_2 = 0$, $H_1 = H_,, H_2 = H'$, (15.6) reduces to (15.5); hence it is sufficient to prove (15.6).

The elements g^*, h^* are cosets $g_1 + G_2$, $h_1 + H_2$. If $g = g_1 + g_2$, $h = h_1 + h_2$ are in the cosets we have $gh = g_1h_1$. Since this product is independent of g_2, h_2 we may take its value as the definition of g^*h^*, and this product is distributive in both factors. Since gh is a continuous function of (g, h) given g_{01}, h_{01} and a neighborhood W of $g_{01}h_{01} = \overset{*}{g_0}\overset{*}{h_0}$ in K, there exist neighborhoods U of g_{01} in G_1 and V of h_{01} in H_1 such that $g_1 \in U$, $h_1 \in V \rightarrow g_1h_1 \in W$. Since the projections π: $G_1 \rightarrow G_1/G_2$, ω: $H_1 \rightarrow H_1/H_2$ are open, πU and ωV are neighborhoods of $\overset{*}{g_0}$ in G_1/G_2 and $\overset{*}{h_0}$ in H_1/H_2 such that $g^* \in \pi U$, $h^* \in \omega V \rightarrow g^*h^* \in W$. Therefore g^*h^* is continuous, and so it is a multiplication behaving as described under (15.6a).

Under the assumption of (15.6b) suppose $g^*h^* = 0$ for given g^* whatever h^*. Then $g_1h_1 = 0$ for a given $g_1 \in g^*$ whatever h_1 and so $g_1 \in G_2$, $g^* = 0$. Similarly with g^*, h^* interchanged and so (15.6b) follows.

16. **Pairing of products and limit-groups.** Let $\{G_\lambda\}$, $\{H_\lambda\}$ be two systems of groups indexed by the same set $\Lambda = \{\lambda\}$ and under the following conditions:

(a) The H_λ are discrete.

(b) For each λ the groups G_λ, H_λ are paired to a fixed group K.

(16.1) *Let* $G = \mathbf{P}G_\lambda$, $H^w = \mathbf{P}^w H_\lambda$, $g = \{g_\lambda\} \in G$, $h = \{h_\lambda\} \in H^w$. *Then* $gh = \sum g_\lambda h_\lambda$ *defines a multiplication pairing* G *and* H^w *to* K. *Furthermore when* G_λ, H_λ *are orthogonal throughout, so are* G, H^w.

Since at most a finite number of the h_λ are different from 0, $gh = \sum g_\lambda h_\lambda$ has a meaning and gh thus defined is distributive in both g and h. Since H^w is discrete, the continuity of gh in (g, h) reduces to that of gh as a function of g alone for h fixed. That gh is continuous under these conditions is obvious since g_λ is a continuous function of g and $g_\lambda h_\lambda$ a continuous function of g_λ.

Suppose G_λ, H_λ orthogonal throughout. Then if η_λ is the injection $H_\lambda \rightarrow H^w$ and if $gh = 0$ for every h we have in particular $g\eta_\lambda(h_\lambda) = g_\lambda h_\lambda = 0$ for every h_λ, hence $g_\lambda = 0$, $g = 0$. Similarly with g, h interchanged. Therefore G, H^w are orthogonal.

(16.2) Definition. *Two inverse and direct systems* $S = \{G_\lambda ; \pi_\mu^\lambda\}$, $S^* = \{H_\lambda ; \overset{*}{\pi}{}_\lambda^\mu\}$ *both indexed by* $\Lambda = \{\lambda; \succ\}$ *are said to be paired to a group* K, *if* G_λ, H_λ *are paired to* K *and there holds the permanence relation*

$$(16.3) \qquad g_\lambda \cdot (\overset{*}{\pi}{}_\lambda^\mu h_\mu) = (\pi_\mu^\lambda g_\lambda) \cdot h_\mu , \qquad \lambda \succ \mu.$$

(16.4) *If* S, S^* *are paired to* K *so are their limit-groups* $'G$, H^* *and this under a multiplication* $'gh^*$ *such that if* g_λ *is a coordinate of* $'g$ *and* h_λ *a representative of* h^*, *then* $'gh^* = g_\lambda h_\lambda$.

We have seen that $'G$ is a closed subgroup of G (13.2) and that $H^* = H^w/L$, where L is the subgroup of H^w generated by the elements $h_\mu - \overset{*}{\pi}{}_\lambda^\mu h_\mu$ $(\lambda \succ \mu)$ (14.1). By (16.1) we also have a pairing of G, H^w to K. I say that $'G$ annuls L.

This merely requires proving that

$$'g\cdot(h_\mu - \pi_\lambda^{*\mu}h_\mu) = 0. \tag{16.5}$$

Now in view of the multiplication between G, H^w:

$$'g\cdot(h_\mu - \pi_\lambda^{*\mu}h_\mu) = 'g_\mu h_\mu - 'g_\lambda \pi_\lambda^{*\mu}h_\mu . \tag{16.6}$$

Since $'g \in 'G$ we have $'g_\mu = \pi_\mu^\lambda\, 'g_\lambda$. By substituting in (16.6) and in view of the relation of permanence (16.3) we find that (16.5) holds.

Since $'G$ annuls L the asserted pairing of $'G$, H^* is a consequence of (15.5).

17. We shall require later in the theory of intersections in nets (VI, 8) generalizations of (16.4) where the pairing of G_λ, H_λ is to a variable group K_λ. There are two cases which must be dealt with separately.

First case. S, S^* are as before and in addition there is a third inverse system $S_0 = \{K_\lambda ; \theta_\mu^\lambda\}$ likewise indexed by Λ and with limit-group $'K$. We assume a relation of permanence

$$\theta_\mu^\lambda(g_\lambda \cdot \pi_\lambda^{*\mu}h_\mu) = \pi_\mu^\lambda g_\lambda \cdot h_\mu , \qquad \lambda > \mu, \tag{17.1}$$

and we have this time:

(17.2) *The same as* (16.4) *except that* $'G$, H^* *are now paired to* $'K$.

It follows readily from (13.3) that if $\mathrm{M} = \{\mu\}$ is cofinal in Λ and $\{k_\mu\}$ is such that $\mu > \mu' \rightarrow \theta_{\mu'}^\mu k_\mu = k_{\mu'}$ there is a unique element $'k \in 'K$ with the coordinates $\{k_\mu\}$. Moreover if every $k_\mu = 0$ then $'k = 0$.

Let now $h^* \in H^*$ have the representative h_ν and let $'g = \{g_\lambda\} \in 'G$. For $\mu > \nu$ define $h_\mu = \pi_\mu^{*\nu}h_\nu$ and set $k_\mu = g_\mu h_\mu$. If $\mu > \mu' > \nu$ then $\pi_\mu^{*\mu'}h_{\mu'} = h_\mu$, and hence from (17.1) $\theta_{\mu'}^\mu k_\mu = k_{\mu'}$. Since $\mathrm{M} = \{\mu\}$ is cofinal in Λ, $\{k_\mu\}$ are coordinates of a unique $'k \in 'K$. If $h_{\nu'}$ is another representative of h^* and yields $'k_1$ then for some $\mu > \nu, \nu'$: $\pi_\mu^{*\nu}h_\nu - \pi_\mu^{*\nu'}h_{\nu'} = 0$, from which readily follows that $'k - 'k_1$ has coordinates zero for every $\mu' > \mu$. Since $\{\mu'\}$ is cofinal in Λ we have $'k = k_1$.

Thus $'k$ depends solely upon $'g$ and h^*. We now define $g'h^* = 'k$ and this function is manifestly distributive.

Regarding continuity, since H^* is discrete, we must prove it only for $'gh^*$ as a function of $'g$. Taking M as above we have $'gh^* = 'k$ where $'k$ has the coordinates $k_\mu = g_\mu h_\mu$ for all $\mu \in \mathrm{M}$ where M is cofinal in Λ and depends solely upon h^*. Let $'g_0 = \{g_{0\lambda}\}$, $'k_0 = 'g_0 h^*$, U_λ any open set of K_λ, $V_\lambda = \{'k \mid k_\lambda \in U_\lambda\}$. Since $\{V_\lambda\}$ is a base for $'K$ (I, 38.2) if V is any neighborhood of $'k_0$ in $'K$ there is a V_λ between $'k_0$ and V. Since θ_λ^μ is continuous and M cofinal in Λ, there is a $\mu > \lambda$ and a $U_\mu \ni k_{0\mu}$ such that $\theta_\lambda^\mu U_\mu \subset U_\lambda$ and hence $'k_0 \subset V_\mu \subset V_\lambda$. Since $g_\mu h_\mu$ for h_μ fixed is continuous in g_μ there is a neighborhood W_μ of $g_{0\mu}$ in G_μ such that $g_\mu \in W_\mu \rightarrow g_\mu h_\mu \in U_\mu$. Therefore $W = \{'g \mid g_\mu \in W_\mu\}$ is a neighborhood of $'g$ in $'G$ such that $'g \in W \rightarrow 'gh^* \in V$. This proves that $'gh^*$ is continuous. Hence $'G$ and H^* are paired to $'K$.

Second case. This time there are three direct systems: $S^* = \{G_\lambda ; \pi_\lambda^{*\mu}\}$, $S_1^* = \{H_\lambda ; \omega_\lambda^{*\mu}\}$, $S_2^* = \{K_\lambda ; \theta_\lambda^{*\mu}\}$, all three directed by $\Lambda = \{\lambda; >\}$ with limit-groups G^*, H^*, K^* and with the relation of permanence

$$\pi_{\lambda}^{*\mu} g_{\mu} \cdot \omega_{\lambda}^{*\mu} h_{\mu} = \theta_{\lambda}^{*\mu}(g_{\mu} h_{\mu}), \qquad \lambda > \mu. \tag{17.3}$$

We now prove:

(17.4) *G^*, H^* are paired to K^* under a multiplication g^*h^* such that if g_λ, h_λ are representatives of g^* and h^* then $g_\lambda h_\lambda$ is a representative of g^*h^*.*

Let g_λ, $h_{\lambda'}$ be representatives of g^*, h^*. Choose any $\mu > \lambda, \lambda'$. Both have representatives for the index μ, and hence for all the elements of some $\mathrm{M} = \{\mu\}$ cofinal in $\{\Lambda\}$. We may show then that $g_\mu h_\mu = k_\mu$ are the representatives of a $k^* \in K^*$ which is independent of M and distributive in g^*, h^*. The details of the proof are essentially like those of the preceding case and so they are omitted.

(17.5) The systems of (13.7), (14.8) give rise to an interesting generalization of (17.4). We will suppose both systems indexed by $\{\lambda; >\}$. The notations of (13.7) remain the same except that the letters H, h are to be replaced everywhere by G'', g''. Thus we will have G''_λ for H''_λ, etc. The notations of (14.8) are modified in that G, g, H, h, τ are replaced by H, h, H'', h'', τ^*.

We suppose then G''_λ, H''_λ paired to K in a multiplication $g''_\lambda h''_\lambda$ which satisfies the permanence relation analogous to (16.3):

$$g''_\lambda \cdot (\omega_\lambda^{*\mu} h''_\mu) = (\omega_\mu^\lambda g''_\lambda) \cdot h''_\mu. \tag{17.6}$$

(17.7) *Under the preceding circumstances the groups $°G$, $°H$ are paired to K in a multiplication gh such that if g_λ is a coordinate of $g \in °G$ and h_λ a representative of $h \in °H$ then $gh = (\tau_\lambda g_\lambda) \cdot (\overset{*}{\tau}_\lambda h_\lambda)$.*

In view of (17.6) gh is a function of g, h alone with values in K and is distributive in both variables. Since g_λ is a continuous function of g, and the operations are continuous gh is continuous in g, hence in both g, h since $°H$ is discrete, and this proves (17.7).

(17.8) Similar extensions may be given for (17.2, 17.4) and they are left to the reader.

§5. CHARACTERS. DUALITY

18. (18.1) Definition. *A character h of a group G is a homomorphism $G \to \mathfrak{P}$ (= group of the reals* mod 1). *Instead of the functional notation $h(g)$ for the value of h at g we denote it by a product gh.*

If h_1, h_2 are two characters we define $h_1 + h_2$ as the character given by $g(h_1 + h_2) = gh_1 + gh_2$, and the character 0 as the one mapping G into 0. Under these definitions $H = \{h\}$ is a group. Following Pontrjagin we topologize H thus: If E is any compact subset of G and P any nucleus of $\mathfrak{P}$ we choose as nuclear base for H the family of all the sets $N(E, P) = \{h \mid gh \in P; g \in E\}$.

(18.2) *The topology assigned to H turns it into a topological group, known as the character-group of G.*

We must verify (a), $\cdots$, (f) of (2.4). The verification of (a), (b) is immediate.

Suppose $h_0 \in N(E, P)$. The image E_1 of E under h_0 is compact and hence closed in $\mathfrak{P}$. Since E_1 and the complement of P in $\mathfrak{P}$ are disjoint closed subsets of $\mathfrak{P}$ there is a real number η between 0 and 1 such that if $p \in E_1$ then $(p \pm \eta) \in P$. Hence if P' is the nucleus $-\eta < p < \eta \bmod 1$ we have $h_0 + N(E, P') \subset N(E, P)$, and this is (2.4c).

From $N(E_1 \cup E_2, P_1 \cap P_2) \subset N(E_1, P_1) \cap N(E_2, P_2)$ follows (2.4d). If $h \neq 0$ there is a g such that $gh \neq 0$, hence a $P \not\ni gh$, and so $h \notin N(g, P)$ which is (2.4e). Finally given P there is a P' such that $P' - P' \subset P$ and hence $N(E, P') - N(E, P') \subset N(E, P)$, which is (2.4f). This proves (18.2).

19. (19.1) *If the group G is compact, discrete or more generally locally compact then its character-group H is, respectively, discrete, compact or locally compact, and in addition the multiplication gh defined in* (18.1) *pairs G and H to $\mathfrak{P}$.*

Although the locally compact case offers no major difficulty it is not needed later and so we shall treat only the other two cases.

(a) *G is compact.* Let P be the nucleus of $\mathfrak{P}$ defined by $-1/4 < p < 1/4 \bmod 1$. Clearly the only closed subgroup of $\mathfrak{P}$ in P is the element 0 (3.3). On the other hand if $h \in N(G, P)$ then $h(G)$ is a closed subgroup of $\mathfrak{P}$ in P and so $h(G) = 0$ or $h = 0$, and finally $N(G, P) = 0$. Since 0 is an open set of H every $h \in H$ is an open set and so H is discrete.

(b) *G is discrete.* Let $\{\mathfrak{P}_g\}$ indexed by G be such that $\mathfrak{P}_g = \mathfrak{P}$ and let $'H = \mathbf{P}\mathfrak{P}_g$. The group H is clearly a subgroup of $'H$ (in the algebraic sense, i.e., except for the topology). We wish to show that it has also the correct relative topology. Let H' be the space with the same elements as H and the relative topology as a subset of $'H$. It is sufficient to show that they have a common nuclear subbase. Since G is discrete its compact subsets are its finite subsets. Hence if $\{P\}$ are the nuclei of $\mathfrak{P}$ the collection $\{N(g, P)\}$ is a nuclear subbase for H. Now it follows from the definition of $N(g, P)$ and the known topology of $'H$ that the same collection is likewise a nuclear subbase for H' thus proving our assertion.

Consider the function on $'H$ to $\mathfrak{P}$ defined by $\varphi_{g,g'}(h) = h_{g+g'} - h_g - h_{g'}$. A n.a.s.c. in order that $h \in H$ is $\varphi_{g,g'}(h) = 0$ for all pairs $g, g' \in G$. Let $F(g, g')$ be the subset of $'H$ defined by $\varphi_{g,g'}(h) = 0$. Assuming g, g' fixed the coordinates $h_{g+g'}, h_g, h_{g'}$ are continuous in h, and since the group operation in $\mathfrak{P}$ is continuous so is $\varphi_{g,g'}$. It follows that $F(g, g')$ is closed in $'H$. Hence $H = \bigcap\{F(g, g') \mid g, g' \in G\}$ is likewise closed in $'H$, and since $'H$ is compact so is H.

(c) *Continuity of gh.* Let $g_0h_0 = p_0$. Given a nucleus P of $\mathfrak{P}$ there exists a symmetrical one P_1 such that $P_1 + P_1 = P_1 - P_1 \subset P$. Since G is locally compact g_0 has a neighborhood U with $\bar{U}$ compact. Since h_0 is continuous g_0 has also a neighborhood $V \subset U$ such that $h_0(V) \subset p_0 + P_1$, and $V \subset U$ implies that $\bar{V}$ is compact. Consequently $g \in V, h \in h_0 + N(\bar{V}, P_1) \rightarrow gh \in p_0 + P$. Therefore gh is continuous.

(19.2) It is clear that if we had interchanged throughout the roles of G and H but *not their order* the argument would go through as before. A similar observation may be made repeatedly in the sequel.

(19.3) *If a locally compact group G has a countable base then its character-group H has one also.*

If E is a compact subset of G and W is an open set of $\mathfrak{P}$, we will write $M(E, W) = \{h \mid g \in E,\ gh \in W\}$ and first prove:

(a) *$M(E, W)$ is an open set of H.*

Let $h_0 \in M(E, W)$. Since h_0 is continuous $A = h_0(E)$ is a compact subset of W. If $p \in A$ there is a nucleus P of $\mathfrak{P}$ such that $p + P \subset W$. Paraphrasing the reasoning in the proof of (5.5) (with A in place of G', loc. cit.) we show that P may be chosen independent of p, i.e., such that the inclusion holds for every $p \in A$. Suppose now $h \in N(E, P)$. We will have for $g \in E$: $g(h_0 + h) \in gh_0 + P \subset A + P \subset W$, and hence $h_0 + h \in M(E, W)$, or $h_0 + N(E, P) \subset M(E, W)$. Since $N(E, P)$ is a nucleus of H this proves (a).

Let now $\{U_\lambda\}$ be a base for G such that the $\overline{U}_\lambda$ are compact (I, 29.2) and let $\{W_n\}$ be a countable base for $\mathfrak{P}$. By (a) $V_{\lambda n} = M(\overline{U}_\lambda, W_n)$ is open in H. We prove:

(b) *$\{V_{\lambda n}\}$ is a subbase for H.*

Given any open set V of H and $h_0 \in V$, some $h_0 + N(E, P) \subset V$. Choose a P' such that $P' + P' \subset P$. If $g \in E$ there exists a $W_n \ni gh_0$ such that $W_n - gh_0 \subset P'$, and then owing to the continuity of h_0, a $U_\lambda \ni g$ such that $gh_0 - h_0(\overline{U}_\lambda) \subset P'$. Since $\{U_\lambda \cap E\}$ is an open covering of the compact set E there is a finite subcovering. Therefore there is a finite set $\{g_i\}$, and for each g_i sets U_{λ_i}, W_{n_i} such that:

(c) $\bigcup U_{\lambda_i} \supset E$.

(d) $W_{n_i} - gh_0 \subset P$ for every $g \in U_{\lambda_i} \cap E$.

Let $V_\rho = \bigcap V_{\lambda_i n_i}$. If $h \in V_\rho$, hence $h \in V_{\lambda_i n_i}$, and $g \in E$ then by (c) some $U_{\lambda_i} \ni g$ and $gh \in W_{n_i} \rightarrow gh - gh_0 \in P \rightarrow h - h_0 \in N(E, P)$. Hence $V_\rho \subset h_0 + N(E, P) \subset V$. Since clearly $h_0 \in V_\rho$, V_ρ is between h_0 and V, and since V_ρ is a finite intersection of sets $V_{\lambda n}$ property (b) is proved.

If G has a countable base we may choose $\{U_\lambda\}$ countable (I, 6.8). Then the related subbase $\{V_{\lambda n}\}$ will also be countable and this implies (19.3).

(19.4) *If G is a compactum* [*countably discrete*] *then its character-group H is countably discrete* [*a compactum*].

If G is a compactum it has a countable base, and so has H by (19.3). Since $\{h\}$ is a base of which no subcollection is a base by (I, 6.8) it must be countable. If G is countably discrete $\{g\}$ is a countable base. Hence H has a countable base by (19.3), and since it is compact it is a compactum (I, 46.4). For examples see (21.2, 21.3, 21.6).

(19.5) Let G, H be paired to $\mathfrak{P}$ under a multiplication gh and let G^*, H^* be their character-groups. If g is kept fixed gh becomes a function φ_g on H to $\mathfrak{P}$, and there is a similar function φ_h on G to $\mathfrak{P}$.

(19.6) $g \rightarrow \varphi_g$ [$h \rightarrow \varphi_h$] *defines a homomorphism* $\chi_g : G \rightarrow H^*$ [$\chi_h : H \rightarrow G^*$].

Clearly χ_g, χ_h are homomorphisms in the algebraic sense, so continuity alone requires proof, and it will be sufficient to give it for χ_g. It reduces at once to showing that given a nucleus $N(E, P)$ of H^* there is a nucleus N of G

such that $\chi_g N \subset N(E, P)$. This reduces in turn to finding N such that $g \in N$, $h \in E \rightarrow gh \in P$. Since gh is continuous, if $h \in E$ there is a nucleus N_h of G and a neighborhood U_h of h in H such that $g \in N_h$, $h \in U_h \rightarrow gh \in P$. Since E is compact the open covering $\{U_h \cap E\}$ of E has a finite subcovering $\{U_{h_i} \cap E\}$ and $N = \bigcap N_{h_i}$ is a nucleus of G behaving as required. This proves (19.6).

(19.7) The two homomorphisms χ_g, χ_h are said to be *induced* by the pairing gh. If both are isomorphisms then G, H are said to be *dually paired* and the pairing is called a *dual* pairing.

(19.8) If H is the character-group of G then the multiplication gh giving the value of h at g (18.1) is known as the *natural* multiplication of the two groups. Similarly with G, H interchanged.

(19.9) *If G, H are dually paired they are orthogonal.*

20. We shall now state the fundamental results of the Pontrjagin-van Kampen duality theory.

(20.1) Duality theorem. *Let G, H be locally compact, and let one of the two be the character-group of the other with gh as the natural multiplication. Then the multiplication gh is a dual pairing.*

An apparently more general but equivalent form of the theorem is:

(20.2) *Let G, H be locally compact and paired to $\mathfrak{P}$ under a multiplication gh. If one of the induced homomorphisms χ_g, χ_h is an isomorphism so is the other, so that the multiplication gh is a dual pairing for the two groups.*

If χ_g is an isomorphism we may identify G with the character-group of H so that we have $g = \chi_g g$ and this reduces (20.2) to (20.1). Since (20.1) is a special case of (20.2) the two are equivalent.

Referring to (19.1, 19.4) we may also state:

(20.3) *In the collection of all locally compact groups there may be set up a one-one correspondence to within isomorphisms such that corresponding groups are the character-groups of one another. This correspondence establishes similar one-one correspondences*: (a) *between the collections of all compact and all discrete groups*; (b) *between the collections of all the groups which are compacta and all the countably discrete groups.*

The full proof of these theorems in their general form will be found in van Kampen [a] and A. Weil [W, VI]. The initial case treated by Pontrjagin [P, V] corresponds to G a compactum and H countably discrete. The only case required in the sequel is that of G compact and H discrete and will be dealt with in (21). Explicitly stated it is:

(20.4) *If G is compact, H discrete and one of them is the character-group of the other with gh as the natural multiplication, then each is the character-group of the other and gh is a dual pairing.*

Before considering (20.4) we shall discuss certain consequences of (20.1).

(20.5) *If G, H are locally compact and dually paired, G' is a closed subgroup of G, H' its annihilator in H, then G' is the annihilator of H' in G.*

We will require the following property, deduced here from (20.2), but actually a step in its derivation. The part required in (21) will be discussed there (21.7).

(20.5a) *If G is locally compact and $g \in G$, $g \neq 0$, there is a character h of G which does not annul g.*

Let H be the character-group of G and gh their natural multiplication. Since we may replace G by an isomorph, by (20.1) we may assume that it is the character-group of H. Since $g \neq 0$ there is an h such that $gh \neq 0$, and h answers the question.

Proof of (20.5). Let $G/G' = G^* = \{g^*\}$ and let π be the projection $G \to G^*$. Under the hypothesis the annihilator G'' of H' in G contains G'. Take now g_0 in G but not in G' and let $g_0^* = \pi g_0$. Then $g_0^* \neq 0$. Since G^* is locally compact (5.6) by (20.5a) there is a character h^* of G^* such that $g_0^* h^* \neq 0$. Hence the function $h(g)$, whose values gh are given by $gh = (\pi g)h^*$, is a character of G. It is clear that h annuls G' and so $h \in H'$. On the other hand $g_0 h = g_0^* h^* \neq 0$, and so $g_0 \notin G''$. Thus $G' \supset G''$ and hence $G' = G''$, which is (20.5).

(20.6) *A n.a.s.c. for a compact and a discrete group to be orthogonal in a pairing to $\mathfrak{P}$ is that the pairing be dual.*

Sufficiency is a consequence of (19.9). Regarding necessity, let G be the compact group and H the discrete group and let H^* be the character-group of G. Then χ_h, as defined in (19.6), is a homomorphism $H \to H^*$ and since H^* is discrete $H_1 = \chi_h H$ is closed in H^*. Furthermore since G and H are orthogonal χ_h is univalent. Therefore χ_h is an isomorphism $H \to H_1$ and G, H_1 are orthogonal under the natural multiplication of G and H^*. Since G, H_1 are orthogonal, the annihilator of H_1 in G is 0, and so H_1 is H^* itself, hence $H \cong H_1 = H^*$. This together with (20.2) yields (20.6).

(20.7) *Let $\{G_\lambda\}$, $\{H_\lambda\}$ be two systems of groups both indexed by $\{\lambda\}$, and such that:* (a) *the G_λ are compact and the H_λ discrete;* (b) *G_λ, H_λ are dually paired under a multiplication $g_\lambda h_\lambda$. Then $G = \mathbf{P}G_\lambda$, $H^w = \mathbf{P}^w H_\lambda$ are dually paired under the multiplication $gh = \sum g_\lambda h_\lambda$.*

This is an immediate consequence of (16.1) and (20.6).

(20.8) *Let $S = \{G_\lambda ; \pi_\mu^\lambda\}$, $S^* = \{H_\lambda ; \pi_\lambda^{*\mu}\}$ be an inverse and a direct system paired to $\mathfrak{P}$ (see 16.2) and such that the G_λ and H_λ are dually paired throughout. (We will call S, S^* "dual systems.") Then the limit-groups $'G$, H^* are also dually paired and this under the multiplication of* (16.4).

Let G, H^w be the same as in (16.1). We have $'G \subset G$ and $H^* = H^w/L$, where L annuls $'G$ (see the proof of 16.4). We shall show that $'G$ is the annihilator of L. In fact let $g \in G$ annul L. If $\lambda > \mu$ and $h_\mu \in H_\mu$ we have $g(h_\mu - \pi_\lambda^{*\mu} h_\mu) = 0 = g_\mu h_\mu - g_\lambda(\pi_\lambda^{*\mu} h_\mu) = (g_\mu - \pi_\mu^\lambda g_\lambda)h_\mu$ by (16.3). Since G_μ, H_μ are orthogonal we must have $g_\mu = \pi_\mu^\lambda g_\lambda$ and so $g \in {'G}$.

We conclude now from (20.5) that L is likewise the annihilator of $'G$ and by (15.5) that $'G$, H^* are orthogonal. Since they are also paired to $\mathfrak{P}$ (20.8) follows from (20.6).

(20.9) As an application of the duality theorems we discuss certain groups needed in the following chapters. Given any group G and an integer m we set:

$G[m]$ = the subgroup consisting of the elements whose order divides m;
$G(m)$ = the subgroup of the elements mg;
$G^*(m) = G/\overline{G(m)}$.

The group $G[m]$ is always closed in G. For if we define $f(g) = mg$, f is a homomorphism $G \to G$ and $G[m]$ is its kernel and hence closed. Not so, however, regarding $G(m)$ and we lay down with Steenrod the

(20.10) DEFINITION. *A division-closure group is a group G such that all the subgroups $G(m)$ are closed.*

(20.11) *Compact groups, discrete groups and fields are division-closure groups* (Steenrod [a]).

This is trivial when G is discrete or a field. When G is compact then $G(m)$ as the continuous image of a compact group must also be compact and therefore closed.

(20.12) *If G and H are the one compact, the other discrete and they are dually paired then $G[m]$ and $H^*(m)$ are likewise dually paired.*

Let g annul $H(m)$. Then $g(mh) = (mg)h = 0$ whatever h and so $mg = 0$, or $g \in G[m]$. Conversely, if $g \in G[m]$ then $g(mh) = 0$ whatever h and so $G[m]$ is the annihilator of $H(m)$. By (20.11) $H(m)$ is closed. By (20.5) $H(m)$ is then also the annihilator of $G[m]$, and so by (15.5) $G[m]$ and $H^*(m)$ are orthogonal to $\mathfrak{P}$. Suppose first G compact and H discrete. Since $G[m]$ is closed in G it is compact and since H is discrete so is $H^*(m)$ (since the natural projection $H \to H^*(m)$ is open). Therefore $G[m]$ and $H^*(m)$ are dually paired by (20.6). Suppose now G discrete and H compact. Then $G[m]$ is also discrete. Since $H(m)$ is closed (20.11) we are justified in considering $H^*(m)$ as a topological group and it is compact (5.5). Therefore by (20.6) $G[m]$ and $H^*(m)$ are dually paired here also.

(20.13) REMARK. One might expect that all groups are of the division-closure type. The following example due to Steenrod shows that this is not the case. G is the subgroup of the additive group of the real numbers consisting of the rational numbers $\{2^{-n}\cdot m \mid n, m$ any integers$\}$. Then $G(3)$ is dense in G and different from G since it does not contain 1/2. We have then: $\overline{G(3)} = G \neq G(3)$ and so $G(3)$ is not closed. Therefore G is not a division-closure group.

Another example pointed out to the author by L. J. Savage and with well known historical significance is the following: G is the multiplicative group of all real positive rational numbers, with the topology of the straight line; $G(2)$ is the subgroup consisting of all the rational squares. Since 2 is not a perfect square $2 \notin G(2)$ and yet $2 \in \overline{G(2)}$ since it is the limit of an increasing sequence of rational squares.

Let $\mathfrak{J}$ be the group of the integers. $\mathfrak{P}[m]$ is the group of fractions n/m mod 1, $\mathfrak{J}^*(m)$ the group of residues mod m, both cyclic of order m. Hence

(20.14)
$$\mathfrak{P}[m] \cong \mathfrak{J}^*(m).$$

21. We will now take up the proof of the duality theorem (20.4). The general argument runs thus: Following Pontrjagin [P, V] we first dispose of the so-called

elementary groups and of certain preliminary results (21.1, $\cdots$, 21.11), after which the proof is brought to a rapid conclusion (21.12, 21.14) by means of a device communicated to the author by van Kampen and Whitney.

Notations. Generally G denotes a compact group, H a discrete group, G^*, $\cdots$, the character-groups. We write

$$\mathfrak{P} = \{p\}, \quad \mathfrak{J} = \{i\}, \quad \mathfrak{P}^* = \{p^*\}, \quad \mathfrak{J}^* = \{i^*\}.$$

(21.1) We begin with an important preliminary observation. Supposing G, H orthogonal in a multiplication gh to $\mathfrak{P}$ we have the homomorphism χ_g $[\chi_h]$: $G \to H^*$ $[H \to G^*]$ whereby g $[h]$ is sent into the character of H $[G]$ whose value at $h[g]$ is gh. In view of orthogonality χ_g, χ_h are univalent. Since G and H are compact and discrete, $\chi_g G \cong G$, $\chi_h H \cong H$. Thus we may identify G with $\chi_g G$ and H with $\chi_h H$, so that G, H will be subgroups of H^*, G^*. This procedure will be followed wherever possible. Under the circumstances to prove that $G = H^*$ $[H = G^*]$ it will be sufficient to show that G $[H]$ contains all the characters of H $[G]$.

(21.2) *$\mathfrak{P}$ and $\mathfrak{J}$ are isomorphic with one another's character-groups and dually paired by the numerical product pi* mod 1.

At all events $\mathfrak{P}$ and $\mathfrak{J}$ are orthogonal in the multiplication in question, and so by (21.1) it is only necessary to show that each contains all the characters of the other. Now if $i^* \in \mathfrak{J}^*$ sends 1 into p it sends i into pi, and the values of i^* on $\mathfrak{J}$ are those of pi. Thus $i^* \in \mathfrak{P}$. Passing to $\mathfrak{P}^*$ if $p^* \in \mathfrak{P}^*$ then p^{*-1} (0) is a closed subgroup of $\mathfrak{P}$, and is by (3.3) $\mathfrak{P}$ itself or a finite cyclic subgroup of $\mathfrak{P}$. In the former case $p^* = 0$; in the latter case on the circumference the points of the subgroup are the vertices of a regular i-sided polygon, one of which is the zero of $\mathfrak{P}$. As a consequence p^* maps the arc $0 < p < 1/i$ topologically on the arc $0 < p < 1$. The mapping may be sense-preserving or sense-reversing. According as one or the other alternative occurs we will have $p^*(m/ni) = \epsilon m/n$, $\epsilon = \pm 1$, m/n a proper positive fraction. Since $\{m/n\}$ is dense on $\mathfrak{P}$ by a standard argument $p^*(p) = \epsilon pi$ for all $p \in \mathfrak{P}$, and hence $p^* \in \mathfrak{J}$. This completes the proof of (21.2).

(21.3) *If G is cyclic of (finite) order n then $G \cong G^*$. More precisely let $G_1 \cong G$ be the subgroup $\{m/n\}$ of $\mathfrak{P}$ and $G_2 \cong G$ the additive group of the integral residues* mod n. *Then G_1, G_2 are dually paired under the ordinary product $g_1 g_2$ taken* mod 1 (proof similar to that of 21.2).

(21.4) *If G_i, H_i $(i = 1, 2)$ are dually paired in a multiplication $g_i h_i$ then $G = G_1 \times G_2$ and $H = H_1 \times H_2$ are dually paired in the multiplication $gh = g_1 h_1 + g_2 h_2$, where $g = (g_1, g_2)$ and $h = (h_1, h_2)$.*

By (16.1) G, H are orthogonal and so applying (21.1) we merely have to show that, say, G contains all the characters of H. Identify G_i with $G_i \times 0$ so that $g_i = (g_i, 0)$ and similarly H_i with $H_i \times 0$. Since every h may be written $h = h_1 + h_2$, we may write $h^* = g_1 + g_2$, $g_i(h_1 + h_2) = h^*(h_i)$. Therefore $H^* = G$, and similarly $G^* = H$, proving (21.4).

(21.5) Definitions. *A compact group is said to be elementary if it is the product of a finite-dimensional toroidal group* (6.3) *by a finite group. A discrete group is said to be elementary if it has a finite number of generators.*

(21.6) *Let* $G = G_1 \times G_0$, $H = H_1 \times G_0$ *where* G_1 *is an n-dimensional toroidal group,* H_1 *a discrete free group of rank n and* G_0 *a finite group. Then each is isomorphic with the character-group of the other.*

Since we may replace G, H by isomorphs we may suppose that $G = G_1 \times \cdots \times G_m$, $H = H_1 \times \cdots \times H_m$, where $H_i = \mathfrak{J}$ and then $G_i = \mathfrak{P}$, or else H_i is cyclic of finite order and then $G_i = H_i^* \cong H_i$. If $g_i h_i = g_i(h_i)$ is the natural multiplication of G_i with H_i corresponding to $G_i = H_i^*$, then $g_i h_i$ is a dual pairing. Hence by repeated application of (21.4) the multiplication $gh = \sum g_i h_i$, $g = (g_1, \cdots, g_m)$, $h = (h_1, \cdots, h_m)$ is a dual pairing for G, H and this proves (21.6).

Rather than (21.6) we shall need later the following closely related property which is a special case of (20.4):

(21.6a) *If* H *is elementary and discrete then* H *and* H^* *are dually paired in their natural multiplication.*

We may assume $H = H_1 \times \cdots \times H_m$ where the notations are as before. Furthermore in accordance with (7.3) we identify H_i with the subgroup $0 \times \cdots \times 0 \times H_i \times 0 \times \cdots \times 0$ of H. We have just shown that G, H are dually paired by gh. As a consequence they are orthogonal and so by (21.1) we may assume $G \subset H^*$, the multiplication gh being then the value of the natural multiplication h^*h when $h^* \in G$. Take now any h^* whatever. We have $h^* \mid H_i = g_i \in G_i$ and $g = (g_1, \cdots, g_m) \in G$ is a character of H such that $gh = h^*(h) = h^*h$. Hence $h^* = g$, and so $H^* = G$, showing that H, H^* are paired in the asserted way.

(21.7) If G is any group, G' a closed subgroup of G, g_1^* a character of $G_1 = G/G'$, π the natural projection $G \rightarrow G_1$, then $g^* = g_1^*\pi$ is a character of G. We call g_1^* the *projection* of g^*.

(21.8) Conversely, in the same notations, if g^* is a character of G which takes the value zero on G' then $g^*(g)$ depends solely upon the coset of g mod G', and by (5.3c) there is a character g_1^* of G_1 which is the projection of g^*.

(21.9) *If* H' *is a subgroup of the discrete group* H *then every character* g' *of* H' *may be extended to a character* g *of* H, *i.e.,* g *exists such that* $g \mid H' = g'$.

Let $\gamma_\lambda = (g_\lambda, H_\lambda)$ where H_λ is a subgroup of H containing H' and g_λ is a character of H_λ such that $g_\lambda \mid H' = g'$. Order the collection $\Gamma = \{\gamma_\lambda\}$ by $\gamma_\lambda \succ \gamma_\mu$ whenever $H_\lambda \supset H_\mu$ and $g_\lambda \mid H_\mu = g_\mu$. If $\{\gamma_\mu\}$ is a simply ordered subsystem set $H_\nu = \sum H_\mu$, and define the character g_ν of H_ν by the condition that if $h \in H_\mu$ then $g_\nu(h) = g_\mu(h)$. It is clear that $\gamma_\nu = (g_\nu, H_\nu) \in \Gamma$ and $\gamma_\mu \prec \gamma_\nu$ for every μ. We may thus apply Zorn's lemma, and it asserts the existence of a maximal $\gamma = (g, H_0)$. Since $g \mid H' = g'$, to prove (21.9) we merely need to show that $H_0 = H$. Suppose this false and let $h_1 \in H$, $h_1 \notin H_0$. If H_1 is the subgroup of H generated by the elements $h_0 + mh_1$, $h_0 \in H_0$, then $H_1 \neq H_0$.

Let q denote the least positive integer if any exists, such that $qh_1 \in H_0$; otherwise set $q = \infty$. Define a character g_1 of H_1 as follows: On H_0 the values of g_1 are those of g; if $q = \infty$ take $g_1(h_1) = 0$, if $q \neq \infty$ take for $g_1(h_1)$ one of the q determinations of $g_1(qh_1)/q$. Clearly $\gamma_1 = (g_1, H_1) \in \Gamma$ and $\gamma_1 > \gamma, \gamma_1 \neq \gamma$, a contradiction. Therefore $H_0 = H$ and (21.9) is proved.

(21.10) *If G is compact or discrete and $g_0 \in G$, $g_0 \neq 0$, then there is a character g^* of G such that $g^*(g_0) \neq 0$.*

The proof for G compact requires an extensive appeal to integration in groups and to the theory of representations and so it is omitted. The proof given by Pontrjagin [P, 146C] for G a compactum is valid for any compact G. See also van Kampen [a, proof of Lemma 3] and Gelfond-Raikov [a].

Suppose now G discrete. The multiples of g_0 generate a cyclic subgroup G_1 of G and by (21.2, 21.3) G_1 has a character g_1^* such that $g_1^*(g_0) \neq 0$. By (21.9) there is a character g^* of G such that $g^* \mid G_1 = g_1^*$ and so $g^*(g_0) \neq 0$. This proves (21.10) for a discrete G.

(21.11) *Let $G = H^*$ (G compact, H discrete) and let H' be a subgroup of H and G' its annihilator in G. Then G' is closed in G* (15.4) *and $G_1 = G/G' = H'^*$. More precisely $\chi_{g_1} G_1 = H'^*$.*

Let gh be the natural multiplication of G, H. Since $G = H^*$ the only g annulling H is $g = 0$. Given $h \in H$ by (21.10) there is a g such that $gh \neq 0$ and hence the only h annulling G is $h = 0$. Thus G, H are orthogonal, and so (15.5b) the compact group G_1 and discrete group H' are orthogonal in the multiplication g_1h' induced by gh in accordance with (15.5a). Thus again by (21.1) we merely need to show that G_1 contains every character h'^* of H'.

Since H is discrete by (21.9) there is a character g of H such that $g \mid H' = h'^*$. If g_1 is the coset of g mod G' then $g_1h' = h'^*(h')$, hence $h'^* = g_1$ and (21.11) follows.

(21.12) *Let G be compact or discrete and G^*, $G^{**} = (G^*)^*$, $G^{***} = (G^{**})^*$ the successive character-groups. If G^{**} is not $\chi_g G$, then G^{***} is not $\chi_{g^*} G^*$.*

Using orthogonality in the appropriate natural multiplications and by reference to (21.1) we may suppose $G \subset G^{**}$ and $G^* \subset G^{***}$. Suppose then $G^{**} \neq G$. Since G^{**}/G is compact or discrete and different from 0, by (21.7) it has a character different from 0 with an extension h to G^{**} which is zero on G but not everywhere. Hence h is an element of G^{***} but not of G^*. This proves (21.12).

(21.13) As a consequence of the preceding result if G is compact [discrete] and different from G^{**} then G^* is discrete [compact] and different from $(G^*)^{**}$. Therefore in proving (20.4) it is sufficient to consider $G = H^*$, G compact.

(21.14) Supposing then G compact, and $G = H^*$ we will prove that $G^* = \chi_h H$. Once more by (21.1) we may suppose $H \subset G^*$ and show that there is no $g^* \notin H$. Suppose such a g^* exists. Take in $\mathfrak{P}$ the nucleus $P = \{p \mid |p| < 1/4 \bmod 1\}$. Thus P contains no closed subgroup different from 0. Since g^* is a continuous function on G to $\mathfrak{P}$, there must exist a nucleus N of G such that $g^*N \subset P$. Since $G = H^*$ we may choose N of form $N(\{h_1, \cdots, h_r\}, P')$. Let H' be the subgroup of H generated by $\{h_1, \cdots, h_r\}$ and G' its annihilator in G. By (21.11) if $G_1 = G/G'$ then $\chi_{g_1} G_1 = H'^*$. Identifying now g_1 with $\chi_{g_1} g_1$, hence

G_1 with H'^*, makes the multiplication g_1h' of (21.11) the natural multiplication. Since H' is elementary, by (21.6a) G_1, H' are dually paired by g_1h'.

Now if g^* annuls G', g^* has for projection a character g_1^* of G_1 (21.8). Since H', G_1 are dually paired by g_1h' there exists an h' such that $g_1h' = g_1^*(g_1)$ for every $g_1 \in G_1$.. Hence whatever $g \in G$ if g_1 is the coset of g mod G' we will have $gh' = g_1h' = g_1^*(g_1) = g^*(g)$. It follows that h' is the character g^* of G and so $g^* \in H$ contrary to assumption. Thus g^* cannot annul G'. As a consequence $g^*(G')$ is a closed subgroup of $\mathfrak{P}$ which is different from 0 and hence $\not\subset P$. Since $G' \subset N$, this is a contradiction and the proof of (20.4) is completed.

§6. VECTOR SPACES

22. While the earlier group-duality theorems utilized in topology (and explicitly contained in the duality theorems for infinite manifolds of [L, VII, §3]) have been eclipsed by the brilliant results of Pontrjagin, they have not been reduced to mere corollaries. They refer essentially to groups with a field as domain of operators, i.e., to vector spaces. We propose to consider these spaces with particular emphasis on duality.

Henceforth we utilize a fixed field Ω which is taken with discrete topology (discrete field). The elements of Ω are usually denoted by $\alpha, \beta, \gamma, \cdots$.

(22.1) Definition. *A vector space over Ω is an additive group $G = \{g\}$ for which there is defined an operation assigning to every pair (α, g) an element of G written αg which is continuous and distributive in both variables and such that $1 \cdot g = g$, $\alpha(\alpha' g) = (\alpha\alpha')g$ for every $\alpha, \alpha' \in \Omega$ and $g \in G$. Notice that since Ω is discrete, αg is continuous in (α, g) when it is continuous in g alone.*

Let G, H be vector spaces over Ω and let τ be a homomorphism $G \to H$. We say that τ is *linear* whenever $\tau(\alpha g) = \alpha(\tau g)$, $\alpha \in \Omega$, $g \in G$. Suppose that τ is an isomorphism and linear in the sense just stated. It means that $\tau g = g' \to \tau(\alpha g) = \alpha g'$, and hence that $g = \tau^{-1} g' \to \alpha g = \tau^{-1}(\alpha g')$, or τ^{-1} is likewise linear. Therefore if an isomorphism τ is linear so is τ^{-1}. This makes it unnecessary to introduce the concept of "bilinear" isomorphism.

Let now G, H, K be vector spaces over Ω such that G, H are paired to K under a multiplication gh. This multiplication is said to be *linear* whenever $\alpha(gh) = (\alpha g)h = g(\alpha h)$, $\alpha \in \Omega$.

(22.2) *Fundamental conventions for vector spaces.* Hereafter unless otherwise stated a homomorphism of one vector space into another, or a multiplication pairing two vector spaces to a third will always be understood to be linear. Furthermore throughout the present chapter (and later also with chain- and related groups) vector spaces will be taken with a topology, likewise called *linear*, and fully described in (25). Until then the questions dealt with are really non-topological or rather independent of the topology.

23. If $A = \{g_a\}$ is a subset of the vector space G so is the set $\{\alpha g_a\}$ and it is denoted by αA. A *subspace* of G is a subgroup H such that $h \in H \to \alpha h \in H$

for every $\alpha \in \Omega$. It is easily seen that this property is equivalent to: $\alpha H = H$ for every $\alpha \in \Omega$ and $\neq 0$.

A finite set of vectors $g_1, \cdots, g_n$ is *linearly independent* whenever $\sum \alpha_i g_i = 0 \rightarrow$ every $\alpha_i = 0$. More generally a subset A of G is *linearly independent* if every finite subset of A has that property.

The intersection of any number of subspaces is clearly a subspace. Therefore all those containing a given set B intersect in a subspace H, the "smallest" subspace containing B, and said to be *spanned* by B. If B is linearly independent it is said to be a *base* for H.

Notice the mild deviations from the meaning previously attached to "linear independence" and "base" in (9). A supplementary mention "relative to Ω" will be used wherever needed to avoid misunderstanding, but this will rarely be necessary.

(23.1) *Let G be a vector space and H a subspace of G. Then:*

(a) *If C is a base for H there exists a base B of G containing C.*

(b) *There is a subspace H' of G such that $G = H + H'$, $H \cap H' = 0$. Consequently every g may be written uniquely in the form $g = h + h'$, $h \in H$, $h' \in H'$.*

If $H = 0$ we have $C = \emptyset$ and hence a special case of (23.1a) is:

(23.2) *Every vector space G has a base.*

Proof of (23.1a). Consider all the linearly independent sets A containing C. By Zorn's theorem there is a maximal set B. If $g \notin B$ then $B \cup g$ is not an A and so there must exist a non-trivial relation $\alpha g = \alpha_1 g_1 + \cdots + \alpha_n g_n$, $g_i \in B$, and we must have $\alpha \neq 0$ since the g_i are linearly independent. Therefore $g = \sum \alpha^{-1} \alpha_i g_i$ and B is a base.

Proof of (23.1b). In the same notations the complement C' of C in B spans an H' behaving as stated.

24. (24.1) *Any two bases of the same vector space have the same cardinal number* (proof by Chevalley).

Let B be a base, A any linearly independent set; $|A|$ and $|B|$ the cardinal numbers of A and B. It is sufficient to prove

$$|A| \leqq |B|. \tag{24.2}$$

Let φ denote a one-one transformation of a subset B_φ of B into a subset A_φ of A such that if A'_φ is the complement of A_φ in A then $B_\varphi \cup A'_\varphi$ is linearly independent. The set $\Phi = \{\varphi\} \neq \emptyset$; for if φ_0 sends the empty subset of B into A then $B_{\varphi_0} = A_{\varphi_0} = \emptyset$, $A'_{\varphi_0} = A$ and $B_{\varphi_0} \cup A'_{\varphi_0} = A$ is linearly independent. Therefore $\varphi_0 \in \Phi \neq \emptyset$.

Order Φ as follows: $\varphi \prec \varphi'$ whenever $B_\varphi \subset B_{\varphi'}$ and $\varphi = \varphi' \mid B_\varphi$. Let Ψ be a simply ordered subset of Φ and set $B^* = \bigcup\{B_\varphi \mid \varphi \in \Psi\}$. We may define a one-one mapping ψ of B^* onto a subset A^* of A such that $\psi \mid B_\varphi = \varphi$ for every $\varphi \in \Psi$ and moreover $A^* = \bigcup\{A_\varphi \mid \varphi \in \Psi\}$. Let $A^{*\prime}$ be the complement of A^* in A and let $g_1, \cdots, g_p, h_1, \cdots, h_q$ be a finite set of vectors of $B^* \cup A^{*\prime}$ where the g_i include all the vectors of the set in B^*, and hence the h_i are in $A^{*\prime}$. Then $\{g_i\} \subset B_\varphi$ for some $\varphi \in \Psi$, and $\{h_i\} \subset A^{*\prime} \subset A'_\varphi$. Therefore $g_1, \cdots, h_q$ are linearly independent. Thus $\psi \in \Phi$ and $\varphi \in \Psi \rightarrow \varphi \prec \psi$.

It follows from Zorn's theorem that Φ contains a maximal element φ_0. I say that $A_{\varphi_0} = A$. If this does not hold then $A'_{\varphi_0} \neq \emptyset$ and so it contains an element h. Let A''_{φ_0} be the set of elements different from h in A'_{φ_0}. Since the vectors of $B_{\varphi_0} \cup A'_{\varphi_0}$ are linearly independent h is not in the space H spanned by $B_{\varphi_0} \cup A''_{\varphi_0}$. On the other hand since B is a base we have $h = \alpha_1 g_1 + \cdots + \alpha_n g_n$, n finite, $g_i \in B$, and at least one of the g_i say $g_1 \notin H$. Therefore $B_{\varphi_0} \cup A''_{\varphi_0} \cup g_1$ is linearly independent. Extend now φ_0 to φ_1 defined by $\varphi_1 \mid B_{\varphi_0} = \varphi_0$, $\varphi_1(g_1) = h$. Clearly $\varphi_1 \in \Phi$ and $\varphi_1 > \varphi_0$ yet $\varphi_1 \neq \varphi_0$, hence φ_0 is not maximal. This contradiction proves that $A = A_{\varphi_0}$.

Now $B_{\varphi_0} \subset B \rightarrow |B_{\varphi_0}| \leqq |B|$. On the other hand since $\varphi_0^{-1} A = B_{\varphi_0}$ and φ_0 is one-one we have $|A| = |B_{\varphi_0}| \leqq |B|$ which proves (24.2) and hence also (24.1).

(24.3) DEFINITION. *The common value of* $|B|$ *for all the bases of* G *is called the dimension of* G.

(24.4) *If* H *is a subspace of* G *then* $\dim H \leqq \dim G$. *Hence*: (a) *if* G *has a countable base so has* H; (b) *if* $\dim H = \dim G$ *is finite then* $H = G$.

In view of (23.1a) we may take a base B for G with a subset C which is a base for H and so $\dim H = |C| \leqq \dim G = |B|$. Property (a) is then obvious, and as for (b) if both dimensions are finite and equal, $C = B$ and hence $G = H$.

(24.5) *If* $\dim G = n$ *is finite then* n *is the maximum number of linearly independent vectors in* G. *Moreover any* n *linearly independent vectors form a base.*

If $\{g_1, \cdots, g_n\}$ is a base the g_i are linearly independent. On the other hand if $g'_1, \cdots, g'_{n+1}$ are any $n + 1$ vectors we have relations

$$g'_i = \sum \alpha_{ij} g_j$$

from which follows at once that there is at least one non-trivial relation $\sum \beta_i g'_i = 0$. Hence the g'_i are not linearly independent and n is maximal.

Suppose now $g_1, \cdots, g_n$ merely linearly independent. Since n is maximal if $g \in G$ there is a relation $\alpha g = \sum \alpha_i g_i$, $\alpha \neq 0$. Hence $g = \sum \alpha^{-1} \alpha_i g_i$, and so $\{g_i\}$ is a base.

(24.6) *If* H *is a subspace of* G *and* $G/H = K$ *(discrete topology) then* $\dim G = \dim H + \dim K$.

Select a base B for G with a subset C as a base for H. If D is the complement of C in B then K is isomorphic with the subspace spanned by D, from which to (24.6) is but a step.

25. **Linear topology.** We have already made it a part of our conventions that vector spaces are topological groups with a specialized topology. This topology is described in the

(25.1) DEFINITIONS. *A vector space* G *is said to be linearly topologized or to have a linear topology if it is a topological group with a nuclear base composed of subspaces. A subspace which is also a nucleus will be called a nuclear subspace. Since an isomorphism transforms a nuclear base into a nuclear base and a subspace into a subspace, it preserves the linearity of the topology. That is to say if* τ *is an isomorphism* $G \to H$ *and* G *has a linear topology the same holds for* H. *Without this property the linear topology would have of course but little value.*

We recall then that under the fundamental convention (22.2) throughout the rest of the chapter all vector spaces are assumed linearly topologized.

Since the intersection of two subspaces is a subspace we have at once:

(25.2) *If* G *is linearly topologized so are its subspaces.*

By means of this property and the definition we may now prove:

(25.3) *Under our fundamental convention* (22.2) *when the operations: closure, taking a factor-group, product, weak product, are applied to vector spaces alone, they yield only vector spaces (understood with a linear topology). Furthermore even with these added restrictions all the results of* (§§1, 3, 4) *continue to hold.*

Closure. Let H be a subspace of G and $\{N\}$ a nuclear base of G composed of subspaces. Then $\bar{H}$ consists of the elements h such that every $h + N$ meets H. As a consequence $\alpha h + \alpha N = \alpha h + N$ meets $\alpha H = H$, hence $\alpha h \in \bar{H}$ and $\bar{H}$ is a subspace.

Factor-group. The notations remaining the same, suppose H closed in G and let $G^* = G/H$. If g^* is the coset of g mod H denote by αg^* the coset of αg mod H. This multiplication obeys the algebraic rules required for vector spaces. Let π denote the natural projection $G \to G^*$. If N^* is a nucleus of G^* then $\pi^{-1}N^*$ is a nucleus of G. Since π is continuous there is an $N \subset \pi^{-1}N^*$. Since π is open πN is a nucleus of G^* and since $\pi N \subset N^*$, $\{\pi N\}$ is a nuclear base for G^*. Evidently $g^* \in \pi N \to \alpha g^* \in \pi N$, and so αg^* is a continuous multiplication. Therefore G^* is a vector space. Since $\pi(\alpha g) = \alpha(\pi g)$ the πN are subspaces. Therefore $\{\pi N\}$ is a nuclear base of G^* composed of subspaces, and the topology of G^* is linear. Thus G^* is a vector space.

Products. Let $\{G_\lambda\}$ be an indexed system of vector spaces and let $G = \mathbf{P}G_\lambda$. If $g = \{g_\lambda\} \in G$ then $\{\alpha g_\lambda\}$ is also an element of G and if we denote it by αg, then $g \in G \to \alpha g \in G$. This multiplication obeys the algebraic rules required for a vector space (22.1). Since the coordinates αg_λ are continuous in g_λ, hence in g, αg is also continuous in g. Let $\{N_\lambda\}$ be a nuclear base of G_λ composed of subspaces. Then if $\{\lambda_1, \cdots, \lambda_k\}$ is any finite subset of $\{\lambda\}$ and μ ranges over the $\lambda \neq \lambda_1, \cdots, \lambda_k$, the products $N_{\lambda_1} \times \cdots \times N_{\lambda_k} \times \mathbf{P}G_\mu$ are subspaces and make up a nuclear base. Therefore G has a linear topology. Thus G is a vector space. Notice also that if $\{\mu\}$ is any subset of $\{\lambda\}$ then the projection $G \to \mathbf{P}G_\mu$ is a linear homomorphism.

The weak product is treated in the same way.

If G is a group of chains over Ω based on $X = \{x_\lambda\}$ and $g = \sum \alpha_\lambda x_\lambda$ then αg is the element $\sum (\alpha\alpha_\lambda)x_\lambda$.

(25.4) DEFINITION. *By an inverse or direct system of vector spaces is meant,* *respectively, inverse or direct systems whose groups are vector spaces over Ω and whose projections are linear.*

Since the limit-groups of inverse and direct systems are defined in terms of products, weak products, subspaces and factor-groups, they are vector spaces and the appropriate results in (§§1, 3, 4) hold.

From the definitions of the operations involved and the preceding considerations there follows immediately:

(25.5) *The natural projection $G \to G/H$, H a closed subspace of G, is a linear open homomorphism. Similarly if $\{\mu\}$ is a subset of $\{\lambda\}$ as regards the natural projection $\mathbf{P}G_\lambda \to \mathbf{P}G_\mu$.*

(25.6) *A finite-dimensional vector space G with linear topology is discrete.*

Since G is a Hausdorff space, 0 is the intersection of all the nuclei. Let $\{N\}$ be a nuclear base composed of subspaces and suppose $N \neq 0$. If $g \in N$, $g \neq 0$, then some $N' \not\ni g$. Hence $N'' = N \cap N' \subset N$ and $N'' \neq N$. Therefore $\dim N'' < \dim N$. Since $\dim N$ is finite after repeating the process a finite number of times we arrive at a nucleus reduced to 0. Therefore G is discrete.

(25.7) *A discrete vector space has a linear topology* (obvious).

(25.8) *Let N be a nuclear subspace of G. Then*

(a) *N is both open and closed;*

(b) *$H = G/N$ is discrete;*

(c) *$G = N + H$, $N \cap H = 0$;*

(d) *every element g may be written uniquely $g = n + h$, $n \in N$, $h \in H$, and more generally if G' is any subspace of G then*

$$G' = N' + H', \qquad N' = N \cap G', \qquad N' \cap H' = 0, \qquad H' = G'/N';$$

(e) *the transformation $\tau: (n + h) \to (n, h)$ is an isomorphism $G \to N \times H$.*

Since $g \in N \to g + N \subset N$, N is open. Since $g \notin N \to (g + N) \cap N = \emptyset$, the complement of N is open, hence N is closed and (a) holds.

Let π be the natural projection $G \to G/N$. Since π is open $\pi N = 0$ is a nucleus of G/N, and so the latter is discrete, which is (b).

By (23.1b): $G = N + H$, $N \cap H = 0$. Therefore 0 is a nucleus of H and so it is discrete. It is obviously $\cong G/N$ in the algebraic sense, and hence topologically also, since both are discrete. This proves (c). As for (d) it is an obvious consequence of (c).

That τ under (e) is an isomorphism in the algebraic sense is immediate. If $\{N_1\}$ is a nuclear base for N it is also one for G. Since H is discrete $\{N_1 \times 0\}$ is a nuclear base for $N \times H$. Since $\tau N_1 = N_1 \times 0$, τ establishes a one-one correspondence between the elements of two nuclear bases, and so it is topological, proving (e).

While the pairing of vector spaces will be taken up later we may prove at the present time a simple property which has often been used in topology in connection with duality.

(25.9) *Let the vector spaces G, H be orthogonal under a multiplication gh pairing them to Ω. Then:*

(a) *If $g_1, \cdots, g_n$ are linearly independent vectors of G, then* $\dim H \geqq n$ *and there may be selected in H linearly independent vectors $h_1, \cdots, h_n$ such that the determinant* $|g_i h_j| \neq 0$.

(b) *More precisely $h_1, \cdots, h_n$ may be chosen such that $\|g_i h_j\|$ is the unit matrix of order n, or in another form such that $g_i h_j = \delta_j^i$ (Kronecker deltas).*

(c) *The two preceding properties hold with G, H interchanged.*

(d) *Both* $\dim G$, $\dim H$ *are finite and equal, or else both are infinite.*

With the situation as in (a) suppose $\dim H = m < n$ and let $\{h_1, \cdots, h_m\}$ be a base for H. The system

$$\alpha_1(g_1 h_j) + \cdots + \alpha_n(g_n h_j) = 0$$

in the α_i has a solution in elements of Ω not all zero. Therefore $g = \alpha_1 g_1 + \cdots + \alpha_n g_n$ is an element of G which is different from 0 and such that $gh_j = 0$, $(j = 1, 2, \cdots, m)$. Since $\{h_j\}$ is a base this implies $gh = 0$ for every $h \in H$ and since orthogonality rules this out we have $m \geqq n$. In particular then if $\dim G$ is infinite: $\dim H \geqq n$ whatever n hence $\dim H$ is infinite also, and conversely. Since G and H may manifestly be interchanged necessarily $\dim H = \dim G$ when one of them is finite and this is (d).

Returning to (a) let G_1 be the subspace of G spanned by $\{g_1, \cdots, g_n\}$ and let H_1 be its annihilator in H. If $H^* = H/H_1$ then G_1 and H^* are orthogonal under a multiplication $g_1 h^*$ such that if $h \in h^*$ then $g_1 h^* = g_1 h$ (15.5b). Furthermore referring to (25.3) H^* is a vector space and $g_1 h^*$ a correct multiplication for G_1, H^*. By the above argument $\dim H^* \geqq n$. If $\{h_1^*, \cdots, h_n^*\}$ are independent then $|g_i h_j^*| \neq 0$, since otherwise we could obtain the same violation of orthogonality as before. Select now for each j an $h_j \in h_j^*$. We have $g_i h_j = g_i h_j^*$ and hence $|g_i h_j| \neq 0$. This relation, or the linear independence of $\{h_j^*\}$ implies the same for $\{h_j\}$. This proves (a). Property (b) is then an elementary consequence of (a) and (c) is obvious.

26. Linear varieties.

(26.1) Definitions. *A linear variety V in G is a coset* mod H, *H a subspace of G. The dimension of V, written* $\dim V$, *is the dimension of H. If $H = 0$ the coset of g is merely the element g itself and* $\dim V = 0$. *Thus the elements may be viewed as the zero-dimensional linear varieties. G and its subspaces are all linear varieties. In fact if* $\dim G = n$ *is finite, G itself is the only n-dimensional linear variety which it contains.*

(26.2) *If G' is a closed subspace of G, π the natural projection $G \to G/G'$, V a linear variety in G, then πV is a linear variety in G/G'.*

For π is a linear homomorphism.

(26.3) *If V is a linear variety so is $\bar{V}$.*

For if $V = g + H$ then $\bar{V} = g + \bar{H}$, and this is a linear variety since $\bar{H}$ is a subspace (25.3).

(26.4) *If $V' \subset V$ both are linear varieties and* $\dim V$ *is finite then* $\dim V' < \dim V$ *or else* $V' = V$.

For if $g \in V'$ then V, V' are cosets $g + H$, $g + H'$, and $V' \subset V \to H' \subset H$ from which to (26.4) is but a step.

27. **Linear compactness.** Various considerations, notably the applications to homology suggest a weakening of the concept of compactness as applied to vector spaces in accordance with the

(27.1) DEFINITION. *A linearly topologized vector space G, and more generally a linear variety V in G, is said to be linearly compact whenever given any family $\{V_a\}$ of linear varieties which are closed in G or V as the case may be and have the finite intersection property then $\bigcap V_a \neq \emptyset$. It is hardly necessary to observe that linear compactness is preserved under an isomorphism: if V is linearly compact in G and τ is an isomorphism $G \to H$ then τV is linearly compact in H.*

Many of the important properties of compactness carry over to linear compactness as we shall now show.

(27.2) *A product of linearly compact vector spaces is linearly compact.*

For the results of (26) make it possible to carry over the proof of (I, 24.1).

(27.3) *If G is linearly compact so is every closed linear variety in G* (obvious).

(27.4) *If G is linearly compact then*: (a) *its image under a homomorphism is also linearly compact*; (b) *if H is a closed subspace of G then G/H is linearly compact.*

The proof of (a) is the same as for (I, 23.2) with closed linear varieties replacing closed sets. As for (b) it is a consequence of (a) plus the fact that the natural projection $G \to G/H$ is a homomorphism.

(27.5) *If the linear variety V is linearly compact in G it is closed in G.*

Let $\{N\}$ be the nuclear subspaces and $g \in \bar{V}$. Since $g + N$ is a neighborhood of g, $(g + N) \cap V = W \neq \emptyset$. Since a finite intersection of sets $g + N$ is a set $g + N$, $\{W\}$ has the finite intersection property. Moreover since N is closed (25.8a) so is $g + N$, and hence W is closed in V. By the compactness condition $\bigcap W \neq \emptyset$. From $\bigcap N = 0$ follows $\bigcap((g + N) \cap V) = \bigcap W = g$, hence $g \in V$, or $\bar{V} = V$, proving (27.5).

Referring to (I, 38, 39) we find that the results just obtained enable us to prove:

(27.6) *Let $S = \{G_\lambda ; \pi_\mu^\lambda\}$ be an inverse system of vector spaces and let V_λ be a linearly compact variety in G_λ such that $\lambda > \mu \to \pi_\mu^\lambda V_\lambda \subset V_\mu$. Then the results of* (I, 38, 39) *are valid for the inverse mapping system $\Sigma = \{V_\lambda ; \pi_\mu^\lambda\}$, with compactness replaced by linear compactness.*

(27.7) *A n.a.s.c. for a discrete G to be linearly compact is that its dimension be finite. Hence every finite-dimensional G is linearly compact* (25.6).

Suppose dim $G = 1$, so that G is discrete. A linear variety V in G is G itself or else an element, so G is linearly compact. If dim $G = n$ is finite and G is discrete then G is the product of n one-dimensional vector spaces and so by (27.2) it is linearly compact. Thus the condition is sufficient.

Suppose now G discrete and linearly compact. The case $G = 0$ is trivial; so we assume $G \neq 0$. The space has then a base $B = \{b\}$ and we have $g = \sum \alpha_b b$

(finite sum). Now $V_b = \{g \mid \alpha_b = 1\}$ is a linear variety and is closed since G is discrete. Furthermore $\{V_b\}$ has the finite intersection property. Hence if G is linearly compact $\bigcap V_b \neq \emptyset$, which is impossible if B is not finite. Since B is finite so is dim G. This proves necessity and hence (27.7).

(27.8) *If τ is a univalent homomorphism of G onto H and G is linearly compact then τ is an isomorphism.*

Since τ is already one-one, all that needs to be proved is that it is open. If N is a nuclear subspace of G we must show then that $N_1 = \tau N$ is one for H. By (25.8) N is closed and G/N discrete. Since G/N is the natural projection of G it is also linearly compact (27.4) and hence finite-dimensional (27.7). Since N is closed in G it is linearly compact and so is N_1. Therefore N_1 is closed in H. Now to τ there corresponds a homomorphism of G/N onto H/N_1. Hence H/N_1 is finite-dimensional and therefore discrete. Since N_1 is the inverse image in H of the nucleus 0 of H/N_1 under the natural projection $H \to H/N_1$, N_1 is a nuclear subspace for H and (27.8) follows.

The natural extension of the notion of local compactness in the direction of linear compactness is manifestly given by the

(27.9) DEFINITION. *The vector space G is said to be locally linearly compact whenever it has a linearly compact nuclear subspace.*

Evidently this property is preserved under an isomorphism. Moreover compact and discrete vector spaces are locally linearly compact.

(27.10) *A n.a.s.c. for G to be locally linearly compact is that $G \cong G_1 \times G_2$, where G_1 is discrete and G_2 linearly compact.*

Necessity is a consequence of (25.8). Conversely, suppose G behaves as stated. We may as well assume $G = G_1 \times G_2$ and then $0 \times G_2$ is a linearly compact nuclear subspace of G, hence G is locally linearly compact.

(27.11) *Essential elements.* This concept of importance in the homology theory of nets is due to Čech [a]. Generally speaking, if $S = \{G_\lambda ; \pi_\mu^\lambda\}$ is for the present any inverse system of groups then an *essential element* of G_μ is an element x_μ such that $\lambda > \mu \to (\pi_\mu^\lambda)^{-1} x_\mu \neq \emptyset$, or which is the same such that $x_\mu \in \pi_\mu^\lambda G_\lambda$. We have from (27.6) and (I, 39.3):

(27.12) *When S is an inverse system of compact groups or an inverse system of linearly compact vector spaces then a n.a.s.c. for x_μ to be essential is that it be a coordinate of an element of the limit-group or limit-space as the case may be.*

A noteworthy property is the following:

(27.13) *Let S be an inverse system of finite-dimensional vector spaces. Then for every μ there is a $\lambda_0 > \mu$ such that all the $\pi_\mu^\lambda x_\lambda$, $\lambda > \lambda_0$, are essential* (Čech).

The essential elements of G_μ are those of the subspace $H_\mu = \bigcap\{\pi_\mu^\lambda G_\lambda \mid \lambda > \mu\}$. Since dim $G_\mu = n$ is finite there exists a finite set $\lambda_1, \cdots, \lambda_r$, such that $H_\mu = \bigcap(\pi_\mu^{\lambda_i} G_{\lambda_i})$. Choose any $\lambda_0 > \lambda_1, \cdots, \lambda_r$. Then $\pi_\lambda^\mu G_\lambda \subset \pi_\mu^{\lambda_i} G_{\lambda_i}$, hence for $\lambda > \lambda_0$: $\pi_\mu^\lambda G_\lambda \subset H$. Since the inclusion may also be reversed $H_\mu = \pi_\mu^\lambda G_\lambda$.

28. **Field characters. Duality.** The preceding developments will enable us

to extend in a significant way the duality theory of Pontrjagin-van Kampen. The basic definitions are:

(28.1) DEFINITIONS. *A field character h of a vector space G over a field Ω is a homomorphism $G \to \Omega$. As in* (18.1) *we denote by gh the value of h at g. If h_1, h_2 are field characters and α_1, $\alpha_2 \in \Omega$ then the relation $gh = \alpha_1 gh_1 + \alpha_2 gh_2$ defines a character which is written $\alpha_1 h_1 + \alpha_2 h_2$. Except for continuity conditions $H = \{h\}$ is thus turned into a vector space. To topologize H if E is any linearly compact subspace of G and $N(E) = \{h \mid gh = 0, g \in E\}$, then we choose $\{N(E)\}$ as a nuclear base for H. Since $h \in N(E) \to \alpha h \in N(E)$, αh is continuous under this topology. Since $N(E)$ is a subspace of H, the topology is linear and so H is a vector space behaving in accordance with* (22.2). *This vector space is known as the field character-group of G, or else also as the character-space of G.*

We shall now prove the analogue of (19.1):

(28.2) *If G is linearly compact, discrete, locally linearly compact then its field character-space H is respectively discrete, linearly compact, locally linearly compact and gh pairs G and H to Ω.*

(a) G *is linearly compact.* Then $N(G) = 0$ is a nucleus of H, and so H is discrete.

(b) *G is discrete.* Let $B = \{b\}$ be a base for G, $\{h_b\}$ a set of symbols such that $b \to h_b$ is one-one, Ω_b a copy of Ω corresponding to b, $H' = \mathbf{P}\Omega_b$. Any element h' of H' may be represented as an infinite chain over $\{h_b\}$

$$h' = \sum \beta_b h_b .$$

On the other hand if $g \in G$ we have

$$g = \sum \alpha_b b \text{ (finite sum)}$$

so that G may be identified with the weak product $\mathbf{P}^w\Omega_b$, the chains of G being thus considered as representations of the weak product by finite chains over $\{b\}$.

If we assign to g the element of Ω:

$$\text{(b}'\text{)} \qquad gh' = \sum \alpha_b \beta_b \text{ (finite sum)}$$

the assignment $g \to gh'$ makes h' a field character of G. The h_b are the particular field characters such that $bh_b = 1$, $b'h_b = 0$, $b \neq b'$, or in terms of Kronecker deltas:

$$b'h_b = \delta_b^{b'}.$$

Now if $h \in H$ sends b into β_b then $h \to h' = \sum \beta_b h_b$ defines an isomorphism τ: $H \to H'$ in the algebraic sense. Let $N(E)$ be a nuclear subspace of H. Since G is discrete so is E, and since E is linearly compact it is finite-dimensional and therefore a subspace of the space spanned by a finite subset $\{b_1, \cdots, b_n\}$ of B. Now $N_1 = \{h \mid h \in H', \beta_{b_1} = \cdots = \beta_{b_n} = 0\}$ is a nucleus of H' and since $\tau N(E) \supset N_1$, τ is open. Therefore τ^{-1} is a univalent homomorphism $H' \to H$

and since H' is linearly compact τ^{-1} is an isomorphism. Since H' is linearly compact so is $H = \tau^{-1}H'$.

(c) *G is locally linearly compact.* We have then $G \cong G_1 \times G_2$, G_1 discrete and G_2 linearly compact. For our purposes we may assume $G = G_1 \times G_2$ and then

$$G = G_1 \times 0 + 0 \times G_2, \qquad (G_1 \times 0) \cap (0 \times G_2) = 0.$$

Let now H, H_i denote the character-spaces of G, G_i. We have $g = (g_1, g_2)$, $g_i \in G_i$. Hence if h_i is a field character of G_i the relation $gh = g_1h_1 + g_2h_2$ defines a univalent homomorphism τ: $H_1 \times H_2 \to H$, in the algebraic sense, whereby $(h_1, h_2) \to h$. Conversely, let $h \in H$. Since $g = (g_1, 0) + (0, g_2)$ the relations $g_1h_1 = (g_1, 0)h$, $g_2h_2 = (0, g_2)h$ define elements $h_i \in H_i$ such that $\tau(h_1, h_2) = h$. Hence τ is an isomorphism in the algebraic sense.

In order to show that τ is topological we prove that τ establishes a one-one correspondence between the elements of nuclear bases for $H_1 \times H_2$ and H. Let $N(E)$ be as before and $N_1(E_1)$ the analogue of $N(E)$ for H_1. The set $\{N(E)\}$ is a nuclear base for H. If we write $E = E_1 \times E_2(E_1 \subset G_1, E_2 \subset G_2)$ E_1 and E_2 must be linearly compact since the projection $g_1 \times g_2 \to g_1 \times 0$ $[g_1 \times g_2 \to 0 \times g_2]$ is continuous. On the other hand every $E = E_1 \times E_2$ is linearly compact, if E_1 and E_2 are linearly compact. Furthermore every $N(E_1 \times E_2)$ contains $N(E_1 \times G_2)$. It follows that $\{N(E_1 \times G_2)\}$ (E_1 any linearly compact subspace of G_1) is a nuclear base for H. But clearly $\tau(N(E_1) \times 0) = N(E_1 \times G_2)$ and any $N(E_1 \times G_2)$ can be obtained in that way, proving τ topological.

Since $H = \tau(H_1 \times H_2)$, and $H_1 \times H_2$ is locally linearly compact so is H.

(d) *gh is a multiplication pairing G, H to Ω.* The algebraic properties (22.2) are easily verified. We take G locally linearly compact and as in (c) since this is the general case. If $E = 0 \times G_2$ then E is both linearly compact and open. Hence $E \times N(E)$ is a nucleus of $G \times H$ mapped by gh into zero. Therefore gh is continuous. This proves (d), and also (28.2).

29. Suppose now G, H paired to Ω with a multiplication gh. As in (19.5, 19.6) we introduce the associated field characters $\varphi_g(h) = gh$, $\varphi_h(g) = gh$ of H and G, and then likewise the *induced homomorphisms* χ_g defined by $g \to \varphi_g$ of G into the character-space of H, and χ_h defined by $h \to \varphi_h$ into the character-space of G. If both χ_g, χ_h are isomorphisms G and H are said to be *dually paired.* We now prove the analogue of the Pontrjagin-van Kampen duality theorem (20.2):

(29.1) Duality theorem for vector spaces. *Let G, H be locally linearly compact vector spaces paired to Ω under a multiplication gh. Then if one of the induced homomorphisms χ_g, χ_h is an isomorphism so is the other. Thus G, H are dually paired.*

Let us assume that χ_h is an isomorphism. If G^* is the character-space of H we must show that χ_g: $G \to G^*$ is an isomorphism.

30. *Suppose first G linearly compact.* Then H is discrete. Take a base $B = \{b\}$ of H, choose a copy Ω_b of Ω for each b and set $'G = \mathbf{P}\Omega_b$. Corresponding to φ_g we have then $\{\varphi_g(b)\} = \{gb\} \in {'G}$ and the result of (28.2b) may be interpreted as showing that $\varphi_g \to \{\varphi_g(b)\}$ defines an isomorphism $G^* \to {'G}$. Therefore to prove that χ_g is an isomorphism it is sufficient in the present instance to prove that the homomorphism τ: $G \to {'G}$ defined by $g \to \{gb\}$ is an isomorphism.

(a) *τ is univalent.* We must prove $g_0 \neq 0 \to \tau g_0 \neq 0$, or that $g_0 b \neq 0$ for some b. There exists a nuclear subspace N of G such that $g_0 \notin N$. By (25.8) we have $G = N + G'$, $N \cap G' = 0$. Let $\{b'\}$ be a base for G'. By (25.8d) every $g \in G$ has a unique representation

$$g = g_N + \sum \alpha_{b'}(g)b',$$

where $g_N \in N$ and the sum is finite. Since $g_0 \notin N$, some $\alpha_{b'}(g_0) \neq 0$. Hence $\alpha_{b'}(g)$ is a character of G mapping the nucleus N into zero and different from 0 at g_0. Hence some b must exist such that $g_0 b \neq 0$, as asserted.

(b) $\tau G = {'G}$. Let $\{b_1, \cdots, b_n\}$ and $\{g_1, \cdots, g_m\}$ be finite subsets of B and G. Consider now the matrix $\| (g_i b_j) \|$ and suppose that whatever $\{g_i\}$ its rank is at most $r < n$. As a consequence we may choose n linearly independent combinations b_i' of the b_i and m vectors g_i such that in $\| (g_i b_j') \|$ the last column consists of zeros. Since the b_i' may replace the b_i in B, we may assume that $\| (g_i b_j) \|$ already behaves in this manner. Since the matrix

$$\left\| \begin{matrix} (g_i b_j) \\ (g b_j) \end{matrix} \right\|$$

is of rank not exceeding r, whatever g, we must have $gb_n = 0$ and hence $b_n = 0$' which is ruled out.

We conclude then that we may choose $g_1, \cdots, g_n$ such that $\| (g_i b_h) \|$ is of rank n. Therefore we may choose $g = \sum \alpha_i(g) g_i$ such that $\{(gb_j)\}$ are n preassigned elements of Ω.

Let now $'g = \{\alpha b\} \in {'G}$ and let U be any neighborhood of $'g$. The sets in $'G$ for which a finite number of the coordinates are zero and the rest arbitrary, form a nuclear base. Therefore there exists a set $\{b_1, \cdots, b_n\}$ such that the points $'g$ whose b_i coordinate is α_i $(i = 1, 2, \cdots, n)$ are all in U. We have seen that among these there is a point τg. Therefore $\overline{\tau G} = {'G}$. However G being linearly compact τG is closed in $'G$ (27.5), and so $\tau G = \overline{\tau G} = {'G}$.

If we combine (a), (b) with (27.8) we find that τ is an isomorphism. This proves (29.1) for the present case.

31. *Suppose now G discrete.* The notations being those of (28.2b), and since in proving (29.1) we may replace H by any isomorph, we may assume $H = \mathbf{P}\Omega_b$, the pairing being given by (28.2b′). Let $g^* \in G^*$. Since g^* is continuous and 0 is a nucleus of Ω, the subset of H mapped by g^* into 0 will contain an element of a nuclear base. Now the sets N of H consisting of the elements having all but a finite number of coordinates zero, and the rest arbitrary form a nuclear

base for H. Hence g^* will take the value zero on a certain set N, say the set for which the coordinates $b_1, \cdots, b_n$ are zero. That is to say if $g^*h_b = \alpha_b$, then $\alpha_b = 0$ for $b \neq b_1, \cdots, b_n$. If $g = \sum \alpha_b b$ then $g \to g^*$ defines χ_g and it is a homomorphism of G onto G^*. Since G, G^* are both discrete, (29.1) merely requires here to show that χ_g is univalent. Suppose $g_0 \neq 0$. By (23.1a) we may assume $g_0 = b \in B$ and since $gh_b = 1 \neq 0$, we have $\chi_g g_0 \neq 0$. Therefore χ_g is univalent and (29.1) holds in the case under consideration.

Suppose finally G locally linearly compact, and let the notations be those of (28.2c). Thus $H \cong H_1 \times H_2$ and we may identify the two vector spaces, so that $H = H_1 \times H_2$. For similar reasons if G_i^* is the character-space of H_i we may assume $G^* = G_1 \times G_2$. If φ_{ig}, χ_{ig} have their obvious meaning, and if $h = (h_1, h_2)$ then $\varphi_g(h) = \varphi_{1g}(h_1) + \varphi_{2g}(h_2)$ and therefore χ_g is defined by $g \to (\varphi_{1g}(h_1), \varphi_{2g}(h_2))$. Coupling this with the fact already proved that the χ_{ig} are isomorphisms we find that the same holds for χ_g. This completes the proof of the duality theorem (29.1).

32. Two results obtained incidentally deserve mention. It is a consequence of the argument of (30) that:

(32.1) Theorem. *Every linearly compact vector space is a product of one-dimensional spaces.*

As a consequence of (30) we also have:

(32.2) *Let $B = \{b\}$ be any set and for each b select a copy Ω_b of Ω. Then $G = \mathbf{P}\Omega_b$ and $H = \mathbf{P}^w\Omega_b$ are dually paired under a multiplication which may be described as follows. Let*

$$g = \sum \alpha_b g_b, \qquad h = \sum \beta_b h_b \tag{32.3}$$

be respective representations of the elements of G, H by infinite and finite chains. Then

$$gh = \sum \alpha_b \beta_b. \tag{32.4}$$

(32.5) *Noteworthy special case: G, H are both one-dimensional. Then each may be identified with Ω and the dual pairing is under a multiplication which is merely the multiplication in Ω.*

33. Once we are in possession of the analogue (29.1) of (20.2) the remaining results: (20.5, $\cdots$, 20.8), of (20) are derived as loc. cit. "Dual systems" in the sense of (20.8) will refer of course to pairing to Ω. There are other minor deviations which the reader will readily supply.

34. **Weak duality.** We shall have occasion in homology theory to consider a couple S, S^* as in (20.8) save that G_λ, H_λ *will both be discrete* spaces orthogonal in a pairing to Ω. We shall then say that we have *weak duality*. If the dimension of one of the groups G_λ, H_λ is finite whatever λ then the same holds for the other and they are in fact equal (25.9). In that case the G_λ are again linearly compact (27.7) and we are back to ordinary vector space duality.

35. **Dimensional complements.** Certain questions which are important in connection with homology will be treated here.

(35.1) *The dimension of the limit-space of a direct system of vector spaces.* Let $S^* = \{H_\lambda ; \pi_\lambda^{*\mu}\}$ be a direct system with limit-space H. Consider any finite set $\{h_{\mu i}\}$, $i = 1; 2, \cdots, t$, of elements of H_μ such that $\{\pi_\lambda^{*\mu} h_{\mu i}\}$ is a linearly independent set for H_λ. Let $\rho(\lambda, \mu) = \sup t$ and set

$$\text{(35.2)} \qquad \rho = \sup_\mu \{\inf_\lambda \rho(\lambda, \mu)\}.$$

In connection with Betti numbers and certain other characters Alexandroff has repeatedly considered numbers analogous to ρ. For this reason we will call ρ the *Alexandroff number* of S^*.

Let us set $\rho(\mu) = \inf_\lambda \rho(\lambda, \mu)$, $r_\mu = \dim H_\mu$, $r = \dim H$, $H''_\mu =$ the subspace of the representatives of the zero of H in H_μ. By (14.4) H''_μ is a subgroup, and since $h''_\mu \in H''_\mu$, $\alpha \in \Omega \rightarrow \alpha h''_\mu \in H''_\mu$, it is also a subspace. By (23.1b) there is a second subspace H'_μ of H_μ such that

$$\text{(35.3)} \qquad H_\mu = H'_\mu + H''_\mu, \qquad H'_\mu \cap H''_\mu = 0.$$

From the definition of these subspaces follow also if $\dim H'_\mu = r'_\mu$:

$$\text{(35.4)} \qquad r'_\mu \leqq r;$$

$$\text{(35.5)} \qquad r'_\mu \leqq \rho(\mu).$$

We will now prove:

(35.6) *If* $\rho(\mu)$ *is finite then* $r'_\mu = \rho(\mu)$.

Choose $\lambda > \mu$ such that $\rho(\lambda, \mu) = \rho(\mu)$ and let $H''_{\mu\lambda}$ be the subspace of the elements $h''_{\mu\lambda}$ of H_μ such that $\pi_\lambda^{*\mu} h''_{\mu\lambda} = 0$. We have again a decomposition

$$H_\mu = H'_{\mu\lambda} + H''_{\mu\lambda}, \qquad H'_{\mu\lambda} \cap H''_{\mu\lambda} = 0.$$

Suppose that there exists an $h'_{\mu\lambda} \in H'_{\mu\lambda} \cap H''_{\mu\nu}$, where also $\nu > \mu$ and $h'_{\mu\lambda} \neq 0$. Take a $\nu' > \nu, \lambda$. Then $h'_{\mu\lambda} \in H''_{\mu\nu'}$ and $\pi_{\nu'}^{*\mu} h'_{\mu\lambda} = 0$. Since every element of H_μ is in $H'_{\mu\lambda} \bmod H''_{\mu\lambda}$, and $H''_{\mu\lambda} \subset H''_{\mu\nu'}$, we will have $\dim H'_{\mu\nu'} < \dim H'_{\mu\lambda} = \rho(\mu)$ which is ruled out, since $\rho(\mu) = \inf_\lambda \dim H'_{\mu\lambda}$. Therefore $H'_{\mu\lambda} \cap H''_{\mu\nu} = 0$. In other words whatever $\nu > \mu$ we have, $\pi_\nu^{*\mu} h'_{\mu\lambda} \neq 0$. Thus $H'_{\mu\lambda} \cap H''_\mu = 0$. It follows that $H''_\mu = H''_{\mu\lambda}$, and since each element of each of H'_μ, $H'_{\mu\lambda}$ is equal to an element of the other mod H''_μ, their dimensions are the same. This is precisely (35.6).

It is a consequence of (35.4), (35.6) that if $\rho(\mu)$ is finite then $r \geqq \rho(\mu)$, and hence also $r \geqq \rho$. Thus if the $\rho(\mu)$ are all finite and ρ is infinite so is r.

Suppose now that H contains s linearly independent elements. Then for some μ the space H'_μ will also contain s linearly independent elements and so $s \leqq r'_\mu \leqq \rho(\mu) \leqq \rho$. Therefore if $r = \sup s = \infty$ likewise $\rho = \infty$, and if r is finite then $\rho \geqq r$. Consequently if all the $\rho(\mu)$ are finite then r and ρ are both finite and equal or else both are infinite.

Suppose some $\rho(\mu) = \infty$. Then $\rho = \infty$ and no comparison is possible. Since $\rho(\mu) \leqq \rho(\lambda, \mu) \leqq r_\lambda$ for every $\lambda > \mu$, this circumstance will certainly fail to occur if the r_λ are all finite or else all the $\rho(\lambda, \mu)$ are finite. Therefore

(35.7) *If ρ is finite then $\rho = r$. More precisely if the $\rho(\lambda, \mu)$ or else the r_λ are all finite then ρ and r are both finite and equal or else both are infinite.*

(35.8) The following example shows that when the r_λ are all infinite (35.7) need not hold. Take a countable set of symbols $\{b_n\}$ as a base for finite chains over Ω and let H_n be the group of the finite chains based on $\{b_n, b_{n+1}, \cdots\}$, and $\{\lambda; >\} = \{n\}$ with $> \leftrightarrow \geqq$. Define π_n^{*m}, $m \leqq n$ as follows: if $h = \alpha_m b_m + \cdots + \alpha_q b_q$ then $\pi_n^{*m} h = \alpha_n b_n + \cdots + \alpha_q b_q$. It is readily seen that r_m, $\rho(n, m)$ are all infinite. Nevertheless the limit-group H reduces to the zero and so $r = 0$. Thus when the r_λ are infinite we may have $r = 0$, $\rho = \infty$.

(35.9) Consider now an inverse system of vector spaces $S = \{G_\lambda; \pi_\mu^\lambda\}$ and let $\rho'(\lambda, \mu)$ be the maximum number of elements of G_λ whose projections by π_μ^λ, $\lambda > \mu$, form a linearly independent set. The number

$$\rho' = \sup_\mu \{\inf_\lambda \rho'(\lambda, \mu)\}$$

is called the *Alexandroff number* of S. We prove:

(35.10) *If S, S^* form a dual system* (20.8, 33) *then their Alexandroff numbers are equal.*

Let $\{h_{\mu i}\}$, $i = 1, 2, \cdots, s$, be such that $\{\pi_\lambda^{*\mu} h_{\mu i}\}$, $\lambda > \mu$, is a linearly independent subset of H_λ. Since G_λ and H_λ are dually paired under the multiplication $g_\lambda h_\lambda$, by (33) and (25.9b) G_λ contains a linearly independent subset $\{g_{\lambda i}\}$, $(i = 1, 2, \cdots, s)$ such that

$$g_{\lambda i}(\pi_\lambda^{*\mu} h_{\mu j}) = \delta_i^j \text{ (Kronecker delta).}$$

Hence by (16.3)

$$(\pi_\mu^\lambda g_{\lambda i}) h_{\mu j} = \delta_i^j .$$

Therefore $\{\pi_\mu^\lambda g_i^\lambda\}$ is a linearly independent subset of G_μ and so $\rho'(\lambda, \mu) \geqq s$, and hence $\rho'(\lambda, \mu) \geqq \rho(\lambda, \mu)$. By interchanging the roles of G, H we find similarly $\rho(\lambda, \mu) \geqq \rho'(\lambda, \mu)$. Therefore $\rho(\lambda, \mu) = \rho'(\lambda, \mu)$ are both finite and equal, or else both infinite. If $\rho(\lambda, \mu)$ and $\rho'(\lambda, \mu)$ are always finite then (35.10) holds, while if they are infinite for some pair (λ, μ) then ρ, ρ' are both infinite and so (35.10) holds again.

36. **Field extension.** Let G be a vector space over Ω and Ω_1 a field which is an extension of Ω. We propose to examine certain consequences for G and certain related spaces, having in mind chiefly an important application to homology. Since we shall only be concerned with questions of dimension we may as well assume all the vector spaces discrete.

(36.1) Let $B = \{b\}$ be a base for G and correspondingly introduce $X = \{x_b\}$ such that $b \leftrightarrow x_b$ is one-one. Thus G is isomorphic with the space of the finite chains over Ω based on X. Let G_1 be the similar space for Ω_1 and the same X.

Suppose now that $C = \{c\}$ is a second base for G, and let $Y = \{y_c\}$, $\bar{G}_1$ be the analogues of X, G_1 for C. We first prove

$$\bar{G}_1 \cong G_1 . \tag{36.2}$$

Since B, C are both bases for G we have relations

$$b = \sum \lambda(b, c)c, \qquad c = \sum \mu(c, b)b, \tag{36.3}$$

where the sums are finite. Since

$$b = \sum_{c,b'} \lambda(b, c)\mu(c, b')b', \qquad c = \sum_{b,c'} \mu(c, b)\lambda(b, c')c',$$

and B, C are bases we must have

$$\begin{aligned} \sum_c \lambda(b, c)\mu(c, b') &= \delta_{b'}^b , \\ \sum_b \mu(c, b)\lambda(b, c') &= \delta_{c'}^c , \end{aligned} \tag{36.4}$$

where $\delta_{b'}^b$, $\delta_{c'}^c$ are the Kronecker deltas.

Now by linear extension

$$x_b \to \sum \lambda(b, c)y_c , \qquad y_c \to \sum \mu(c, b)x_b ,$$

define in the obvious way homomorphisms $\tau_1 : G_1 \to \bar{G}_1$, $\tau_2 : \bar{G}_1 \to G_1$. Using (36.4) we find at once $\tau_1\tau_2 = 1$, $\tau_2\tau_1 = 1$. Hence τ_1 is an isomorphism and (36.2) follows.

(36.5) In view of (36.2) we may as well denote G_1 by $\Omega_1 G$ without reference to the special base used in its construction. It is clear that G_1 may be identified with the additive group of the finite linear combinations of the elements b of B with coefficients in Ω_1. Thus every linear function on G (= homomorphism into another vector space over Ω) has a unique extension to G_1.

(36.6) Let now G' be a subspace of G, and let there be given a set of linear functions $\{f_\alpha(g)\}$ on G to Ω. The system

$$f_\alpha(g) = 0 \tag{36.7}$$

determines a subspace G'' of G. It is assumed that $G' \subset G''$ and so we form $G^* = G''/G'$ which is again a vector space over Ω. Let now $\Omega_1 G = G_1$, $\Omega_1 G' = G_1'$. By linear extension $f_\alpha(g)$ becomes a linear function on G_1 to Ω_1 and so (35.7) determines a vector space G_1'' over Ω_1. Clearly $G_1' \subset G_1'' \subset G_1$, so that we may again introduce $G_1^* = G_1''/G_1'$. We prove:

(36.8) $G_1^* \cong \Omega_1 G^*$ *and in particular* $\dim G_1^* = \dim G^*$.

We first prove that $G_1'' = \Omega_1 G''$. Since Ω_1 may be considered as a vector space over Ω, as such it has a base $\{\omega_\lambda\}$. Hence any $g \in G_1''$ is of the form $\sum \omega_\lambda g_\lambda$ where the sum is finite and $g_\lambda \in G$. We have then $f_\alpha(g) = 0 \to \sum \omega_\lambda f_\alpha(g_\lambda) = 0 \to f_\alpha(g_\lambda) = 0$ since $\{\omega_\lambda\}$ is a base over Ω and $f_\alpha(g_\lambda) \in \Omega$. Thus $g_\lambda \in G''$ and so $G_1'' \subset \Omega_1 G''$. Since the converse is obvious our assertion follows.

Let now C be a base for G' and extend it to a base $C \cup D$, $C \cap D = \emptyset$, for G''. Define $\Omega_1 G'$ and $\Omega_1 G''$ using this base. Then it is clear that G^*, G_1^* are, respectively, isomorphic with the vector spaces over Ω, Ω_1 spanned by D and this means precisely that (36.8) holds.

37. **Kronecker products.**

(37.1) Let $G = \{g\}$, $H = \{h\}$ be two discrete vector spaces over Ω. Choose two bases $B = \{b\}$, $C = \{c\}$ for G, H and introduce new symbols $\{b \otimes c\}$. The finite linear forms $\sum \alpha_{bc} b \otimes c$ (finite chains) give rise by addition and multiplication by elements of Ω to a new vector space K over Ω. If we have

$$g = \sum \beta_b b\,, \qquad h = \sum \gamma_c c \text{ (finite sums)}$$

then $\sum \beta_b \gamma_c b \otimes c$ is a definite element of K written $g \otimes h$. The operation $\otimes$ thus defined between the elements of G and H is distributive and we have $(\alpha g) \otimes h = g \otimes (\alpha h) = \alpha(g \otimes h)$,

(37.2) If we replace B or C, and hence both, by new bases B', C' and form the analogue K' of K we show as in (36) that there are two homomorphisms $\tau_1 : K \to K'$, $\tau_2 : K' \to K$ such that $\tau_1\tau_2 = 1$, $\tau_2\tau_1 = 1$. Hence τ_1 is an isomorphism. If $\otimes'$ is the analogue of $\otimes$ relative to K', we readily verify that the identification of $g \otimes h$ with $g \otimes' h$ turns τ_1 into the identity; this identification is assumed henceforth and so we will have $K = K'$. We have thus arrived at a unique new discrete vector space K depending solely upon G and H. It is known as the *Kronecker product* of G, H and denoted by $G \otimes H$.

(37.3) *When G, H are both finite-dimensional so is $G \otimes H$ and* $\dim G \otimes H = \dim G \dim H$. *When one of G, H is infinite-dimensional and the other is not zero then $G \otimes H$ is likewise infinite-dimensional.*

(37.4) *If G admits the decomposition in subspaces*:

$$G = G' + G'', \qquad G' \cap G'' = 0,$$

then $G \otimes H$ admits the analogous decomposition

$$G \otimes H = G' \otimes H + G'' \otimes H, \tag{37.4a}$$

$$G' \otimes H \cap G'' \otimes H = 0. \tag{37.4b}$$

Similarly with G, H interchanged.

All but (37.4b) are obvious, and (37.4b) is a consequence of the fact that if B', B'' are bases for G', G'' then $B' \cup B''$ is one for G.

(37.5) *Under the same circumstances as for* (37.4) *if $g \otimes h \in G' \otimes H$ and $h \neq 0$ then $g \in G'$.*

For if $g \otimes h \neq 0$ then it is of the form $g' \otimes h$, $g' \in G'$, and if $g \otimes h = 0$, $h \neq 0$, then $g = 0$ and again $g \in G'$.

(37.6) Extensive generalizations and indeed a full treatment of the Kronecker product, under the name "tensor product" will be found in Whitney [f], to which the reader is referred for further details on this topic.

CHAPTER III

COMPLEXES

A complex is a particular type of partially ordered set with complementary properties designed to carry an algebraic superstructure, its homology theory. Complexes thus appear as the tool par excellence for the application of algebraic methods to topology.

For the present we shall deal chiefly with finite complexes and give a complete treatment of their homology and cohomology groups and duality theory. Polyhedral and Euclidean complexes are discussed as special examples. Infinite complexes are likewise considered as well as a special class, the simple complexes, introduced by A. W. Tucker, and may be said to have all the main algebraic attributes of the polyhedral type. It is for simple complexes that an intersection theory is developed in (V), and the combinatorial manifolds of (V) are also simple complexes.

Summation notation. It is the same as in tensor calculus: non-dimensional indices (usually clear from the context) repeated up and down are to be summed unless an explicit statement is made to the contrary. Thus $g^i x_i$ stands for $\sum_i g^i x_i$.

Kronecker deltas. They are the well known numbers defined by $\delta^i_j = 0$ for $i \neq j$, $\delta^i_j = 1$ for $i = j$.

Designations for some special groups. We will write as in (II): $\mathfrak{J}$ = the group of the integers, $\mathfrak{J}_m$ = the group of the residues mod m, $\mathfrak{P}$ = the group of the reals mod 1, $\mathfrak{R}$ = the additive group of the rational numbers (rational group).

If $G = \{g\}$ is any group then $\{g\alpha\}$ is a group under the composition law $g\alpha - g'\alpha = (g - g')\alpha$, and this group is written $G\alpha$. The designations $G(m)$, $G^*(m)$, $G[m]$ are as in (II, 20.9).

The function $\beta(p)$. Convenient in many calculations it is defined by

$$\beta(p) = (-1)^{\frac{p(p-1)}{2}},$$

and we notice the useful relations:

$$\beta(-p) = (-1)^{\frac{p(p+1)}{2}},$$

$$\beta(p)\beta(q) = (-1)^{pq}\beta(p + q).$$

General references: Alexander [b, c], Alexandroff [f], Alexandroff-Hopf [A-H, Part 2], Hopf [a], Lefschetz [L, I, VII; L_1], Mayer [a, c], Poincaré [b], Seifert-Threlfall [S-T], Steenrod [a], Tucker [a], Veblen [V], Whitney [d].

§1. COMPLEXES. DEFINITIONS AND EXAMPLES

1. (1.1) DEFINITIONS. *A complex X is a set $\{x\}$ of elements ordered by a proper reflexive ordering relation $<$ (I, 4) together with two associated functions of the elements and element pairs whose values are integers: one,* dim x, *the dimension of x, also denoted by means of an index as x^p, the element then being called p-dimensional or a p-element, and the other, $[x:x']$, the incidence number of x and x', subject to the following conditions:*

K1. $x' < x \rightarrow \dim x' \leqq \dim x$;

K2. $[x:x'] = [x':x]$;

K3. $[x:x'] \neq 0 \rightarrow x < x'$ or $x' < x$, *and* $|\dim x - \dim x'| = 1$.

K4. *For every pair of elements x, x'' whose dimensions differ by two there is at most a finite number of x' such that $[x:x'][x':x''] \neq 0$ and then*

$$\sum_{x'} [x:x'][x':x''] = 0. \tag{1.2}$$

When the complex is finite K4 *may be replaced by the simpler condition:*

K4′. *For every pair of elements x, x'' whose dimensions differ by two, the relation* (1.2) *holds.*

The dimension of X, written dim X is sup dim x. When its value n is finite X is sometimes called an *n-complex.*

(1.3) Let $\alpha(x)$ be a function of x whose values are ± 1. If $[x:x']$ is replaced by $\alpha(x)\alpha(x')[x:x']$ conditions K1234 are fulfilled and so we still have a complex, say X'. In conformity with the usual conventions we agree to consider X' as identical with X. Thus the function $[\ :\ ]$ for a given X is to be considered as not unique but only given to within a factor $\alpha(x)\alpha(x')$. The different sets of incidence numbers thus arising are said to be *admissible*, the passage from one to the other is described as *reorienting* X. The function $\alpha(x)$ is known as an *orientation function* and we say that x has been reoriented if $\alpha(x) = -1$, and that it has *preserved its orientation* otherwise.

(1.4) REMARK. The definition of complexes adopted here is essentially Tucker's [a] and differs from his only in that: (a) the dimensions are not restricted to being greater than or equal to 0 which will be of importance in dealing with duality; (b) the complexes need not be finite. Indeed persistent attention to infinite complexes will characterize our treatment.

The complexes which we have just introduced are often called "abstract complexes." Other general types have been considered in the literature notably by M. H. A. Newman [a] and W. Mayer [a]. Newman's type is designed chiefly to preserve as many as possible of the properties of polyhedra and for many purposes it is decidedly too "geometric." In Mayer's type on the other hand only the properties which flow from the incidence numbers are preserved and the type is thus too "algebraic." Tucker's type may be said to occupy a reasonable intermediate position.

2. There are three important sets associated with any element $x \in X$: the *star* of x, written St x, the *closure* of x, written Cl x and the *boundary* of x, written $\mathfrak{B}x$. Their defining relations are

$$\mathrm{St}\, x = \{x' \mid x < x'\}, \qquad \mathrm{Cl}\, x = \{x' \mid x' < x\},$$

$$\mathfrak{B}x = \mathrm{Cl}\, x - x = \{x' \mid x' < x, x' \neq x\}.$$

An analogue $\mathrm{St}\, x - x$ of $\mathfrak{B}x$ may be formally introduced but will not be needed in the sequel.

We say that x' is a *face* of x when $x' < x$ (a *proper face* when $x' \neq x$), also that x and x' are *incident* when $x' >$ or $< x$. By the *incidence relations* in x we shall mean the incidences $<$ together with all the incidence numbers.

The notions of star, closure and boundary of a single element may be generalized as follows: if Y is any subaggregate of X the *star* of Y, written $\mathrm{St}\, Y$, is the union of the stars of all the elements of Y. Similarly the *closure* of Y, written $\mathrm{Cl}\, Y$, is the union of the closures of all the elements of Y. $\mathrm{St}\, Y$ is the union of all the elements $>$ some element of Y, while $\mathrm{Cl}\, Y$ is the union of all the elements $<$ some element of Y. The *boundary* of an open subcomplex Y is $\mathfrak{B}Y = \mathrm{Cl}\, Y - Y$.

A *subcomplex* of X is a subaggregate $Y = \{x'\}$ of X such that with the same dimensions and incidence relations as in X, conditions K1234 hold in Y alone. It is clear that the verification of the complex conditions for Y merely requires the verification of K4 alone.

We say that the subcomplex Y of X is

open whenever $\mathrm{St}\, Y = Y$, or $x \in Y \to \mathrm{St}\, x \subset Y$;

closed whenever $\mathrm{Cl}\, Y = Y$, or $x \in Y \to \mathrm{Cl}\, x \subset Y$.

Immediate consequences are:

(2.1) *If one of the sets Y, $X - Y$ is an open subcomplex the other is a closed subcomplex, and conversely.*

(2.2) *Any union or intersection of open or closed subcomplexes is, respectively, an open or a closed subcomplex.*

The aggregates $\mathrm{St}\, Y$, $\mathrm{Cl}\, Y$, $\mathfrak{B}x$, are subcomplexes of X. The proof merely requires that we verify K4. Let us do so for the first. If $x, x'' \in \mathrm{St}\, Y$, the only significant contribution to $\sum [x:x'][x':x'']$ occurs say when $x < x''$ and from elements x' between both. But in that case $x' \in \mathrm{St}\, Y$ also, so that the relation in question holds in $\mathrm{St}\, Y$ alone. Evidently $\mathrm{St}\, x$ is an open subcomplex, and $\mathrm{Cl}\, x$ a closed subcomplex; since $\mathrm{Cl}\, x$ is closed so is $\mathfrak{B}x$.

The *p-section* X^p of X is the set of all the elements of X whose dimension does not exceed p. X^p is likewise a closed subcomplex of X. For the union of all $\mathrm{St}\, x$, $\dim x > p$, is an open subcomplex and X^p is its complement.

When the dimensions of the elements of X are greater than or equal to 0, the zero-dimensional elements are frequently called the *vertices* of X.

(2.3) *Connectedness and components.* The component of any element x is the set of all x' such that there exists a finite collection $x = x_1, \cdots, x_r = x'$ in which any two consecutive elements are incident. We will then say briefly that x, x' are in the relation R. It is not difficult to see that:

(2.4) *Properties* (I, 17.1, $\cdots$, 17.4) *hold for X and the present definition of components.*

Furthermore we also prove readily:

(2.5) *The relation* R *is equivalent to each of the following*:

(a) St x, St x' *are in the same component of* $\{\text{St } x\}$ *in the sense of* (I, 17);

(b) Cl x, Cl x' *are in the same component of* $\{\text{Cl } x\}$ *in the same sense.*

It is also a consequence of the definition that

(2.6) *The component* X' *of* x *contains both* Cl x *and* St x. *Hence* X' *is both an open and a closed subcomplex of* X.

The complex X is said to be *connected* whenever it consists of a single component, i.e., when any two elements are in the relation R.

3. The complex X is said to be *star-finite*, or *closure-finite* whenever every St x or Cl x is finite, and to be *locally finite* when it has both properties. Notice these properties:

(3.1) *Finiteness* $\rightarrow$ *local finiteness.*

(3.2) *When* X *is star- or closure-finite there is at most a finite number of elements between* x *and* x'' *and hence* K4 *may then be replaced by the simpler condition* K4$'$.

(3.3) *Every component of a locally finite complex* X *is countable.*

Let Y be a component of X and x any element of Y. Consider the sequence $Y_1 = x, Y_2, \cdots$, where $Y_{n+1} = \text{St Cl } Y_n$. Every Y_n is finite and since $\bigcup Y_n = Y$, Y is countable.

4. Let $X_1 = \{x_1\}$, $X_2 = \{x_2\}$ be two complexes and suppose that there exists a one-one, order-preserving transformation T: $\{x_1\} \rightarrow \{x_2\}$ such that: (a) dim $Tx_1 = \dim x_1 + k$ where k is a fixed integer; (b) the numbers $[x_1 : x_1']$ are appropriate incidence numbers for Tx_1, Tx_1'. Whenever $k = 0$ we call T an *isomorphic* transformation or *isomorphism* $X_1 \leftrightarrow X_2$ and say that X_1 and X_2 are *isomorphic.* When $k \neq 0$ we say that we have a *weak* isomorphism and refer to X_1, X_2 as *weakly* isomorphic.

It is clear that these two types of isomorphisms give rise to equivalence classes but we will not refer to them particularly in the sequel as they are not sufficiently broad for the applications.

Dual complex. Given the complex $X = \{x\}$, let us introduce a new set of elements $X^* = \{x^*\}$ such that $x \leftrightarrow x^*$ is a one-one correspondence with the following properties: (a) $x \prec x' \leftrightarrow x'^* \prec x^*$; (b) $\dim x^* = -\dim x$; (c) $[x^* : x'^*] = [x : x']$. We verify immediately that conditions K1234 continue to be fulfilled, so that X^* is also a complex. It is known as the *dual* of X. Clearly $X^{**} = (X^*)^* \cong X$. If we agree to choose as the elements x^{**} the elements x themselves: $x^{**} = x$, then we will have $X^{**} = X$. *Thus* X, X^* *will be dual to one another.*

By way of notation if x_i^p is any element of X its image x_i^{p*} in X^* is conveniently denoted by x_p^i. Thus when the dimensional index of an element is a subscript it denotes the *negative* of the true dimension of the element.

The reader will not have missed the fact that our definitions have been so couched as to continue to give free play to the dualism which permeates the general theory of ordered sets. This is the chief justification for imposing *symmetry* in the incidence numbers and for introducing *negative* dimensions. The so-called dual complexes hitherto considered in topology had of course only positive dimensions, and also in fact nonsymmetrical incidence numbers. Wherever they occurred these complexes were weak isomorphs of X^* with dimensions raised to make them greater than or equal to 0, and with a certain reorientation that need not be described at the present time.

5. Simplicial complexes. As we shall see later (VII, VIII) this is the dominant type wherever complexes occur in topology.

(5.1) We must first define the *simplex*. A *p-simplex* σ^p is merely any set of $p + 1$ objects $\{A_0, \cdots, A_p\}$, known as the *vertices* of σ^p. It will generally be assumed that they are assigned a definite order modulo an even permutation and we will write accordingly $\sigma^p = A_0 \cdots A_p$, the specified order being the one in which the A_i are named. The simplex σ with its vertices ordered as just stated is said to be *oriented.* The number p is the *dimension* of σ^p. The simplexes whose vertices are among those of σ^p are known as the *faces* of σ^p, more precisely its *q-faces* for those of dimension q. In particular σ^p has $p + 1$ zero-faces, the vertices A_i, and a single p-face, namely itself.

If $\sigma_1 = A_0 \cdots A_q$, $\sigma_2 = A_{q+1} \cdots A_p$, where $A_0, \cdots, A_p$ are all distinct, we write $\sigma^p = \sigma_1\sigma_2$, call σ^p the *join* of σ_1, σ_2 and also σ_1, σ_2 *opposite* faces of σ^p. The symbol $\sigma_1 \cdots \sigma_r$ is defined by recurrence.

(5.2) We come now to the simplicial complex. The elements of a simplicial complex K are simplexes. They make up a set $\{\sigma\}$ such that if σ is in the set then every face of σ is likewise in the set. The dimension of σ is as defined in (5.1). The relation $\sigma' \prec \sigma$ means that σ' is a face of σ in the sense of (5.1). The incidence numbers are defined as follows:

(d) *If σ_1 is the face opposite the vertex A in σ set $\epsilon = \pm 1$ according as σ is or is not ordered like $A\sigma_1$; set $\epsilon = 0$ in all the other cases (σ_1 not a face opposite a vertex of σ nor the other way around). Then $[\sigma:\sigma_1] = [\sigma_1:\sigma] = \epsilon$.*

EXAMPLE. $\sigma = A_1A_2A_3$, $\sigma_1 = A_1A_3$. Since $A_1A_2A_3$ is not $A_2A_1A_3$ modulo an even permutation we have $[\sigma:\sigma_1] = -1$. On the other hand for instance $[\sigma:A_1] = 0$.

It remains to verify that K is a complex. Since K123 are manifestly satisfied it is only necessary to verify K4. It reduces here to:

$$\sum_{\sigma_1} [\sigma:\sigma_1][\sigma_1:\sigma_2] = 0. \tag{5.3}$$

The verification is trivial unless the situation may be so arranged that σ_2 is the opposite face of a one-simplex AB of σ, and σ_1 is then, except for ordering σ_1, one of the simplexes $A\sigma_2$, $B\sigma_2$. Furthermore if the order is changed in any simplex occurring in (5.3) the left-hand side will at most change sign. Therefore the order of the vertices may be chosen as specified by the above symbols. And now (5.3) reduces to

$$[AB\sigma_2:A\sigma_2][A\sigma_2:\sigma_2] + [AB\sigma_2:B\sigma_2][B\sigma_2:\sigma_2] = 0.$$

Therefore K is a complex.

(5.4) If the ordering of the vertices is changed the effect upon the incidence numbers is the same as applying to K an orientation function $\alpha(\sigma) = (-1)^\nu$, where ν is 0 or 1 according as the permutation of the vertices of σ is even or odd. Under our conventions this does not modify K.

To *orient* a simplicial complex $K = \{\sigma\}$ is to orient every $\sigma \in K$. The order assigned to the vertices of each σ is the one which is to serve in calculating the incidence numbers in (5.2). A convenient mode often utilized in orienting K is to range its vertices $\{A_i\}$ in a definite order, then to orient every σ as $\sigma = A_i \cdots A_j, i < \cdots < j$.

(5.5) *Special terms.* A one-dimensional complex is sometimes called a *linear graph*, or merely a *graph*. If L is a closed subcomplex of the simplicial complex K then L is also a simplicial complex. The complement $K - L$ is known as an *open* simplicial complex. By contrast K itself is sometimes called a *closed* simplicial complex.

The boundary $S^n = \mathfrak{B}\sigma^{n+1}$ of a σ^{n+1} is a closed simplicial n-complex, sometimes called an *n-sphere*. The zero-sphere S^0 consists of two vertices.

(5.6) While K may be infinite, it is clearly closure-finite but need not be star-finite.

(5.7) *An alternate scheme for the incidence numbers.* It is convenient on occasion to define the incidence numbers in the following way: If $\sigma = \sigma' A$ then the new incidence numbers, denoted temporarily by $[\ :\ ]'$ are $[\sigma:\sigma']' = [\sigma':\sigma]' = 1$ and all the other incidence numbers are zero. Clearly $[\sigma^p:\sigma^{p-1}]' = (-1)^p[\sigma^p:\sigma^{p-1}]$. That is to say, the new incidence numbers correspond to reorientation by means of $\alpha(\sigma^p) = \beta(-p)$, where $\beta(p)$ is as in the Introduction. Thus they are admissible incidence numbers for K.

(5.8) Remark. Unless otherwise stated the incidence numbers will always be selected in accordance with (5.2).

(5.9) *Duals.* The dual K^* of the simplicial complex K is defined as for any complex. Its elements are denoted by σ_p^i, and in particular the dual of A_i is written A^i. If $\sigma_i^p = A_j \cdots A_k$ we write $\sigma_p^i = A^j \cdots A^k$, and call $A^j, \cdots, A^k$ the *vertices* of σ_p^i. If $\sigma^p = A_h\sigma^{p-1}$ or $\sigma^p = \sigma^q\sigma^r$ then we also write $\sigma_p = A^h\sigma_{p-1}$ or $\sigma_p = \sigma_q\sigma_r$, as the case may be. The incidence numbers may be defined directly in K^* as in K by the rule $[A\sigma:\sigma] = [\sigma:A\sigma] = 1$ in the case of (5.2), $[\sigma A:\sigma] = [\sigma:\sigma A] = 1$ in the case of (5.7), and all the other $[\ :\ ]$ zero. We also have $\dim A^h = 0$, $\dim \sigma_p = -p$ and $\sigma_p \prec \sigma_q$ signifies that the set of vertices of σ_p *contains* the set of vertices of σ_q. In other words, the passage from K to K^* consists essentially: (a) in ordering the subsets of $\{A^i\}$ by the inclusions of their complements; (b) in replacing the dimensions by their negatives.

6. **Polyhedral complexes.** We will consider polyhedral complexes in an Euclidean space $\mathfrak{E}^n$ and indicate the extension to those in the Hilbert parallelotope (6.14).

(6.1) Definitions. *A polyhedral complex in $\mathfrak{E}^n$ is a countable locally finite complex $\Pi = \{E\}$ with the following properties:*

(a) *a p-dimensional element E^p is a p-cell which is a bounded convex region of some $\mathfrak{E}^p$ of $\mathfrak{E}^n$;*

(b) *the cells are disjoint;*
(c) *the union of the cells of* Cl E^p *is* $\bar{E}^p$;
(d) *If* $\varphi(E)$ *is the union of the cells* $E' \notin$ St E, *then* $\overline{\varphi(E)} \cap E = \emptyset$.
The set $\bigcup E$ *is known as a polyhedron, written* $|\Pi|$.

Notice that (c) implies that $E' < E$ when and only when it is E or else a cell $\subset \bar{E} - E$. In other words, $E' < E$ is equivalent to: $E' = E$ or else E' is a face of E in the commonly accepted sense.

Evidently (d) holds automatically when the polyhedral complex is finite. Its purpose is to eliminate certain topological complications which are foreign to the structure of complexes (see 6.2).

The incidence numbers are described below but they necessitate an extensive discussion of Euclidean coordinates.

(6.2) EXAMPLES. The regular solids are well known finite polyhedral complexes. The subdivision of the plane by the lines $x, y = 0, \pm 1, \pm 2, \cdots$, is a good example of an infinite polyhedral complex. On the other hand the set of segments:

$$l_0 : 0 \leqq x \leqq 1, y = 0; \qquad l_n : 0 \leqq x \leqq 1, y = \frac{1}{n}, \qquad n = 1, 2, \cdots,$$

is not a polyhedral complex since $\varphi(l_0) = \bigcup l_n$ is such that $\overline{\varphi(l_0)} \supset l_0$.

It may be noticed that under our definition a given point set A may admit of a decomposition in disjoint cells in two distinct ways, one of which gives rise to a polyhedral complex, and the other fails to do so. In particular (c) may cease to hold. Thus let λ be the set $0 < u \leqq 1$. The interval together with $u = 1$ is not a polyhedral complex. However, the intervals $1/(n+1) < u < 1/n$ together with their end points decompose λ into the cells of a polyhedral complex.

(6.3) The incidence numbers in Π will be described in terms of auxiliary coordinate systems in the spaces of the cells. As is well known, an Euclidean space $\mathfrak{E}^n$ may be viewed as a linear variety L in a real vector space $\mathfrak{V}$, i.e., a vector space over the field of reals. An $\mathfrak{E}^p$ in $\mathfrak{E}^n$ is a linear p-dimensional variety L' contained in L. The points of $\mathfrak{E}^p$ may be represented as vectors $a + x$, where $\{x\}$ spans a p-subspace $\mathfrak{V}_1$ of $\mathfrak{V}$. If $\{b^1, \cdots, b^p\}$ is a base for $\mathfrak{V}_1$ then we have $x = x_i b^i$, and $\{x_1, \cdots, x_p\}$ is a *coordinate system* for $\mathfrak{E}^p$. The point a is known as the *origin* of the system.

We will suppose once for all that every $\mathfrak{E}^p$ has been assigned a definite coordinate system called its *basic* coordinate system. Let $\{x_1, \cdots, x_p\}$ be the one of $\mathfrak{E}^p$. Then if $\{y_1, \cdots, y_p\}$ is any coordinate system for $\mathfrak{E}^p$ we will have:

$$\text{(6.3a)} \qquad y_i = a_i^j x_j + a_i, \qquad \alpha = |a_i^j| \neq 0.$$

The number $\epsilon^p = \alpha/|\alpha| = \pm 1$ is known as the *characteristic number* of the coordinate system $\{y_i\}$.

An $\mathfrak{E}^{p-1} \subset \mathfrak{E}^p$ partitions $\mathfrak{E}^p$ into two convex regions $\mathfrak{E}'^p, \mathfrak{E}''^p$. We may choose a coordinate system $\{x_i\}$ for $\mathfrak{E}^p$ such that $x_p = 0$ represents $\mathfrak{E}^{p-1}$. The two regions $\mathfrak{E}'^p$, $\mathfrak{E}''^p$ are then the two sets $x_p > 0$, $x_p < 0$. Since we may choose a new coordinate system in which x_p is replaced by $-x_p$, and the other coordinates are unchanged, we may assume $\{x_i\}$ such that say $\mathfrak{E}'^p$ is the region $x_p > 0$. The coordinates $\{x_1, \cdots, x_{p-1}\}$ of any point of $\mathfrak{E}^{p-1}$ define a coordinate system for $\mathfrak{E}^{p-1}$, with characteristic number say ϵ^{p-1}. We will introduce incidence numbers

$$(6.3b) \qquad [\mathfrak{E}'^p:\mathfrak{E}^{p-1}] = -[\mathfrak{E}''^p:\mathfrak{E}^{p-1}] = \epsilon^p\epsilon^{p-1}.$$

Let the basic coordinate system $\{x_i\}$ of $\mathfrak{E}^p$ serving to determine the characteristic numbers be replaced by another $\{\bar{x}_i\}$. We have then relations

$$(6.3c) \qquad \bar{x}_i = m_i^j x_j + n_i, \qquad \mu = | m_i^j | \neq 0.$$

If $\mu > 0$ the characteristic numbers are unchanged, if $\mu < 0$ they are all changed in sign. To *orient* $\mathfrak{E}^p$ is to assign to it a coordinate system $\{x_i\}$ modulo a transformation (6.3c) with $\mu > 0$. It is said to have its orientation *reversed* if the basic coordinate system undergoes a transformation (6.3c) with $\mu < 0$.

(6.4) We are now ready for the incidence numbers of Π. Let $\mathfrak{E}_i^p$ denote the subspace of the space $\mathfrak{E}^n$ of Π containing E_i^p and suppose $E_j^{p-1} < E_i^p$. The subspace $\mathfrak{E}_j^{p-1}$ divides $\mathfrak{E}_i^p$ into two regions one of which, say $\mathfrak{E}_i'^p$ contains E_i^p, and we define $[E_i^p:E_j^{p-1}] = [E_j^{p-1}:E_i^p] = [\mathfrak{E}_i'^p:\mathfrak{E}_j^{p-1}] = \pm 1$. All the other incidence numbers which are not determined by this rule are set equal to zero. Thus K123 hold and we merely have to verify K4.

(6.5) Suppose $E_h^{p-2} < E_i^p$. In $\mathfrak{E}_i^p$ let $\mathfrak{E}^2$ be a plane meeting $\mathfrak{E}_h^{p-2}$ at a single point $A \in E_h^{p-2}$. The intersection of $\mathfrak{E}^2$ with $\bar{E}_i^p$ is a closed convex plane polygonal region with the vertex A. In such a region each vertex is incident with exactly two edges. Hence E_h^{p-2} is the common face of exactly two $(p - 1)$-faces E_j^{p-1}, E_k^{p-1} of E_i^p. There are now two possibilities:

(a) $\mathfrak{E}_j^{p-1} = \mathfrak{E}_k^{p-1}$. Let $\mathfrak{E}_j'^{p-1}$, $\mathfrak{E}_k''^{p-1}$ be the two regions of the partition of $\mathfrak{E}_j^{p-1}$ by $\mathfrak{E}_h^{p-2}$. Since $E_j^{p-1} \neq E_k^{p-1}$ and they have the common face E_h^{p-2}, there must be one in each of the two regions, say $E_j^{p-1} \subset \mathfrak{E}_j'^{p-1}$, $E_k^{p-1} \subset \mathfrak{E}_j''^{p-1}$. Hence under our definition of the incidence numbers:

$$[E_j^{p-1}:E_h^{p-2}] = -[E_k^{p-1}:E_h^{p-2}] = \pm 1.$$

On the other hand:

$$[E_i^p:E_j^{p-1}] = [E_i^p:E_k^{p-1}] = [\mathfrak{E}_i'^p:\mathfrak{E}_j^{p-1}] = \pm 1.$$

Hence

$$(6.6) \qquad \sum_m [E_i^p:E_m^{p-1}][E_m^{p-1}:E_h^{p-2}] = 0$$

and K4 holds.

(b) $\mathfrak{E}_j^{p-1} \neq \mathfrak{E}_k^{p-1}$. Choose a coordinate system $\{x_1, \cdots, x_p\}$ for $\mathfrak{E}_i^p$ such that $\mathfrak{E}_j^{p-1}$, $\mathfrak{E}_k^{p-1}$ are, respectively, represented by $x_{p-1} = 0$ and $x_p = 0$ and that on $E_i^p : x_{p-1} > 0$, $x_p > 0$. As a consequence on $E_j^{p-1}:x_p > 0$ and on E_k^{p-1}:

$x_{p-1} > 0$. For we must have, respectively, $x_p \geqq 0$, $x_{p-1} \geqq 0$ and equality is excluded. To proceed further let us introduce the characteristic numbers $\epsilon_i^p, \epsilon_j^{p-1}, \epsilon_k^{p-1}, \epsilon_h^{p-2}$ of $\mathfrak{E}_i^p, \mathfrak{E}_j^{p-1}, \mathfrak{E}_k^{p-1}, \mathfrak{E}_h^{p-2}$ with respect to the coordinate systems $\{x_1, \cdots, x_p\}$, $\{x_1, \cdots, x_{p-2}, x_p\}$, $\{x_1, \cdots, x_{p-1}\}$, $\{x_1, \cdots, x_{p-2}\}$. We have then

$$[E_i^p : E_j^{p-1}] = -\epsilon_i^p \epsilon_j^{p-1}; \qquad [E_i^p : E_k^{p-1}] = \epsilon_i^p \epsilon_k^{p-1};$$

$$[E_j^{p-1} : E_h^{p-2}] = \epsilon_j^{p-1} \epsilon_h^{p-2}; \qquad [E_k^{p-1} : E_h^{p-2}] = \epsilon_k^{p-1} \epsilon_h^{p-2}.$$

That the last three incidence numbers have the correct value is clear. Regarding the first, to determine it we may utilize for $\mathfrak{E}_i^p$ the coordinate system $\{x_1, \cdots, x_{p-2}, x'_{p-1}, x'_p\}$ with $x'_p = x_{p-1}$, $x'_{p-1} = x_p$. This system has the characteristic $\epsilon_i'^p = -\epsilon_i^p$. The space $\mathfrak{E}_j^{p-1}$ is now represented by $x'_p = 0$, and has the coordinate system $\{x_1, \cdots, x_{p-2}, x'_{p-1}\}$ with the same characteristic ϵ_j^{p-1} as before. Hence

$$[E_i^p : E_j^{p-1}] = \epsilon_i'^p \epsilon_j^{p-1} = -\epsilon_i^p \epsilon_j^{p-1}.$$

Substituting the incidence numbers in (6.6) we find that this relation holds here also. Thus K4 is fulfilled in all cases and so Π is a complex.

(6.7) The characteristic numbers ϵ_i^p have been defined throughout relative to fixed orientations of the spaces $\mathfrak{E}_i^p$. If these are modified one will obtain new characteristic numbers $\epsilon_i'^p$. Setting $\alpha(E_i^p) = \epsilon_i^p \epsilon_i'^p$ we find then that the effect upon the incidence numbers $[E_i^p : E_j^{p-1}]$ is equivalent to applying the orientation function $\alpha(E)$. Thus the polyhedron *as a complex* is independent of the basic coordinate systems which serve to determine the ϵ_i^p and hence the incidence numbers.

(6.8) *A polyhedral complex is countable and locally finite.*

This is an immediate consequence of the definition (6.1).

(6.9) *Euclidean complexes.* We must first define *Euclidean simplexes*, the constituent parts of Euclidean complexes.

We will consider again a fixed $\mathfrak{E}^n$ and as in (6.3) take a representation of the space as a subset of a real vector space $\mathfrak{V}$. Let then σ^p be a simplex whose vertices $\{a_i\}$ are independent points of $\mathfrak{E}^n$, that is to say, contained in no $\mathfrak{E}^{p-1}$ of $\mathfrak{E}^n$ or equivalently contained in a unique $\mathfrak{E}^p$ of $\mathfrak{E}^n$. We associate with σ^p a set σ_e^p, known as an *Euclidean p-simplex*, and composed of the points of $\mathfrak{E}^n$ given by:

$$(6.10) \quad x = y^i a_i, \quad 0 < y^i < 1, \quad \sum y^i = 1, \quad p > 0; \quad x = a_0, \quad p = 0.$$

The y^i are the *barycentric coordinates* of x. To the face $\sigma' = a_i \cdots a_j$ of σ^p there corresponds the set of points obtained by replacing above $0 < y^k$ by $0 = y^k$ for $k \neq i, \cdots, j$. It is the σ'_e associated with σ' and is known as a *face* of σ_e^p. We transfer to σ_e^p and all its faces the terminology and concepts introduced for σ^p and thus we have notably the incidences, and incidence numbers of (5).

(6.11) *σ_e^p is a p-cell; its boundary $\mathfrak{B}\sigma_e^p$ is a topological $(p-1)$-sphere and $\bar{\sigma}_e^p$ is a closed p-cell.*

Let $\mathfrak{E}^p$ be the space of σ_e^p (i.e., the $\mathfrak{E}^p \supset \sigma_e^p$). It is an immediate consequence of (6.10) that: (a) σ_e^p is convex; (b) if $A \in \sigma_e^p$, any ray AL issued from A in $\mathfrak{E}^p$ intersects $\bar{\sigma}_e^p$ in a segment AB, $B \neq A$, $B \in \mathfrak{B}\sigma_e^p$. Now σ_e^p is contained in some parallelotope P of $\mathfrak{E}^n$. Since $\mathfrak{B}\sigma_e^p$ is thus a closed subset of the compactum P it is likewise a compactum. Hence $A \notin \mathfrak{B}\sigma_e^p \to d(A, \mathfrak{B}\sigma_e^p) > \rho > 0$ (I, 45.2) and therefore $\mathfrak{S}(A, \rho) \cap \mathfrak{E}^p \subset \sigma_e^p$. Thus A is an interior point of σ_e^p in $\mathfrak{E}^p$ and so σ_e^p is a convex region of $\mathfrak{E}^p$. Hence (6.11) is a consequence of (I, 12.9).

(6.12) Let now $K = \{\sigma\}$ be a countable locally finite simplicial complex whose vertices $\{a_i\}$ are points of an $\mathfrak{E}^n$ and such that:

(a) the vertices of any $\sigma^p \in K$ are independent and therefore determine an Euclidean simplex σ_e^p of $\mathfrak{E}^n$;

(b) $\sigma \neq \sigma' \to \sigma_e \cap \sigma_e' = \emptyset$;

(c) if $\varphi(\sigma_e)$ is the union of all the $\sigma_e' \notin \operatorname{St} \sigma_e$, then $\overline{\varphi(\sigma_e)} \cap \sigma_e = \emptyset$.

If we transfer to $\{\sigma_e\}$ the dimensions, incidences "is a face of," and incidence numbers prevailing in K, it becomes a complex $K_e \cong K$, known as an *Euclidean complex*. We also speak of K_e as an *Euclidean realization* of K, of K as an *antecedent* of K_e.

It follows from the definition of K_e that every $x \in \sigma_e \in K_e$ satisfies a relation

$$x = y^i a_i, \qquad \sum y^i = 1,$$

where if $x \in \sigma_e$, and σ_e has the vertices $a_j, \cdots, a_k$, then $y^j, \cdots, y^k \neq 0$, and all the other y^h are zero. The y^i are the *barycentric coordinates* of x and are uniquely determined by the point.

(6.13) *K_e is a polyhedral complex.*

Let K^q be the q-section of K. Since (6.13) is trivial for K_e^0 we may assume it for K_e^{p-1} and prove it for K_e^p. At all events all the requisite conditions except those referring to the incidence numbers are fulfilled. Thus we merely have to show that admissible incidence numbers of K are suitable for K_e as a polyhedral complex.

Let the vertices $\{a_i\}$ be ranged in some fixed order and let σ^p, σ_e^p have the common vertices $a_{i_0}, \cdots, a_{i_p}$, $i_0 < \cdots < i_p$. If $\mathfrak{E}^p$ is the space of σ^p, we agree to choose as its basic coordinate system $\{x_1, \cdots, x_p\}$ a system with origin at a_{i_0} and such that a_{i_h}, $h > 0$, has the coordinates δ_i^h (Kronecker deltas). Let σ^{p-1}, σ_e^{p-1} be the faces of σ, σ_e^p with the vertices $a_{i_0}, \cdots, a_{i_{q-1}}, a_{i_{q+1}}, \cdots, a_{i_p}$ and let the incidence numbers in K be defined by taking the vertices in the increasing order of the subscripts and in accordance with (5.7). Then

$$[\sigma^p : \sigma^{p-1}] = (-1)^{p-q} = [\sigma_e^p : \sigma_e^{p-1}].$$

Therefore (6.13) is proved.

(6.14) *Polyhedral complexes in the Hilbert parallelotope.* Let P^ω be referred to the coordinates $\{x_1, x_2, \cdots\}$, $0 \leqq x_n \leqq 1/n$. Consider a real vector space $\mathfrak{B}$ defined as follows: its elements are all the ordered countable sets of real numbers $y = \{y_1, y_2, \cdots\}$; if a is real then $ay = \{ay_n\}$; if $y' = \{y_n'\}$ then $y + y' = \{y_n + y_n'\}$. We may identify P^ω with a subset of a real variety L in $\mathfrak{B}$, where

L is so chosen that it contains no proper linear subvariety $\supset P^\omega$. If we interpret everywhere an $\mathfrak{E}^p$ as being a set $L' \cap P^\omega$, where L' is a linear p-dimensional variety contained in L, then all the preceding considerations are applicable. We will thus obtain polyhedral complexes, Euclidean simplexes and complexes in P^ω which have exactly the same properties as before.

(6.15) If $K_e = \{\sigma_e\}$ then in accordance with (6.1) the set $\bigcup \sigma_e$ is designated by $| K_e |$. Similarly if L_e is a closed subcomplex of K_e then the union of the $\sigma_e \in K_e - L_e$ is written $| K_e - L_e |$.

§2. HOMOLOGY THEORY OF FINITE COMPLEXES

(a) Generalities

7. The group-theoretic role of the algebraic structure imposed upon a complex receives its full significance through the medium of the chain-groups and certain associated subgroups and factor-groups. The groups related to finite complexes are to be investigated first and to the full. They will serve as a fundamental pattern for all later developments.

(7.1) Let then $X = \{x\}$ be a finite complex, and G an additive group. In the terminology of (II, 8) we use $\{x_i^p\}$ as a base to form the p-chains *over* G, or chains

$$C^p = g^i x_i^p , \qquad g^i \in G.$$

These are all finite since X is finite, and their group $\mathbf{P}(Gx_i^p)$ is denoted by $\mathfrak{C}^p(X, G)$. Instead of "C^p is a chain of X," we will also say more simply "C^p is contained in X," written $C^p \subset X$.

(7.2) If there are no p-elements it is convenient to introduce formally a group $\mathfrak{C}^p(X, G)$ consisting solely of zero.

(7.3) The set of all the x_i^p appearing in C^p with a coefficient $g^i \neq 0$ together with all their faces is a closed subcomplex of X denoted by $| C |$.

(7.4) Instead of "chain over $\mathfrak{J}$, $\mathfrak{J}_m$, $\mathfrak{P}$, over the rational group," we will say "integral, mod m, mod 1, rational chain" and similarly later for related entities (cycles, homology groups, etc.), the meaning being clear from the context.

(7.5) Notice that in dealing with finite complexes the groups of chains arising out of $\mathbf{P}^w$ may be considered as merely those arising out of $\mathbf{P}$ over a discrete G, and the distinction between the two possible types of chains of (II, 8) disappears.

(7.6) When G is a discrete field the chain-groups over G and the related groups introduced below will all be vector spaces over G and will conform with the basic convention (II, 22.2) for such spaces. That the homomorphisms and multiplications which will arise are always linear will generally be obvious.

8. The *chain-boundary*, or merely *boundary* of C^p is the $(p-1)$-chain

$$\text{(8.1)} \qquad \mathrm{F}C^p = \sum_j g^i [x_i^p : x_j^{p-1}] x_j^{p-1} .$$

The boundary operator F is thus defined simultaneously *for all groups G whatever.* When $\mathrm{F}C^p = 0$, C^p is said to be a *p-cycle of X over G.* Notice that $\mathrm{F}x^p$ is a chain of Cl x^p, and hence $\mathrm{F}C^p$ is a chain of Cl $| C^p |$. Identifying for convenience x_i^p with the integral p-chain $1.x_i^p$, F defines the integral boundary of x_i^p as

$$\mathrm{F}x_i^p = \sum_j [x_i^p : x_j^{p-1}] x_j^{p-1}. \tag{8.2}$$

Since the $[x_i^p : x_j^{p-1}]$ are integers, the finite sum $g^i[x_i^p : x_j^{p-1}]$ is an element of G so $\mathrm{F}C^p$ is an element of $\mathfrak{C}^{p-1}$. We notice that when x_i^p has no $(p-1)$-faces, $\mathrm{F}x_i^p = 0$ for every G. In particular if q is the lowest dimension of all the elements of X we have $\mathrm{F}x_i^q = 0$ and hence $\mathrm{F}C^q = 0$ whatever C^q: all the chains of the lowest dimension are cycles. This is not an exception as it corresponds merely to the fact that $\mathfrak{C}^{q-1} = 0$.

According to (II, 8.4) the operation F determines a homomorphism $\mathfrak{C}^p \to \mathfrak{C}^{p-1}$. The transformed group $\mathfrak{F}^p = \mathrm{F}\mathfrak{C}^{p+1}$ is known as the *group of the bounding p-chains over G.* The homomorphism $\mathrm{F}: \mathfrak{C}^p \to \mathfrak{C}^{p-1}$ has a kernel $\mathfrak{Z}^p$ in $\mathfrak{C}^p$ whose elements are the p-cycles over G, and $\mathfrak{Z}^p$ is known as *the group of the p-cycles over G.* If there are no p-elements then as in (7.2) we define formally $\mathfrak{Z}^p = \mathfrak{F}^p = 0$.

(8.3) As a consequence of K4 (K4′ of 1 in fact since X is finite) we have $\mathrm{FF}C^p = 0$ whatever C^p and whatever G. This relation is generally expressed in the operator form

$$\mathrm{FF} = 0. \tag{8.3a}$$

In fact condition K4′ is strictly equivalent to (8.3a) for $G = \mathfrak{J}$ and it is at least as frequently expressed in this form. Our formulation offered the formal advantage of being wholly divested of any connection with the chain-groups.

(8.4) It is a consequence of (8.3) that $\mathfrak{F}^p$ is a subgroup not merely of $\mathfrak{C}^p$ but actually of $\mathfrak{Z}^p$. Or in words: *every boundary is a cycle.* For $G = \mathfrak{J}$ this is merely another formulation of K4′.

9. Since $\mathfrak{Z}^p = \mathrm{F}^{-1}\bar{\omega}_{p-1}$, where $\bar{\omega}_q$ is the zero of $\mathfrak{C}^q$, $\mathfrak{Z}^p$ is a closed subgroup of $\mathfrak{C}^p$. Since $\mathfrak{Z}^p$ is closed $\bar{\mathfrak{F}}^p \subset \mathfrak{Z}^p$. Thus $\bar{\mathfrak{F}}^p$, which is also a subgroup (II, 3.2), is actually a closed subgroup of $\mathfrak{Z}^p$. In accordance with (II, 5) we may therefore introduce the factor-group:

$$\mathfrak{H}^p(X, G) = \mathfrak{Z}^p(X, G)/\bar{\mathfrak{F}}^p(X, G). \tag{9.1}$$

It is known as the *pth homology group* of X over G, and its elements as the *pth homology classes* of X over G. If $C^p - D^p \in \bar{\mathfrak{F}}^p$, we then write with Poincaré the homology:

$$C^p \sim D^p. \tag{9.2}$$

It is hardly necessary to point out that the homologies (9.2) combine like linear equations with integral coefficients, i.e., like arithmetical congruences.

The homology groups have various important characteristic numbers, notably their rank. The rank R^p of the pth integral homology group is known as the *pth Betti* number of the complex. The calculation of these numbers will be illustrated in some of the examples.

(9.3) The reader will have no difficulty in proving also

$$\mathfrak{C}^p(X, G)/\mathfrak{Z}^p(X, G) \cong \mathfrak{F}^{p-1}(X, G). \tag{9.4}$$

However, while of some interest, (9.4) rarely occurs in the applications.

10. **Complementary remarks.**

(10.1) *Reorientation convention for chains.* We shall agree that if X is reoriented by the orientation function $\alpha(x)$, the element gx^p of $\mathfrak{C}^p$ is to be replaced by $\alpha(x^p)gx^p$. In other words the chain-groups are to undergo a simultaneous automorphism α that may be described as:

$$\alpha\colon x \to \alpha(x)\, x.$$

Referring to (8.1) we find that $\alpha\mathrm{F} = \mathrm{F}\alpha$ (α commutes with F). It follows that $\mathfrak{F}^p$, $\mathfrak{Z}^p$ are unchanged by α, and hence the same holds regarding $\mathfrak{H}^p$. In other, words our convention merely introduces isomorphisms of all the groups $\mathfrak{C}$ $\mathfrak{F}$, $\mathfrak{Z}$, $\mathfrak{H}$.

(10.2) *Influence of isomorphisms upon the different groups.* An isomorphism $X \to X'$ induces likewise isomorphisms of the groups $\mathfrak{C}^p(X, G), \cdots, \mathfrak{H}^p(X, G)$, with the corresponding groups of X'. A weak isomorphism $X \to X'$ raising dimensions n units induces isomorphisms of the groups $\mathfrak{C}^p(X, G), \cdots$, with the groups $\mathfrak{C}^{p+n}(X, G), \cdots$. A similar remark applies to all the groups of complexes introduced later and will not be repeated.

(10.3) *Separation of dimensions.* While we are separating dimensions throughout, this is not absolutely necessary. We could have defined a chain over G as any expressions $C = g^ix_i$, $g^i \in G$, with resulting groups $\mathfrak{C}, \cdots, \mathfrak{H}$ related as before. Evidently $\mathfrak{C} = \mathbf{P}\mathfrak{C}^p, \cdots$. However, the more interesting parts of the theory of complexes arise precisely from the comparison of certain dimensions, and so the scheme which we are following is preferable.

(10.4) *Simplicial complexes and their duals.* Let the notations be as in (5), notably as in (5.9) regarding the dual. If $C^p = \sum g_i\, \sigma_i^p$ is a chain of K, then AC^p denotes the chain $\sum g_i(A\sigma_i^p)$, where if either $A\sigma_i^p$ is not a simplex of K, or A is a vertex of σ_i^p then the term $g_i(A\sigma_i^p)$ is to be set equal to zero. A similar convention is adopted for K^* and its chains.

It is convenient to have the explicit expression of $\mathrm{F}\sigma^p$, $\mathrm{F}\sigma_p$ under both schemes (5.2, 5.7):

(a) under the scheme (5.2)

$$\mathrm{F}A_0 \cdots A_p = \sum (-1)^q A_0 \cdots A_{q-1}\, A_{q+1} \cdots A_p, \qquad p > 0;$$

(b) under the scheme (5.7) ($[\sigma A : \sigma] = 1$):

$$\mathrm{F}A_0 \cdots A_p = \sum (-1)^{p-q} A_0 \cdots A_{q-1}\, A_{q+1} \cdots A_p, \qquad p > 0;$$

(c) under both schemes $\mathrm{F}\sigma^0 = 0$.

In K^*, we have if $\gamma_0 = \sum A^i$, then:

(d) under the scheme (5.2):

$$\mathrm{F}\sigma_p = \gamma_0\sigma_p\,; \qquad \mathrm{F}C_p = \gamma_0 C_p\,;$$

(e) under the scheme (5.7):

$$\mathrm{F}\sigma_p = \sigma_p\gamma_0; \qquad \mathrm{F}C_p = C_p\gamma_0.$$

We verify directly $\mathrm{F}\gamma_0 = 0$, so γ_0 is a zero-cycle of K^*. This zero-cycle plays the role of a boundary operator for K^*: multiplied to the left under (d), and to the right under (e).

The usual scheme is (d), and so γ_0 will generally act as a left multiplier.

(10.5) *Boundary relations in polyhedral complexes.* Let $\Pi = \{E\}$ be a polyhedral complex and let $E^n \in \Pi$ have as its $(n-1)$-faces $\{E_i^{n-1}\}$. Then $[E^n : E_i^{n-1}] = \epsilon^i = \pm 1$ (6.4), and so:

$$\mathrm{F}E^n = \epsilon^i E_i^{n-1}.$$

Therefore:

(10.6) *Every $(n-1)$-face of E^n appears in $\mathrm{F}E^n$ with a coefficient* ± 1.

Since E^n is a bounded region in an $\mathfrak{E}^n$, $\mathfrak{B}E^n$, $n > 0$, cannot be contained in a finite set of $\mathfrak{E}^p$'s, $p < n-1$. Therefore E^n has at least one $(n-1)$-face and hence:

(10.7) E^n, $n > 0$, *is not a cycle.*

In Π we have in more general form:

$$\mathrm{F}E_i^n = \eta_i^j(n-1)E_j^{n-1}, \tag{10.8}$$

where $\eta_i^j(n-1) = \pm 1$ if E_j^{n-1} is a face of E_i^n, and $\eta_i^j(n-1) = 0$ otherwise.

§3. HOMOLOGY THEORY OF FINITE COMPLEXES

(b) Integral groups

11. Further progress is contingent upon a full investigation of the integral groups. They are assumed taken with discrete topology and it is to these that the symbols $\mathfrak{C}$, $\mathfrak{Z}$, $\mathfrak{F}$, $\mathfrak{H}$ shall refer in the present section.

Since $\mathfrak{C}^p$ is a free group on a finite number of generators, we shall naturally utilize properties of such groups as given in (II, §2).

Set for convenience $[x_i^{p+1} : x_j^p] = \eta_i^j(p)$, and denote by $\eta(p)$ the matrix of these numbers, or *pth incidence matrix* of X. The group $\mathfrak{Z}^p$ of the p-cycles is the subgroup of the elements $\gamma^p = g^i x_i^p$ of $\mathfrak{C}^p$ which satisfy the relations $\mathrm{F}\gamma^p = 0$ or

$$g^i \eta_i^j(p-1) = 0. \tag{11.1}$$

Since $\mathfrak{Z}^p$ is a subgroup of a free group of finite rank it is itself a free group of finite rank (II, 10.1). Since the topology is discrete $\bar{\mathfrak{F}}^p = \mathfrak{F}^p$. The latter is the group generated by the elements $\mathrm{F}x_i^{p+1} = \eta_i^j(p)x_j^p$. The group $\mathfrak{H}^p$ is the factor-group of $\mathfrak{Z}^p$ by $\mathfrak{F}^p$ and so it is isomorphic with the group of the elements $g^i x_i^p$ such that (11.1) holds and with the relations

$$\eta_i^j(p)x_j^p = 0. \tag{11.2}$$

Thus $\mathfrak{H}^p$ is a group on a finite number of generators and so from the basic reduction theorem for such groups (II, 12.8) we conclude:

(11.3) THEOREM. *The pth integral homology group of a finite complex X satisfies a relation*:

$$\mathfrak{H}^p \cong \mathfrak{B}^p \times \mathbf{P}\mathfrak{T}_i^p \tag{11.4}$$

where: (a) $\mathfrak{B}^p$ *is a free group on a number of generators equal to the pth Betti number* $R^p(X)$; (b) *the* $\mathfrak{T}_i^p$ *are cyclic groups in finite number whose orders* t_i^p *are finite and such that* t_i^p *divides* t_{i+1}^p.

The t_i^p are the *pth torsion coefficients* of X and $\mathfrak{B}^p$ its *pth Betti group*. From the reduction theorem we have also the complementary result:

(11.5) *The torsion coefficients* t_i^p *are the invariant factors greater than* 1 *of the incidence matrix* $\eta(p)$.

The group $\mathfrak{T}^p = \mathbf{P}\mathfrak{T}_i^p$ is also known as the *pth torsion group* of X and we have:

(11.6) $\mathfrak{H}^p$ *is isomorphic with the product of the pth Betti group and the pth torsion group.*

12. Pursuing our investigation we shall obtain a simultaneous reduction of the chain-groups to a form clearly exhibiting their mutual relations.

The group $\mathfrak{C}^p$ of the integral p-chains is a free group on the x_i^p with the subgroups $\mathfrak{Z}^p$, $\mathfrak{F}^p$, where $\mathfrak{F}^p \subset \mathfrak{Z}^p$. By (II, 12.6) a base may be chosen for $\mathfrak{C}^p$ consisting of elements A_i^p, B_i^p such that certain multiples $s_i A_i^p$ make up a base for $\mathfrak{Z}^p$.

Now $t\gamma^p = tg^i x_i^p$ is a cycle when and only when the tg^i satisfy (11.1). But in that case the g^i themselves satisfy also these relations and so γ^p is likewise a cycle. In other words, $t\gamma^p \in \mathfrak{Z}^p \rightarrow \gamma^p \in \mathfrak{Z}^p$. Applying this to the A_i^p we deduce that $A_i^p \in \mathfrak{Z}^p$, which can only be if the s^i are unity. Thus:

(12.1) *A base* $\{A_i^p, B_i^p\}$ *may be chosen for* $\mathfrak{C}^p$ *such that* $\{A_i^p\}$ *is a base for the group* $\mathfrak{Z}^p$ *of the integral p-cycles.*

Regarding the B_i^p no element based on them is a cycle, i.e., no $g^i B_i^p \in \mathfrak{Z}^p$. unless every $g^i = 0$.

The same reduction may be carried out for all dimensions. Since the A_i^{p+1} are cycles, $\mathrm{F}A_i^{p+1} = 0$ and so $\mathfrak{F}^p$ has for generators the chains $\mathrm{F}B_i^{p+1}$ which are in fact p-cycles (8). Again by (II, 12.6) a base may be chosen for $\mathfrak{Z}^p$ consisting of cycles $A_i'^p$, c_i^p such that certain multiples $r_i^p A_i'^p$ form a base for $\mathfrak{F}^p$. Furthermore r_i^p divides r_{i+1}^p throughout.

Notice that the replacement of the A_i^p by the $A_i'^p$, c_i^p as free generators of $\mathfrak{C}^p$, is a change of base in $\mathfrak{C}^p$ which leaves the B_i^p undisturbed.

Since $\mathfrak{H}^p = \mathfrak{Z}^p/\mathfrak{F}^p$ it is isomorphic with the group on the generators $A_i'^p$, c_i^p with the relations:

$$r_i^p A_i'^p = 0. \tag{12.2}$$

We now divide the $A_i'^p$ into two sets. The first will consist of the elements, denoted by a_i^p, such that the corresponding $r_i^p = 1$, the second of those, denoted by b_i^p, such that the corresponding r_i^p, henceforth written t_i^p are greater than 1. The notations are so chosen that t_i^p divides t_{i+1}^p. The group $\mathfrak{H}^p$ is now isomorphic with the group on the generators a_i^p, b_i^p, c_i^p with the relations:

$$a_i^p = 0, \tag{12.3a}$$

$$t_i^p b_i^p = 0. \tag{12.3b}$$

Evidently the a_i^p may be suppressed among the generators so that $\mathfrak{H}^p$ is in fact isomorphic with the group on the b_i^p, c_i^p with the relations (12.3b). We have then

$$\mathfrak{H}^p \cong \mathbf{P}(\mathfrak{J}c_i^p) \times \mathbf{P}(\mathfrak{J}^*(t_i^p)b_i^p). \tag{12.4}$$

By comparing with (11.4) we have then:

$$\mathfrak{B}^p \cong \mathbf{P}(\mathfrak{J}c_i^p),\ \mathfrak{T}^p = \mathbf{P}\mathfrak{T}_i^p \cong \mathbf{P}(\mathfrak{J}^*(t_i^p)b_i^p). \tag{12.5}$$

Therefore: (a) the rank R^p of $\mathfrak{B}^p$, or of $\mathfrak{H}^p$, is the number of c_i^p; (b) the t_i^p are the orders of the $\mathfrak{T}_i^p$, i.e., they are the pth torsion coefficients of X.

13. Since $\{a_i^p, t_i^p b_i^p\}$ are free generators for $\mathfrak{F}^p$, by (II, 12.3) they must be reducible to the set of free generators $\{\mathrm{F}B_i^{p+1}\}$ by a unimodular transformation. That is to say there exist relations:

$$t_i^p b_i^p = \lambda_i^j \mathrm{F} B_j^{p+1} = \mathrm{F}(\lambda_i^j B_j^{p+1})$$
$$a_i^p = \mu_i^j \mathrm{F} B_j^{p+1} = \mathrm{F}(\mu_i^j B_j^{p+1})$$

with a unimodular matrix

$$\left\| \begin{matrix} \lambda_i^j \\ \mu_i^j \end{matrix} \right\|.$$

It follows that if we set

$$d_i^{p+1} = \lambda_i^j B_j^{p+1}, \qquad e_i^{p+1} = \mu_i^j B_j^{p+1},$$

the d_i^{p+1}, e_i^{p+1} may replace the free generators B_i^{p+1} in the set of free generators $\{A_i^{p+1}, B_i^{p+1}\}$ for $\mathfrak{C}^{p+1}$. Notice that this substitution does not affect $\{A_i^{p+1}\} = \{a_i^{p+1}, b_i^{p+1}, c_i^{p+1}\}$.

14. Let us suppose now that the dimensions in X run from q to r. We reduce the bases for $\mathfrak{C}^q$, $\mathfrak{C}^{q+1}$, $\cdots$ in succession as follows:

Group $\mathfrak{C}^q$. Here $\mathfrak{C}^{q-1} = 0$ so $\mathfrak{C}^q = \mathfrak{Z}^q$. The first reduction yields the base $\{A_i^q\}$ and the second the base $\{a_i^q, b_i^q, c_i^q\}$. There are no d_i^q, e_i^q.

Group $\mathfrak{C}^{q+1}$. The first reduction yields the base $\{A_i^{q+1}, B_i^{q+1}\}$, and the second $\{a_i^{q+1}, b_i^{q+1}, c_i^{q+1}, B_i^{q+1}\}$. The a_i^q, b_i^q determine the e_i^{q+1}, d_i^{q+1} and as already observed the choice of these in place of the B_i^{q+1} as generators is tantamount to a change of base for $\mathfrak{C}^{q+1}$. We thus have a third reduction to a final base $\{a_i^{q+1}, \cdots, e_i^{q+1}\}$ for $\mathfrak{C}^{q+1}$.

Group $\mathfrak{C}^p$. The same reduction may be applied step by step for $p = q + 1$ $\cdots, r$. We have thus proved the

(14.1) THEOREM. *The bases for the integral chain-groups of a finite complex may be chosen to consist for each dimension p of five sets of elements* $a_i^p, \cdots, e_i^p$ *with the boundary relations*:

$$\begin{aligned} \mathrm{F}e_i^{p+1} &= a_i^p, & \mathrm{F}a_i^p &= 0,\\ \mathrm{F}d_i^{p+1} &= t_i^p b_i^p, & \mathrm{F}b_i^p &= 0,\\ & & \mathrm{F}c_i^p &= 0,\\ & & \mathrm{F}d_i^p &= t_i^{p-1}b_i^{p-1},\\ & & \mathrm{F}e_i^p &= a_i^{p-1}. \end{aligned} \tag{14.2}$$

The number of c_i^p *is the pth Betti number* R^p, *and the t's are the torsion coefficients.*

The bases in the reduced form described in the theorem are said to be *canonical*.

15. There remains the explicit calculation of the R^p. Let α^p denote as before the number of elements x_i^p, and let ρ^p be the rank of the incidence matrix $\eta(p)$.

Since $\mathfrak{Z}^p$ is the subgroup of the elements $g^i x_i^p$ for which (11.1) holds, its rank is $\alpha^p - \rho^{p-1}$. Since $\mathfrak{H}^p \cong$ the factor-group of $\mathfrak{Z}^p$ by the subgroup of the elements which satisfy the relations (11.2), its rank is $\alpha^p - \rho^{p-1} - \rho^p$, or

$$R^p = \alpha^p - \rho^{p-1} - \rho^p. \tag{15.1}$$

Since $\eta(p)$ exists only for $q \leqq p \leqq r - 1$, to make (15.1) hold formally for all dimensions we define $\rho^s = 0$, for $s < q$ or $s > r - 1$. Thus (15.1) provides an explicit expression for the Betti numbers in terms of the incidence matrices.

If we multiply both sides in (15.1) by $(-1)^p$ and add there comes the classical *Euler-Poincaré relation* for finite complexes:

$$\sum (-1)^p \alpha^p = \sum (-1)^p R^p. \tag{15.2}$$

The common value $\chi(X)$ of these two sums is known as the *Euler characteristic of* X. The relation (15.2) is often convenient for computing Betti numbers, notably when the range of the dimensions in X is small.

(15.3) *The Poincaré polynomial.* We understand thereby the polynomial

$$P(t, X) = \sum R^p t^p \tag{15.4}$$

whose coefficients are the Betti numbers of X. Under certain circumstances (formation of the product, IV) this polynomial obeys very convenient formal rules. Moreover it may be used to advantage in describing the Betti numbers of certain simple complexes. Thus for the boundary of the $(n + 1)$-simplex we shall find later (22.4) that in substance $P(t, X) = 1 + t^n$.

§4. HOMOLOGY THEORY OF FINITE COMPLEXES

(c) Arbitrary groups of coefficients

16. Let again G be an arbitrary topological group and let this time $\mathfrak{C}, \mathfrak{Z}, \mathfrak{F}, \mathfrak{H}$ refer to the groups over G. Following Steenrod [a] we shall give a complete analysis of these groups.

By definition (II, 8)

$$\mathfrak{C}^p = \mathbf{P}(Gx_i^p).$$

Let us designate temporarily by $x_i'^p$ the elements $a_i^p, \cdots, e_i^p$ of (14.1), where we have a relation

(16.1) $$x_i'^p = \lambda_i^j(p)x_j^p, \qquad \lambda(p) = \|\lambda_i^j(p)\| \text{ unimodular.}$$

Consider now the group

$$\mathfrak{C}'^p = \mathbf{P}(Gx_i'^p).$$

Referring to (II, 8.4):

$$g^i x_i'^p \to g^i \lambda_i^j(p) x_j^p$$

defines a homomorphism τ: $\mathfrak{C}'^p \to \mathfrak{C}^p$. Since $\lambda(p)$ is unimodular, it has an inverse

$$\mu(p) = \lambda^{-1}(p) = \|\mu_i^j(p)\|,$$

and so

$$g^i x_i^p \to g^i \mu_i^j(p) x_j'^p$$

defines a homomorphism θ: $\mathfrak{C}^p \to \mathfrak{C}'^p$. It is an elementary matter to verify $\tau\theta = 1$, $\theta\tau = 1$; hence τ, θ are isomorphisms (II, 4.5), and consequently $\mathfrak{C}^p \cong \mathfrak{C}'^p$.

From the result just obtained we infer that we may represent every element $C^p \in \mathfrak{C}^p$ in the form:

(16.2) $$C^p = g^h a_h^p + y^i b_i^p + z^j c_j^p + u^k d_k^p + v^l e_l^p$$

where the coefficients $\in G$. From (14.2) follows now:

(16.3) $$\mathrm{F}C^p = u^k t_k^{p-1} b_k^{p-1} + v^l a_l^{p-1}.$$

Hence C^p is a cycle when and only when $v^l = 0$ and $t_k^{p-1} u^k = 0$ (k unsummed). Therefore in the group symbolism of the introduction to the present chapter:

(16.4) $$\mathfrak{Z}^p = \mathbf{P}(Ga_h^p) \times \mathbf{P}(Gb_i^p) \times \mathbf{P}(Gc_j^p) \times \mathbf{P}(G[t_k^{p-1}]d_k^p).$$

On the other hand $\mathfrak{F}^p$ is the set of all chains

$$C^p = g^h a_h^p + y^i t_i^p b_i^p . \tag{16.5}$$

Since $y^i t_i^p$ (i unsummed) is merely any element of $G(t_i^p)$ we have:

$$\mathfrak{F}^p = \mathbf{P}(Ga_h^p) \times \mathbf{P}(G(t_i^p)b_i^p). \tag{16.6}$$

The topology of the factor Ga_h^p is governed by that of G and the isomorphism $Ga_h^p \cong G$. The topology of $G(t_i^p)b_i^p$ is its relative topology as a subgroup of $Gb_i^p \cong G$. Therefore $\bar{\mathfrak{F}}^p$ is obtained by merely replacing $G(t_i^p)b_i^p$ by its closure $\overline{G(t_i^p)b_i^p}$ as a subgroup of Gb_i^p, and this is the same as $\overline{G(t_i^p)}b_i^p$ (closure in G). With this meaning of the symbols clearly before us we have then

$$\bar{\mathfrak{F}}^p = \mathbf{P}(Ga_h^p) \times \mathbf{P}(\overline{G(t_i^p)}b_i^p). \tag{16.7}$$

If we combine with (16.4) and recall that $\mathfrak{H} = \mathfrak{Z}/\bar{\mathfrak{F}}$, we have:

$$\mathfrak{H}^p(X, G) \cong \mathbf{P}((G/\overline{G(t_i^p)})b_i^p) \times \mathbf{P}(Gc_j^p) \times \mathbf{P}(G[t_k^{p-1}]d_k^p), \tag{16.8}$$

or finally in equivalent form:

$$\mathfrak{H}^p(X, G) = \mathbf{P}(G^*(t_i^p)b_i^p) \times \mathbf{P}(Gc_j^p) \times \mathbf{P}(G[t_k^{p-1}]d_k^p). \tag{16.9}$$

We have thus obtained a basic decomposition of the homology groups over any coefficient group G.

Some simple conclusions may immediately be drawn from the relations just obtained notably:

(16.10) *If there are no torsion coefficients for the dimension p then a p-cycle ~ 0 is a bounding cycle. Hence if there are no torsion coefficients "~ 0" and "bounding" are equivalent.* (See 9.)

For the second product at the right in (16.6) is then absent, hence $\bar{\mathfrak{F}}^p = \mathfrak{F}^p$, which is (16.10).

(16.11) *If there are no torsion coefficients for the dimensions p and $p - 1$ the pth homology group over G reduces to the "Betti" part $\mathbf{P}(Gc_j^p)$. Hence if there are no torsion coefficients all the homology groups reduce to their Betti parts.*

(16.12) *If in X: $p \leqq \dim x \leqq q$ then*: (a) *no q-cycle ~ 0 unless it is zero*; (b) *every p-chain is a cycle.*

Noteworthy special case: *In a simplicial complex or in a polyhedral complex, every zero-chain is a zero-cycle.*

Since there are no $(q + 1)$-chains different from 0, we have $\bar{\mathfrak{F}}^{q+1}(X, G) = 0$ which is (a). Owing to the absence of $(p - 1)$-chains different from 0 we have $FC^p = 0$ whatever C^p and this is (b).

17. **Some noteworthy coefficient-groups.**

(17.1) *Division-closure groups.* For these groups the $G(m)$ are all closed (II, 20.9) and so from (16.7):

(17.2) *When G is a division-closure group, $\bar{\mathfrak{F}}^p = \mathfrak{F}^p$, and hence*: (a) "$\sim 0$" $\leftrightarrow$ "*bounding*;" (b) $\mathfrak{H}^p = \mathfrak{Z}^p/\mathfrak{F}^p$ (Steenrod [a]).

Noteworthy special cases: *G is compact or discrete.*

(17.3) *G is a discrete field.* The groups $\mathfrak{C}^p(X, G), \cdots, \mathfrak{H}^p(X, G)$ are then finite-dimensional vector spaces over G and hence discrete (II, 25.6). The dimension $R^p(X, G)$ of $\mathfrak{H}^p(X, G)$ is known as the *pth Betti number over G*. Let π be the characteristic of G. It will be recalled that π is such that $\pi g = 0$ for every $g \in G$, and is a prime number or zero. Among the special fields of characteristic $\pi > 0$ is found $\mathfrak{J}_\pi$, the field of the residues mod π, and it is a subfield of every field G of characteristic π. The corresponding chains, $\cdots$ are known as *chains*, $\cdots$, mod π and the associated groups and Betti numbers are written $\mathfrak{C}^p(X, \pi), \cdots, R^p(X, \pi)$. We shall show in substance that for most purposes $\mathfrak{J}_\pi$ may replace G.

We will make use of (II, 36). In the notations there utilized and since $\mathfrak{J}_\pi$ is a subfield of G we recognize immediately that $\mathfrak{C}^p(X, G) = G\mathfrak{C}^p(X, \pi)$. Furthermore $\mathfrak{Z}^p(X, \pi)$ is defined by means of $\mathrm{F}C^p = 0$, where the coefficients of $\mathrm{F}C^p$ are reduced mod π. The group $\mathfrak{Z}^p(X, G)$ is defined by the same relation save that the values $\mathrm{F}C^p$ for the chains over G are obtained by linear extension from the values for the chains mod π. Both the groups $\mathfrak{F}^p(X, G)$ and $\mathfrak{F}^p(X, \pi)$ are spanned by the chains $\mathrm{F}E_i^{p+1}$ taken mod π (i.e., with coefficients reduced mod π). We have therefore the exact situation of (II, 36.8). By that result then $\mathfrak{H}^p(X, G)$ is isomorphic with the vector space over G spanned by $\mathfrak{H}^p(X, \pi)$, and $\mathfrak{H}^p(X, G)$, $\mathfrak{H}^p(X, \pi)$ have the same dimension, or

$$(17.4) \qquad R^p(X, G) = R^p(X, \pi).$$

We have obtained (17.4) without utilizing the reduction (16.9). We may also use the latter for the same purpose, and it will lead to an expression for $R^p(X, \pi)$ in terms of the integral Betti numbers and torsion coefficients.

Referring to (16.9), $R^p(X, G)$ is equal to the number of products effectively present at the right:

(a) When π does not divide t_i^p, $G(t_i^p) = G$, and the corresponding term is absent. When π divides t_i^p, $G(t_i^p) = 0$ and there is a term Gb_i^p. Therefore the first product at the right in (16.9) consists of θ_π^p isomorphs of G, where θ_π^p is the number of t_i^p with π as a prime factor.

(b) The second product in (16.9) consists of R^p isomorphs of G.

(c) When π divides t_k^{p-1}, $G[t_k^{p-1}] \cong G$, otherwise it is zero. Therefore the third product in (16.9) consists of θ_π^{p-1} isomorphs of G.

Thus $\mathfrak{H}^p(X, G)$ is the product of $R^p + \theta_\pi^{p-1} + \theta_\pi^p$ isomorphs of G. Hence

$$(17.5) \qquad R^p(X, G) = R^p + \theta_\pi^{p-1} + \theta_\pi^p .$$

Since this value depends only on π, (17.4) follows. We write explicitly

$$(17.6) \qquad R^p(X, \pi) = R^p + \theta_\pi^{p-1} + \theta_\pi^p .$$

The case $\pi = 0$, i.e., G of characteristic zero, is not exceptional. The field $\mathfrak{J}_\pi$ is then to be replaced by the rational field $\mathfrak{R}$. The corresponding chains, $\cdots$, are said to be *rational*. We verify directly that in (16.9) the second product alone remains, thus yielding for G of characteristic zero:

(17.7) $$R^p(X, G) = R^p(X, \mathfrak{R}) = R^p(X).$$

This shows that the integral Betti numbers themselves may also be defined as the dimensions of the homology groups over a field, namely the rational field $\mathfrak{R}$, or for that matter any field of characteristic zero. An incidental and frequently convenient result is that $\{c_i^p\}$ is a base for the rational p-cycles with respect to homology.

To sum up we have proved for finite complexes, a theorem which will recur in a number of instances later, and is formulated for later reference in a more general form than immediately required:

(17.8) UNIVERSAL THEOREM FOR FIELDS. *The homology groups over a field G of characteristic π are vector spaces over G, and their dimensions, the Betti numbers over G, are equal to the corresponding Betti numbers* mod π.

COMPLEMENTARY RESULT. *A maximal set of p-cycles* mod π *independent with respect to homology is likewise a maximal independent set for any field of characteristic π.*

These properties hold likewise for $\pi = 0$, the cycles mod π, *and their Betti numbers being then the rational cycles and Betti numbers.*

Complements (for finite complexes only): (a) *The Betti numbers are all finite, and the rational Betti numbers are the numbers R^p (integral Betti numbers) previously defined.* (b) *The Euler-Poincaré formula holds for all fields G:*

(17.9) $$\sum (-1)^p R^p(X, G) = \sum (-1)^p \alpha^p.$$

(Immediate consequence of (15.2) and (17.6).)

Historical note. The special theory mod 2 played an important role in earlier topology, as a reference to Veblen [V] will show.

(17.10) $G = \mathfrak{P}$, *the group of the reals* mod 1. This time $\mathfrak{P}(m) = \mathfrak{P}$, hence $\mathfrak{P}^*(m) = 0$. Therefore

(17.11) $$\mathfrak{H}^p(X, \mathfrak{P}) \cong \mathbf{P}(\mathfrak{P}c_j^p) \times \mathbf{P}(\mathfrak{P}[t_k^{p-1}]d_k^p).$$

The first term is a toroidal group, which is the direct product of R^p isomorphs of $\mathfrak{P}$. Since $\mathfrak{P}[t_k^{p-1}] \cong \mathfrak{J}^*(t_k^{p-1})$, (II, 20.14) the second term in (17.11) $\cong \mathfrak{T}^{p-1}(X)$. Therefore

$$\mathfrak{H}^p(X, \mathfrak{P}) \cong \mathbf{P}(\mathfrak{P}c_j^p) \times \mathfrak{T}^{p-1}(X).$$

This proves:

(17.12) *The pth homology group of X* mod 1 *is $\cong$ the product of an R^p-dimensional toroidal group by the $(p - 1)$-dimensional torsion group of X (which is a finite group).*

18. **Universal coefficient-groups.** A group G_0 is called a universal coefficient-group for X, whenever given the full set $\{\mathfrak{H}^p(X, G_0)\}$ and an arbitrary G it is possible to determine in terms of the groups of this set and of G all the groups $\{\mathfrak{H}^p(X, G)\}$.

When the integral homology groups and G are known, so are the Betti numbers and torsion coefficients and hence also an isomorph of the product at the right in (16.9), and finally $\mathfrak{H}^p(X, G)$ itself. Similarly, given all the homology groups mod 1 and the group G, we learn from (17.11) the values of the numbers R^p, and also the $\mathfrak{T}^{p-1}(X)$, hence the torsion coefficients. Consequently, we may again determine the terms in (16.9) and hence the groups over G. Therefore

(18.1) *The groups* $\mathfrak{J}$, $\mathfrak{P}$ *of the integers and of the reals* mod 1 *are both universal coefficient-groups for finite complexes.*

(18.2) *A n. a. s. c. for two finite complexes to have the same homology groups over every G is that they have the same integral homology groups, or else the same homology groups* mod 1, *and so, in the last analysis, that they have the same Betti numbers and torsion coefficients.*

19. The following properties which are often useful, are ready consequences of the general theory:

(19.1) *If* $X = \bigcup X_i$, *where the* X_i *are disjoint complexes, then*:

$$\mathfrak{H}^p(X, G) = \mathbf{P}\mathfrak{H}^p(X_i, G) \qquad \text{(every } G\text{)}, \tag{19.2}$$

$$R^p(X, G) = \sum R^p(X_i, G) \qquad (G \text{ a field}). \tag{19.3}$$

The second relation follows from the first, so we merely prove (19.2). Every chain C^p of X over G is of the form

$$C^p = \sum C_i^p, \tag{19.4}$$

where C_i^p is a chain of X_i over G. N. a. s. c. for C^p to be a cycle, or to be ~ 0, are, respectively, that every C_i^p be a cycle, or be ~ 0. From this to (19.2) is but a step.

(19.5) *If* X^{p+1} *is the* $(p + 1)$*-section of* X *then for every* $r \leqq p$:

$$\mathfrak{H}^r(X, G) = \mathfrak{H}^r(X^{p+1}, G) \qquad \text{(every } G\text{)} \tag{19.6}$$

$$R^r(X, G) = R^r(X^{p+1}, G) \qquad (G \text{ a field}). \tag{19.7}$$

For $\mathfrak{H}^r(X, G)$ depends solely on the elements of X^{p+1}.

§5. APPLICATION TO SOME SPECIAL COMPLEXES

20. **Simplicial complexes.** Let $K = \{\sigma\}$ be a finite simplicial complex. The vertices of K will be designated by A with possible supplementary indices. If $\sigma^p = A_0 \cdots A_p \in K$ then (10.4ac):

$$\mathrm{F}\sigma^p = \sum (-1)^q A_0 \cdots A_{q-1}A_{q+1} \cdots A_p, \qquad p > 0; \tag{20.1}$$

$$\mathrm{F}A = 0. \tag{20.2}$$

In particular for $p = 1$

$$\mathrm{F}\sigma^1 = A_1 - A_0 . \tag{20.3}$$

(20.4) DEFINITION. *If* $C^0 = g^i A_i$ *is a zero-chain then* $\sum g^i$ *is a function of* C^0 *known as the Kronecker index of* C^0 *and denoted by* $\mathrm{KI}(C^0)$. *As we shall see later* (28, 46) *this number is a special case of a more general numerical function with the same designation.*

We will now prove a series of properties relating connectedness and the zero-cycles. They are so interlocked that they will have to be proved more or less together. The following notations will be used:

$\{K_i\}$ = the components of K;

$\{A_{ih}\}$ = the vertices of K_i ; however, the vertex A_{i1} will be written A_i .

(20.5) *A n. a. s. c. for two vertices* A, A' *to be in the same component is* $A \sim A'$.

(20.6) $C^0 \sim 0 \rightarrow \mathrm{KI}(C^0) = 0$.

(20.7) *Every zero-cycle over any* G *satisfies a relation*

$$C^0 \sim g^i A_i , \qquad g^i \in G. \tag{20.7a}$$

Moreover a relation

$$g^i A_i \sim 0, \qquad g^i \in G, \tag{20.7b}$$

implies that every $g^i = 0$.

(20.8) *For every group* G:

$$\mathfrak{H}^0(K, G) \cong \mathbf{P}(GA_i).$$

In particular when $G = \mathfrak{J}$, *the group of the integers, then* $\mathfrak{H}^0(K, \mathfrak{J})$ *is isomorphic with the free group on the generators* A_i. *This group has no elements of finite order and so there are no torsion coefficients for the dimension zero.*

(20.9) *The number of components of* K *is the zero-dimensional Betti number* $R^0(K, G)$, G *any field. Hence* $R^0(K, G)$ *is independent of* G *and it will be designated by* $R^0(K)$.

(20.10) *If* K *is connected and* A *any vertex then every* C^0 *satisfies the relation*:

$$C^0 \sim \mathrm{KI}(C^0)A.$$

Therefore when K *is connected* $C^0 \sim 0 \leftrightarrow \mathrm{KI}(C^0) = 0$.

(a) *Suppose first* $G = \mathfrak{J}$. Let A, A' be any two vertices of K_i . Since K_i is a component there is a finite set of elements of $K_i : A = \sigma_1 , \cdots , \sigma_r = A'$ in which any two consecutive elements are incident (2.3). If σ_j is not a vertex it contains a vertex A'_j of σ_{j-1} and a vertex A''_j of σ_{j+1} , hence also the one-simplex $A'_j A''_j$. It follows that in the sequence joining A to A' we may replace σ_j by the set of simplexes: A'_j , $(A'_j A''_j)$, A''_j, and still have consecutive elements incident. Proceeding thus we will arrive at a set of the same type and of form $A_{ih_1} = A, \sigma_1^1 , A_{ih_2} , \cdots , A_{ih_r} = A'$. Hence $\mathrm{F}\sigma_k^1 = A_{ih_{k+1}} - A_{ih_k}$ and so if $C^1 = \sum \sigma_j^1$ then $\mathrm{F}C^1 = A' - A \sim 0$. This proves (20.5) as regards necessity.

If $C^0 \sim 0$ and $G = \mathfrak{J}$ we have $C^0 = \mathrm{F}(g^h \sigma_h^1)$, hence $\mathrm{KI}(C^0) = g^h \mathrm{KI}(\mathrm{F}\sigma_h^1) = 0$ by (20.3). This is (20.6) for the present situation.

If C^0 is an integral chain we have $C^0 = \sum C_i^0$, $C_i^0 \subset K_i$, and $C_i^0 = g^h A_{ih}$. By the necessity part of (20.5) for $G = \mathfrak{J}$ we have $A_{ih} \sim A_i$, hence $C_i^0 \sim (\sum g^h)A_i = \mathrm{KI}(C_i^0)A_i$ from which (20.7a) follows. Furthermore this also proves (20.10) for integral chains.

Suppose that (20.7b) holds with the g^i integers and not all zero. Then $g^i A_i = \mathrm{F}C^1 = \mathrm{F} \sum C_i^1$, $C_i^1 \subset K_i$. Since chains in different components cannot cancel out we have $\mathrm{F}C_i^1 = g^i A_i$ (i unsummed). Hence by (20.6) already proved for $G = \mathfrak{J}$: $\mathrm{KI}(\mathrm{F}C_i^1) = 0 = g^i$. This proves (20.7b) for $G = \mathfrak{J}$. From this follows also the sufficiency proof of (20.5). For suppose $A \sim A'$ and A, A' in different components say K_i, K_j, $i \neq j$. The two vertices may then be chosen as A_i, A_j and we would have $A_i \sim A_j$ which is ruled out. Therefore A, A' are in the same component, and the proof of (20.5) is completed.

It follows from (20.7) for $G = \mathfrak{J}$ that $\mathfrak{H}^0(K, \mathfrak{J})$ is isomorphic with the free group on the generators $\{A_i\}$ and this is merely another way of stating (20.8) for $G = \mathfrak{J}$. This is as much as may be obtained for $G = \mathfrak{J}$.

(b) *Suppose now that G is any group.* We first notice that (20.9) is a consequence of (20.8) and so requires no further consideration. For all but (20.8) the only property required is that "~ 0" $\leftrightarrow$ "bounding" for the zero-cycles, and this follows from (16.10) and the fact that there are no t_i^0 (20.8 for $G = \mathfrak{J}$). Regarding (20.8), if we go back to the derivation of (12.4) we verify that, since there are no t_i^0, $\{c_i^0\}$ is merely any set of zero-cycles such that $\mathfrak{H}^0(K, \mathfrak{J})$ is isomorphic with the free group on the generators c_i^0. Therefore we may choose $\{c_i^0\} = \{A_i\}$, and so (20.8) follows from (16.11). This completes the proofs of all our propositions.

21. **Complexes with cyclic and acyclic properties.** An important and simple property of many noteworthy complexes is to have all the homology groups for certain dimensions vanishing or merely isomorphic with the coefficient-groups. Among these are found, for example, simplexes and their boundaries.

(21.1) DEFINITIONS. *The complex X is said to be cyclic* [*acyclic*] *in the dimension p over G if $\mathfrak{H}^p(X, G) \cong G\,[= 0]$. It is said to be $(p, \cdots, q)$-cyclic* [*to be $(p, \cdots, q)$-acyclic, acyclic*] *over G if it is cyclic over G in the dimensions $p, \cdots, q$ and acyclic over G in the other dimensions* [*acyclic over G in the dimensions $p, \cdots, q$, in all dimensions*]. *If X say, is cyclic in the dimension p over G for every G it is merely said to be "cyclic in the dimension p," and similarly for the other properties.*

As in similar instances when $G = \mathfrak{J}, \mathfrak{J}_m, \mathfrak{R}$ we will say "integrally acyclic," "acyclic mod m" or "rationally acyclic," and similarly for the other concepts.

We have at once from (16.9):

(21.2) *A n.a.s.c. for a finite complex to be acyclic is that all the Betti numbers and torsion coefficients vanish.*

(21.3) *A n.a.s.c. for a finite complex to be $(p, \cdots, q)$-cyclic is that all the torsion coefficients vanish, that the Betti number $R^s = 1$ for $s = p, \cdots, q$ and $R^s = 0$ otherwise.*

A convenient result is:

(21.4) *If* $\dim X = n + 1 \geqq 2$, *and X is zero-cyclic and with a single $(n + 1)$-element x^{n+1}, then its n-section $X^n = X - x^{n+1}$ is $(0, n)$-cyclic, and all its n-cycles are of the form $g\delta^n$, $\delta^n = \mathrm{F}x^{n+1}$.*

By (19.5) it is only necessary to show that X^n is cyclic in the dimension n, and this will follow if we can show that every n-cycle of X is of the form $g\delta^n$, $\delta^n = \mathrm{F}x^{n+1}$. Referring to the canonical bases (14), since there is only one $(n + 1)$-element x^{n+1} it is the single e_i^{n+1} on hand and there are no d_i^{n+1}. Therefore at the right in $(16.6)_n$ only the first product is present and it is $G\delta^n$. By (16.7) we find then $\mathfrak{F}^n(X, G) = \bar{\mathfrak{F}}^n(X, G) = G\delta^n$. Since X is acyclic in the dimension n we must have $\mathfrak{Z}^n(X, G) = \bar{\mathfrak{F}}^n(X, G) = G\delta^n$. Hence every n-cycle is a $g\delta^n$, and (21.4) follows.

(21.5) *A closed connected simplicial complex is cyclic in the dimension zero* (20.10).

22. **The simplex, its closure and boundary.** Let σ^n be an n-simplex. We will consider the groups of σ^n, $\mathrm{Cl}\,\sigma^n$, $\mathfrak{B}\sigma^n$. It is often convenient to call σ^n an *open simplex*, $\mathrm{Cl}\,\sigma^n$ *a closed simplex*.

(22.1) *Groups of* σ^n. The simplex σ^n itself is (trivially) n-cyclic.

(22.2) *Groups of* $\mathrm{Cl}\,\sigma^n$. We will show that it is zero-cyclic. For $n = 0$ this is the same as (22.1) so we assume $n > 0$. If A, A' are any two vertices then AA' is a one-simplex of $\mathrm{Cl}\,\sigma^n$ and so $\mathrm{Cl}\,\sigma^n$ is connected and hence cyclic in the dimension zero (21.5). Let now $\sigma^n = A\sigma^{n-1}$, and $\mathrm{Cl}\,\sigma^{n-1} = \{\sigma_i^{p-1}\}$. By (20.1): $\mathrm{F}A\sigma_i^{p-1} = \sigma_i^{p-1} - A\mathrm{F}\sigma_i^{p-1}$, $p > 1$, and $\mathrm{F}A\sigma_i^0 = \sigma_i^0 - A$. Hence

$$(22.3)\qquad \begin{cases} \mathrm{F}AC^{p-1} = C^{p-1} - A\mathrm{F}C^{p-1}, & p > 1; \\ \mathrm{F}AC^0 = C^0 - \mathrm{KI}(C^0)\cdot \mathrm{A}. \end{cases}$$

Let now γ^p, $p > 0$, be a cycle of $\mathrm{Cl}\,\sigma^n$, and suppose first $p > 1$. We have $\gamma^p = AC^{p-1} + C^p$, C^{p-1} and $C^p \subset \mathrm{Cl}\,\sigma^{n-1}$. Hence $\mathrm{F}\gamma^p = C^{p-1} - A\mathrm{F}C^{p-1} + \mathrm{F}C^p = 0$. Since the middle term alone contains the vertex A we have $\mathrm{F}C^{p-1} = 0$, $C^{p-1} = -\mathrm{F}C^p$ and hence $\gamma^p = \mathrm{F}AC^p$. When $p = 1$ we obtain: $\mathrm{F}\gamma^1 = C^0 - \mathrm{KI}(C^0)A + \mathrm{F}C^1 = 0$, hence $\mathrm{KI}(C^0) = 0$, $C^0 = -\mathrm{F}C^1$, and the conclusion is again the same. Therefore $\gamma^p \sim 0$ for every $p > 0$ and so $\mathrm{Cl}\,\sigma^n$ is zero-cyclic.

(22.4) *Groups of* $\mathfrak{B}\sigma^n$. By (21.4) when $n > 1$, $\mathfrak{B}\sigma^n$ is $(0, n - 1)$-cyclic. Since $\mathfrak{B}\sigma^1$ consists of two points A, B its sole homology group different from 0 is the one for the dimension zero and it is the product for any given G of two isomorphs of G.

23. **Dissection of a complex. Relative cycles.**

(23.1) Let X_1 be a closed subcomplex of X and $X_0 = X - X_1$ its open complement. The pair (X_0, X_1) in the order named will be referred to as a *dissection* of X. Our present purpose is to compare the groups of the X_i with those of

X itself. We will denote by F_i the boundary operator for X_i. If C is a chain of X we have $C = C_0 + C_1$, $C_i \subset X_i$, and we will call C_i : the chain C reduced mod X_j $(j \neq i)$, or merely "C mod X_j." If $\varphi(C)$ is a function whose range and values are chains of X, we have

$$\varphi(C) = \varphi_0(C) + \varphi_1(C), \qquad \varphi_i(C) \subset X_i,$$

and we will call $\varphi_i(C)$: the function φ reduced mod X_j $(j \neq i)$.

(23.2) *The groups of* X_1. If $x \in X_1$ then $\mathfrak{B}x \subset X_1$, and hence $\mathrm{F}_1 = \mathrm{F} \mid X_1$. It follows that a cycle γ^p of X_1 is also a cycle of X and that if C_1 bounds in X_1 it also bounds in X.

Since the elements of X_1 are among those of X we have an injection $\mathfrak{C}^p(X_1, G) \rightarrow \mathfrak{C}^p(X, G)$ in the sense of (II, 8.6). We may consider this as a simultaneous operation on all the chain-groups of X_1 into those of X, and denote it briefly as $\eta: X_1 \rightarrow X$, as if it were an operation on X_1 to X. This operation is called an *injection* of X_1 into X. We notice the obvious property: $\mathrm{F}\eta = \eta\mathrm{F}_1$. As a consequence of this, η maps, respectively, $\mathfrak{Z}^p(X_1, G), \mathfrak{F}^p(X_1, G)$ into $\mathfrak{Z}^p(X, G), \mathfrak{F}^p(X, G)$, and since η is continuous also $\bar{\mathfrak{F}}^p(X_1, G)$ into $\bar{\mathfrak{F}}^p(X, G)$. Hence (II, 5.4) η induces a homomorphism $\bar{\eta}$: $\mathfrak{H}^p(X_1, G) \rightarrow \mathfrak{H}^p(X, G)$. We will set $\eta\mathfrak{F}^p(X_1, G) = \mathfrak{F}_1^p(X_1, G)$, $\bar{\eta}\mathfrak{H}^p(X_1, G) = \mathfrak{H}_1^p(X_1, G)$.

(23.3) Since $\eta\mathfrak{Z}^p(X_1, G) = \mathfrak{Z}^p(X_1, G)$, the group $\mathfrak{H}_1^p(X_1, G) = \mathfrak{Z}^p(X_1, G)/\mathfrak{Z}^p(X_1, G) \cap \bar{\mathfrak{F}}^p(X, G)$ may be viewed as the group of the cycles of X_1 *as to bounding in* X. Taking now integral chains it will be seen that the considerations of (12, 13) are still applicable when the p-chains are restricted to X_1 and the $(p + 1)$-chains are still chains of X. They will lead to a system (14.2) *for the particular dimension* p (but not simultaneously for all dimensions), with the $(p + 1)$-chains d_i^{p+1}, e_i^{p+1} chains of X and the rest chains of X_1. The reductions and the other results of (16), of (17) (all but 17.8b), and of (18), follow automatically.

(23.4) *The groups of* X mod X_1. The important operation is now the reduction π of the chains of X mod X_1. That is to say, if $C = C_0 + C_1$, $C_i \subset X_i$, then $\pi C = C_0$. Since $\mathfrak{C}^p(X, G) = \mathfrak{C}^p(X_0, G) \times \mathfrak{C}^p(X_1, G)$, π is a collection of open homomorphisms: $\mathfrak{C}^p(X, G) \rightarrow \mathfrak{C}^p(X_0, G)$, (II, 6.2). We call π the *projection of* X *into* X_0 and denote it symbolically also as $\pi: X \rightarrow X_0$. We verify here: $\pi\mathrm{F}C = \mathrm{F}_0\pi C$ or $\pi\mathrm{F} = \mathrm{F}_0\pi$, and we conclude as before that π induces a collection of homomorphisms $\bar{\pi}: \mathfrak{H}^p(X, G) \rightarrow \mathfrak{H}^p(X_0, G)$. This time a cycle γ^p of X_0 is not a cycle of X but merely a chain of X_0 whose boundary is in X_1, and $\gamma^p \sim 0$ in X_0 means that γ^p + a chain of $X_1 \sim 0$ in X. For this reason γ^p is called a *relative cycle* or *a cycle of* X mod X_1, and the groups of X_0 are correspondingly written $\mathfrak{C}^p(X, X_1, G)$, $\cdots$. The term *absolute* cycle is sometimes applied to the cycles of X itself. Thus the cycles of X_1 are absolute cycles, those of X_0 are relative cycles.

(23.5) We will now make certain identifications in accordance with (II, 8.6, $\cdots$, 8.9). First $\mathfrak{C}^p(X_1, G)$ is identified with $\eta\mathfrak{C}^p(X_1, G)$ in accordance with (II, 8.8) and thus becomes a closed subgroup of $\mathfrak{C}^p(X, G)$. We will say that two chains C^p, $C_1^p \in \mathfrak{C}^p(X, G)$ are *congruent* mod X_1 if they are congruent mod $\mathfrak{C}^p(X_1, G)$ in the sense of (II, 5.1). We will now identify all the chains congruent to a given chain mod X_1, first with their coset mod $\mathfrak{C}^p(X_1, G)$ (II, 5.1), then with the representative of the coset in $\mathfrak{C}^p(X_0, G)$, thus obtaining in particular the topological identification of $\mathfrak{C}^p(X, G)/\mathfrak{C}^p(X_1, G)$ with $\mathfrak{C}^p(X_0, G)$ (II, 8.9). *All these identifications will be assumed throughout the sequel in all similar instances.*

Hereafter a chain mod X_1 over G is then merely a chain given to within a chain of X_1 over G. The identification of the chains causes the identification of $\mathfrak{Z}^p(X_0, G)$ with the group of the cosets of $\mathfrak{C}^p(X, G)$ mod $\mathfrak{C}^p(X_1, G)$ consisting of the chains γ^p over G such that $\mathrm{F}\gamma^p \subset X_1$. Under our identification such a chain is also to be described henceforth as a *cycle* mod X_1 and it is known only to within a chain of X_1.

Similarly $\mathfrak{F}^p(X_0, G)$ is identified henceforth with the group of the cosets of the chains δ^p of $\mathfrak{C}^p(X, G)$ mod $\mathfrak{C}^p(X_1, G)$ such that $\delta^p = \mathrm{F}C^{p+1} + D^p, D^p \subset X_1$. The chain δ^p is also to be described as a *bounding cycle* mod X_1 and is again only known to within a chain in X_1. For the same reasons $\delta^p \sim 0$ mod X_1 is now understood to mean that $\delta^p - D_p \in \bar{\mathfrak{F}}^p(X_0, G)$, $D^p \subset X_1$.

(23.6) Remark. We have temporarily denoted by F_i the boundary operator of X_i. However if we return to our previous custom and designate by F the boundary operator of any complex whatever then we have $\eta\mathrm{F} = \mathrm{F}\eta$, $\pi\mathrm{F} = \mathrm{F}\pi$. Thus both η and π commute with F. The general class of the operations with this property will come strongly to the fore in the next chapter under the designation of "chain-mapping" (IV, 9).

24. **Circuits.** An absolute *n-circuit* or merely an *n-circuit* is an n-complex X with the following properties: (a) $\Gamma^n = \sum x_i^n$ is an n-cycle mod 2; (b) no proper closed subcomplex of X possesses such a cycle, i.e., X is irreducible with respect to (a). Property (b) implies in particular: $X = |\Gamma^n|$ (notation of 7.3).

When X is simplicial (a) means that every σ^{n-1} is the face of an even number of σ^n.

(24.1) *If an n-circuit X has integral n-cycles different from* 0 *then*: (a) *their group is infinite cyclic*; (b) *if D^n is any integral n-cycle different from* 0, *then* $|D^n| = X$.

Let $D^n = a^i x_i^n \neq 0$ be an integral n-cycle. We show first that $|D^n| = X$. If the a^i have a common factor p, $D'^n = (1/p)D^n$ is likewise an n-cycle and as $|D'^n| = |D^n|$, D'^n may replace D^n. Therefore we may suppose the a^i relatively prime and hence one of them, say a^1, to be odd. Let $b^i = 0, 1$ according as a^i is even or odd. Then $b^i x_i^n$ is a cycle mod 2 and hence $b^i x_i^n = \Gamma^n$, $b^i = 1$. Therefore every element of D^n is also an element of Γ^n and hence $|D^n| = |\Gamma^n| = X$.

Suppose that there are integral n-cycles different from 0. Their group $\mathfrak{Z}^p(X)$ is free (11) and its dimension $d > 0$. If $d > 1$ a suitable combination D^n of the base elements will lack some x^n, which contradicts $|D^n| = X$. Therefore $d = 1$ and $\mathfrak{Z}^p(X)$ is infinite cyclic.

Under the same conditions $\mathfrak{Z}^p(X)$ will have a base consisting of a single Δ^n, called a *basic* n-cycle. The only other basic n-cycle is $-\Delta^n$.

(24.2) The n-circuit is called *orientable* when it possesses integral n-cycles, *non-orientable* otherwise.

A *simple* n-circuit (sometimes called an *n-pseudo-manifold*) is an n-circuit in which every $(n - 1)$-element is a face of precisely two n-elements. The simple n-circuit may be orientable or not.

If Y is a closed subcomplex of X and $X - Y$ is an n-circuit, X is called a *relative* n-circuit, or an n-circuit mod Y. This may be combined with "orientability" or the "simple-circuit" property. In the relative circuits the n-cycles Γ^n, Δ^n are cycles of X mod Y.

Let X denote an n-circuit (absolute or relative). From the definition we infer that its nth homology group $\mathfrak{H}^n(X, 2)$ is cyclic, i.e., consists of 0 and a single element Γ^n. Then X is orientable whenever its integral homology group $\mathfrak{H}^n(X)$ is cyclic, non-orientable when $\mathfrak{H}^n(X) = 0$. When the circuit is relative, the homology groups are those of X mod Y.

(24.3) Examples. $\mathfrak{B}\sigma^{n+1}$, $n \geqq 1$, is an absolute orientable n-circuit. Take a rectangle $ABCD$, match A with C, B with D, AB with CD. There results the so-called Möbius strip. If $[ABCD:AB] = 2$, the resulting complex is a non-orientable n-circuit mod $(AD \cup BC)$.

(24.4) *Simplicial simple n-circuit.* For these important circuits the defining properties may be given a more elementary form in accordance with:

(24.5) *If $K - L$ is simplicial, n.a.s.c. for K to be a simple n-circuit* mod L *are*:

(α) *every simplex $\sigma \in K - L$ is a face of a σ^n;*

(β) *every σ^{n-1} is a face of two and only two σ^n;*

(γ) *the set M of the σ^{n-1} and σ^n of $K - L$ is connected.*

Notice that M is the complement of the $(n - 2)$-section of $K - L$ and so it is an open subcomplex of $K - L$.

When $K - L$ is a simple n-circuit both (α) and (β) hold by definition. As for (γ) if $\{M_i\}$, $i = 1, 2, \cdots, r$, are the components of M then $\Gamma^n = \sum \Gamma_i^n$, $\Gamma_i^n \subset M_i$, and Γ_i^n is an n-cycle mod $(L, 2)$. Hence if $K - L$ is an n-circuit we must have $r = 1$, or (γ) holds. Conversely, suppose that (α), (β), (γ) hold. In view of (β), $\sum \sigma_i^n = \Gamma^n$ is a cycle mod $(L, 2)$ so property (a) holds. Suppose that a proper closed subcomplex of $K - L$ contained another such cycle Γ'^n. Owing to (α), Γ'^n must lack at least one n-simplex say σ_1^n. Let σ_r^n be present in Γ'^n. Since $K - L$ is connected in view of (α) there is a sequence which under proper labelling may be put in the form $\sigma_1^n, \sigma_2^{n-1}, \sigma_3^n, \cdots, \sigma_r^n$, where consecutive terms are incident. Now owing to (β), and since Γ'^n is a cycle mod $(L, 2)$, if σ_h^n is a face of Γ'^n, so must σ_{h-1}^{n-1} be, and hence likewise σ_{h-2}^n. Consequently σ_1^n must be a face of Γ'^n, and this contradiction proves that Γ'^n cannot exist, or property (b) holds also. Therefore $K - L$ is an n-circuit, and in view of (β) it is simple. This proves (24.5).

(24.6) Example. *The sphere $S^n = \mathfrak{B}\sigma^{n+1}$, $n > 0$, is a simplicial, simple, orientable n-circuit.*

By (21.4, 22.2) if $\gamma^n = \text{F}\sigma^{n+1}$, then: (a) γ^n is an integral n-cycle of S^n, hence also a cycle mod 2; (b) there are no other n-cycles mod 2 different from 0. Since $|\gamma^n| = S^n$, the two basic circuit conditions (a, b) are fulfilled. It is an elementary matter to show that every $\sigma^{n-1} \in S^n$ is the face of just two σ^n, and so S^n is a simplicial simple n-circuit. Since it contains the integral n-cycle γ^n it is orientable. Thus S^n has all the properties asserted in (24.6).

§6. DUALITY THEORY FOR FINITE COMPLEXES

25. (25.1) Let $X = \{x_i^p\}$ be a finite complex and $X^* = \{x_p^i\}$ its dual. We shall compare the various groups of the two complexes.

Since X^* is a finite complex it has all the general properties of finite complexes. However, it is convenient to adopt a terminology referring the relations in X^* back to X. A $(-p)$-chain or $(-p)$-cycle of X^* is called a *p-cochain* or *p-cocycle* of X, and denoted by C_p, γ_p. Their groups are written $\mathfrak{C}_p(X, G)$, $\mathfrak{Z}_p(X, G)$, those of the bounding cocycles $\mathfrak{F}_p(X, G)$. The $(-p)$-dimensional homology groups of X^* are called the *p-dimensional cohomology* groups of X, written $\mathfrak{H}_p(X, G)$, and the corresponding Betti numbers and torsion coefficients are written $R_p(X, \pi)$, t_p^i. For reasons of euphony we will sometimes say: *dual* Betti numbers, groups, etc. In substance then in the notations the dimension $(-p)$ in X^* is indicated by the subscript p.

All the necessary modifications are obvious enough and need not be discussed. Notice, for later reference, that the basic boundary relations for the cochains are

$$\text{F}g_i x_p^i = \sum_j g_i[x_p^i : x_{p+1}^j]x_{p+1}^j. \tag{25.2}$$

The boundary of C_p is then a C_{p+1} whose dimension is that of C_p decreased by one. The "dimensional" behavior is thus the same as for chains.

Referring to (23), and in the same notations, we may also introduce new types of absolute or relative cocycles. They are: the *absolute cocycles of* X_0 and the *cocycles of* X mod X_0 (cycles of X^* mod X_0^*).

(25.3) Let K be simplicial. The notations being those of (10.4) we will call $\gamma_0 = \sum A^i$ the *fundamental zero-cocycle* of K (27.7a). The coboundary relations are then $\text{F}C_p = \gamma_0 C_p$, under the usual incidence number scheme (5.2), and $\text{F}C_p = C_p\gamma_0$, under the scheme of (5.7). This is a mere restatement of (10.4de) in the "co-terminology."

26. Instead of considering the "co-theory" as a theory of a different collection of elements from those of X, some authors prefer to view it as a theory of the elements of X with $\prec$ reversed. It is then necessary to introduce, side by side with F, a second operator F*, the *coboundary operator*, defined by

$$\text{F}^*(g^i x_i^p) = \sum_j g^i[x_i^p : x_j^{p+1}]x_j^{p+1}, \tag{26.1}$$

which raises the dimensions by one unit, instead of lowering them like F. Thus Whitney proceeds in that manner and writes ∂, δ for F, F*. The operator F* is a homomorphism $\mathfrak{C}^p \to \mathfrak{C}^{p+1}$ with similar properties to those of F, the cocycles are the chains of X whose coboundary vanishes, etc.

In the present work we shall definitely consider the elements of X^* *as distinct from those of* X *with the notations and terminology indicated in* (25).

To justify our choice we may anticipate and consider cartesian products of chains and cochains as in (IV, §2). If we write down expressions such as $C^p \times C^q$, $C^p \times C_q$, we know by inspection the rules for calculating the appropriate boundary chains $\mathrm{F}(C^p \times C^q)$, $\mathrm{F}(C^p \times C_q)$, the operator F being the same throughout. However, if we adopted the alternate procedure with F, F*, we should have to write all these expressions $C^p \times C^q$, and choose each time one of *four* possible operators. Three factors would impose a choice between *eight* operators.

It may be pointed out also that our convention merely represents adherence to those employed for many years in projective geometry and related doctrines, whereby *covariant* and *contravariant* elements are represented by distinct symbols. This is in keeping with the fact that they undergo distinct transformations.

27. It is evident that all the results of (§3) are applicable to X^*, i.e., to cochains, etc. Let α_p, $\cdots$ have the same meaning for X^* as α^p, $\cdots$, for X. Evidently $\alpha_p = \alpha^p$, and from (25.1) follows that if $\eta^*(p)$ is the $(-p)$th incidence matrix for X^*, then $\eta^*(p+1) = (\eta(p))'$ (the prime means the transpose). Therefore $\rho_{p+1} = \rho^p$ and the torsion coefficients t^i_{p+1} are the same as the t^p_i. Since the subscripts are the negatives of the dimensions we obtain in place of (15.1)

$$R_p = \alpha_p - \rho_{p+1} - \rho_p = \alpha^p - \rho^p - \rho^{p-1} = R^p.$$

Hence $\mathfrak{T}_{p+1}$, $\mathfrak{B}_p$ are abstractly the same as $\mathfrak{T}^p$, $\mathfrak{B}^p$. This proves the following theorem which is the analogue of Poincaré's initial duality theorem for manifolds (V, 33.1) and as far as Betti groups go, is the duality theorem of [L, 286] (duality theorem for pseudo-cycles):

(27.1) FIRST DUALITY THEOREM. The *pth Betti and dual Betti groups are isomorphic, and likewise the pth torsion and* $(p+1)$*st dual torsion groups, and*

$$R_p = R^p, \qquad t^i_{p+1} = t^p_i. \tag{27.2}$$

We state also explicitly the convenient property:

(27.3) *When X is torsion-free so is X^* and the integral pth homology and cohomology groups are isomorphic with one another as well as with the pth Betti group of X.*

(27.4) *The Betti numbers and torsion coefficients of a finite complex determine all its homology and cohomology groups.*

Let us define X as *p-cocyclic*, $\cdots$ whenever X^* is $(-p)$-cyclic, $\cdots$. Then we have by (21.3):

(27.5) *Whenever X is $(p, \cdots, q)$-cyclic or acyclic it is also $(p, \cdots, q)$-cocyclic or acocyclic, and conversely.*

(27.6) *Let $X = \{x\}$ be such that $p \leqq \dim x \leqq q$. Then*: (a) *no p-cocycle ~ 0 unless it is zero*; (b) *every q-cochain is a cocycle* (16.12).

(27.7) *Let $K = \{\sigma\}$ be a simplicial complex with vertices $\{A_i\}$ and duals $\{A^i\}$. Then*:

(a) $\gamma_0 = \sum A^i$ *is a cocycle*;

(b) *if K is connected every zero-cocycle is of the form* $g\gamma_0$;

(c) *if K is a simple n-circuit then every σ_n^i is an n-cocycle and $\sigma_n^i \sim \pm \sigma_n^j$ for all i, j; hence every n-cocycle $\sim g\sigma_n^i$.*

PROOF OF (a). We have $\mathrm{F}\sigma_i^1 = \eta_i^j A_j$, where $\sum_j \eta_i^j = 0$. Hence $\mathrm{F}A^j = \eta_i^j \sigma_1^i$, and so $\mathrm{F}\gamma_0 = \mathrm{F}\sum A^j = \sum_i \sigma_1^i \sum_j \eta_i^j = 0$. Therefore γ_0 is a cocycle.

PROOF OF (b). When K is connected then $R_0 = R^0 = 1$ and there are no t_0^i. It follows that K is cocyclic in the dimension 0. Consequently every zero-cocycle is of the form $g\delta_0$ where δ_0 is an integral cocycle. Suppose $\delta_0 = x_i A^i$, and let $\gamma_0 = y\delta_0$. We have then $\sum A^i = yx_i A^i$, and so $yx_i = 1$. Hence $x_i = y = \pm 1$, $\delta_0 = \pm\gamma_0$, from which (b) follows.

PROOF OF (c). Let K be a simple n-circuit and let K^{n-2} be its $(n-2)$-section. By (24.5γ) $K - K^{n-2}$ is connected. It follows that if σ^n, σ'^n are any two n-simplexes of K there is a sequence $\sigma^n = \sigma_1^n, \sigma_1^{n-1}, \sigma_2^n, \cdots, \sigma_{r-1}^{n-1}, \sigma_r^n = \sigma'^n$, in which consecutive elements are incident. Consequently this holds equally regarding $\sigma_n^1, \sigma_{n-1}^1, \cdots, \sigma_n^r$. Since K is a simple circuit: $[\sigma_{n-1}^i : \sigma_n^i] = \pm 1 = \pm[\sigma_{n-1}^i : \sigma_n^{i+1}]$. Since the only elements of $\{\sigma_n\}$ incident with σ_{n-1}^i are σ_n^i and σ_n^{i+1}, we have $\mathrm{F}\sigma_{n-1}^i = \pm(\sigma_n^i \pm \sigma_n^{i+1}) \sim 0$, or $\sigma_n^i \sim \pm\sigma_n^{i+1}$, and so finally $\sigma_n \sim \pm\sigma_n'$.

(27.8) EXAMPLE. Consider the sphere $S^n = \mathfrak{B}\sigma^{n+1}, \sigma^{n+1} = A_0 \cdots A_{n+1}, n > 0$. Since S^n is $(0, n)$-cyclic it is also $(0, n)$-cocyclic. Its zero-cocycles are all of the form $g \sum A^i$. We have seen (24.6) that S^n is a simple n-circuit and so its n-cocycles are all $\sim gA^0 \cdots A^n$.

28. **Kronecker index of chains and cochains.** Further progress will rest upon an extension of the concept of Kronecker index. The connection with the earlier concept will be made in (46).

(28.1) DEFINITION. *Let $\beta(p)$ be as in the Introduction. Then the Kronecker index of the couple x_i^p, x_p^j is the number*

(28.2) $$\mathrm{KI}(x_i^p, x_p^j) = \beta(p)\delta_j^i \text{ (Kronecker delta)},$$

and the Kronecker index of x_p^j, x_i^p is

(28.3) $$\mathrm{KI}(x_p^j, x_i^p) = \beta(-p)\delta_i^j = (-1)^p \mathrm{KI}(x_i^p, x_p^j).$$

(28.4) We have just specified values for the Kronecker index whenever X, X^* are so oriented that $[x_i^p : x_j^{p-1}] = [x_{p-1}^j : x_p^i]$. In order to allow for arbitrary reorientations we add the convention that if X, X^* are reoriented by means of orientation functions $\alpha(x_i^p)$, $\alpha^*(x_p^i)$ then

$$\mathrm{KI}(x_i^p, x_p^j) = \alpha(x_i^p)\alpha^*(x_p^j)\beta(p)\delta_j^i$$
$$= (-1)^p \mathrm{KI}(x_p^j, x_i^p).$$

REMARK. In [L, 165] the analogous definition of the index was given by means of (28.2) but without the factor $\beta(p)$, thus causing dissymetry under dualization. To pass from the present to the earlier definition it is merely necessary to reorient X^* by $\alpha(x_p) = \beta(-p)$.

(28.5) We shall now choose two groups G, H paired to a third J and with a multiplication gh. We define $hg = gh$, so that H, G are formally paired to J

with the same multiplication gh. Thus the two groups G, H are paired to J in one or the other order and with a multiplication independent of the order of the pairing. We shall briefly describe this relationship by the statement "G, H are *commutatively paired* to J."

Suppose that we have a chain and cochain over G and H:

$$C^p = g^i x_i^p, \quad g^i \in G; \qquad C_p = h_i x_p^i, \quad h_i \in H.$$

The Kronecker index of C^p, C_p, written $\mathrm{KI}(C^p, C_p)$ is an element of J defined by:

$$\text{(28.6)} \qquad \mathrm{KI}(C^p, C_p) = g^i h_j \mathrm{KI}(x_i^p, x_p^j) = \beta(p) g^i h_i .$$

Similarly with the terms in reverse order we define

$$\text{(28.7)} \qquad \mathrm{KI}(C_p, C^p) = \beta(-p) g^i h_i .$$

(28.8) *Interpretation.* A noteworthy interpretation, very close to the initial reason for introducing the index, is to consider that the dual elements x_i^p, x_p^j *cross one another* when $i = j$, and *do not cross one another* when $i \neq j$. It will be convenient to say that C^p, C_p have a *crossing* at x_i^p, x_p^i whenever both $g^i \neq 0$, $h_i \neq 0$. We agree to count this crossing with the weight $\beta(p) g^i h_i$ (i unsummed) and so the index (28.6) may be interpreted as a mode of counting the crossings suitably weighted. If $G = H = J =$ the ring of the integers, the weights become multiplicities in a reasonable sense. Viewed in this manner the index has for example played an important role in the author's work on Algebraic Geometry. (See [L, VIII, §4].)

29. The Kronecker index will now be utilized as a basis for deriving the duality relations between the chain- and cochain-groups. We shall use the following notations:

The chains and cochains over G and H are denoted by C^p, C_p, the cycles and cocycles over G and H by γ^p, γ_p and their homology and cohomology classes by Γ^p, Γ_p. We shall also denote by $\mathfrak{C}^p$, $\mathfrak{Z}^p$, $\mathfrak{F}^p$, $\mathfrak{H}^p$ the groups of chains, cycles, bounding cycles and homology groups over G, and by $\mathfrak{C}_p$, $\cdots$ the same for the cochains, $\cdots$ over H. *The group H is assumed discrete.*

As a preliminary we prove the important relation:

$$\text{(29.1)} \qquad \mathrm{KI}(\mathrm{F}C^{p+1}, C_p) = (-1)^p \, \mathrm{KI}(C^{p+1}, \mathrm{F}C_p)$$

which is the analogue of Formula 20 of [L, 169]. If $C^{p+1} = x_j^{p+1}$, $C_p = x_p^k$, both sides of (29.1) become $\beta(p)\,[x_j^{p+1} : x_k^p]$, so (29.1) holds. Since the two sides are bilinear in x_j^{p+1}, x_p^k, (29.1) holds in all cases.

(29.2) *The index obeys the commutation rule*

$$\text{(29.3)} \qquad \mathrm{KI}(C_p, C^p) = (-1)^p \, \mathrm{KI}(C^p, C_p).$$

(29.4) $\mathrm{KI}(C^p, C_p)$ *is a group multiplication for* $\mathfrak{C}^p$, $\mathfrak{C}_p$ *which pairs them to* J.

Since Gx_i^p, Hx_p^i are respective isomorphs of G, H they are paired to J with the multiplication $\beta(p)gh$. Since $\mathfrak{C}_p$ is discrete

$$\text{(29.5)} \qquad \mathfrak{C}^p = \mathbf{P} G x_i^p, \qquad \mathfrak{C}_p = \mathbf{P}^w H x_p^i,$$

(29.4) follows from (II, 16.1).

(29.6) $$\gamma^p \text{ or } \gamma_p \sim 0 \rightarrow \mathrm{KI}(\gamma^p, \gamma_p) = 0.$$

An equivalent formulation is

(29.7) $$\mathfrak{Z}_p, \mathfrak{Z}^p \text{ annul } \bar{\mathfrak{F}}^p, \mathfrak{F}_p.$$

It is only necessary to prove the property of the pair $\bar{\mathfrak{F}}^p$, $\mathfrak{Z}_p$, and hence (II, 15.4) that $\mathfrak{F}^p$ annuls $\mathfrak{Z}_p$, or that $\gamma^p = \mathrm{F}C^{p+1}$ annuls $\mathfrak{Z}_p$, and this follows immediately from $(29.1)_p$, since it yields:

$$\mathrm{KI}(\gamma^p, \gamma_p) = (-1)^p\mathrm{KI}(C^{p+1}, 0) = 0.$$

(29.8) $\mathrm{KI}(\gamma^p, \gamma_p)$ *depends solely upon the classes* Γ^p, Γ_p (29.6).

(29.9) Definition. *The fixed value of* $\mathrm{KI}(\gamma^p, \gamma_p)$ *under* (29.8) *is called the Kronecker index of the classes* Γ^p, Γ_p *written* $\mathrm{KI}(\Gamma^p, \Gamma_p)$.

(29.10) $\mathrm{KI}(\Gamma^p, \Gamma_p)$ *is a group multiplication for* $\mathfrak{H}^p$, $\mathfrak{H}_p$, *and obeys the commutation rule* (29.3), (*with* Γ *in place of* C).

Except for commutation (29.10) is a consequence of (29.4), and (II, 15.6), while the commutation rule follows from (29.2).

(29.11) *If* G, H *are* J*-orthogonal so are* $\mathfrak{C}^p$, $\mathfrak{C}_p$.

For Gx_i^p, Hx_p^i are then J-orthogonal and so (29.11) follows from (29.5) and (II, 16.1).

(29.12) *If* G, H *are* J*-orthogonal,* $\mathfrak{Z}_p$ *is the annihilator of* $\bar{\mathfrak{F}}^p$ *and likewise* $\mathfrak{Z}^p$ *of* $\mathfrak{F}_p$.

It is sufficient to prove the property of the pair $\mathfrak{Z}_p$, $\bar{\mathfrak{F}}^p$. We have just shown that every γ_p annuls $\bar{\mathfrak{F}}^p$, so it is only necessary to prove the converse, or that if γ_p annuls $\bar{\mathfrak{F}}^p$ it is a cocycle. If γ_p annuls $\bar{\mathfrak{F}}^p$ it annuls $\mathfrak{F}^p$ and so by (29.1):

$$\mathrm{KI}(\mathrm{F}C^{p+1}, \gamma_p) = (-1)^p\mathrm{KI}(C^{p+1}, \mathrm{F}\gamma_p) = 0.$$

Thus $\mathrm{F}\gamma_p \in \mathfrak{C}_{p+1}$ annuls $\mathfrak{C}^{p+1}$, and so by $(29.11)_{p+1}$, $\mathrm{F}\gamma_p = 0$ or γ_p is a cocycle.

30. **Duality theorems.** The situation which will now be faced will recur again and again in a more or less similar form. It is therefore best to introduce at the outset a systematic terminology designed especially to avoid undue repetition later.

(30.1) Definition. *The pair* (G, H) *in the order named, is said to form a normal couple whenever one of the following two possibilities arises:*

(a) G *is compact,* H *is discrete and they are dually paired by a commutative multiplication* gh *to* $\mathfrak{P}$. *In particular then they are orthogonal and each* $\cong$ *the character-group of the other.*

(b) $G = H = J =$ *a discrete field and the multiplication* gh *is merely the multiplication of the field* J. *Notice that* G *may be viewed as a linearly compact (one-dimensional) vector space over* J, *and* H *as a (one-dimensional) discrete vector space over* J, *dually paired under the multiplication* gh, *which is merely the multiplication of the field* J (II, 32.5).

We may now state the

(30.2) SECOND DUALITY THEOREM FOR FINITE COMPLEXES. *If G, H is a normal couple then the pth homology and cohomology groups $\mathfrak{H}^p(X, G)$, $\mathfrak{H}_p(X, H)$ are dually paired (to $\mathfrak{P}$ or to the discrete field J when $G = H = J$) and with the class Kronecker index as the group multiplication.*

This is the duality theorem of [L, 286] with the all-important Pontrjagin group duality complement.

Since G, H are dually paired to $\mathfrak{P}$ or J as the case may be, so are their isomorphs Gx_i^p, Hx_p^i and with the Kronecker index as the multiplication. Hence the same holds for $\mathfrak{C}^p$, $\mathfrak{C}_p$ (II, 20.7, 33). Since G, H are dually paired, $\mathfrak{Z}_p$ and $\bar{\mathfrak{F}}^p$ are one another's annihilators in $\mathfrak{C}_p$, $\mathfrak{C}^p$ (29.12; II, 20.5, 33), and likewise for $\mathfrak{Z}^p$, $\mathfrak{F}_p$ $(= \bar{\mathfrak{F}}_p)$. Therefore $\mathfrak{H}^p = \mathfrak{Z}^p/\bar{\mathfrak{F}}^p$ and $\mathfrak{H}_p = \mathfrak{Z}_p/\bar{\mathfrak{F}}_p$ are likewise orthogonal to $\mathfrak{P}$ or J as the case may be (II, 15.6) and hence dually paired (II, 20.6, 33). Since Γ^p, Γ_p are merely the cosets of γ^p, γ_p mod $\bar{\mathfrak{F}}^p$, $\bar{\mathfrak{F}}_p$, the multiplication of the dual pairing is the one described under (II, 15.5a) and it is precisely the class Kronecker index. This proves the theorem.

31. **Dual categories.** The preceding theorem will serve as a pattern for a number of similar theorems occurring later. In order to facilitate their description and minimize repetition, we introduce the convenient concept of dual categories.

Let A, B be two collections of cycles and cocycles of all the different dimensions and over various groups of coefficients. For the missing dimensions the groups are taken to be zero. Let it be possible to define the groups $\mathfrak{F}$ and hence the homology and cohomology groups $\mathfrak{H} = \mathfrak{Z}/\bar{\mathfrak{F}}$, likewise the Kronecker index $\mathrm{KI}(\gamma^p, \gamma_p)$ with the same properties other than orthogonality as in (29). When the coefficient-group is a field J it is assumed that the corresponding groups $\mathfrak{Z}$, $\mathfrak{F}$ are vector spaces over J, and in particular satisfy the basic convention (II, 22.2). Under our assumptions one may define a class index $\mathrm{KI}(\Gamma^p, \Gamma_p)$. If (G, H) is any normal couple and $\mathfrak{H}^p(G)$, $\mathfrak{H}_p(H)$ are the corresponding homology and cohomology groups, we shall say that the cycles of A and the cocycles of B [the cocycles of B and the cycles of A] are:

dual categories whenever the groups $\mathfrak{H}^p(G)$, $\mathfrak{H}_p(H)$ [$\mathfrak{H}_p(G)$, $\mathfrak{H}^p(H)$] are dually paired (to $\mathfrak{P}$ or to the discrete field J when $G = H = J$) and with the class Kronecker index as the group multiplication;

weak dual categories whenever the groups are defined only for $G = H = J =$ a discrete field, and are vector spaces orthogonal to J with the class Kronecker index as the multiplication. Whenever the dimensions of the paired spaces are finite their pairing is again a full dual pairing of vector spaces (II, 34).

Since orthogonality to $\mathfrak{P}$ or a discrete field J results in each case in the dual pairing (II, 20.6, 33) we may say that: (a) the characteristic property of dual categories is orthogonality to $\mathfrak{P}$ or J; (b) weak dual categories are those where only orthogonality to J may take place.

In the terminology just introduced (30.2) assumes the form:

(31.1) *The cycles and cocycles of a finite complex in one or the other order are dual categories.*

As a further application we also have:

(31.2) *Let* X, X_0, X_1 *be as in* (23). *Then the cycles of* X mod X_1 *and the cocycles of* X_0 *in one or the other order are dual categories.*

A similar statement may be made for the cocycles of X mod X_0 and the cycles of X_1, but it is merely the expression of (31.1) for X_1 itself. Notice also that when $X_0 = \emptyset$ and $X = X_1$, (31.2) reduces to (31.1).

32. Several noteworthy properties of dual categories are immediate consequences of properties of vector spaces.

We suppose then that A, B are dual categories of any sort and take the groups over a discrete field J. The formulation is given so as to include possible infinite-dimensional groups which may occur later. The Betti and dual Betti numbers have their usual significance of dimension of the homology and cohomology groups.

(32.1) *The pth Betti and dual Betti numbers over* J *are finite and equal or else both infinite* (II, 25.9d).

When these numbers are finite, in particular for a finite complex, (32.1) gives the full content of (30.2) for the groups over the field J.

(32.2) *If the cycles* γ_i^p, $(i = 1, 2, \cdots, r)$ *are independent with respect to homology, there can be selected cocycles* γ_p^j, $(j = 1, 2, \cdots, r)$ *such that*

$$\text{(32.3)} \qquad \mathrm{KI}(\gamma_i^p, \gamma_p^j) = \delta_i^j .$$

For this is true for the classes (II, 25.9b), and so by (29.8) for γ^p, γ_p.

(32.4) *If the Betti numbers are finite and* $\{\gamma_i^p\}$, $\{\gamma_p^j\}$ *are maximal independent sets (with respect to* $\sim$*), then*

$$\text{(32.5)} \qquad | \mathrm{KI}(\gamma_i^p, \gamma_p^j) | \neq 0.$$

Since the Γ_i^p are independent, by (II, 25.9a) classes Γ_p^j may be chosen such that

$$| \mathrm{KI}(\Gamma_i^p, \Gamma_p^j) | \neq 0,$$

which, in view of (29.8) yields (32.4).

33. Returning to the duality theorem for finite complexes, in view of its importance, and also for later purposes, we shall indicate another proof based on the comparison of canonical bases (14).

Let us pass from the bases $\{x_i^p\}$ for the integral chains to new bases $\{\epsilon_i^p\}$ by simultaneous transformations

$$\text{(33.1)} \qquad x_i^p = \lambda_i^{pj} \epsilon_j^p , \qquad \lambda^p = \| \lambda_i^{pj} \| \text{ unimodular.}$$

It will be convenient to designate by $\lambda_p = \| \lambda_{pj}^i \|$ the matrix $(\lambda^p)^{-1}$, i.e., such that $\lambda_j^{pi} \lambda_{pk}^j = \delta_k^i$ (p unsummed). Since λ_p is also unimodular,

$$\text{(33.2)} \qquad x_p^i = \lambda_{pj}^i \epsilon_p^j$$

is a simultaneous transformation from the bases $\{x_p^i\}$ to the new bases $\{\epsilon_p^i\}$. If we let $[x_i^{p+1}:x_j^p] = \eta_i^{pj}$, then:

$$\text{(33.3)} \qquad \mathrm{F}x_i^{p+1} = \eta_i^{pj} x_j^p ,$$

$$\text{(33.4)} \qquad \mathrm{F}x_p^i = \eta_j^{pi} x_{p+1}^j ,$$

the matrix in (33.4) being the transpose of the matrix in (33.3). Using (33.1) and (33.2) we now obtain

$$\text{(33.5)} \qquad \mathrm{F}\epsilon_i^{p+1} = \zeta_i^{pj} \epsilon_j^p ,$$

$$\text{(33.6)} \qquad \mathrm{F}\epsilon_p^i = \zeta_j^{pi} \epsilon_{p+1}^j ,$$

where $\zeta_i^{pj} = \lambda_{p+1,i}^k \eta_k^{ph} \lambda_h^{pj}$; *again the matrix of* (33.6) *is the transpose of the matrix of* (33.5).

By an elementary calculation:

$$\text{(33.7)} \qquad \mathrm{KI}(\epsilon_i^p , \epsilon_p^j) = \beta(p)\delta_i^j .$$

In other words the index is invariant under simultaneous application of (33.1) and (33.2).

Suppose in particular that (33.1) is the transformation to the canonical bases $\{a_i^p , \cdots , e_i^p\}$ of (14) and let the corresponding new bases $\{\epsilon_p^i\}$ for the cocycles be $\{a_p^i , \cdots , e_p^i\}$. In other words if $\epsilon_i^p = a_i^p , \cdots$ then $\epsilon_p^i = a_p^i , \cdots$. Formula (14.2) specified the form of the diagonal matrix ζ_j^{pi} of (33.5) so that

$$\text{(33.8)} \qquad \mathrm{F}a_i^{p+1} = 0, \quad \mathrm{F}b_i^{p+1} = 0, \quad \mathrm{F}c_i^{p+1} = 0,$$
$$\mathrm{F}d_i^{p+1} = t_i^p b_i^p , \quad \mathrm{F}e_i^{p+1} = a_i^p .$$

Applying (33.6) we have immediately

$$\text{(33.9)} \quad \mathrm{F}a_p^i = e_{p+1}^i , \quad \mathrm{F}b_p^i = t_{p+1}^i d_{p+1}^i , \quad \mathrm{F}c_p^i = 0, \quad \mathrm{F}d_p^i = 0, \quad \mathrm{F}e_p^i = 0,$$

where $t_{p+1}^i = t_i^p$, and i is not summed in (33.8), (33.9).
Furthermore

$$\text{(33.10)} \qquad \mathrm{KI}(a_i^p , a_p^i) = \cdots = \mathrm{KI}(e_i^p , e_p^i) = \beta(p),$$

and all the other indices will be zero. Thus we have proved:

(33.11) *At the same time as the bases for the chains are reduced to the canonical form* (14.2) *those for the cochains may be reduced to the canonical form* (33.9) *with indices related as stated. Notice that in* (33.9) *the analogues of* $a_i^p , \cdots , e_i^p$ *are* $e_p^i , \cdots , a_p^i$.

34. The application to the duality theorem is immediate. Suppose that we have two groups of coefficients G and H for the homology and cohomology groups, respectively. Then the direct decomposition (16.9) and the result of (33.9) yield:

$$\text{(34.1)} \qquad \mathfrak{H}^p(X, G) \cong \mathbf{P}(G^*(t_i^p)b_i^p) \times \mathbf{P}(Gc_i^p) \times \mathbf{P}(G[t_i^{p-1}]d_i^p),$$

$$(34.2) \qquad \mathfrak{H}_p(X, H) \cong \mathbf{P}(H[t^i_{p+1}]b^i_p) \times \mathbf{P}(Hc^i_p) \times \mathbf{P}(H^*(t^i_p)d^i_p).$$

Referring to (II, 20.12), or else directly if $G = H = J$, a field, we find that when (G, H) is a normal couple then $G^*(t^p_i)b^p_i$ and $H[t^i_{p+1}]d^i_p$ are dually paired with the Kronecker index as the group multiplication. Similarly each group in (34.1) is dually paired with one and only one of the groups in (34.2). Hence (II, 20.7, 33) $\mathfrak{H}^p(X, G)$ and $\mathfrak{H}_p(X, H)$ are likewise dually paired with the Kronecker index as the group multiplication, and this is (30.2).

§7. LINKING COEFFICIENTS. DUALITY IN THE SENSE OF ALEXANDER

35. The Kronecker index may be considered as the algebraic analogue of the intuitive concept of "multiplicity of intersection," for instance of two plane curves, in geometry. Another closely related geometric concept is that of *linking coefficient,* of two curves C, D in a three-space $\mathfrak{E}^3$, which describes the "algebraic" number of times each twists around the other. We shall show that under certain conditions such numbers may be introduced in complexes, and as we shall see later (VII, 9) in certain topological spaces.

Much of the argument will refer to *finite complexes which are* $(p - 1, p)$-*acyclic.* Let X be such a complex, and (G, H) a normal couple. If γ^{p-1} is a cycle over G and γ_p a cocycle over H, we have $\gamma^{p-1} \sim 0$ and so since G, H are division-closure groups $\gamma^{p-1} = \mathrm{F}C^p$ (17.2). Suppose also that $\gamma^{p-1} = \mathrm{F}C'^p$. Therefore $\mathrm{F}(C^p - C'^p) = 0$ and $C^p - C'^p$, being a p-cycle, is ~ 0. Hence by (29.6):

$$(35.1) \qquad \mathrm{KI}(C^p, \gamma_p) = \mathrm{KI}(C'^p, \gamma_p).$$

Thus the index at the left is independent of the C^p bounded by γ^{p-1}, and its value is known as the *linking coefficient* of γ^{p-1}, γ_p written $\mathrm{Lk}(\gamma^{p-1}, \gamma_p)$. *One must keep in mind that it is only defined for* γ^{p-1}, γ_p *over a normal couple* G, H.

Since cycles and cocycles are dual categories X is also $(p - 1, p)$-acocyclic. This enables us to interchange their role and so define a linking coefficient $\mathrm{Lk}'(\gamma^{p-1}, \gamma_p)$. However (29.3) yields at once $\mathrm{Lk}' = (-1)^{p-1}\,\mathrm{Lk}$, so except for a fixed change in sign, their values are equal.

36. The duality theorems which have been given so far relate merely the groups of a complex to one another. The linking coefficients will enable us to give full expression to duality theorems of a different type introduced by J. W. Alexander. They may be described at this stage, as relating under certain conditions the groups of a closed subcomplex to those of the complement. What is commonly known as *Alexander's duality theorem* is a duality theorem for topological complexes immersed in spheres. However the general intent is always the same, and we shall refer to the whole class of similar propositions as "duality theorems of the type of Alexander."

37. (37.1) THEOREM. *Let X be $(p - 1, p)$-acyclic and let X_1 be a closed subcomplex of X and G compact or a field. Then there subsists the isomorphism*

$$(37.2) \qquad \mathfrak{H}^{p-1}(X_1, G) \cong \mathfrak{H}^p(X, X_1, G).$$

If γ^p is a cycle of X mod X_1, $\delta^{p-1} = \mathrm{F}\gamma^p$ is an absolute cycle of X_1. Here F is the boundary operator for X itself. Thus F induces a homomorphism: $\mathfrak{Z}^p(X, X_1, G) \rightarrow \mathfrak{Z}^{p-1}(X_1, G)$. Suppose $\gamma^p \sim 0$ mod X_1. Since G has the division-closure property, $\gamma^p = \mathrm{F}C^{p+1} + D^p$, $D^p \subset X_1$, and hence $\delta^{p-1} = \mathrm{F}D^p$, or $\delta^{p-1} \sim 0$ in X_1. Therefore F maps $\bar{\mathfrak{F}}^p(X, X_1, G) \rightarrow \mathfrak{F}^{p-1}(X_1, G)$ and hence (II, 5.4) F induces a homomorphism φ: $\mathfrak{H}^p(X, X_1, G) \rightarrow \mathfrak{H}^{p-1}(X_1, G)$. To prove (37.1) we merely need to show that φ is an isomorphism.

(a) *φ is a mapping of $\mathfrak{H}^p(X, X_1, G)$ onto $\mathfrak{H}^{p-1}(X_1, G)$.* Since X is acyclic in the dimension $p - 1$ and G is a division-closure group every δ^{p-1} is an $\mathrm{F}\gamma^p$, so (a) holds.

(b) *φ is univalent.* It is required to show that $\mathrm{F}\gamma^p = \delta^{p-1} \sim 0$ in $X_1 \rightarrow \gamma^p \sim 0$ mod X_1 in X. Since G is a division-closure group if $\delta^{p-1} \sim 0$ in X_1 there is a D^p in X_1 such that $\delta^{p-1} = \mathrm{F}D^p$ and as a consequence $\gamma^p - D^p$ is a cycle of X. Since X is acyclic in the dimension p we have $\gamma^p - D^p \sim 0$ in X or $\gamma^p \sim 0$ mod X_1 in X which proves (b).

The group G may be compact or else a discrete field. Suppose G compact. The groups $\mathfrak{H}$ over G are then compact also. By (a), (b) φ is a mapping which is an isomorphism in the algebraic sense of one compact group into another. It follows that φ is topological and hence it is an isomorphism. When G is a discrete field the groups $\mathfrak{H}$ are finite-dimensional vector spaces, hence discrete and so φ is again an isomorphism. This proves (37.1).

38. Let again G, H be a normal couple and (X_0, X_1) a dissection of X. If δ^{p-1} is a cycle of X_1, we have $\delta^{p-1} \backsim \mathrm{F}\gamma^p$, γ^p a cycle of X_0. We may therefore introduce $\mathrm{Lk}(\delta^{p-1}, \gamma_p)$ and we have

$$\mathrm{KI}(\gamma^p, \gamma_p) = \mathrm{Lk}(\delta^{p-1}, \gamma_p) = \mathrm{Lk}(\mathrm{F}\gamma^p, \gamma_p). \tag{38.1}$$

It is obvious that, Lk takes a fixed value when δ^{p-1}, γ_p vary in fixed classes Δ^{p-1}, Γ_p of X_1 and X_0, and this value is by definition the *class linking* coefficient $\mathrm{Lk}(\Delta^{p-1}, \Gamma_p)$. From (37.2) we deduce:

$$\mathrm{KI}(\Gamma^p, \Gamma_p) = \mathrm{Lk}(\Delta^{p-1}, \Gamma_p) = \mathrm{Lk}(\mathrm{F}\Gamma^p, \Gamma_p), \tag{38.2}$$

where $\mathrm{F}\Gamma^p$ denotes the homology class of $\mathrm{F}\gamma^p$ in X_1. From the duality theorem (30.2) and (38.2) follows then:

(38.3) DUALITY THEOREM. *Let X be $(p - 1, p)$-acyclic, and let (X_0, X_1) be a dissection of X, with X_0 open and X_1 closed. Given any normal couple (G, H) the groups $\mathfrak{H}^{p-1}(X_1, G)$ and $\mathfrak{H}_p(X_0, H)$ are dually paired with the class linking coefficient as the group multiplication.*

(38.4) OBVIOUS REMARK. In (38.3) the two groups G, H may be interchanged.

Coupling (38.3, 38.4) with (30.2) for X_1 we find:

(38.5) *Under the same conditions as in* (38.3) *we have $\mathfrak{H}_{p-1}(X_1, G) \cong \mathfrak{H}_p(X_0, G)$ for any G which is compact, discrete or a field* (Kolmogoroff [b]; see Alexandroff [f]).

In the special case where $G = H =$ a field of characteristic π we have:

(38.6) *Under the same conditions as in* (38.3):

(38.6a) $$R^{p-1}(X_1, \pi) = R_p(X_0, \pi).$$

39. We shall now consider certain important related special cases.

(39.1) *X is acyclic.* (38.3) holds then for all p.

(39.2) *X is simplicial and zero-cyclic.* Then (38.3) holds for $p > 1$. Since X is zero-cyclic, it is connected, and a zero-cycle $\gamma^0 \sim 0$ in X when and only when $\mathrm{KI}(\gamma^0) = 0$ (20.10).

The homology classes of the zero-cycles in X_1 which are ~ 0 in X form a subgroup $'\mathfrak{H}^0(X_1, G)$ of $\mathfrak{H}^0(X_1, G)$, and the same argument goes through as before provided that $\mathfrak{H}^0(X_1, G)$ is replaced by $'\mathfrak{H}^0(X_1, G)$. Now if γ^0 is any zero-cycle of X_1, and if A_i are vertices one on each component of X_1, then (20.7a):

$$\gamma^0 \sim g^i A_i = \mathrm{KI}(\gamma^0)A_1 + g^i(A_i - A_1) = \mathrm{KI}(\gamma^0)A_1 + \delta^0,$$

where $\mathrm{KI}(\delta^0) = 0$. Since $A_1 \nsim 0$ in X, if Γ^0 is the class of A_1 then

$$'\mathfrak{H}^0(X_1, G) \cong \mathfrak{H}^0(X_1, G)/G\Gamma^0.$$

Thus in (38.3) in the present instance $\mathfrak{H}^0(X_1, G)$ must be replaced by $\mathfrak{H}^0(X_1, G)/G\Gamma^0$. In these and similar expressions later $G\Gamma^0$ represents the subgroup of the classes of the cycles gA_1.

(39.3) *X is simplicial, n-dimensional, and* (0, n)*-cyclic.* Suppose first $n > 1$. For $1 < p < n$, the situation is as under (39.1), and for $p = 1$ as under (39.2). Let $p = n$. Since dim $X = n$ and X is n-cyclic: $\mathfrak{H}^n(X, G) \cong G\gamma_0^n$, where γ_0^n s a basic integral n-cycle and so (37b) must be replaced by

(b′) $$(\delta^{n-1} \sim 0 \text{ in } X) \rightarrow (\gamma^n \sim g\gamma_0^n \bmod X_1 \text{ in } X).$$

As a consequence in place of (37.2) we have, if Γ_0^n is the class of γ_0^n (basic class):

(39.4) $$\mathfrak{H}^n(X, X_1, G)/G\Gamma_0^n \cong \mathfrak{H}^{n-1}(X_1, G),$$

and the factor-group at the left must replace $\mathfrak{H}^n(X, X_1, G)$.

Finally if $n = 1$, we must combine the operation under (39.2) with the one just described and as they cancel, (38.3) is applicable as it stands.

To sum up we may state:

(39.5) *Theorem* (38.3) *is valid when X is*: (a) *acyclic for all dimensions p*; (b) *zero-cyclic and simplicial for all p, provided that $\mathfrak{H}^0(X_1, G)$ is replaced by $\mathfrak{H}^0(X_1, G)/G\Gamma^0$, where Γ^0 is the class of a vertex of X_1*; (c) (0, n)*-cyclic and simplicial for all n provided that $\mathfrak{H}^0(X_1, G)$, $\mathfrak{H}^n(X, X_1, G)$ are replaced by $\mathfrak{H}^0(X_1, G)/G\Gamma^0$, $\mathfrak{H}^n(X, X_1, G)/G\Gamma_0^n$, except that* (38.3) *applies as it stands for $n = 1$.*

(39.6) The explicit Betti number relations are:

(a) *X acyclic:* (38.6a) for all p;

(b) *X simplicial and zero-cyclic*

$$R^{p-1}(X_1, \pi) = R_p(X_0, \pi) + \delta_1^p.$$

(c) *X simplicial,* (0, n)*-cyclic and* dim $X = n$:

$$R^{p-1}(X_1, \pi) = R_p(X_0, \pi) + \delta_1^p - \delta_n^p .$$

For Betti numbers mod 2 the last formula is the analogue for complexes of Alexander's original result for manifolds.

(39.7) EXAMPLES. An augmented closed n-simplex, in the sense defined later in (42) is acyclic (42.6), and so (38.3) is valid for such a complex and all p. The ordinary closed n-simplex is zero-cyclic and falls under (39.2) (second case of 39.5). Finally $\mathfrak{B}\sigma^{n+1}$, $n > 0$, is $(0, n)$-cyclic and n-dimensional, thus falling under (39.3) (third case of 39.5).

§8. HOMOLOGY THEORY OF INFINITE COMPLEXES

40. In endeavoring to carry over to infinite complexes the theory developed so far, serious difficulties arise in defining groups $\mathfrak{Z}$, $\mathfrak{F}$, of any sort, unless the complexes are at least star- or closure-finite. The simplest situation is of course when they are locally finite. Fortunately these types include all the types of interest in topology and certainly all those for which any general results are known. We shall therefore confine our attention to *star-, closure-, and locally finite complexes.*

Let then $X = \{x\}$ be infinite and of one of the three types just mentioned. This time we may introduce two kinds of chain- or cochain-groups:

$\mathfrak{C}^p(X, G) = \mathbf{P}(Gx_i^p)$, the group of the *infinite* chains over any G;

$\mathfrak{C}_f^p(X, G) = \mathbf{P}^w(Gx_i^p)$, the group of the *finite* chains over a discrete G; and the similar cochain groups $\mathfrak{C}_p(X, G)$, $\mathfrak{C}_p^f(X, G)$.

Referring now to (II, 8.4) we have the following situations.

(a) *X is star-finite.* Then F defines for every p and G a chain-homomorphism $\mathfrak{C}^p(X, G) \to \mathfrak{C}^{p-1}(X, G)$. When G is discrete F defines in addition homomorphisms $\mathfrak{C}_p^f(X, G) \to \mathfrak{C}_{p+1}^f(X, G)$.

(b) *X is closure-finite.* The situation is the same for X^* as previously for X, i.e., with cycles and cocyles interchanged. We have then homomorphisms of the groups of finite chains over a discrete G and in addition F defines homomorphisms $\mathfrak{C}_p(X, G) \to \mathfrak{C}_{p+1}(X, G)$ (any G).

(c) *X is locally finite.* Then F defines the four types of homomorphisms considered under (a) and (b).

We notice also that when G is a discrete field all the groups $\mathfrak{C}$ under discussion are vector spaces and so they fall under the fundamental convention (II, 22.2) for such spaces.

In any one of the three cases just considered the groups $\mathfrak{Z}$ may be defined as in (7, 8, 9) and likewise the groups $\mathfrak{H}$ as the factor-groups $\mathfrak{H} = \mathfrak{Z}/\bar{\mathfrak{F}}$ ($\bar{\mathfrak{F}} = \mathfrak{F}$ for the groups of finite chains). We may therefore state the comprehensive

(40.1) THEOREM. *When X is star-finite [closure-finite] the homology [cohomology] groups of the infinite cycles [cocycles] of X over any G may be introduced in the same manner as for finite complexes. When X is locally finite this holds for both the infinite cycles and cocycles. In all three cases this holds also for the finite cycles and cocycles over a discrete G.*

Complementary remarks. (40.2) We call attention to the fact that many definitions given for finite complexes are directly applicable to certain infinite complexes. In particular:

(a) When X is star-finite we may introduce as before the following concepts: dissections and related groups for infinite cycles (23), the *cyclic* or *acyclic* types of (21) corresponding to infinite cycles, and also the *circuits* of various kinds (24) which are now described in terms of groups of infinite n-cycles. infinite n-cycles only.

(b) When X is closure-finite the dissections and the cocyclic and acocyclic types may be introduced.

(c) When X is locally finite there may be introduced all the concepts mentioned under (a) and (b).

(40.3) *Suppose G compact.* Then if X is star-finite the groups of infinite chains $\mathfrak{C}^p(X, G)$, $\mathfrak{Z}^p(X, G)$, $\mathfrak{F}^p(X, G)$ which may then be introduced are all compact: the second as a closed subgroup of the first, and the third as the image of $\mathfrak{C}^{p+1}(X, G)$ under F. As a consequence $\mathfrak{F}^p(X, G)$ is closed in $\mathfrak{C}^p(X, G)$, or $\bar{\mathfrak{F}}^p(X, G) = \mathfrak{F}^p(X, G)$. Hence $\mathfrak{H}^p(X, G) = \mathfrak{Z}^p(X, G)/\mathfrak{F}^p(X, G)$ and it is also compact. Similarly of course for a closure-finite X and the groups $\mathfrak{C}_p$, $\mathfrak{Z}_p$, $\mathfrak{F}_p$, $\mathfrak{H}_z$

(40.4) *When G is a linearly compact field* the groups, $\mathfrak{C}$, $\mathfrak{Z}$, $\mathfrak{F}$ are linearly compact and the same argument goes through as is seen by reference to (II, 27.2, $\cdots$, 27.5). The groups $\mathfrak{H} = \mathfrak{Z}/\mathfrak{F}$ are found this time to be linearly compact.

(40.5) *Betti numbers.* They are defined in the same way as before, as the dimensions of the vector spaces $\mathfrak{H}^p(X, J)$ or $\mathfrak{H}_p(X, J)$, J a discrete field. We may notice here and now that the *universal theorem for fields* (17.8) *is valid for the case under consideration.* For $\mathfrak{Z}^p(X, J)$, $\mathfrak{F}^p(X, J)$ are spanned here also by $\mathfrak{Z}^p(X, \pi)$, $\mathfrak{F}^p(X, \pi)$ and so the asserted property is a direct consequence of (II, 36.8).

(40.6) *Alternate definition of the homology groups.* Let X be star-finite. Besides the topologized homology group $\mathfrak{H}^p = \mathfrak{Z}^p/\bar{\mathfrak{F}}^p$ one may consider the purely formal algebraic factor-group $\mathfrak{H}^p = \mathfrak{Z}^p/\mathfrak{F}^p$ (or even more generally $\mathfrak{H}^p = \mathfrak{Z}^p/'\mathfrak{F}^p$ where $'\mathfrak{F}^p$ is a subgroup of $\mathfrak{Z}^p$ such that $\mathfrak{F}^p \subset {}'\mathfrak{F}^p \subset \bar{\mathfrak{F}}^p$). This would amount to taking $\mathfrak{C}^p$ untopologized. As stated in (40.3) and (40.4) the two concepts are algebraically equivalent when G is compact or a field. In other cases, however, (for instance for integral chains) it may very well happen that $\mathfrak{F}^p \neq \bar{\mathfrak{F}}^p$ and that also $\mathfrak{Z}^p/\bar{\mathfrak{F}}^p$, $\mathfrak{Z}^p/\mathfrak{F}^p$ are essentially different. The latter and likewise the group $\bar{\mathfrak{F}}^p/\mathfrak{F}^p$ (taken discrete) have been considered recently to advantage by Eilenberg [a] and Steenrod [b] (Appendix A).

(40.7) *Universal coefficient-groups.* It has been proved by Čech [d] that the group of the integers is universal for the homology groups of the finite cycles of a locally finite complex. A complete description of all the groups of such complexes has just been obtained by Eilenberg and MacLane [a] (Appendix A).

(40.8) The analogue of the question considered in (23.3) is of interest later. We suppose X infinite, Y a finite closed subcomplex and consider the groups of the cycles of Y over a discrete G reduced with respect to bounding in X. Here again we readily arrive at (14.2) for a single dimension p, except that the $(p + 1)$-chains in (14.2) are to be replaced by finite cycles of X mod Y. Let $\sum_p$ denote the system thus obtained. Let also M be a finite closed subcomplex of X which includes Y and all the d_i^{p+1}, e_i^{p+1}. If we reduce the cycles of Y with respect to bounding in M we still obtain $\sum_p$, for we have already utilized all the relations of bounding in M. Similarly if M is replaced by any other closed finite subcomplex $M_1 \supset M$. Since the groups in question for any G depend solely upon $\sum_p$, (23.3), the groups of Y reduced with respect to bounding in M and M_1 must be the same. From this we conclude that the groups relative to bounding in X and M are the same. For otherwise Y must contain a cycle $\gamma^p \sim 0$ in X but $\nsim 0$ in M. Hence if $\gamma^p = FC^{p+1}$, C^{p+1} finite, and M_1 is any finite closed subcomplex containing M and C^{p+1}, the reductions relative to bounding in M and M_1 cannot yield the same groups, a contradiction proving our statement. We conclude then:

(40.9) *If Y is a finite closed subcomplex of the complex X, then the homology groups of the cycles of Y reduced relative to finite bounding in X are the same as those reduced relative to bounding in a certain finite closed subcomplex M containing Y. Hence in particular the remarks of* (23.3) *are applicable to the groups in question. Thus they have finite Betti numbers and the group of the integers is a universal coefficient-group* (18).

41. **Duality.** Let X be star-finite, and G, H commutatively paired to J. We consider the group $\mathfrak{C}^p$ of the infinite chains over G. Since X is star-finite we may introduce the infinite cycles over G; they form a subgroup $\mathfrak{Z}^p$ of $\mathfrak{C}^p$, likewise the infinite *bounding* cycles over G with group $\mathfrak{F}^p \subset \mathfrak{Z}^p$. Therefore the homology groups of the infinite cycles over G are $\mathfrak{H}^p = \mathfrak{Z}^p/\mathfrak{F}^p$.

Regarding the cocycles, since X need not be closure-finite, only finite cocycles may be allowed, and groups over a discrete H: $\mathfrak{C}_p^f$, $\mathfrak{Z}_p^f$, $\mathfrak{F}_p^f$, $\mathfrak{H}_p^f = \mathfrak{Z}_p^f/\mathfrak{F}_p^f$.

It is hardly necessary to observe that the index $\mathrm{KI}(C^p, C_p)$ may be defined as in (28). Indeed it may even be defined when both C^p, C_p are infinite (H being then any topological group) *provided that they have a finite number of crossings.*

When X is closure-finite the situation is the same with cycles and cocycles interchanged.

We are now in position to state

(41.1) *The properties of the Kronecker index given in* (29) *are valid for infinite cycles* [*cocycles*] *and finite cocycles* [*cycles*] *in a star-finite* [*closure-finite*] *complex X.*

For the proofs loc. cit. apply without modification.

We may now repeat for X, and also for X^* when X is closure-finite, the argument of (30) and thus obtain

(41.2) Duality theorem for star- or closure-finite complexes. *When X is star-finite* [*closure-finite*] *the infinite cycles* [*cocycles*] *and the finite cocycles* [*cycles*] *are dual categories. When X is locally finite both types of dual categories are present.*

(41.3) *Linking coefficients.* The full argument and definitions of (§7) may be extended to locally finite complexes, and in particular:

(41.4) Theorem. *The duality theorems* (38.3, 39.5) *of the Alexander type, hold for locally finite complexes.*

REMARK. We shall return to infinite complexes in (VI, §6) when we shall apply to them the powerful method of nets and webs.

§9. AUGMENTABLE AND SIMPLE COMPLEXES

42. Let $K = \{\sigma\}$ be a simplicial complex, $\{A_i\}$ its vertices, $\{A^i\}$ their duals. Upon examining the argument in (5.1) we readily verify that K does not cease to be a complex if we increase it by a new *null-simplex* ϵ such that: (a) ϵ is a face of every σ; (b) $\dim \epsilon = -1$; (c) $[A_i:\epsilon] = [\epsilon:A_i] = 1$, $[\sigma^p:\epsilon] = [\epsilon:\sigma^p] = 0$ for $p > 0$. The complex $K_a = K \cup \epsilon$ thus obtained is said to be K *augmented* (A. W. Tucker [a]). The chief differences between K and K_a are embodied in the properties:

(42.1) *A finite zero-chain C^0 is a cycle of K_a when and only when its Kronecker index* $\mathrm{KI}(C^0) = 0$.

For in K_a we have $\mathrm{F}C^0 = \mathrm{KI}(C^0)\epsilon$.

As a noteworthy special case:

(42.2) *The differences $A_i - A_j$ are integral zero-cycles of K_a but A_i is not.*

Let $\{K_i\}$ be the components of K and B_i a vertex of K_i. A one-chain C^1 of K is likewise one of K_a, and whether considered as in K or K_a its boundary $\mathrm{F}C^1$ is the same. It follows that (20.7) holds for K_a and finite chains. If γ^0 is a zero-cycle of K_a we have $\mathrm{KI}(\gamma^0) = 0$ and hence by (20.7a):

$$(42.3) \qquad \gamma^0 \sim g^i(B_i - B_1) + \mathrm{KI}(\gamma^0)B_1 \sim g^i(B_i - B_1).$$

By (20.7b) also

$$g^i(B_i - B_1) \sim 0 \rightarrow g^i = 0.$$

From this we deduce the analogue of (20.8) for K_a:

$$(42.4) \qquad \mathfrak{H}^0(K_a, G) \cong \mathfrak{H}^0(K, G)/GB_1,$$

where for simplicity B_1 is identified with its class. As a special case of (42.3) if K is connected $\gamma^0 \sim 0$, and hence:

(42.5) *If K is connected then K_a is acyclic in the dimension zero.*

(42.6) $(\mathrm{Cl}\ \sigma^n)_a$ *is acyclic and* $(\mathfrak{B}\sigma^n)_a$ *is* $(n-1)$*-cyclic* (22.2, 22.4, 42.5).

$$(42.7) \qquad R^0(K) = R^0(K_a) + 1, \ (42.4).$$

43. Let now $X = \{x\}$ be any closure-finite complex with $\dim x \geqq 0$. Is it possible to "augment" X, i.e., to increase it by a (-1)-dimensional element ϵ, which is to be a face of every x, and with incidence numbers $\lambda_i = [x_i^0:\epsilon] = [\epsilon:x_i^0]$ not all zero and $[x^p:\epsilon] = [\epsilon:x^p] = 0$ for $p > 0$? If so X is said to be *augmentable* and the new complex $X_a = X \cup \epsilon$ is called X *augmented.* In order that X_a be a complex it must fulfill conditions K1234 of (1), or which is equivalent, its dual X_a^* must fulfill them. The first three are automatically satisfied and so only K4 is in question. By (8.3) it reduces to requiring that if ϵ^* is the dual (one-dimensional) of ϵ, then $\mathrm{FF}\epsilon^* = 0$. Since $[x_i^0:\epsilon] = [\epsilon^*:x_0^i] = \lambda_i$, this

is equivalent to requiring $F\lambda_i x_0^i = 0$, i.e., that $\gamma_0 = \lambda_i x_0^i$ be an integral zero-cocycle, a result due to Tucker [a]. The particular zero-cocycle arising in the augmentation is called the *fundamental* zero-cocycle, and we shall say that "X is augmentable, or augmented, with fundamental zero-cocycle γ_0."

44. All the possible modes of augmenting X correspond to its different non-trivial integral zero-cocycles, i.e., to the nonzero elements of $\mathfrak{Z}_0(X)$. Since $\dim X^* \leqq 0$: $\mathfrak{Z}_0(X) = \mathfrak{H}_0(X)$ (27.6a). Hence the integral zero-cocycles form a free group of rank $R_0 = R^0$. Therefore

(44.1) *A n. a. s. c. for augmentability is that the Betti number $R^0 \neq 0$. Each mode of augmenting is uniquely determined by an integral zero-cocycle $\gamma_0 = \lambda_i x_0^i$, and in $X_a = X \cup \epsilon$, $\lambda_i = [x_0^i : \epsilon]$.*

An equivalent condition of augmentability is $FFC^1 = 0$ for every finite chain C^1 over G in X_a. If $FC^1 = g^i x_i^0$ this yields as n. a. s. c.:

$$\lambda_i g^i = \mathrm{KI}(FC^1, \gamma_0) = 0, \tag{44.2}$$

where the index is an element of G. Therefore

(44.3) *The n. a. s. c. for augmentability in* (44.1) *is equivalent to requiring the existence of a non-trivial integral zero-cocycle γ_0 such that $\mathrm{KI}(FC^1, \gamma_0) = 0$ for all finite C^1.*

45. Suppose X augmented and with the fundamental zero-cocycle γ_0. If Y is a closed subcomplex of X we may write $\gamma_0 = \gamma_0' + \gamma_0''$ where γ_0' is in Y^* and γ_0'' has no element in Y^*. We shall say that Y^* *meets* γ_0 when $\gamma_0' \neq 0$, and we shall call γ_0' for the present the *intersection* of Y^* with γ_0. Since γ_0' is a cocycle of Y we have:

(45.1) *Let X be augmentable with fundamental cocycle γ_0. Then every closed subcomplex Y such that Y^* meets γ_0 is also augmentable and with a fundamental cocycle γ_0' which is the intersection of γ_0 with Y^*.*

When Y is finite and augmentable as stated it may be augmented with γ_0' as fundamental cocycle. We shall denote in any case by Y_a the new augmented complex $Y \cup \epsilon$, when $\gamma_0' \neq 0$, and Y itself when $\gamma_0' = 0$. Notice that Y_a may depend *a priori* upon the cocycle γ_0 chosen as fundamental for X. In fact the significance of the choice of γ_0 as fundamental cocycle lies in a sense in that it provides a uniform procedure for augmenting the finite subcomplexes of X.

46. Returning for a moment to the simplicial complex K, let $\delta_0 = \sum A^i$. If $C^0 = g^i A_i^0$, we have

$$\sum g^i = \mathrm{KI}(C^0, \delta_0). \tag{46.1}$$

Therefore the Kronecker index as a sum of coefficients is in fact also a "chain-cochain" index. Furthermore if $FC^1 = g^i A_i^0$, C^1 finite,

$$\sum g^i = 0 = \mathrm{KI}(FC^1, \delta_0). \tag{46.2}$$

Therefore K is augmentable with $\gamma_0 = \delta_0$, i.e., with unity as the new incidence numbers $[A_i : \epsilon]$. This is precisely the way in which K_a has been obtained in (42). We also know that if K is connected, $R^0 = 1$, so that all the zero-cocycles are of the form $g\delta_0$. Therefore *when K is finite and connected it can only be augmented in essentially one way and with incidence numbers λ_i all equal.* Their common value λ is the only "indeterminate" in the augmentation.

Referring also to (28.8) we have the following noteworthy interpretation for $\mathrm{KI}(C^0)$: it is the number of weighted crossings of C^0 with the cocycle δ_0.

47. The preceding properties suggest the following extension of simplicial complexes:

(47.1) DEFINITION. *A complex $X = \{x\}$ is said to be simple whenever*: (a) *X is closure-finite*; (b) *X is augmentable and this with a fundamental zero-cocycle which in a suitable orientation of X is given by $\gamma_0 = \sum x_0^i$*; (c) *every* $(\mathrm{Cl}\, x)_a$ *is acyclic.*

We agree first of all to orient X so that $\gamma_0 = \sum x_0^i$. Then in X the Kronecker index $\mathrm{KI}(C^0, \gamma_0)$, C^0 finite, is the sum of the coefficients of C^0, and it will be denoted once more by $\mathrm{KI}(C^0)$.

Notice that when X is simplicial with vertices $\{A_i\}$ then $\gamma_0 = \sum A^i$. Thus for a simplicial complex the *fundamental cocycle* in the sense of (25.3) is the same as the fundamental cocycle of (47.1).

If X is simple and Y is a closed subcomplex of X, then $X - Y$ is called an *open* simple complex. By contrast X or Y are also called *closed* simple complexes.

How close the approximation is to simplicial complexes is attested by the following properties:

(47.2) *Simplicial and polyhedral complexes are simple.*

For simplicial complexes it is a consequence of (44). For polyhedral complexes the proof will be given later (IV, 28.2).

(47.3) *Every closed subcomplex of a simple complex is simple.*

(47.4) *When X is simple every p-element has at least one* $(p - 1)$*-face, hence at least one vertex, and every one-face has two vertices* (Whitney [d]).

Suppose $\mathrm{F}x^p = 0$, $p > 0$. Then x^p is a p-cycle of $(\mathrm{Cl}\, x)_a$, hence $x^p \sim 0$ in $(\mathrm{Cl}\, x)_a$, and since both have the dimension p, $x^p = 0$, which is absurd. Therefore $\mathrm{F}x^p \neq 0$, so that x^p must have at least one $(p - 1)$-face, and therefore step by step it is shown to have at least one vertex.

Suppose that x^1 has only one vertex x^0 so that $\mathrm{F}x^1 = gx^0$. Since $(\mathrm{Cl}\, x^1)_a$ is augmented, $\mathrm{KI}(\mathrm{F}x^1) = g = 0$, hence again x^1 must be a cycle which we have just ruled out. Suppose on the other hand that x^1 has three vertices x_1^0, x_2^0, x_3^0. Then $x_1^0 - x_2^0$ and $x_1^0 - x_3^0$ are cycles of $(\mathrm{Cl}\, x^1)_a$ and hence $x_1^0 - x_2^0 = g\mathrm{F}x^1$, $x_1^0 - x_3^0 = h\mathrm{F}x^1$, where g, h are distinct nonzero integers. As a consequence $h(x_1^0 - x_2^0) = g(x_1^0 - x_3^0)$, $g = h$, $x_2^0 = x_3^0$, a contradiction.

(47.5) *When X is finite and satisfies* (47.1b) *then "X is zero-cyclic" and "X_a is acyclic" are equivalent. Hence in* (47.1) *"every* $(\mathrm{Cl}\, x)_a$ *is acyclic" may be replaced by "every* $\mathrm{Cl}\, x$ *is zero-cyclic."*

Let X be zero-cyclic and let $C^0 = g^i x_i^0$ be a zero-cycle of X_a. Then in X_a: $\mathrm{F}C^0 = 0 = \mathrm{KI}(C^0)$. Since X is zero-cyclic $x_1^0 - gx_i^0 \sim 0$ for some g. Hence $\mathrm{KI}(x_1^0 - gx_i^0) = (1 - g) = 0, g = 1, x_1^0 - x_i^0 \sim 0$. Hence $C^0 = C^0 - \mathrm{KI}(C^0)x_1^0 = g^i(x_i^0 - x_1^0) \sim 0$ in X and hence also in X_a. Therefore X_a is acyclic. The converse is immediate.

(47.6) *A one-dimensional simple complex X^1 is simplicial.*

By (47.4) X^1 has the ordering relations of a simplicial complex. Let A, B be the vertices of x^1. If $\mathrm{F}x^1 = gA + hB$ we have $\mathrm{KI}(\mathrm{F}x^1) = g + h = 0$, $g = -h$, $\mathrm{F}x^1 = g(A - B)$. Also by (47.5) $A \sim kB$, or $A - kB = \mathrm{F}mx^1 = mg(A - B)$. Hence $mg = 1$, $g = \pm 1$. Therefore $[x^1:A] = -[x^1:B] = \pm 1$. Thus x^1 has the incidence numbers of a simplicial one-complex and (47.6) follows.

(47.7) *Properties* (20.5, $\cdots$, 20.10) *hold for a simple complex, provided that the zero-dimensional homology groups and Betti numbers are those of the finite zero-cycles.*

For the elements of X are all connected with those of its one-section X^1 which is simplicial, and the proof of (20.5) refers solely to X^1 and its finite zero-cycles.

(47.8) *When X is connected and simple all the modes of augmenting X are essentially unique, in the sense that all the possible fundamental cocycles are merely the multiples of a single cocycle* (Whitney [d]).

For by (41.2) $R_0 = R_f^0 = 1$. It follows that every integral cocycle $\delta_0 = \sum \lambda_i x_0^i$ is a rational multiple of $\gamma_0 = \sum x_0^i$, and hence it is an *integral* multiple: $\delta_0 = \lambda \sum x_0^i$.

(47.9) *Properties* (42.1, $\cdots$, 42.5, 42.7) *hold for any simple complex X.*

For they depend solely upon the one-section X^1 of X and X^1 is simplicial (47.6).

(47.10) *When X is simple, $(\mathfrak{B}x^p)_a$, $p \geqq 1$, is $(p-1)$-cyclic, and all its $(p-1)$-cycles are of the form $g\mathrm{F}x^p$.*

For $p > 1$ this follows from (21.4, 47.5), and for $p = 1$ from the fact that x^1 is a simplex (47.6).

CHAPTER IV

COMPLEXES: PRODUCTS. TRANSFORMATIONS. SUBDIVISIONS

The title gives sufficient indication of the ground covered in the chapter. There are two basic types of transformations: those of a complex as a set of elements, and certain homomorphisms of the chain-groups, the chain-mappings. The latter are of fundamental importance in the sequel, and their properties are similar in many ways to those of point set mappings. Thus one may introduce a very useful concept of chain-homotopy, classify chain-mappings with respect to this relation, etc.

While the treatment is mainly developed for finite complexes, the modifications required for infinite complexes are discussed in full in (§6).

General references: Alexandroff-Hopf [A-H], Hopf [a], Künneth [a], Lefschetz [L, I, V; e, f], Tucker [a, c].

§1. PRODUCTS OF COMPLEXES

1. Let X, Y be two complexes. With Tucker [a] we will turn the product $\{x_i^p\} \times \{y_j^q\}$ into a complex to be called the *product* of X *by* Y, written $X \times Y$, in the following manner. For convenience the elements of $\{x_i^p\} \times \{y_j^q\}$ are written $x_i^p \times y_j^q$. Then:

(a) $x' \times y' \prec x \times y \leftrightarrow x' \prec x$ and $y' \prec y$;

(b) $\dim(x \times y) = \dim x + \dim y$;

(c) $[x \times y : x' \times y] = [x : x']$; $\quad [x \times y : x \times y'] = (-1)^{\dim x}[y : y']$;

the incidence numbers that are not of one of these two forms are all zero, by definition.

Of the four basic properties K1 $\cdots$ 4 of (III, 1), all but the last are trivially verified; so K4 alone requires proof. Given $x \times y$, $x'' \times y''$, there is at most a finite number of elements $x' \times y'$ (we do not exclude $x' = x$ or x'', $y' = y$ or y'') such that

$$[x \times y : x' \times y'] \cdot [x' \times y' : x'' \times y''] \neq 0. \tag{1.1}$$

We may assume in fact that $x'' \times y'' \prec x \times y$, and then the only two cases requiring verification are:

(α) $x'' = x$, or else $y'' = y$. The verification of K4 reduces then to the same for X or Y.

(β) $\dim x'' = \dim x - 1$, $\dim y'' = \dim y - 1$. The only elements $x' \times y'$ to be considered are $x \times y''$ and $x'' \times y$. The sum of the corresponding expressions (1.1) must be zero, i.e., we must have:

$$[x \times y : x \times y''][x \times y'' : x'' \times y''] + [x \times y : x'' \times y][x'' \times y : x'' \times y''] = 0.$$

By (c), together with $\dim x'' = \dim x - 1$, this reduces to $(-1)^{\dim x}[y : y''][x : x''] + (-1)^{\dim x - 1}[x : x''][y : y''] = 0$. Therefore $X \times Y$ is a complex.

(1.2) REMARK. The choice of the factor $(-1)^{\dim x}$ in (c) is solely on grounds of expediency. By examining the rectangle as product of two segments the reader will quickly convince himself that some such factor is needed if the orientations of the elements are to behave in accordance with the rather natural rules for polyhedra given in (III, 6.4). This will be brought out more fully in connection with the product of polyhedra (3).

(1.3) *Product of any finite number of factors.* Consider first $(X_1 \times X_2) \times X_3$ and $X_1 \times (X_2 \times X_3)$. Their elements are the same and so are their ordering relations and incidence numbers. Hence the two products coincide and are isomorphic. We identify them (a convenient procedure followed in similar cases later) and designate the complex thus obtained by $X_1 \times X_2 \times X_3$. The product $X_1 \times \cdots \times X_n$ is defined by an obvious recurrence.

Let now X, Y be disjoint and compare $X \times Y$ with $Y \times X$. Their elements are once more the same and with the same ordering relations. If we set

$$\alpha(x \times y) = (-1)^{\dim x \dim y}, \tag{1.4}$$

then we verify that the incidence numbers in $Y \times X$ are such that it is merely $X \times Y$ reoriented by $\alpha(x \times y)$.

(1.5) *Identical factors.* The product of n factors equal to X is written sometimes X^n. It is to be noted that while the product of n elements $x_1 \times \cdots \times x_n$ of X is associative it is not generally commutative. Thus generally $x_1 \times x_2 \neq x_2 \times x_1$.

2. The following properties are direct consequences of the definitions:

$$\operatorname{St}(x \times y) = \operatorname{St} x \times \operatorname{St} y; \tag{2.1}$$

$$\operatorname{Cl}(x \times y) = \operatorname{Cl} x \times \operatorname{Cl} y; \tag{2.2}$$

$$\mathfrak{B}(x \times y) = \mathfrak{B}x \times \operatorname{Cl} y \cup \operatorname{Cl} x \times \mathfrak{B}y. \tag{2.3}$$

(2.4) *The product of two open [closed] subcomplexes of X, Y is an open [a closed] subcomplex of $X \times Y$* (2.1, 2.2).

(2.5) *When X, Y are both star-finite [closure-finite and hence when they are both locally finite] so is $X \times Y$* (2.1, 2.2).

(2.6) *If X, Y are reoriented by the orientation functions $\alpha(x)$, $\alpha'(y)$ then $X \times Y$ is reoriented by the orientation function $\alpha''(x \times y) = \alpha(x)\alpha'(y)$.*

(2.7) *Duals.* Let us denote temporarily by x^*, y^*, $(x \times y)^*$ the duals of x, y, $x \times y$ in X^*, Y^*, $(X \times Y)^*$. From the incidences we find at once that $(x \times y)^* \to x^* \times y^*$ defines an isomorphism of $(X \times Y)^*$ with $X^* \times Y^*$. It will be convenient therefore to identify the two complexes so that $(x \times y)^* = x^* \times y^*$ and we will then have the relation

$$(X \times Y)^* = X^* \times Y^*. \tag{2.8}$$

Thus the dual of $x_i^p \times y_j^q$ will be $x_p^i \times y_q^j$, and $(X \times Y)^* = \{x_p^i \times y_q^j\}$.

(2.9) *The components of $X \times Y$ are the products of those of X and Y. Hence if X and Y are connected so is $X \times Y$.*

Referring to (III, 2.3) and to (1a) a n. a. s. c. for $x \times y$ and $x' \times y'$ to be in the relation R, in the sense of (III, 2.3), is that this holds for x, x' and for y, y'. In other words, $x \times y$ and $x' \times y'$ are in the same component of $X \times Y$ when and only when x, x' are in the same component of X and y, y' in the same component of Y. This is essentially (2.9).

3. **Products of polyhedral complexes.** It will be sufficient to consider a product of two polyhedral complexes $\Pi_i = \{E^p_{ij}\}$, $i = 1, 2$. We have on the one hand the topological product of polyhedra $|\Pi_1| \times |\Pi_2|$, on the other hand the product of complexes $\Pi_1 \times \Pi_2$. We prove:

(3.1) *$\Pi_1 \times \Pi_2$ is a polyhedral complex whose polyhedron* $|\Pi_1 \times \Pi_2| = |\Pi_1| \times |\Pi_2|$.

This provides the best possible justification for the product convention of (1).

Proof of (3.1).

(3.2) *If Ω, Ω' are bounded convex regions in the Euclidean spaces $\mathfrak{E}^m$, $\mathfrak{E}^n$ then $\Omega \times \Omega'$ is a bounded convex region in* $\mathfrak{E}^m \times \mathfrak{E}^n$ (proof elementary).

$$\text{(3.3)} \qquad \mathfrak{B}(\Omega \times \Omega') = \mathfrak{B}\Omega \times \bar{\Omega}' \cup \bar{\Omega} \times \mathfrak{B}\Omega' \qquad \text{(I, 12.5).}$$

If we combine (2.3) and (3.3) we find by an elementary induction on the dimension that

(3.4) *$\{E^p_{1i} \times E^q_{2j}\}$ are the cells of a polyhedral complex Π such that* $|\Pi| = |\Pi_1| \times |\Pi_2|$.

The complexes Π, $\Pi_1 \times \Pi_2$ consist of the same elements with the same assignment of dimensions and incidences. Hence to prove (3.1) it is sufficient to show that

(3.5) *The incidence numbers of the elements in Π and $\Pi_1 \times \Pi_2$ are the same.*

(3.6) Let $\mathfrak{E}^m$ be an Euclidean space referred to the coordinate system $\{x_1, \cdots, x_m\}$ and let its characteristic be ϵ^m (III, 6.3). The subspace $\mathfrak{E}^{m-1}$: $x_p = 0$ has the coordinate system $\{x_1, \cdots, x_{p-1}, x_{p+1}, \cdots, x_m\}$ and with a characteristic, say ϵ^{m-1}. Let $\mathfrak{E}'^m$, $\mathfrak{E}''^m$ be the two regions $x_p > 0$, $x_p < 0$. We modify the convention of (III, 6.3) in that we now define the incidence numbers as:

$$[\mathfrak{E}'^m : \mathfrak{E}^{m-1}] = -[\mathfrak{E}''^m : \mathfrak{E}^{m-1}] = (-1)^{p-1} \epsilon^m \epsilon^{m-1}.$$

Under our original convention we should introduce a new coordinate system $\{x'_1, \cdots, x'_m\}$ for $\mathfrak{E}^m$ such that: $x'_i = x_i$, $i < p$; $x'_i = x_{i+1}$, $i > p$; $x'_m = x_p$, whose characteristic for $\mathfrak{E}^m$ is $(-1)^{m-p+1} \epsilon^m$. Hence we would have $[\mathfrak{E}'^m : \mathfrak{E}^{m-1}] = (-1)^{m-p+1} \epsilon^m \epsilon^{m-1}$. One sees readily, however, that this does not modify the argument of (III, 6.5), and the present convention is more convenient for our immediate purpose.

(3.7) Let $\mathfrak{E}^n$ be a second Euclidean space with coordinates $\{y_1, \cdots, y_n\}$ and characteristic ϵ^n. Then $\mathfrak{E}^m \times \mathfrak{E}^n$ is an $\mathfrak{E}^{m+n}$ with coordinates $\{x_1, \cdots, x_m, y_1, \cdots, y_n\}$ and characteristic $\epsilon^m \epsilon^n$. Let also $\mathfrak{E}^{n-1}$, $\mathfrak{E}'^n$ be analogues of $\mathfrak{E}^{m-1}$, $\mathfrak{E}'^m$ relative to y_q. Then again $\mathfrak{E}^{m-1} \times \mathfrak{E}^n$, $\mathfrak{E}'^m \times \mathfrak{E}^n$ and $\mathfrak{E}^m \times \mathfrak{E}^{n-1}$, $\mathfrak{E}^m \times \mathfrak{E}'^n$ are similar pairs for x_p, y_q relative to $\mathfrak{E}^m \times \mathfrak{E}^n$. Their incidence numbers are given by:

$$[\mathfrak{E}'^m \times \mathfrak{E}^n : \mathfrak{E}^{m-1} \times \mathfrak{E}^n] = [\mathfrak{E}'^m : \mathfrak{E}^{m-1}], \tag{3.8}$$

$$[\mathfrak{E}^m \times \mathfrak{E}'^n : \mathfrak{E}^m \times \mathfrak{E}^{n-1}] = (-1)^m [\mathfrak{E}'^n : \mathfrak{E}^{n-1}]. \tag{3.9}$$

The incidence numbers for Π calculated by means of (3.8, 3.9) and the rule of (III, 6.4) being those of $\Pi_1 \times \Pi_2$, (3.5) follows and so (3.1) is proved.

4. **The joins of simplicial complexes.** It is readily seen that the product of two simplicial complexes is only simplicial in the trivial case when their dimensions are zero. This is already apparent in the property that the product of two segments is a rectangle and not a triangle. A substitute operation, the join, will have the advantage of preserving simpliciality.

(4.1) Let $K = \{\sigma\}$, $L = \{\zeta\}$ be two simplicial complexes and let K_a, L_a be K, L augmented by the same (-1)-element ϵ (the null-simplex). It is convenient to introduce also the formal joins $\sigma\epsilon$, $\epsilon\sigma$ which we define as $\sigma\epsilon = \epsilon\sigma = \sigma$. Similarly $\zeta\epsilon = \epsilon\zeta = \zeta$.

Consider now the following sets of joins:

$$KL = \{\sigma\zeta\}, K_aL = \{\sigma\zeta\} \cup L,$$

$$KL_a = \{\sigma\zeta\} \cup K, K_aL_a = \{\sigma\zeta\} \cup K \cup L \cup \epsilon.$$

We call $KL, \cdots$ the join of K and $L, \cdots$. If we set $M = \{\sigma\zeta\} \cup K \cup L$, then M is a closed simplicial complex, and hence KL, K_aL, KL_a are open simplicial complexes, while $K_aL_a = M_a =$ an augmented simplicial complex.

(4.2) Let $'K, \cdots$ denote the weak isomorphs of $K, \cdots$ obtained by raising all the dimensions one unit. Then we quickly verify: $'(KL) \cong {'K} \times {'L}, \cdots$, $'(K_aL_a) \cong {'K_a} \times {'L_a}$. Thus if we raised all dimensions one unit we could replace the joins by products.

(4.3) Let K_e, L_e be finite Euclidean complexes situated, respectively, in $\mathfrak{E}^m$, $\mathfrak{E}^n$ and let K, L be the simplicial companions with the same vertices such as in (III, 6.12). If $x \in K_e$, $y \in L_e$ the segments $\overline{xy}$ in $\mathfrak{E}^m \times \mathfrak{E}^n$ generate an Euclidean complex whose simplicial companion augmented is K_aL_a. This provides a good geometric illustration for the join K_aL_a. Similar configurations related to K_aL, KL_a, KL are generated by $\overline{xy} - x$, $\overline{xy} - y$, $\overline{xy} - x - y$.

§2. PRODUCTS OF CHAINS AND CYCLES

5. In this section the restriction is imposed that *all* the factor complexes $X, Y, \cdots$ are star-finite. As a consequence their products will be likewise star-finite (2.5), and so the boundary operator F will have free scope throughout.

The natural definition of chain-products for two complexes X, Y requires that the chains of X, Y be taken over two groups G, H commutatively paired to a third J. Given then two chains

$$\xi^p = g^i x_i^p, \quad g^i \in G; \qquad \eta^q = h^j y_j^q, \quad h^j \in H, \tag{5.1}$$

their product is by definition the $(p + q)$-chain of $X \times Y$ over J

$$\xi^p \times \eta^q = g^i h^j x_i^p \times y_j^q . \tag{5.2}$$

Since G, H are paired to J their isomorphs Gx_i^p, Hy_j^q are paired to $Jx_i^p \times y_j^q \cong J$ under a multiplication such that the product of gx_i^p by hy_j^q is $ghx_i^p \times y_j^q$. Hence by (II, 16.1) the chain-groups $\mathfrak{C}^p(X, G)$ and $\mathfrak{C}^q(Y, H)$ are paired to $\mathfrak{C}^{p+q}(X \times Y, J)$ by the multiplication given loc. cit. If the product of ξ^p by η^q is written $\xi^p \times \eta^q$, this multiplication is precisely (5.2). Therefore

(5.3) *The product of chains defined by* (5.2) *is a multiplication pairing the chain-groups* $\mathfrak{C}^p(X, G)$, $\mathfrak{C}^q(Y, H)$ *to* $\mathfrak{C}^{p+q}(X \times Y, J)$.

From the assigned incidence numbers there follows:

$$\mathrm{F}(x_i^p \times y_j^q) = (\mathrm{F}x_i^p) \times y_j^q + (-1)^p x_i^p \times (\mathrm{F}y_j^q) \tag{5.4}$$

and therefore:

(5.5) *The boundary relation for the product is*:

$$\mathrm{F}(\xi^p \times \eta^q) = (\mathrm{F}\xi^p) \times \eta^q + (-1)^p \xi^p \times (\mathrm{F}\eta^q). \tag{5.6}$$

Denote cycles of X, Y by γ, δ and their classes by Γ, Δ. We have at once from (5.6):

(5.7) $\gamma^p \times \delta^q$ *is a* $(p + q)$*-cycle*.

Thus in other words the multiplication (5.3) pairs $\mathfrak{Z}^p(X, G)$, $\mathfrak{Z}^q(Y, H)$ to $\mathfrak{Z}^{p+q}(X \times Y, J)$.

(5.8) γ^p *or* $\delta^q \sim 0 \rightarrow \gamma^p \times \delta^q \sim 0$.

For if say $\gamma^p = \mathrm{F}\xi^{p+1}$, by (5.6): $\gamma^p \times \delta^q = \mathrm{F}(\xi^{p+1} \times \delta^q)$, and hence $\gamma^p \times \delta^q$ for δ^q fixed, maps $\mathfrak{F}^p(X, G) \rightarrow \mathfrak{F}^{p+q}(X \times Y, J)$. Since the multiplication is continuous it maps also $\bar{\mathfrak{F}}^p(X, G) \rightarrow \bar{\mathfrak{F}}^{p+q}(X \times Y, J)$. Similarly with X, Y interchanged, and this is (5.8).

(5.9) $\gamma^p \in \Gamma^p$, $\delta^q \in \Delta^q \rightarrow \gamma^p \times \delta^q$ *lies in a fixed homology class called the product of* Γ^p, Δ^p, *written* $\Gamma^p \times \Delta^q$, *and this product defines a multiplication pairing* $\mathfrak{H}^p(X, G)$, $\mathfrak{H}^q(Y, H)$ *to* $\mathfrak{H}^{p+q}(X \times Y, J)$ (5.8; II, 15.6a).

(5.10) While we have allowed infinite chains everywhere, everything that precedes continues to hold if the chains are restricted to being finite. The groups are then groups of finite chains throughout.

(5.11) Assuming X, Y disjoint let $X \times Y$ be replaced by $Y \times X$. Under our convention (1.3) this is equivalent to reorienting $X \times Y$ by means of (1.4). Let us agree to designate by $\eta^q \times \xi^p$ the chain corresponding to $\xi^p \times \eta^q$ of $X \times Y$ after commutation of the factors X, Y. Referring to (III, 10.1) we will have:

$$\eta^q \times \xi^p = (-1)^{pq} \xi^p \times \eta^q. \tag{5.12}$$

This may be described as the commutation rule for the chains of the product.

If we have a product $X_1 \times \cdots \times X_n$ then we may commute consecutive *disjoint factors* X_i, X_{i+1} and treat each time the chain-products in accordance with (5.12).

(5.13) *Joins.* Let the notations be those of (4). If $\xi^p = g^i \sigma_i^p$, $\eta^q = h^j \zeta_j^q$, then the $(p + q + 1)$-chain $\xi^p\eta^q = g^i h^j \sigma_i^p \zeta_j^q$ is known as the join of ξ^p and η^q. We readily verify:

$$\text{(5.14)} \qquad \mathrm{F}(\xi^p \eta^q) = (\mathrm{F}\xi^p)\eta^q + (-1)^{p+1}\xi^p(\mathrm{F}\eta^q).$$

The same relation holds relative to $K_a L_a$ provided that we also allow $\xi = \eta = \epsilon$, it being understood that we are to write ξ, η for $\xi\epsilon$, $\eta\epsilon$ and $\epsilon\xi$, $\epsilon\eta$. In particular (5.14) holds also for p or $q = 0$ provided that we replace $\mathrm{F}\xi^0$, $\mathrm{F}\eta^0$ by $\mathrm{KI}(\xi^0)$, $\mathrm{KI}(\eta^0)$.

(5.15) Remark. There is an interesting connection between chain-products and the Kronecker product of (II, 37). To simplify matters suppose $G = H =$ a field, and let the chain-groups be those of the finite chains. Then we readily verify the relation $\mathfrak{C}^r(X \times Y, G) = \sum_{p+q=r} \mathfrak{C}^p(X, G) \otimes \mathfrak{C}^q(Y, G)$. If X, Y are both finite a similar relation holds for the homology groups.

6. **Homology groups of finite products.** Let X, Y be two finite complexes. The homology groups of $X \times Y$ are uniquely determined by the integral groups and so we concentrate our efforts primarily upon these groups. The results given below are essentially due to Künneth [a]. See also [L, V], [A-H, 299].

(6.1) *Notations.* $a_i^p, \cdots, e_i^p$ are the same for X as in (III, 14) for K; $\alpha_i^q, \cdots, \epsilon_i^q$ are the analogues for Y; t_i^p, $\theta_i^q =$ the torsion coefficients of X, Y; $\xi^p =$ a chain of X, $\eta^q =$ a chain of Y; $\zeta^s =$ a chain of $X \times Y$.

Let us fix our attention upon a particular dimension s. A change of bases for the p-chains of X and the q-chains of Y, for all p, q such that $p + q = s$, induces a change of bases for the s-chains of $X \times Y$. Hence $\{a_i^p \times \alpha_j^q, a_i^p \times \beta_j^q, \cdots, e_i^p \times \epsilon_j^q\}$, for all $p + q = s$, is a base for the s-chains of $X \times Y$. We will say that a chain or cycle of $X \times Y$ is a *reduced* chain or cycle if it is a sum of terms containing only factors b, c, d, and β, γ, δ.

The basic boundary relations for X, Y are:

$$\text{(6.2)} \qquad \mathrm{F}e_i^{p+1} = a_i^p, \qquad \mathrm{F}d_i^{p+1} = t_i^p b_i^p; \qquad \mathrm{F}c_i^p = 0;$$

$$\text{(6.3)} \qquad \mathrm{F}\epsilon_j^{q+1} = \alpha_j^q, \qquad \mathrm{F}\delta_j^{q+1} = \theta_j^q \beta_j^q; \qquad \mathrm{F}\gamma_j^p = 0.$$

The chains a, b, c, α, β, γ are cycles. The boundary relations in $X \times Y$ are obtained by means of (5.6) and need not be written down. We find from them immediately that a reduced chain has a reduced boundary. Therefore if $\mathfrak{C}_r^s$, $\mathfrak{Z}_r^s$ denote the groups of the integral reduced chains and cycles then $\mathrm{F}\mathfrak{C}_r^{s+1} = \mathfrak{F}_r^s \subset \mathfrak{Z}_r^s$, and so we may form the reduced integral homology group $\mathfrak{H}_r^s = \mathfrak{Z}_r^s/\mathfrak{F}_r^s$. If $\mathfrak{H}^s(X \times Y)$ is the integral s-dimensional homology group of $X \times Y$ we prove:

$$\text{(6.4)} \qquad \mathfrak{H}^s(X \times Y) \cong \mathfrak{H}_r^s.$$

Let ζ^s be a given integral cycle of $X \times Y$. If η^q is an integral cycle of Y, then $\mathrm{F}(e_i^{p+1} \times \eta^q) = a_i^p \times \eta^q \sim 0$. Hence we may suppress in ζ^s the terms $a \times \alpha$, $a \times \beta$, $a \times \gamma$ and similarly the terms $b \times \alpha$, $c \times \alpha$, without modifying its homology class. Suppose this already done. The terms in ζ^s with a_i^p as a factor make up a chain $a_i^p \times \eta_i^q$. From

$$\mathrm{F}(e_i^{p+1} \times \eta_i^q) = a_i^p \times \eta_i^q + (-1)^{p+1} e_i^{p+1} \times \mathrm{F}\eta_i^q \sim 0,$$

we conclude that we may suppress $a_i^p \times \eta_i^q$ in ζ^s and replace it by terms $e \times \eta$. We will thus have

$$\zeta^s \sim \text{a reduced chain} + \sum e_i^p \times \eta_i^q + \sum \xi_i^p \times \alpha_i^q + \sum \xi_i'^p \times \epsilon_i^q + \sum x_{pq}^{ij} e_i^p \times \epsilon_j^q,$$

where η_i^q contains no ϵ term, and ξ, ξ' contain no a, e terms. We must have $\mathrm{F}\zeta^s = 0$. Since reduced chains have reduced boundaries this yields

$$\begin{aligned}\sum (a_i^{p-1} \times \eta_i^q + (-1)^p e_i^p \times \mathrm{F}\eta_i^q) + \sum (\mathrm{F}\xi_i^p) \times \alpha_i^q \\ + \sum ((\mathrm{F}\xi_i'^p) \times \epsilon_i^q + (-1)^p \xi_i'^p \times \alpha_i^{q-1}) \\ + \sum x_{pq}^{ij} (a_i^{p-1} \times \epsilon_j^q + (-1)^p e_i^p \times \alpha_j^{q-1}) = 0.\end{aligned}$$

Since the terms $e \times \alpha$ occur only in the last sum we have $x_{pq}^{ij} = 0$. Then a^{p-1} terms occur only in the first sum and so $\eta_i^q = 0$. The coefficient of α_i^{q-1} is $(\mathrm{F}\xi_i^{p+1} + (-1)^p \xi_i'^p)$, and as it must be zero we have $\xi_i'^p = (-1)^{p+1}\mathrm{F}\xi_i^{p+1}$. Then $\mathrm{F}\xi_i'^p = 0$, and so $(\mathrm{F}\xi_i'^p) \times \epsilon_i^q = 0$. Thus ultimately:

$$\zeta^s \sim \text{a reduced chain} + \sum (-1)^{p+1}\mathrm{F}(\xi_i^{p+1} \times \epsilon_i^q) \sim \text{a reduced chain}.$$

Thus every $\Gamma^s \in \mathfrak{H}^s(X \times Y)$ contains a reduced cycle.

Consider now any $\Delta^s \in \mathfrak{H}_r^s$. The cycles in Δ^s are in a unique Γ^s and $\Delta^s \to \Gamma^s$ defines a homomorphism τ of $\mathfrak{H}_r^s$ onto $\mathfrak{H}^s$ as just shown. Moreover τ is univalent. For suppose that $\zeta^s \in \Delta^s$ bounds in $X \times Y$, so that $\zeta^s = \mathrm{F}\zeta^{s+1}$. We may write $\zeta^{s+1} = \zeta_1^{s+1} + \zeta_2^{s+1}$, where ζ_1^{s+1} is reduced and each term of ζ_2^{s+1} contains one of the factors a, α, e, ϵ. Thus $\mathrm{F}\zeta_2^{s+1} = \zeta^s - \mathrm{F}\zeta_1^{s+1} =$ a reduced chain. By direct computation we find then that this implies $\mathrm{F}\zeta_2^{s+1} = 0$. Therefore $\zeta^s = \mathrm{F}\zeta_1^{s+1}$, or $\zeta^s \sim 0$ in $X \times Y$, ζ^s reduced $\to \zeta^s \in \mathfrak{F}_r^s$. Therefore τ is univalent. Hence it is an isomorphism, and (6.4) follows.

Several simple but very useful conclusions may already be drawn from (6.4).

(6.5) *The homology groups of a product of finite complexes are uniquely determined by those of the factors.*

It is clearly sufficient to consider $X \times Y$. Suppose that X', Y' have the same homology groups as X, Y and let $b_i'^p, \cdots, \delta_j'^q$ have their obvious meaning. Then $b_i^p \times \beta_j^q \to b_i'^p \times \beta_j'^q, \cdots$ defines an isomorphism of the groups of reduced chains $\mathfrak{C}_r^s$, $\mathfrak{C}_r'^s$ of $X \times Y$, $X' \times Y'$, under which the related groups $\mathfrak{Z}_r^s$, $\mathfrak{Z}_r'^s$ and $\mathfrak{F}_r^s$, $\mathfrak{F}_r'^s$ correspond to one another. Hence the reduced homology groups of $X \times Y$, $X' \times Y'$ are isomorphic, and therefore also by (6.4) the integral homology groups, and finally by (III, 18.2) all the homology groups of the two complexes. This proves (6.5).

(6.6) *A product of finite acyclic complexes is acyclic.*

For if X, Y are acyclic there are no b, c, d, β, γ, δ, hence no reduced cycles and so by (6.4) all the integral homology groups are zero. It follows that the product has no torsion coefficients and that all its Betti numbers vanish. Hence it is acyclic (III, 21.3).

(6.7) *If X, Y have no torsion coefficients the same holds regarding $X \times Y$ and $\{c_i^p \times \gamma_j^q\}$, $p + q = s$, is a base for a group $\cong \mathfrak{H}^s(X \times Y)$.*

For in the absence of torsion coefficients there are no b, β, d, δ, and hence all the reduced chains are cycles and none is bounding.

(6.8) *Betti numbers.* Consider the groups mod π, a prime, or the rational group ($\pi = 0$). The basic system for the chains and their boundary relations in X is still (III, 14.2). However if π does not divide t_i^p, writing now d_i^{p+1} for $s_i^p d_i^{p+1}$, $s_i^p t_i^p = 1 \bmod \pi$, the second relation of the system becomes $\mathrm{F} d_i^{p+1} = b_i^p$, and so of the same form as the first, while if π divides t_i^p, d_i^{p+1} is merged with the c_i^{p+1}. Hence we may suppress the b, d. The system (III, 14.2) retains its form but with chains b, d absent. The argument may then proceed as before as if there were no such terms, i.e., as if there were no torsion coefficients. The $\{c_i^p\}$, $\{\gamma_j^q\}$ are now merely bases for groups isomorphic with the groups $\mathfrak{H}^p(X, \pi)$, $\mathfrak{H}^q(Y, \pi)$. We still obtain the analogue of (6.7) and so $\{c_i^p \times \gamma_j^q\}$, $p + q = s$, is a base for a group isomorphic with $\mathfrak{H}^s(X \times Y, \pi)$. Since we are dealing with vector spaces over $\mathfrak{J}_\pi$ the dimensions are the corresponding Betti numbers and also the numbers of elements in the bases. Hence

$$R^s(X \times Y, \pi) = \sum_{p+q=s} R^p(X, \pi) R^q(Y, \pi). \tag{6.9}$$

In particular for the ordinary (rational) Betti numbers:

$$R^s(X \times Y) = \sum_{p+q=s} R^p(X) R^q(Y). \tag{6.10}$$

This relation is equivalent to the following noteworthy relation between the Poincaré polynomials (III, 15.3):

$$P(t; X \times Y) = P(t; X) P(t; Y). \tag{6.11}$$

We may also introduce in the obvious way the Poincaré polynomial mod π (i.e., whose coefficients are the Betti numbers mod π), say $P_\pi(t; X)$, and then (6.9) is equivalent to

$$P_\pi(t; X \times Y) = P_\pi(t; X) P_\pi(t; Y). \tag{6.12}$$

(6.13) *If X is p-cyclic and Y is q-cyclic then $X \times Y$ is $(p + q)$-cyclic* (6.7, 6.10; III, 21.3).

(6.13a) *If ξ^p, η^q are cycles of X, Y* mod π *then $\xi^p \times \eta^q \sim 0$, $\xi^p \nsim 0 \rightarrow \eta^q \sim 0$, and similarly with ξ, η interchanged.*

In the notations of (6.8) $\xi^p \sim g^i c_i^p$, $\eta^q \sim h^j \gamma_j^q$, $\xi^p \times \eta^q \sim g^i h^j c_i^p \times \gamma_j^q \sim 0$. By the argument in (6.8) the last homology implies that every $g^i h^j = 0$, and since not every $g^i = 0$, we must have $h^j = 0$ for every j, and hence $\eta^q \backsim 0$.

(6.14) *Application.* Let l be a segment considered as a polyhedron. Its homology groups are those of the closed one-simplex and so l is zero-cyclic. Hence the Euclidean parallelotope P^n (product of n segments) is also zero-cyclic. Since P^n has a single n-element, when $n > 1$, the boundary sphere $\mathfrak{B}l^n$ is $(0, n - 1)$-cyclic (III, 21.4).

The preceding results lead to a rapid proof of:

(6.15) *The product of a finite number of simple complexes is simple.*

It is sufficient to consider a product of two simple factors X, Y.

(a) $\dim x \times y \geqq 0$.

(b) Let ξ_0, η_0 be the fundamental zero-cocycles of X, Y. We have

$$\xi_0 = \sum x_0^i, \qquad \eta_0 = \sum y_0^i.$$

Since ξ_0, η_0 are zero-cycles of X^*, Y^*, $\xi_0 \times \eta_0$ is a zero-cycle of $X^* \times Y^*$, hence a zero-cycle of $(X \times Y)^*$ (2.8) and it is

$$\xi_0 \times \eta_0 = \sum x_0^i \times y_0^j$$

which is the sum of the duals of the vertices of $X \times Y$. Therefore $X \times Y$ is augmentable and with a fundamental zero-cocycle equal to the sum of the duals of its vertices.

(c) $\mathrm{Cl}(x \times y)$ is finite and zero-cyclic. This follows from (2.2) and (6.13). Since $X \times Y$ is augmentable, $(\mathrm{Cl}(x \times y))_a$ is acyclic (III, 47.5).

Referring now to (III, 47.1), property (6.15) is a consequence of (a, b, c).

(6.16) *Betti and torsion groups.* In view of (6.4) to determine these groups we only need to consider reduced chains. Among their generators all those not containing a factor d or δ are cycles. Moreover a chain ζ containing terms $d \times \delta$ is readily seen not to be a cycle. Suppose

$$\zeta^s = ub_i^p \times \delta_j^q + vd_i^{p+1} \times \beta_j^{q-1} + \cdots.$$

From the relation $\mathrm{F}\zeta^s = 0$ we find $u\theta_j^{q-1} = (-1)^{p+1}vt_i^p$. Let generally $T_{ij}^{pq} = \text{h. c. f. } (t_i^p, \theta_j^q)$. Then

$$u = \frac{\lambda}{T_{ij}^{p,q-1}} t_i^p, \qquad v = (-1)^{p+1} \frac{\lambda}{T_{ij}^{p,q-1}} \theta_j^{q-1},$$

and therefore

$$\zeta^s = \frac{\lambda}{T_{ij}^{p,q-1}} \mathrm{F}(d_i^{p+1} \times \delta_j^q) + \cdots.$$

Thus the group of reduced cycles contains also the generator $\zeta_{ij}^s(p, q) = \mathrm{F}(d_i^{p+1} \times \delta_j^q)/T_{ij}^{p,q-1}$. Hence a full set of generators for $\mathfrak{Z}_r^s$ consists of the cycles of the type just written together with the products $b_i^p \times \beta_j^q$, $b_i^p \times \gamma_j^q$, $c_i^p \times \beta_j^q$, $c_i^p \times \gamma_j^q$. The basic boundary relations are:

$$\mathrm{F}(d_i^{p+1} \times \delta_j^q) = T_{ij}^{p,q-1}\zeta_{ij}^s(p, q),$$

$$\mathrm{F}(d_i^{p+1} \times \beta_j^q) = t_i^p b_i^p \times \beta_j^q; \qquad \mathrm{F}((-1)^p b_i^p \times \delta_j^{q+1}) = \theta_j^q b_i^p \times \beta_j^q;$$

$$\mathrm{F}(d_i^{p+1} \times \gamma_j^q) = t_i^p b_i^p \times \gamma_j^q; \qquad \mathrm{F}((-1)^p c_i^p \times \delta_j^{q+1}) = \theta_j^q c_i^p \times \beta_j^q.$$

We may express T_{ij}^{pq} in terms of t_i^p, θ_j^q as $T_{ij}^{pq} = t_i^p t_i'^p + (-1)^p \theta_j^q \theta_j'^q$, and this enables us to replace the second and third basic boundary relations by the unique relation

$$F(t_i'^p d_i^{p+1} \times \beta_j^q + \theta_j'^q b_i^p \times \delta_j^q) = T_{ij}^{pq} b_i^p \times \beta_j^q .$$

Since the $c \times \gamma$ are the only products which occur in no boundary, the Betti group $\mathfrak{B}^s(X \times Y)$ is isomorphic with the free group on $\{c_i^p \times \gamma_j^q\}$, $p + q = s$.

From the basic boundary relations we also verify that the torsion group is isomorphic with a product of cyclic groups of orders $T_{ij}^{p,q-1}$, T_{ij}^{pq}, t_i^p, θ_j^q, $p + q = s$. Thus the torsion coefficients are the invariant factors greater than 1 of a diagonal matrix with all these terms in the diagonal.

(6.17) *Application to joins.* Let K, L be as in (4) and finite. Let also $a, \cdots, e$ and $\alpha, \cdots, \epsilon$ be the elements of canonical bases for K and L. Then from (4.2) and the appropriate results for the product we deduce:

(6.18) $\{c_i^p \gamma_j^q\}$, $p + q = s - 1$, *is a base for the Betti group* $\mathfrak{B}^s(KL)$.

$$R^s(KL, \pi) = \sum_{p+q=s-1} R^p(K, \pi) R^q(K, \pi). \tag{6.19}$$

(6.20) *When K, L have no torsion coefficients this is also the case for KL.*

(6.21) *The same results hold for the joins $K_a L$, KL_a, $K_a L_a$, provided that the basic elements c_i^0, γ_j^0 are properly chosen.*

As an application of the preceding results or else directly one may also prove the useful property:

(6.22) *Let K be a finite simplicial complex and A a point. Then the pth homology groups of AK [of AK_a] are the same as the $(p - 1)$st of K [of K_a].*

§3. SET-TRANSFORMATIONS

7. The dual nature, algebraic and structural, of complexes, reflects itself in their transformations. We shall have to consider separately those of X as an ordered space, then the homomorphisms on the chain-groups, and ultimately combine the two types. The following treatment of set-transformations is based on Tucker [b].

(7.1) Definition. *Given two complexes X, Y, we shall understand by a set-transformation $t: X \to Y$ a transformation of the set X into the set Y (for each x, tx is a set of elements of Y). We call t:*

closed when $t \text{ Cl} = \text{Cl } t$;

weakly closed when $\text{Cl } t \text{ Cl} = \text{Cl } t$ (Cl *commutes with* Cl t);

open when $t \text{ St} = \text{St } t$;

weakly open when $\text{St } t \text{ St} = \text{St } t$;

simple when Y is simple and every $(\text{Cl } tx)_a$ *is acyclic.*

Convenient formal relations to keep in mind in connection with these definitions are:

$$\text{Cl Cl} = \text{Cl}, \qquad \text{St St} = \text{St}. \tag{7.2}$$

We also notice the following properties of set-transformations:

(7.3) *When t is closed [open] it is weakly closed [open].*

(7.4) *When t is weakly closed, $x' < x \rightarrow \text{Cl}\, tx' \subset \text{Cl}\, tx$. Similarly for the obvious dual situation.*

(7.5) *The product $t't$ of two closed, weakly closed, open, weakly open set-transformations $t:X \rightarrow Y$, $t':Y \rightarrow Z$, is a set-transformation $X \rightarrow Z$ of the same type as the factors.*

(7.6) *The identity $X \rightarrow X$ is both open and closed.*

Properties (7.3, 7.6) are obvious. Regarding (7.4): $x' < x \rightarrow x' \in \text{Cl}\, x \rightarrow tx' \subset t\, \text{Cl}\, x \rightarrow \text{Cl}\, tx' \subset \text{Cl}\, t\, \text{Cl}\, x = \text{Cl}\, tx$, which disposes of the weakly closed case. The dualization is obvious. Passing to (7.5) it is only necessary to consider the weakly closed case. If we apply $\text{Cl}\, t'$ to the two sides of the relation $\text{Cl}\, t\, \text{Cl} = \text{Cl}\, t$ and if we remember $\text{Cl}\, t'\, \text{Cl} = \text{Cl}\, t'$ we have in view of (7.1): $\text{Cl}\, t'\, \text{Cl}\, t\, \text{Cl} = \text{Cl}\, t't\, \text{Cl} = \text{Cl}\, t'\, \text{Cl}\, t = \text{Cl}\, (t't)$ which disposes of the case in question, and similarly for the rest.

Examples (7.7). Take a simplicial complex $K = \{\sigma\}$ and the join with a point A, $A_a K_a = L$. Then $\sigma \rightarrow A\sigma$ defines a set-transformation $t:K \rightarrow L$. Now $t\, \text{Cl}\, \sigma = A\, \text{Cl}\, \sigma$, while $\text{Cl}\, t\sigma = \text{Cl}\, A\sigma = A\, \text{Cl}\, \sigma \cup \text{Cl}\, \sigma \cup A$, and so $t\, \text{Cl} \neq \text{Cl}\, t$. On the other hand $\text{Cl}\, t\, \text{Cl}\, \sigma = \text{Cl}\, (A\, \text{Cl}\, \sigma) = A\, \text{Cl}\, \sigma \cup \text{Cl}\, \sigma \cup A = \text{Cl}\, t\sigma$. Thus t is weakly closed but not closed.

(7.8) Let K consist of the closed simplex $A_0 \cdots A_p$ and let t be the set-transformation $K \rightarrow K$ whereby $A_i A_j \cdots A_k$ $(i < j < \cdots < k)$ is transformed into $A_{i+n} A_{j+n} \cdots A_{k+n}$, where the indices are understood mod $(p+1)$. Then $t\, \text{Cl} = \text{Cl}\, t$ is immediately verified and so t is closed.

(7.9) *Simplicial set-transformations.* Let $K = \{\sigma\}$, $L = \{\zeta\}$ be two simplicial complexes with respective vertices $\{A_i\}$, $\{B_j\}$. A simplicial set-transformation $t:K \rightarrow L$ is one which sends every vertex A_i of K into a vertex B_j of L and sends $\sigma = A_i \cdots A_j \in K$ into $\zeta = t\sigma = (tA_i) \cdots (tA_j) \in L$. We notice that:

(a) *t is single-valued;*

(b) *t is closed;*

(c) *the two preceding properties are sufficient to characterize t.*

Property (a) is obvious. Regarding (b), clearly $\sigma' < \sigma \rightarrow t\sigma' < t\sigma$. Moreover if $\zeta' = B_h \cdots B_k < t\sigma$ there are indices $h', \cdots, k'$ such that $tA_{h'} = B_h, \cdots, tA_{k'} = B_k$, where $A_{h'}, \cdots, A_{k'}$ are vertices of σ. Thus $\sigma' = A_{h'} \cdots A_{k'} < \sigma$ and $t\sigma' = \zeta'$. Hence $t\, \text{Cl} = \text{Cl}\, t$, which is (b). Suppose now that t has properties (a), (b). Since $\text{Cl}\, t\, A_i = t\, \text{Cl}\, A_i = tA_i$, $\text{Cl}\, tA_i$ is a simplex and so it is a point. Suppose that we have shown that every $t\sigma^q$, $q < p$, is the simplex of L whose vertices are the transforms of those of σ^q. Consider now $\sigma^p = A_0 \cdots A_p$. By hypothesis $t\mathfrak{B}\sigma^p$ has no other vertices than $tA_0, \cdots, tA_p$. Suppose $\zeta = t\sigma^p$ has a vertex B not among these. Since $\text{Cl}\, t\sigma^p = \text{Cl}\, \zeta = t\, \text{Cl}\, \sigma^p$, B must be in some $t\sigma'$, $\sigma' < \sigma^p$, $\sigma' \neq \sigma^p$, contrary to assumption. Therefore $\zeta = (tA_0) \cdots (tA_p)$, t is simplicial and (c) is proved.

(7.10) An interesting generalization is the following. The notations remaining the same, t is closed and such that every $t\sigma$ is a $\text{Cl}\, \zeta$. Thus t is a single-valued transformation $\{\sigma\} \rightarrow \{\text{Cl}\, \zeta\}$. We will call t a *generalized simplicial set-transformation.*

8. **Inverse and dual set-transformations.**

(8.1) The definition of the inverse is as usual: t^{-1} is a set-mapping $Y \to X$ such that $x \in t^{-1}y$ when and only when $y \in tx$. The dual t^* of t will be defined as a set-transformation $Y^* \to X^*$ such that $x^* \in t^*y^* \leftrightarrow y \in tx$. It is convenient to introduce also an auxiliary set-transformation $t_*: X^* \to Y^*$ defined by: $y^* \in t_*x^* \leftrightarrow y \in tx$. However it is definitely t^* that we require later. We note the following properties:

(8.2) $$t^{**} = t, \qquad t_{**} = t, \qquad t^* = (t^{-1})_* ;$$

(8.3) *if t is closed [open] then t^{-1} and t_* are open [closed];*

(8.4) *if one of t, t^* is closed so is the other.*

Property (8.2) is obvious and (8.4) is a consequence of (8.3) and the third of (8.2) so that (8.3) alone requires proof. The assertion as to t, t_* is again obvious so that only the one concerning t, t^{-1} requires proof. Suppose t closed. Given then $x \in t^{-1}\,\mathrm{St}\, y$, there is a $y' > y$ such that $x \in t^{-1}y'$ and so $y' \in tx$ and $y \in \mathrm{Cl}\, y' \subset \mathrm{Cl}\, tx = t\, \mathrm{Cl}\, x$. Hence there is an $x' < x$ such that $y \in tx'$ or $x' \in t^{-1}y$. Therefore $x \in \mathrm{St}\, t^{-1}y$, and hence $t^{-1}\,\mathrm{St}\, y \subset \mathrm{St}\, t^{-1}y$. Conversely, let $x \in \mathrm{St}\, t^{-1}y$. There is an $x' < x$ such that $x' \in t^{-1}y$. Since $x' \in \mathrm{Cl}\, x$ we have $\mathrm{Cl}\, tx = t\, \mathrm{Cl}\, x \supset tx' \ni y$. Hence there is a $y' \in tx$ and $> y$. Hence $y' \in \mathrm{St}\, y$ and $x \in t^{-1}y' \subset t^{-1}\,\mathrm{St}\, y$. Therefore $\mathrm{St}\, t^{-1}y \subset t^{-1}\,\mathrm{St}\, y$. Thus $t^{-1}\,\mathrm{St} = \mathrm{St}\, t^{-1}$ or t^{-1} is open. By dualizing the proof we find that t open implies t^{-1} closed; so (8.3) is proved.

§4. CHAIN-MAPPINGS

9. Let X, Y be finite complexes and let $\| a_i^j(p) \|$, $p = 0, \pm 1, \cdots$ be matrices of integers. Referring to (II, 8.4) the relations

(9.1) $$\tau x_i^p = a_i^j(p) y_j^p$$

define a system of homomorphisms of the integral chain-groups $\tau: \mathfrak{C}^p(X) \to \mathfrak{C}^p(Y)$. The operation τ is known as a *chain-transformation*. When in addition τ commutes with F:

(9.2) $$\tau \mathrm{F} = \mathrm{F} \tau,$$

(for integral chains) then τ is called a *chain-mapping* $X \to Y$.

(9.3) If G is any coefficient-group whatever, a chain-mapping τ induces a homomorphism $\mathfrak{C}^p(X, G) \to \mathfrak{C}^p(Y, G)$ defined by the relations

(9.4) $$\tau(g^i x_i^p) = a_i^j(p) g^i y_j^p , \qquad g^i \in G,$$

which still commutes with F. Furthermore owing to (9.2) τ maps $\mathfrak{Z}^p(X, G) \to \mathfrak{Z}^p(Y, G)$, $\mathfrak{F}^p(X, G) \to \mathfrak{F}^p(Y, G)$, and hence (II, 5.4) τ also induces homomorphisms: $\mathfrak{H}^p(X, G) \to \mathfrak{H}^p(Y, G)$. Thus:

(9.5) THEOREM. *A chain-mapping $\tau: X \to Y$ induces homomorphisms of the groups $\mathfrak{C}^p, \cdots, \mathfrak{H}^p$ of X over any G into the corresponding groups of Y.*

Let the boundary relations in X, Y be

(9.6) $$\mathrm{F} x_i^p = \alpha_i^j(p) x_j^{p-1}, \qquad \mathrm{F} y_i^p = \beta_i^j(p) y_j^{p-1}.$$

Notice that for X^*, Y^* they are

$$\text{(9.7)} \qquad \mathrm{F}x^i_{p-1} = \alpha^i_j(p)x^j_p\,, \qquad \mathrm{F}y^i_{p-1} = \beta^i_j(p)y^j_p\,.$$

A simple calculation yields

$$\text{(9.8)} \qquad (\tau\mathrm{F} - \mathrm{F}\tau)g^i x^p_i = \{\alpha^j_i(p)a^k_j(p-1) - a^j_i(p)\beta^k_j(p)\}g^i y^{p-1}_k.$$

A n. a. s. c. for (9.2) to hold is that this last expression vanish identically. Therefore (9.2) and

$$\text{(9.9)} \qquad \alpha^j_i(p)a^k_j(p-1) - a^j_i(p)\beta^k_j(p) = 0$$

are equivalent.

(9.10) *Simple chain-mapping.* Let X, Y be simple. The chain-mapping $\tau{:}X \to Y$ is said to be *simple* whenever it preserves the Kronecker indices of integral zero-chains: $\mathrm{KI}(\tau\xi^0) = \mathrm{KI}(\xi^0)$. It implies that if X, Y are augmented to X_a, Y_a with (-1)-elements ϵ, η then τ may be extended to $\tau_a{:}X_a \to Y_a$ such that $\tau_a\epsilon = \eta$, where $\tau_a = \tau$ on X, $\tau_a\mathrm{F}\xi^0 = \mathrm{F}\tau_a\xi^0$. Explicit n. a. s. c. for τ to be simple are

$$\mathrm{KI}(x^0_i) = \mathrm{KI}(a^j_i(0)y^0_j),$$

which are equivalent to:

$$\text{(9.11)} \qquad \sum_j a^j_i(0) = 1$$

for every i.

(9.12) *Simplicial chain-mapping.* Let $X = \{\sigma\}$, $Y = \{\zeta\}$ be simplicial with vertices A_i, B_i, and let t be a simplicial set-transformation $X \to Y$. Set $tA_i = B_i$ where the B's need not be distinct. Define a chain-mapping $\tau{:}X \to Y$ by

$$\text{(9.13)} \qquad \tau\sigma = \tau A_{i_0} \cdots A_{i_p} = \begin{cases} B_{i_0} \cdots B_{i_p} \text{ when the } B_{i_n} \text{ are distinct,} \\ 0 \text{ otherwise.} \end{cases}$$

To prove commutation with F it is sufficient to prove $\tau\mathrm{F}\sigma = \mathrm{F}\tau\sigma$. Assume the labels so chosen that $A_{i_h} = A_h$. Then if the B_i are distinct:

$$\begin{aligned} \tau\mathrm{F}\sigma &= \tau \sum (-1)^i A_0 \cdots A_{i-1}A_{i+1} \cdots A_p \\ &= \sum (-1)^i B_0 \cdots B_{i-1}B_{i+1} \cdots B_p = \mathrm{F}\tau\sigma. \end{aligned}$$

If the B's are not distinct the labels may be so chosen that $B_0 = B_1$, and then the first two terms in the second sum cancel and the rest vanish. Hence $\tau\mathrm{F}\sigma = \mathrm{F}\tau\sigma$ in any case, so τ is a chain-mapping. Since $\mathrm{KI}(A_h) = \mathrm{KI}(B_h) = 1$, τ is also simple. A chain-mapping such as τ is said to be *simplicial.*

(9.14) It may be noticed that while t determines τ uniquely the converse is also true. For this reason we will often write $\tau\sigma$ for $t\sigma$ and this will cause no ambiguity.

(9.15) *Product of chain-mappings: If τ is a chain-mapping $X \to Y$ and τ' is a chain-mapping $Y \to Z$ then $\tau'\tau$ is a chain-mapping $X \to Z$.*

(9.16) *Chain-mappings and reorientation.* The effect of applying to X an orientation function $\alpha(x)$ is the same as subjecting X to the chain-mapping $\alpha : x \to \alpha(x)x$. Similarly to reorient Y by means of $\alpha_1(y)$ is the same as subjecting Y to the chain-mapping $\alpha_1 : y \to \alpha_1(y)y$. Therefore if X, Y both undergo the preceding reorientations the equations of τ given by (9.1) become instead

$$\tau x_i^p = \alpha(x_i^p)a_i^j(p)\alpha_1(y_j^p)y_j^p \, . \tag{9.17}$$

(9.18) *Notation.* If X' is a closed subcomplex of X and τ a chain-mapping $X \to Y$, the values of τ on the chains of X' define a chain-mapping $\tau' : X' \to Y$ which we will conveniently denote by $\tau \mid X'$. This is a slight but very natural deviation from the $\mid$ notation of (I, 2).

(9.19) Definition. *Let $\{G_\lambda\}$, $\{H_\lambda\}$ be two systems of groups indexed by the same set $\Lambda = \{\lambda\}$ and let τ be an operation upon the groups of the first system such that τ is a homomorphism, an open homomorphism, $\cdots : G_\lambda \to H_\lambda$. We will call τ a simultaneous homomorphism, $\cdots$, of $\{G_\lambda\}$ into [onto or with, if need be] $\{H_\lambda\}$, written symbolically as usual $\tau : G_\lambda \to H_\lambda$. Frequently when the meaning is otherwise clear we shall drop "simultaneous" and still call τ a homomorphism, $\cdots$.*

Examples: A chain-mapping $\tau : X \to Y$ is a simultaneous homomorphism of the groups $\mathfrak{C}^p(X, G) \to \mathfrak{C}^p(Y, G)$. Here $\Lambda = \{(p, G)\}$. Other simultaneous homomorphisms are: the projections and injections corresponding to a dissection of a complex, the set of homomorphisms of (9.5). Very frequently the indexing system will be of the form $\{(p, G)\}$ but other types will also occur.

10. Side by side with τ we may consider the *dual* mapping $\tau^* : Y^* \to X^*$ defined by

$$\tau^* y_p^j = a_i^j(p)x_p^i \, . \tag{10.1}$$

If we compute $\tau^*\mathrm{F} - \mathrm{F}\tau^*$ we find again that its vanishing is equivalent to (9.9). Therefore

(10.2) *If one of τ, τ^* is a chain-mapping so is the other.*

We also have, with $\beta(p)$ as in (III, Introduction),

$$\mathrm{KI}(\tau x_i^p \, , y_p^k) = \beta(p)a_i^k(p) = \mathrm{KI}(x_i^p \, , \tau^* y_p^k), \tag{10.3}$$

and hence, since the index is distributive:

(10.4) *Invariance of the Kronecker index. If G, H are commutatively paired to J, and ξ^p, η_p are, respectively, over G, H then:*

$$\mathrm{KI}(\tau\xi^p, \eta_p) = \mathrm{KI}(\xi^p, \tau^*\eta_p). \tag{10.5}$$

Conversely, let it be known that τ, τ^* are so related that (10.5) is satisfied, where τ is as before and τ^* is given by

$$\tau^* y_p^i = a_j'^i(p)x_p^j \, .$$

From (10.5) follows for $\xi^p = x_i^p$, $\eta_p = y_p^j : a_j'^i(p) = a_j^i(p)$ and so τ^* is the dual of τ. Therefore:

(10.6) *A n. a. s. c. for* τ, τ^* *to be dual is that* (10.5) *hold for integral* ξ^p, η_p.

Suppose now that X, Y, τ are simple, and let ξ_0, η_0 be the fundamental zero-cocycles. By (10.4; III, 47.1) and since τ is simple we have:

$$\text{(10.7)} \qquad \mathrm{KI}(\tau\xi^0, \eta_0) = \mathrm{KI}(\xi^0, \xi_0) = \mathrm{KI}(\xi^0, \tau^*\eta_0).$$

Select $\xi^0 = x_i^0$ and let $\xi_0 - \tau^*\eta_0 = \lambda_j \cdot x_0^j$. From (10.7) follows

$$\mathrm{KI}(x_i^0, \lambda_j x_0^j) = \lambda_i = 0$$

or $\xi_0 = \tau^*\eta_0$. Conversely, if $\xi_0 = \tau^*\eta_0$, from (10.5) follows the first equality in (10.7), and hence τ is simple. Therefore:

(10.8) *When X, Y are simple a n. a. s. c. for* τ *to be simple is that* τ^* *map the fundamental zero-cocycle of Y into the same for X.*

Since τ^* is a chain-mapping, we have from (9.5):

(10.9) THEOREM. *The dual* τ^* *of* τ *induces homomorphisms of the groups* $\mathfrak{C}_p, \cdots, \mathfrak{H}_p$ *of Y into the corresponding groups for X.*

(10.10) *If* τ *induces isomorphisms of the homology groups of X with the corresponding groups of Y, then* τ^* *induces isomorphisms of the cohomology groups of Y with the corresponding groups of X.*

The proof rests upon

(10.11) *If* τ *induces isomorphisms of the integral homology groups, then it induces isomorphisms of those over every G.*

Let there be given sets $\{b_i^p, c_i^p, d_i^p\}$ of chains of X such that: (a) they satisfy the same relations $\mathrm{F}d_i^{p+1} = t_i^p b_i^p$, $\mathrm{F}c_i^p = 0$ as the elements b, c, d of canonical bases (III, 14.2); (b) every integral p-cycle $\sim$ a combination of the b_i^p, c_i^p. Upon examining the proof of (III, 14.1) it is found that the sets may be completed by suitable chains a_i^p, e_i^p to canonical bases $\{a_i^p, b_i^p, \cdots, e_i^p\}$.

Let now $\tau b_i^p = \beta_i^p$, $\tau c_i^p = \gamma_i^p$, $\tau d_i^p = \delta_i^p$. Under our assumptions X, Y have the same Betti numbers and torsion coefficients, and furthermore (b) will hold for Y and the β_i^p, γ_i^p. Since τ commutes with F we have as a consequence of (a): $\mathrm{F}\delta_i^{p+1} = t_i^p\beta_i^p$, $\mathrm{F}\gamma_i^p = 0$, which is (a) for Y and the β, γ, δ. Therefore the β, γ, δ may be completed to canonical bases $\{\alpha_i^p, \beta_i^p, \cdots, \epsilon_i^p\}$ for Y.

Referring now to the explicit expression (III, 16.9) for the $\mathfrak{H}^p(X, G)$, and to the same for Y, property (10.11) becomes obvious.

PROOF OF (10.10). In view of the result just obtained, clearly we only need to prove the asserted property for the integral groups. By (III, 30.2) $\mathfrak{H}^p(X, \mathfrak{P})$ and $\mathfrak{H}_p(X, \mathfrak{J})$, likewise $\mathfrak{H}^p(Y, \mathfrak{P})$ and $\mathfrak{H}_p(Y, \mathfrak{J})$ are dually paired with the Kronecker index as the multiplication. Let Γ^p, Γ_p denote the elements of the above homology and cohomology groups of X, and Δ^p, Δ_p the same for Y. By (10.5) and (III, 29.8) we have:

$$\mathrm{KI}(\tau\Gamma^p, \Delta_p) = \mathrm{KI}(\Gamma^p, \tau^*\Delta_p).$$

This relation states in substance that the character $\tau^*\Delta_p$ of $\mathfrak{H}^p(X, \mathfrak{P})$ takes the same value at Γ^p as the character Δ_p of $\mathfrak{H}^p(Y, \mathfrak{P})$ at $\tau\Gamma^p$. Since τ is an isomorphism the relationship between the characters implies that τ^* is a univalent homomorphism onto, and so that it is an isomorphism.

(10.12) *If* $\tau : X \to Y$, $\tau' : Y \to Z$ *are chain-mappings then the chain-mapping* $\tau'\tau : X \to Z$ *has for dual* $(\tau'\tau)^* = \tau^*\tau'^*$.

This is an immediate consequence of the definition of the duals by (10.1).

(10.13) *Application to dissections.* Let (X_0, X_1) be a dissection of X, π the projection $X \to X_0$, η the injection $X_1 \to X$ (III, 23). Then (X_1^*, X_0^*) is a dissection of X^* and there are a related projection

$$\eta^* : X^* \to X_1^*$$

and an injection

$$\pi^* : X_0^* \to X^*.$$

(10.14) (π, π^*) *and* (η, η^*) *are two pairs of dual chain-mappings.*

The proof that they are chain-mappings has already been given in (III, 23). The equations of π, π^* are

$$\pi x_{0i}^p = x_{0i}^p, \qquad \pi x_{1i}^p = 0,$$
$$\pi^* x_p^{0i} = x_p^{0i},$$

which proves that they are dual. Similarly, of course, for η, η^*.

11. The consideration of the graphs is particularly significant for chain-transformations. Let us associate with τ as *chain-graph*, or merely *graph*, the zero-chain of $X^* \times Y$ given by

$$(11.1) \qquad \begin{aligned} \Gamma &= \sum \beta(p) x_p^i \times \tau x_i^p = \sum_p (-1)^{\frac{p(p-1)}{2}} a_i^j(p) x_p^i \times y_j^p \\ &= \sum_p \beta(p) a_i^j(p) x_p^i \times y_j^p . \end{aligned}$$

Except for the (-1) factor this chain is strongly suggested by the symbolism, and the choice of the supplementary factor will be justified in a moment. Furthermore Γ *is also the graph of* τ^*. We have in fact

$$\begin{aligned} \Gamma &= \sum \beta(-p) y_j^p \times \tau^* y_p^j = \sum_p \beta(p)(-1)^p a_i^j(p) y_j^p \times x_p^i \\ &= \sum_p \beta(-p) a_i^j(p) y_j^p \times x_p^i , \end{aligned}$$

which shows that it bears the same relation to both τ, τ^*. In other words, the image of Γ in $Y \times X^*$ is the graph of τ^*. We may thus consider Γ as the graph of both τ, τ^*.

The boundary of Γ is:

$$\begin{aligned}\mathrm{F}\Gamma &= \sum_p \beta(p)a_i^j(p)\{(\mathrm{F}x_p^i) \times y_j^p + (-1)^p x_p^i \times \mathrm{F}y_j^p\}\\ &= \sum_p \beta(p)a_i^j(p)\{\alpha_h^i(p+1)x_{p+1}^h \times y_j^p + (-1)^p \beta_j^k(p)x_p^i \times y_k^{p-1}\}\\ &= \sum_p \beta(-p)\{a_i^j(p)\beta_j^k(p) - \alpha_i^j(p)a_j^k(p-1)\}x_p^i \times y_k^{p-1}.\end{aligned}$$

Therefore a n. a. s. c. for Γ to be a cycle is that (9.9) hold, or which is the same, that (9.2) hold. Since Γ and (9.1) determine one another uniquely we have:

(11.2) *The graph Γ of a chain-mapping $\tau:X \to Y$ is a zero-cycle of $X^* \times Y$. Conversely, every cycle such as Γ determines τ uniquely. Moreover Γ is also the graph of the dual τ^** (Tucker [a]).

Let unimodular transformations yield new bases:

$$x_i^p = \lambda_i^{pj}\bar{x}_j^p\,; \qquad y_i^p = \mu_i^{pj}\bar{y}_j^p\,. \tag{11.3}$$

Referring to (III, 33), the dual transformations are of the same form but with the matrices λ^p, μ^p replaced by their inverses λ_p, μ_p (notations loc. cit.). If we set

$$\bar{a}_i^j(p) = \lambda_{pi}^h a_h^k(p)\mu_k^{pj}, \tag{11.4}$$

we find that τ, τ^*, Γ have the same expressions as before with $\bar{x}$, $\bar{y}$, $\bar{a}$ in place of x, y, a. Therefore

(11.5) *If the x_i^p, y_i^p undergo unimodular transformations to the $\bar{x}_j^p$, $\bar{y}_j^p$, the formal relations between τ, τ^*, Γ remain the same.*

(11.6) *Application.* Suppose that one of X, Y is *torsion-free* (without torsion coefficients) or for that matter that X [Y] has no torsion coefficients t_i^p for the dimensions p of the elements of X [Y]. If we apply the argument of (6.16) to $Y \times X^*$ and Γ we have then

$$\Gamma \sim \sum \Gamma_p(\tau), \qquad \Gamma_p(\tau) = B_i^p \times \beta_p^i + C_i^p \times \gamma_p^i\,,$$

where B_i^p, C_i^p are linear integral combinations of the b_i^p, c_i^p. These cycles may be identified with elements of the integral homology group $\mathfrak{H}^p(X)$, and hence $\Gamma_p(\tau)$ with a cocycle of Y over $\mathfrak{H}^p(X)$. Thus there is associated with τ a set $\{\Gamma_p(\tau)\}$, where $\Gamma_p(\tau)$ is a cocycle of the type just mentioned.

(11.7) EXAMPLE. X is simplicial, $Y = S^n = \mathfrak{B}\sigma^{n+1}$, $n > 0$, and τ is simplicial. If γ_0, γ_n are the fundamental cocycles of S^n, then $\Gamma_0(\tau) = B^0 \times \gamma_0$, $\Gamma_n(\tau) = B^n \times \gamma_n$, where B^0, B^n are homology classes of X.

12. Complements.

(12.1) If τ_1, τ_2 are chain-mappings $X \to Y$ we will denote by $\tau_1 \pm \tau_2$ the chain-mapping $X \to Y$ defined by $x \to \tau_1 x \pm \tau_2 x$, and by 0 the chain-mapping defined by $x \to 0$.

We will write $\tau_1 \sim \tau_2$ over G if $\tau_1\gamma^p \sim \tau_2\gamma^p$ for every cycle γ^p over G. We say: *τ_1 is homologous to τ_2 over G*, also for $G = \mathfrak{J}$, $\mathfrak{P}$ or $\mathfrak{J}_m$: *τ_1 is integrally ho-*

mologous to τ_2, *or* τ_1 *is homologous to* τ_2 mod 1, mod m ($\tau_1 \sim \tau_2$ mod 1, mod m). If $\tau_1 \sim \tau_2$ over every G then we write $\tau_1 \sim \tau_2$, and say τ_1 *is homologous to* τ_2.

(12.2) *A n. a. s. c. for* $\tau_1 \sim \tau_2$ *or equivalently for* $\tau = \tau_1 - \tau_2 \sim 0$, *is that if* γ^p *is any cycle of* X *then* $\tau\gamma^p$ *bounds.*

The condition is manifestly sufficient. It is also necessary. For a cycle γ^p over a topological group G is also a cycle over the isomorph G_0 of G in the algebraic sense with the discrete topology. Since for $G_0: \sim 0 \leftrightarrow$ bounding, $\tau\gamma^p$ must bound a chain C^{p+1} over G_0. Since C^{p+1} is also a chain over G, $\tau\gamma^p$ bounds in the asserted way.

(12.3) Definition. *It is clear that* $\tau \sim \tau'$ *is a relation of equivalence. We say therefore that two chain-mappings which are homologous are in the same chain-mapping class. If* τ, τ' *are in fixed classes so are* $\tau \pm \tau'$. *If* 0 *denotes the class of* $\tau = 0$, *the classes generate by addition the group of the chain-mapping classes* $H(X, Y)$. *The classes of the chain-mappings* $X \to X$ *give rise to a similar group* $H(X, X)$.

(12.4) Definition. *Two complexes* X, Y *are said to be homologous, written* $X \sim Y$, *whenever there exist chain-mappings* $\tau: X \to Y$, $\theta: Y \to X$ *such that* $\theta\tau \sim 1$, $\tau\theta \sim 1$.

This is again a relation of equivalence and so we have *classes of homologous complexes.*

(12.5) $$X \sim Y \leftrightarrow X^* \sim Y^* \qquad \text{(obvious).}$$

(12.6) *If* $X \sim Y$ *then the homology and cohomology groups of* X *are isomorphic with the corresponding groups of* Y.

For θ, τ of (12.4) induce homomorphisms $\bar{\theta}$, $\bar{\tau}$ of the homology groups and we have: $\bar{\theta}\bar{\tau} = 1$, $\bar{\tau}\bar{\theta} = 1$. Hence $\bar{\tau}$ is an isomorphism, which is (12.6) for the homology groups. The same result for the cohomology groups is then a consequence of (12.5).

(12.7) *Let* $X \sim Y$, $X_1 \sim Y_1$ *with* τ, θ *as in* (12.4) *and* τ_1, θ_1 *the analogues for* X_1, Y_1. *The chain-mappings* $\sigma: X \to X_1$, $\zeta: Y \to Y_1$ *are said to be congruent if* $\zeta \sim \tau_1\sigma\theta$ *and hence* $\sigma \sim \theta_1\zeta\tau$. *The relation* $\zeta \to \sigma$ *defines an isomorphism of* $H(X, X_1)$ *with* $H(Y, Y_1)$ (proof elementary).

(12.8) *Weak chain-mapping.* Let τ_0 be a chain-mapping $X \to Y_0$, and τ_1 a weak isomorphism $Y_0 \to Y$. Then $\tau = \tau_1\tau_0$ is known as a *weak* chain-mapping $X \to Y$. Its basic equations are:

(12.9) $$\tau x_i^p = a_i^j(p + n)y_j^{p+n}$$

where n is the increase in dimensions caused by τ_1. We merely note that for the weak chain-mapping (9.1, 10.1) read as before except that for τ the homomorphisms are $\mathfrak{C}^p(X, G) \to \mathfrak{C}^{p+n}(Y, G), \cdots$, while for τ^* they are $\mathfrak{C}_p(Y, G) \to \mathfrak{C}_{p-n}(X, G)$.

(12.10) Example. Let $K = \{\sigma\}$ be a simplicial complex. Then $\sigma \rightarrow A\sigma$ defines a weak chain-mapping $\tau : K \rightarrow AK$, since $\tau F\sigma^p = -AF\sigma^p = F\tau\sigma^p$.

13. **Carriers of chain-mappings.** In this capacity set-transformations appear in their true role as regards the present work. A carrier of a chain-mapping $\tau : X \rightarrow Y$ is a set-transformation $t : X \rightarrow Y$ such that τx is a chain of Cl tx.

(13.1) Example. X, Y are simplicial and t, τ are simplicial and related as in (9.12). Then t is a carrier of τ.

(13.2) *If τ, τ' are chain-mappings $X \rightarrow Y$, $Y \rightarrow Z$ with carriers t, t', then $\tau'\tau$ is a chain-mapping $X \rightarrow Z$ with carrier $t't$.*

§5. CHAIN-HOMOTOPY

14. As a natural parallel with set-homotopy we shall introduce (following essentially Lefschetz [e]) a chain-homotopy, and its value will rapidly justify itself.

Let first $\xi = ab$ be a closed one-simplex. For convenience we also designate by ξ the integral chain ξ whose boundary is

$$F(\xi) = b - a.$$

Given any complex X consider the product $\xi \times X$. By (5.6) we have for every chain C of X:

$$F(\xi \times C) = b \times C - a \times C - \xi \times FC. \tag{14.1}$$

(14.2) Definition. *Two chain-mappings $\tau_1, \tau_2 : X \rightarrow Y$ are called chain-homotopic whenever there exists a chain-mapping $\tau : \xi \times X \rightarrow Y$ such that $\tau(a \times x) = \tau_1 x$, $\tau(b \times x) = \tau_2 x$. The chain-mapping τ_2 is called a chain-deformation whenever $\tau_1 = 1$.*

Let us set: $\mathfrak{D}C = \tau(\xi \times C)$. Since τ commutes with F we have from (14.1) by applying τ to both sides:

$$F\mathfrak{D}C = \tau_2 C - \tau_1 C - \mathfrak{D}FC, \tag{14.3}$$

or in equivalent operator form

$$F\mathfrak{D} + \mathfrak{D}F = \tau_2 - \tau_1, \tag{14.4}$$

it being understood that the operators are all applied to the chains of X. We call $\mathfrak{D}$ the *homotopy operator* for τ (*deformation operator* when τ is a deformation). We will refer sometime to (14.4) as the fundamental chain-homotopy relation.

(14.5) *A n. a. s. c. for τ_1, τ_2 to be chain-homotopic is the existence of a simultaneous collection $\mathfrak{D}$ of homomorphisms $\mathfrak{C}^p(X) \rightarrow \mathfrak{C}^{p+1}(Y)$ such that* (14.4) *holds.*

Necessity has just been proved. Suppose that $\mathfrak{D}$ exists as stated. Define the chain-transformation $\tau : \xi \times X \rightarrow Y$ by

$$\tau(a \times x) = \tau_1 x, \qquad \tau(b \times x) = \tau_2 x, \qquad \tau(\xi \times x) = \mathfrak{D}x.$$

To prove that τ is a chain-mapping we merely have to show that it commutes with F. Since τ_1, τ_2 are chain-mappings we have:

$$(\tau\mathrm{F} - \mathrm{F}\tau)(a \times x) = (\tau_1\mathrm{F} - \mathrm{F}\tau_1)x = 0,$$

$$(\tau\mathrm{F} - \mathrm{F}\tau)(b \times x) = (\tau_2\mathrm{F} - \mathrm{F}\tau_2)x = 0.$$

Regarding the elements $\xi \times x$ we find:

$$\begin{aligned}(\tau\mathrm{F} - \mathrm{F}\tau)(\xi \times x) &= \tau(b \times x - a \times x - \xi \times \mathrm{F}x) - \mathrm{F}\tau(\xi \times x)\\ &= (\tau_2 - \tau_1 - \mathfrak{D}\mathrm{F} - \mathrm{F}\mathfrak{D})x = 0,\end{aligned}$$

and so (14.5) holds.

15. (15.1) *Chain-homotopy is symmetric, reflexive and transitive.*

We may therefore introduce chain-homotopy equivalence and classes in the usual way. We prove

symmetry by interchanging τ_1 and τ_2, and replacing $\mathfrak{D}$ by $-\mathfrak{D}$ in (14.4);

reflexivity by taking $\mathfrak{D} = 0$, $\tau_1 = \tau_2$ in the same relation, and observing that (14.4) continues to hold.

As for *transitivity* if τ_1 is chain-homotopic to τ_2 and τ_2 to τ_3 with operators $\mathfrak{D}_1$, $\mathfrak{D}_2$ we have:

$$\mathrm{F}\mathfrak{D}_1 + \mathfrak{D}_1\mathrm{F} = \tau_2 - \tau_1, \qquad \mathrm{F}\mathfrak{D}_2 + \mathfrak{D}_2\mathrm{F} = \tau_3 - \tau_2.$$

Hence if we set $\mathfrak{D} = \mathfrak{D}_1 + \mathfrak{D}_2$, then

$$\mathfrak{D}\mathrm{F} + \mathrm{F}\mathfrak{D} = \tau_3 - \tau_1.$$

Therefore τ_1 and τ_3 are chain-homotopic.

(15.2) *If τ_1, τ_2 are chain-homotopic mappings $X \to Y$ then $\tau_1 \sim \tau_2$. Hence if τ is a chain-deformation $X \to X$ then $\tau \sim 1$, that is to say, τ does not change the homology groups.*

For if γ is a cycle (14.3) yields $\mathrm{F}\mathfrak{D}\gamma = (\tau_2 - \tau_1)\gamma \sim 0$, and so $\tau_1 \sim \tau_2$.

(15.3) *Dualization.* Supposing τ_1, τ_2 chain-homotopic mappings $X \to Y$, the relations defining $\mathfrak{D}$ are:

$$\mathfrak{D}x_i^p = A_i^j(p)y_j^{p+1}. \tag{15.4}$$

Now the relations

$$\mathfrak{D}^*y_{p+1}^j = A_i^j(p)x_p^i \tag{15.5}$$

define simultaneous homomorphisms $\mathfrak{C}_{p+1}(Y) \to \mathfrak{C}_p(X)$, and it is readily seen that the matrices of $\mathfrak{D}\mathrm{F}$ and $\mathrm{F}\mathfrak{D}^*$, likewise those of $\mathrm{F}\mathfrak{D}$ and $\mathfrak{D}^*\mathrm{F}$, are the transverses of one another. Therefore

$$\mathfrak{D}^*\mathrm{F} + \mathrm{F}\mathfrak{D}^* = \overset{*}{\tau_2} - \overset{*}{\tau_1}. \tag{15.6}$$

The operator $\mathfrak{D}^*$ is known as the *dual* of $\mathfrak{D}$. Evidently $(\mathfrak{D}^*)^* = \mathfrak{D}^{**} = \mathfrak{D}$. In view of (15.6) we have:

(15.7) *When τ_1, τ_2 are chain-homotopic so are their duals $\overset{*}{\tau_1}$, $\overset{*}{\tau_2}$ and their homotopy-operators $\mathfrak{D}$, $\mathfrak{D}^*$ are dual to one another. When τ is a chain-deformation so is τ^*.*

(15.8) DEFINITION. *The two chain-mappings τ_1, $\tau_2:X \to Y$ with a common weakly closed carrier t are said to be contiguous in t, whenever they are chain-homotopic and with an operator $\mathfrak{D}$ such that $\mathfrak{D}x \subset \mathrm{Cl}\ tx$ for every x. Otherwise stated, for every x the homotopy takes place over* $\mathrm{Cl}\ tx$ (Tucker [c]).

Under the same conditions if t is closed so is t^* (8.4). In the notations of (15.3) if y_j^{p+1} is an element of $\mathfrak{D}x_i^p$, by assumption $y_j^{p+1} < y_h^q \in tx_i^p$. Hence $x_p^i \in t^*y_q^h \subset \mathrm{Cl}\ t^*y_q^h \subset \mathrm{Cl}\ t^*y_{p+1}^j$, which proves finally: $\mathfrak{D}^*y_{p+1}^j \subset \mathrm{Cl}\ t^*y_{p+1}^j$. Therefore

(15.9) *If the τ_i are contiguous in a closed carrier t then their duals are contiguous in the dual t^* (which is also closed).*

(15.10) *Chain-retraction.* This concept is carried over from retraction (I, 47.6). Explicitly:

(15.11) DEFINITION. *The closed subcomplex Y of X is said to be a chain-retract of X if there exists a chain-mapping $\rho:X \to Y$ such that $\rho \mid Y = 1$. We call ρ a chain-retraction. If ρ is a chain-deformation then Y is known as a chain-deformation retract of X and ρ as a chain-deformation retraction.*

(15.12) *If ρ is a chain-deformation retraction $X \to Y$ then ρ induces an isomorphism of the homology groups of X with the corresponding groups of Y.*

Let η be the injection $Y \to X$. If γ^p is a cycle of X then $\rho\gamma^p \sim \gamma^p$ in X and $\eta\rho\gamma^p \sim \rho\gamma^p \sim \gamma^p$ in X. Similarly if γ_1^p is a cycle of Y then $\eta\gamma_1^p = \rho\eta\gamma_1^p = \gamma_1^p$. Hence $\eta\rho \sim 1$, $\rho\eta = 1 \sim 1$, and (15.12) follows.

16. **A noteworthy chain-homotopy in simplicial complexes.** The situation to be considered below is essentially the one from which the concept of chain-homotopy arose (see [L, 78]). It is also in close relation with a particularly simple decomposition of prisms into simplexes discussed later (VIII, 22.1).

(16.1) Let $K = \{\sigma\}$, $L = \{\zeta\}$ be simplicial complexes, τ_1 and τ_2 simplicial chain-mappings $K \to L$, t_1 and t_2 their simplicial carriers (9.12, 13.1). We say that τ_1, τ_2 are *prismatically related* whenever there exists a simple ordering $\{A_i\}$ of the vertices of K such that if $t_1A_i = B_i$, $t_2A_i = C_i$, and

(16.1a) $$\sigma = A_{i_0} \cdots A_{i_p}, \qquad i_0 < \cdots < i_p,$$

is a simplex of K then every simplex

(16.1b) $$\zeta = B_{i_0} \cdots B_{i_q}C_{i_q} \cdots C_{i_p}$$

is a simplex of L. Clearly:

(16.2) *A sufficient condition for τ_1, τ_2 to be prismatically related (independently of the ordering of the vertices of K) is that if $\sigma \in K$ then the join $(t_1\sigma)(t_2\sigma) \in L$. We will then say that τ_1, τ_2 are prismatically related "in the strong sense."*

We will now prove:

(16.3) *If τ_1, τ_2 are prismatically related then they are chain-homotopic and with homotopy operator $\mathfrak{D}$ given by*

(16.3a) $$\mathfrak{D}\sigma = \mathfrak{D}A_{i_0} \cdots A_{i_p} = \sum (-1)^q B_{i_0} \cdots B_{i_q}C_{i_q} \cdots C_{i_p}.$$

This implies in particular that: (a) *the vertices of the simplexes of* $\mathfrak{D}\sigma$ *are among those of* $t_1\sigma$ *and* $t_2\sigma$; (b) *if* τ_1, τ_2 *are prismatically related in the strong sense then* $\mathfrak{D}\sigma$ *is a chain of* $\mathrm{Cl}(t_1\sigma)(t_2\sigma)$.

(16.4) A chain-homotopy such as described in (16.3) will be called *prismatic*.

PROOF OF (16.3). Suppose first that t_1, t_2 are isomorphisms of K with disjoint subcomplexes K_1, K_2 of L. Thus ζ in (16.1b) will always be $(p + 1)$-dimensional, and the simplexes in the expression (16.3a) of $\mathfrak{D}\sigma$ will all be non-degenerate. Under the circumstances an elementary calculation yields $\mathrm{F}\mathfrak{D}\sigma = (\tau_2 - \tau_1 - \mathfrak{D}\mathrm{F})\sigma$ for every $\sigma \in K$. Thus the fundamental chain-homotopy relation (14.4) holds and (16.3) is proved for the present case.

Passing to the general case, introduce for each vertex A_i of K two new vertices B_i', C_i' and for each ζ as in (16.1b) the simplex $\zeta' = B_{i_0}' \cdots B_{i_q}'C_{i_q}' \cdots C_{i_p}'$. If M is the simplicial complex made up of the ζ' and all their faces, then $A_i \to B_i'$, $A_i \to C_i'$ define simplicial chain-mappings τ_1', $\tau_2':K \to M$ which are manifestly prismatically chain-homotopic, and this with respect to the adopted ordering of the vertices of K. We are clearly in the situation already considered, and so if $\mathfrak{D}'$ is the homotopy operator we have

$$\mathrm{F}\mathfrak{D}' + \mathfrak{D}'\mathrm{F} = \tau_2' - \tau_1'. \tag{16.5}$$

Now $B_i' \to B_i$, $C_i' \to C_i$ define a simplicial chain-mapping $\theta:M \to L$ such that $\tau_h = \theta\tau_h'$, $(h = 1, 2)$, $\mathfrak{D} = \theta\mathfrak{D}'$. Since $\theta\mathrm{F}\mathfrak{D}' = \mathrm{F}\theta\mathfrak{D}' = \mathrm{F}\mathfrak{D}$, applying θ to both sides of (16.5) we obtain (14.4) and (16.3) follows.

17. **Comparison of homologous and chain-homotopic chain-mappings.**

(17.1) Let the designations a_i^p, $\cdots$, for the canonical bases of X be as in (6), and let A_i^p, $\cdots$, designate integral chains of Y. We will first endeavor to characterize more completely the chain-mappings: $X \to Y$.

A chain-transformation $\tau:X \to Y$ is uniquely defined by relations

$$\tau a_i^p = A_i^p, \cdots, \tau e_i^p = E_i^p.$$

In order that τ be a chain-mapping we must have $\tau\mathrm{F} = \mathrm{F}\tau$. This yields here:

$$\mathrm{F}C_i^p = 0 \quad (C_i^p \text{ is a cycle}); \tag{17.1a}$$

$$\mathrm{F}D_i^{p+1} = t_i^p B_i^p \quad (D_i^{p+1} \text{ is a cycle mod } t_i^p); \tag{17.1b}$$

$$\mathrm{F}E_i^{p+1} = A_i^p. \tag{17.1c}$$

The C, D, E thus determine the A, B and hence τ. We may therefore state:

(17.2) *Each chain-mapping defines uniquely and is uniquely defined by the following elements*: (a) *a set of integral cycles* $\{C_i^p\}$; (b) *a set* $\{D_i^{p+1}\}$, D_i^{p+1} *a cycle* mod t_i^p; (c) *a set of chains* $\{E_i^{p+1}\}$. *The range of* p *is that of the dimensions of the elements of* X.

(17.3) THEOREM. *A n. a. s. c. in order that two chain-mappings* τ_1, $\tau_2:X \to Y$ *be chain-homotopic is that they be homologous. In other words chain-homotopy and homology are equivalent properties of chain-mappings.*

Necessity is implicit in (15.2). Suppose $\tau_1 \sim \tau_2$ or $\tau = \tau_1 - \tau_2 \sim 0$. To prove sufficiency we must show that τ is chain-homotopic to zero, i.e., exhibit an operator $\mathfrak{D}$ such that

$$(17.4) \qquad \mathfrak{D}\mathrm{F} + \mathrm{F}\mathfrak{D} = \tau.$$

The notations being as before $\tau \sim 0 \rightarrow A_i^p = \mathrm{F}A_{1i}^{p+1}$, $B_i^p = \mathrm{F}B_{1i}^{p+1}$ where $A_{1i}^{p+1}, \cdots$, are suitable integral chains of Y, and so (17.1bc) become

$$(17.5)_p \qquad \mathrm{F}E_i^{p+1} = \mathrm{F}A_{1i}^{p+1}, \qquad \mathrm{F}D_i^{p+1} = t_i^p \mathrm{F}B_{1i}^{p+1}.$$

Furthermore

$$(17.6)_p \qquad C_i^p = \mathrm{F}C_{1i}^{p+1}.$$

To prove the existence of τ we merely have to find chains $\mathfrak{D}a_i^p, \cdots$, such that (17.4) is satisfied when both sides are applied to $a_i^p, \cdots$. We choose

$$(17.7) \qquad \mathfrak{D}a_i^p = A_{1i}^{p+1} + A_{2i}^{p+1}, \qquad \mathfrak{D}b_i^p = B_{1i}^{p+1} + B_{2i}^{p+1}, \qquad \mathfrak{D}c_i^p = C_{1i}^{p+1},$$

where A_{2i}^{p+1}, B_{2i}^{p+1} are cycles to be specified in a moment. The relation for determining $\mathfrak{D}d_i^p$ reduces to:

$$t_i^{p-1}\mathfrak{D}b_i^{p-1} + \mathrm{F}\mathfrak{D}d_i^p = D_i^p,$$

or by (17.7) to:

$$\mathrm{F}\mathfrak{D}d_i^p = D_i^p - t_i^{p-1}(B_{1i}^p + B_{2i}^p).$$

From the basic relations for the canonical bases (III, 14.4) we infer that d_i^p is a cycle mod t_i^{p-1}. Hence $\tau \sim 0$ implies that τd_i^p bounds mod t_i^{p-1} or:

$$(17.8) \qquad D_i^p = \mathrm{F}D_{1i}^{p+1} + t_i^{p-1}D_{2i}^p.$$

Therefore

$$\mathrm{F}\mathfrak{D}d_i^p = \mathrm{F}D_{1i}^{p+1} + t_i^{p-1}(D_{2i}^p - B_{1i}^p - B_{2i}^p).$$

From (17.8) and the second relation of $(17.5)_{p-1}$ we find $\mathrm{F}D_{2i}^p = \mathrm{F}B_{1i}^p$. Hence $D_{2i}^p - B_{1i}^p$ is a cycle and so we may choose

$$B_{2i}^p = D_{2i}^p - B_{1i}^p, \qquad \mathfrak{D}d_i^p = D_i^{p+1}.$$

Similarly the relation for $\mathfrak{D}e_i^p$ reduces to:

$$\mathrm{F}\mathfrak{D}e_i^p = E_i^p - (A_{1i}^p + A_{2i}^p),$$

where $\mathrm{F}E_i^p = \mathrm{F}A_{1i}^p$. Thus, $E_i^p - A_{1i}^p$ being a cycle, we may choose

$$\mathfrak{D}e_i^p = 0, \qquad A_{2i}^p = E_i^p - A_{1i}^p.$$

Therefore the required operator τ exists and (17.3) is proved.

(17.9) Observe that in proving the existence of $\mathfrak{D}$ we have merely utilized the following properties:

(17.9a) $\tau \sim 0$ integrally;

(17.9b) $\tau d_i^p = D_i^p \sim 0 \bmod t_i^{p-1}$.

Coupling this with (17.2) we obtain then:

(17.10) *The classes of chain-homotopic (or equivalently of homologous) chain-mappings* $X \to Y$ *are in one-one correspondence with the following sets of simultaneous homomorphisms*: (a) *a simultaneous homomorphism of the integral homology groups* $\mathfrak{H}^p(X) \to \mathfrak{H}^p(Y)$; (b) *a simultaneous homomorphism* $\mathfrak{H}^p(X; t_i^p) \to \mathfrak{H}^p(Y; t_i^p)$. *The range of p is that of the dimensions of the elements of* X.

If X is torsion-free (a) alone remains. Coupling this with (15.7) we find:

(17.11) *If* X [Y] *is torsion-free then a n.a.s.c. for* τ_1, τ_2 *to be chain-homotopic is that they* [*that* $\overset{*}{\tau_1}$, $\overset{*}{\tau_2}$] *be integrally homologous.*

(17.12) *Application.* Let $Y = S^n = \mathfrak{B}\sigma^{n+1}$, $n > 0$, and let X, τ_1, τ_2 be simplicial. Applying the designations $\alpha_i^p, \cdots$, of (6) to Y, there are no β_i^p, δ_i^p and only two cycles $\gamma_i^p : \gamma^0, \gamma^n$. Moreover since τ_1, τ_2 are simplicial we readily find: $\overset{*}{\tau_1}\gamma_0 \sim \overset{*}{\tau_2}\gamma_0$. Hence $\tau_1 \sim \tau_2$ if and only if $\overset{*}{\tau_1}\gamma_n \sim \overset{*}{\tau_2}\gamma_n$. In other words corresponding to each chain-mapping $\tau : X \to Y$ there is a *characteristic n*-cocycle $\gamma_n(\tau)$ of X. Referring to (17.2) one may show that the chain-homotopy classes $X \to S^n$ are in one-one correspondence with the nth integral cohomology classes of X, i.e., with the elements of the group $\mathfrak{H}_n(X)$.

18. **Uniqueness of certain chain-mappings.** To what extent does a carrier determine a chain-mapping? This question was raised recently by Tucker [c] and answered in part in an important theorem of which we give the following special case, more than ample however for the sequel.

(18.1) THEOREM. *Let* X, Y *be two complexes, with* Y *simple, and let* X_0 *be a closed subcomplex of* X *such that all the elements of* $X - X_0$ *are of positive dimension. Consider also chain-mappings* τ, τ_1, $\tau_2 : X \to Y$ *with a common weakly closed carrier t such that* $(\mathrm{Cl}\ tx)_a$, $x \in X - X_0$, *is acyclic. Then:*

(a) *if* $\tau_0 = \tau \mid X_0$ *is assigned and is such that* $\mathrm{KI}(\tau_0 \mathrm{F} x^1) = 0$, $x^1 \in X - X_0$, τ_0 *has an extension* $\tau : X \to Y$, *and* τ *is unique to within contiguity in t. Moreover when* $\dim tx \leqq \dim x$, $x \in X - X_0$, *then* τ *is unique;*

(b) *if* $\tau_1 \mid X_0$, $\tau_2 \mid X_0$ *are contiguous in* $t \mid X_0$ *then* τ_1, τ_2 *are contiguous in t.*

Noteworthy for its interest is the following special case obtained from (b) by taking for X_0 the zero-section of X:

(18.2) THEOREM. *If* X, Y *are simplicial and* τ_1, τ_2 *are simplicial chain-mappings* $X \to Y$ *with the same generalized simplicial carrier t* (*see* 7.10) *then* τ_1, τ_2 *are contiguous in t.*

PROOF OF (18.1a). Suppose that the elements of $X - X_0$ have dimensions not less than $q > 0$, and take $x^q \in X - X_0$. Since $\mathrm{F}x^q \in X_0$, $\tau_0 \mathrm{F} x^q$ is defined and is a cycle γ^{q-1} of $(\mathrm{Cl}\ tx^q)_a$. If $q > 1$, γ^{q-1} bounds in $(\mathrm{Cl}\ tx^q)_a$. If $q = 1$, by hypothesis $\mathrm{KI}(\gamma^0) = \mathrm{KI}(\tau_0 \mathrm{F} x^1) = 0$, and again γ^0 bounds in $(\mathrm{Cl}\ tx^q)_a$. Therefore in any case the latter contains a chain ξ^q such that $\mathrm{F}\xi^q = \gamma^{q-1}$. We define $\tau x^q = \xi^q$, add x^q to X_0 and proceed likewise for all $x^q \in X - X_0$. The situation is then as before, with a new X_0 and with q replaced by $q + 1$, and so step by step τ is extended to the whole of X. By assumption $\tau \mathrm{F} = \mathrm{F} \tau$ holds on X_0, and τx, $x \in X - X_0$, is defined at each step so as to preserve this relation. Therefore τ commutes with F throughout. It is also chosen so as to have the carrier t. It follows that τ fulfills all the required conditions under (18.1a). Its uniqueness to within contiguity in t will be a consequence of (18.1b).

Suppose that dim $tx \leqq x$ throughout. To show uniqueness it is sufficient to prove ξ^q unique. Assume the existence of a second ξ^q, say ξ'^q. Since $F(\xi^q - \xi'^q) = 0$, $\xi^q - \xi'^q$ is a cycle of $(\text{Cl } tx)_a$ and of the dimension of that complex. Therefore $\xi^q - \xi'^q \sim 0$ and hence is equal to 0, or $\xi^q = \xi'^q$. Thus all that is left to complete the proof of (18.1a), the uniqueness to within contiguity in t, is reduced to (18.1b), whose proof now follows.

Proof of (18.1b). Let again dim $x \geqq q$, $x \in X - X_0$. By (14.5, 15.8) we require an operator $\mathfrak{D}$ such that

(α) $\mathfrak{D}x^p \subset \text{Cl } tx^p$;

(β) $F\mathfrak{D}x^p = (\tau_2 - \tau_1 - \mathfrak{D}F)x^p$.

By the assumption of the contiguity of τ_1, τ_2 in t on X_0, $\mathfrak{D}$ is already suitably defined (i.e., to satisfy (α), (β)) for $x^p \in X_0$, and in particular for $p < q$. Suppose now $p \geqq q$, and assume that we have obtained all the $\mathfrak{D}x^r$, $q \leqq r < p$, such that (α), (β) hold for dimensions less than p. Setting

$$\zeta^p = (\tau_2 - \tau_1 - \mathfrak{D}F)x^p, \tag{18.3}$$

we find

$$F\zeta^p = F(\tau_2 - \tau_1 - \mathfrak{D}F)x^p = (\tau_2 - \tau_1 - F\mathfrak{D})Fx^p. \tag{18.4}$$

Since by assumption (β) holds for dimensions less than p, we find from (18.4): $F\zeta^p = \mathfrak{D}FFx^p = 0$, so ζ^p is a cycle. Clearly $(\tau_2 - \tau_1)x^p \subset \text{Cl } tx^p$. Also by the hypothesis of the induction: $\mathfrak{D}Fx^p \subset \text{Cl } t(\mathfrak{B}x^p) \subset \text{Cl } t \text{ Cl } x^p = \text{Cl } tx^p$, since t is weakly closed. Therefore in view of (18.3), $\zeta^p \subset \text{Cl } tx^p$. Since $(\text{Cl } tx^p)_a$ is acyclic and $p > 0$, ζ^p bounds in $\text{Cl } tx^p$. There exists then in $\text{Cl } tx^p$, a chain $\mathfrak{D}x^p$ such that $F\mathfrak{D}x^p = \zeta^p$. Thus we have found a $\mathfrak{D}x^p$ which satisfies (α), (β), and (18.1b) is proved.

Noteworthy complements are:

(18.5) *The results of* (18.1) *also hold under the following conditions*: X, Y, τ, τ_1, τ_2, t *are simple,* dim $X_0 = 0$, *and* t *is a weakly closed carrier of* τ, τ_1, τ_2.

For the Kronecker indices of finite zero-chains exist and are preserved by τ, τ_1, τ_2. Since X, τ are simple $\text{KI}(Fx^1) = 0 = \text{KI}(\tau F x^1)$; so (18.1a) follows. For similar reasons $\text{KI}(\tau_1 x^0) = \text{KI}(\tau_2 x^0)$, and so $\tau_1 x^0 - \tau_2 x^0$ bounds in $(\text{Cl } tx^0)_a$. This means that the latter contains a chain $\mathfrak{D}x^0$ such that $F\mathfrak{D}x^0 = \tau_1 x^0 - \tau_2 x^0$. Since in the present instance $X_0 = X^0$, the zero-section of X, $\tau_1 \mid X_0$ and $\tau_2 \mid X_0$ are contiguous in $t \mid X_0$, so (18.1b) holds also.

(18.6) *If* τ_1, τ_2 *under* (18.1b) *coincide on* X_0 *we may choose* $\mathfrak{D} = 0$ *as the definition of* $\mathfrak{D}$ *on* X_0.

Since X_0 is closed in X this is implicit in the proof of reflexivity in (15.1). Explicitly also (14.4) is manifestly satisfied for the elements of X_0 alone, when $\mathfrak{D} = 0$ on X_0.

(18.7) *Given* X, Y, t *as in* (18.1) *and* $X_0 = X^0$, (X^q = *the q-section of* X), *if* $\mathrm{KI}(\tau_1 x^0) = \mathrm{KI}(\tau_2 x^0)$, *then* τ_1, τ_2 *are contiguous in* t.

Since the dimensions in Y are greater than or equal to 0, both τ_1 and τ_2 map the elements of X^{-1} into zero, so we may choose $\mathfrak{D} = 0$ on X^{-1}. Under the stated conditions we can find again $\mathfrak{D}x^0$ such that $\mathrm{F}\mathfrak{D}x^0 = \tau_1 x^0 - \tau_2 x^0$ and $\mathfrak{D}$ will satisfy (14.4) for all of X^0. This makes $\tau_1 \mid X^0$, $\tau_2 \mid X^0$ contiguous in $t \mid X^0$ and so (18.7) follows from (18.1b) with $X_0 = X^0$.

§6. COMPLEMENTS

19. Transformations of infinite complexes.

(19.1) *Locally finite complexes.* Let X, Y be locally finite complexes and let $\| a_i^j(p) \|$, $p = 0, \pm 1, \pm 2, \cdots$ be matrices of integers. Let

$$\tau x_i^p = a_i^j(p) y_j^p .$$

When for a given p and i there is only a finite number of j's such that $a_i^j(p) \neq 0$ then τx_i^p is a finite chain of Y and therefore τ defines simultaneous homomorphisms of the *finite* integral chain-groups

$$\tau : \mathfrak{C}_f^p(X) \to \mathfrak{C}_f^p(Y).$$

When τ commutes with F: $\tau\mathrm{F} = \mathrm{F}\tau$, we shall say that τ is an *f-chain-mapping* $X \to Y$.

When for a given p and j there is only a finite number of i's such that $a_i^j(p) \neq 0$ then for every infinite integral chain

$$\xi^p = g^i x_i^p$$

the expression

$$\tau \xi^p = a_i^j g^i y_j^p$$

is an infinite chain of Y and therefore τ defines simultaneous homomorphisms of the *infinite* integral chain-groups (II, 8.4):

$$\tau : \mathfrak{C}^p(X) \to \mathfrak{C}^p(Y).$$

When τ commutes with F we shall say that τ is an *i-chain-mapping* $X \to Y$.

When τ is both an f-mapping and an i-mapping we shall say that τ is an *fi-chain-mapping*.

With τ there is associated a dual mapping $\tau^* : Y^* \to X^*$ defined by

$$\tau^* y^j = a_i^j x^i.$$

As in (9, 10) we have

(19.2) τ *is an f-chain- [i-chain- or fi-chain-] mapping if and only if* τ^* *is an i-chain- [f-chain- or fi-chain-] mapping.*

(19.3) *An f-chain- [i-chain- or fi-chain-] mapping* $\tau : X \to Y$ *induces homomorphisms of the groups* $\mathfrak{C}^p, \cdots, \mathfrak{H}^p$ *of the finite [infinite or both] chains of* X *over any* G *into the corresponding groups for* Y. *The dual* $\tau^* : Y^* \to X^*$ *induces then* *homomorphisms of the groups* $\mathfrak{C}_p, \cdots, \mathfrak{H}_p$ *of the infinite [finite or both] cochains of* Y *into the corresponding groups of* X.

(19.4) *The product of two chain-mappings of a given type* (f, i *or* fi) *is a chain-mapping of the same type.*

The modifications required in (9, 10) (all but (10.10) which is left out of consideration) are clear enough. We notice explicitly that ξ^0 in $\mathrm{KI}(\xi^0)$, and one of ξ^p, η_p in $\mathrm{KI}(\xi^p, \eta_p)$ must be finite.

The graph Γ is still the zero-chain of $X^* \times Y$ given by (11.1). If τ is an f- [or i-] chain-mapping then every factor x_p^i or y_j^p occurs in at most a finite number of terms of Γ. If we bear in mind this restriction concerning Γ (11.2) still holds.

The remarks of (13) may be repeated throughout with the sole restriction that a carrier of an f-mapping $\tau : X \to Y$ must be a finite valued set-transformation $t : X \to Y$. If τ is an i-mapping t^{-1} must be finite-valued.

(19.5) *Star- or closure-finite complexes.* Let X, Y be star-finite and let the meaning of f-, $\cdots$ transformations be as before. In that sense the boundary operator F is an i-transformation from the groups of p-chains to those of $(p-1)$-chains. Therefore if τ is an i-transformation in the same sense as before (from groups of p-chains to groups of p-chains) $\mathrm{F}\tau - \tau\mathrm{F}$ will have a meaning and so τ may be defined as an i-chain-mapping in the same way as previously. Similarly for an fi-chain-mapping. Likewise for closure-finite X, Y and f- and fi-chain-mappings. And naturally also (19.2, 19.3, 19.4) continue to hold.

20. Chain-homotopy in infinite complexes. Let X, Y be both star-finite or both closure-finite. Two f-chain- [i-chain- or fi-chain-] mappings $\tau_1, \tau_2 : X \to Y$ are called *f-chain* [*i-chain or fi-chain-*] *homotopic* whenever in the terminology of (14) there exists an f-chain- [i-chain- or fi-chain-] mapping $\tau : \xi \times X \to Y$, such that $\tau(a \times X) = \tau_1 X$, $\tau(b \times X) = \tau_2 X$. If $\tau_1 = 1$, τ_2 is called an *f-chain-* [*i-chain- or fi-chain-*] *deformation.* It is understood throughout that the f-type [i-type] are considered only when X, Y are closure- [star-] finite, while the fi-type may be considered in both cases.

As in (14) it may be proved that the existence of an f-, i- or fi-chain-homotopy is equivalent to the existence of a simultaneous homomorphism

$$\mathfrak{D} : \mathfrak{C}^p(X) \to \mathfrak{C}^{p+1}(Y)$$

such that

$$\mathrm{F}\mathfrak{D} + \mathfrak{D}\mathrm{F} = \tau_2 - \tau_1$$

given by formulas

$$\mathfrak{D}x_i^p = A_i^j(p) y_j^{p+1}$$

where the matrices of integers $\| A_i^j(p) \|$ are subject to the same type of restrictions as the matrices $\| a_i^j(p) \|$ defining τ_1 and τ_2.

As in (14) we may define the dual $\mathfrak{D}^*$ of $\mathfrak{D}$ and prove that

(20.1) τ_1, τ_2 *are f-homotopic [i-homotopic or fi-homotopic] if and only if* τ_1^*, τ_2^* *are i-homotopic [f-homotopic or fi-homotopic].*

(20.2) *If τ_1, τ_2 are f-homotopic [i-homotopic] then $\tau_1\xi^p \sim \tau_2\xi^p$ for every finite [infinite] cycle ξ^p of X.*

(20.3) Similarly we have three types of homologous complexes (12.4). It is readily seen that:

(20.4) *X, Y are f- [i- or fi-] homologous when and only when X^*, Y^* are i- [f- or fi-] homologous.*

(20.5) *If X, Y are f- [i-] homologous then the finite [infinite] homology groups and the infinite [finite] cohomology groups of X are isomorphic with the corresponding groups of Y.*

(20.6) The considerations of (17) hold for X finite, Y closure-finite and τ, τ_i, $\mathfrak{D}$ of f-type. It is understood that all the chains and cycles under consideration are finite and wherever homologous or chain-homotopic chain-mappings occur they are f-homologous or f-chain-homotopic.

(20.7) In the remainder of (§5) the required modifications are quite straightforward and may safely be left to the reader. In particular, everything may be carried out for transformations of the *fi*-type.

For the sake of simplicity we shall assume throughout the rest of the chapter that *all complexes are locally finite and all chain-mappings are of the fi-type.*

21. **Transformations of products.** Let t, τ be a set-transformation and a chain-mapping $X \to Y$, and t', τ' the same: $X' \to Y'$. Define $t \times t'$ and $\tau \times \tau'$ by the relations

$$(21.1)\qquad \begin{aligned}(t \times t')(x \times x') &= (tx) \times (t'x'),\\ (\tau \times \tau')(x \times x') &= (\tau x) \times (\tau' x').\end{aligned}$$

Evidently $t \times t'$ is a set-transformation $X \times X' \to Y \times Y'$. From the commutation property of τ, τ' with F we find also $(\tau \times \tau')\mathrm{F} = \mathrm{F}(\tau \times \tau')$, so that $\tau \times \tau'$ is a chain-mapping $X \times X' \to Y \times Y'$. We prove readily

(21.2) *If both t, t' are open, closed, weakly or otherwise, so is $t \times t'$.*

(21.3) *If t is the carrier of τ and t' the carrier of τ' then $t \times t'$ is a carrier of $\tau \times \tau'$.*

(21.4) *The chain-graph of $\tau \times \tau'$ is the product of the chain-graphs of τ and τ'.*

(21.5) *Let τ_1, τ_2 be chain-homotopic chain-mappings $X \to Y$, and τ_1', τ_2' the same: $X' \to Y'$. Then $\tau_1 \times \tau_1'$ and $\tau_2 \times \tau_2'$ are chain-homotopic chain-mappings $X \times X' \to Y \times Y'$.*

It is manifestly sufficient to prove that the two product chain-mappings considered are both chain-homotopic to $\tau_1 \times \tau_2'$. We have an operator $\mathfrak{D}$ on X' such that

$$(21.6)\qquad \mathfrak{D}\mathrm{F} + \mathrm{F}\mathfrak{D} = \tau_2' - \tau_1'.$$

If we set

$$(21.7)\qquad \Delta(x \times x') = (-1)^p(\tau_1 x) \times (\mathfrak{D}x'),\ p = \dim x,$$

we find directly

$$(21.8)\qquad (\Delta\mathrm{F} + \mathrm{F}\Delta)(x \times x') = (\tau_1 \times \tau_2' - \tau_1 \times \tau_1')(x \times x'),$$

which is, in fact, (14.4) for Δ relative to $\tau_1 \times \tau_1'$ and $\tau_1 \times \tau_2'$. We prove similarly the chain-homotopy of the latter with $\tau_2 \times \tau_2'$ and with operator Δ'. Hence (15.1) $\tau_1 \times \tau_1'$ and $\tau_2 \times \tau_2'$ are chain-homotopic and with operator $\Delta + \Delta'$.

(21.9) *The situation being as in* (21.5) *if* τ_1, τ_2 *are contiguous in* t, *and* τ_1', τ_2' *are contiguous in* t' *then* $\tau_1 \times \tau_1'$ *and* $\tau_2 \times \tau_2'$ *are contiguous in* $t \times t'$.

For $\Delta(x \times x') \subset \text{Cl}\,(t \times t')(x \times x')$, and likewise for Δ', hence also for $\Delta + \Delta'$.

(21.10) *The preceding properties hold for any finite product.*

22. **Induced chain-mappings.** Let (X_0, X_1), (Y_0, Y_1) be dissections of X, Y (III, 23) and let π, ω be the projections $X \to X_0$, $Y \to Y_0$ and η, ζ the injections $X_1 \to X$, $Y_1 \to Y$. Suppose now that we have a chain-mapping τ and a set-transformation $t: X \to Y$, such that $\tau X_1 \subset Y_1$, $tX_1 \subset Y_1$. Let $\tau_0 = \omega\tau \mid X_0$, and define t_0: $X_0 \to Y_0$ as the set-transformation such that $t_0 x = tx \cap Y_0$, $x \in X_0$. Set also $\tau_1 = \tau \mid X_1$, $t_1 = t \mid X_1$. We say that t_i, τ_i are induced by t, τ.

(22.1) (a) τ_i *is a chain-mapping* $X_i \to Y_i$; (b) *if* t *is a closed carrier of* τ *then* t_i *is a closed carrier of* τ_i.

Only (a) for τ_0 requires proof. We must prove $\omega\tau\pi\mathrm{F} = \omega\mathrm{F}\omega\tau$. Since $\omega^2 = \omega$ and F commutes with ω and τ we have $\omega\mathrm{F}\omega\tau = \omega\mathrm{F}\tau$. Clearly also $\omega\tau(1 - \pi) = 0$, and hence $\omega\tau = \omega\tau\pi$. From this follows $\omega\tau\pi\mathrm{F} = \omega\tau\mathrm{F} = \omega\mathrm{F}\tau = \omega\mathrm{F}\omega\tau$, proving the assertion regarding τ_0.

Let τ' analogous to τ induce τ_i', and suppose τ, τ' chain-homotopic with operator $\mathfrak{D}$ such that $\mathfrak{D}X_1 \subset Y_1$. If $\mathfrak{D}_0 = \omega\mathfrak{D}$, $\mathfrak{D}_1 = \mathfrak{D} \mid X_1$, we find as above $\mathfrak{D}_0\pi = \omega\mathfrak{D}\pi = \omega\mathfrak{D}$, and so, using $\mathfrak{D}\mathrm{F} + \mathrm{F}\mathfrak{D} = \tau - \tau'$, we verify that $\mathfrak{D}_i\mathrm{F} + \mathrm{F}\mathfrak{D}_i = \tau_i' - \tau_i$. Thus τ_i, τ_i' are chain-homotopic with operator $\mathfrak{D}_i$ said to be *induced* by $\mathfrak{D}$. If τ, τ' are contiguous in t, and if Cl_i designates closures in Y_i, then $\mathfrak{D}_i x \subset \text{Cl}_i(t_i x)$, $x \in X_i$, and so τ_i, τ_i' are contiguous in t_i. Thus:

(22.2) *If* τ, τ': $X \to Y$ *are chain-homotopic with operator* $\mathfrak{D}$ *and* τ, τ', $\mathfrak{D}$ *send* X_1 *into* Y_1 *then the induced chain-mappings* τ_i, τ_i' : $X_i \to Y_i$ *are also chain-homotopic and their homotopy operators are the operators* $\mathfrak{D}_0 = \omega\mathfrak{D}$, $\mathfrak{D}_1 = \mathfrak{D} \mid X_1$ *induced by* $\mathfrak{D}$. *Furthermore if* τ, τ' *are contiguous in* t *such that* $tX_1 \subset Y_1$, *then* τ_i, τ_i' *are contiguous in the set-transformation* t_i: $X_i \to Y_i$ *induced by* t.

§7. SUBDIVISION. DERIVATION. PARTITION

23. Subdivision may be viewed as a method for carrying over to complexes the geometric process of partitioning a polyhedron Π into arbitrarily small pieces.

We continue to adhere to locally finite complexes, and to take all chain-mappings of the *fi*-type. In particular homologous complexes are also to be understood in the *fi*- sense throughout.

The basic definition, due essentially to Tucker [a, c] is:

(23.1) Definitions. *The complex* Y *is said to be a subdivision of* X *whenever there exists a set-transformation* $S: X \to Y$, *and chain-mappings* $\sigma: X \to Y$, τ: $Y \to X$ *with the following properties*:

Sd1. *S is finite-valued and closed;*

Sd2. *S^{-1} is single-valued;*

Sd3. *σ has the carrier S, τ the carrier S^{-1}, $\tau\sigma^{\cdot} = 1$, and $\sigma\tau$ is contiguous to the identity in SS^{-1}.*

We call S set-subdivision or also subdivision, σ a chain-subdivision, τ a reciprocal of σ.

(23.2) $S^{-1}S = 1$. *However $SS^{-1} = 1$ only when S is one-one.*

(23.3) *S^{-1} is weakly closed.*

This is equivalent to the relation:

$$\mathrm{Cl}\, S^{-1}\, \mathrm{Cl}\, y = \mathrm{Cl}\, S^{-1}y. \tag{23.4}$$

At all events from $y \in \mathrm{Cl}\, y$ follows $S^{-1}\, \mathrm{Cl}\, y \supset S^{-1}y$ or $\mathrm{Cl}\, S^{-1}\, \mathrm{Cl}\, y \supset \mathrm{Cl}\, S^{-1}y$. Let now $x = S^{-1}y$ and $x' \in \mathrm{Cl}\, S^{-1}\, \mathrm{Cl}\, y$. We have then $x' \in \mathrm{Cl}\, S^{-1}\, \mathrm{Cl}\, Sx = \mathrm{Cl}\, S^{-1}S\, \mathrm{Cl}\, x = \mathrm{Cl}\, x = \mathrm{Cl}\, S^{-1}y$ and hence $\mathrm{Cl}\, S^{-1}\mathrm{Cl}\, y \subset \mathrm{Cl}\, S^{-1}y$. Since each side of (23.4) contains the other, (23.4) follows.

$$\tau\sigma \sim 1, \qquad \sigma\tau \sim 1 \qquad (15.8,\ 15.2). \tag{23.5}$$

24. (24.1) Theorem. *When Y is a subdivision of X then X and Y are homologous* (12.4, 23.5).

(24.2) Theorem. *When Y is a subdivision of X then X and Y have the same homology and cohomology groups. More precisely σ [its dual σ^*] induces isomorphisms of the homology [cohomology] groups of X [of Y] with the corresponding groups of Y [of X]* (12.6, 24.1). *Similarly for τ with X, Y interchanged.*

(24.3) *Let (X_0, X_1) be a dissection of X and $Y = SX$ a subdivision of X, with σ, τ as in* (23.1). *Then*

(a) *(Y_0, Y_1), $Y_i = SX_i$, is a dissection of Y;*

(b) *$S_i = S \mid X_i$ is a set-subdivision of X_i;*

(c) *the corresponding σ, τ are $\sigma_i: X_i \to Y_i$ induced by σ and $\tau_i: Y_i \to X_i$ induced by τ. Explicitly and with π, ω as in* (22) *we have: $\sigma_0 = \omega\sigma \mid X_0$, $\tau_0 = \pi\tau \mid Y_0$, $\sigma_1 = \sigma \mid X_1$, $\tau_1 = \tau \mid Y_1$.*

Let $Sx = Z$. By Sd 1: $S\, \mathrm{Cl}\, x = \mathrm{Cl}\, Z$. Since X_1 is closed: $x \in X_1 \to \mathrm{Cl}\, x \subset X_1 \to \mathrm{Cl}\, Z \subset Y_1$. On the other hand an element y of Y_1 is in some Z and hence $\mathrm{Cl}\, y \subset \mathrm{Cl}\, Z \subset Y_1$. Thus Y_1 is closed and therefore Y_0 is open, and so (Y_0, Y_1) is a dissection of Y, which is (a). Since $SX_1 = Y_1$ we also have $\sigma X_1 \subset Y_1$, $S^{-1}Y_1 = S^{-1}SX_1 = X_1$ (23.2), and $\tau Y_1 \subset X_1$. Coupling these properties with (22.1, 22.2) it is but a step to (b, c).

(24.4) *Sx and $S\, \mathrm{Cl}\, x$ are subdivisions of X and $\mathrm{Cl}\, x$* (24.3).

(24.5) *Let S be a subdivision of X into Y, and S' a subdivision of Y into Z, with σ, τ related as before to S, and σ', τ' the analogues for S'. Then $S'S$ is a subdivision of X into Z with $\sigma'\sigma$ and $\tau\tau'$ as the related σ, τ.*

(24.6) *Let S, σ, τ, X, Y be as in* (23.1), *and let S', σ', τ', X', Y' be a second* *similar set. Then* $S \times S'$ *is a subdivision of* $X \times X'$ *into* $Y \times Y'$ *with* $\sigma \times \sigma'$, $\tau \times \tau'$ *as the related* σ, τ (21).

(24.7) *The situation being as in* (24.6) *if* θ *is a chain-mapping* $X \to X'$ *then* $\theta' = \sigma'\theta\tau$ *is a chain-mapping* $Y \to Y'$. *Furthermore if* $\bar{\sigma}$ *is the isomorphism of the homology groups of* X *with those of* Y *induced by* σ *and* $\bar{\sigma}'$ *the same for* σ', *and if* $\bar{\theta}$, $\bar{\theta}'$ *are the homomorphisms in the homology groups of* X, X' *induced by* θ, θ' *then* $\bar{\theta}' = \bar{\sigma}'\bar{\theta}\bar{\sigma}^{-1}$.

(24.8) *Let* X, Y *be simple, and* S *a finite-valued closed set-transformation* $X \to Y$ *such that*: (a) S^{-1} *is single-valued*; (b) Sx^0 *is a vertex*; (c) $(\mathrm{Cl}\ Sx)_a$ *is acyclic. Then* S *is a set-subdivision of* X *into* Y *and the related* σ, τ *may be chosen simple.*

Conditions Sd12 of (23.1) being already satisfied there remains to exhibit suitable σ, τ satisfying Sd3.

Define σ on X^0, the zero-section of X, as $\sigma x^0 = Sx^0$. This makes the Kronecker index of zero-chains invariant under σ. As a consequence $\mathrm{KI}(\mathrm{F}\sigma x^1) = \mathrm{KI}(\mathrm{F}x^1) = 0$, since X, Y are simple. By (18.1a) σ may be extended to the whole of X with S as carrier, and it is clearly simple.

We proceed in similar manner as regards τ and S^{-1}. Define τ on Y^0, the zero-section of Y, as follows. Take any vertex x^0 of $S^{-1}y^0$ and choose $\tau y^0 = x^0$. This makes the Kronecker index of zero-chains invariant under τ. Hence $\mathrm{KI}(\mathrm{F}\tau y^1) = \mathrm{KI}(\mathrm{F}y^1) = 0$. Moreover $(\mathrm{Cl}\ S^{-1}y)_a$ is a $(\mathrm{Cl}\ x)_a$, and hence it is acyclic since X is simple. Therefore by (18.1a) τ may be extended to Y with the carrier S^{-1}, and it is clearly simple.

Consider now $\tau\sigma : X \to X$. We have $\tau\sigma \mid X^0 = 1$ and since σ, τ leave $\mathrm{KI}(\xi^0)$ invariant, this holds also for $\tau\sigma$, from which we conclude as above that $\mathrm{KI}(\mathrm{F}\tau\sigma x^1) = 0$. Since $\tau\sigma$ has the carrier $S^{-1}S$ which contains the chain-mapping 1, and $(\mathrm{Cl}\ S^{-1}Sx) = (\mathrm{Cl}\ x)_a$ is acyclic, $\tau\sigma$ is contiguous to 1 in $S^{-1}S$. Since $S^{-1}S = 1$ (23.2), and so does not raise dimensions, by (18.1a) $\tau\sigma = 1$.

The same argument holds for $\sigma\tau$ and SS^{-1}. We have $(\mathrm{Cl}\ SS^{-1}y)_a = (\mathrm{Cl}\ Sx)_a$, and so it is acyclic. As before $\mathrm{KI}(\mathrm{F}\sigma\tau y^1) = 0$, and since SS^{-1} is the carrier of both $\sigma\tau$ and the chain-mapping 1, they are contiguous in SS^{-1}.

Thus Sd123 are satisfied and (24.8) is proved.

25. **Derived complexes.** There is a noteworthy process for constructing new complexes out of given complexes and which applied to simple and simplicial complexes, or polyhedra yields subdivisions. It is described in the

(25.1) DEFINITIONS. *Let* X *be any complex and choose as simplexes the ordered collections of distinct elements*

$$\sigma = \{x_0, \cdots, x_p\}, \qquad x_0 \prec x_1 \prec \cdots \prec x_p. \tag{25.2}$$

Every face of σ *is a collection of this nature, hence* $X' = \{\sigma\}$ *is a simplicial complex. It is known as the first derived of* X. *Similarly the second derived* X'' *of* X *is the first derived* $(X')'$ *of* X', $\cdots$, *the* $(n + 1)$*st derived* $X^{(n+1)}$ *of* X *is* $(X^{(n)})'$. *Any* $X^{(n)}$ *is called a derived of* X. *In general however "the derived" shall refer to* X'.

For the sake of clarity it is convenient to designate x as a vertex of X' by a new symbol $'x$. Or if we prefer we may consider the $'x$'s as new vertices in one-one correspondence with the elements x of X. The general simplex σ of X' is then described by

$$\sigma = \{'x_i \cdots 'x_j\}, \qquad x_i < \cdots < x_j. \tag{25.3}$$

For every $\sigma \in X'$ the representation (25.3) is unique and in this form $'x_i$, $'x_j$ are called the *first* and *last* vertex of σ.

(25.4) *Weakly isomorphic complexes have isomorphic derived complexes.*

(25.5) $X' \cong X^{*\prime}$, *the isomorphism being such that the orientations of* $X^{*\prime}$ *are those of* X' *after applying the orientation function* $\alpha(\sigma^p) = \beta(-p)$.

Let $'x^*$ denote x^* as a vertex of $X^{*\prime}$. Then to $\sigma \in X'$ given by (25.3) there corresponds $\sigma_0 \in X^{*\prime}$ given by

$$\sigma_0 = 'x_j^* \cdots 'x_i^*, \qquad x_j^* < \cdots < x_i^*,$$

and clearly $'x \to 'x^*$ defines an isomorphism of X' with $X^{*\prime}$ which conforms with (25.5).

26. **Derived of simplicial complexes.** Let $K = \{\sigma\}$ be our customary simplicial complex and let $'\sigma$ be the new vertex of its derived K' associated with σ. Thus the elements of K', now written ζ, assume the form

$$\zeta = '\sigma_i \cdots '\sigma_j, \qquad \sigma_i < \cdots < \sigma_j. \tag{26.1}$$

We will now introduce the following operations:

(26.2a) *Derivation D.* This is a set-transformation $K \to K'$ defined recursively by

$$D\sigma^0 = '\sigma^0 = \sigma^0; \qquad D\sigma^p = '\sigma^p(D\mathfrak{B}\sigma^p)_a, \qquad p > 0.$$

If K is augmented to K_a with the (-1)-simplex ϵ, we add the convention

$$D\epsilon = \epsilon.$$

(26.2b) *Chain-derivation.* This is a chain-transformation defined recursively by:

$$\delta\sigma^0 = \sigma^0; \qquad \delta\sigma^p = '\sigma^p\delta\mathrm{F}\sigma^p, \qquad p > 0.$$

If we set $[\sigma_i^q : \sigma_j^{q-1}] = \lambda_{ij}^q$, then

$$p > 0: \delta\sigma_i^p = '\sigma_i^p \sum \lambda_{ik}^p \delta\sigma_k^{p-1},$$

and so step by step:

$$\begin{aligned} p > 0: \delta\sigma_i^p &= \sum \lambda_{ik_1}^p \lambda_{k_1k_2}^{p-1} \cdots \lambda_{k_{p-1},k_p}^1 {'\sigma_i^p}\, {'\sigma_{k_1}^{p-1}} \cdots {'\sigma_{k_p}^0} \\ &= \sum \beta(-p)\lambda_{ik_1}^p \cdots \lambda_{k_{p-1},k_p}^1 {'\sigma_{k_p}^0} \cdots {'\sigma_i^p}, \end{aligned}$$

the summation being extended to all $k_1, \cdots, k_p$ such that $\sigma_{k_p}^0 < \sigma_{k_{p-1}}^1 < \cdots < \sigma_{k_1}^{p-1} < \sigma_i^p$.

(26.2c) *A reciprocal τ of chain-derivation.* Assign to each $'\sigma$ a vertex $\tau\,'\sigma$ of the simplex σ. It follows then from the expression (26.1) for the simplex ζ of K', that $\tau\,'\sigma_i, \cdots, \tau\,'\sigma_j$ are all vertices of σ_j. Hence $'\sigma \to \tau\,'\sigma$ defines a simplicial chain-mapping $K' \to K$ which we continue to call τ. This τ is precisely our *reciprocal* of σ (see 26.8).

(26.3) (a) *D is a closed set-transformation*; (b) $D\sigma^p = \{'\sigma_i \cdots '\sigma_j\,'\sigma^p \mid \sigma_i < \cdots < \sigma_j < \sigma^p\}$; (c) *$D^{-1}$ is single-valued.*

PROOF OF (a). Evidently $\mathrm{Cl}\, D\sigma^0 = D\, \mathrm{Cl}\, \sigma^0$. For $p > 0$, we have then

$$\mathrm{Cl}\, D\sigma^p = D\sigma^p \cup D\mathfrak{B}\sigma^p = D\, \mathrm{Cl}\, \sigma^p.$$

PROOF OF (b). It is trivial for $p = 0$, so we may assume $p > 0$. We find then from (26.2a) that $D\sigma^p$ consists of $'\sigma^p$ together with all the joins $\zeta\,'\sigma^p$, $\zeta \in D\mathfrak{B}\sigma^p$. Under the hypothesis of the induction if $\sigma_j \in \mathfrak{B}\sigma^p$ then $D\sigma_j$ consists of those and only those $\zeta \in K'$ which terminate with $'\sigma_j$. Therefore $\{\zeta \mid \zeta \in D\mathfrak{B}\sigma^p\} = \{'\sigma_i \cdots '\sigma_j \mid \sigma_i < \cdots < \sigma_j < \sigma^p\}$, and from this to (b) is but a step.

PROOF OF (c). From (b) follows that $\zeta = '\sigma_i \cdots '\sigma_j'\sigma^p$ can occur only in $D\sigma^p$, and so $D^{-1}\zeta = \sigma^p$, which proves (c).

(26.4) *δ is a chain-mapping $K \to K'$ with the carrier D and τ a chain-mapping $K' \to K$ with the carrier D^{-1}.*

The carrier properties are immediate and it is already known that τ is a simplicial chain-mapping. There remains to show that δ is a chain-mapping, or that $\mathrm{F}\delta\sigma^p = \delta\mathrm{F}\sigma^p$. For $p = 0$ this property is trivial and so we assume it for $p - 1$ and prove it for p. Now $\mathrm{F}\delta\sigma^p = \delta\mathrm{F}\sigma^p - {'\sigma^p}\mathrm{F}\delta\mathrm{F}\sigma^p = \delta\mathrm{F}\sigma^p - {'\sigma^p}(\delta\mathrm{F}\mathrm{F}\sigma^p) = \delta\mathrm{F}\sigma^p$, which is the asserted property, and so (26.4) is proved.

(26.5) $$\tau\delta = 1.$$

We must show that $\tau\delta\sigma^p = \sigma^p$ for every p. This is obvious for $p = 0$; so we assume it again for $p - 1$, $p > 0$, and prove it for p. Suppose $\tau\,'\sigma^p = A$ and $\sigma^p = \epsilon A\sigma^{p-1}$, $\epsilon = \pm 1$. Then

$$\tau\delta\sigma^p = \epsilon\tau\,'\sigma^p\delta(\sigma^{p-1} - A\mathrm{F}\sigma^{p-1}).$$

By hypothesis $\tau\delta\sigma^{p-1} = \sigma^{p-1}$, $\tau\delta A\mathrm{F}\sigma^{p-1} = A\mathrm{F}\sigma^{p-1}$. Since $\tau'\sigma^p = A$, and A is a vertex of every element of the chain $A\mathrm{F}\sigma^{p-1}$, we have $\tau\delta\,'\sigma^p A\mathrm{F}\sigma^{p-1} = 0$. Hence $\tau\delta\sigma^p = \epsilon A\sigma^{p-1} = \sigma^p$, and (26.5) is proved.

(26.6) THEOREM. *Every derived of a simplicial complex is a subdivision.*

From this and (24.2) there will follow

(26.7) THEOREM. *Derivation alters neither the homology nor the cohomology groups.*

Proof of (26.6). It is sufficient to show that K' is a subdivision. To this effect we will show by means of (24.8) that D is a set-subdivision $K \to K'$. In view of (26.3) all that is left to do is to show that $(D \text{ Cl } \sigma)_a$ is acyclic.

At all events $(D \text{ Cl } \sigma^0)_a = (\sigma^0)_a$ is acyclic. Moreover if σ^1 is the simplex AB then $D \text{ Cl } \sigma^1$ consists of two one-simplexes AC, CB ($C = {}'\sigma^1$) with their end points. An elementary calculation shows that it has no one-cycles. It is connected and hence zero-cyclic, and so $(D \text{ Cl } \sigma^1)_a$ is also acyclic.

Consider now $(D \text{ Cl } \sigma^p)_a$, $p > 1$. We may assume it proved acyclic for dimensions less than p. As a consequence (26.6) and (26.7) will hold for those dimensions. Hence $D\mathfrak{B}\sigma^p$, like $\mathfrak{B}\sigma^p$, is $(0, p - 1)$-cyclic (III, 21.4), $(D\mathfrak{B}\sigma^p)_a$ is $(p - 1)$-cyclic, and finally $(D \text{ Cl } \sigma^p)_a = {}'\sigma_a^p(D\mathfrak{B}\sigma^p)_a$ is acyclic. We conclude then from (24.8) that *D is a subdivision.* This proves (26.6).

We will complete (26.6) by:

(26.8) *δ is a chain-subdivision and τ is a reciprocal of δ.*

The carrier properties have been proved in (26.4). Referring also to (24.8) we find that δ, τ assume on the zero-sections of K, K' the values specified there for a chain-subdivision and its reciprocal. Then the argument of (24.8) yields (26.8).

(26.9) *If K is an n-circuit, a simple n-circuit, or an orientable n-circuit, so are all its respective derived* (proof elementary).

(26.10) *Derivation does not modify the dimension of a simplicial complex* (proof elementary).

27. **Derived of simple complexes.** Everything that has just been said up to (26.9), and which does not involve the description of (26.2c) carries over to simple complexes. In particular D, δ still have the forms (26.2ab), and since the proof of (26.6) does not utilize (26.2c) it continues to hold, and so does (26.7). By (24.8) δ and τ are both simple. We state explicitly for reference:

(27.1) *Derivation of a simple complex is a subdivision and so it alters neither the homology nor the cohomology groups.*

It is a direct consequence of the definition of the derived (25.1) that $\dim Dx^p \leqq p$. On the other hand $\tau\delta x^p = x^p \to \delta x^p \neq 0$, and so Dx^p contains the nonzero p-chain δx^p. It follows that $\dim Dx^p \geqq p$, and so $\dim Dx^p = p$. Thus as for simplicial complexes (26.10):

(27.2) *Derivation does not alter the dimension of a simple complex.*

Finally we have the following complementary result:

(27.3) *If X is finite,* $\dim x \geqq 0$ *and* $(\text{Cl } x)_a$ *is acyclic then* (27.1) *still holds.*

For no other properties than those stated are needed in the proof.

28. **Partitions of polyhedral complexes.** Let $\Pi = \{E\}$, $\Pi_1 = \{E_1\}$ be polyhedral complexes. We say that Π_1 is a *partition* of Π whenever every E_1 is contained in some E and every E is the union of a finite set of cells E_1. If Π_1 is an Euclidean complex the partition is said to be *simplicial.*

(28.1) Theorem. *A partition Π_1 of a polyhedral complex is a subdivision of Π and hence Π, Π_1 have the same homology and cohomology groups.*

To prove (28.1) it will also be necessary to derive the following result which is interesting for its own sake:

(28.2) *A polyhedral complex is simple.*

Let the partition operation S be defined as the replacement of each cell E by the set of cells E_1 whose union is E. To prove (28.1) it is sufficient to show that S is a set-subdivision.

Referring to (III, 6.1, 6.14) Π may be in an Euclidean space or in the Hilbert parallelotope P^ω. In any case, however, it is identified with a subset of a linear variety L in a real vector space $\mathfrak{B}$. Each E^n will then span a definite subspace $\mathfrak{B}'$ and $\mathfrak{B}' \cap L$ will be a unique linear n-dimensional variety $\mathfrak{E}^n$ determined by E^n. We refer to $\mathfrak{E}^n$ as the *space* of E^n. When Π is in an Euclidean space then $\mathfrak{E}^n$ is a subspace of that space.

Let then $\mathfrak{E}^n$ be the space of a given cell E^n of Π. If $\mathfrak{E}^{n-1} \subset \mathfrak{E}^n$ is a subspace meeting E^n, it decomposes the latter into two convex n-cells E'^n, E''^n and an $(n-1)$-cell E^{n-1}. It decomposes also similarly $\mathfrak{B}E^n$ into a polyhedral complex Π^{n-1}. The replacement of Cl E^n by $\Pi^{n-1} \cup \{E'^n, E''^n, E^{n-1}\}$ is a partition S_0 of Π. Suppose $\mathfrak{E}^{n-1}$ is such that $\mathfrak{B}E^n = \Pi^{n-1}$. Then S_0 consists merely in replacing E^n by $\{E'^n, E''^n, E^{n-1}\}$. Such a partition operation will be called *elementary*.

(28.3) *Every partition* Π_1 *of* Π *may be further partitioned to* Π_2 *which may be obtained from* Π *or* Π_1 *by a succession of elementary partition operations.*

In the space $\mathfrak{E}^n$ of E^n take the intersection of Cl E^n with all the subspaces of all the $E^q \subset E^n$, and repeat the process for all $E \in \Pi$. Let S' be the combined operation which is a partition operation and let $\Pi_2 = S'\Pi$. If Π^q is the q-section of Π, and $\Pi_2^q = S'\Pi^q$, it is clear that we can pass from Π^0 to Π_2^0 by a succession of elementary partition operations. Suppose this also proved regarding the passage from Π^{n-1} to Π_2^{n-1}. It is evident from the construction that we can then pass from Π^n to Π_2^n, by a succession of elementary partition operations applied to the n-cells of Π^n one at a time. The resulting operation on Π is S'.

At the same time as Cl E^n is being partitioned by the subspaces of the $E_1 \subset E^n$, these E_1 themselves are also partitioned and so the combined operation on Π_1 is likewise a partition operation $S_1' : \Pi_1 \rightarrow \Pi_2$. As above Π_2 is shown to be obtainable by a succession of elementary partition operations from Π_1 and so (28.3) follows.

(28.4) *An elementary partition* T *is a set-subdivision.*

The proof is by means of (24.8). We notice that:

(28.5) Π *is augmentable and with a zero-cocycle* $\gamma_0 = \sum A^i$, *where* $\{A_i\}$ *are the vertices.*

For these properties depend solely upon the one-section, which possesses them since it is an Euclidean complex and hence isomorphic with a simplicial complex.

Among the conditions which (24.8) requires, (24.8ab) are clearly fulfilled and the rest will follow if we can prove (28.2) and

(28.6) $(T\text{Cl } E)_a$ *is acyclic.*

Let $(28.2)_n$ denote (28.2) for dim $\Pi = n$, and $(28.6)_n$ denote (28.6) for $E = E^n$. Since $(28.2)_n$, $(28.6)_n$ are trivial for $n = 0$ we may assume them for $n - 1$ and prove them for n.

In order to prove $(28.2)_n$ all that is lacking, in view of (28.5), is to show that $(\text{Cl } E^n)_a$ is acyclic. Since this is obvious for $n = 0, 1$ we may assume $n > 1$. Then the asserted property is equivalent to: Cl E^n is zero-cyclic (III, 47.5). It will also follow from:

(28.7) *$\mathfrak{B}E^n$, $n > 1$, is $(0, n - 1)$-cyclic.*

For if (28.7) holds $\mathfrak{B}E^n$ has an integral $(n - 1)$-cycle γ^{n-1} such that every other is a multiple of γ^{n-1}. In particular $\text{F}E^n = t\gamma^{n-1}$. Let $\{E_i^{n-1}\}$ be the $(n-1)$-faces of E^n. By definition $[E^n : E_i^{n-1}] = \epsilon^i = \pm 1$, hence $\text{F}E^n = \epsilon^i E_i^{n-1} \neq 0$ since $\mathfrak{B}E^n$ is $(0, n - 1)$-cyclic and so must have $(n - 1)$-faces. We also have $\gamma^{n-1} = \eta^i E_i^{n-1}$. Hence say $\epsilon^i = t\eta^i = \pm 1$, and so $t = \pm 1$. It follows that $\gamma^{n-1} = \pm \text{F}E^n$, and so the $(n - 1)$-cycles of Cl E^n, which are the same as those of $\mathfrak{B}E^n$, are all ~ 0. Since $\text{F}tE^n = \pm t\gamma^{n-1} \neq 0$, tE^n cannot be a cycle. For dimensions less than $n - 1$ the homology groups of Cl E^n and $\mathfrak{B}E^n$ are the same and so Cl E^n is zero-cyclic. Thus (28.2) is reduced to (28.7).

Take a point $A \in E^n$ and choose an Euclidean simplex σ_e^n (III, 6.9) such that $A \in \sigma_e^n \subset E^n$. Let $\mathfrak{E}^n$ be the space of E^n. Each face of σ_e^n or of E^n defines together with A a certain subspace of $\mathfrak{E}^n$. The set of subspaces thus obtained, whose number is finite, is readily shown to cause isomorphic partitions of $\mathfrak{B}\sigma_e^n$ and $\mathfrak{B}E^n$. Therefore the two partitions have the same homology groups. Under the hypothesis of the induction these groups are the same as those of $\mathfrak{B}\sigma_e^n$ and of $\mathfrak{B}E^n$. Hence the latter is $(0, n - 1)$-cyclic like $\mathfrak{B}\sigma_e^n$. This proves (28.7) and hence (28.2).

There remains to prove $(28.6)_n$. We assume then E^n replaced by E^{n-1} together with two other convex cells E_1^n, E_2^n and so we have to show that if $K_1 = (\text{Cl } E_1^n)_a$, $K_2 = (\text{Cl } E_2^n)_a$, $K = K_1 \cup K_2$, then K is acyclic. Since K is a finite complex it is sufficient to show that all the integral homology groups of K are zero, i.e., that every integral cycle γ^p bounds. Now we may write $\gamma^p = C_1^p + C_2^p$, $C_i^p \subset K_i$. We have then $\gamma^{p-1} = \text{F}C_1^p = -\text{F}C_2^p \subset K_1 \cap K_2 = (\text{Cl } E^{n-1})_a$. Since the latter is acyclic $\gamma^{p-1} = \text{F}C^p$, $C^p \subset K_1 \cap K_2$. Hence $\gamma^p = \gamma_1^p + \gamma_2^p$, $\gamma_1^p = C_1^p - C^p$, $\gamma_2^p = C_2^p + C^p$, γ_i^p a cycle of K_i. Since K_i is acyclic $\gamma_i^p = \text{F}C_i^{p+1}$, $C_i^{p+1} \subset K_i$. Hence $\gamma^p = \text{F}(C_1^{p+1} + C_2^{p+1})$. Thus K is acyclic, and $(28.6)_n$ follows. This completes the proof of (28.6) and also of (28.4).

(28.8) There remains to show that any partition operation S is a subdivision. We apply again (24.8). All that it requires is to prove:

(28.9) *$(S \text{ Cl } E)_a$ is acyclic.*

By (28.3) there are two products of elementary partitions T_1, T_2 such that $T_1 S \text{ Cl } E = T_2 \text{ Cl } E$. By (28.4) and (24.5) both T_1, T_2 are set-subdivisions and so they do not modify the homology groups. Therefore those of $(\text{Cl } E)_a$ and $(S \text{ Cl } E)_a$ are the same. Since $(\text{Cl } E)_a$ has just been proved acyclic (28.9) follows and (28.1) is proved.

29. **Barycentric subdivision.** Let $K_e = \{\sigma_e\}$ be a finite Euclidean complex

and let $'\sigma_e$ be the centroid of σ_e. Define an operation D on K_e formally as follows:

(29.1) $$D\sigma_e^0 = '\sigma_e^0 = \sigma_e^0 ; \qquad D\sigma_e^p = '\sigma_e^p(DB\sigma_e^p)_a , \qquad p > 0.$$

Thus D is essentially the analogue of derivation applied to K_e.

(29.2) *D is a simplicial partition of K_e.*

Let K_e^p denote the p-section of K_e. Clearly (29.2) holds for K_e^0 ; so we assume it for K_e^{p-1} and prove it for K_e^p. Since the simplexes of $D\sigma_e^p$ are disjoint from those of DK_e^{p-1}, all that is required is to show that:

(a) each $x \in \sigma_e^p$ is in a simplex of $D\sigma_e^p$;

(b) the simplexes of $D\sigma_e^p$ are disjoint.

Since (a) holds for $'\sigma_e^p$ we may assume $x \neq '\sigma_e^p$. Since σ_e^p is convex the segment $\overline{'\sigma_e^p x}$ meets $| \mathfrak{B}\sigma_e^p |$ in a point x'. Under the hypothesis of the induction $x' \in | D\mathfrak{B}\sigma_e^p |$ and so $x' \in \sigma_{ei}^q \subset | D\mathfrak{B}\sigma_e^p |$, $x \in | '\sigma_e^p D\mathfrak{B}\sigma_e^p |$, which is (a).

Since D induces a partition of $\mathfrak{B}\sigma_e^p$, the simplexes σ_{ei}^q of $D\mathfrak{B}\sigma_e^p$ are disjoint and hence this holds also for the set $\{'\sigma_e^p , '\sigma_e^p\sigma_{ei}^q\}$, which proves (b) and therefore (29.2).

(29.3) The analogy of D with the operation of (26.2a) is obvious, and so we call it *barycentric subdivision*. If K is a simplicial antecedent of K_e (III, 6.12) then K' is an antecedent of DK_e. Owing to this DK_e is called a *barycentric derived* of K_e, and denoted by K_e'. The nth barycentric derived $K_e^{(n)}$ is defined as D^nK_e. As a partition D is a subdivision and the related chain-subdivision and reciprocal are δ given by (26.2b) with σ_e in place of σ, and τ defined in (26.2c).

(29.4) The fact that $'\sigma_e$ has been chosen as the centroid of σ_e played no role whatever, and it could equally be any other point of σ_e. The partition then obtained is merely called a *derived* of K_e. However, unless otherwise stated "derived" will be understood to refer to the barycentric derived.

(29.5) *Choose the points $\{'\sigma_e^p\}$ such that $'\sigma_e^p \in \bar{\sigma}_e^p$ and also that $\sigma_e^q < \sigma_e^p \rightarrow$ $'\sigma_e^q = '\sigma_e^p$. Then D defined by* (29.1) *is still a simplicial partition of K_e.*

The proof is the same as before and need not be repeated.

Finally the same proof yields:

(29.6) *Let $\Pi = \{E\}$ be a polyhedral complex and $'E$ the centroid of E. Then D defined by* (29.1) *with E in place of σ, defines a simplicial partition of $\Pi \cong \Pi'$ (= the derived of Π), and called the barycentric derived of Π.*

For convenience $D\Pi$ is identified henceforth with Π'.

30. Application to the duals of simplicial complexes.

(30.1) Let first $\sigma^{n+1} = A_0 \cdots A_{n+1}$, $n > 1$, and let $S^n = \mathfrak{B}\sigma^{n+1} = \{\sigma\}$, where we suppose that in S^n the incidence numbers are as in (III, 5.7) ($[\sigma A:\sigma] = 1$). For convenience if A, σ^n are opposite one another then we suppose σ^n so oriented that $\sigma^nA = \sigma^{n+1}$.

If $\sigma_i^p \in S^n$ we denote by ζ_i^{n-p} the face opposite σ_i^p so oriented that $\sigma_i^p\zeta_i^{n-p} = \sigma^{n+1}$. By the convention just made this merely yields as the ζ_i^0 the vertices A_i. Let $S_0^n = \{\zeta_i^{n-p}\}$ with the incidence numbers defined in the usual way ($[A\zeta:\zeta] = 1$). Thus S_0^n is simply S^n reoriented in a definite way.

(30.2) The dual of S^n is designated as usual by $(S^n)^*$ and its elements by $\{\sigma_p^i\}$. Now it is an elementary matter to show that $\sigma_p^i \rightarrow \zeta_i^{n-p}$ preserves the incidences in $(S^n)^*$ and merely raises dimensions n units. Therefore it defines a weak isomorphism $\theta:(S^n)^* \rightarrow S_0^n$ with dimensions raised n units. Thus S^n *is weakly isomorphic with its own dual, the dimensions being raised n units.*

(30.3) Let now K be any finite simplicial complex and let its vertices be augmented by at least two thus producing a set $\{A_i\}$, $i = 0, \cdots, n + 1$, where the first $r \leqq n - 1$ are the vertices of K. We construct $\sigma^{n+1} = A_0 \cdots A_{n+1}$, then S^n as before and so we have K immersed as a subcomplex in S^n. We adopt in K the incidences of S^n, i.e., it is to be a subcomplex of S^n itself. Then again $\theta K^* = K_0$ is a subcomplex of S_0^n (i.e., with the incidence numbers of S_0^n) which is a weak isomorph of K^* with dimensions raised n units. Thus:

(30.4) *Given any finite simplicial complex K, and an integer n above a certain value, then S^n contains an isomorph of K and also a weak isomorph K_0 of K^* with dimensions raised n units.*

(30.5) We shall adopt naturally enough the same incidence numbers of (III, 5.7) for the derived of S^n, and also for those of K, as in S^n and K. The only change required in (26) is in the replacement of (26.2b) as the definition of δ by:

$$\delta\sigma^0 = \sigma^0, \qquad \delta\sigma^p = (\delta \mathrm{F}\sigma^p)'\sigma^p, \qquad p > 0. \tag{30.5a}$$

It is easily seen that this affects no ulterior argument in (26).

(30.6) Take now the first derived $(S^n)'$ of S^n. The vertex mapping $'\sigma_i^p \rightarrow {'\zeta_i^{n-p}}$ defines an isomorphism t of $(S^n)'$ with $(S_0^n)'$ under which $'\sigma_i \cdots {'\sigma_j}$, $\sigma_i < \cdots < \sigma_j$, of $(S^n)'$ goes into $'\zeta_j \cdots {'\zeta_i}$, $\zeta_j < \cdots < \zeta_i$, of $(S_0^n)'$. In other words the set of simplexes beginning with $'\sigma_i$ goes into the set of those ending with $'\zeta_i$. Under t there corresponds to the set of simplexes of the derived $(\zeta_i^{n-p})'$ of ζ_i^{n-p} (i.e., with $'\zeta_i^{n-p}$ as last vertex) the set η_i^{n-p} of the simplexes beginning with $'\sigma_i^p$. Since t is an isomorphism, if we introduce in $\{\eta_i^{n-p}\}$ the dimensions and incidence relations of S_0^n, there is obtained an isomorph S_1^n of S_0^n such that $(S_1^n)' = (S_0^n)'$. By (30.4) S_1^n is a weak isomorph of $(S^n)^*$ with dimensions raised n units and in S_1^n the image of σ_p^i has for derived the set of simplexes of $(S^n)'$ beginning with $'\sigma_i^p$.

Notice that $(\sigma_i^p)'$ and η_j^{n-q} have common simplexes when and only when there are simplexes whose first vertex is $'\sigma_j^q$ and last vertex $'\sigma_i^p$, i.e., when and only when $\sigma_j^q < \sigma_i^p$, and all such simplexes are common to both. Thus we have:

(30.7) *$(\sigma_i^p)'$ and η_j^{n-q} have common simplexes (elements of $(S^n)'$) when and only when $\sigma_j^q < \sigma_i^p$, and their common simplexes consist then of all those beginning with $'\sigma_j^q$ and ending with $'\sigma_i^p$.*

(30.8) Suppose now K immersed as a subcomplex in S^n. Then $K_1 = \{\eta_i^{n-p} \mid \sigma_i^p \in K\}$ is a weak isomorph of K^* with dimensions raised n units and (30.7) continues to hold as between the $\sigma_i^p \in K$ and the corresponding $\eta_i^{n-p} \in K_1$.

Since η_i consists of all the simplexes of $(S^n)'$ which begin with the vertex $'\sigma_i^p \in K'$ we have:

$$K_1' = \text{St } K', \text{ (star in } (S^n)').$$

31. Let us take now an Euclidean $(n + 1)$-simplex, and to simplify matters let us preserve the notations of (30). Thus the Euclidean simplex will be written σ^{n+1}, and we write throughout $\sigma^p, \cdots$ instead of $\sigma_e^p, \cdots$ for the Euclidean elements. It is then an elementary matter to verify:

(31.1) *Let* $\mathfrak{E}^{n+1}$ *be the space of* σ^{n+1} *and if* $\sigma_i^p < \sigma^{n+1}$, $p \leqq n$, *let* $\mathfrak{E}_i^{p+1}$ *be the subspace of* $\mathfrak{E}^{n+1}$ *defined by* σ_i^p *and* $'\sigma^{n+1}$. *Then* $(S^n)'$ *is the partition of* S^n *caused by its intersections with the* $\mathfrak{E}_i^{p+1}$.

(31.2) η_i^{n-p} *consists of all the simplexes of* $(S^n)'$ *with the vertex* $'\sigma_i^p$ *and otherwise exterior to* σ_i^p. *The set* η_i^{n-p} *is in the space defined by* $'\sigma_i^p$ *and the complement* ζ_i^{n-p} *of* σ_i^p *in* σ^{n+1}.

Owing to this property one may also describe η_i^{n-p} as a cell *transverse* to σ^p. Notice that if σ^{n+1} is regular (in the sense of elementary geometry) then η_i^{n-p} is in the $\mathfrak{E}^{n-p+1}$ orthogonal to the space of σ_i^p in $\mathfrak{E}^{n+1}$ at $'\sigma_i^p$. From (30.7, 30.8) we deduce also:

(31.3) $\sigma_i^p \cap \eta_i^{n-q} \neq \emptyset$ *when and only when* $\sigma_j^q < \sigma_i^p$ *and the intersection is the union of the Euclidean simplexes of* $(S^n)'$ *whose extreme vertices are* $'\sigma_j^q$ *and* $'\sigma_i^p$. *The same property holds for a subcomplex* K *of* S^n *as regards the* $\sigma_i^p \in K$ *and the related elements* $\eta_i^{n-p} \in K_1$ *(related to* K *as in* 30).

This property will provide valuable indications regarding chain-intersections (V, 4).

CHAPTER V

COMPLEXES: MULTIPLICATIONS AND INTERSECTIONS. FIXED ELEMENTS. MANIFOLDS

The present chapter brings to a close the general theory of complexes. Multiplications for finite complexes are introduced in a general form and then specialized to intersections. The complements required by infinite complexes will be found in (3, 18). The algebraic treatment of fixed elements of chain-mappings and the related question of coincidences follow and the chapter concludes with an extensive treatment of combinatorial manifolds.

General references: Alexander [c, d, e], Alexandroff [f], Čech [f], Kolmogoroff [b, c], Lefschetz [a; L, III–VI], Newman [b], Pontrjagin [d], Tucker [a], Veblen [V], Whitehead [a], Whitney [d].

§1. MULTIPLICATIONS

1. (1.1) We shall be dealing for the time being with three basic complexes X, Y, Z which are assumed finite unless otherwise stated. In addition we shall have two groups G, H commutatively paired to a third group J. Occasionally also we may have $G = H = J = \rho$, a commutative ring.

(1.2) Let μ be a chain-mapping $X \times Y \rightarrow Z$ given by $\mu(x_i^p \times y_j^q) = \mu_{i\,j}^{h}(p, q)z_h^r$, $r = p + q$. If we have chains $\xi^p = g^i x_i^p$, $\eta^q = h^j y_j^q$ of X, Y over G, H then

$$(1.3) \qquad \mu(\xi^p \times \eta^q) = g^i h^j \mu_{i\,j}^{h}(p, q) z_h^r = \zeta^r,$$

is a chain of Z over J, also denoted by

$$(1.4) \qquad \zeta^r = \xi^p \times_\mu \eta^q.$$

The operation $\times_\mu$ thus defined is known as a *multiplication of* X, Y *to* Z, or merely a *multiplication* whenever X, Y, Z are clear from the context.

Let the notations be those of (IV, 5). From the fact that μ is a chain-mapping together with (IV, 5.3, $\cdots$, 5.9) we deduce:

(1.5) *μ is a multiplication pairing $\mathfrak{C}^p(X, G)$, $\mathfrak{C}^q(Y, H)$ to $\mathfrak{C}^{p+q}(Z, J)$.*

$$(1.6) \qquad \mathrm{F}(\xi^p \times_\mu \eta^q) = (\mathrm{F}\xi^p) \times_\mu \eta^q + (-1)^p \xi^p \times_\mu \mathrm{F}\eta^q.$$

(1.7) *$\gamma^p \times_\mu \delta^q$ is a $(p + q)$-cycle.*

(1.8) *γ^p or $\delta^q \sim 0 \rightarrow \gamma^p \times_\mu \delta^q \sim 0$.*

(1.9) *If $\gamma^p \in \Gamma^p$ and $\delta^q \in \Delta^q$ then $\gamma^p \times_\mu \delta^q$ is in a fixed homology class written $\Gamma^p \times_\mu \Delta^q$ and this operation is a multiplication pairing $\mathfrak{H}^p(X, G)$, $\mathfrak{H}^q(Y, H)$ to $\mathfrak{H}^{p+q}(Z, J)$.*

(1.10) *Multiplication ring.* Consider the special case $X = Y = Z$ and $G = H = J = \rho$. Then we only have cycles γ over ρ and their classes Γ. Let

$\mathfrak{H}(X, \rho) = \mathbf{P}\mathfrak{H}^p(X, \rho)$. The group $\mathfrak{H}(X, \rho)$ may be identified with the additive group generated by all the homology classes over ρ. Between the generating elements $\{\Gamma^p\}$ we have relations of the form $\Gamma^p \times_\mu \Gamma^q = \Gamma^{p+q}$. They may be considered as defining a distributive *but not necessarily associative* operation of multiplication between the elements of $\mathfrak{H}(X, \rho)$, thus turning the group into a ring $R(X, \rho, \mu)$. Such a ring was introduced (in connection with intersections, (8.8c) by J. W. Alexander [d] and Gordon [a]. See also the Pontrjagin ring of (VIII, 53.1).

(1.11) Suppose that there is merely given a chain-transformation $\mu: X \times Y \to Z$ and we wish to find if it is a multiplication. This requires that we show that $\mu\mathrm{F} = \mathrm{F}\mu$, or by (IV, 5.5) that:

$$\mathrm{F}\mu(\xi^p \times \eta^q) = \mu\mathrm{F}(\xi^p \times \eta^q) = \mu\{(\mathrm{F}\xi^p) \times \eta^q + (-1)^p\xi^p \times \mathrm{F}\eta^q\}.$$

If we write μ as a product $\times_\mu$ this reduces to (1.6). Therefore:

(1.12) *A n. a. s. c. that a chain-transformation $\mu:X \times Y \to Z$ be a multiplication is that it satisfy the boundary relation* (1.6).

(1.13) An example. The following noteworthy example is due to W. W. Flexner [c]. It will be of interest in another connection later. We choose X locally finite but otherwise arbitrary and define a multiplication φ of X, X^* to X', the derived of X (IV, 25) as follows. Set

$$[x_i^p : x_j^{p-1}] = \lambda_{ij}^p = [x_{p-1}^j : x_p^i],$$

and let the designations for the simplexes of X' be those of (IV, 25). Then the relations defining φ are:

$$\varphi(x_i^p \times x_p^j) = \beta(p)\delta_i^j\,'x_i^p\,; \tag{1.14}$$

$$p < q\colon \varphi(x_i^p \times x_q^j) = 0; \tag{1.14a}$$

$$\varphi(x_i^p \times x_{p-1}^j) = \begin{cases} \beta(p-1)\lambda_{ij}^p\,'x_i^p\,'x_j^{p-1}, \text{ where } x_j^{p-1} \prec x_i^p, \\ 0 \text{ otherwise;} \end{cases} \tag{1.15}$$

$$q < p - 1\colon \varphi(x_i^p \times x_q^j) = \sum \beta(q)\lambda_{ik_1}^p\lambda_{k_1k_2}^{p-1} \cdots \lambda_{k_rj}^{q+1}\,'x_i^p\,'x_{k_1}^{p-1} \cdots\,'x_j^q, \tag{1.16}$$

where $r = p - q - 1$ and the summation is over all $k_1, \cdots, k_r$ such that $x_i^p \succ x_{k_1}^{p-1} \succ \cdots \succ x_j^q$. We must verify (1.6). It is trivial for (1.14) and immediate for (1.15). The relation FF = 0 yields:

$$\sum_{k_s} \lambda_{k_{s-1}k_s}^{p-s+1}\lambda_{k_sk_{s+1}}^{p-s} = 0.$$

From this follows for $q < p - 1$:

$$\begin{aligned} \mathrm{F}(x_i^p \times_\varphi x_q^j) = &\sum \beta(q)\lambda_{ik_1}^p \cdots \lambda_{k_rj}^{q+1}\,'x_{k_1}^{p-1} \cdots\,'x_j^q \\ &+ (-1)^{p-q}\sum \beta(q)\lambda_{ik_1}^p \cdots \lambda_{k_rj}^{q+1}\,'x_i^p \cdots\,'x_{k_r}^{q+1}. \end{aligned} \tag{1.17}$$

The term in the first sum is at once reduced to $(\mathrm{F}x_i^p) \times_\varphi x_q^j$. The term in the second sum is

$$(-1)^{p-q}\beta(q)\beta(q+1)\{\beta(q+1)\lambda^p_{ik_1}\cdots\lambda^{q+1}_{k_r j}{}'x^p_i\cdots{}'x^{q+1}_{k_r}\} = (-1)^p x^p_i \times_\varphi \mathrm{F}x^j_q.$$

Substituting in (1.17) we obtain (1.6) for φ, and so φ is a multiplication.

2. Let again μ be as in (1.1). Since it is a chain-mapping $X \times Y \to Z$, μ has a chain-graph Γ which is a zero-cycle of $X^* \times Y^* \times Z$ (IV, 11). We propose to utilize Γ to show that there are in general certain new multiplications related to μ. Referring to (IV, 11) we have:

$$\Gamma = \sum_{p,q} \beta(r)\mu^h_{ij}(p,q)x^i_p \times y^j_q \times z^r_h, \qquad r = p+q. \tag{2.1}$$

By permuting, if possible, the factors X^*, Y^*, Z we shall obtain new expressions for Γ and hence the new multiplications. These permutations are only possible when the factors to be permuted are disjoint and so we have to examine several cases.

(a) *X, Y are disjoint.* Then $X^* \cap Y^* = \emptyset$ and permutation of the two factors X^*, Y^* is the same as applying to $X^* \times Y^* \times Z$ an orientation function $\alpha(x^i_p \times y^j_q \times z^r_h) = (-1)^{pq}$ (IV, 1.4). Therefore by (III, 10.1) Γ as a cycle of $Y^* \times X^* \times Z$ is to be written:

$$\Gamma = \sum_{p,q} (-1)^{pq}\beta(r)\mu^h_{ij}(p,q)y^j_q \times x^i_p \times z^r_h. \tag{2.2}$$

Consequently (IV, 11) Γ determines a chain-mapping $Y \times X \to Z$, written likewise as a μ-multiplication, and given by:

$$y^q_j \times_\mu x^p_i = (-1)^{pq}\mu^h_{ij}(p,q)z^r_h. \tag{2.3}$$

That is to say,

$$y^q_j \times_\mu x^p_i = (-1)^{pq}x^p_i \times_\mu y^q_j,$$

and therefore for a product of chains

$$\eta^q \times_\mu \xi^p = (-1)^{pq}\xi^p \times_\mu \eta^q. \tag{2.4}$$

(b) *Y^*, Z are disjoint.* We may then write Γ as a cycle of $X^* \times Z \times Y^*$:

$$\Gamma = \sum \beta(-q)\{\beta(p)\mu^h_{ij}(p,q)\}x^i_p \times z^r_h \times y^j_q. \tag{2.5}$$

From this expression of Γ we infer that there exists a chain-mapping $X \times Z^* \to Y^*$, which written as a μ-multiplication, is given by:

$$x^p_i \times_\mu z^h_r = \beta(p)\mu^h_{ij}(p,q)y^j_q. \tag{2.6}$$

(c) *Z disjoint from X^*, Y^*.* Proceeding as under (a) we obtain a multiplication $\zeta_r \times_\mu \xi^p$ given by:

$$\zeta_r \times_\mu \xi^p = (-1)^{pr}\xi^p \times_\mu \zeta_r, \tag{2.7}$$

which may be viewed as giving the commutation rule for the operation (2.6).

(d) *X and $Y \cup Z^*$ are disjoint.* Then Γ may be written as a cycle of $Y^* \times Z \times X^*$:

(2.8) $$\Gamma = \sum \beta(-p)\{\beta(p)\beta(r)\mu^{h}_{i\,j}(p, q)\}y^{j}_{q} \times z^{r}_{h} \times x^{i}_{p}\,.$$

This yields the chain-mapping $Y \times Z^* \to X^*$ which written as a multiplication reads:

(2.9) $$y^{q}_{j} \times_{\mu} z^{h}_{r} = \beta(p)\beta(r)\mu^{h}_{i\,j}(p, q)x^{i}_{p}\,.$$

If in addition Y, Z^* *are disjoint* there is obtained for this last multiplication the commutation rule

(2.10) $$\zeta_r \times_{\mu} \eta^{q} = (-1)^{qr}\eta^{q} \times_{\mu} \zeta_r\,.$$

(2.11) To sum up then, in addition to the initial multiplication μ there may exist two more with possible commutation of the factors under the circumstances which have just been described.

(2.12) In similar manner we may introduce a multiplication of $n - 1$ complexes $X_1, \cdots, X_{n-1}$ to yield chains of an nth complex X_n. It will be a chain-mapping $X_1 \times \cdots \times X_{n-1} \to X_n$ with a graph which is a zero-cycle in $X^*_1 \times \cdots \times X^*_{n-1} \times X_n$.

3. (3.1) The situation is particularly interesting when the three factors X, Y, Z coincide or else are one of X, X^*. The complexes which are geometrically significant are the simple complexes or their duals, and for these $\dim x \geqq 0$ or else $\leqq 0$. Hence on dimensional grounds we may exclude chain-mappings of type $X^2 \to X^*$ as of little significance. This leaves the types $X^2 \to X$ and $X^{*2} \to X^*$. They are really the same with X, X^* interchanged and so we will describe them as *dual to one another*. The results obtained for the one may be applied to the other provided that all the elements are replaced by their duals. It will be sufficient therefore to concentrate upon the second type. This is the one which interests us primarily since it includes intersections.

It may be observed that $X \cap X^* = \emptyset$, and so the disjunction properties required in (2) are easily verified here.

(3.2) We shall then discuss principally the multiplication of X^*, X^* to X^*. The associated graph is a zero-cycle Γ of $X^2 \times X^*$. The elements of the complex are the triples $(x^{q}_{j}, x^{r}_{h}, x^{i}_{p})$, written $x^{q}_{j} \times x^{r}_{h} \times x^{i}_{p}$, and ordered with respect to x^{q}_{j}, x^{r}_{h}. To be precise the element just written is *distinct* from $x^{r}_{h} \times x^{q}_{j} \times x^{i}_{p}$. On the other hand if we write the complex as $X \times X^* \times X$ we are supposed to replace $x^{q}_{j} \times x^{r}_{h} \times x^{i}_{p}$ by $x^{q}_{j} \times x^{i}_{p} \times x^{r}_{h}$, the effect being the same as applying an orientation function whose value on the element under consideration is $(-1)^{pr}$. Thus

(3.3) $$x^{q}_{j} \times x^{i}_{p} \times x^{r}_{h} = (-1)^{pr}x^{q}_{j} \times x^{r}_{h} \times x^{i}_{p}\,.$$

Similarly as regards replacing $X \times X^* \times X$ by $X^* \times X^2$. Thus the commutation rules of (2) will only be applicable to the admissible commutations, namely of X^* with one of the other factors. As a consequence we have only two basic distinct equivalent forms for Γ. In the first form it appears as a cycle of $X^* \times X^2$:

$$\Gamma = \sum_{q,r} (-1)^p \mu_i^{jh}(q, r) x_p^i \times x_j^q \times x_h^r, \qquad q + r = p. \tag{3.4}$$

In the second form Γ appears as a cycle of $X^2 \times X^*$:

$$\Gamma = \sum_{q,r} \mu_i^{jh}(q, r) x_j^q \times x_h^r \times x_p^i. \tag{3.5}$$

The expression of Γ as a cycle of $X \times X^* \times X$ is not needed since it merely yields a commutation rule that may be written down directly.

As in (2) we read off the multiplications from the expressions of Γ. Since $X^* \times X^2 = (X \times X^*)^* \times X$, (3.4) yields the chain-mapping $\mu_1: X \times X^* \to X$ defined by:

$$\mu_1(x_i^p \times x_q^j) = (-1)^p \beta(r) \mu_i^{jh}(q, r) x_h^r. \tag{3.6}$$

Since X, X^* can be commuted, if μ_1 is written as an operation $\times_{\mu_1}$ we have

$$\xi_q \times_{\mu_1} \xi^p = (-1)^{pq} \xi^p \times_{\mu_1} \xi_q. \tag{3.7}$$

Similarly from (3.5) there comes the chain-mapping $\mu_2^*: X^{*2} \to X^*$ defined by:

$$\mu_2^*(x_q^j \times x_r^h) = \beta(-p) \mu_i^{jh}(q, r) x_p^i. \tag{3.8}$$

This time there is no commutation rule such as (3.7).

(3.9) *The graph* Γ *determines two multiplications* $\mu_1: X \times X^* \to X$ *and* $\mu_2^*: X^{*2} \to X^*$, *and each of* μ_1, μ_2^*, Γ *determines the other two uniquely.*

(3.10) A pair of multiplications $\mu = (\mu_1, \mu_2^*)$ such as just considered, determined by the same graph Γ, will be called a *multiplication* in X, and μ_1, μ_2^* will be referred to as the *first* and *second* component of μ.

(3.11) Side by side with μ_2^* we shall also consider on occasion its dual μ_2 which is a chain-mapping $X \to X^2$.

(3.12) *Multiplications in infinite complexes.* Going back to the situation of (1) we may consider multiplications of X, Y to Z of f-, i-, or fi-types in the sense of (IV, 19) and there is no difficulty in extending to them the results of (1, 2). We may also restrict the chains of X or Y or both to being finite with appropriate modifications of the paired groups.

(3.13) A more interesting situation concerns multiplications in a complex X. Suppose first X *closure-finite.* Only finite chains are admissible and so μ_1 must be of type f. We learn from (3.6) that when q, r, i, j are given then $\mu_i^{jh}(q, r) \neq 0$ for at most a finite number of values of h. Suppose that we have a μ_1 with the property just asserted. In order to have a μ_2^* which is an i-chain-mapping we find from (3.8) that given i, q, r then $\mu_i^{jh}(q, r) \neq 0$ for at most a finite number of values of j, h. Therefore in order that we have a multiplication $\mu = (\mu_1, \mu_2^*)$ in X the matrices $\| \mu_i^{jh}(q, r) \|$ (i, q, r fixed) must have at most a finite number of terms different from 0 in each row or column.

The treatment of a star-finite X is the same with chains and cochains interchanged and otherwise obvious modifications. The locally finite type partakes of the properties of both and its discussion is safely omitted.

§2. INTERSECTIONS

4. We shall now assume that the basic complex X is finite and simple and, by specializing the multiplication μ, obtain *intersections* in X. The concentration upon simple complexes is amply justified on geometric grounds since they include simplicial and polyhedral complexes. The extension to infinite complexes will be taken up later (18).

From our earlier intersection theory (see notably [L, IV]) we are already led to associating intersections of chains with certain bilinear functions of chains, i.e., with multiplications. On the other hand returning to the situation and notations of (IV, 31) we have seen that σ_i^p and the transverse η_j^{n-q} of σ_j^q intersect when and only when $\sigma_j^q \prec \sigma_i^p$. If we think of η_j^{n-q} as an image of the dual σ_q^j then we may expect an intersection of σ_i^p with σ_q^j only when $\sigma_j^q \prec \sigma_i^p$, and its dimension will be less than or equal to $p - q$ (IV, 31.3).

On the strength of the preceding remarks we introduce for the finite simple complex X a multiplication $\mu = (\mu_1, \mu_2^*)$, written as a dot-product: $x^p \cdot x_q$, $x_p \cdot x_q$, $x_p \cdot x^q$, and satisfying the following two conditions:

(4.1) $x^p \cdot x_q$ = *a* $(p - q)$*-chain of* Cl x^p *which is zero whenever* $x^q \not\prec x^p$;

(4.2) $x^p \cdot x_p = \beta(p)x^0$, x^0 *a vertex of* x^p.

(4.3) The important condition is (4.1), and it is patterned after a "local" condition due to Whitney [d]. Condition (4.2) has an interesting consequence as regards the Kronecker index. We have already defined in (III, 28): $\mathrm{KI}(x^p, x_p) = \beta(p)$. On the other hand as a zero-chain $x^p \cdot x_p$ has likewise a Kronecker index which is

$$\mathrm{KI}(\beta(p)x^0) = \beta(p) = \mathrm{KI}(x^p, x_p).$$

Hence from (4.2) for any pair ξ^p, η_p:

$$\mathrm{KI}(\xi^p, \eta_p) = \mathrm{KI}(\xi^p \cdot \eta_p). \tag{4.4}$$

(4.5) It is to be noted that $x^p \cdot x_p$ is not completely determined by (4.2) when $p > 0$, since x^0 is an unspecified vertex of x^p. A similar remark will apply regarding the vertices present in certain expressions below.

(4.6) Conditions (4.1, 4.2) refer to the component μ_1 alone. Suppose that there has been found a μ_1 satisfying (4.1, 4.2). Writing this operation μ_1 as a dot-product we will have (3.6):

$$x_i^p \cdot x_q^j = (-1)^p \beta(r) \mu_i^{jh}(q, r) x_h^r, \qquad q + r = p. \tag{4.7}$$

For the particular value $q = p$ condition (4.2) yields explicitly

$$x_i^p \cdot x_p^j = \beta(p) \delta_i^j x_h^0, \qquad x_h^0 \prec x_i^p, \tag{4.8}$$

and is equivalent to:

$$\mu_i^{jh}(p, 0) = \begin{cases} (-1)^p \beta(p) & \text{for one } h \text{ such that } x_h^0 \prec x_i^p; \\ 0 & \text{otherwise.} \end{cases} \tag{4.9}$$

The operation $\overset{*}{\mu}_2$ written as a dot-product is:

$$x_q^j \cdot x_r^h = \beta(-p)\mu_i^{jh}(q, r)x_p^i, \qquad p = q + r, \tag{4.10}$$

and (4.1, 4.2) are equivalent to the following conditions on $\overset{*}{\mu}_2$:

(4.11) *$x_q \cdot x_r$ is a $(q + r)$-cochain of* St $x^q \cap$ St x^r.

(4.12) *$x_p \cdot x_0 = x_p$ for a certain $x_0 > x_p$, and $= 0$ for all other x_0.*

We recall that the fundamental cocycle of X is $\gamma_0 = \sum x_0^i$. A consequence of (4.12) is the noteworthy relation:

$$x_p \cdot \gamma_0 = x_p. \tag{4.13}$$

5. Since (4.1, 4.2) and (4.11, 4.12) are equivalent we introduce the

(5.1) Definition. *By a chain-intersection or merely intersection in a simple complex X is meant a multiplication $\mu = (\mu_1, \overset{*}{\mu}_2)$ written as a dot-product, which satisfies any one of the two sets of conditions* (4.1, 4.2) *or* (4.11, 4.12), *and therefore both sets. More explicitly, an intersection in X is a pair of chain-mappings $\mu_1 : X \times X^* \to X$, $\overset{*}{\mu}_2 : X^{*2} \to X^*$ related as in* (3) *by their common graph and which satisfy, respectively, the equivalent pairs of conditions* (4.1, 4.2) *and* (4.11, 4.12).

(5.2) In everything that precedes, the basic convention prevails that X^* is so oriented that $[x_i^p : x_j^{p-1}] = [x_{p-1}^j : x_p^i]$. We shall find it convenient at times to adopt a different orientation for the dual complex. If it is subjected to an orientation function $\alpha(x_p^i)$ then $X \times X^*$ will be subjected to the orientation function $\alpha'(x_i^p \times x_q^j) = \alpha(x_q^j)$ and so we will have to replace $x^p \cdot x_q$ and x_q by $\alpha(x_q)x^p \cdot x_q$, $\alpha(x_q)x_q$. Incidentally we recall that $\mathrm{KI}(x^p, x_p)$ will then also go into $\alpha(x_p)\mathrm{KI}(x^p, x_p)$ (III, 28.4).

6. Before proceeding we require suitable carriers for the components μ_1, $\overset{*}{\mu}_2$ of intersection.

(6.1) *$x^p \times x_q \to x^p$ defines a set-transformation $T_1 : X \times X^* \to X$ which is a closed carrier for μ_1.*

The carrier property is obvious. To prove T_1 closed, observe that $(x^p \times x_q < x^r \times x_s \to x^p < x^r) \to T_1 \mathrm{Cl}(x^r \times x_s) \subset \mathrm{Cl}\, T_1(x^r \times x_s)$. On the other hand $\mathrm{Cl}\, T_1(x^r \times x_s) = \mathrm{Cl}\, x^r = \{x^p \mid x^p < x^r\} \subset T_1\{x^p \times x_q \mid x^p < x^r, x_s > x_q\} = T_1 \mathrm{Cl}(x^r \times x_s)$. Hence $T_1 \mathrm{Cl} = \mathrm{Cl}\, T_1$, or T_1 is closed.

(6.2) *$x \to \mathrm{Cl}(x \times x)$ defines a set-transformation $T_2 : X \to X^2$ which is a closed carrier for μ_2. Hence* (IV, 8.4) *$\overset{*}{T}_2$ is a closed carrier for $\overset{*}{\mu}_2$.*

The proof of $T_2 \mathrm{Cl} = \mathrm{Cl}\, T_2$ is elementary. By (4.11) $\overset{*}{\mu}_2$ maps $x_r \times x_s$ into a cochain of elements x_p such that $x^r, x^s < x^p$. Therefore μ_2 maps x^p into a chain of $\mathrm{Cl}(x^p \times x^p)$ and so it has the carrier T_2.

7. (7.1) Existence and uniqueness theorem. *A chain-intersection may be introduced in every simple complex and is unique to within contiguity in the carriers of the components* (Whitney [d]).

We first observe that μ_1 is to be a chain-mapping $X \times X^* \rightarrow X$ which vanishes on the negative-dimensional elements and also on the set $Y_1 = \{x^p \times x_q \mid x^q \nless x^p\}$. Now $x^p \times x_q < x^r \times x_s \rightarrow x^p < x^r, x^s < x^q$. Therefore $x^s \nless x^r \rightarrow x^q \nless x^r \rightarrow x^q \nless x^p$, or Y_1 is a closed subcomplex of $X \times X^*$. Since the zero-section Y^0 of $X \times X^*$ is likewise a closed subcomplex, the same holds for $Y = Y^0 \cup Y_1$. Thus the requirements on μ_1 may be formulated as: (a) it is to have the carrier T_1; (b) it is to be zero on all but the zero-elements of Y, and on the zero-elements it is to receive values in accordance with (4.1, 4.2). Since $(\mathrm{Cl}\ T_1 x^p \times x_q)_a = (\mathrm{Cl}\ x^p)_a$ is acyclic, and the elements of $X \times X^* - Y$ are positive-dimensional, and X is simple, by (IV, 18.1a) there will exist a suitable μ_1 provided that the values on the zero-dimensional elements satisfy

$$\text{(7.2)} \qquad \mathrm{KI}(\mu_1 \mathrm{F}(x_i^p \times x_{p-1}^j)) = 0.$$

This relation reduces at once to

$$\text{(7.3)} \qquad \mathrm{KI}(\mu_1((\mathrm{F}x_i^p) \times x_{p-1}^j)) + (-1)^p \mathrm{KI}(\mu_1(x_i^p \times \mathrm{F}x_{p-1}^j)) = 0.$$

By (4.1) the indices in (7.3) have the respective values $\beta(p-1)[x_i^p : x_j^{p-1}]$, and $(-1)^p \beta(p)[x_{p-1}^j : x_p^i]$, and as they are equal and opposite in signs (7.2) holds. Thus μ_1, and hence a chain-intersection $\mu = (\mu_1, \mu_2^*)$, does exist in X.

Suppose that there are two chain-intersections $\mu = (\mu_1, \mu_2^*)$, $\mu' = (\mu_1', \mu_2'^*)$. We have to prove that μ_1 and μ_1' are contiguous in T_1 and μ_2^*, $\mu_2'^*$ are contiguous in T_2^*. Since the contiguity of μ_2^*, $\mu_2'^*$ in T_2^* is equivalent to that of μ_2, μ_2' in T_2, in the last analysis it is to be shown that μ_i, μ_i' $(i = 1, 2)$ are contiguous in T_i. Since $(\mathrm{Cl}\ T_1 x^p \times x_q)_a$, $(\mathrm{Cl}\ T_2 x)_a$ are acyclic and μ_i, μ_i' map the negative-dimensional elements into zero, the required result will follow by (IV, 18.7), if given any finite zero-chain of the complex upon which μ_i operates we have

$$\text{(7.4)} \qquad \mathrm{KI}(\mu_i \xi^0) = \mathrm{KI}(\mu_i' \xi^0).$$

Take first μ_1, μ_1'. By (4.2) the value of $\mathrm{KI}(\mu_1(x_i^p \times x_p^j))$ is the same for all intersections, hence also the same for μ_1, μ_1'. Therefore (7.4) holds and so μ_1, μ_1' are contiguous in T_1.

Referring to (4.11, 4.12) the transforms of the zero-cochains under μ_2^* are defined by:

$$\text{(7.5)} \qquad \mu_2^*(x_0^i \times x_0^j) = 0, \quad i \neq j; \qquad \mu_2^*(x_0^i \times x_0^i) = x_0^i.$$

Hence the transforms of the zero-chains under μ_2 are defined by

$$\text{(7.6)} \qquad \mu_2 x_i^0 = x_i^0 \times x_i^0.$$

Once more $\mathrm{KI}(\mu_2 x_i^0) = 1$ and so it is independent of μ_2. Thus (7.4) holds for $i = 2$ also. It follows that μ_2, μ_2' are contiguous in T_2 and the proof of (7.1) is completed.

8. **Multiplicative properties of intersections.** They are first of all those of intersections as multiplications. In addition, however, there are important complementary properties which will require separate treatment.

(8.1) *Notations.* We will consider groups G, H, J and the ring ρ as in (1). We write:

$$\begin{aligned} \xi &= \text{a chain or cochain over } G; \\ \gamma &= \text{a cycle or cocycle over } G; \\ \Gamma &= \text{the class of } \gamma; \\ \eta, \delta, \Delta & \quad \text{the same as } \xi, \gamma, \Gamma \text{ over } H; \end{aligned}$$

if $G = H = J = \rho$ the only designations will be ξ, γ, Γ.

Let then $\mu = (\mu_1, \overset{*}{\mu_2})$ be an intersection in X. Since μ_1 and $\overset{*}{\mu_2}$ are, respectively multiplications of X, X^* to X and X^*, X^* to X^*, we have from (1.5, $\cdots$, 1.9, 3.7):

(8.2) *Intersections pair* $\mathfrak{C}^p(X, G)$, $\mathfrak{C}_q(X, H)$ *to* $\mathfrak{C}^{p-q}(X, J)$, *and* $\mathfrak{C}_p(X, G)$, $\mathfrak{C}_q(X, H)$ *to* $\mathfrak{C}_{p+q}(X, J)$.

(8.3) $\eta_q \cdot \xi^p = (-1)^{pq} \xi^p \cdot \eta_q$.

(8.4) $\mathrm{F}(\xi \cdot \eta) = (\mathrm{F}\xi) \cdot \eta + (-1)^{\dim \xi} \xi \cdot \mathrm{F}\eta$.

(8.5) $\gamma^p \cdot \delta_q$, $\gamma_p \cdot \delta^q$ *are* $(p - q)$*-cycles over* J *and* $\gamma_p \cdot \delta_q$ *is a* $(p + q)$*-cocycle over* J.

(8.6) γ *or* $\delta \sim 0 \rightarrow \gamma \cdot \delta \sim 0$.

(8.7) *The class of* $\gamma \cdot \delta$ *depends solely upon the classes* Γ, Δ *of* γ, δ. *It is written* $\Gamma \cdot \Delta$ *and called the class intersection of* Γ, Δ.

With class intersections at our disposal we are in position to give the following comprehensive theorem:

(8.8) THEOREM. *The class intersection* $\Gamma^p \cdot \Delta_q$ *pairs* $\mathfrak{H}^p(X, G)$, $\mathfrak{H}_q(X, H)$ *to* $\mathfrak{H}^{p-q}(X, J)$ *and the class intersection* $\Gamma_p \cdot \Delta_q$ *pairs* $\mathfrak{H}_p(X, G)$, $\mathfrak{H}_q(X, H)$ *to* $\mathfrak{H}_{p+q}(X, J)$. *Furthermore:*

(a) (*Commutation rule*) $\Delta \cdot \Gamma = (-1)^{\dim \Gamma \cdot \dim \Delta} \Gamma \cdot \Delta$.

(b) *If the classes are over a ring* ρ *then their intersections are associative.*

(c) *If the cocycles are over a ring* ρ *then* $\Gamma_p \cdot \Gamma_q$ *considered as a formal product generates an associative but not necessarily commutative ring* $R(X, \rho)$.

We call $R(X, \rho)$ the *cohomology ring over* ρ, or merely the *cohomology ring*, written $R(X)$, when ρ is the ring of the integers. (Regarding this ring, see Alexander [c, d, e], Gordon [a], also Freudenthal [a].)

(8.9) AN EXAMPLE. Consider the sphere $S^n = \mathfrak{B}\sigma^{n+1}$, $\sigma^{n+1} = A_0 \cdots A_{n+1}$. Referring to (III, 22.4, 27.5) S^n is $(0, n)$-cocyclic and if $\gamma_0 = \sum A^i$, $\gamma_n = A^0 \cdots A^n$ then every zero-cocycle is of the form $g\gamma_0$, and every n-cocycle $\sim g\gamma_n$ (III, 27.7). The only intersections of the elements γ_0, γ_n which are $\nsim 0$ are $\gamma_0 \cdot \gamma_0$ and $\gamma_0 \cdot \gamma_n = \gamma_n \cdot \gamma_0$, and we shall calculate these. We have from (4.11): $A^i \cdot A^i = A^i$, $A^i \cdot A^j = 0$ for $j \neq i$. Hence

$$\gamma_0 \cdot \gamma_0 = \sum A^i \cdot A^i = \gamma_0 .$$

We also have $\sigma_n \cdot \sigma^n = \beta(-n) A_i$ and $\gamma_0 \cdot \gamma_n \sim \lambda \sigma_n$. Hence

$$\gamma_0 \cdot \gamma_n \cdot \gamma_n \sim \lambda \gamma_0 \cdot (\sigma_n \cdot \sigma^n) = \lambda \beta(-n) \gamma_0 \cdot A_i = \lambda \beta(-n) A_i ;$$

$$\gamma_0 \cdot \gamma_n \cdot \gamma^n = \sum A^i \cdot (\gamma_n \cdot \gamma^n) = \beta(-n) A_i .$$

Hence $\lambda\beta(-n)A_i = \beta(-n)A_i$, and so $\lambda = 1$. Therefore $\gamma_0\cdot\gamma_n \sim \gamma_n$. Thus if Γ_0, Γ_n are the classes of γ_0, γ_n the class intersections are fully determined by the relations: $\Gamma_0\cdot\Gamma_0 = \Gamma_0$, $\Gamma_0\cdot\Gamma_n = \Gamma_n\cdot\Gamma_0 = \Gamma_n$.

(8.10) *Outline of the proof of* (8.8). The pairing properties are a consequence of (8.7) and the ring properties follow from (1.10) and (8.8b). There remain (8.8ab) which we will prove separately by means of certain auxiliary chain-mappings.

9. **Proof of** (8.8a). For $\Gamma^p\cdot\Delta_q$, that is to say for μ_1, it is a consequence of (8.3), so we only have to consider $\Gamma_p\cdot\Delta_q$, i.e., $\overset{*}{\mu_2}$. This operation has been defined as a chain-mapping $X^{*2} \to X^*$ which satisfies (4.11, 4.12). The condition that $\overset{*}{\mu_2}$ be a chain-mapping is $\overset{*}{\mu_2}\mathrm{F} = \mathrm{F}\overset{*}{\mu_2}$. If the resulting relations are written down for the relations (3.8) they are still found to be verified when $\mu_i^{jh}(q, r)$ is replaced by $(-1)^{qr}\mu_i^{jh}(q, r)$ and $x_q^j \times x_r^h$ by $x_r^h \times x_q^j$. It follows that

$$\bar{\mu}_2^*(x_q^j \times x_r^h) = (-1)^{qr}\beta(-p)\mu_i^{hj}(r, q)x_p^i$$

is a chain-mapping $\bar{\mu}_2^*: X_2^* \to X^*$. The conditions (4.11, 4.12) are readily verified for $\bar{\mu}_2^*$ and so it is a second component of an intersection in X. Hence by (7.1) $\bar{\mu}_2^* \sim \bar{\mu}_2$ as a chain-mapping $X^{*2} \to X^*$. Therefore if γ_q, γ_r are cocycles, we have $\overset{*}{\mu_2}(\gamma_q \times \gamma_r) \sim \bar{\mu}_2^*(\gamma_q \times \gamma_r)$, or finally

$$\gamma_q\cdot\gamma_r \sim (-1)^{qr}\gamma_r\cdot\gamma_q,$$

and this is (8.8a).

10. **Proof of** (8.8b) (**associativity**). This property, so it must be kept in mind, applies only to cycles and cocycles, and to within homology. It requires that we prove the four relations:

$$(10.1)\qquad (\gamma^p\cdot\gamma_q)\cdot\gamma_r \sim \gamma^p\cdot(\gamma_q\cdot\gamma_r);$$

$$(10.2)\qquad (\gamma_q\cdot\gamma^p)\cdot\gamma_r \sim \gamma_q\cdot(\gamma^p\cdot\gamma_r);$$

$$(10.3)\qquad (\gamma_p\cdot\gamma_q)\cdot\gamma_r \sim \gamma_p\cdot(\gamma_q\cdot\gamma_r);$$

$$(10.4)\qquad (\gamma_q\cdot\gamma_r)\cdot\gamma^p \sim \gamma_q\cdot(\gamma_r\cdot\gamma^p).$$

The last reduces immediately to the first by permutation of the elements. For by (10.1) and (8.1a) we have:

$$(\gamma_q\cdot\gamma_r)\cdot\gamma^p = (-1)^{p(q+r)}\gamma^p\cdot(\gamma_q\cdot\gamma_r) \sim (-1)^{pq+pr+qr}\gamma^p\cdot(\gamma_r\cdot\gamma_q),$$

$$\gamma_q\cdot(\gamma_r\cdot\gamma^p) = (-1)^{pq+pr+qr}(\gamma^p\cdot\gamma_r)\cdot\gamma_q \sim (\gamma_q\cdot\gamma_r)\cdot\gamma^p.$$

Thus we need only to prove (10.1), (10.2), (10.3).

11. To prove (10.1) introduce the two chain-mappings τ_1, $\tau_2: X \times X^{*2} \to X$ defined by $\tau_1(x^p \times x_q \times x_r) = (x^p\cdot x_q)\cdot x_r$, $\tau_2(x^p \times x_q \times x_r) = x^p\cdot(x_q\cdot x_r)$. They have the common carrier $t: X \times X^{*2} \to X$ such that $t(x^p \times x_q \times x_r) = x^p$. We prove as for T_1 in (6.1) that t is closed and we notice that $(\mathrm{Cl}\ t(x^p \times x_q \times x_r))_a = (\mathrm{Cl}\ x^p)_a$ is acyclic. Since both τ_1, τ_2 map the negative-dimensional elements into zero, by (IV, 18.7) they are contiguous in t provided that:

$$\text{(11.1)} \qquad \mathrm{KI}(\tau_1(x_i^p \times x_q^j \times x_r^h)) = \mathrm{KI}(\tau_2(x_i^p \times x_q^j \times x_r^h)),$$

or that

$$\text{(11.2)} \qquad \mathrm{KI}((x_i^p \cdot x_q^j), x_r^h) = \mathrm{KI}(x_i^p, (x_q^j \cdot x_r^h)).$$

Referring to (4.7, 4.8, 4.10) the two sides of (11.2) are found to be

$$\mathrm{KI}((-1)^p \beta(r) \mu_i^{jk}(q, r) x_k^r, x_r^h) = (-1)^p \mu_i^{jh}(q, r),$$

$$\mathrm{KI}(x_i^p, \beta(-p) \mu_k^{jh}(q, r) x_p^k) = (-1)^p \mu_i^{jh}(q, r).$$

This proves (11.2) and hence (10.1).

The same argument reduces (10.2) to:

$$\text{(11.3)} \qquad \mathrm{KI}(x_q^j, (x_i^p \cdot x_r^h)) = \mathrm{KI}((x_q^j \cdot x_i^p), x_r^h).$$

Here again the two sides are found to be

$$\mathrm{KI}(x_q^j, (-1)^p \beta(q) \mu_i^{hk}(r, q) x_k^q) = (-1)^r \mu_i^{hj}(r, q),$$

$$\mathrm{KI}((-1)^{pq+p} \beta(r) \mu_i^{jk}(q, r) x_k^r, x_r^h) = (-1)^r \mu_i^{hj}(r, q),$$

which proves (11.3), and hence (10.2).

12. A slightly different argument is required for (10.3). The chain-mappings θ_1^*, θ_2^* to be proved contiguous will be $X^{*^3} \to X^*$, and are defined by:

$$\theta_1^*(x_p^i \times x_q^j \times x_r^k) = (x_p^i \cdot x_q^j) \cdot x_r^k = \nu_{m1}^{ijk}(p, q, r) x_s^m,$$

$$\nu_{m1}^{ijk} = \beta(-p - q) \beta(-s) \mu_h^{ij}(p, q) \mu_m^{hk}(p + q, r),$$

$$\theta_2^*(x_p^i \times x_q^j \times x_r^k) = x_p^i \cdot (x_q^j \cdot x_r^k) = \nu_{m2}^{ijk}(p, q, r) x_s^m,$$

$$\nu_{m2}^{ijk} = \beta(-s) \beta(-q - r) \mu_h^{jk}(q, r) \mu_m^{ih}(p, q + r),$$

where $s = p + q + r$, and $\nu = 0$ unless $x_p^i, x_q^j, x_r^k < x_s^m$. Therefore the duals are chain-mappings $\theta_i : X \to X^3$ with the common carrier T such that $Tx = x^3$. It is immediately seen that T is simple. If we can show that θ_ρ ($\rho = 1, 2$) are contiguous in T, the θ_ρ^* will be proved contiguous in T^* (IV, 15.9) and (10.3) will follow. As before, everything reduces to showing that the transforms of any x_i^0 by the θ_ρ have equal Kronecker indices. Now we have:

$$\theta_\rho x_m^s = \sum_{p+q+r=s} \nu_m^{ijk}(p, q, r) x_i^p \times x_j^q \times x_k^r,$$

and hence

$$\theta_\rho x_m^0 = \nu_{m\rho}^{ijk}(0, 0, 0) x_i^0 \times x_j^0 \times x_k^0.$$

From the expression of the ν's and since, in (4.9) as applied here we have $h = i$, there comes

$$\theta_\rho x_i^0 = x_i^0 \times x_i^0 \times x_i^0.$$

Hence $\mathrm{KI}(\theta_\rho x_i^0)$, is the same for $\rho = 1$ and $\rho = 2$. Therefore the θ_ρ are contiguous in T and (10.3) follows. This completes the proof of associativity and also of Theorem (8.8).

13. **Induced intersections.** The notations being as in (4, 6), let X_1 be a closed subcomplex of X and $X_0 = X - X_1$. The product $X_1 \times X^* = Y_1$ is closed in $X \times X^*$ and we set $Y_0 = X \times X^* - Y_1 = X_0 \times X^*$. We verify immediately that μ_1, T_1 send Y_1 into X_1. In fact $T_1 Y_1 = X_1$. To indicate explicitly the complex in which $\mu_1, \cdots$ operate we shall designate these operations by $\mu_1(X) \cdots$. The operations $\mu_1(X)$, T_1 induce $\mu_1(X, X_i)$, $T_{1i}: Y_i \to X_i$ (IV, 22) and in particular $\mu_1(X, X_1)$ is a suitable $\mu_1(X_1)$ for X_1, and T_{1i} is the analogue of T_1 for X_1.

A slightly different argument is required for the second multiplication $\overset{*}{\mu_2}$. Here also it is more convenient to pass to μ_2, $T_2: X \to X^2$. We define separately μ_{20}, μ_{21}. Since X_1, X_1^2 are closed in X, X^2 and both μ_2, T_2 transform the first into the second, there are corresponding induced elements $\mu_2(X, X_1)$, T_{21}. Since X_0^2 is open in X^2, its complement $Z_1 = X^2 - X_0^2$ is closed and it contains X_1^2. It follows that μ_2, T_2 transform X_1 into Z_1 and so we may define induced transformations T_{20}, $\mu_{20}(X, X_0): X_0 \to X_0^2$. The duals are then taken, thus leading to T_{2i}^*, $\overset{*}{\mu_2}(X, X_i)$. The couples

$$\mu(X, X_i) = \{\mu_1(X, X_i), \overset{*}{\mu_2}(X, X_i)\}$$

are called the chain-intersections *induced* in the X_i by the intersection $\mu(X)$. We notice the following properties:

(13.1) *The induced intersections possess properties* (4.1, 4.11) *it being understood that* Cl, St *are taken relative to* X_i *in each case.*

(13.2) $\mu_1(X, X_1) = \mu_1(X)$ on X_1;

(13.3) $\mu_1(X, X_0) = \mu_1(X) \bmod X_1$ on X_0;

(13.4) $\overset{*}{\mu_2}(X, X_1) = \overset{*}{\mu_2}(X) \bmod X_0^*$ on X_1;

(13.5) $\overset{*}{\mu_2}(X, X_0) = \overset{*}{\mu_2}(X)$ on X_0.

From these relations follow immediately:

(13.6) *Property* (4.2) *is unchanged for* X_1, *but for* X_0 *it is to be replaced by*

$$x^p \cdot x_p = \beta(p) x^0 \bmod X_1, \qquad x^0 < x^p.$$

(13.7) *Property* (4.12) *is unchanged for* X_0, *but for* X_1 *it is replaced by*

$$x_p \cdot x_0 = x_p \bmod X_0^* \text{ for some } x^0 < x^p.$$

Observe that since X_1 is closed in X it is simple. In view of (13.1, 13.6) $\mu_1(X, X_1)$ possesses the properties (4.1, 4.2), and so by (7.1):

(13.8) $\mu_1(X, X_1)$ *is the component* $\mu_1(X_1)$ *of an intersection* $\mu(X_1)$ *in* X_1.

14. **Chain-mappings and intersections.** Let X, Y be simple complexes and let there be adopted a chain-intersection for each, denoted by a dot-product. Let Γ, Δ designate the classes in X, Y. We prove:

(14.1) THEOREM. *Let τ be a simple chain-mapping $X \to Y$ with a simple carrier t and let τ^* be the dual of τ. Denote also for simplicity by τ, τ^* the induced homomorphisms in the homology and cohomology groups. Then:*

(a) $$\tau(\Gamma^p \cdot \tau^*\Delta_q) = (\tau\Gamma^p)\cdot\Delta_q .$$

(b) $$\tau^*\Delta_p \cdot \tau^*\Delta_q = \tau^*(\Delta_p \cdot \Delta_q).$$

(c) *τ^* is both an additive and multiplicative homomorphism of the cohomology ring $R(Y, \rho)$ into the cohomology ring $R(X, \rho)$.*

(d) *The preceding properties hold when X, Y, τ are simplicial.*

(e) *Let (X_0, X_1) and (Y_0, Y_1) be dissections of X and Y. Then if t, τ both send X_1 into Y_1 the relations* (a, b) *still hold for the induced intersections provided of course that τ, τ^* are replaced by the induced chain-mappings τ_i, τ_i^** (IV, 22).

Property (c) is implicit in (b); (d) is obvious; (e) is a consequence of (a, b) and (13). There remains then to prove (a, b).

Consider the chain-mappings $X \times Y^* \to Y$ defined by:

$$\tau_1 x^p \times y_q = \tau(x^p \cdot \tau^* y_q), \qquad \tau_2 x^p \times y_q = (\tau x^p)\cdot y_q$$

with the common carrier s defined by: $s(x^p \times y_q) = tx^p$. It is an elementary matter to show that s is closed and simple. We first prove:

(14.2) *The chain-mappings τ_1, τ_2 are contiguous in the carrier s.*

Since the dimensions in X are greater than or equal to 0, by (IV, 18.7) we merely need to show that:

$$\text{(14.3)} \qquad \mathrm{KI}(\tau(x_i^p, \tau^* y_p^j)) = \mathrm{KI}(\tau x_i^p, y_p^j).$$

Since τ is simple, the first index is equal to $\mathrm{KI}(x_i^p, \tau^* y_p^j)$; so (14.3) may be reduced to:

$$\text{(14.4)} \qquad \mathrm{KI}(x_i^p, \tau^* y_p^j) = \mathrm{KI}(\tau x_i^p, y_p^j).$$

Let the explicit equations of τ, τ^* be:

$$\text{(14.5)} \qquad \tau x_i^p = a_i^j(p) y_j^p , \qquad \tau^* y_p^j = a_i^j(p) x_p^i .$$

By an elementary computation there comes:

$$\mathrm{KI}(x_i^p, \tau^* y_p^j) = \beta(p) a_i^j(p) = \mathrm{KI}(\tau x_i^p, y_p^j);$$

so (14.4) is proved, hence also (14.2). Then (14.1a) is a direct consequence of (14.2).

15. Consider now the two chain-mappings $Y^{*2} \to X^*$ defined by

$$\text{(15.1)} \qquad \theta_1^*(y_q^j \times y_r^h) = \tau^* y_q^j \cdot \tau^* y_r^h ; \qquad \theta_2^* y_q^j \times y_r^h = \tau^*(y_q^j \cdot y_r^h),$$

and let u be the simple set-transformation $X \to Y^2$ defined by $ux = tx \times tx$. We prove:

(15.2) *θ_1^*, θ_2^* have the carrier u^* and are contiguous in this carrier.*

It is clear that (14.1b) will be a consequence of (15.2). To prove (15.2) it is sufficient (IV, 15.9) to show that the dual chain-mappings $\theta_i : X \to Y^2$ have the simple carrier u and are contiguous in u.

Let chain-intersection in Y be denoted by the same notations as in X, but with $\bar{\mu}$ in place of μ throughout. The defining relations of θ_1, θ_2 are found to be:

(15.3) $$\theta_1 x_i^p = a_l^j(q) a_m^h(r) \mu_i^{lm}(q, r) y_j^q \times y_h^r$$
$$= \mu_i^{lm}(q, r)(\tau x_l^q) \times (\tau x_m^r);$$

(15.4) $$\theta_2 x_i^p = a_i^l(p) \bar{\mu}_l^{jh}(q, r) y_j^q \times y_h^r .$$

Since τx_l^q, $\tau x_m^r \subset \mathrm{Cl}\, t x_i^p$, θ_1 has the asserted carrier property. Regarding θ_2, $\bar{\mu}_l^{jh}(q, r) \neq 0$ only when y_j^q, $y_h^r < y_l^p$, so the coefficient of $a_i^l(p)$ is a chain of $\mathrm{Cl}\, y_l^p \times \mathrm{Cl}\, y_l^p$. Since $\mathrm{Cl}\, y_l^p \subset t\, \mathrm{Cl}\, x_i^p \subset \mathrm{Cl}\, t\, \mathrm{Cl}\, x_i^p = \mathrm{Cl}\, t\, x_i^p$, $\theta_2 x_i^p$ is a chain of $\mathrm{Cl}\,(t x_i^p \times t \bar{x}_i^p)$ so θ_2 has likewise the carrier u. Since u is simple the contiguity of θ_1, θ_2 in u and hence the proof of (14.3), is reduced again to showing that θ_1, θ_2 transform the indices of zero-chains in the same way, or in the last analysis to proving:

(15.5) $$\mathrm{KI}(\theta_1 x_i^0) = \mathrm{KI}(\theta_2 x_i^0).$$

Now $p = 0 \to q = r = 0$, and

$$\mu_i^{lm}(0, 0) = \begin{cases} 1 \text{ when } l = m = i, \\ 0 \text{ otherwise}; \end{cases}$$

$$\bar{\mu}_l^{jh}(0, 0) = \begin{cases} 1 \text{ when } j = h = l, \\ 0 \text{ otherwise}. \end{cases}$$

Furthermore since τ is simple it preserves the Kronecker index and hence

$$1 = \mathrm{KI}(x_i^0) = \mathrm{KI}(\tau x_i^0) = \mathrm{KI}(a_i^j(0) y_j^0) = \sum_j a_i^j(0).$$

If we calculate the indices in (15.5) we find then

$$\mathrm{KI}(\theta_1 x_i^0) = 1 = \mathrm{KI}(\theta_2 x_i^0);$$

so (14.1b) is proved.

16. **Application to subdivisions.** Let X, Y be simple, with Y a subdivision of X. Then chain-subdivision σ and its reciprocal τ are likewise simple and so (14.1) is applicable to the chain-mapping σ. By (IV, 24.2) if Γ, Δ denote the classes in X, Y then $\sigma\Gamma^p = \Delta^p$ and $\sigma^*\Delta_q = \Gamma_q$ are unique classes of Y^*, X^*, and $\{\Delta^p\}$, $\{\Gamma_q\}$ include all the elements of the pth homology and qth cohomology groups. If we choose σ as the τ of (14.1) we obtain:

(16.1) $$\sigma(\Gamma^p \cdot \Gamma_q) = \Delta^p \cdot \Delta_q, \qquad \Gamma_p \cdot \Gamma_q = \sigma^*(\Delta_p \cdot \Delta_q).$$

If we combine this with (8.8) we find:

(16.2) *Let X, Y be simple and Y a subdivision of X, with σ as the chain-subdivision. Then σ, σ^* set up isomorphisms between the class intesections in X,*

Y and furthermore σ^* maps the ring $R(Y, \rho)$ isomorphically on $R(X, \rho)$. In particular these properties are true when X is simplicial and Y is any derived of X. Furthermore they hold equally for the induced intersections in open or closed subcomplexes of X and the (induced) subdivisions.

17. **Intersections in a product.** Let X, Y be simple with intersections (denoted by dot-products), which are multiplications $\mu = (\mu_1, \overset{*}{\mu_2})$ in X and $\nu = (\nu_1, \overset{*}{\nu_2})$ in Y. We will set $Z = X \times Y$. We assume X, X^* disjoint from Y, Y^*.

Since μ_1, ν_1 are chain-mappings $X \times X^* \to X$, $Y \times Y^* \to Y$, by (IV, 21.1) $\mu_1 \times \nu_1 = \rho_1$ is a chain-mapping $X \times X^* \times Y \times Y^* \to X \times Y$ defined by:

$$\rho_1(x^p \times x_q \times y^r \times y_s) = (x^p \cdot x_q) \times (y^r \cdot y_s). \tag{17.1}$$

The effect of permuting X^* and Y in $X \times X^* \times Y \times Y^*$ (a permissible operation since X^* and Y are disjoint) is the same as applying the orientation function $\alpha(x^p \times x_q \times y^r \times y_s) = (-1)^{qr}$. Therefore, by (IV, 9.17), (17.1) yields the chain-mapping $\rho_1 : Z \times Z^* \to Z$ defined by

$$\rho_1(x^p \times y^r \times x_q \times y_s) = (-1)^{qr}(x^p \cdot x_q) \times (y^r \cdot y_s). \tag{17.2}$$

Since $x^p \cdot x_q = 0$ when $x^q \not< x^p$ and $y^r \cdot y_s = 0$ when $y^s \not< y^r$, we have $\rho_1(x^p \times y^r \times x_q \times y_s) = 0$ whenever $x^q \times y^s \not< x^p \times y^r$. Moreover dim $(x^p \cdot x_q) \times (y^r \cdot y_s) = p - q + r - s = p + r - (q + s)$. Thus ρ_1 satisfies (4.1) relative to Z. We also have from (4.2) for X, and Y, and (17.2):

$$\begin{aligned} \rho(x^p \times y^r \times x_p \times y_r) &= (-1)^{pr}\beta(p)\beta(r)x^0 \times y^0 \\ &= \beta(p + r)x^0 \times y^0 \end{aligned} \tag{17.3}$$

where $x^0 \times y^0$ is a vertex of $x^p \times y^r$. Therefore μ_1 satisfies also (4.2). Thus it is the first component of a chain-intersection in $X \times Y$. Let this intersection in $X \times Y$ be also designated by a dot-product. We will then have from (17.2):

$$(x^p \times y^r) \cdot (x_q \times y_s) = (-1)^{qr}(x^p \cdot x_q) \times (y^r \cdot y_s). \tag{17.4}$$

The relation (17.4) expresses the identity of two chain-mappings $Z \times Z^* \to Z$. Therefore the induced mappings on the homology classes are the same. Let Γ, Δ denote the classes in X, Y. It is to be understood throughout that the coefficient groups are to be such that the indicated chain-multiplications have a meaning. For instance this will be the case if all the coefficient groups are the same commutative ring ρ. From (17.4) there comes then:

$$(\Gamma^p \times \Delta_r) \cdot (\Gamma_q \times \Delta_s) = (-1)^{qr}(\Gamma^p \cdot \Gamma_q) \times (\Delta^r \cdot \Delta_s). \tag{17.5}$$

A wholly similar argument yields

$$(\Gamma_p \times \Delta_r) \cdot (\Gamma_q \times \Delta_s) = (-1)^{qr}(\Gamma_p \cdot \Gamma_q) \times (\Delta_r \cdot \Delta_s). \tag{17.6}$$

We may also combine (17.5), (17.6) as:

$$(\Gamma \times \Delta) \cdot (\Gamma' \times \Delta') = (-1)^{\dim\Delta \dim\Gamma'}(\Gamma \cdot \Gamma') \times (\Delta \cdot \Delta'). \tag{17.7}$$

The importance of these formulas lies in the fact that they express intersections in a product in terms of those in the factors.

(17.8) *Kronecker indices in a product.* By means of (III, 28.1), first applied to products of elements then to products of chains and cochains, with $\xi \subset X$, $\eta \subset Y$, we find in $X \times Y$:

$$\mathrm{KI}(\xi^p \times \eta^q, \xi_p \times \eta_q) = (-1)^{pq}\mathrm{KI}(\xi^p, \xi_p)\mathrm{KI}(\eta^q, \eta_q), \tag{17.9}$$

$$\mathrm{KI}(\xi^p \times \eta^q, \xi_r \times \eta_s) = 0, \quad p \neq r, p + q = r + s. \tag{17.10}$$

Both results are valid *for any pair of complexes* X, Y (not necessarily simple) provided only that in any index of a chain and cochain in (17.9) or (17.10) one of the two is finite.

18. **Intersections in infinite complexes.** The extension of the results obtained so far offers no serious difficulty and few novel features.

Suppose first that X is an arbitrary simple complex and hence merely closure-finite. Then the groups of chains and cycles must be finite and hence discrete and everything said so far may be preserved provided that we add to (4.12) the condition that $x_q \cdot x_r$ *is to be finite.*

Suppose now X locally finite. If a multiplication μ_1 of X, X^* to X satisfies (4.1) it will then be of *fi*-type, and if a $\overset{*}{\mu}_2$ satisfies (4.11) it will be likewise of *fi*-type. In view of this it is an elementary matter to show that the situation of (4) carries over without modifications and all the results obtained are directly applicable to locally finite complexes. It is merely to be observed that wherever in (8) there are paired groups, if one of them is a group of finite chains or cycles, the intersections are finite also. As a consequence there are two kinds of rings over ρ: the ring $R(X, \rho)$ of the infinite cocycles, and the ring $R_f(X, \rho)$ of the finite cocycles.

19. **A second method for intersections.** The multiplication of W. W. Flexner considered in Example (1.13) may be utilized to advantage for defining intersections. For if X is simple then we have a simple chain-mapping $\tau: X' \to X$, the reciprocal of chain-derivation (IV, 27). As a consequence $\mu_1 = \tau\phi$ is a chain-mapping $X \times X^* \to X$. Referring to the relations (1.14, 1.15, 1.16) defining φ, the simplexes of $\varphi(x^p \times x_q)$ have as their extreme vertices $'x^p$, $'x^q$. Therefore $\varphi(x^p \times x_q) = 0$ when $x^q \not< x^p$, and when $\varphi(x^p \times x_q) \neq 0$ it consists of simplexes of $(\mathrm{Cl}\ x^p)'$, since these include all the simplexes of X' with $'x^p$ as their last vertex. Let, on the other hand, D be set-derivation in X (IV, 26.2a, 27). The chain-mapping τ has the carrier D^{-1} and from the relations defining D we verify that $D^{-1}(\mathrm{Cl}\ x)' = \mathrm{Cl}\ x$. Hence $\tau(\mathrm{Cl}\ x)' \subset \mathrm{Cl}\ x$, and so $\tau\varphi(x^p \times x_q)$ is zero whenever $x^q \not< x^p$, and $\tau\varphi(x^p \times x_q)$ is a chain of $\mathrm{Cl}\ x^p$ otherwise. In other words, μ_1 satisfies (4.1).

Regarding (4.2), we have $\tau' x^p = x^0$, a vertex of x^p, and so from (1.14): $\mu_1(x_i^p \times x_p^j) = \beta(p)\delta_i^j x^0$, $x^0 < x_i^p$. Thus (4.2) holds likewise and so:

(19.1) $\mu_1 = \tau\varphi$ *is a first component of a chain-intersection in* X.

As we know (3.9) μ_1 will determine the second component $\overset{*}{\mu}_2$, and hence μ_1 *determines an intersection in* X. It is not difficult to recognize in this type of intersection the analogue of those developed for a manifold in [L, IV].

20. Special method for intersections in simplicial complexes (third method). In recent years Čech [f] and Whitney [d], perfecting a scheme due to Alexander and Kolmogoroff have introduced a special type of intersections for simplicial complexes which is particularly convenient for computations.

Let then $K = \{\sigma\}$ be any simplicial complex whatever and let $\{A_\alpha\}$ be its vertices ranged in some simple order. The simplexes of K are oriented by naming their vertices in the increasing order of the indices and they will be so designated throughout. The duals of the vertices are written A^α and the duals of the simplexes are likewise named in the increasing order of the indices. We will introduce the two components μ_1, $\overset{*}{\mu}_2$ of an intersection in K. However following essentially Whitney we will denote μ_1, by $\frown$ and $\overset{*}{\mu}_2$ by $\smile$ ("cap" and "cup" products).

(20.1) The operation $\smile$ is defined as follows:

(a) if $\sigma_p = \sigma_{p-1}A^\alpha$, $\sigma_q = A^\alpha\sigma_{q-1}$, then

$$\sigma_p \smile \sigma_q = \sigma_{p-1}A^\alpha\sigma_{q-1};$$

(b) in all other cases $\sigma_p \smile \sigma_q = 0$;

(c) if G, H are commutatively paired to J and $\xi_p = g_i\sigma_p^i$, $\eta_q = h_j\sigma_q^j$ then

$$\xi_p \smile \eta_q = g_ih_j\sigma_p^i \smile \sigma_q^j.$$

Similarly $\frown$ is defined as follows:

(d) if $\sigma_q = \sigma_{q-1}A^\alpha$, $\sigma^p = \sigma^{q-1}A_\alpha\sigma^{p-q-1}$, then

$$\sigma_q \frown \sigma^p = \beta(-q)A_\alpha\sigma^{p-q-1};$$

(e) in all other cases $\sigma_q \frown \sigma^p = 0$;

(f) analogous to (c).

(Notice that our $\frown$ differs from Whitney's by the presence of the factor $\beta(-q) = \pm 1$, but this is immaterial.)

We shall now show that $\smile$, $\frown$ are chain-multiplications. Given α let $\{A_\beta\}$ be the vertices such that $\beta < \alpha$ and $\{A_\gamma\}$ those such that $\gamma > \alpha$. We have then under (a) (III, 10.4):

$$\begin{aligned}
\mathrm{F}(\sigma_p \smile \sigma_q) &= \left(\sum A^\beta + \sum A^\gamma\right)\sigma_{p-1}A^\alpha\sigma_{q-1} \\
&= \left(\sum A^\beta\sigma_{p-1}A^\alpha\right)\sigma_{q-1} + (-1)^{p+1}\sigma_{p-1}A^\alpha\left(\sum A^\gamma\sigma_{q-1}\right) \\
&= (\mathrm{F}\sigma_p) \smile \sigma_q + (-1)^{p+1}\sigma_{p-1}A^\alpha \smile A^\alpha \sum A^\gamma\sigma_{q-1} \\
&= \mathrm{F}\sigma_p \smile \sigma_q + (-1)^p\sigma^p \smile \mathrm{F}\sigma_q.
\end{aligned}$$

Thus in the present instance:

$$\mathrm{F}(\sigma_p \smile \sigma_q) = \mathrm{F}\sigma_p \smile \sigma_q + (-1)^p\sigma_p \smile \mathrm{F}\sigma_q. \tag{20.2}$$

Consider now (b) and let $\sigma_p = \sigma_{p-1}A^\beta$, $\sigma_q = A^\gamma\sigma_{q-1}$. If $\beta > \gamma$ then all the terms in (20.2) are zero and so it holds identically.

Suppose $\beta < \gamma$. Then

$$(\mathrm{F}\sigma_p) \smile \sigma_q = (-1)^{p+1}\sigma_{p-1}A^\beta A^\gamma \smile A^\gamma \sigma_{q-1} = (-1)^{p+1}\sigma_{p-1}A^\beta A^\gamma \sigma_{q-1};$$

$$(-1)^p \sigma_p \smile \mathrm{F}\sigma_q = (-1)^p \sigma_{p-1}A^\beta \smile A^\beta A^\gamma \sigma_{q-1} = (-1)^p \sigma_{p-1}A^\beta A^\gamma \sigma_{q-1},$$

and since $\mathrm{F}(\sigma_p \smile \sigma_q) = 0$, (20.2) holds here also.

Thus by (1.12) $\smile$ is a multiplication of X^*, X^* to X^*. In other words $\smile$ is an operation $\overset{*}{\mu}_2$ in the sense of (3). We verify immediately that it satisfies (4.11, 4.12), and so $\smile$ is the second component of an intersection $\mu = (\mu_1, \overset{*}{\mu}_2)$ in K. To find the corresponding μ_1 it is best to write down the graph (3.5). In a convenient form for our present purpose it is with $p = q + r$:

$$\Gamma = \sum \beta(-p)\sigma^{q-1}A_\alpha \times A_\alpha \sigma^{r-1} \times \sigma_{q-1}A^\alpha \sigma_{r-1}$$
$$= \sum \beta(r)\beta(-q)\sigma^{q-1}A_\alpha \times \sigma_{q-1}A^\alpha \sigma_{r-1} \times A_\alpha \sigma^{r-1}.$$

Hence the operation μ_1 is given by:

$$\mu_1 \sigma_{q-1}A^\alpha \times \sigma^{q-1}A_\alpha \sigma^{r-1} = \beta(-q)A_\alpha \sigma^{r-1};$$

$\mu_1 \sigma_q \times \sigma^p = 0$ in all other cases.

Written as $\frown$ it satisfies (d, e, f). This proves:

(20.3) *$\frown$, $\smile$ are respectively the components μ_1, $\overset{*}{\mu}_2$ of an intersection in K.*

(20.4) In view of (20.3) one may introduce the products $\frown$, $\smile$ for the classes. Since the results are the same for all intersections they are the same for $\frown$, $\smile$ as for the dot-product. In particular the same rings are obtained and generally the basic theorem (8.8) is applicable. An important consequence is that *the class products $\frown$, $\smile$ are independent of the ordering of the vertices.* To be precise if $\frown'$, $\smile'$ are a similar system corresponding to a different ordering of the vertices $\{A_\alpha\}$ then

$$\gamma_p \smile' \gamma_q \sim \gamma_p \smile \gamma_q,$$
$$\gamma_q \frown' \gamma^p \sim \gamma_q \frown \gamma^p.$$

(20.5) Example. Consider again the sphere S^n of (8.9) and let the notations remain those of (8.9). We have at once

$$\gamma_0 \smile \gamma_0 = \sum A^i \smile \sum A^j = \sum A^i = \gamma_0;$$
$$\gamma_0 \smile \gamma_n = \left(\sum A^i\right) \smile A^0 \cdots A^n = A^0 \smile A^0 \cdots A^n = \gamma_n.$$

Hence as loc. cit.:

$$\Gamma_0 \cdot \Gamma_0 = \Gamma_0, \qquad \Gamma_0 \cdot \Gamma_n = \Gamma_n \cdot \Gamma_0 = \Gamma_n.$$

(20.6) Remark. Strictly speaking Whitney's operations $\smile$, $\frown$ are applied to the simplexes themselves: where we write $\sigma_p \smile \sigma_q$, $\sigma_p \frown \sigma^q$, Whitney writes $\sigma^p \smile \sigma^q$, $\sigma^p \frown \sigma^q$, and applies on the other hand his two boundary and coboundary operators ∂, δ. Thus in his notation

$$\delta(\sigma^p \smile \sigma^q) = \delta\sigma^p \smile \sigma^q + (-1)^p \sigma^p \smile \delta\sigma^q, \tag{20.6a}$$

(20.6b) $$\partial(\sigma^p \frown \sigma^q) = \delta\sigma^p \frown \sigma^q + (-1)^p \sigma^p \smile \partial\sigma^q.$$

See again in this connection the remarks of (III, 26).

§3. COINCIDENCES AND FIXED ELEMENTS

21. To simplify matters we will suppose again all complexes finite, although a much more general situation could well be considered.

Let first σ be a chain-transformation $X \to X$ given by

(21.1) $$\sigma x_i^p = a_i^j(p) x_j^p .$$

We say that x_i^p is a *fixed element* of σ if it actually occurs in σx_i^p, i.e., if $a_i^i(p) \neq 0$ (i unsummed). More generally, consider two chain-mappings $\tau: X \to Y$, $\theta: Y \to X$ given by

(21.2) $$\tau x_i^p = a_i^j(p) y_j^p , \qquad \theta y_j^p = b_j^i(p) x_i^p .$$

If y_j^p occurs in τx_i^p and at the same time x_i^p occurs in θy_j^p, i.e., if $a_i^j(p) b_j^i(p) \neq 0$ (i, j unsummed) then the pair (x_i^p, y_j^p) is said to be a *coincidence* of τ, θ. If we introduce the dual $\theta^*: X^* \to Y^*$,

(21.3) $$\theta^* x_p^i = b_j^i(p) y_p^j$$

then we may also define a coincidence (x_i^p, y_p^j) of τ, θ^* by the condition that $y_j^p \subset \tau x_i^p$ at the same time as $y_p^i \subset \theta^* x_p^i$. The condition for its occurrence is still that $a_i^j(p) b_j^i(p) \neq 0$ (i, j unsummed). Clearly the coincidences of τ, θ and τ, θ^* are thus in one-one correspondence.

We propose to derive expressions which yield important information regarding the coincidences and fixed elements. Notice first that the search for coincidences and fixed elements are equivalent problems. For if we choose $Y = X$, and $\theta = 1$, the coincidences of σ, 1 are the fixed elements of σ. And on the other hand the coincidences of τ, θ are (in many-one) correspondence with the fixed elements of $\theta\tau: X \to X$.

22. **First method.** Referring to (IV, 11) the graphs of τ, θ are:

(22.1) $$\tau:\Gamma = \sum_p \beta(p) a_i^j(p) x_p^i \times y_j^p, \qquad \text{in } X^* \times Y;$$

(22.2) $$\theta:\Gamma' = \sum_p \beta(p) b_j^i(p) y_p^j \times x_i^p, \qquad \text{in } Y^* \times X.$$

Observe now that $Y^* \times X$ is merely $(X^* \times Y)^*$ reoriented by the orientation function $\alpha(x_i^p \times y_q^j) = (-1)^{pq}$. Therefore

$$\mathrm{KI}(x_p^i \times y_j^p, y_p^h \times x_k^p) = (-1)^p \delta_k^i \delta_j^h,$$

and consequently:

(22.3) $$\mathrm{KI}(a_i^j(p) x_p^i \times y_j^p, b_h^k(p) y_p^h \times x_k^p) = (-1)^p a_i^j(p) b_h^k(p) \delta_k^i \delta_j^h,$$

(no summation on i, j, h, k).

It follows that the crossings of Γ, Γ' (III, 28) correspond to $k = i$, $h = j$, and $a_i^j(p)$, $b_j^i(p)$ not both zero, that is to say, *the crossings are in one-one correspondence with the coincidences.* Thus $\mathrm{KI}(\Gamma, \Gamma')$ is the weighted number of crossings in the sense of (III, 28). It is natural to take this number as a measure for what may be termed the *weighted number of coincidences.* Its value is then by (22.3):

$$(22.4)\qquad \varphi(\tau, \theta) = \sum_p (-1)^p a_i^j(p) b_h^k(p) \delta_k^i \delta_j^h = \sum_p (-1)^p a_i^j(p) b_j^i(p).$$

The weight of the coincidence (x_i^p, y_j^p):

$$(-1)^p a_i^j(p) b_j^i(p) = \mathrm{KI}(a_i^j(p) x_p^i \times y_j^p, b_j^i(p) y_p^j \times x_i^p), \qquad (i, j \text{ unsummed})$$

is also called the *algebraic multiplicity* or merely *multiplicity* of the coincidence. If we introduce the matrices

$$(22.5)\qquad a^p = \| a_i^j(p) \|, \qquad b^p = \| b_j^i(p) \|,$$

we have from (22.4)

$$(22.6)\qquad \varphi(\tau, \theta) = \sum (-1)^p \operatorname{trace} (a^p b^p).$$

This expression is known as the *algebraic number of coincidences* of τ and θ.

Since θ^* has the same graph as θ, if we define the coincidences of τ, θ through the intermediary of θ^*, we are led again to $\mathrm{KI}(\Gamma, \Gamma')$ as an enumeration for them. As a function of τ, θ^* it is to be designated by $\varphi_1(\tau, \theta^*)$, but clearly $\varphi_1(\tau, \theta^*) = \varphi(\tau, \theta)$.

We may introduce $\psi(\sigma) = \varphi(\sigma, 1)$ to measure the number of fixed elements. Denote by Γ the graph of σ and by Γ_0 the graph of 1. Then $\psi(\sigma) = \mathrm{KI}(\Gamma, \Gamma_0)$. Since $b^p = 1$, the algebraic multiplicity of the "coincidence" (x_i^p, x_i^p) for σ, 1 is $(-1)^p a_i^i(p)$ and we define it as the *algebraic multiplicity* or merely *multiplicity* of x_i^p as a fixed element. The algebraic number of fixed elements is then obtained from (22.6) with $b^p = 1$, and it yields:

$$(22.7)\qquad \psi(\sigma) = \sum (-1)^p \operatorname{trace} a^p.$$

(22.8) (a) *If* $\varphi(\tau, \theta) \neq 0$ *then* τ, θ *have at least one coincidence.*

(b) *Similarly if* $\varphi_1(\tau, \theta^*) \neq 0$ *then* τ, θ^* *have at least one coincidence.*

(c) *If* $\psi(\sigma) \neq 0$ *then* σ *has at least one fixed element.*

Clearly if in (a) $\varphi(\tau, \theta) \neq 0$ then not every product $a_i^j(p) b_j^i(p)$ (i, j unsummed) may be zero and hence there must be a coincidence. Similarly in (b). In (c), $\psi(\sigma) \neq 0$ implies that not every $a_i^i(p)$ (i unsummed) is zero and so there must be a fixed element.

$$(22.9)\qquad \varphi(\tau, \theta) = \psi(\theta\tau) = \psi(\tau\theta).$$

(22.10) Let us change the bases for the chains of X, Y to the new set $\{\bar{x}_i^p\}$, $\{\bar{y}_i^p\}$ related to the old set by

$$\bar{x}_i^p = \pi_i^j(p) x_j^p, \qquad \bar{y}_i^p = \omega_i^j(p) y_j^p.$$

The equations of τ, θ in terms of the new bases have the same form with new matrices expressed in obvious notations in the form:

$$\bar{a}(p) = \pi(p)a(p)\omega^{-1}(p), \qquad \bar{b}(p) = \omega(p)b(p)\pi^{-1}(p).$$

Hence trace $\bar{a}(p)\bar{b}(p)$ = trace $\pi(p)a(p)b(p)\pi^{-1}(p)$ = trace $a^p b^p$. Similarly for σ and the change of bases in X alone. Therefore:

(22.11) *The numbers trace a^p, trace $a^p b^p$, $\varphi(\tau, \theta)$, $\varphi(\tau, \theta^*)$, $\psi(\sigma)$ depend solely upon τ, θ, θ^*, σ but not upon the choice of bases for the complexes.*

23. **Second method.** Consider again $\sigma: X \to X$ given by (21.1). We have defined x_i^p as a fixed element whenever $a_i^i(p) \neq 0$. Therefore we may choose $(-1)^p a_i^i(p)$ as a measure of the weight or multiplicity of x_i^p as a fixed element. The factor $(-1)^p$ which came naturally before is now justified on grounds of expediency. The sum $\psi(\sigma)$ of all the multiplicities is chosen as an algebraic measure of the fixed elements and it is directly found to be (22.7).

If we have τ, θ and consider their coincidences, we use the property of (22), that the algebraic multiplicity of the coincidences corresponds to that of the fixed elements of $\theta\tau$. Since $a^p b^p$ are the transformation matrices of the latter, we obtain an algebraic estimate of the coincidences by means of $\psi(\theta\tau)$, and if this is denoted by $\varphi(\tau, \theta)$ it yields (22.4). As for $\varphi_1(\tau, \theta^*)$ its coincidences are the same as those of τ, θ, hence (22.4) is again its expression.

24. **Application to chain-mappings.** When τ, θ, σ are chain-mappings we will find that φ, φ_1, ψ may be expressed in terms of the induced homomorphisms on the *rational* homology groups. The reduction will be by means of the canonical bases. Certain preliminary observations are first necessary.

(24.1) Generally speaking an integral cycle or cocycle γ in a complex X is said to be a *torsion cycle* or *torsion cocycle* if a multiple $m\gamma \sim 0$, i.e., if the class of γ is of finite order. We will also write in that case $\gamma \approx 0$, or $\gamma \approx \delta$ when $\gamma - \delta \approx 0$. We notice that if $\gamma^p \approx 0$ so that $m\gamma^p \sim 0$, then $\mathrm{KI}(\gamma^p, \delta_p) = (1/m)\mathrm{KI}(m\gamma^p, \delta_p) = 0$ whatever the cocycle δ_p, and similarly with γ, δ interchanged.

(24.2) According to (III, 33.11) we may apply unimodular transformations on the bases in X, Y, X^*, Y^*, reducing them simultaneously to the canonical form. By (IV, 11.5) the relation between the chain-mappings and their graphs continue to remain the same after the transformations. Let then $a_i^p, \cdots, e_i^p$ and $\alpha_i^p, \cdots, \epsilon_i^p$ be the canonical bases for X and Y, and let the corresponding dual bases be $a_p^i, \cdots$ and $\alpha_p^i, \cdots$. One must bear in mind that the torsion cycles in the bases are the b_i^p, β_i^p and the torsion cocycles the d_p^i, δ_p^i (III, 14.2, 33.9). It follows that the reduced zero-cycles of $X^* \times Y$ (IV, 6.1) are the products $c_p \times \gamma^p$, $c_p \times \beta^p$, $d_p \times \gamma^p$, $d_p \times \beta^p$, and of these all but $c_p \times \gamma^p$ are torsion cycles (IV, 6.16). Hence Γ expressed in terms of the canonical bases is of the form:

$$\Gamma = \sum \beta(p)\tau_i^j(p)c_p^i \times \gamma_j^p + \Delta, \tag{24.3}$$

where Δ is a torsion cycle. Thus

(24.4) $$\Gamma \approx \sum \beta(p)\tau_i^j(p)c_p^i \times \gamma_j^p .$$

We have:

$$\Delta = \sum c_p^i \times \eta_i^p + \Delta_1 ,$$

where Δ_1 contains no term in $c_p \times \gamma^p$. From the relation between a chain-mapping and its graph there results:

$$\tau c_i^p = \tau_i^j(p)\gamma_j^p \pm \eta_i^p ,$$

where η_i^p is the difference of two cycles and hence a cycle; moreover it contains no γ^p and so it is a torsion cycle. Therefore

(24.5) $$\tau c_i^p \approx \tau_i^j(p)\gamma_j^p .$$

Similarly regarding Γ' and θ we will have:

(24.6) $$\Gamma' \approx \sum \beta(p)\theta_j^i(p)\gamma_p^j \times c_i^p .$$

(24.7) $$\theta\gamma_j^p \approx \theta_j^i(p)c_i^p .$$

If we set

(24.8) $$\tau^p = \| \tau_i^j(p) \|, \qquad \theta^p = \| \theta_j^i(p) \|,$$

the same calculation as in (22) yields now:

(24.9) $$\varphi(\tau, \theta) = \sum (-1)^p \operatorname{trace} \tau^p\theta^p.$$

If $\sigma: X \to X$ is as before, then we will have:

(24.10) $$\sigma c_i^p \approx \sigma_i^j(p)c_j^p ,$$

and hence as in (22) we derive from (24.9) with $\tau = \sigma$, $\theta = 1$ and $\sigma^p = \| \sigma_i^j(p) \|$:

(24.11) $$\psi(\sigma) = \sum (-1)^p \operatorname{trace} \sigma^p.$$

(24.12) The $\{c_i^p\}$, $\{\gamma_i^p\}$ are bases for the *rational* p-cycles of X, Y (III, 17). By the same argument as in (22.9) one shows now that the traces in (24.9, 24.11) and the numbers φ, φ_1, ψ depend solely upon the chain-mappings and not upon the special bases selected for the rational groups. That is to say, if $\{c_i^p\}$, $\cdots$ are replaced by other homology bases and the corresponding relations (24.5, 24.7, 24.10) written, the expressions (24.9, 24.11) preserve their form.

(24.12a) We shall indicate an alternate and more direct way of deriving (24.11), and hence by a previous argument, (24.9). In the canonical bases let us lump together the terms a_i^p and b_i^p, likewise the terms d_i^p and e_i^p. The basic relations of (III, 14) assume then the form:

$$\mathrm{F}d_i^{p+1} = t_i^p b_i^p, \qquad \mathrm{F}c_i^p = 0, \qquad \mathrm{F}d_i^p = t_i^{p-1}b_i^{p-1},$$

where the t's are torsion coefficients, or unity. The defining relations of σ assume the form:

$$\sigma b_i^p = \beta_i^j(p) b_j^p ,$$
$$\sigma c_i^p = \cdots + \gamma_i^j(p) c_j^p ,$$
$$\sigma d_i^p = \cdots + \delta_i^j(p) d_j^p .$$

The first two have the form indicated since σ maps a cycle [bounding cycle] into a cycle [bounding cycle]. Since $\psi(\sigma)$ is independent of the choice of the bases, we have:

$$\psi(\sigma) = \sum (-1)^p (\beta_i^i(p) + \gamma_i^i(p) + \delta_i^i(p)).$$

Since σ commutes with F we also have $\sigma \mathrm{F} d_i^p = \mathrm{F} \sigma d_i^p$. Identifying the coefficients of b_i^{p-1} on both sides we find $\beta_i^i(p-1) = \delta_i^i(p)$. Hence the terms in β, δ cancel in the sum and so in view of $\gamma_i^j(p) = \sigma_i^j(p)$:

$$\psi(\sigma) = \sum (-1)^p \gamma_i^i(p) = \sum (-1)^p \sigma_i^i(p)$$

which is (24.11).

To sum up then we may state:

(24.13) THEOREM. (a) *Let X, Y be finite complexes and let τ, θ be chain-mappings $X \to Y$ and $Y \to X$, with matrices of transformations τ^p, θ^p of a set of bases for the rational cycles, or of bases for the Betti groups. Then if $\varphi(\tau, \theta)$ given by* (24.9) *is not zero, τ and θ have coincidences.*

(b) *The same properties hold as regards the coincidences of τ and θ^* relative to $\varphi_1(\tau, \theta^*)$ defined in terms of $\varphi(\tau, \theta)$ as in* (21).

(c) *Let σ be a chain-mapping $X \to X$ with σ^p as the matrices of the transformations of the bases for the rational cycles or of those for the Betti groups. Then if $\psi(\sigma)$ given by* (24.11) *is not zero σ has fixed elements.*

Evidently also (IV, 12.3):

(24.14) *The numbers $\varphi(\tau, \theta)$ and $\psi(\sigma)$ depend solely upon the homology classes of τ, θ or σ.*

(24.15) *Let $X \sim Y$, $X_1 \sim Y_1$. If τ, τ_1 are chain-mappings $X \to X_1$, $X_1 \to X$ and θ, θ_1 congruent chain-mappings $Y \to Y_1$, $Y_1 \to Y$* (IV, 12.7) *then $\varphi(\tau, \tau_1) = \varphi(\theta, \theta_1)$.*

(24.16) *If $X \sim Y$ and τ, θ are congruent chain-mappings $X \to X$, $Y \to Y$ then $\psi(\tau) = \psi(\theta)$* (proof elementary).

Properties (24.15, 24.16) may be summarized in the statement: $\varphi(\tau, \tau_1)$ and $\psi(\tau)$ are congruence invariants.

(24.17) APPLICATION (Tucker). Let X, X_1 be simple with X', X_1' as their derived. If δ, τ are chain-derivation and a reciprocal in X, and δ_1, τ_1 analogues for X_1 then $\sigma : X \to X_1$, $\sigma' : X' \to X_1'$, $\sigma' = \delta_1 \sigma \tau$, are congruent. If ζ, ζ' are another congruent pair of this type then

$$\varphi(\sigma, \zeta) = \varphi(\sigma', \zeta')$$

and if $X_1 = X$ then $\psi(\sigma) = \psi(\sigma')$.

25. Examples. (25.1) Let τ be the identity mapping of the finite complex X into itself, and let α^p be the number of p-elements. Each p-element has the multiplicity $(-1)^p$ as a fixed element. Therefore

$$\psi(1) = \sum (-1)^p \alpha^p.$$

On the other hand in (24.11) every $\nu^p = 1$, hence trace $\nu^p =$ order $\nu^p = R^p$. Therefore

$$\psi(1) = \sum (-1)^p \alpha^p = \sum (-1)^p R^p.$$

This is the Euler-Poincaré formula (III, 15.2) for which we have thus obtained a new proof.

(25.2). Let X be p-cyclic and σ a chain-mapping $X \to X$. The only matrix ν^r is ν^p, and it consists of a single term, ν. It corresponds to the sole equation of transformation

$$\sigma c^p = \nu c^p.$$

The number ν is the *order* of σ and we have $\psi(\sigma) = (-1)^p \nu$. *Therefore a chain-mapping of a p-cyclic finite complex into itself whose order is different from* 0 *always has a fixed element.*

(25.3) Let X be (p, q)-cyclic and σ again a rational chain-mapping $X \to X$. Here we have two equations

$$\sigma c^p = \nu c^p, \qquad \sigma c^q = \nu' c^q$$

and hence

$$\psi(\sigma) = (-1)^p \nu + (-1)^q \nu',$$

so that no simple result may be stated in the present instance.

(25.4) Let X be $(0, p)$-cyclic and σ a chain-mapping $X \to X$ which preserves the Kronecker index of the zero-chains. Then

$$\sigma c^0 = c^0, \qquad \sigma c^p = \nu c^p$$

are the only equations and hence

$$\psi(\sigma) = 1 + (-1)^p \nu.$$

ν is called the *degree* of σ (Brouwer) and we see that *if p is even and $\nu \geqq 0$, σ possesses a fixed element.*

The results obtained in (25.2, 25.4) are at the root of Brouwer's fixed point theorems (see VIII, 30).

§4. COMBINATORIAL MANIFOLDS

26. Combinatorial manifolds arise out of the search for analogues among complexes of structures with the local smoothness of an Euclidean space. Our basic definition will be "in the large," by means of a comparison with the dual. A local equivalent frequently more convenient, if less elegent, will also be given afterward.

Special references for this topic: Lefschetz [L], Tucker [a], Whitney [d].

(26.1) Definitions. *Let $X = Y - Z$ be an open simple complex, where Y is closed simple and Z is a closed subcomplex of Y. We say that X is an orientable combinatorial manifold whenever the following two conditions are fulfilled.*

μ1. *The dual X^* of X has a closed simple weak isomorph $\bar{X}$.*

μ2. *If x, x' are distinct elements of X then* $\mathrm{Cl}\, x \cap \mathrm{St}\, x'$ *is acyclic or void.*

It will be noticed that if Y is *increased* by any closed simple complex Y_1 disjoint from X, X continues to satisfy $\mu 12$. In other words $\mu 12$ depend strictly upon Cl X (closure in Y) and nothing more. In particular if Cl $X = X$, or again if Z may be chosen null the manifold is said to be *absolute*; otherwise it is said to be *relative*, or Y is called a *manifold* mod Z. The complex $\mathfrak{B}X =$ Cl $X - X$ is known as the *boundary* of the manifold, and the latter (when $\mathfrak{B}X \neq \emptyset$) is also called a manifold *with boundary*.

The role of the two conditions $\mu 12$ may be described thus: $\mu 1$ is the source of the special duality properties of manifolds, while $\mu 2$ causes invariance under subdivision, that is to say, it marks out manifolds as geometrically significant.

(26.2) Let $\bar{X}$, in $\mu 1$, be obtained by raising the dimensions in X^* by n units, and let $\bar{x}$ be the image in $\bar{X}$ of x^*, the dual of x.

We have then $\dim \bar{x} = \dim x^* + n = -\dim x + n$. Since $\bar{X}$ is simple, $\dim \bar{x} \geqq 0$, and some $\dim \bar{x} = 0$. Therefore $0 \leqq \dim x \leqq n$, and some $\dim x = n$. In other words, n is precisely $\dim X$, and in particular X is *finite-dimensional*. We call X an *n-dimensional manifold*, or more simply *n-manifold*, with the generic designation M^n.

Suppose now that X is an absolute M^n. Then likewise $\dim x \geqq 0$ and some $\dim x = 0$. Therefore $0 \leqq \dim \bar{x} \leqq n$ and both extremes are reached. In particular, $\bar{X}$ is likewise a simple n-complex.

(26.3) Remark. Non-orientable manifolds will be considered in (34).

27. **Digression: Reciprocal complexes.** Let X denote for the present any n-complex. There corresponds to X^* a unique weak isomorph $\bar{X}$ obtained by raising by n units the dimensions in X^*. We call $\bar{X}$ the *reciprocal* of X. To $x_i^p \in X$ there corresponds x_p^i in X^*, and to x_p^i an $(n-p)$-element in $\bar{X}$, the *reciprocal* of x_i^p, denoted by $\bar{x}_i^{n-p}$.

(27.1) *The correspondence* $x_i^p \leftrightarrow \bar{x}_i^{n-p}$ *is one-one, incidence reversing and such that the sum of the dimensions of any two corresponding elements is* n.

Let $'x$, $'\bar{x}$ denote the vertices of the derived X', $\bar{X}'$ associated with x, $\bar{x}$. Then:

(27.2) $'x \to '\bar{x}$ *defines an isomorphism with* $\bar{X}'$ *of* X' *reoriented by* $\alpha(x^p) = \beta(-p)$.

By (IV, 25.5): $'x \to 'x^*$ defines an isomorphism with $X^{*\prime}$ of X' reoriented by $\alpha(x^p) = \beta(-p)$, while $'x^* \to '\bar{x}$ defines an isomorphism $X^{*\prime} \to \bar{X}'$, and from this to (27.2) is but a step.

(27.3) *If* $\dim x = 0$ *for some* x, *hence if* X *is an absolute* M^n, *then* $\bar{\bar{X}} \cong X$.

For under the circumstances $\dim X = n$, $0 \leqq \dim x \leqq n$, the extreme values being reached, and the asserted isomorphism is obvious.

Notice that when X has no vertices (27.3) need not hold. Thus if $X = \sigma^n$, $n > 0$, then $\bar{X}$ is a point and so is $\bar{\bar{X}}$. Hence the latter is not isomorphic with X.

(27.4) Given two finite-dimensional complexes X, Y with $\dim X = n$, we readily verify that $\bar{X} \times \bar{Y}$ reoriented by $\alpha(x^p \times y^q) = (-1)^{nq}$ is in an orientation preserving isomorphism with $\overline{X \times Y}$, which clearly exists under the circumstances. For convenience the two are henceforth identified, so that $\overline{X \times Y} = \bar{X} \times \bar{Y}$ reoriented as stated.

(27.5) It is clear that *reciprocation* may serve in place of *dualization.* Its mechanism is substantially the same save that instead of passing from the dimension p to the dimension $-p$, we must pass to the dimension $n - p$. This will be more fully developed in connection with the Kronecker index and intersections.

(27.6) Referring again to the comparison of X and $\bar{X}$, when X is an M^n, we may say that the absolute orientable M^n is characterized by the fact that the complex X and its reciprocal are both closed simple and n-dimensional.

(27.7) REMARK. In view of (27.1), in dealing with a pair of complexes X, $\bar{X}$, we shall adopt the summation notation relative to repeated lower indices. Thus $a_{ij}x_i^p \times \bar{x}_j^q$ shall stand for $\sum_{i,j} a_{ij}x_i^p \times \bar{x}_j^q$, etc.

28. The characterization of X as an M^n given in definition (26.1) has the disadvantage of not being intrinsic, and also of masking, as it were, the important local properties of manifolds. We shall now give a characterization free from these defects.

Since $\bar{X}$ is simple: (a) every Cl $\bar{x}$ is zero-cyclic; (b) the sum of the duals of its zero-faces is a zero-cocycle of $\bar{X}$. Since the homology groups of Cl $\bar{x}^p$ are the cohomology groups of St x^{n-p} by (III, 21.3) (a) is equivalent to the condition that every St x is n-cyclic. Since the system of the p-cochains of $\bar{X}$ is in an isomorphic boundary preserving correspondence with the system of the $(n - p)$-chains of X, (b) is equivalent to the condition that the sum of the n-elements of X is an n-cycle. Coupling this with the fact that X itself is to be simple (open or closed), we replace $\mu12$ by the equivalent conditions for a manifold expressed in terms of X alone:

Mn1. *X is an open or a closed simple complex.*

Mn2. *Every* St x *is n-cyclic.*

Mn3. *In* Cl X *every* Cl $x \cap$ St x', $x \neq x'$ *is acyclic or void.*

Mn4. *The sum of the n-faces suitably oriented is an integral n-cycle.*

The four conditions together characterize X as an orientable M^n, or equivalently Y as an orientable M^n mod Z. If only the first three hold then X is called a *non-orientable* M^n, or equivalently Y a *non-orientable* M^n mod Z. *Unless otherwise stated it will be understood that the manifold is orientable.* Non-orientable manifolds are discussed in (34, 35).

Conditions Mn24 state that $\bar{X}$ is simple while Mn3 is $\mu2$ of (26.1). We notice also that when X is a relative manifold the n-cycle in Mn4 is a relative cycle (cycle of Y mod Z).

In the applications one may frequently require a certain regularity of the boundary $\mathfrak{B}X$. It is said to be *regular* whenever the following supplementary conditions hold:

Mn5. *When* $x \in \mathfrak{B}X$ *then* St x *is n-cyclic* mod Z.

Mn6. $\mathfrak{B}X$ *is an absolute* M^{n-1}.

SOME EXAMPLES. (28.1) X is an n-simplex σ^n, $Y = \text{Cl}\,\sigma^n$. X^* has for weak isomorph a point so $\mu1$ is fulfilled. Regarding $\mu2$ the sets different from $\emptyset$ occurring in it are all of the form $\sigma'(\mathrm{Cl}\ \sigma'')_a$, where σ', σ'' have no common vertices. By (IV, 6.21) all such sets are acyclic and so $\mu 2$ holds. Therefore σ^n is an M^n mod $\mathfrak{B}\sigma^n$.

(28.2) $X = \mathfrak{B}\sigma^{n+1}$. We have seen (IV, 30.2) that X^* is weakly isomorphic with X, dimensions being raised by n, i.e., $\bar{X} \cong X$. Therefore X is an absolute n-manifold.

(28.3) The "reciprocal" of a regular convex polyhedron in 3-space (in the sense of elementary geometry) is a regular convex polyhedron. Therefore the polyhedral complexes defined by such convex polyhedra are absolute 2-manifolds.

(28.4) The decomposition of a plane into equal squares furnishes an example of an absolute infinite M^2.

(28.5) Clearly σ^n, $n > 0$, is an M^n with regular boundary. Take the boundaries of two tetrahedra in 3-space intersecting in a polygon Z and thus giving rise to a polyhedral 2-complex Y. Then $X = Y - Z$ is an M^2 with the boundary Z. If Z is an edge with its end points neither Mn5 nor Mn6 holds, while if Z is the perimeter of a triangle Mn6 will hold but not Mn5.

29. We shall now consider a series of properties of orientable combinatorial manifolds.

(29.1) *N.a.s.c. for X to be an absolute orientable manifold is that both X and $\bar{X}$ be simple and that* Mn3 *holds. Hence if one of X, $\bar{X}$ is an absolute orientable M^n so is the other.*

This follows immediately from $\mu 12$ of (26).

Suppose X simplicial (open or closed) and n-dimensional. Then Mn13 are fulfilled. This is clear for Mn1. Regarding Mn3 it asserts here that the stars in a closed simplex are acyclic, and this has been shown to hold in (28.1). Therefore

(29.2) *For a simplicial complex the manifold conditions* Mn1234 *reduce to* Mn24.

It is an elementary matter to prove:

(29.3) *Every component of an M^n [of an absolute M^n] is an M^n [an absolute M^n].*

From Mn2 follows:

(29.4) *Every element of an M^n is the face of an n-element. Hence we find once more* $\dim X = n$.

We also have:

(29.5) *Every $(n - 1)$-element of an M^n is the face of exactly two n-elements.*

For the incidences between the $(n - 1)$- and the n-elements of X are the same as between the zero- and one-elements of $\bar{X}$, and since $\bar{X}$ is simple, (29.5) is a consequence of (III, 47.4).

(29.5a) Important remark. When M^n is simplicial (29.5) is a consequence of Mn2 alone. For let σ^{n-1} be a face of $\sigma_1^n, \cdots, \sigma_r^n$. We may assume the orientations such that $[\sigma_i^n : \sigma^{n-1}] = 1$. Then the integral n-cycles of St σ^{n-1} are the linear combinations $a^i(\sigma_i^n - \sigma_1^n)$, and St σ^{n-1} can only be n-cyclic if $r = 2$.

(29.6) *An orientable M^n is locally finite.*

For Y and $\bar{X}$ are simple and hence closure-finite. Since X is a subcomplex of Y and weakly isomorphic with the reciprocal of $\bar{X}$ it is both closure- and star-finite, hence locally finite.

(29.7) *The incidence numbers in* Cl X *are as follows: if $x^{p-1} \prec x^p$ then $[x^p : x^{p-1}] = [x^{p-1} : x^p] = \pm 1$; all the other incidence numbers are zero.*

Since the one-section of Cl X is simplicial the property is true for $p = 1$, so we assume it for $p - 1 > 0$ and prove it for p. Let $x^{p-1} < x^p \in \mathrm{Cl}\, X$. By (III, 47.4) there is an $x^{p-2} < x^{p-1}$ and, in view of Mn3, St x^{p-2} in Cl x^p is acyclic. As in (29.5a) we show that there are only two $(p - 1)$-elements in the star. Thus in $\mathfrak{B}x^p$ each x^{p-2} is the common face of exactly two elements. Hence $\mathfrak{B}x^p$ is the union of a finite set of simple $(p - 1)$-circuits $X_1, \cdots, X_r$, where $X_i \cap X_j$ is of dimension less than $p - 2$. However by (III, 47.10) $\mathfrak{B}x^p$ is $(0, p - 1)$-cyclic and so consists of a single simple circuit. Furthermore, (loc. cit.) its p-cycles are of the form $g\mathrm{F}x^p$ and also, as in a simple circuit, of the form $g(\epsilon^i x_i^{p-1})$, where $\{x_i^{p-1}\}$ are all the $(p - 1)$-faces of x^p and $\epsilon^i = \pm 1$. From this follows $\pm \mathrm{F}x^p = \epsilon^i x_i^{p-1}$ and finally $[x^p : x_i^{p-1}] = \pm 1$ which suffices to prove (29.7).

(29.8) *If* $x^p \in \mathrm{Cl}\, X$ *then* $\mathfrak{B}x^p$ *is an absolute orientable* M^{p-1}.

Since Y is simple Mn1 holds in $\mathfrak{B}x^p$. Then Mn3 in X implies Mn23 in $\mathfrak{B}x^p$ while Mn4 in $\mathfrak{B}x^p$ is a consequence of (29.7), the $(p - 1)$-cycle being $\mathrm{F}x^p$.

(29.9) *A connected orientable n-manifold* mod Z *is a simple orientable n-circuit* mod Z. *When the manifold is absolute so is the circuit* (29.4, 29.5; III, 40.2a, 24).

(29.10) *If* Y *is an* M^n mod Z *then* $Y^{(k)}$ *(kth derived) is an* M^n mod $Z^{(k)}$.

It is evidently sufficient to prove that Y' is an M^n mod Z'. In view of (29.2) we merely have to verify Mn24.

Consider first Mn2. By (IV, 25.2) any simplex of X' is of the form:

$$\sigma = {}'x^{p_1}, \cdots, {}'x^{p_r}, \quad x^{p_1} < \cdots < x^{p_r}, x^{p_i} \in X.$$

Therefore the star of σ in Y' is:

$$\mathrm{St}\ \sigma = \{\sigma_1 \mid \sigma_1 = {}'x^{p_{01}} \cdots {}'x^{p_{0s_0}}\, {}'x^{p_1}\, {}'x^{p_{11}} \cdots {}'x^{p_r}\, {}'x^{p_{r1}} \cdots {}'x^{p_{rs_r}}, \quad x^{p_{01}} < \cdots < x^{p_{rs_r}}\}.$$

Consider the following sets of simplexes of St σ:

$$Y_0 = \{{}'x^{p_{01}} \cdots {}'x^{p_{0s_0}}\}, \qquad Y_r = \{{}'x^{p_{r1}} \cdots {}'x^{p_{rs_r}}\}, \qquad Y_i = \{{}'x^{p_{i1}} \cdots {}'x^{p_{is_i}}\}.$$

From the expressions just written it is clear that Y_0, Y_r, Y_i are closed simplicial complexes and we have the expression of St σ as a join (IV, 4):

$$\mathrm{St}\ \sigma = Y_{0a} \cdots Y_{ra}\, {}'x^{p_1} \cdots {}'x^{p_r}.$$

Referring to (IV, 6.17) we readily verify that if we can show that whenever the complexes under consideration below are different from $\emptyset$ then:

(a) Y_{0a} is $(p_1 - 1)$-cyclic;

(b) Y_{ia}, $0 < i < r$, is $(p_{i+1} - p_i - 2)$-cyclic;

(c) Y_{ra} is $(n - p_r - 1)$-cyclic,

there will follow that St σ is n-cyclic, which will prove Mn2.

Proof of (a). Y_0 consists of all the simplexes ${}'x_i \cdots {}'x_j$, $x_i < \cdots < x_j < x^{p_1}$, $x_j \neq x^{p_1}$, or $Y_{0a} = (\mathfrak{B}x^{p_1})'_a$. Hence Y_{0a} has the homology groups of $(\mathfrak{B}x^{p_1})_a$ (IV, 26.7) and so it is $(p_1 - 1)$-cyclic (III, 47.10).

Proof of (b). By Mn3, St x^{p_i} in Cl $x^{p_{i+1}}$ is acyclic. Since St $x^{p_i} \cap \mathrm{Cl}\, x^{p_{i+1}}$

contains a single p_{i+1}-element a paraphrase of the proof of (III, 21.4) will show that St $x^{p_i} \cap$ Cl $x^{p_{i+1}} - x^{p_{i+1}}$ is $(p_{i+1} - 1)$-cyclic. That is to say, the star of x^{p_i} in $\mathfrak{B}x^{p_{i+1}}$, which we will denote by $\mathrm{St}_1\, x^{p_i}$, is $(p_{i+1} - 1)$-cyclic. If we lower all dimensions in $\mathrm{St}_1\, x^{p_i}$ by p_{i+1} units it becomes a complex Z_i such that Z_{ia} is $(p_{i+1} - p_i - 2)$-cyclic. The closures of the elements of Z_i are obtained by the same process from sets Cl $x \cap$ St x^{p_i} and so they are acyclic (Mn3). Hence (IV, 27.3) Z'_{ia} is likewise $(p_{i+1} - p_i - 2)$-cyclic. It is easily seen however that $Y_{ia} \cong Z'_{ia}$, and so (b) follows.

Proof of (c). This time Z_r is St x^{p_r} in X with dimensions lowered $p_r + 1$ units. Hence by Mn2, Z_r is $(n - p_r - 1)$-cyclic, and the rest is as in the proof of (b).

We have thus shown that Mn2 is fulfilled by $Y' - Z'$.

Consider now the operation δ (chain-derivation) of (IV, 26.2b, 27). By means of (IV, 26.2b) and (29.7) we may show that δx^p is the sum of the p-elements of $(x^p)'$ each taken with a coefficient ± 1. Since $\sum x^n = \Gamma^n$ is an n-cycle of X, $\sum \delta x^n$ will be an n-cycle of X'. Therefore the sum of the n-elements of X' suitably oriented is an n-cycle. This proves Mn4 for $Y' - Z'$ and hence (29.10).

(29.11) *Under the same conditions as in* (29.8) *if Y is an M^n* mod Z *with regular boundary then $Y^{(k)}$ is likewise an M^n* mod $Z^{(k)}$ *with regular boundary.*

We may here also merely consider the case $k = 1$. Furthermore we may assume $Y =$ Cl X and then Z is the boundary. It is to be shown this time that Mn56 are preserved under derivation. Regarding Mn6 this is a consequence of (29.10) and for Mn5 the proof is essentially like the proof just given that Mn2 is preserved, and so we omit it.

(29.12) *If X, X_1 are both open subcomplexes of Y with $X_1 \subset X$, and X is an M^n so is X_1* (obvious).

(29.13) *The product of an M^p by an M^q is an M^{p+q} and if the factors are absolute so is the product.*

Let $X = M^p$, $X_1 = M^q$. Since $\bar{X}$, $\bar{X}_1$ are simple so is $\bar{X} \times \bar{X}_1$ and hence also $\overline{X \times X_1}$ (27.4). If X, X_1 are simple so is $X \times X_1$. Hence by (29.1) all that is required is to verify Mn3 (or $\mu 2$) for $X \times X_1$. And this follows from Mn3 for X, X_1 together with: $(\mathrm{Cl}\, x \times x_1) \cap (\mathrm{St}\, x' \times x'_1) = (\mathrm{Cl}\, x \cap \mathrm{St}\, x') \times (\mathrm{Cl}\, x_1 \cap \mathrm{St}\, x'_1)$.

30. **Elementary manifolds.** We refer thereby to the well known finite absolute simplicial manifolds. The following considerations are valid for both orientable and non-orientable manifolds (see 34). Since it is sufficient to characterize the components, by (29.3) we may suppose the manifolds connected. Consider first a connected M^1. By (29.4, 29.5) we find readily that it has the structure of a partition of the circumference.

Take now an M^2. By (29.5) every edge of M^2 is a face of two triangles. From this follows that about every vertex the triangles and edges fall into circular systems consisting of alternating edges and triangles. Then by means of Mn2 we may show that about each vertex there is just one circular system. With this preliminary characterization as a basis, one may proceed to a complete classification of the absolute M^2 into types by means of the homology characters (Betti numbers and torsion coefficients). Each type corresponds to a unique topological class of polyhedra, and the polyhedra of distinct types are topologically distinct. For details see Veblen [V, II].

31. **Kronecker index.** For the proper formulation of the duality relations we require an extension of the Kronecker index. This is done chiefly by replacing everywhere dual elements by reciprocal elements.

For the dimensions p, $n - p$ automatically:

$$\mathrm{KI}(\xi^p, \bar{\eta}^{n-p}) = \mathrm{KI}(\xi^p, \eta_p), \tag{31.1}$$

where η_p is the $(-p)$-chain of X^* (p-cochain of X) corresponding to $\bar{\eta}^{n-p}$ in the weak isomorphism $X^* \leftrightarrow \bar{X}$. From (III, 28.2) we have explicitly,

$$\mathrm{KI}(x_i^p, \bar{x}_j^{n-p}) = \beta(p)\delta_{ij} \tag{31.2}$$

where $\delta_{ij} = \delta_i^j$.

If we consider $\bar{X}$ as the basic complex we have from (31.2)

$$\mathrm{KI}(\bar{x}_j^{n-p}, x_i^p) = \beta(n - p)\delta_{ij},$$

and therefore

$$\mathrm{KI}(x^p, \bar{x}^{n-p}) = \beta(n)(-1)^{p(n-p)}\mathrm{KI}(\bar{x}^{n-p}, x^p).$$

The commutation rule is thus

$$\mathrm{KI}(\xi^p, \bar{\eta}^{n-p}) = \beta(n)(-1)^{p(n-p)}\mathrm{KI}(\bar{\eta}^{n-p}, \xi^p). \tag{31.3}$$

We observe now that under the weak isomorphism $X^* \to \bar{X}$, η_p and $\mathrm{F}\eta_p$ go over into $\bar{\eta}^{n-p}$ and $\mathrm{F}\bar{\eta}^{n-p}$. Hence from (III, 29.1) and under the assumption that *one of* ξ^{p+1}, $\bar{\eta}^{n-p}$ *is finite* we have [L, 169]:

$$\mathrm{KI}(\mathrm{F}\xi^{p+1}, \bar{\eta}^{n-p}), = (-1)^p\mathrm{KI}(\xi^{p+1}, \mathrm{F}\bar{\eta}^{n-p}). \tag{31.4}$$

It is a consequence of our definitions that the new Kronecker index may serve as a group multiplication under the same conditions as the earlier index (III, 29, 40). Moreover with (31.4) at our disposal we may introduce wherever need be a class index $\mathrm{KI}(\Gamma^p, \bar{\Gamma}^{n-p})$ and related multiplication of the corresponding groups. This is the multiplication implicit in the statements of the duality theorems given below.

(31.5) *Linking coefficients.* Whenever M^n is $(p - 1, p)$-acyclic a linking coefficient $\mathrm{Lk}\,(\gamma^{p-1}, \bar{\gamma}^{n-p})$ and related class linking coefficient $\mathrm{Lk}\,(\Gamma^{p-1}, \bar{\Gamma}^{n-p})$ may be defined directly by mere paraphrase of (III, 35). They may also be defined indirectly in terms of $\mathrm{Lk}\,(\gamma^{p-1}, \gamma_p)$ in the same way as we have just introduced the Kronecker index.

32. **Duality theorems.** The duality theorems for manifolds will all be obtained from those for complexes by transfer from pth cohomology groups to $(n - p)$th homology groups. The situation being simpler for absolute manifolds we will consider them first.

Suppose then that X is an absolute M^n. Since the weak isomorphism of X^* with $\bar{X}$ raises dimensions n units, the $(-p)$th homology groups of X^*, or the

pth cohomology groups of X are isomorphic with the $(n - p)$th homology groups of $\bar{X}$ of the *same kind*. By this we mean that if the cohomology groups are those of the finite [infinite] cocycles over G, then the homology groups are understood to be those of the finite [infinite] cycles over G and *vice versa*. Since X, $\bar{X}$ are simple and $X' \cong \bar{X}'$ reoriented, X and $\bar{X}$ have the same homology groups and so:

(32.1) *The pth cohomology groups of an absolute M^n are isomorphic with the $(n - p)$th homology groups of the same kind.*

Let now Y be an M^n mod Z with $X = Y - Z$. The best formulation is in terms of the derived X', Y', Z'. Since $X' \cong \bar{X}'$ and $\bar{X}$ is simple closed, X' is a closed simplicial complex. The open simplicial complex $Y' - X'$ consists of all the simplexes $'x_i \cdots 'x_j$ $(x_i \prec \cdots \prec x_j)$ of Y' such that one of $x_i, \cdots, x_j \in Z$, hence such that $x_i \in Z$. These elements of Y' make up St Z' in Y', so that $X' = Y' -$ St Z'. Since $\bar{X}$ is closed and simple its homology groups are isomorphic with those of its derived $\bar{X}'$ and hence with those of $X' = Y' -$ St Z'. On the other hand, owing to the weak isomorphism between X^* and $\bar{X}$, the pth cohomology groups of X are the isomorphs of the corresponding $(n - p)$th homology groups of $\bar{X}$. Therefore

(32.2) *If Y is an M^n* mod Z *then the pth absolute cohomology group of X is isomorphic with the $(n - p)$th homology group of $X' = Y' -$* St Z' *which is of the same kind.*

The analogues of dual categories (III, 31) are given by:

(32.3) Definition. *Two categories A, B of cycles of M^n are said to be quasi-dual whenever if (G, H) is a normal couple, $\mathfrak{H}^p(A, G)$ and $\mathfrak{H}^{n-p}(B, H)$ are dually paired and with the class Kronecker index for multiplication.*

33. In view of (32.1, 32.2) the duality theorems for complexes (III, 30, 31, 32, 38, 39, 41) yield here:

(33.1) Duality theorem of Poincaré. *When M^n is finite the Betti groups for the dimensions p and $n - p$, and likewise the torsion groups for the dimensions p and $n - p - 1$ are isomorphic. As a consequence the Betti numbers and torsion coefficients satisfy:*

(33.1a) $$R^p = R^{n-p};$$

(33.1b) $$t_i^p = t_i^{n-p-1}.$$

(33.2) *The following are quasi-dual categories:*

(a) *Finite absolute M^n. The cycles repeated (one category repeated).*

(b) *Infinite absolute M^n. The infinite and the finite cycles.*

(c) *Finite relative M_n. Y is an M^n* mod Z, $X = Y - Z$: *The cycles of Y* mod Z *and the absolute cycles of X'.*

(d) *Infinite relative M^n. In the same notations the infinite cycles of X' [of Y* mod Z] *and the finite cycles of Y* mod Z *[of X'].* (Lefschetz [L, 142, 314].)

(33.3) *If A, B are quasi-dual categories in M_n and J a field of characteristic π then:*

(a)
$$R^p(A, \pi) = R^{n-p}(B, \pi).$$

(b) *If $\{\gamma_i^p\}$ is an independent set of r-cycles of A over J there may be selected a set of r independent cycles $\{\delta_i^{n-p}\}$ of B over J such that*

$$\mathrm{KI}(\gamma_i^p, \delta_j^{n-p}) = \delta_{ij}.$$

(c) *If the Betti numbers for dimensions p and $n - p$ are finite and $\{\gamma_i^p\}$, $\{\delta_i^{n-p}\}$ are maximal independent sets then:*

$$|\mathrm{KI}(\gamma_i^p, \delta_j^{n-p})| \neq 0$$

[L, 178, 314].

(33.4) *Alexander duality.* The duality theorems of this type hold when Y is simple with a dissection (X, Z) such that Y is an M^n mod Z, the transfer being always:

pth cohomology group of $X \to (n - p)$th homology group of X'.

We state explicitly the following simple case:

(33.5) *If Y is finite $(p - 1, p)$-acyclic and (G, H) is a normal couple, then the groups $\mathfrak{H}^{p-1}(Z, G)$ and $\mathfrak{H}^{n-p}(X', H)$ are dually paired with the class linking coefficient as the group multiplication* (Alexander [a]).

(33.6) *Weak manifolds.* The proofs of (33.1, 33.2ab) rest solely upon the symmetry between X and $\bar{X}$ and do not depend upon $\mu 2$ of (26.1), (or which is the same, upon Mn3). If we call *weak absolute* M^n a complex satisfying Mn 124, we have:

(33.7) *Properties* (33.1, 33.2ab) *hold for a weak absolute manifold.*

34. **Non-orientable manifolds.** For simplicity we will only discuss the simplicial types. Thus X is now an open simplicial complex which merely satisfies Mn2 but not Mn4. In an obvious sense M^n may be *absolute, relative, with regular boundary* (*orientable or otherwise*).

(34.1) *All the properties of* (29) *except* (29.9) *hold for a non-orientable M^n.*

For they do not involve Mn4. In place of (29.9) we now have:

(34.2) *A connected non-orientable M^n is a simple non-orientable n-circuit.*

(34.3) *The duality theorems for groups* mod 2 *hold also for non-orientable manifolds.*

35. In view of (34.3) one could expect to lose in a non-orientable M^n all the homology properties except those mod 2. The loss is restored by means of a device due to De Rham [c]. Corresponding to the elements $\{x_i^p\}$ of every Cl x^n introduce two new sets $\{x_{1i}^p\}$, $\{x_{2i}^p\}$ where in each set the order and the incidence numbers not involving x_2^n, are the same as in Cl x^n, and take $[x_2^n : x_{2i}^{n-1}] = -[x^n : x_i^{n-1}]$. Suppose that x^n, x'^n have a common face x^{n-1}, and let $x_{1i}'^p$, $x_{2i}'^p$ have the obvious meaning for x'^n. There are then two images x_1^{n-1}, x_2^{n-1} for x^{n-1} as a face of x^n, and similarly $x_1'^{n-1}$, $x_2'^{n-1}$ for x^{n-1} as a face of x'^n. If $[x^n : x^{n-1}] = -[x'^n : x^{n-1}]$, we identify Cl x_h^{n-1} with Cl $x_h'^{n-1}$ $(h = 1, 2)$,

otherwise Cl x_1^{n-1} with Cl $x_2'^{n-1}$ and Cl x_2^{n-1} with Cl $x_1'^{n-1}$. The result is readily shown to be a new orientable (simplicial) n-manifold M_0^n. It is known as the *doubly-covering manifold* of M^n. The same process applied to an orientable M^n yields an M_0^n which consists merely of two isomorphs of M^n. For many purposes M_0^n may very well replace its non-orientable companion. In particular its homology and intersection properties may be viewed as properties of M^n itself.

36. **Intersections.** Owing to the possible transfer from cochains to chains there may be defined intersections of chains by chains in an M^n. We will first introduce those for an absolute, orientable n-manifold X, and consider relative manifolds afterward.

(36.1) Several important auxiliary operations will be required. Let δ, τ be chain-derivation and a reciprocal in X, and $\bar{\delta}$, $\bar{\tau}$ the same for $\bar{X}$. We will conveniently identify the vertices $'x$, $'\bar{x}$ of X', $\bar{X}'$, so that $\bar{X}'$ is now X' reoriented by $\alpha(x'^p) = \beta(-p)$, $(x'^p \in X')$. Let this operation be considered as a chain-mapping $\alpha:X'$ or $\bar{X}' \to \bar{X}'$. Clearly $\alpha^2 = 1$. The operations $\eta = \bar{\tau}\alpha\delta$, $\bar{\eta} = \tau\alpha\bar{\delta}$ are, respectively, chain-mappings $X \to \bar{X}$, $\bar{X} \to X$. From the known relations $\delta\tau \sim 1$, $\tau\delta \sim 1$, $\bar{\delta}\bar{\tau} \sim 1$, $\bar{\tau}\bar{\delta} \sim 1$ (IV, 23.5) there follows:

$$\eta\bar{\eta} \sim 1, \qquad \bar{\eta}\eta \sim 1. \tag{36.2}$$

We may even state a more precise result. For if D, $\bar{D}$ are derivations in X, $\bar{X}$ (IV, 26.2a), then $\delta\tau$ is contiguous to the identity in DD^{-1} (IV, 26.2c, 23.1) and similarly for $\bar{\delta}\bar{\tau}$ in $\bar{D}\bar{D}^{-1}$. From this follows readily that: $\eta\bar{\eta}$ is contiguous to the identity in $\bar{D}^{-1}DD^{-1}\bar{D}$. Furthermore we also verify that if Δ is the set-transformation $x \to$ Cl St Cl x then $\bar{D}^{-1}DD^{-1}\bar{D}x \subset \Delta x$. Hence $\eta\bar{\eta}$ is contiguous to the identity in Δ. Thus if $\bar{\Delta}$ is the analogue of Δ for $\bar{X}$ we have:

(36.3) *$\eta\bar{\eta}$, $\bar{\eta}\eta$ are contiguous to the identity in Δ, $\bar{\Delta}$.*

(36.4) In addition to the preceding operations we shall need an operation ζ whose effect on *any complex* merely consists in lowering all dimensions by n units. Thus $\zeta\bar{X} = X^*$, $\zeta X = \bar{X}^*$, etc. Evidently ζ is a weak isomorphism.

(36.5) Consider the multiplication φ of (1.13), of X, X^* to X'. We recall that by (19.1), if τ is a reciprocal of chain-derivation in X then $\tau\varphi = \mu_1$ is the first component of an intersection in X. We may thus freely assume that we have an intersection $\mu = (\mu_1, \mu_2^*)$ in X which we will also denote by a dot-product. Thus

$$x^p \cdot x_q = \tau(x^p \times_\varphi x_q).$$

Let us suppose the x_i^n so oriented that the basic n-cycle of M^n, whose class Γ^n is the *basic class*, is:

$$\gamma^n = \beta(-n) \sum x_i^n . \tag{36.6}$$

From (1.14, 1.15, 1.16) we have:

$$\gamma^n \times_\varphi x_n^i = \beta(-n)\beta(n)\,'x_i^n ,$$

$$p < n: \gamma^n \times_\varphi x_p^i = \beta(-n)\beta(p) \sum \lambda_{k_1 k_2}^n \cdots \lambda_{k_{n-p} i}^{p+1} {}' x_{k_1}^n \cdots {}' x_i^p .$$

Referring to the expressions of the λ's as incidence numbers we verify by means of (IV, 26.2b, 27):

$$x_p^i \times_\varphi \gamma^n = \beta(p - n)\bar{\delta}\bar{x}_i^{n-p}.$$

Hence

(36.7) $$\tau(x_p^i \times_\varphi \gamma^n) = x_p^i \cdot \gamma^n = \bar{\eta}\bar{x}_i^{n-p}.$$

The relation just written may also be put in the form

$$(\zeta \bar{x}_i^{n-p}) \cdot \gamma^n = \bar{\eta}\bar{x}_i^{n-p}.$$

Hence for any chain $\bar{\xi}^{n-p}$ of $\bar{X}$:

(36.8) $$(\zeta \bar{\xi}^{n-p}) \cdot \gamma^n = \bar{\eta}\bar{\xi}^{n-p}.$$

If ξ^p is a chain of X over G then $\eta\xi^p$ will be a chain of $\bar{X}$ over G, and so from (36.8):

(36.9) $$(\zeta\eta\xi^p) \cdot \gamma^n = \bar{\eta}\eta\xi^p.$$

(36.10) Let $\xi'_{n-p} = \zeta\eta\xi^p$. Since η is a chain-mapping and ζ is a weak isomorphism, if ξ^p is a cycle γ^p in a fixed homology class Γ^p over G then $\gamma'_{n-p} = \zeta\eta\gamma^p$ is a cocycle in a fixed cohomology class Γ'_{n-p} over G. From (36.2, 36.9) there results:

(36.11) $$\gamma'_{n-p} \cdot \gamma^n \sim \gamma^p; \qquad \Gamma'_{n-p} \cdot \Gamma^n = \Gamma^p.$$

(36.12) Since η is an isomorphism (36.2) and ζ is a weak isomorphism, $\Gamma^p \to \Gamma'_{n-p}$ defines an isomorphism $\omega: \mathfrak{H}^p(X, G) \to \mathfrak{H}_{n-p}(X, G)$. It follows that ω^{-1} is likewise an isomorphism. Thus

(36.12a) *Given* Γ'_{n-p}, *there is a unique* $\Gamma^p = \omega^{-1}\Gamma'_{n-p}$ *such that* $\Gamma'_{n-p} \cdot \Gamma^n = \Gamma^p$.

For later purposes we also require some information regarding what may be described as a *carrier* of $\zeta\eta$. We recall that δx consists of simplexes with $'x$ as last vertex, i.e., of form $\sigma = {}'x_i \cdots {}'x_j {}'x$, $x_i \prec \cdots \prec x_j \prec x$. When σ is considered as a simplex of $\bar{X}'$ its last vertex is $'x_i = {}'\bar{x}_i$. Hence $\eta\sigma = \bar{\tau}\alpha\sigma$ is an element of $\mathrm{Cl}\, \bar{x}_i = \overline{\mathrm{St}\, x_i} \subset \overline{\mathrm{St}\, \mathrm{Cl}\, x}$ (the bar denotes reciprocation). It follows that $\eta x = \bar{\tau}\alpha\delta x \subset \overline{\mathrm{St}\, \mathrm{Cl}\, x}$ and hence $\zeta\eta x \subset (\mathrm{St}\, \mathrm{Cl}\, x)^*$. Therefore

(36.13) $\zeta\eta x$ *is a cochain of* $\mathrm{St}\, \mathrm{Cl}\, x$.

In an evident sense we could also say that the set-transformation $t: X \to X^*$ defined by $x \to \mathrm{St}\, \mathrm{Cl}\, x$ is a *carrier* of $\zeta\eta$.

Observe finally that since ω is an isomorphism, we have:

(36.14) $$\Gamma^p = 0 \to \Gamma'_{n-p} = 0.$$

37. We are now ready for the chain-chain-intersections. Two types will be considered, a direct one which operates upon the chains of X alone, and another which is more suitable for relative manifolds. We will deal at length with the

finite absolute M^n and then complete the treatment with a few remarks about the general case.

We assume then that X is a finite, absolute, orientable M^n and take G, H, J, ρ as in (1.1).

A. *Intersections of the first type.* Given two chains ξ^p, ξ^q over G, H we define as their intersection the $(p + q - n)$-chain over J, denoted by $\xi^p \circ \xi^q$ and given by:

$$\xi^p \circ \xi^q = \xi'_{n-p} \cdot \xi^q. \tag{37.1}$$

From (8.2, $\cdots$, 8.7) there results then:

(37.2) *The operation $\circ$ pairs $\mathfrak{C}^p(X, G)$, $\mathfrak{C}^q(X, H)$ to $\mathfrak{C}^{p+q-n}(X, J)$.*

$$\mathrm{F}(\xi^p \circ \xi^q) = (\mathrm{F}\xi^p) \circ \xi^q + (-1)^{n-p} \xi^p \circ \mathrm{F}\xi^q. \tag{37.3}$$

(37.4) *$\gamma^p \circ \gamma^q$ is a $(p + q - n)$-cycle over J.*

(37.5) *The class of $\gamma^p \circ \gamma^q$ depends solely upon the classes Γ^p, Γ^q of γ^p, γ^q. It is written $\Gamma^p \circ \Gamma^q$ and called the class intersection of Γ^p, Γ^q.*

We have clearly by (36.8):

$$\xi'_{n-p} \cdot \xi'_{n-q} \cdot \gamma^n = \xi^p \circ \bar{\eta} \eta \xi^q,$$

and therefore by (36.2):

$$\gamma'_{n-p} \cdot \gamma'_{n-q} \cdot \gamma^n \sim \gamma^p \circ \gamma^q,$$

or equivalently

$$\Gamma'_{n-p} \cdot \Gamma'_{n-q} \cdot \Gamma^n = \Gamma^p \circ \Gamma^q. \tag{37.6}$$

In terms of the operation ω of (36.12) the last relation may also be written

$$\Gamma^p \circ \Gamma^q = \omega^{-1} (\omega \Gamma^p \cdot \omega \Gamma^q). \tag{37.6a}$$

By combining (37.6) with (8.8) we have the comprehensive

(37.7) THEOREM. *The class intersection $\Gamma^p \circ \Gamma^q$ in the finite absolute orientable n-manifold X pairs*

$$\mathfrak{H}^p(X, G),\ \mathfrak{H}^q(X, H) \text{ to } \mathfrak{H}^{p+q-n}(X, J).$$

Furthermore:

(a) *(Commutation rule)* $\Gamma^p \circ \Gamma^q = (-1)^{(n-p)(n-q)} \Gamma^q \circ \Gamma^p$.

(b) *If the classes are over a ring ρ then their intersections are associative.*

(c) *Under the same conditions as in* (b), *$\Gamma^p \circ \Gamma^q$ considered as a formal product generates an associative but not necessarily commutative ring $R(X, \rho)$, the homology ring of the manifold over ρ.*

(37.8) We may define the Kronecker index of ξ^p, ξ^{n-p} as

$$\mathrm{KI}(\xi^p, \xi^{n-p}) = \mathrm{KI}(\xi'_{n-p}, \xi^{n-p}),$$

and we have from (4.4):

$$\mathrm{KI}(\xi^p, \xi^{n-p}) = \mathrm{KI}(\xi^p \circ \xi^{n-p})$$

where the last index is the index of the zero-chain $\xi^p \circ \xi^{n-p}$.

Given two classes Γ^p, Γ^{n-p} over G, there are two possible ways of defining the class index. First in the notations of (36.10) we may set

(37.9) $$\mathrm{KI}(\Gamma^p, \Gamma^{n-p}) = \mathrm{KI}(\Gamma'_{n-p} \cdot \Gamma^{n-p}).$$

Second we may choose any $\gamma^p \in \Gamma^p$. Then $\eta\gamma^p$ is in a fixed class $\bar{\Gamma}^p$ of $\bar{X}$ and we define:

(37.10) $$\begin{aligned} \mathrm{KI}(\Gamma^p, \Gamma^{n-p}) &= \mathrm{KI}(\bar{\Gamma}^p, \Gamma^{n-p}) \\ &= \beta(n)(-1)^{p(n-p)}\mathrm{KI}(\Gamma^{n-p}, \bar{\Gamma}^p), \end{aligned}$$

where the last two indices are the class indices as defined in (31). It is only necessary, however, to refer to the definitions of the index in (31) to show that the two values of $\mathrm{KI}(\Gamma^p, \Gamma^{n-p})$ are in fact the same. Furthermore (37.6) yields

(37.11) $$\mathrm{KI}(\Gamma^p, \Gamma^{n-p}) = \mathrm{KI}(\gamma^p \circ \gamma^{n-p}), \qquad \gamma^p \in \Gamma^p, \gamma^{n-p} \in \Gamma^{n-p}.$$

(37.12) It is also a consequence of (37.10) that $\mathrm{KI}(\Gamma^p, \Gamma^{n-p})$ may be chosen as the group multiplication of (33.2a) and $\mathrm{KI}(\gamma^p, \gamma^{n-p})$ in the sense of (37.8) as the index of (33.3) (in place of the indices $\mathrm{KI}(\gamma^p, \delta^{n-p})$ there considered).

B. *Intersections of the second type.* This time we only have the intersections of a chain $\bar{\xi}^p$ of $\bar{X}$ over G with a chain ξ^q of X over H. It will be once more a $(p + q - n)$-chain of X over J, written $\bar{\xi}^p * \xi^q$ and given by

(37.13) $$\bar{\xi}^p * \xi^q = \xi_{n-p} \cdot \xi^q, \ \xi_{n-p} = \zeta\bar{\xi}^p.$$

This new operation is related to $\circ$ by:

(37.14) $$\xi^p \circ \xi^q = \eta\xi^p * \xi^q.$$

We notice explicitly that:

(37.15) *Properties* (37.2, $\cdots$, 37.5) *hold for* $*$ *with the appropriate* (*and obvious*) *modifications.*

We also have with the Kronecker index defined as in (31.1):

(37.16) $$\mathrm{KI}(\bar{\xi}^p, \xi^{n-p}) = \mathrm{KI}(\bar{\xi}^p * \xi^{n-p}).$$

(37.17) *Intersections in a product.* If X, Y are finite absolute orientable manifolds of dimensions n, n' then $X \times Y$ is an absolute orientable $M^{n+n'}$ (29.13). Let the previous notations: ξ, γ, Γ apply to X, and let η, δ, Δ be the analogues for Y. To simplify matters all chains are over a ring ρ. We have now by (37.1, 17.4)

$$\begin{aligned} (\xi^p \times \eta^r)\circ(\xi^q \times \eta^s) &= (\xi'_{n-p} \times \eta'_{n'-r})\cdot(\xi^q \times \eta^s) \\ &= (-1)^{q(n'-r)}(\xi'_{n-p}\cdot\xi^q) \times (\eta'_{n'-r}\cdot\eta^s), \end{aligned}$$

and so

(37.18) $$(\xi^p \times \eta^r)\circ(\xi^q \times \eta^s) = (-1)^{q(n'-r)}(\xi^p \circ \xi^q) \times (\eta^r \circ \eta^s),$$

(37.19) $$(\Gamma^p \times \Delta^r)\circ(\Gamma^q \times \Delta^s) = (-1)^{q(n'-r)}(\Gamma^p \circ \Gamma^q) \times (\Delta^r \circ \Delta^s).$$

Assuming, in particular, $n = n'$ we have from (37.16) the following relation on the Kronecker index which is required below:

$$(37.20) \quad \mathrm{KI}(\xi^p \times \eta^r, \xi^{n-p} \times \eta^{n-r}) = (-1)^{(n-p)(n-r)}\mathrm{KI}(\xi^p, \xi^{n-p})\mathrm{KI}(\eta^r, \eta^{n-r}).$$

38. So much for the finite absolute orientable M^n; for the other cases a few complementary remarks will suffice.

(38.1) *Infinite absolute orientable* M^n. For this case all the considerations of (37) may be repeated, with the complement that if one of the intersecting chains is finite so is the intersection. The ring property (37.7c) is only preserved as regards intersections between classes of finite cycles or between classes of infinite cycles.

(38.2) *Relative orientable* M^n. Suppose $X = Y - Z =$ an orientable M^n mod Z. Then we may only define the operation $*$ of (37.13), it being understood that the intersections at the right in (37.14) are the intersections induced in X by those in Y in the sense of (13). Properties (37.2, $\cdots$, 37.5) hold with the obvious modifications. We merely notice that as regards the cycles we are taking the intersection of an absolute p-cycle of X with a q-cycle of X mod Z and obtain an absolute $(p + q - n)$-cycle of X as the intersection. The index is as defined in (31.1) and satisfies (37.16).

APPLICATIONS. (38.3) Take as orientable M^1 the complex X obtained from the subdivision of the real line $L: -\infty < u < +\infty$ by the points $A_i : u = i$, the elements being the vertices A_i and the one-simplexes A_iA_{i+1}. Let B_i denote the midpoint of A_iA_{i+1} or vertex of X' in A_iA_{i+1}. We may then identify $\bar{X}$ with the analogue of X corresponding to the subdivision points B_i. However we will orient $\bar{X}$ so that $\bar{A}_i = B_iB_{i-1}$, which corresponds to reorienting it by means of $\alpha(\bar{x}^{n-p}) = \beta(n - p)$. The operation τ is defined by $B_i \to A_{i+1}$, $A_i \to A_i$.

The vertices A_i are homologous finite integral zero-cycles of X and we will denote by A their common class (element of the homology group of the finite integral zero-chains). The cycle $\sum A_iA_{i+1}$ is basic for X and its class (element of the homology group of the infinite integral one-chains), which is a basic class, will also be written L for convenience. Using the above data we find readily from (36, 37) the class intersections in X:

$$(38.4) \qquad A\cdot A = 0, \qquad A\cdot L = L\cdot A = A, \qquad L\cdot L = L.$$

(38.5) Take now n real lines $\{L_i\}$, $i = 1, 2, \cdots, n$; $L_i : -\infty < u_i < +\infty$, and turn L_i into a complex X_i such as just considered. The complex $X = X_1 \times \cdots \times X_n$ is an orientable M^n. Let L_i be the basic class of X_i and A_i the class of its vertices. Now $\Gamma^p = L_1 \times \cdots \times L_p \times A_{p+1} \times \cdots \times A_n$ and $\Gamma^{n-q} = A_1 \times \cdots \times A_q \times L_{q+1} \times \cdots \times L_n$ are the basic classes for the submanifolds:

$$X^p_{1,\cdots,p} = X_1 \times \cdots \times X_p \times A_{p+1} \times \cdots \times A_n,$$

$$X^{n-q}_{q+1,\cdots,n} = A_1 \times \cdots \times A_q \times X_{q+1} \times \cdots \times X_n.$$

Applying (37.19) we find if $q < p$:

$$(38.6) \quad \Gamma^p\circ\Gamma^{n-q} = \Gamma^{p-q} = A_1 \times \cdots \times A_q \times L_{q+1} \times \cdots \times L_p \times A_{p+1} \times \cdots \times A_n.$$

Therefore:

(38.7) *The intersection of the basic homology classes of* $X_1 \times \cdots \times X_p \times A_{p+1} \times \cdots \times A_n$ *and* $A_1 \times \cdots \times A_q \times X_{q+1} \times \cdots \times X_n$ *is the basis homology class of*

$$A_1 \times \cdots \times A_q \times X_{q+1} \times \cdots \times X_p \times A_{p+1} \times \cdots \times A_n .$$

The analogy of this intersection rule with the initial rules of Lefschetz [a] (for convex polyhedral cells) and of [L, IV] is obvious enough. It will become completely obvious in the topological applications in (VIII, 27.10, 47).

If $p + q = n$ the intersection is $A_1 \times \cdots \times A_n$ and so

$$\mathrm{KI}(\Gamma^p, \Gamma^{n-p}) = 1. \tag{38.8}$$

It is hardly necessary to point out that the preceding results have obvious extensions to the case where the X_i are arbitrarily scattered among the A_i, but as the explicit forms are not required we will not derive them here.

39. Chain-mappings and their graphs. Let again X, Y be absolute orientable manifolds. We will suppose them this time of the same dimension n. A chain-mapping $\tau : X \to Y$ will be written

$$\tau x_i^p = a_{ij}^p y_j^p \tag{39.1}$$

with summation on the lower indices. We wish to associate with τ a chain-graph which will be an n-cycle Γ^n in $\bar{X} \times Y$. We first write the usual graph as a cycle of $Y \times X^*$

$$\Gamma = \sum \beta(-p) a_{ij}^p y_j^p \times x_p^i .$$

Under the weak isomorphism $Y \times X^* \to Y \times \bar{X}$ raising dimensions n units Γ goes into an n-cycle of $Y \times \bar{X}$:

$$\Gamma^n = \sum \beta(-p) a_{ij}^p y_j^p \times \bar{x}_i^{n-p} = \sum \beta(-p)(\tau x_i^p) \times \bar{x}_i^{n-p}. \tag{39.2}$$

It is clear that τ and Γ^n determine one another uniquely.

Consider now a second chain-mapping $\bar{\theta} : \bar{X} \to \bar{Y}$ given by

$$\bar{\theta}\bar{x}_i^p = b_{ij}^p \bar{y}_j^p , \tag{39.3}$$

with its graph in $\bar{Y} \times X$

$$\bar{\Gamma}'^n = \sum \beta(-p) b_{ij}^p \bar{y}_j^p \times x_i^{n-p}. \tag{39.4}$$

Written as a cycle of $\overline{Y \times \bar{X}}$ by (27.4) $\bar{\Gamma}'^n$ assumes the form:

$$\bar{\Gamma}'^n = \sum \beta(-p) b_{ij}^p (-1)^{n(n-p)} \overline{y_j^{n-p} \times \bar{x}_i^p} . \tag{39.5}$$

This form will be utilized in a moment.

40. Coincidences.

(40.1) The results of (§3) are of course applicable to manifolds. However, by utilizing the transfer from cocycles to cycles it will be found possible to extend them to a pair of chain-mappings τ, θ both proceeding in the *same direction* (from X to Y).

(40.2) Consider first the two chain-mappings τ, $\bar{\theta}$. By mere transfer from (21) (for the τ, θ^* of 21), we define a coincidence of τ, $\bar{\theta}$ as a pair (x_i^p, y_j^p) such that y_j^p is found in τx_i^p and $\bar{y}_j^{n-p}$ in $\bar{\theta}\bar{x}_i^{n-p}$. The number of signed coincidences is then naturally defined as

(40.3) $$\chi(\tau, \bar{\theta}) = \text{KI}(\Gamma^n, \bar{\Gamma}'^n).$$

Since $Y \times \bar{X}$ is an absolute orientable M^{2n} (29.13) we have from (31):

$$\text{KI}(\overline{y_j^p \times \bar{x}_i^{n-p}}, y_j^p \times \bar{x}_i^{n-p}) = \beta(n),$$

and whenever other pairs of elements are involved:

$$\text{KI}(\overline{y_j^p \times \bar{x}_i^{n-p}}, y_h^q \times \bar{x}_k^{n-q}) = 0.$$

Referring to the properties of $\beta(p)$ (Introduction to III) we find then:

$$\chi(\tau, \bar{\theta}) = \sum (-1)^p a_{ij}^{n-p} b_{ij}^p .$$

If we set

$$a^p = \| a_{ij}^p \|, \qquad b^p = \| b_{ij}^p \|,$$

there comes:

$$\chi(\tau, \bar{\theta}) = \sum (-1)^p \text{ trace } a^{n-p}(b^p)'.$$

Let $\{c_i^p\}$, $\{\bar{c}_i^p\}$ be bases for the rational homology groups of X, $\bar{X}$ and $\{d_i^p\}$, $\{\bar{d}_i^p\}$ the same for Y, $\bar{Y}$. The chain-mappings τ, $\bar{\theta}$ induce homomorphisms on the groups in question represented by:

(40.4) $$\tau c_i^p \approx \tau_{ij}^p d_j^p , \qquad \tau^p = \| \tau_{ij}^p \|,$$

(40.5) $$\bar{\theta}\bar{c}_i^p \approx \bar{\theta}_{ij}^p \bar{d}_j^p , \qquad \bar{\theta}^p = \| \bar{\theta}_{ij}^p \|.$$

The same argument as in (24) yields then:

(40.6) $$\chi(\tau, \bar{\theta}) = \sum (-1)^p \text{ trace } \tau^{n-p}(\bar{\theta}^p)'.$$

The expression $\chi(\tau, \bar{\theta})$ is *by definition* the number of signed coincidences of τ, $\bar{\theta}$.

(40.7) Consider now two chain-mappings τ, $\theta : X \to Y$, where we suppose τ as before and θ such that

(40.8) $$\theta c_i^p \approx \theta_{ij}^p d_j^p , \qquad \theta^p = \| \theta_{ij}^p \|.$$

Let η, $\bar{\eta}$ be the same as in (36.1) for X and η', $\bar{\eta}'$ the analogues for Y. The chain-mapping $\bar{\eta}$ induces an isomorphism λ on the rational homology groups given by

$$\lambda \bar{c}_i^p = \lambda_{ij}^p c_j^p , \qquad \lambda^p = \| \lambda_{ij}^p \|,$$

and by (36.2), η induces λ^{-1}. Similarly η' induces μ given by

$$\mu d_i^p = \mu_{ij}^p \bar{d}_j^p , \qquad \mu^p = \| \mu_{ij}^p \|,$$

and $\bar{\eta}'$ induces μ^{-1}.

To $\theta: X \to Y$ there corresponds the congruent chain-mapping (in the sense of IV, 12.7) $\bar{\theta}: \bar{X} \to \bar{Y}$ given by $\bar{\theta} = \eta'\theta\bar{\eta}$. This chain-mapping induces the homomorphisms on the rational homology groups defined by:

$$\bar{\theta}\bar{c}_i^p = \bar{\theta}_{ij}^p \bar{d}_j^p, \qquad \bar{\theta}^p = \| \bar{\theta}_{ij}^p \| = \mu^p \theta^p \lambda^p.$$

In the light of (24.15) it is rather natural to define directly the *number of signed coincidences* of τ, θ as $\omega(\tau, \theta) = \chi(\tau, \bar{\theta})$. We have at once from (40.6) after replacing the product of matrices by its transverse:

$$\omega(\tau, \theta) = \sum (-1)^p \operatorname{trace} \mu^p \theta^p \lambda^p (\tau^{n-p})'.$$

Since X, $\bar{X}$ have the same Betti numbers, (33.1a) yields $R^p(X) = R^{n-p}(\bar{X})$, and similarly for Y, $\bar{Y}$. Hence the $\bar{c}$, $\bar{d}$ may be chosen such that

$$\mathrm{KI}(c_i^p * \bar{c}_j^{n-p}) = \delta_{ij}, \qquad \mathrm{KI}(\bar{d}_i^p * d_j^{n-p}) = \delta_{ij}.$$

Now from (37.12, 37.14) and (31) there follow:

$$\mathrm{KI}(\bar{c}_i^p \circ \bar{c}_j^{n-p}) = \lambda_{ih}^p \mathrm{KI}(c_h^p * \bar{c}_j^{n-p}) = \beta(n-p)\lambda_{ij}^p,$$
$$\mathrm{KI}(d_i^p \circ d_j^{n-p}) = \mu_{ih}^p \mathrm{KI}(\bar{d}_h^p * d_j^{n-p}) = \beta(n-p)\mu_{ij}^p,$$

where the indices are taken, respectively, in $\bar{X}$ and Y. Hence if we set

$$\alpha^p = \| \mathrm{KI}(\bar{c}_i^p, \bar{c}_j^{n-p}) \|, \qquad \beta^p = \| \mathrm{KI}(d_i^p, d_j^{n-p}) \|,$$

then

$$(40.9) \qquad \omega(\tau, \theta) = \sum (-1)^p \operatorname{trace} \beta^p \theta^p \alpha^p (\tau^{n-p})'.$$

This expression is wholly similar to the coincidence formula (24) of [L, 269].

(40.10) *Fixed elements.* If $Y = X$ and σ is a chain-mapping $X \to X$ then $\chi(\sigma, 1) = \psi(\sigma)$ where ψ is as in (24). Thus the number of signed fixed elements is the same as for any finite complex and this is as might be expected.

CHAPTER VI

NETS OF COMPLEXES

The passage from finite complexes to infinite complexes or topological spaces necessitates some limiting process, and the theory of nets will provide the necessary mechanism. In its general form it may be viewed as abstracted from the Čech homology theory for topological spaces (Čech [a]), which will be adopted as the basic theory in (VII). An important special type of net, the sequential spectrum, was already utilized by Alexandroff [a], for compacta, likewise for infinite complexes and compacta in [L, VII].

A close parallel will be found between nets and finite complexes and we shall have here also open and closed subnets, their projections and injections. In the general net there are no chains and so the operations bear directly upon the cycles. However for a spectrum chains may again be introduced, the similarity with complexes being greatly increased thereby.

By combining the subnets there will be obtained a noteworthy complementary mechanism, the web, which will have important applications in (VII).

The general theory of nets and webs will be applied to infinite complexes, and in particular to a type which we have termed *metric*. Such complexes will be shown to have a special "metric" homology theory, which includes the well known Vietoris theory for compacta, but has other applications as well.

General references: Alexandroff [a, f], Čech [a, b], Chevalley [a], Freudenthal [b], Lefschetz [L_1, XVII], Steenrod [a].

§1. DEFINITION OF NETS AND THEIR GROUPS

1. A net is a collection of complexes with special relations. A good point of departure is therefore a suitable type of infinite product of complexes. Since these infinite products have as yet but few applications we will not dwell upon them very long.

(1.1) Consider then a system $\{X_\lambda\}$ of finite complexes indexed by $\Lambda = \{\lambda\}$ and let $\mathfrak{X} = \mathbf{P}X_\lambda$ be the product of the complexes as sets of elements. It is not our purpose to turn X into a complex in the sense of (III, 1)—such a complex would be, in fact, irrelevant here. We may, however, introduce the groups of chains, $\cdots$ of $\mathfrak{X}$ over a given coefficient group G in the following way. Write $\mathfrak{C}^p_\lambda$, $\mathfrak{Z}^p_\lambda$, $\mathfrak{F}^p_\lambda$, $\mathfrak{H}^p_\lambda$ for $\mathfrak{C}^p(X_\lambda, G)$, $\cdots$, where the groups are as in (III, 7, 8). Define now the groups $\mathfrak{C}^p(\mathfrak{X}, G)$, $\mathfrak{Z}^p(\mathfrak{X}, G)$, $\mathfrak{F}^p(\mathfrak{X}, G)$ as

$$\mathfrak{C}^p(\mathfrak{X}, G) = \mathbf{P}\mathfrak{C}^p_\lambda, \qquad \mathfrak{Z}^p(\mathfrak{X}, G) = \mathbf{P}\mathfrak{Z}^p_\lambda, \tag{1.2}$$

$$\mathfrak{F}^p(\mathfrak{X}, G) = \mathbf{P}\mathfrak{F}^p_\lambda.$$

By (I, 12.5) we have then

$$\bar{\mathfrak{F}}^p(\mathfrak{X}, G) = \mathbf{P}\bar{\mathfrak{F}}_\lambda^p ,$$

and we readily show that

$$\mathfrak{H}^p(\mathfrak{X}, G) = \mathfrak{Z}^p(\mathfrak{X}, G)/\bar{\mathfrak{F}}^p(\mathfrak{X}, G) = \mathbf{P}\mathfrak{H}_\lambda^p . \tag{1.3}$$

There is an obvious parallel development in the direction of "weak" products whose details are omitted.

2. The next step is essentially analogous to the passage from products of groups to inverse or direct systems (II, 13, 14).

(2.1) Definition. *A net X is a system of finite complexes $\{X_\lambda\}$ indexed by a directed set $\Lambda = \{\lambda; >\}$ and with the following properties:*

N1. *When $\lambda > \mu$ there exist one or more chain-mappings, also called "projections," $\pi_\mu^\lambda: X_\lambda \to X_\mu$.*

N2. *When $\lambda > \mu > \nu$ and π_μ^λ, π_ν^μ are projections so is $\pi_\nu^\mu \pi_\mu^\lambda$.*

N3. *Any two projections π_μ^λ, $\pi_\mu^{\prime\lambda}$, $\lambda > \mu$, are homologous ($\pi_\mu^\lambda \gamma_\lambda^p \sim \pi_\mu^{\prime\lambda} \gamma_\lambda^p$ for every cycle γ_λ^p of X_λ).*

(2.2) Let N*i denote Ni with $>$ replaced by $<$. If X, still indexed by $\Lambda = \{\lambda; >\}$, has properties N*i it is known as a *conet*. In one or the other case a convenient designation is $X = \{X_\lambda ; \pi_\mu^\lambda\}$. Unless otherwise stated, in such a designation X will be understood to be a net.

(2.3) Let X_λ^* denote as usual the dual of X_λ, and $\pi_\lambda^{*\mu}: X_\mu^* \to X_\lambda^*$, the dual of π_μ^λ. Then if one of $X = \{X_\lambda ; \pi_\mu^\lambda\}$, $X^* = \{X_\lambda^* ; \pi_\lambda^{*\mu}\}$ is a net the other is a conet.

(2.4) *Special designations.* The net X is called

simplicial if the X_λ and the π_μ^λ are simplicial;

simple if the X_λ and the π_μ^λ are simple and in addition the π_μ^λ have simple carriers;

sequential if $\Lambda = \{1, 2, \cdots\}$;

a *spectrum* if the π_μ^λ are unique;

a *sequential spectrum* if the net is both sequential and a spectrum.

Evidently *a simplicial net is simple.*

The *dimension* of X, written dim X, is sup dim X_λ.

3. Since there may occur multiple projections π_μ^λ, $\pi_\lambda^{*\mu}$, it will be necessary for the groups $\mathfrak{Z}$, $\mathfrak{F}$ to have recourse to the mechanism of (II, 13.7, 14.8). Not so, however, as we shall see, for the groups $\mathfrak{H}$. Since the cycles and cocycles, and not their classes, are the elements usually arising in the applications their properties will be examined in some detail.

As a consequence of N3 we have:

(3.1) *Any two projections π_μ^λ, $\pi_\mu^{\prime\lambda}$ $[\pi_\lambda^{*\mu}, \pi_\lambda^{\prime*\mu}]$, $\lambda > \mu$, induce the same simultaneous homomorphism $\bar{\pi}_\mu^\lambda: \mathfrak{H}_\lambda^p \to \mathfrak{H}_\mu^p$ $[\bar{\pi}_\lambda^{*\mu}: \mathfrak{H}_p^\mu \to \mathfrak{H}_p^\lambda]$ so that there is a unique inverse* [*direct*] *system* $\{\mathfrak{H}_\lambda^p ; \bar{\pi}_\mu^\lambda\}$ $[\{\mathfrak{H}_p^\lambda ; \bar{\pi}_\lambda^{*\mu}\}]$ *in the sense of* (II, 13.1) [(II, 14.1)].

We have then limit-groups of the two systems and we lay down:

(3.2) DEFINITIONS. *$\mathfrak{H}^p = \lim \{\mathfrak{H}^p_\lambda ; \bar{\pi}^\lambda_\mu\}$ and $\mathfrak{H}_p = \lim \{\mathfrak{H}^\lambda_p ; \bar{\pi}^{*\mu}_\lambda\}$ are, respectively, the pth homology and pth cohomology groups of the net over G. The elements Γ^p, Γ_p of the two groups are the homology and cohomology classes over G; the terms will be justified presently.*

An element of $\mathfrak{H}^p$ is a collection $\Gamma^p = \{\Gamma^p_\lambda\}$ where $\Gamma^p_\lambda \in \mathfrak{H}^p_\lambda$, and $\lambda > \mu \rightarrow \bar{\pi}^\lambda_\mu \Gamma^p_\lambda = \Gamma^p_\mu$. The Γ^p_λ are the *coordinates* of Γ^p. Furthermore referring to (I, 38.2) if u_λ is any open set in $\mathfrak{H}^p_\lambda$, and $U_\lambda = \{\Gamma^p \mid \Gamma^p_\lambda \in u_\lambda\}$, the aggregate $\{U_\lambda\}$ is a base for $\mathfrak{H}^p$.

(3.3) Let now a *p-cycle γ^p of X over G* be defined as follows: $\gamma^p = \{\gamma^p_\lambda\}$, where $\gamma^p_\lambda \in \mathfrak{Z}^p_\lambda$ and $\lambda > \mu \rightarrow \pi^\lambda_\mu \gamma^p_\lambda \sim \gamma^p_\mu$ in X_μ. The γ^p_λ are the *coordinates* of γ^p. Define $\gamma^p = 0$ when every $\gamma^p_\lambda = 0$, and if $\gamma'^p = \{\gamma'^p_\lambda\}$, set $\gamma^p + \gamma'^p = \{\gamma^p_\lambda + \gamma'^p_\lambda\}$. As a consequence $\mathfrak{Z}^p = \{\gamma^p\}$ is a group, the *group of the p-cycles over G*. The topology in $\mathfrak{Z}^p$ is assigned by means of a subbase as follows: take any open set U_λ in $\mathfrak{Z}^p_\lambda$ and let $V_\lambda = \{\gamma^p \mid \gamma^p_\lambda \in U_\lambda\}$. The collection $\{V_\lambda\}$ is chosen as a subbase for $\mathfrak{Z}^p$.

(3.4) *$\mathfrak{Z}^p$ is a closed subgroup of $\mathfrak{Z}^p(\mathfrak{X}, G) = \mathbf{P}\mathfrak{Z}^p_\lambda$, the group of* (1.2) (II, 13.7a).

Among the cycles of the net are found the collections $\delta^p = \{\delta^p_\lambda \mid \delta^p_\lambda \sim 0\}$ known as *bounding cycles*. The terminology is justified by (3.5). If $\gamma^p - \gamma'^p$ is a bounding cycle we write the usual homology $\gamma^p \sim \gamma'^p$. Evidently $\mathfrak{F}^p = \{\delta^p\} = \bar{\mathfrak{F}}^p(\mathfrak{X}, G)$ (group of 1.2) is a subgroup of $\mathfrak{Z}^p$.

(3.5) *$\mathfrak{F}^p$ is closed in $\mathfrak{Z}^p$.*

For $\mathfrak{F}^p$ is closed in $\mathfrak{Z}^p(\mathfrak{X}, G) \supset \mathfrak{Z}^p$.

(3.6) $\mathfrak{H}^p \cong \mathfrak{Z}^p/\mathfrak{F}^p$ (II, 13.7c).

Henceforth $\mathfrak{H}^p$ is identified with $\mathfrak{Z}^p/\mathfrak{F}^p$, Γ^p being identified with the coset of γ^p mod $\mathfrak{F}^p$. It is also called the homology class of γ^p.

(3.7) *When G is compact so are $\mathfrak{Z}^p$, $\mathfrak{F}^p$ and $\mathfrak{H}^p$.*

For $\mathfrak{Z}^p_\lambda(X, G)$ is then compact and (3.7) is a consequence of (II, 13.7d).

(3.8) *Let G be a division-closure group and G_0 a discrete group isomorphic with G in the algebraic sense. Then the groups $\mathfrak{Z}^p$, $\mathfrak{F}^p$, $\mathfrak{H}^p$ over G are isomorphic in the algebraic sense with the corresponding groups over G_0. In other words, when G is a division-closure group its topology may be disregarded without modifying $\mathfrak{Z}^p$, $\mathfrak{F}^p$, $\mathfrak{H}^p$ algebraically.*

Let $\gamma^p = \{\gamma^p_\lambda\}$ be a cycle over G. Owing to the division-closure property (III, 17.2): (a) $\lambda > \mu \rightarrow \pi^\lambda_\mu \gamma^p_\lambda - \gamma^p_\mu = FC^{p+1}_\mu$; (b) $\gamma^p \sim 0 \leftrightarrow \gamma^p_\lambda = FC'^{p+1}_\lambda$; the chains C^{p+1}_μ, C'^{p+1}_λ are chains of X_μ, X_λ over G. Since chains over G are likewise chains over G_0, γ^p is a cycle over G_0 and conversely, if γ^p is a cycle over G_0 it is likewise a cycle over G. If $\gamma^p \sim 0$ as a cycle over one of G, G_0 then also $\gamma^p \sim 0$ as a cycle over the other. Since the identification of γ^p as a cycle over G with itself as a cycle over G_0 manifestly defines an isomorphism in the algebraic sense of the groups $\mathfrak{Z}^p$, $\mathfrak{F}^p$ over G with the corresponding groups over G_0, (3.8) follows.

(3.9) *If $\gamma^p = \{\gamma^p_\lambda\}$ is a cycle and $\gamma^p_\nu \sim 0$ for all elements of $\{\nu\}$ cofinal in $\{\lambda\}$ then $\gamma^p \sim 0$.*

Given λ there is a $\nu > \lambda$ and so $\gamma^p_\lambda \sim \pi^\nu_\lambda \gamma^p_\nu \sim 0$. Hence $\gamma^p \sim 0$.

(3.10) *If* $\dim X = n$ *is finite then all the groups* $\mathfrak{H}^p$, $p > n$, *are zero* (2.4, 3.9)

(3.11) *Essential cycles.* Following Čech, γ_μ^p is defined as *essential* for X_μ whenever $\lambda > \mu$ implies the existence in X_λ of a γ_λ^p such that $\pi_\mu^\lambda \gamma_\lambda^p \sim \gamma_\mu^p$ in X_μ. Under these conditions Γ_μ^p is an essential element of $\mathfrak{H}_\mu^p$ in the sense of (II, 27.11).

(3.12) *If G is a field then for every* μ *there is a* λ_0 *such that if* $\gamma_{\lambda_0}^p \in \mathfrak{Z}_{\lambda_0}^p$ *then* $\pi_\mu^{\lambda_0} \gamma_{\lambda_0}^p$ *is an essential cycle for* X_μ *and hence the* μ *coordinate of a cycle of* X (Čech [a]; II, 27.13).

By (II, 27.13) there exists for given μ and every $p \leqq \dim X_\mu = n$ an index $\lambda_p > \mu$ such that $\bar{\pi}_\mu^{\lambda_p} \Gamma_{\lambda_p}^p$ is essential, and hence such that $\pi_\mu^{\lambda_p} \gamma_{\lambda_p}^p$ is essential. Since n is finite we may choose a $\lambda_0 > \lambda_1, \cdots, \lambda_n$, and then every $\pi_p^{\lambda_0} \gamma_{\lambda_0}^p$ (all p) will be essential.

(3.13) *Betti and Alexandroff numbers.* If G is a field the pth Betti number of X over G is $R^p(X, G) = \dim \mathfrak{H}^p(X, G)$, when the latter is finite, and $R^p(X, G) = \infty$ otherwise.

Since the X_λ are finite so are their Betti numbers. Hence the Alexandroff numbers of X (i.e., of the $\mathfrak{H}^p(X, G)$) are equal to the corresponding Betti numbers of the net.

4. (4.1) We now pass to the cocycles γ_p and their classes Γ_p over a discrete group G. The groups $\mathfrak{H}_p$ over G have already been defined. Any Γ_p is a collection of elements Γ_p^λ, the *representatives* of Γ_p, such that if Γ_p^μ and Γ_p^ν exist then for some $\lambda > \mu, \nu$ we have $\bar{\pi}_\lambda^{*\mu} \Gamma_p^\mu = \bar{\pi}_\lambda^{*\nu} \Gamma_p^\nu$. A *p-cocycle* γ_p over G is now a maximal collection of cocycles γ_p^λ over G, the *representatives* of γ_p, such that if γ_p^μ and γ_p^ν exist then for some $\lambda > \mu, \nu$ we have $\pi_\lambda^{*\mu} \gamma_p^\mu = \pi_\lambda^{*\nu} \gamma_p^\nu$ in X_λ. The cocycles $\gamma_p^\lambda = 0$ are representatives of a unique γ_p denoted by 0. If $\{\gamma_p^\mu\}$, $\{\gamma_p'^\nu\}$ are the representatives of γ_p, γ_p' choose corresponding to μ, ν any $\lambda > \mu, \nu$. Then one may show that the cocycles $\{\pi_\lambda^{*\mu} \gamma_p^\mu + \pi_\lambda^{*\nu} \gamma_p'^\nu\}$ are representatives of a unique cocycle written $\gamma_p + \gamma_p'$. Under these conditions $\mathfrak{Z}_p = \{\gamma_p\}$ is a group, the group of the cocycles over G, and it is taken discrete. The cocycles δ_p which have a representative δ_p^μ such that for some $\lambda > \mu$ we have $\pi_\lambda^{*\mu} \delta_p^\mu \sim 0$ are known as *bounding cocycles.* If $\gamma_p - \gamma_p'$ is a bounding cocycle we write the usual homology $\gamma_p \sim \gamma_p'$. Evidently $\mathfrak{F}_p = \{\delta_p\}$ is a subgroup of $\mathfrak{Z}_p$.

(4.2) $$\mathfrak{H}_p \cong \mathfrak{Z}_p / \mathfrak{F}_p \qquad \text{(II, 14.8a).}$$

We may now identify $\mathfrak{H}_p$ with $\mathfrak{Z}_p/\mathfrak{F}_p$ and thus consider Γ_p as the coset of γ_p mod $\mathfrak{F}_p$. It is also called the *cohomology class* of γ_p.

(4.3) If G is a field we define the dual Betti numbers as $R_p(X, G) = \dim \mathfrak{H}_p(X, G)$, when the latter is finite, and $R_p(X, G) = \infty$ otherwise. The "dual" Alexandroff numbers of X (i.e. those of the $\mathfrak{H}_p(X, G)$) are, as in (3.13), the same as the dual Betti numbers.

(4.4) *If* $\dim X = n$ *is finite then all the groups* $\mathfrak{H}_p$, $p > n$, *are zero* (2.4).

(4.5) γ_p *or* Γ_p *have representatives for some* $\{\mu\}$ *cofinal in* $\{\lambda\}$.

For if say Γ_p^μ is a representative of Γ_p then so is every $\bar{\pi}_\lambda^{*\mu} \Gamma_p^\mu$, $\lambda > \mu$.

(4.6) *Conets.* All the results obtained for nets carry over to conets with cycles and cocycles interchanged. In particular *in conets the discrete groups are those of the cycles.*

5. **Application to connectedness in simplicial nets.** We shall find here again the same general relations between connectedness and the zero-dimensional groups as in simplicial and simple complexes (III, 20, 47.7). In point of fact we could develop the same considerations for simple nets, but the simplicial case is sufficient for the topological applications.

Let then $X = \{X_\lambda ; \pi_\mu^\lambda\}$ be simplicial and let $\{X_{\lambda i}\}$ be the components of X_λ. Since each contains a vertex and homologous vertices are mapped by π_μ^λ into homologous vertices $\pi_\mu^\lambda X_{\lambda i}$ is connected and hence in some $X_{\mu j}$. We shall say that the component $X_{\mu j}$ is *essential* whenever if $\lambda > \mu$ there exists an $X_{\lambda i}$ such that $\pi_\mu^\lambda X_{\lambda i} \subset X_{\mu j}$. A *component* of X is a collection $X' = \{X'_\lambda\}$, where X'_λ is a component of X_λ, (one of the sets $X_{\lambda i}$), and where $\lambda > \mu \rightarrow \pi_\mu^\lambda X'_\lambda \subset X'_\mu$. Evidently X' itself is a net, (a subnet of X in the terminology of 12). We shall denote by ρ the cardinal number of the components of X.

(5.1) *The component X' is uniquely determined by its coordinates X'_ν for $\{\nu\}$ cofinal in $\{\lambda\}$.*

For X'_λ is uniquely determined by X'_ν, $\nu > \lambda$.

(5.2) *If $X^1, \cdots, X^r$ are distinct components of X (r finite), with $X^i = \{X^i_\lambda\}$, then for some μ the X^i_μ are distinct components of X_μ.*

Take any pair X^i, X^j, $i \neq j$. For some index $\mu(i, j)$ we must have $X^i_{\mu(i,j)} \neq X^j_{\mu(i,j)}$. Therefore whatever $\nu > \mu(i, j)$ necessarily $X^i_\nu \neq X^j_\nu$, for otherwise their projections $X^i_{\mu(i,j)}$, $X^j_{\mu(i,j)}$ in $X_{\mu(i,j)}$ would coincide. Since the number of indices $\mu(i, j)$ is finite there is a $\mu >$ every $\mu(i, j)$, and the X^i_μ will then all be distinct.

(5.3) *The Betti numbers $R^0(X, G)$, over any field G, are all equal to ρ when ρ is finite, and infinite otherwise.*

Their common value is designated by $R^0(X)$ or R^0, and known as the *zero-dimensional Betti number* of the net.

The notations remaining the same choose a vertex $A_{\lambda i}$ on $X^i_{\lambda i}$. Since $\pi_\mu^\lambda A_{\lambda i}$ is a vertex of X_μ in the same component as $A_{\mu i}$, we have $\pi_\mu^\lambda A_{\lambda i} \sim A_{\mu i}$. Therefore $\gamma_i^0 = \{A_{\lambda i}\}$ is an integral zero-cycle of X^i and hence of X. Similarly $\{gA_{\lambda i}\}$, $g \in G$, is a zero-cycle of X^i over G and it is denoted by $g\gamma_i^0$.

The γ_i^0 are independent. For suppose $g^i\gamma_i^0 \sim 0$ and choose μ as in (5.2). We must have $g^iA_{\mu i} \sim 0$ in X_μ, which implies $g^i = 0$, since the $A_{\mu i}$ are in distinct components of X_μ. It follows that $R^0(G) \geqq \rho$, so if $\rho = \infty$ likewise $R^0(G) = \infty$. There remains then to dispose of the finite case.

We assume then ρ finite and $r = \rho$, so that $\{X^i\}$ is now a maximal set of distinct components of X. Choose again μ as in (5.2), so that now the ρ components X^i_μ of X_μ are distinct, and of course essential. Moreover clearly no other component of X_μ may be essential. It follows that for $\nu > \mu$ there are exactly ρ essential components of X_ν. For there can be no more, and there are at least that many, else there would be fewer than ρ in X_μ. The ρ essential components of X_ν must be the X^i_ν and they are thus distinct. If $\delta^0 = \{\delta^0_\lambda\}$ is a zero-cycle we have then $\delta^0_\nu \sim g^i_\nu A_{\nu i}$ and $\pi^\nu_\mu g^i_\nu A_{\nu i} \sim g^i_\nu A_{\mu i} \sim g^i_\mu A_{\mu i}$. Since the $A_{\mu i}$ are independent $g^i_\nu = g^i_\mu = g^i$ is independent of ν. Thus $\delta^0_\nu \sim g^i A_{\nu i}$. Therefore $\delta^0 - g^i \gamma^0_i$ has its ν coordinate ~ 0, and so (3.9), $\delta^0 \sim g^i \gamma^0_i$. It follows that $R^0(G) \leqq \rho$, and finally

$$R^0(G) = \rho.$$

We have also proved the complementary property:

(5.4) *When R^0 is finite $\{\gamma^0_i\}$, where $\gamma^0_i = \{\gamma^0_{i\lambda}\}$ is a zero-cycle of X, is a base for the zero-cycles over any group G and so*

$$\mathfrak{H}^0(X, G) = \mathbf{P}(G\gamma^0_i). \tag{5.5}$$

More generally:

(5.6) *When R^0 is finite,*

$$\mathfrak{H}^p(X, G) = \mathbf{P}\mathfrak{H}^p(X^i, G).$$

For if $\gamma^p = \{\gamma^p_\lambda\}$ is any cycle over G we have $\gamma^p_\nu = \sum \gamma^p_{\nu i}$, $\gamma^p_{\nu i} \subset X^i_\nu$, and $\gamma^p_\nu \sim 0 \rightarrow \gamma^p_{\nu i} \sim 0$ in X^i_ν. From this and the fact that $\{\nu\}$ is cofinal in $\{\lambda\}$ follows readily that $\gamma^p_i = \{\gamma^p_{\nu i}\}$ is a cycle of X^i with $\gamma^p = \sum \gamma^p_i$, and also that $\gamma^p \sim 0$ in X when and only when every $\gamma^p_i \sim 0$ in X^i. From this to (5.6) is but a step.

§2. DUALITY AND INTERSECTIONS

6. For the duality relations a Kronecker index is required. Keeping nets and conets together for the present let the groups G, H be commutatively paired to J, and let Γ^p, Γ_p be over G, H. If $\lambda > \mu$ and Γ_p in the net, Γ^p in the conet have λ, μ representatives we find from (IV, 10.5) with $\tau = \pi^\lambda_\mu$, the *permanence relation for the index*:

$$\mathrm{KI}(\Gamma^p_\lambda, \Gamma^\lambda_p) = \mathrm{KI}(\Gamma^p_\mu, \Gamma^\mu_p). \tag{6.1}$$

Their common value is defined as the class index $\mathrm{KI}(\Gamma^p, \Gamma_p)$. One may likewise introduce $\mathrm{KI}(\Gamma_p, \Gamma^p)$ and one finds from the commutation rule (III, 29.3) for the index in X_λ:

$$\mathrm{KI}(\Gamma^p, \Gamma_p) = (-1)^p \mathrm{KI}(\Gamma_p, \Gamma^p). \tag{6.2}$$

If $\gamma^p \in \Gamma^p$, $\gamma_p \in \Gamma_p$ we define

$$\mathrm{KI}(\gamma^p, \gamma_p) = (-1)^p \mathrm{KI}(\gamma_p, \gamma^p) = \mathrm{KI}(\Gamma^p, \Gamma_p). \tag{6.3}$$

Clearly $\mathrm{KI}(\gamma^p, \gamma_p) = \mathrm{KI}(\gamma^p_\lambda, \gamma^\lambda_p)$ for the λ for which γ_p has a representative when X is a net, or γ^p has one when X is a conet.

Referring to (II, 16.4, 17.7), and the known pairing of the groups $\mathfrak{H}^p_\lambda$, $\mathfrak{H}^\lambda_p$ and $\mathfrak{Z}^p_\lambda$, $\mathfrak{Z}^\lambda_p$ (III, 29.4, 29.10) we find;

(6.4) $\mathrm{KI}(\Gamma^p, \Gamma_p)$ *is a multiplication pairing* $\mathfrak{H}^p(X, G)$, $\mathfrak{H}_p(X, H)$ *to* J, *and similarly for* $\mathrm{KI}(\gamma^p, \gamma_p)$ *and* $\mathfrak{Z}^p(X, G)$, $\mathfrak{Z}_p(X, H)$.

Referring now to (II, 20.7, 33) and (III, 31.1) we have the basic

(6.5) DUALITY THEOREM. *The cycles and cocycles of a net [cocycles and cycles of a conet] are dual categories.*

7. **Application to the Betti numbers.** It follows from (6.5) that for any field J we have the duality relation for the Betti numbers:

$$R^p(X, J) = R_p(X, J). \tag{7.1}$$

We extend the universal theorem for fields (III, 17.8) to nets and prove:

(7.2) *The Betti numbers depend solely upon the characteristic of the field.*

Owing to (7.1) it is sufficient to consider only one of R^p, R_p. We will assume that X is a net and consider R_p. For a conet the reasoning would be the same with R^p and cycles in place of R_p and cocycles. To prove (7.2) we merely need to show that if J_1 is an extension of J we have

$$R_p(X, J_1) = R_p(X, J). \tag{7.3}$$

Let then $\{\Gamma^i_p\}$ be a base for $\mathfrak{H}_p(X, J)$ and $\{\Gamma^{i_h}_p\}$ a maximal subset of elements of this base with linearly independent representatives $\{\Gamma^{\lambda i_h}_p\}$ for a given fixed λ. Since every $\Gamma^\lambda_p \in \mathfrak{H}_p(X_\lambda, J)$ represents a cohomology class of X linearly dependent upon the $\{\Gamma^{i_h}_p\}$, a base may be obtained for $\mathfrak{H}_p(X_\lambda, J)$ consisting of $\{\Gamma^{\lambda i_h}_p\}$ and of a set $\{'\Gamma^{\lambda j}_p\}$ whose elements are representatives of the zero of $\mathfrak{H}_p(X, J)$. It is now a consequence of (III, 17.8) that if $\Delta_p \in \mathfrak{H}_p(X, J_1)$ has the representative Δ^λ_p then

$$\Delta^\lambda_p = \alpha_h \Gamma^{\lambda i_h}_p + \beta_j {'\Gamma^{\lambda j}_p}; \qquad \alpha_h, \beta_j \in J_1.$$

Since the $'\Gamma^{\lambda j}_p$ are representatives of the zero of $\mathfrak{H}_p(X, J)$ and finite in number, there exists a $\nu > \lambda$ such that $\bar{\pi}^{*\lambda}_\nu {'\Gamma^{\lambda j}_p} = 0$ and hence $\bar{\pi}^{*\lambda}_\nu(\beta_j \Gamma^{\lambda j}_p) = \beta_j \bar{\pi}^{*\lambda}_\nu \Gamma^{\lambda j}_p$ is in the zero of $\mathfrak{H}_p(X, J_1)$, or

$$\Delta_p = \alpha_h \Gamma^{i_h}_p.$$

Thus every Δ_p depends upon the Γ^i_p. On the other hand a non-trivial relation of the form

$$\alpha_h \Gamma^{i_h}_p = 0, \qquad \alpha_h \in J_1,$$

means that for some λ we have

$$\alpha_h \Gamma^{\lambda i_h}_p = 0.$$

By (III, 17.8) the $\Gamma^{\lambda i_h}_p$ are linearly dependent elements of $\mathfrak{H}_p(X_\lambda, J)$. Hence the $\Gamma^{i_h}_p$ are linearly dependent elements of $\mathfrak{H}_p(X, J)$, which is ruled out since they are elements of a base for this vector space. This contradiction proves (7.3) and hence (7.2).

Since $R^p(X, J)$ depends solely upon the characteristic π we designate henceforth by $R^p(X, \pi)$ its value for all fields of characteristic π and similarly for

$R_p(X, J)$ and $R_p(X, \pi)$. For the Betti numbers over the rationals (characteristic zero) we write as usual $R^p(X)$, $R_p(X)$.

8. **Intersections.** The only permanence relations available are those based upon (V, 14.1) for simple complexes. *We must therefore assume X to be simple, or as a special case simplicial.* The groups being as before let $\Gamma^p = \{\Gamma^p_\lambda\}$, $\Gamma_q = \{\Gamma^\lambda_q\}$ be over G, H. Then (V, 14.1ab) yield the *permanence relations for intersections*:

$$\Gamma^p_\mu \cdot \Gamma^\mu_q = \bar{\pi}^\lambda_\mu(\Gamma^p_\lambda \cdot \Gamma^\lambda_q), \tag{8.1}$$

$$\Gamma^\lambda_p \cdot \Gamma^\lambda_q = \bar{\pi}^{*\mu}_\lambda(\Gamma^\mu_p \cdot \Gamma^\mu_q). \qquad \lambda > \mu, \tag{8.2}$$

It is understood of course that λ, μ are such that, wherever need be, representatives are available and this may always be assumed to be the case for some cofinal subset of $\{\lambda\}$.

Suppose now that X is a net and let H be discrete. By (II, 17.2) and (8.1) the set $\{\Gamma^p_\lambda \cdot \Gamma^\lambda_q\}$ defines a class of $(p - q)$-cycles of X over J which is by definition the intersection of Γ^p, Γ_q and is written $\Gamma^p \cdot \Gamma_q$. Similarly when G, H, J are discrete by (II, 17.4) and (8.2) $\{\Gamma^\lambda_p \cdot \Gamma^\lambda_q\}$ defines a class of $(p + q)$-cocycles of X over J which is the intersection of Γ_p, Γ_q and is written $\Gamma_p \cdot \Gamma_q$. There are obvious modifications for conets which we leave to the reader. From these definitions follows immediately:

(8.3) *Theorem* (V, 8.8) *holds for intersections in simple and a fortiori in simplicial nets and conets.*

§3. FURTHER PROPERTIES OF NETS

9. **Partial nets.** Let $X = \{X_\lambda;\ \pi^\lambda_\mu\}$ be a net [conet] and let $\{\mu\} \subset \{\lambda\}$. Then $X' = \{X_\mu;\ \pi^\mu_{\mu'}\}$ is likewise a net [conet], said to be a *part* of X or a *partial* net [conet] of X. When $\{\mu\}$ is cofinal in $\{\lambda\}$ then X' is said to be *cofinal* in X. In the statement to follow we shall take cycles over G, cocycles over H, where G, H are commutatively paired to J. In the net case the groups of cycles are topologized and those of the cocycles discrete, while in the conet case it is the other way around. The classes in X, X' are, respectively, denoted by Γ, Γ'.

(9.1) *If X' is a partial net* [*conet*] *of X then there are the simultaneous homomorphisms of* (II, 13.3, 14.5):

$$\tau: \mathfrak{H}^p(X, G) \rightarrow \mathfrak{H}^p(X', G), \tag{9.2}$$

$$\tau^*: \mathfrak{H}_p(X', H) \rightarrow \mathfrak{H}_p(X, H), \tag{9.3}$$

and there subsists the permanence relation for the index:

$$\mathrm{KI}(\tau\Gamma^p, \Gamma'_p) = \mathrm{KI}(\Gamma^p, \tau^*\Gamma'_p). \tag{9.4}$$

Furthermore if there are intersections in X and hence in X′ then:

$$(\tau\Gamma^p) \cdot \Gamma'_q = \tau(\Gamma^p \cdot \tau^*\Gamma'_q), \tag{9.5}$$

$$\tau^*\Gamma'_p \cdot \tau^*\Gamma'_q = \tau^*(\Gamma'_p \cdot \Gamma'_q). \tag{9.6}$$

(9.7) *It is a consequence of* (9.6) *that* τ^* *induces a homomorphism of the cohomology rings of* X' *into the corresponding rings for* X.

(9.8) *When* X' *is cofinal in* X *then* X *and* X' *have the same homology theory.*

(9.9) Remark. Let X, X' be nets and G, H a normal couple. Since $\mathfrak{H}^p(X, G)$, $\mathfrak{H}_p(X, H)$ are then dually paired with the Kronecker index as the group multiplication (6.5), the class $\tau^*\Gamma'_p$ is uniquely determined by the values of $\mathrm{KI}(\Gamma^p, \tau^*\Gamma'_p)$ for all Γ^p. Therefore in view of (9.4), τ^* *is uniquely determined by* τ. Similarly in the conet case τ is uniquely determined by τ^*. Thus as regards the homology and cohomology groups τ and τ^* are related like dual chain-mappings of complexes. For this reason we shall say that τ, τ^* are *dual*.

The relations (9.2), (9.3), (9.4) are the analogues of the permanence relations for the index and intersections under chain-mappings (IV, 10.4; V, 14.1).

The proofs are very simple. Assume X to be a net. Then if $\Gamma^p = \{\Gamma^p_\lambda\} \in \mathfrak{H}^p(X, G)$ the subcollection $\{\Gamma^p_\mu\}$ of $\{\Gamma^p_\lambda\}$ consists of the coordinates of a $\Gamma'^p \in \mathfrak{H}^p(X', G)$ and by (II, 13.3) $\Gamma^p \to \Gamma'^p$ defines a simultaneous homomorphism τ (9.2). Similarly if $\Gamma'_p \in \mathfrak{H}_p(X', G)$ has for representatives $\{\Gamma'^\mu_p\}$ then the latter are also representatives of a $\Gamma_p \in \mathfrak{H}_p(X, G)$ and by (II, 14.5) $\Gamma'_p \to \Gamma_p$ defines a simultaneous homomorphism τ^* (9.3). Thus the coordinates of Γ^p include those of $\tau\Gamma^p$ and the representatives of Γ'_p are also representatives of $\tau^*\Gamma'_p$. Since the Kronecker index $\mathrm{KI}(\Gamma^p, \Gamma_p)$ in X is defined by its values for any coordinate λ such that Γ^λ_p exists we may choose a coordinate μ for which Γ'^μ_p exists. Then the two sides in (9.4) become both equal to $\mathrm{KI}(\Gamma^p_\mu, \Gamma^\mu_p)$ and so they are equal. A similar argument applies to (9.5), (9.6). This proves (9.1) for nets. The modifications needed for conets are obvious.

If $\{\mu\}$ is cofinal in $\{\lambda\}$, by (II, 13.3, 14.5) τ and τ^* become isomorphisms and (9.8) follows.

Application. Suppose that $\{\lambda\}$ has an upper bound λ_0, that is to say, there is a *last* complex X_{λ_0} in X. Then λ_0 is in fact cofinal in $\{\lambda\}$. The homology theory of X_{λ_0} as a "net" is obviously the same as its homology theory as a complex. Therefore (9.8) yields here:

(9.10) *If* $\{\lambda\}$ *has an upper bound* $\{\lambda_0\}$, *the homology theory of* X *is the same as the homology theory of its last complex* X_{λ_0}.

10. Augmentation.

(10.1) Let our usual net be *simple*. If $\gamma^0 = \{\gamma^0_\lambda\}$ then since π^λ_μ, $\lambda > \mu$, is simple we have $\mathrm{KI}(\pi^\lambda_\mu\gamma^0_\lambda) = \mathrm{KI}(\gamma^0_\lambda)$ and since $\pi^\lambda_\mu\gamma^0_\lambda \sim \gamma^0_\mu$, also $\mathrm{KI}(\pi^\lambda_\mu\gamma^0_\lambda) = \mathrm{KI}(\gamma^0_\mu)$. Thus $\mathrm{KI}(\gamma^0_\lambda) = \mathrm{KI}(\gamma^0_\mu)$. If μ, ν are any two indices there is a $\lambda > \mu$, ν and then $\mathrm{KI}(\gamma^0_\mu) = \mathrm{KI}(\gamma^0_\lambda) = \mathrm{KI}(\gamma^0_\nu)$. Thus $\mathrm{KI}(\gamma^0_\lambda)$ is independent of λ and its fixed value is defined as the index $\mathrm{KI}(\gamma^0)$.

(10.2) Let γ^λ_0 be the fundamental zero-cocycle of X_λ (the sum of the duals of the vertices of X_λ). According to (IV, 10.8) the constancy of $\mathrm{KI}(\gamma^\lambda_0)$ implies for $\lambda > \mu$: $\pi^{*\mu}_\lambda\gamma^\mu_0 = \gamma^\lambda_0$. Hence $\gamma_0 = \{\gamma^\lambda_0\}$ is a zero-cocycle of X. We call it the *fundamental* zero-cocycle of the net. From $\mathrm{KI}(\gamma^0) = \mathrm{KI}(\gamma^0_\lambda)$, and $\mathrm{KI}(\gamma^0, \gamma_0) = \mathrm{KI}(\gamma^0_\lambda, \gamma^\lambda_0)$, follows $\mathrm{KI}(\gamma^0) = \mathrm{KI}(\gamma^0, \gamma_0)$.

(10.3) Let us now augment X_λ to $X_{a\lambda}$ by the addition of a (-1)-element ϵ_λ, then extend π^λ_μ, $\lambda > \mu$, to a chain-mapping $\pi^\lambda_{a\mu}: X_{a\lambda} \to X_{a\mu}$ by imposing $\pi^\lambda_{a\mu}\epsilon_\lambda = \epsilon_\mu$ (IV, 9.10). It is an elementary matter to verify that properties Ni of (2.1) continue to hold and so $X_a = \{X_{a\lambda}; \pi^\lambda_{a\mu}\}$ is still a net. It is known as X *augmented.*

(10.4) We observe that a zero-cycle of $X: \gamma^0 = \{\gamma^0_\lambda\}$ is also a zero-cycle of X_a when and only when γ^0_λ is a cycle of $X_{a\lambda}$, i.e., when and only when $\mathrm{KI}(\gamma^0_\lambda) = \mathrm{KI}(\gamma^0) = 0$.

(10.5) It is an elementary matter to verify that the properties of components developed in (5) for simplicial complexes are also valid for simple complexes. If $\{X'_\lambda\}$ is a component of X and A_λ is a vertex of X'_λ then $\gamma^0 = \{A_\lambda\}$ is a zero-cycle of X such that $\mathrm{KI}(\gamma^0) = 1$. From this follows readily as in (III, 42), with Γ^0 as the class of γ^0:

(10.6) $$\mathfrak{H}^0(X_a, G) \cong \mathfrak{H}^0(X, G)/G\Gamma^0,$$

(10.7) $$R^0(X) = R^0(X_a) + 1.$$

(10.8) To sum up, we may say that *as regards augmentation in simple nets the situation is the same as for simple complexes* (see notably (III, 42.1, $\cdots$, 42.5, 42.7)).

11. **Products of nets.** Let $X = \{X_\lambda; \pi^\lambda_{\lambda'}\}$, $Y = \{Y_\mu; \omega^\mu_{\mu'}\}$ be two nets and $\Lambda = \{\lambda; >\}$, $\mathrm{M} = \{\mu; >\}$ their indexing sets. The product $\Lambda \times \mathrm{M} = \{(\lambda, \mu)\}$ ordered by $(\lambda, \mu) > (\lambda', \mu') \leftrightarrow \lambda > \lambda', \mu > \mu'$ is directed by $>$. Hence $\{X_\lambda \times Y_\mu; \pi^\lambda_{\lambda'} \times \omega^\mu_{\mu'}\}$ is a net known as the product of X, Y and denoted by $X \times Y$.

(11.1) *If X, Y are both simple, spectra or countable spectra, so is $X \times Y$. However, if X, Y are both simplicial and of dimension greater than* 0 *then $X \times Y$ is not simplicial.*

(For simple nets (V, 17.8) is needed.)

(11.2) *Suppose now that for dimensions not exceeding s both X, Y have finite Betti numbers* mod π, *$R^p(X, \pi)$, $R^q(Y, \pi)$, p, $q \leqq s$, and that the dimensions of the elements of the X_λ, Y_μ are above a certain fixed t. Then*:

(11.3) $$R^s(X \times Y, \pi) = \sum_{p+q=s} R^p(X, \pi)R^q(Y, \pi).$$

If one of the numbers $R^p(X, \pi)$, $R^q(Y, \pi)$, p, $q \leqq s$, is infinite so is $R^s(X \times Y, \pi)$. Thus the relations for the Betti numbers are the same as for products of finite complexes (IV, 6.9, 6.10). If all the Betti numbers mod π of the two nets are finite we may write down formal Poincaré power series (analogous to the Poincaré polynomials) and (IV, 6.11, 6.12) will hold here also. All that is necessary, however, is to prove (11.3).

If $\gamma^p = \{\gamma^p_\lambda\}$, $\delta^p = \{\delta^p_\mu\}$ are cycles of X, Y mod π, we verify at once that $\{\gamma^p_\lambda \times \delta^p_\mu\}$ is a $(p + q)$-cycle of $X \times Y$ mod π which we call the product of γ^p, δ^q and denote by $\gamma^p \times \delta^q$. Moreover $\gamma^p \times \delta^q \sim 0 \to$ every $\gamma^p_\lambda \times \delta^q_\mu \sim 0$. If say $\gamma^p \nsim 0$ then some $\gamma^p_{\lambda_0} \nsim 0$ and hence every $\delta^q_\mu \sim 0$ (IV, 6.13a), or $\delta^q \sim 0$. Thus $\gamma^p \times \delta^q \sim 0 \to$ one of γ^p, $\delta^q \sim 0$. The same remarks apply with obvious modifications to cocycles.

(11.4) Assuming then that the Betti numbers at the right in (11.3) are finite we can find maximal linearly independent sets of p-cycles and p-cocycles of X mod π, $\{\gamma_i^p\}$, $\{\gamma_p^i\}$ where $\gamma_i^p = \{\gamma_{i\lambda}^p\}, \gamma_p^i = \{\gamma_p^{i\lambda}\}$ and $i = 1, 2, \cdots, R^p(X, \pi)$. Furthermore (III, 32.3) the sets may be so chosen that

$$\text{(11.5)} \qquad \| \mathrm{KI}(\gamma_i^p, \gamma_p^j) \| = 1.$$

The substitutions $X \to Y$, $\gamma \to \delta$, $p \to q$, $\lambda \to \mu$ yield similar elements for Y. Hence (IV, 6.7, 6.8; V, 17.9, 17.10)

$$\text{(11.6)} \qquad \{\gamma_i^p \times \delta_j^q\}, \qquad \{\gamma_p^i \times \delta_q^j\}$$

are sets such that

$$\text{(11.7)} \qquad | \mathrm{KI}(\gamma_i^p \times \delta_j^q, \gamma_{p'}^h \times \delta_{q'}^l) | = \pm 1, \qquad p + q = p' + q' = s.$$

From this we conclude that the elements in the respective sets (11.6) are independent. Hence if ρ^s is the sum at the right in (11.3) we have $R^s(X \times Y) \geqq \rho^s$.

If say $R^p(X, \pi) = \infty$, $p \leqq s$, the same argument with i running to any integer m shows that $R^s(X \times Y, \pi) \geqq m$, and $R^s(X \times Y, \pi) = \infty$ also.

Returning now to the case where the Betti numbers at the right in (11.3) are all finite, suppose that there exists an s-cocycle d_s of $X \times Y$ mod π independent of those in the second set (11.6). This cocycle will have a representative $d_s^{\lambda\mu}$, a cocycle of $X_\lambda \times Y_\mu$ mod π, which will be $\not\sim 0$ mod π. From the definition of the γ_p^i we infer that a maximal linearly independent set of p-cocycles of X_λ mod π is obtained by augmenting $\{\gamma_p^{i\lambda}\}$ by a set $\{\bar{\gamma}_p^{h\lambda}\}$ consisting of representatives of the zero pth cohomology class of X. Thus for each h, and hence for all together since their number is finite, there exists a $\lambda_0 > \lambda$ such that $\pi_{\lambda_0}^{*\lambda}\bar{\gamma}_p^{h\lambda} \sim 0$. Similarly a maximal linearly independent set for Y_μ is obtained by augmenting $\{\delta_q^{j\mu}\}$ by a set $\{\bar{\delta}_q^{k\mu}\}$ such that for some $\mu_0 > \mu$ we have $\omega_{\mu_0}^{*\mu}\bar{\delta}_q^{k\mu} \sim 0$.

Now by (IV, 6.8) we have

$$d_s^{\lambda\mu} \sim^{\cdot} \sum_{p+q=s} a_{ij}(p, q)\gamma_p^{i\lambda} \times \delta_q^{j\mu} + \text{terms } \gamma \times \bar{\delta}, \bar{\gamma} \times \delta, \bar{\gamma} \times \bar{\delta}.$$

From this follows

$$\pi_{\lambda_0}^{*\lambda} \times \omega_{\mu_0}^{*\mu}(d_s^{\lambda\mu} - \sum a_{ij}(p, q)\gamma_p^{i\lambda} \times \delta_q^{j\mu}) \sim 0,$$

and so finally

$$d_s \sim \sum a_{ij}(p, q)\gamma_p^i \times \delta_q^j.$$

Consequently $R^s(X \times Y) \leqq \rho^s$, hence both are equal. This proves (11.3).

12. **Open or closed subnets. Dissections.**

(12.1) The various concepts centering around dissections in complexes (III, 23) will now be extended to nets. The only limitation, caused by the fact that projections need not be unique, is that the operations loc. cit. will not have to bear upon the cycles. Moreover the subnets can only be introduced for nets which satisfy in place of N3 of (2.1) the more restricted condition:

N3′. *if* π_μ^λ, $\pi_\mu'^\lambda$, $\lambda > \mu$, *are distinct projections then they are chain-homotopic.*

That N3′ → N3 is a consequence of (IV, 15.2). Notice also that if N*3′ is the analogue of N3′ with $>$ replaced by $<$, then N*3′ for $\pi_\lambda^{*\mu}$ is a consequence of N3′ (IV, 15.7) and N*3′ → N*3 (IV, 15.2).

Suppose then that X satisfies N3′, hence also N*3′ (relative to the $\pi_\lambda^{*\mu}$) and let $(X_{0\lambda}, X_{1\lambda})$ be a dissection of X_λ, ($X_{0\lambda}$ open, $X_{1\lambda}$ closed) such that:

(a) $\lambda > \mu \rightarrow \pi_\mu^\lambda X_{1\lambda} \subset X_{1\mu}$;

(b) if π_μ^λ, $\pi_\mu'^\lambda$, $\lambda > \mu$, are distinct projections (hence chain-homotopic) then the related homotopy operator $\mathfrak{D}_\mu^\lambda$ is such that $\mathfrak{D}_\mu^\lambda X_{1\lambda} \subset X_{1\mu}$.

Referring to (IV, 22) there may be introduced the induced operations $\pi_{i\mu}^\lambda$, $\mathfrak{D}_{i\mu}^\lambda$ and their duals $\pi_{i\lambda}^{*\mu}$, $\mathfrak{D}_{i\lambda}^{*\mu}$. Then (IV, 22), $\pi_{i\mu}^\lambda$, $\pi_{i\lambda}^{*\mu}$, $\mathfrak{D}_{i\mu}^\lambda$, $\mathfrak{D}_{i\lambda}^{*\mu}$ are related like π_μ^λ, $\cdots$. It follows that $X_i = \{X_{i\lambda}; \pi_{i\mu}^\lambda\}$ is a net which satisfies N123′. We call X_0 an *open subnet* of X, X_1 a *closed subnet* of X, and the pair (X_0, X_1) (in that order) a *dissection* of X. The two subnets X_0, X_1 are also said to be *complementary.*

We are thus in position to introduce the cycles and cocycles of the X_i, their groups and Kronecker indices.

(12.2) Suppose that in addition to (a, b) we have:

(c) there is a simple set-transformation $t_\mu^\lambda : X_\lambda \rightarrow X_\mu$, $\lambda > \mu$, such that $t_\mu^\lambda X_{1\lambda} \subset X_{1\mu}$, and that any two projections π_μ^λ, $\pi_\mu'^\lambda$ are contiguous in t_μ^λ.

Under these conditions t_μ^λ induces the set-transformations $t_{i\mu}^\lambda : X_{i\lambda} \rightarrow X_{i\mu}$ and their duals $t_{i\lambda}^{*\mu}$. Referring then to (V, 13, 14) one may repeat the considerations of (8) regarding the intersections in X_i and derive (8.3) for these intersections. In point of fact X_1 is a simple net and the intersections induced in X_1 are recognized, by reference to (V, 13), to coincide with its intersections as a simple net.

(12.3) Since a cycle of $X_{1\lambda}$ [cocycle of $X_{0\lambda}$] is absolute and a cycle of $X_{0\lambda}$ [cocycle of $X_{1\lambda}$] is a cycle of X_λ mod $X_{1\lambda}$ [cocycle of X_λ mod $X_{0\lambda}$] we shall naturally describe the cycles of X_1 [cocycles of X_0] as *absolute*, and the cycles of X_0 [cocycles of X_1] as *cycles of* X mod X_1 [*cocycles of* X mod X_0], or also as *relative* cycles [cocycles].

Since the X_i are nets (6.5) yields

(12.4) *The cycles of* X mod X_1 [*of* X_1] *and the cocycles of* X_0 [*of* X mod X_0] *are dual categories and they have intersections when* (c) *holds. The universal theorem for fields* (7.2) *holds for both types.*

(12.5) Let (X_0', X_1') be a second dissection of X and let $X_{i\lambda}'$, $\cdots$ have their obvious meaning. If $X_{1\lambda}' \subset X_{1\lambda}$ throughout we will say that X_1' is *contained* in X_1, written $X_1' \subset X_1$. Suppose this to be the case. Since $\pi_{1\mu}^\lambda = \pi_\mu^\lambda \mid X_{1\lambda}$, we have $\pi_\mu^\lambda X_{1\lambda}' = \pi_{1\mu}^\lambda X_{1\lambda}' \subset X_{1\mu}'$, and since $X_{1\lambda}'$ is a closed subcomplex of $X_{1\lambda}$, X_1' is in fact a closed subnet of X_1. Similarly if $X_{0\lambda} \subset X_{0\lambda}'$ throughout we say that X_0 is *contained* in X_0', written $X_0 \subset X_0'$. The same argument shows that X_0 is then an open subnet of X_0' also.

(12.6) *If* X *is a spectrum or a sequential spectrum so are* X_0, X_1, *while if* X *is simplicial or simple so is* X_1.

13. The comparison of the groups of X with those of the subnets will yield certain important sets of homomorphisms associated with (X_0, X_1).

(13.1) We choose a fixed coefficient group G and denote by $\mathfrak{Z}^p, \cdots, \mathfrak{Z}_i^p, \cdots, \mathfrak{Z}_\lambda^p, \cdots, \mathfrak{Z}_{i\lambda}^p, \cdots$ the groups of the cycles of $X, X_i, X_\lambda, \cdots, X_{i\lambda}$ and by $\mathfrak{Z}_p, \cdots$ the same for the cocycles. In the case of the latter G is assumed discrete. We will set also:

$$\tau_\lambda = \text{the projection } X_\lambda \to X_{0\lambda};$$

$$\theta_\lambda = \text{the injection } X_{1\lambda} \to X_\lambda;$$

$$\tau_\lambda^* = \text{the injection } X_{0\lambda}^* \to X_\lambda^*;$$

$$\theta_\lambda^* = \text{the projection } X_\lambda^* \to X_{1\lambda}^*.$$

As we recall (IV, 10.14) $\tau_\lambda, \tau_\lambda^*$, likewise $\theta_\lambda, \theta_\lambda^*$ are dual chain-mappings.

Each of the chain-mappings $\pi_\mu^\lambda, \tau_\lambda, \cdots$ induces a simultaneous homomorphism in the corresponding homology or cohomology groups which will be denoted by $\bar{\pi}_\mu^\lambda, \bar{\tau}_\lambda, \cdots$.

(13.2) If $\gamma_1^p = \{\gamma_{1\lambda}^p\}$ is a cycle of X_1 then it is also a cycle of X. The identical transformation $\gamma_1^p \to \gamma_1^p$ defines a simultaneous homomorphism (imbedding) in the algebraic sense $\theta: \mathfrak{Z}_1^p \to \mathfrak{Z}^p$. If U_λ is open in $\mathfrak{Z}_\lambda^p$ and $V_\lambda = \{\gamma^p \mid \gamma_\lambda^p \in U_\lambda\}$ $W_\lambda = \{\gamma_1^p \mid \gamma_{1\lambda}^p \in \theta_\lambda U_\lambda\}$ then $W_\lambda = \theta V_\lambda$ and since $\{V_\lambda\}, \{W_\lambda\}$ are subbases for $\mathfrak{Z}^p, \mathfrak{Z}_1^p$, θ is a simultaneous isomorphism (of each $\mathfrak{Z}_1^p$ with a subgroup of the corresponding $\mathfrak{Z}^p$). Now $\gamma_1^p \sim 0$ in $X_1 \to \gamma_{1\lambda}^p \sim 0$ in $X_{1\lambda} \to \gamma_{1\lambda}^p \sim 0$ in $X_\lambda \to \gamma_1^p \sim 0$ in $X \to \theta\mathfrak{F}_1^p \subset \mathfrak{F}^p$. Therefore θ induces a simultaneous homomorphism $\bar{\theta}: \mathfrak{H}_1^p \to \mathfrak{H}^p$. We call θ the *injection* $X_1 \to X$. Its analogy with the injection in the case of complexes (III, 23.2) is obvious.

Similarly the mapping $\overset{0}{\gamma}_p \to \gamma_p$ of each cocycle of X_0 on itself generates a simultaneous homomorphism $\tau^*: \overset{0}{\mathfrak{Z}}_p \to \mathfrak{Z}_p$ called the *injection* $X_0^* \to X^*$ which induces a simultaneous homomorphism $\bar{\tau}^*$ of the cohomology groups: $\overset{0}{\mathfrak{H}}_p \to \mathfrak{H}_p$.

(13.3) For the proper treatment of the projections it is advisable to deal with the cycles of X_0 and cocycles of X_1 in a manner similar to (III, 23.5).

As in loc. cit. we first identify a cycle $\gamma_{0\lambda}^p$ with the chains $\gamma_{0\lambda}^p + C_\lambda^p$, $C_\lambda^p \subset X_{1\lambda}$. Thus a cycle mod $X_{1\lambda}$, or cycle of $X_{0\lambda}$ is now merely any chain with boundary in $X_{1\lambda}$. A cycle of X mod X_1 is then defined as a collection $\gamma_0^p = \{\gamma_{0\lambda}^p\}$, where $\gamma_{0\lambda}^p$ is a chain with boundary in $X_{1\lambda}$ such that $\lambda > \mu \to \pi_\mu^\lambda\gamma_{0\lambda}^p - \gamma_{0\mu}^p \sim 0 \bmod X_{1\mu}$. That is to say, $\pi_\mu^\lambda\gamma_{0\lambda}^p - \gamma_{0\mu}^p +$ a chain of $X_{1\mu}$ is in $\mathfrak{F}_\mu^p$. Since $\pi_{0\mu}^\lambda = \pi_\mu^\lambda \bmod X_{1\mu}$, it is easy to identify the groups of the cycles of X mod X_1 with those of the cycles of X_0 (as a net) and this justifies the appellation "cycles mod X_1" for the cycles of X_0 (12.3).

(13.4) It is an elementary matter to verify that

$$\lambda > \mu \to \tau_\mu\pi_\mu^\lambda \sim \pi_{0\mu}^\lambda\tau_\lambda \to \bar{\tau}_\mu\bar{\pi}_\mu^\lambda = \bar{\pi}_{0\mu}^\lambda\bar{\tau}_\lambda.$$

Hence (II, 13.5) if $\Gamma^p = \{\Gamma_\lambda^p\} \in \mathfrak{H}^p$ then $\Gamma_0^p = \{\bar{\tau}_\lambda\Gamma_\lambda^p\} \in \mathfrak{H}_0^p$ and $\Gamma^p \to \Gamma_0^p$ defines a homomorphism $\bar{\tau}: \mathfrak{H}^p \to \mathfrak{H}_0^p$.

Let now $\gamma^p = \{\gamma_\lambda^p\} \in \Gamma^p$ and set $\gamma_0^p = \{\tau_\lambda \gamma_\lambda^p\}$. Since $\tau_\lambda \gamma_\lambda^p \in \bar{\tau}_\lambda \Gamma_\lambda^p$, we have $\gamma_0^p \in \Gamma_0^p$. Thus $\gamma^p \to \gamma_0^p$ defines a homomorphism τ, the *projection* $X \to X_0$ (II, 13.5). We notice that $\gamma^p \sim 0 \to \gamma_\lambda^p \sim 0 \to \tau_\lambda \gamma_\lambda^p \sim 0 \to \tau\gamma^p \sim 0$. Thus $\tau\mathfrak{F}^p \subset \mathfrak{F}_0^p$. It follows in particular that τ induces $\bar{\tau}$.

(13.5) Similar considerations are valid, with all topological arguments omitted, regarding the cocycles mod X_0, the projection being this time $\theta^*: X^* \to X_1^*$.

(13.6) Suppose now G, H commutatively paired to J and let the cycles be over G, the cocycles over H, and the groups of cocycles taken discrete. In particular H is assumed discrete. If γ_p^0 has the representative $\gamma_p^{0\lambda}$ then by (IV, 10.5):

$$\mathrm{KI}(\tau_\lambda \gamma_\lambda^p, \gamma_p^{0\lambda}) = \mathrm{KI}(\gamma_\lambda^p, \tau_\lambda^* \gamma_p^{0\lambda}),$$

and hence from the definition of the indices in nets we obtain the relation of permanence:

$$\mathrm{KI}(\tau\gamma^p, \gamma_p^0) = \mathrm{KI}(\gamma^p, \tau^*\gamma_p^0). \tag{13.7}$$

Therefore τ, τ^* are dual. We prove similarly that θ, θ^* are dual with the permanence relation

$$\mathrm{KI}(\theta\gamma^p, \gamma_p^0) = \mathrm{KI}(\gamma^p, \theta^*\gamma_p^0). \tag{13.8}$$

If intersections are present the relations of permanence (V, 14.1ab) yield the analogous relations here and we shall not repeat them.

It is hardly necessary to observe that the same situation prevails regarding conets.

To sum up then we have proved:

(13.9) THEOREM. *With the dissection* (X_0, X_1) *of a net or conet* X (X_0 *open,* X_1 *closed) there are associated the following simultaneous homomorphisms of the groups of cycles*:

(a) *a projection* $\tau: X \to X_0$, *or reduction* mod X_1 *of the cycles of* X;

(b) *an injection* $\theta: X_1 \to X$ *or mapping into themselves of the groups of the cycles of* X_1;

(c) *a projection* $\theta^*: X^* \to X_1^*$ *or reduction* mod X_0 *of the cocycles of* X;

(d) *an injection* $\tau^*: X_0^* \to X^*$ *or mapping into themselves of the groups of the cocycles of* X_0;

(e) *each of* $\tau, \cdots, \theta^*$ *maps groups* $\mathfrak{F}$ *into groups* $\mathfrak{F}$ *and hence they induce simultaneous homomorphisms* $\bar{\tau}, \cdots, \bar{\theta}^*$ *of the appropriate homology or cohomology groups into one another*;

(f) *each of* τ, τ^*, *likewise each of* θ, θ^* *is dual to the other*;

(g) *the permanence relations* (13.7, 13.8) *hold, and in case there are intersections, the same relations as* (V, 14.1ab) *for* τ, τ^* *and* θ, θ^* *are fulfilled.*

14. **Special properties of subnets when G is compact or a field.** In this special case it will be found possible to restore the chain-cochain relations

existing in complexes. For convenience we will denote by C, γ, Γ chains, cycles, classes related to X_0 (thus γ_p^λ for a cocycle of $X_{0\lambda}$, $\cdots$) and D, δ, Δ the same when related to X_1. The other notations are as in (13.1) with the addition of $\mathfrak{C}$ for groups of chains.

(14.1) *If* $\gamma^p = \{\gamma_\lambda^p\}$ *is a cycle of* X mod X_1, *then* $\delta^{p-1} = \{\mathrm{F}\gamma_\lambda^p\}$ *is an (absolute) cycle of* X_1, *and it is denoted by* $\delta^{p-1} = \mathrm{F}\gamma^p$. *This holds for any division-closure group* G.

Let $\delta^{p-1} = \{\delta_\lambda^{p-1}\}$, $\delta_\lambda^{p-1} = \mathrm{F}\gamma_\lambda^p$. Suppose $\lambda > \mu$. We have $\gamma_\mu^p \sim \pi_\mu^\lambda \gamma_\lambda^p$ mod $X_{1\mu}$ in X_μ, and since G is a division-closure group, X_μ contains C_μ^{p+1} such that

$$\mathrm{F}C_\mu^{p+1} = \gamma_\mu^p - \pi_\mu^\lambda \gamma_\lambda^p - D_\mu^p .$$

Since the right-hand side is an absolute cycle and π_μ^λ commutes with F, we have:

$$\mathrm{F}\gamma_\mu^p - \pi_\mu^\lambda \mathrm{F}\gamma_\lambda^p = \delta_\mu^{p-1} - \pi_\mu^\lambda \delta_\lambda^{p-1} = \mathrm{F}D_\mu^p ,$$

or $\delta_\mu^{p-1} \sim \pi_\mu^\lambda \delta_\lambda^{p-1}$ in $X_{1\mu}$. Therefore δ^{p-1} is a cycle of X_1.

(14.2) *If* $\delta^{p-1} = \{\delta_\lambda^{p-1}\}$ *is a cycle of* X_1 *which is* ~ 0 *in* X *then* δ^{p-1} *is an* $\mathrm{F}\gamma^p$ *in the sense of* (14.1).

The condition on δ^{p-1} means that $\delta_\lambda^{p-1} \sim 0$ in X_λ, and so as before $\delta_\lambda^{p-1} = \mathrm{F}\gamma_{0\lambda}^p$, where $\gamma_{0\lambda}^p$ is a cycle of X_λ mod $X_{1\lambda}$. It follows that $P_\lambda = \mathrm{F}^{-1}\delta_\lambda^{p-1} \neq 0$. Let also $Q_\lambda = \{\Gamma_\lambda^p \mid \gamma_\lambda^p \in P_\lambda\}$ and denote by t_λ the natural projection $\mathfrak{Z}_\lambda^p \rightarrow \mathfrak{H}_\lambda^p$. When G is compact so is $\mathfrak{C}_\lambda^p$ and since F is continuous P_λ is likewise compact (I, 23.1). From the continuity of t_λ follows then that Q_λ is likewise compact (I, 23.2) and it is different from $\emptyset$. Furthermore if $\{\bar{\pi}_\mu^\lambda\}$ are the induced projections of the homology groups then $\lambda > \mu \rightarrow \bar{\pi}_\mu^\lambda Q_\lambda \subset Q_\mu$. Thus $\{Q_\lambda ; \bar{\pi}_\mu^\lambda\}$ is an inverse system of compact spaces, and so it has a limit-element $\Gamma^p = \{\Gamma_\lambda^p\}$. When G is a field $\mathfrak{C}_\lambda^p$ is finite-dimensional and hence linearly compact (II, 27.7). We find then by reference to (II, 27.6) that the same conclusion may be reached.

By assumption the class Γ_λ^p contains a γ_λ^p such that $\mathrm{F}\gamma_\lambda^p = \delta_\lambda^{p-1}$ and $\gamma^p = \{\gamma_\lambda^p\}$ behaves as required.

(14.3) $\gamma^p \sim 0$ mod $X_1 \rightarrow \mathrm{F}\gamma^p \sim 0$ *in* X_1.

For $\gamma^p \sim 0$ mod $X_1 \rightarrow \gamma_\lambda^p \sim 0$ mod $X_{1\lambda} \rightarrow \gamma_\lambda^p - D_\lambda^p = \mathrm{F}C_\lambda^{p+1} \rightarrow \mathrm{F}\gamma_\lambda^p \sim 0$ in $X_{1\lambda} \rightarrow \mathrm{F}\gamma^p \sim 0$ in X_1.

The same propositions may be obtained for the cocycles. This time G must be discrete.

(14.4) *If* $\delta_p = \{\delta_p^\lambda\}$ *is a cocycle of* X mod X_0 *then* $\gamma_{p+1} = \{\mathrm{F}\delta_p^\lambda\}$ *is an (absolute) cocycle of* X_0 *and it is denoted by* $\mathrm{F}\delta_p$.

(14.5) *If* $\gamma_{p+1} = \{\gamma_{p+1}^\lambda\} \sim 0$ *in* X *then it is an* $\mathrm{F}\delta_p$.

(14.6) $\delta_p \sim 0$ mod $X_0 \rightarrow \mathrm{F}\delta_p \sim 0$ *in* X_0.

The proofs are very similar to those of (14.1, 14.2, 14.3) and are omitted.

15. **Net duality in the sense of Alexander.** We shall extend to nets the results of (III, 38, 39) for complexes. The terms "cyclic, $\cdots$" are to receive the same meaning as in (III, 21.1). The notations remain those of (13, 14).

(15.1) *If* X *is* $(p - 1, p)$*-acyclic and* G *is compact or a field then*

$$\mathfrak{H}^{p-1}(X_1, G) \cong \mathfrak{H}^p(X, X_1, G) = \mathfrak{H}^p(X_0, G).$$

It is to be kept in mind that the compared homology groups are those of X^1 itself at the left and the groups of X mod X_1 at the right. The latter are also the groups of the net X_0. In the notations of (13) the groups are also written less explicitly $\mathfrak{H}_1^{p-1}$, $\mathfrak{H}_0^p$.

To prove (15.1) we first need

(15.2) *Under the conditions of* (15.1) $\delta^{p-1} = \mathrm{F}\gamma^p \sim 0$ *in* $X_1 \rightarrow \gamma^p \sim 0 \bmod X_1$.

Let first G be compact. Then $\mathfrak{C}_{1\lambda}^p$ (group of chains of $X_{1\lambda}$) is compact. Owing to the continuity of F and of the group operations, $R_\lambda = \gamma_\lambda^p - (\mathrm{F}^{-1}\delta_\lambda^{p-1}) \cap \mathfrak{C}_{1\lambda}^p$ is likewise compact and is a collection of absolute cycles. If θ_λ is the natural mapping $\mathfrak{Z}_\lambda^p \rightarrow \mathfrak{H}_\lambda^p$, then $S_\lambda = \theta_\lambda R_\lambda$ is again a compact set. If $'\Gamma_\lambda^p \in S_\lambda$ and $'\gamma_\lambda^p \in {}'\Gamma_\lambda^p$, then

$$'\gamma_\lambda^p \sim \gamma_\lambda^p - D_\lambda^p$$

and so

$$'\gamma_\lambda^p = \gamma_\lambda^p - D_\lambda^p + \mathrm{F}D_\lambda^{p+1}.$$

Suppose $\lambda > \mu$. Since $\pi_\mu^\lambda X_{1\lambda} \subset X_{1\mu}$, and γ^p is a cycle of X_0, we have

$$\pi_\mu^\lambda {}'\gamma_\lambda^p = \gamma_\mu^p - D_\mu^p.$$

Since π_μ^λ commutes with F, the right-hand side is a cycle and so $\mathrm{F}D_\mu^p = \mathrm{F}\gamma_\mu^p = \delta_\mu^{p-1}$ and consequently $'\Gamma_\mu^p = \bar{\pi}_\mu^\lambda {}'\Gamma_\lambda^p \in S_\mu$ ($\bar{\pi}_\mu^\lambda$ as in 13.1). In other words $\bar{\pi}_\mu^\lambda S_\lambda \subset S_\mu$. Once more $\{S_\lambda ; \bar{\pi}_\mu^\lambda\}$ is an inverse mapping system of compact spaces, and so it possesses a limit-element $\{'\Gamma_\lambda^p\} = {}'\Gamma^p$. This limit-element is such that $'\Gamma^p$ contains a representative (absolute) cycle $'\gamma_\lambda^p = \gamma_\lambda^p + D_\lambda^p$. Since X is p-acyclic, $'\gamma^p \sim 0$ and hence $'\gamma_\lambda^p \sim 0$, $\gamma_\lambda^p \sim 0 \bmod X_{1\lambda}$, and finally $\gamma^p \sim 0 \bmod X_1$. This proves (15.2) when G is compact. When G is a field the proof is the same with compactness replaced by linear compactness.

Proof of (15.1). Since X is $(p - 1)$-cyclic (14.1) and (14.2) together assert that $\mathrm{F}\mathfrak{Z}_0^p = \mathfrak{Z}_1^{p-1}$, then (14.3) and (15.2) that this mapping induces an isomorphism $\mathfrak{F}_0^p \rightarrow \mathfrak{F}_1^{p-1}$. Since the two $\mathfrak{F}$ groups are closed in the corresponding $\mathfrak{Z}$ groups (3.5) by (II, 5.4) F induces a homomorphism $\tau : \mathfrak{H}_0^p \rightarrow \mathfrak{H}_1^{p-1}$ which by (15.2) is univalent. Thus τ is continuous and one-one. Since the two $\mathfrak{H}$ groups are compact or linearly compact τ is an isomorphism (I, 32.4; II, 27.8) and (15.1) is proved.

(15.3) *Linking coefficient.* Let again X be $(p - 1, p)$-acyclic and let (G, H) be a normal couple (III, 30.1). Take a cycle δ^{p-1} of X_1 over G which is ~ 0, and a cocycle γ_p of X_0 over H. By (14.2) we have $\delta^{p-1} = \mathrm{F}\gamma^p$, where γ^p is a cycle mod X_1, and we define the linking coefficient of δ^{p-1}, γ_p as:

$$\mathrm{Lk}(\delta^{p-1}, \gamma_p) = \mathrm{KI}(\gamma^p, \gamma_p). \tag{15.4}$$

It is shown to have the same properties as for complexes (III, 35, 38), and in particular there is a class linking coefficient $\mathrm{Lk}(\Delta^{p-1}, \Gamma_p)$ given by

$$(15.5) \qquad \mathrm{Lk}(\Delta^{p-1}, \Gamma_p) = \mathrm{KI}(\Gamma^p, \Gamma_p),$$

where $\delta^{p-1} = \mathrm{F}\gamma^p, \gamma^p \in \Gamma^p$. From the known properties of the index we infer then:

(15.6) Lk *defines a group multiplication for* $\mathfrak{H}^{p-1}(X_1 ; G)$ *and* $\mathfrak{H}_p(X_0 ; H)$.

(15.7) From this point to the duality theorem is but a step. We continue to assume X to be $(p - 1, p)$-acyclic. By (12.4) $\mathfrak{H}^p(X_0, G)$ and $\mathfrak{H}_p(X_0, H)$ are dually paired with the Kronecker index as the multiplication. Coupling this with (15.1, 15.6) we obtain (III, 38.3) for nets. It is now possible to repeat the argument leading to the results of (III, 39.1) with simple nets instead of simplicial complexes, and with the vertices, loc. cit., replaced by the cycles γ^0 of (10.5) associated with the components of the net. We may thus state explicitly:

(15.8) Theorem. *The duality theorems of Alexander's type for complexes* (III, 38.3, 39) *are valid for nets, it being understood that simplicial complexes are to be replaced by simple nets.*

§4. SPECTRA

16. We have already defined spectra (2.4). The unique projections characteristic of a spectrum offer the advantage that chains and chain-groups may be introduced, thus bringing the homology theory of a spectrum one step nearer to the prototype, the homology theory of complexes. However, there is no guarantee that the homology groups based on the chains are the same as the net groups, and this question will be our major problem as regards spectra. Notice that since cocycles are determined by a single coordinate they will not concern us seriously in this connection.

A "cospectrum" may of course be introduced, but will be dispensed with as not really useful in the sequel.

Let then $X = \{X_\lambda ; \pi_\mu^\lambda\}$ be a spectrum. Since π_μ^λ is unique if $\mathfrak{C}_\lambda^p$, $\mathfrak{Z}_\lambda^p$, $\mathfrak{F}_\lambda^p$, $\mathfrak{H}_\lambda^p$ are the usual groups of X_λ over a fixed G, the collection $\{\mathfrak{C}_\lambda^p ; \pi_\mu^\lambda\}$ is an inverse system and it has a limit-group $\mathbf{C}^p$ whose elements $c^p = \{c_\lambda^p\}$ are known as the *projective p-chains over G*. If $c^{p+1} = \{c_\lambda^{p+1}\} \in \mathbf{C}^{p+1}$ then $\{\mathrm{F}c_\lambda^{p+1}\} \in \mathbf{C}^p$ and is written $\mathrm{F}c^{p+1}$. The boundary operator F thus defined is a homomorphism $\mathbf{C}^{p+1} \to \mathbf{C}^p$ (II, 13.5) and clearly $\mathrm{FF} = 0$. The *projective cycles, bounding* projective cycles and their groups $\mathbf{Z}^p$, $\mathbf{F}^p$ are defined as for complexes. The *projective pth homology group* over G is of course $\mathbf{H}^p = \mathbf{Z}^p/\overline{\mathbf{F}}^p$.

The co-elements are treated in the same way save that all the systems are direct, and hence all groups discrete.

It follows immediately from the definitions that $\mathbf{H}_p$ is the ordinary group of X as a net, and so the projective groups of cocycles will not require any special considerations.

(16.1) *The mapping* $\tau_\lambda : \mathbf{C}^p \to \mathfrak{C}_\lambda^p$ *defined by* $c^p \to c_\lambda^p$ *is a homomorphism which commutes with* F *and so maps* $\mathbf{Z}^p \to \mathfrak{Z}_\lambda^p$, $\mathbf{F}^p \to \mathfrak{F}_\lambda^p$ (II, 13.4).

(16.2) *$\mathbf{Z}^p$ is closed in $\mathbf{C}^p$.*

(16.3) *When G is compact so are the projective groups $\mathbf{C}^p$, $\mathbf{Z}^p$, $\mathbf{F}^p$, $\mathbf{H}^p$, and when G is a field they are linearly compact vector spaces over G.*

When G is compact so are the groups $\mathfrak{C}_\lambda^p$, and hence also $\mathbf{C}^p$, from which follows the same for $\mathbf{Z}^p$, $\mathbf{F}^p$, and finally for $\mathbf{H}^p$ by (II, 5.5). When G is a field the groups $\mathfrak{C}_\lambda^p$ are finite-dimensional and hence linearly compact. The rest is then the same with "compact" replaced by "linearly compact."

From (II, 13.3) there follow now:

(16.4) *If $\{\mu; >\} \subset \{\lambda; >\}$ then $X' = \{X_\mu\}$ is likewise a spectrum said to be "a part of X." If $c^p = \{c_\lambda^p\}$ is a chain of X then $c'^p = \{c_\mu^p\}$ is a chain of X' and $c^p \to c'^p$ defines a homomorphism τ of the projective groups, $\mathbf{C}^p$, $\mathbf{Z}^p$, $\mathbf{F}^p$, $\mathbf{H}^p$ of X into the corresponding groups of X'.*

(16.5) *Under the same conditions if $\{\mu\}$ is cofinal in $\{\lambda\}$, τ is an isomorphism between the groups $\mathbf{C}^p$, $\mathbf{Z}^p$, $\mathbf{F}^p$, $\mathbf{H}^p$ of the two spectra.*

17. Let $\mathfrak{Z}^p$, $\mathfrak{F}^p$, $\mathfrak{H}^p$ denote the net groups (as defined in §1), all groups in question being over a fixed G. When do we have $\mathbf{H}^p = \mathfrak{H}^p$? Sufficient conditions, covering all requirements later, are given by the

(17.1) THEOREM. *In a spectrum the projective homology groups over a group G which is compact or a field are isomorphic with the corresponding net groups.*

Since in the duality theorems for a net no other types of homology groups occur we have:

(17.2) COROLLARY. *As regards the duality theorems for a spectrum of finite complexes the groups of cycles may be chosen projective throughout.*

Clearly also:

(17.3) *In a spectrum the group of bounding projective cycles over a group G which is compact or a field is closed, i.e., we have $\mathbf{F}^p = \bar{\mathbf{F}}^p$ and hence $\mathbf{H}^p = \mathbf{Z}^p/\mathbf{F}^p$.*

(17.4) Before we proceed with the proofs we shall show that

(α) *Every class $\Gamma^p \in \mathfrak{H}^p$ contains a projective cycle $'\gamma^p$.*

(β) $\Gamma^p = 0 \to {'\gamma^p} \in \mathbf{F}^p$.

(17.5) *Suppose first G compact.* The groups $\mathfrak{C}_\lambda^p$, $\mathfrak{Z}_\lambda^p$, $\mathfrak{F}_\lambda^p$ are then all compact. Let $\gamma^p = \{\gamma_\lambda^p\} \in \Gamma^p$ and set $\zeta_\lambda^p = \{\delta_\lambda^p \mid \delta_\lambda^p \sim \gamma_\lambda^p\}$. Evidently $\pi_\mu^\lambda \zeta_\lambda^p \subset \zeta_\mu^p$ for $\lambda > \mu$. Since $\mathfrak{F}_\lambda^p$ is compact so is $\zeta_\lambda^p = \gamma_\lambda^p + \mathfrak{F}_\lambda^p$. Hence $\{\zeta_\lambda^p ; \pi_\mu^\lambda\}$ is an inverse mapping system of compact spaces. By (I, 39.1) there exists an element $'\gamma^p = \{\gamma_\lambda^p\}$ in the limit-space of the system. Clearly $'\gamma^p$ is a projective cycle and $'\gamma^p \in \Gamma^p$. This proves (α) for G compact.

Suppose now $\Gamma^p = 0$. The projective cycle $'\gamma^p$ just obtained will then have the property that $'\gamma_\lambda^p \sim 0$ for every λ. Since $\bar{\mathfrak{F}}_\lambda^p = \mathfrak{F}_\lambda^p$ there exist in X_λ chains c_λ^{p+1} such that $\mathrm{F}c_\lambda^{p+1} = {'\gamma_\lambda^p}$. Let $\zeta_\lambda^{p+1} = \{c_\lambda^{p+1} \mid \mathrm{F}c_\lambda^{p+1} = {'\gamma_\lambda^p}\}$. If $c_{0\lambda}^{p+1}$ and $c_{1\lambda}^{p+1}$ are elements of ζ_λ^{p+1} then $c_{0\lambda}^{p+1} - c_{1\lambda}^{p+1}$ is a $(p + 1)$-cycle. Therefore

$\zeta_\lambda^{p+1} = c_{0\lambda}^{p+1} + \mathfrak{Z}_\lambda^{p+1}$. Since $\mathfrak{Z}_\lambda^{p+1}$ is compact, ζ_λ^{p+1} is likewise compact. As before $\{\zeta_\lambda^{p+1}; \pi_\mu^\lambda\}$ is an inverse system of compact spaces and there is an element $c^{p+1} = \{c_\lambda^{p+1}\}$ in its limit-space. Clearly c^{p+1} is a projective chain and $\mathrm{F}c^{p+1} = {}'\gamma^p$. This proves ($\beta$) for G compact.

(17.6) *Suppose now G a field.* Since the groups $\mathfrak{C}_\lambda^p$, $\cdots$ are finite-dimensional vector spaces over G, they are linearly compact, and we find by reference to (II, 27.6) that the proofs just given for (α, β) are still valid in the present instance.

(17.7) Proof of (17.1). Let τ denote the homomorphism $\mathbf{Z}^p \to \mathfrak{Z}^p$ in the algebraic sense which is the identity on $\mathbf{Z}^p$. It is clear that $\tau\mathbf{F}^p \subset \mathfrak{F}^p$. Furthermore here $\mathbf{F}^p$ is compact or linearly compact and hence closed in $\mathbf{Z}^p$ (I, 32.1; II, 27.5) and $\mathfrak{F}^p$ is always closed in $\mathfrak{Z}^p$ (3.5). Therefore τ induces a homomorphism $\bar{\tau}:\mathbf{H}^p \to \mathfrak{H}^p$ in the algebraic sense and by (α, β) this is an isomorphism in the algebraic sense. Since $\lim \mathfrak{Z}_\lambda^p = \mathbf{Z}^p$, $\lim \mathfrak{F}_\lambda^p = \mathbf{F}^p$, $\lim \mathfrak{H}_\lambda^p = \mathbf{H}^p$, we find by (II, 13.6) that $\bar{\tau}$ is open and hence it is an isomorphism, proving (17.1).

18. **Sequential spectra.** In a sequential spectrum $\{X_n; \pi_n^m\}$ (2.4) the π_n^{n+1} determine all the projections since N3 of (2) yields

$$\pi_n^{n+k} = \pi_n^{n+1}\pi_{n+1}^{n+2} \cdots \pi_{n+k-1}^{n+k}. \tag{18.1}$$

(18.2) *The homology theory of a countable net is either that of a single complex of the net or else that of a sequential spectrum.*

If $\{\lambda\}$ has an upper bound λ_0 then the homology theory of X is that of X_{λ_0} (9.10). In the contrary case (I, 4.4) $\{\lambda\}$ has a cofinal sequence and so we may assume $X = \{X_1, X_2, \cdots\}$. Select now a definite π_n^{n+1} for each n and determine π_n^{n+k}, $k > 1$, by means of (18.1). The resulting net with these projections is a sequential spectrum whose homology theory is the same as for the initial net.

19. For certain sequential spectra one may strengthen (17.1) as follows:

(19.1) *If the sequential spectrum $X = \{X_n; \pi_n^{n+1}\}$ is such that every π_n^{n+1} is a mapping onto for the chain-groups, then* (17.1) *holds for every division-closure group G.*

We first prove:

(α) *Every class $\Gamma^p \in \mathfrak{H}^p$ contains a projective cycle ${}'\gamma^p$.*

(β) $\Gamma^p = 0 \to {}'\gamma^p \in \bar{\mathbf{F}}^p$.

The first is the same as (17.4α), the second differs from (17.4β) in that $\mathbf{F}^p$ is replaced by its closure. The proofs will be different from those of (17.4$\alpha\beta$).

Proof of (α). Let $\Gamma^p \in \mathfrak{H}^p$ and $\gamma^p = \{\gamma_n^p\} \in \Gamma^p$. Suppose that we have found ${}'\gamma_r^p$, $1 \leqq r \leqq n$, such that

$$\begin{aligned} {}'\gamma_r^p &= \pi_r^{r+1}\,{}'\gamma_{r+1}^p, && r < n; \\ {}'\gamma_s^p &= \gamma_s^p && \text{for some } s, 1 \leqq s \leqq n; \\ {}'\gamma_r^p &\sim \gamma_r^p, && 1 \leqq r \leqq n. \end{aligned} \tag{19.2}$$

I say that a similar set may be found for $n + 1$, with the same s. We have in fact ${}'\gamma_n^p \sim \pi_n^{n+1}\gamma_{n+1}^p$ in X_n. Since X_n is finite and G is a division-closure group X_n contains a chain c_n^{p+1} such that $\mathrm{F}c_n^{p+1} = {}'\gamma_n^p - \pi_n^{n+1}\gamma_{n+1}^p$. Since π_n^{n+1} is a mapping "onto" there exists a chain c_{n+1}^{p+1} such that $\pi_n^{n+1} c_{n+1}^{p+1} = c_n^{p+1}$. If we replace therefore γ_{n+1}^p by $'\gamma_{n+1}^p = \gamma_{n+1}^p + Fc_{n+1}^{p+1}$ we have $'\gamma_n^p = \pi_n^{n+1} {'\gamma_{n+1}^p}$; $'\gamma_{n+1}^p \sim \gamma_{n+1}^p$ in X_{n+1}, and so the same situation as before for the set $'\gamma_1^p, \cdots, '\gamma_{n+1}^p$. This gives an inductive construction for a projective cycle $'\gamma^p$. We have from (19.2), $'\gamma_n^p \sim \gamma_n^p$ for $n \geqq s$ and also for $n < s$: $\pi_n^s {'\gamma_s^p} = {'\gamma_n^p} = \pi_n^s \gamma_s^p \sim \gamma_n^p$ and so $'\gamma^p \sim \gamma^p$. In other words we have obtained a projective cycle $'\gamma^p \in \Gamma^p$ and having in addition an assigned coordinate in common with a given $\gamma^p \in \Gamma^p$. In particular this proves (α).

Passing now to (β), let $\Gamma^p = 0$. Then $'\gamma_n^p \sim 0$ in X_n, and so as before X_n contains c_n^{p+1} such that $Fc_n^{p+1} = {'\gamma_n^p}$. Choose now $c_r^{p+1} = \pi_r^n c_n^{p+1}$, $r < n$. Suppose also that $c_1^{p+1}, \cdots, c_m^{p+1}$, $m \geqq n$, have been found such that $c_r^{p+1} = \pi_r^{r+1} c_{r+1}^{p+1}$, $1 \leqq r \leqq m - 1$. Since π_m^{m+1} is a mapping "onto," there exists a chain c_{m+1}^{p+1} such that $c_m^{p+1} = \pi_m^{m+1} c_{m+1}^{p+1}$. We thus obtain a chain $c^{p+1} = \{c_m^{p+1}\}$ of X itself such that $Fc_r^{p+1} = {'\gamma_r^p}$, $1 \leqq r \leqq n$.

Let U_n be any nucleus of $\mathfrak{Z}_n^p$, and let $'U_n$ be the nucleus of $\mathbf{Z}^p$ consisting of the $'\gamma_0^p = \{'\gamma_{0n}^p\}$ such that $'\gamma_{0n}^p \in U_n$. It is clear that $\{'U_n\}$ is a base for the nuclei of $\mathbf{Z}^p$. We have just shown that for every n there is a chain c^{p+1} such that $'\gamma^p - Fc^{p+1} \in {'U_n}$. Therefore $'\gamma^p$ is in $\bar{\mathbf{F}}^p$. This proves (β).

Referring to (17.7) we find that the mapping $\tau: \mathbf{Z}^p \to \mathfrak{Z}^p$ there considered induces an isomorphism $\bar{\tau}: \mathbf{H}^p \to \mathfrak{H}^p$ in the algebraic sense. Let now U_n be open in $\mathfrak{Z}_n^p$, and let V_n be the aggregate of the Γ^p containing a $\gamma^p = \{\gamma_n^p\}$ with $\gamma_n^p \in U_n$, $'V_n$ the aggregate of the Γ^p containing a projective cycle with the same property. The sets $\{V_n\}$, $\{'V_n\}$ form bases for $\mathfrak{H}^p$, $\mathbf{H}^p$. Since there is a $'\gamma^p \in \Gamma^p$ with the same coordinate γ_n^p as any particular $\gamma^p \in \Gamma^p$ we have $V_n = {'V_n}$ and so $\bar{\tau}$ is topological. This completes the proof of (19.1).

§5. APPLICATION TO INFINITE COMPLEXES

20. Here as in (III, §8) significant results are only obtained when the complex $X = \{x\}$ is star- or closure-finite.

Suppose first X star-finite and let $\{X_\lambda\}$ be the finite open subcomplexes of X. We order $\{\lambda\}$ by the inclusion of the X_λ and so $\{\lambda; \succ\}$ is directed. Denote by π_μ^λ the projection $X_\lambda \to X_\mu$, $\lambda \succ \mu$, and by $\pi_\lambda^{*\mu}$ its dual the injection $X_\mu^* \to X_\lambda^*$ (IV, 10.13). The verification of the conditions Ni, N*i of (2) is elementary and since the projections are unique $\Sigma = \{X_\lambda ; \pi_\mu^\lambda\}$ is a spectrum.

Let G be a topological group and H a discrete group. Let $\mathfrak{C}^p(X, G)$, $\mathfrak{C}^p(X_\lambda, G)$ be the groups of the infinite chains of X and of the chains of X_λ (necessarily finite) over G, both topologized. Let $\mathfrak{C}_p^f(X, H)$, $\mathfrak{C}_p^\lambda(X_\lambda, H)$ be the discrete groups of the finite cochains of X and of the cochains of X_λ (necessarily finite) over H. From the definition of chain-groups (II, 8) we infer at once that $\mathfrak{C}^p(X, G)$ is the limit of the inverse system $\{\mathfrak{C}^p(X_\lambda, G); \pi_\mu^\lambda\}$ and $\mathfrak{C}_p^f(X, H)$ the limit of the direct system $\{\mathfrak{C}_p(X_\lambda, H); \pi_\lambda^{*\mu}\}$.

Similar results are obtained when X is closure-finite by passing to X^*. Therefore:

(20.1) *When X is star-finite [closure-finite] the collection of its finite open [closed] subcomplexes generates a spectrum [cospectrum] Σ whose projective groups of the cycles [cocycles] over G and of the cocycles [cycles] over a discrete H are the same as the groups of the infinite cycles [cocycles] of the complex X itself over G and of the finite cocycles [cycles] of X over H.*

(20.2) When X is simple and star-finite [closure-finite] the available intersections between classes of infinite cycles [cocycles] and finite cocycles [cycles] as defined in (V, §2) are readily identified with those obtained from the spectrum [cospectrum].

(20.3) If we combine (6.5) and (17.2) with (20.1) we obtain a *second proof of the duality theorem* (III, 41.2) *for infinite complexes.*

(20.4) From (17.3) we deduce immediately that *in a star-finite complex and when G is compact or a field then $\bar{\mathfrak{F}}^p = \mathfrak{F}^p$ and hence $\mathfrak{H}^p = \mathfrak{Z}^p/\mathfrak{F}^p$.*

(20.5) *When G is a division-closure group and X is star-finite, the net groups $\mathfrak{H}^p(\Sigma, G)$ of the spectrum Σ and the groups $\mathfrak{H}^p(X, G)$ of the infinite cycles of X over G are isomorphic* (Steenrod [a]).

Since the groups of X are the direct sums of those of its components, we may replace X by any component and so assume it countable. This being the case $\Sigma = \{X_\lambda\}$ is then countable. Let Σ' be the corresponding elementary spectrum constructed in the proof of (18.2) and with the same homology theory. All the projections of Σ existing between the complexes of Σ' are now projections of Σ' also, since Σ is a spectrum. Therefore Σ' is cofinal in Σ. As far as (20.5) is concerned Σ' may clearly replace Σ. Since the projections in Σ' are all "mappings onto," (20.5) becomes a consequence of (19.1).

(20.6) The dual of (20.5) obtained by passing to X^*, is readily stated and left to the reader.

§6. WEBS

21. In the sequel we shall require a weaker analogue of the lattice, the *web* which arises out of the relations of inclusion between sets, complexes, or nets.

(21.1) *Web of sets.* Let $\mathfrak{R}$ be a topological space. Changing slightly our standard notations we designate its open sets by A, and its closed sets by B, with complementary indices, as A_λ, B', etc.

An *open web* of $\mathfrak{R}$ is a collection $\mathfrak{A} = \{A_\lambda\}$ of open sets such that given A_λ, A_μ there exist A_ν, A_ρ such that $A_\nu \subset A_\lambda$, A_μ and $A_\rho \supset A_\lambda$, A_μ. In other words (in an obvious sense) both $\{A_\lambda ; \subset\}$ and $\{A_\lambda ; \supset\}$ are directed. A *closed web* $\mathfrak{B} = \{B_\lambda\}$ is defined in the same way with B replacing A throughout.

If $\mathfrak{A} = \{A_\lambda\}$ is an open web and $B_\lambda = \mathfrak{R} - A_\lambda$ then $\mathfrak{B} = \{B_\lambda\}$ is a closed web. Each of $\mathfrak{A}$, $\mathfrak{B}$ is said to be the *complement* of the other.

A *partial web* $\mathfrak{A}'$ or $\mathfrak{B}'$ of $\mathfrak{A}$ or $\mathfrak{B}$ (partial system in the sense of (I, 40)) is one whose sets make up a subcollection of those of $\mathfrak{A}$ or $\mathfrak{B}$. We say that $\mathfrak{A}' = \{A'_\mu\}$ is *cofinal* [*coinitial*] in $\mathfrak{A}$ if every A_λ is contained in [contains] an A'_μ. If both occur then $\mathfrak{A}'$ is said to be *coterminal* with $\mathfrak{A}$. Similarly with $\mathfrak{B}$ and $\mathfrak{B}'$.

(21.2) *Web of complexes.* Replace $\mathfrak{R}$ by a complex K and subsets by subcomplexes of K. There result, by manifest analogy, the *open* and *closed webs of subcomplexes* of K, or merely *webs of complexes*, and everything said in (21.1) carries over to this case.

(21.3) *Web of nets.* Let $X = \{X_\lambda ; \pi_\mu^\lambda\}$ be a net. In (12.5) there have been defined the relations of inclusion between the open subnets, those between the closed subnets, as well as complementation for open or closed subnets. Therefore we may paraphrase the definitions of (21.1) and introduce *webs of open subnets* or of *closed subnets of* X, (more simply called *open* or *closed webs of nets*). Explicitly an open web of nets is a collection $\mathfrak{A} = \{A_\lambda\}$ of open subnets of X such that given A_λ, A_μ there is an $A_\nu \subset A_\lambda$, A_μ as well as an $A_\rho \supset A_\lambda, A_\mu$ and similarly for closed webs. Complementary webs and the other terms are defined as in (21.1).

(21.4) *Direct and inverse webs.* A web $\mathfrak{A} = \{A_\lambda\}$ of any one of our types is said to be *direct* if $\{\lambda\}$ is ordered by the inclusions of the $A_\lambda : \lambda > \mu \leftrightarrow A_\lambda \supset A_\mu$, and to be *inverse* if $\{\lambda\}$ is ordered in the opposite way.

(21.5) *Ideal elements.* These structures chiefly designed for a closer analysis of the behavior of the cycles of a complex "at infinity" were already considered in [L, 295]. Their description in terms of webs is very simple. An *open ideal element* in a set or complex is an open web $\mathfrak{A} = \{A_\lambda\}$ such that $\bigcap A_\lambda = \emptyset$, $\bigcup A_\lambda = \mathfrak{R}$ or K as the case may be. Similarly for a *closed* ideal element. If $\mathfrak{A}$, $\mathfrak{B}$ are complementary and one of them is an ideal element so is the other. The associated homology groups fall under the general category of the groups of webs to be taken up presently.

22. **Homology theories.** As we shall see, the direct and inverse webs resulting from a given web of complexes or nets generally give rise to two distinct homology theories.

(22.1) An important role is played by certain sets of homomorphisms which we shall now examine. Suppose first that we are dealing with subnets of a given net X. Let A, A' be open subnets of X and let $B = X - A, B' = X - A'$. If $B' \subset B$ then $A \subset A'$. As we have seen (12.5) A is an open subnet of A' and B' a closed subnet of B. Referring then to (13.9) we have the following operations:

a projection $\pi : A' \to A$ or reduction mod $(A' - A)$ of the cycles of A'; this is the same as a reduction mod B of the cycles of X mod B';

an injection $\eta : B' \to B$ or mapping into themselves of the groups of the cycles of B'.

The duals are

an injection $\pi^* : A^* \to A'^*$,

a projection $\eta^* : B^* \to B'^*$,

with their obvious interpretations in terms of the cocycles.

If we are dealing with subcomplexes of a given complex the same operations are to be understood in the sense of (III, 23). This common terminology will enable us to consider together the groups of webs of complexes and nets.

(22.2) Let $\mathfrak{A} = \{A_\lambda\}$, $\mathfrak{B} = \{B_\lambda\}$ be complementary webs of subcomplexes of a fixed complex K or of subnets of a given net X. We suppose $\mathfrak{A}$ open and $\mathfrak{B}$ closed. A fixed coefficient group G is chosen and will not be indicated in the notations for the groups of cycles. The groups $\mathfrak{Z}^p(B_\lambda), \cdots, \mathfrak{Z}_p(A_\lambda), \cdots$ are groups of absolute cycles and cocycles, while $\mathfrak{Z}^p(A_\lambda), \cdots$ are groups of cycles mod B_λ, and $\mathfrak{Z}_p(B_\lambda), \cdots$ are groups of cocycles mod A_λ. We will write:

$$\left.\begin{aligned} \pi_\mu^\lambda &= \text{a projection} \\ \eta_\mu^\lambda &= \text{an injection} \end{aligned}\right\} A_\lambda \to A_\mu \text{ or } B_\lambda \to B_\mu .$$

(22.3) Consider first the direct web $\mathfrak{A}(\lambda > \mu \leftrightarrow A_\lambda \supset A_\mu)$. We have then (22.1) a projection $\pi_\mu^\lambda : A_\lambda \to A_\mu$ and its dual operation an injection $\pi_\lambda^{*\mu} : A_\mu^* \to A_\lambda^*$. Suppose for the present the group of the cycles of the A_λ topologized and those of the cocycles discrete. Referring to (13.9) the system of the cycles and cocycles of the A_λ under consideration constitute two dual categories with the following properties analogous to Ni, N*i of (2.1):

HN1. If $\lambda > \mu$ there exists a unique simultaneous homomorphism $\pi_\mu^\lambda : \mathfrak{Z}^p(A_\lambda) \to \mathfrak{Z}^p(A_\mu)$, and $\pi_\mu^\lambda \mathfrak{F}^p(A_\lambda) \subset \mathfrak{F}^p(A_\mu)$.

HN2. $\lambda > \mu > \nu \to \pi_\nu^\lambda = \pi_\nu^\mu \pi_\mu^\lambda$.

HN*12 the same as HN12 with $\pi_\lambda^{*\mu}$ instead of π_μ^λ, cocycles in place of cycles and $>$ replaced by $<$.

HN3. The relation (6.1) for permanence of the index holds.

It may happen also that in addition we have:

HN4. The dual categories under consideration possess intersections with the properties of (8) and in particular the permanence relations (8.1, 8.2) hold.

(22.4) Definition. *A system of dual categories with the properties* HN123 *and* HN*12 *will be called an H-net (H abridged for homology). When in addition* HN4 *holds, the H-net is said to be with intersection. If the homomorphisms proceed the other way around there is obtained an H-conet.*

Under certain circumstances the dual categories admit only of weak duality (III, 31: only discrete groups and only field duality). We will then say that we have a *weak* H-net or H-conet as the case may be.

We observe now that HN123, HN*12 are the only properties required in developing the theory of cycles, cocycles, homology groups, duality theorems in (3, 4, 6). Moreover the only supplementary property required for intersections is HN4. We have therefore:

(22.5) *There may be defined cycles, cocycles and the groups* $\mathfrak{Z}, \mathfrak{F}, \mathfrak{H}$ *for H-nets or H-conets in the same way as for nets and conets and they have the properties* (3.4, $\cdots$, 3.7, 4.2, 4.5, 6.1, $\cdots$, 6.5). *Furthermore in the H-net [H-conet] the resulting cycles and cocycles [cocycles and cycles] form dual categories and they have intersections when* HN4 *holds. In the weak H-net [H-conet] the same statement holds but with weak dual categories.*

(22.5a) SUPPLEMENTARY REMARK. Since projections are unique, the true analogy is with spectra, and so there may be introduced the same projective groups **C**, **Z**, **F**, **H** as in (§4). We will not stop, however, to compare them with the groups of the H-nets or conets.

(22.6) Returning to the direct web $\mathfrak{A}$ of (22.3) the groups of cycles and cocycles of the A_λ with the projections π_μ^λ and injections $\pi_\lambda^{*\mu}$ make up an H-net which we designate by $\{A_\lambda ; \pi_\mu^\lambda\}$. The resulting limit-groups are called *groups* of the direct web $\mathfrak{A}$ over G, written $\mathfrak{Z}_d^p(\mathfrak{A}, G), \cdots$. In particular if $\bar{\pi}_\mu^\lambda$ is the simultaneous homomorphism in the groups $\mathfrak{H}^p(A_\lambda)$ induced by π_μ^λ, then $\{\mathfrak{H}^p(A_\lambda); \bar{\pi}_\mu^\lambda\}$ is an inverse system whose limit-group $\mathfrak{H}_d^p(\mathfrak{A}, G)$ is the pth homology group of the direct web $\mathfrak{A}$ over G. Similarly for the cohomology group $\mathfrak{H}_p^d(\mathfrak{A}, G)$ (G discrete).

(22.7) Suppose more particularly that $\mathfrak{A}$ is a web of infinite open subcomplexes of K and that the $\mathfrak{Z}^p(A_\lambda)$ are the groups of finite cycles of K mod B_λ. Then these groups must be taken discrete and $\{A_\lambda ; \pi_\mu^\lambda\}$ is a weak H-net. Our notation is not well designed to separate all the numerous possibilities, but the particular case under consideration will generally be clear from the context.

(22.8) If $\mathfrak{A}$ is taken inverse the situation is essentially the same except that the H-net is replaced by an H-conet, with injections for the cycles, projections for the cocycles, and groups $\mathfrak{Z}_i^p(\mathfrak{A}, G), \cdots$. Finally $\mathfrak{B}$ presents the same two possibilities with cycles and cocycles, projections and injections interchanged. The following table summarizes the situation.

(22.9) TABLE

(a): groups of cocycles discrete;
(b): groups of cycles discrete.
$\lambda > \mu$ throughout.

Web		a	b
I. $\mathfrak{A}$ direct..........	$\{A_\lambda ; \pi_\mu^\lambda\}$	H-net	Weak H-net
II. $\mathfrak{A}$ inverse.........	$\{A_\lambda ; \pi_\lambda^\mu\}$	Weak H-conet	H-conet
III. $\mathfrak{B}$ direct..........	$\{B_\lambda ; \eta_\lambda^\mu\}$	Weak H-conet	H-conet
IV. $\mathfrak{B}$ inverse.........	$\{B_\lambda ; \eta_\mu^\lambda\}$	H-net	Weak H-net

(22.10) *If a web $\mathfrak{A}'$ is cofinal [coinitial] in a web $\mathfrak{A}$ then they have the same direct [inverse] groups* (II, 13.3, 14.5).

(22.11) *The Betti numbers of the groups of the table over a given field depend solely upon the characteristic of the field.*

Since this is true for nets or complexes the proof of (7.2) applies here also.

(22.12) *Duality.* The table describes in substance eight dual categories corresponding to each of the H-nets there occurring.

EXAMPLES (22.13) Let K be a locally finite complex and $\mathfrak{A}$ the open ideal element consisting of the finite open subcomplexes of K including $\emptyset$. Taking $\mathfrak{A}$ direct we have Type Ia of the table and the resulting H-net is merely the spectrum of (20) corresponding to the infinite cycles with topologized groups and finite cocycles with discrete groups.

(22.14) The complex remaining the same, let $\mathfrak{B}$ denote the closed ideal element consisting of all the finite closed subcomplexes of K including $\emptyset$ and let $\mathfrak{A}_1$ be the complement of $\mathfrak{B}$. Taking $\mathfrak{A}_1$ inverse we have Type IIa. Since the groups are those of an H-conet, a cycle γ^p of $\mathfrak{A}_1$ inverse is defined by a single coordinate. Therefore it is merely a cycle of $K \bmod B_\lambda$, and $\gamma^p \sim 0$ signifies that $\gamma^p \sim 0 \bmod B_\nu$, for some $\nu > \lambda$, i.e., for some $B_\nu \supset B_\lambda$. Since we are dealing with discrete cycles, "~ 0" and bounding are equivalent. Therefore γ^p is merely any chain with a finite boundary and $\gamma^p \sim 0$ means that $\gamma^p +$ a finite chain bounds. Let $\mathfrak{C}_f^p(K, G)$ be the group of the finite p-chains over a discrete G. Then the homology groups here considered are given by $\mathfrak{H}^p(K, G) = \mathfrak{Z}^p(K, G)/(\mathfrak{F}^p(K, G) + \mathfrak{C}_f^p(K, G))$.

(22.15) We have already defined (21.1) a *partial web* $\mathfrak{A}_1$ of $\mathfrak{A}$. If $\mathfrak{N}$ is any H-net associated with $\mathfrak{A}$ then the analogue $\mathfrak{N}_1$ for $\mathfrak{A}_1$ is a *partial H-net* of $\mathfrak{N}$ in the sense of (9), and it is readily shown that the arguments of (9) carry over to $\mathfrak{N}, \mathfrak{N}_1$.

23. Under certain conditions groups of a web of subnets may be replaced by those of a single net. The resulting properties have interesting topological applications (VII, 14, 15; VIII, 13.4).

(23.1) Let $X = \{X_\lambda ; \pi_\mu^\lambda\}$ be our customary net and $\mathfrak{B}$ a closed web of subnets of X. It will be more in keeping with the prevailing notations of the chapter to write $\mathfrak{B} = \{X_\alpha\}$, $X_\alpha = \{X_{\alpha\lambda} ; \pi_\mu^\lambda\}$. We denote by $\mathfrak{A}$ the complement of $\mathfrak{B}$.

Let $Y_\lambda = \bigcap_\alpha X_{\alpha\lambda}$. Since $\lambda > \mu \rightarrow \pi_\mu^\lambda X_{\alpha\lambda} \subset X_{\alpha\mu}$, we also have $\pi_\mu^\lambda Y_\lambda \subset Y_\mu$. Therefore $Y = \{Y_\lambda ; \pi_\mu^\lambda\}$ is a closed subnet of X.

The two basic properties which we wish to prove are:

(23.2) *The homology theory of the web $\mathfrak{B}$ taken inverse is the same as that of the closed subnet Y.*

(23.3) *The homology theory of the web $\mathfrak{A}$ taken direct is the same as that of the dual categories of the cycles of X* mod Y *and cocycles of the open subnet $X - Y$.*

As usual we denote cycles and cocycles by γ, δ and their classes by Γ, Δ affected with the same indices. The H-nets or conets defined by the inverse and direct webs will be denoted by $\mathfrak{A}_i, \mathfrak{A}_d, \cdots$.

Proof of (23.2). It is clear that $'\mathfrak{B} = \mathfrak{B} \cup Y$ is likewise a closed web. Since $'\mathfrak{B}$ has Y as initial element the theory of Y and the theory of $'\mathfrak{B}_i$ are the same. Therefore (23.2) reduces to proving:

(23.4) *$\mathfrak{B}_i$ and $'\mathfrak{B}_i$ have the same homology theory.*

Every cycle of $'\mathfrak{B}_i$ is of the form $'\gamma^p = \gamma^p \cup \delta^p$, where γ^p is a cycle of $\mathfrak{B}_i$ and δ^p a cycle of Y. The cycle γ^p is merely the set of coordinates of $'\gamma^p$ in the elements of $\mathfrak{B}_i$, and δ^p is its coordinate in Y. It is clear that $'\gamma^p \rightarrow \gamma^p$ defines a simultaneous homomorphism τ of the groups of cycles of $'\mathfrak{B}_i$ into those of $\mathfrak{B}_i$. In its turn τ induces a simultaneous homomorphism $\bar{\tau}$ in the homology groups (the same as τ of (9) for the H-net $'\mathfrak{B}_i$ and its partial net $\mathfrak{B}_i$). As already observed we must show that:

(23.5) *$\bar{\tau}$ is an isomorphism.*

Since X_λ is finite it has at most a finite number of subcomplexes $X_{\alpha\lambda}$ and so there is a smallest $X_{\alpha_0\lambda}$. Given any $X_{\alpha\lambda}$ there is an $X_{\beta\lambda} \subset X_{\alpha\lambda}, X_{\alpha_0\lambda}$. Hence $X_{\beta\lambda} = X_{\alpha_0\lambda} \subset X_{\alpha\lambda}$, which implies $X_{\alpha_0\lambda} = \bigcap X_{\alpha\lambda} = Y_\lambda$.

We now come to the cycles. Since we have the last case in the table (22.9) the basic operations in $\mathfrak{B}_i$ are injections η_β^α, $\alpha > \beta$. In $'\mathfrak{B}_i$ there are in addition the injections $\eta_\alpha : Y \to X_\alpha$.

Now a cycle of X_α is a collection $\gamma_\alpha^p = \{\gamma_{\alpha\lambda}^p\}$, where $\gamma_{\alpha\lambda}^p$ is a cycle of $X_{\alpha\lambda}$ and $\lambda > \mu \to \pi_\mu^\lambda \gamma_{\alpha\lambda}^p \sim \gamma_{\alpha\mu}^p$ in $X_{\alpha\mu}$. Hence a cycle of $\mathfrak{B}_i$ is a collection $\gamma^p = \{\gamma_\alpha^p\}$, where $\alpha > \beta \to \eta_\beta^\alpha \gamma_\alpha^p \sim \gamma_\beta^p$ in X_β. From the definition of homologous cycles in a net (here X_β) there comes: $\gamma_{\alpha\lambda}^p \sim \gamma_{\beta\lambda}^p$ in $X_{\beta\lambda}$. In particular if $\beta = \alpha_0$ then $X_{\beta\lambda} = Y_\lambda$. Therefore Y_λ contains a cycle δ_λ^p $(= \gamma_{\alpha_0\lambda}^p)$ such that if α, β are replaced by α_0, α then

$$\text{(23.6)} \qquad \delta_\lambda^p \sim \gamma_{\alpha\lambda}^p \text{ in } X_{\alpha\lambda}.$$

Let $\lambda > \mu$ and denote by α_1 the analogue of α_0 for μ. Since $\gamma_{\alpha_1}^p$ is a cycle of the closed subnet X_{α_1}, (23.6) yields:

$$\pi_\mu^\lambda \delta_\lambda^p \sim \pi_\mu^\lambda \gamma_{\alpha_1\lambda}^p \sim \gamma_{\alpha_1\mu}^p = \delta_\mu^p \text{ in } X_{\alpha_1\mu} = Y_\mu.$$

Therefore $\delta^p = \{\delta_\lambda^p\}$ is a cycle of Y. By (23.6) $\delta^p \sim \gamma_\alpha^p$ in X_α, which may also be written $\eta_\alpha \delta^p \sim \gamma_\alpha^p$ in X_α. Hence $'\gamma^p = \gamma^p \cup \delta^p$ is a cycle of $'\mathfrak{B}_i$ such that $\tau' \gamma^p = \gamma^p$. Or explicitly:

(23.7) *Corresponding to every cycle* $\gamma^p = \{\gamma_\alpha^p\}$ *of* $\mathfrak{B}$ *inverse there is a cycle* δ^p *of* Y *such that* $'\gamma^p = \gamma^p \cup \delta^p$ *is a cycle of* $'\mathfrak{B}$ *inverse, or equivalently such that* $\gamma_\alpha^p \sim \delta^p$ *in* X_α.

Since $\tau' \gamma^p = \gamma^p$, the mapping τ, hence also $\bar{\tau}$, is onto.

Suppose that $'\gamma^p = \gamma^p \cup \delta^p$ is a given cycle of $'\mathfrak{B}_i$. We have: $\tau' \gamma^p \sim 0$ in $\mathfrak{B}_i \leftrightarrow \gamma^p \sim 0$ in $\mathfrak{B}_i \to \gamma_{\alpha_0}^p \sim 0$ in $X_{\alpha_0} \to \gamma_{\alpha_0\lambda}^p \sim 0$ in $X_{\alpha_0\lambda} \to \delta_\lambda^p \sim 0$ in $Y_\lambda \to \delta^p \sim 0$ in $Y \to {}'\gamma^p \sim 0$ in $'\mathfrak{B}_i$. Therefore $\bar{\tau}$ is univalent.

Let $U_{\alpha\lambda}$ be open in $\mathfrak{H}^p(X_{\alpha\lambda}, G)$ and set

$$V_{\alpha\lambda} = \{\Gamma^p \mid \Gamma_{\alpha\lambda}^p \in U_{\alpha\lambda}\},$$
$$'V_{\alpha\lambda} = \{'\Gamma^p \mid {}'\Gamma^p = \Gamma^p \cup \Delta^p, \Gamma_{\alpha\lambda}^p \in U_{\alpha\lambda}\}.$$

Since the Y_λ are among the $X_{\alpha\lambda}$, it is readily seen that $\{V_{\alpha\lambda}\}$, $\{'V_{\alpha\lambda}\}$ are bases for $\mathfrak{H}^p(\mathfrak{B}_i, G)$, $\mathfrak{H}^p('\mathfrak{B}_i, G)$, and since $V_{\alpha\lambda} = \bar{\tau} V_{\alpha\lambda}$, $\bar{\tau}$ is open. This completes the proof of (23.5).

The representatives of a cocycle γ_p of $\mathfrak{B}_i$ are likewise those of a cocycle $'\gamma_p$ of $'\mathfrak{B}_i$, and $\gamma_p \to {}'\gamma_p$ defines a simultaneous homomorphism τ^* of the groups of the cocycles of $\mathfrak{B}_i$ into those of $'\mathfrak{B}_i$. In its turn τ^* induces a simultaneous homomorphism $\bar{\tau}^*$ of the cohomology groups (the analogue of τ^* of (9) for the two H-nets).

(23.8) $\bar{\tau}^*$ *is an isomorphism.*

Since the cohomology groups are discrete and over a discrete group H or a field J, and (9.4) holds for $\mathfrak{B}_i$, $'\mathfrak{B}_i$, the same character-group argument as for (IV, 10.10) with $\mathfrak{J}$, $\mathfrak{P}$ replaced by H and its character-group or both by J, will yield (23.8).

Referring now to (9), property (23.4) will follow from the remaining relations of permanence. As already observed this proves (23.2).

PROOF OF (23.3). It is essentially obtained by "dualizing" the proof of (23.2). If $'\mathfrak{A} = \mathfrak{A} \cup (X - Y)$, then $X - Y$ is cofinal in $'\mathfrak{A}_d$, and so the dual categories in (23.3) have the same homology theory as $'\mathfrak{A}_d$. This reduces (23.3) to:

(23.9) $\mathfrak{A}_d$ *and* $'\mathfrak{A}_d$ *have the same homology theory.*

Let $\theta, \theta^*, \bar{\theta}, \bar{\theta}^*$ be the analogues of $\tau, \cdots, \bar{\tau}^*$.

(23.10) $\bar{\theta}$ *is an isomorphism.*

Since we have the first case of the table the basic operations in $\mathfrak{A}_d$ are projections ω_β^α, where $\alpha > \beta \leftrightarrow (X - X_\alpha) \supset (X - X_\beta) \leftrightarrow X_\alpha \subset X_\beta$. In $'\mathfrak{A}_d$ there are in addition the projections $\omega_\alpha : X \to (X - X_\alpha)$, or reduction mod X_α of the cycles of X.

Now a cycle of X mod X_α is a collection $\gamma_\alpha^p = \{\gamma_{\alpha\lambda}^p\}$, where $\gamma_{\alpha\lambda}^p$ is a cycle of X_λ mod $X_{\alpha\lambda}$, and $\lambda > \mu \to \pi_\mu^\lambda \gamma_{\alpha\lambda}^p \sim \gamma_{\alpha\mu}^p$ in X_μ mod $X_{\alpha\mu}$. Hence a cycle of $\mathfrak{A}_d$ is a collection $\gamma^p = \{\gamma_\alpha^p\}$ where $\alpha > \beta \to \omega_\beta^\alpha \gamma_\alpha^p \sim \gamma_\beta^p$ in X mod $X_\beta \to \gamma_{\alpha\lambda}^p \sim \gamma_{\beta\lambda}^p$ in X_λ mod $X_{\beta\lambda}$. In particular choosing $\alpha = \alpha_0$ and replacing β by α, we have a cycle δ_λ^p $(= \gamma_{\alpha_0\lambda}^p)$ of X_λ mod Y_λ $(= X_{\alpha_0\lambda})$ such that:

$$\delta_\lambda^p \sim \gamma_{\alpha\lambda}^p \text{ in } X_\lambda \bmod X_{\alpha\lambda}. \tag{23.11}$$

Let now $\lambda > \mu$ and α_0, α_1 as before. Since $\gamma_{\alpha_1}^p$ is a cycle of X mod X_{α_1}, by combining with (23.10) we find:

$$\pi_\mu^\lambda \delta_\lambda^p \sim \pi_\mu^\lambda \gamma_{\alpha_1\lambda}^p \sim \gamma_{\alpha_1\mu}^p \sim \delta_\mu^p \text{ in } X_\mu \bmod X_{\alpha_1\mu} = Y_\mu .$$

Therefore $\delta^p = \{\delta_\lambda^p\}$ is a cycle of X mod Y. By (23.11) again $\delta^p \sim \gamma_\alpha^p$ in X mod X_α. Therefore $'\gamma^p = \gamma^p \cup \delta^p$ is a cycle of $'\mathfrak{A}_d$ such that $\theta\, '\gamma^p = \gamma^p$. Or explicitly:

(23.12) *Corresponding to every cycle* $\gamma^p = \{\gamma_\alpha^p\}$ *of* $\mathfrak{A}$ *direct there is a cycle* δ^p *of* X mod Y *such that* $'\gamma^p = \gamma^p \cup \delta^p$ *is a cycle of* $'\mathfrak{A}$ *direct, or equivalently such that* $\gamma_\alpha^p \sim \delta^p$ *in* X mod X_α.

Thus θ, hence also $\bar{\theta}$, is onto.

If $'\gamma^p = \gamma^p \cup \delta^p$ is a given cycle of $'\mathfrak{A}$, we have: $\theta\, '\gamma^p \sim 0$ in $\mathfrak{A}_d \leftrightarrow \gamma^p \sim 0$ in $\mathfrak{A}_d \to \gamma_{\alpha_0}^p \sim 0$ in X mod $X_{\alpha_0} \to \gamma_{\alpha_0\lambda}^p \sim 0$ in X_λ mod $X_{\alpha_0\lambda} \to \delta_\lambda^p \sim 0$ in X_λ mod $Y_\lambda \to \delta^p \sim 0$ in X mod $Y \to {}'\gamma^p \sim 0$ in $'\mathfrak{A}_d$. Thus $\bar{\theta}$ is univalent.

The proof that $\bar{\theta}$ is open is the same as for $\bar{\tau}$ and (23.10) follows.

From this point on, the conclusion of the proof of (23.9) is the same as for (23.4) and so (23.3) follows.

§7. METRIC COMPLEXES

24. Frequently complexes consist of elements represented by subsets of a metric space. The metric relationships thus arising may be utilized to advantage to introduce new significant webs and related homology groups. The following definition covers all the interesting types thus arising.

(24.1) DEFINITIONS. *A metric complex is a closure-finite complex* $X = \{x\}$ *such that there exists a real-valued function of the element* x, *called the diameter of* x, *written* diam x, *and subjected to the sole condition*: $x' < x \to \operatorname{diam} x' \leqq \operatorname{diam} x$.

If Y is a subcomplex of X we may now define mesh $Y = \sup \{\text{diam } x \mid x \in Y\}$, *and hence for a chain C: mesh* $C =$ *mesh* $|C|$.

(24.2) In the applications metric complexes will usually occur as follows: With each x there will be associated a bounded subset $|x|$ of a certain metric space $\mathfrak{R}$ such that $x' < x \rightarrow \overline{|x'|} \subset \overline{|x|}$, and then diam $x =$ diam $|x|$ will turn X into a metric complex. If Y is a subcomplex of X and C a chain of X, we write $|Y| = \bigcup\{|x| \mid x \in Y\}$ and $\|C\| = |(|C|)|$.

(24.3) A subcomplex Y of X is made metric in the obvious way by assigning to its elements the same diameters as in X.

The subcomplex Y is said to be *essential* if there is an $\epsilon > 0$ such that diam $x < \epsilon \rightarrow x \in Y$.

(24.4) Examples. Geometric and Euclidean complexes (VIII, §1), Vietoris complexes (VII, §5), singular complexes (VIII, 24) are important types of metric complexes.

25. V-cycles.

(25.1) An obvious web related to the metric complex X is readily defined. Given any $\epsilon > 0$, set $B_\epsilon = \{x \mid \text{diam } x < \epsilon\}$. In view of (24.1) B_ϵ is a closed subcomplex of X. If $\epsilon' \leqq \epsilon$ then

$$B_\epsilon \cup B_{\epsilon'} = B_\epsilon , \qquad B_\epsilon \cap B_{\epsilon'} = B_{\epsilon'} ,$$

and so $\mathfrak{B} = \{B_\epsilon\}$ is a closed web of complexes. The only interesting ordering is evidently $\{\epsilon; <\}$. Furthermore since X and hence the B_ϵ, are merely closure-finite, only finite (absolute) cycles are admissible for them. Therefore the appropriate web homology theory is that of the inverse closed web and IVb of the table (22.9). The cycles, $\cdots$ are known as *V-cycles*, $\cdots$. The "V" is abridged for "Vietoris," as the prototype of this homology theory is the classical Vietoris theory for compacta (Vietoris [a]; VII, §5). Since we are dealing here with a weak H-net only discrete coefficient groups are admissible. There are two different approaches to the V-cycles, each with its advantages, and so both will be given here. The resulting homology groups are of course the same.

(25.2) First definition. *Let $\{\epsilon_n\}$ be a positive sequence tending to* 0. *Since it is coinitial in $\{\epsilon\}$, it may replace it in the definition of the groups under consideration. A V-cycle γ^p over a discrete G will then be a collection $\gamma^p = \{\gamma^p_n\}$, where γ^p_n is a finite cycle of B_{ϵ_n} and $n \geqq m \rightarrow \gamma^p_n \sim \gamma^p_m$ in B_{ϵ_m}. The cycle $\gamma^p \sim 0$ whenever $\gamma^p_n \sim 0$ in B_{ϵ_n} for every n. The operations on the cycles are defined in the natural way, and we have the usual groups of the cycles and bounding cycles written $\mathfrak{Z}^p_v(X, G)$, $\mathfrak{F}^p_v(X, G)$. Their topology is defined as follows: The cycles γ^p with a given coordinate γ^p_n form an open set U_n in $\mathfrak{Z}^p_v(X, G)$ and $\{U_n\}$ is a base for the group. $\mathfrak{F}^p_v(X, G)$ receives the relative topology. Since "~ 0" $\leftrightarrow$ "bounding" for the finite cycles in a complex we show as for* (2.8) *that $\mathfrak{F}^p_v(X, G)$ is closed in*

$\mathfrak{Z}_v^p(X, G)$ and so we define $\mathfrak{H}_v^p(X, G) = \mathfrak{Z}_v^p(X, G)/\mathfrak{F}_v^p(X, G)$. From the general theory it is known that $\mathfrak{H}_v^p(X, G)$ remains isomorphic with itself if $\{\epsilon_n\}$ is replaced by any other similar sequence.

(25.3) SECOND DEFINITION. *Homologies between finite chains in B_ϵ are conveniently denoted by $\sim_\epsilon$ and known as ϵ homologies. Thus $C^p \sim_\epsilon C'^p$ means that $C^p - C'^p = \mathrm{F}C^{p+1}$ where all the chains are finite and of mesh less than ϵ.*

A p-dimensional V-cycle over a discrete group G is a countable collection of finite cycles, $\gamma^p \doteq \{\gamma_n^p\}$ such that

(a) *mesh $\gamma_n^p \to 0$;*

(b) *$\gamma_{n+1}^p \sim_{\epsilon_n} \gamma_n^p$, where $\{\epsilon_n\} \to 0$. The cycle γ^p bounds, or is a bounding cycle, written $\gamma^p \sim 0$, whenever*

(c) *$\gamma_n^p \sim_{\eta_n} 0$, where $\{\eta_n\} \to 0$.*

The group of the V-cycles $\mathfrak{Z}_v^p(K, G)$ is obtained by defining $\{\gamma_n^p\} \pm \{\gamma_n'^p\} = \{\gamma_n^p \pm \gamma_n'^p\}$, $0 = \{\gamma_n^p \mid \gamma_n^p = 0\}$, and is taken discrete. The bounding V-cycles form a subgroup $\mathfrak{F}_v^p(X, G)$. The homology group is defined as $\mathfrak{H}_v^p(X, G) = \mathfrak{Z}_v^p(X, G)/\mathfrak{F}_v^p(X, G)$.

Equivalent formulations, frequently useful, are: (a) as before and (b), (c) replaced by:

(b′) for every ϵ there is an m such that $n, n' \geqq m \to \gamma_n^p \sim_\epsilon \gamma_{n'}^p$;

(c′) for every ϵ there is an m such that $n \geqq m \to \gamma_n^p \sim_\epsilon 0$;

or else also by:

(b″) there is a sequence of finite chains $\{C_n^{p+1}\}$ such that

$$\mathrm{F}C_n^{p+1} = \gamma_{n+1}^p - \gamma_n^p, \text{ mesh } C_n^{p+1} \to 0;$$

(c″) there is a sequence of finite chains $\{C'^{p+1}\}$ such that

$$\mathrm{F}C_n'^{p+1} = \gamma_n^p, \text{ mesh } C_n'^{p+1} \to 0.$$

Under the definition just given we have:

(25.4) *If $\gamma^p = \{\gamma_n^p\}$ is a V-cycle and $\gamma'^p = \{\gamma_n'^p\}$ is a set of finite cycles such that $\gamma_n^p \sim_{\epsilon_n} \gamma_n'^p$ where $\{\epsilon_n\} \to 0$, then γ'^p is also a V-cycle and $\gamma'^p \sim \gamma^p$.*

From (25.4) follows readily (still under the second definition):

(25.5) *If $\gamma^p = \{\gamma_n^p\}$ is a V-cycle then $\gamma'^p = \{\gamma_{m_n}^p\}$ is likewise a V-cycle and $\gamma'^p \sim \gamma^p$.*

26. (26.1) We shall now compare the two definitions. For convenience let the cycles, classes and homology groups under the first definition be denoted by $'\gamma$, $'\Gamma$, $'\mathfrak{H}$ and those under the second by γ, Γ, $\mathfrak{H}$. It is not difficult to prove the following properties: (a) all the $'\gamma^p \in {}'\Gamma^p$ are also elements of a fixed Γ^p and $'\Gamma^p \to \Gamma^p$ defines a homomorphism in the algebraic sense $\tau: {}'\mathfrak{H}_v^p \to \mathfrak{H}_v^p$; (b) each $\gamma^p = \{\gamma_n^p\} \in \Gamma^p$ has a subsequence $\{\gamma_m^p\}$ which is a $'\gamma^p$ in a fixed $'\Gamma^p$ and $\Gamma^p \to {}'\Gamma^p$ defines a homomorphism in the algebraic sense $\theta: \mathfrak{H}_v^p \to {}'\mathfrak{H}_v^p$; (c) $\theta\tau = 1$, $\tau\theta = 1$. Hence τ is an isomorphism in the algebraic sense. Thus the two homology groups differ only in their topology. We shall refer to $\mathfrak{H}_v^p$ as the *homology group of the V-cycles*, and to $'\mathfrak{H}_v^p$ as the *homology group of the V-cycles with topology.* For practically all purposes $\mathfrak{H}_v^p$ is quite sufficient.

(26.2) *V-cycles around* Y. Let us suppose that X is metrized as indicated in (24.2) so that we have spheroids $\mathfrak{S}(|Y|, \epsilon)$ and let Y be a closed subcomplex of X. Let the notations be those of (25.3b″c″), except that we impose the following supplementary requirement: given any $\epsilon > 0$, then all but a finite number of γ_n^p, C_n^{p+1} are in $\mathfrak{S}(|Y|, \epsilon)$. This may also be expressed as: $\{\gamma_n^p\}$, $\{C_n^{p+1}\} \rightarrow Y$. Otherwise everything is as before. The new V-cycles are said to be *around* Y. The web interpretation is the same as before except that this time $B_\epsilon = \{x \mid \text{diam } x < \epsilon, x \subset \mathfrak{S}(|Y|, \epsilon)\}$.

(26.3) *V-cocycles.* They must be determined in terms of the general theory of H-nets. A V-cocycle is thus defined by a cocycle in some B_ϵ mod $(X - B_\epsilon)$ (where B_ϵ is as in 25.1), i.e., it is a cochain γ_p such that $\mathrm{F}\gamma_p$ has no elements of diameter less than ϵ. That is to say γ_p is any cochain such that $\inf \{\text{diam } x^{p+1} \mid x_{p+1} \text{ in } \mathrm{F}\gamma_p\} > 0$. The relation $\gamma_p \sim 0$ means that $\gamma_p = \mathrm{F}C_{p-1} + D_p$, where D_p is in some $X - B_\epsilon$, or again that $\inf \{\text{diam } x^p \mid x_p \text{ in } D_p\} > 0$. The groups $\mathfrak{C}_p$, $\mathfrak{Z}_p$, $\mathfrak{F}_p$ are discrete and the cohomology group $\mathfrak{H}_p(X, G) = \mathfrak{Z}_p/\mathfrak{F}_p$. Cocycles may also be introduced for the other types considered but we will not discuss them here.

27. (27.1) Definition. *Let* $X = \{x\}$, $X_1 = \{x_1\}$ *be metric complexes and* t *a set-transformation* $X \rightarrow X_1$. *Then* t *is said to be metric if* diam $tx \rightarrow 0$ *uniformly with* diam x. *An isomorphism* $t{:}X \rightarrow X_1$ *is said to be metric if both* t *and* t^{-1} *are metric. If such a* t *exists we shall say that* X *and* X_1 *are metrically isomorphic. Similarly a chain-mapping* $\tau{:}X \rightarrow X_1$ *is said to be metric if* diam $\|\tau x\| \rightarrow 0$ *uniformly with* diam x.

From the definition of metric isomorphism there follows readily:

(27.2) *A metric isomorphism* $\tau{:}X \rightarrow X_1$ *induces an isomorphism of the* $\mathfrak{Z}_v$, $\mathfrak{F}_v$, $\mathfrak{H}_v$ *groups of* X *with the corresponding groups of* X_1.

(27.3) Application. *Let* X *be made a metric complex in two ways with two distinct functions* diam x, diam$_1$ x. *Then, if each approaches* 0 *uniformly whenever the other approaches* 0, *the two metrics define the same groups* $\mathfrak{Z}_v$, $\mathfrak{F}_v$, $\mathfrak{H}_v$.

(27.4) *A metric chain-mapping* $\tau{:}X \rightarrow X_1$ *induces homomorphisms of the groups* $\mathfrak{Z}_v$, $\mathfrak{F}_v$, $\mathfrak{H}_v$ *of* X *into the same for* X_1.

By (IV, 9) τ induces homomorphisms of the groups of finite chains $\mathfrak{C}$, $\mathfrak{Z}$, $\mathfrak{F}$, $\mathfrak{H}$ of X into the same for X_1. Hence if $\{C_n\}$ are finite chains such that mesh $C_n \rightarrow 0$, then the chains $\{\tau C_n\}$ have the same property. Coupling this with (25.3) there is but a step to (27.4).

(27.5) Let τ_1, τ_2 be metric chain-mappings $X \rightarrow X_1$ which are chain-homotopic, the associated $\mathfrak{D}$ operator being of the f-type (every $\mathfrak{D}x$ finite). If mesh $\mathfrak{D}x \rightarrow 0$ uniformly with diam x then τ_1, τ_2 are said to be *metrically chain-homotopic*. When $\tau_1 = 1$ then τ_2 is said to be a *metric chain-deformation*

(27.6) If X_1 is a chain-retract [chain-deformation retract] of X under a chain-retraction [chain-deformation retraction] which is metric, then X_1 is naturally called a *metric chain-retract* [*metric chain-deformation retract*] of X.

The general properties of chain-homotopy, chain-deformation, chain-retraction, derived in (IV, 14–16) hold here also and with unimportant modifications in the proofs. We note in particular the following:

(27.7) *Metric chain-homotopy is an equivalence* (IV, 15.1).

(27.8) *Metric chain-homotopic mappings* τ_1, τ_2 $X \to X_1$ *induce the same homomorphisms in the V-homology groups. A metric chain-deformation does not alter the homology groups of the V-cycles* (IV, 15.2).

This means in particular that if γ^p is a V-cycle of X then $\tau_1\gamma^p \sim \tau_2\gamma^p$. Also that if τ_2 is a chain-deformation, then $\tau_1\gamma^p \sim \gamma^p$ in X_1.

(27.9) *If* X *is a metric chain-deformation retract of* X_1 *then their V-homology groups are the same* (IV, 15.12).

CHAPTER VII

HOMOLOGY THEORY OF TOPOLOGICAL SPACES

The nerves (in the sense of Alexandroff) of the totality of the open coverings of $\mathfrak{R}$ make up a net whose homology theory is a typical Čech theory. A similar theory arises out of the closed coverings. Let us refer to them as the $\mathfrak{U}$- and $\mathfrak{F}$-theories. The $\mathfrak{U}$-theory is chosen as our basic homology theory and will be discussed at length. For normal spaces both are the same. In addition to the preceding there are in existence:

for compacta: theories due to Vietoris, Alexandroff and Lefschetz;

for general spaces: a theory due to Kurosch, and another due to (or rather patterned after one due to) Alexander and Kolmogoroff.

It is not difficult to reduce the Alexandroff and Lefschetz theories to that of Vietoris (see notably [L, 332]). The latter will be examined more fully and reduced to the $\mathfrak{U}$-type. A complete discussion will also be given of the Kurosch and Alexander-Kolmogoroff types and they will be proved equivalent to the $\mathfrak{F}$-type. Thus at last a certain unity will have been brought into this group of questions.

Noteworthy additional topics treated in the chapter: duality in the sense of Alexander, multiple applications of webs, relations between homology and connectedness, generalization of Mayer's "union" theorem for complexes.

General references: Alexander [d, e], Alexandroff [a, f], Čech [a, b, c, d], Chevalley [a], Kline [a], Kolmogoroff [a, b, c], Kurosch [a], Lefschetz [L, VII], Steenrod [a], Vietoris [a].

§1. HOMOLOGY THEORY: FOUNDATIONS AND GENERAL PROPERTIES

1. The nets which are to carry the homology theory are composed of simplicial complexes, the *nerves*, related to coverings and introduced by Alexandroff. We first define and discuss the nerves and the projections which give rise to the nets.

(1.1) DEFINITIONS. *Let* $\mathfrak{A} = \{A_\alpha\}$ *be any finite aggregate of sets. A simplicial complex* Φ_a *is associated with* $\mathfrak{A}$ *in the following manner. The* A_α *are the vertices of* Φ_a. *A simplex* $\sigma = A_{\alpha_0} \cdots A_{\alpha_p}$ *is a simplex of* Φ_a *when and only when the intersection* $A_{\alpha_0} \cap \cdots \cap A_{\alpha_p}$ *is non-void. If* $\sigma' = A_{\alpha_0'} \cdots A_{\alpha_q'} < \sigma$ *then also*

$$A_{\alpha_0'} \cap \cdots \cap A_{\alpha_q'} \neq \emptyset,$$

and so $\sigma' \in \Phi_a$. *Thus* Φ_a *is indeed a simplicial complex. It is known as the nerve of the aggregate* $\mathfrak{A}$. *In a certain sense it may be said to constitute the "intersection pattern" of the sets of* Φ_a. *Notice that the dimension of* Φ_a *is the order of* $\mathfrak{A}$ (I, 14).

The intersection $\bigcap A_{\alpha_i}$ corresponding to the simplex σ of Φ_a is called the *kernel* of σ and denoted by $[\sigma]$. Thus we may write $\Phi_a = \{\sigma = A_{\alpha_0} \cdots A_{\alpha_p} \mid [\sigma] \neq \emptyset\}$.

REMARK. It is important to bear in mind that while the *simplexes σ of the nerve Φ_a are aggregates of sets, their kernels $[\sigma]$ are actual point sets.*

(1.2) AN EXAMPLE. Let $K = \{\sigma\}$ be a finite simplicial complex with vertices $\{P_\alpha\}$ and let $\mathfrak{A} = \{\text{St } P_\alpha\}$. Then $\text{St } P_{\alpha_0} \cap \ldots \cap \text{St } P_{\alpha_p} = \text{St } P_{\alpha_0} \ldots P_{\alpha_p} \neq \emptyset$ when and only when $P_{\alpha_0} \ldots P_{\alpha_p} \in K$. Therefore $P_\alpha \to \text{St } P_\alpha$ establishes an isomorphism $K \leftrightarrow \Phi_a$, so that we may identify the two complexes. Thus K will become the nerve of the aggregate of the stars of its vertices.

(1.3) Let $\mathfrak{B} = \{B_\beta\}$ be a second finite aggregate of sets and Φ_b its nerve. Suppose that $\mathfrak{B}$ refines $\mathfrak{A}$, so that every B_β is contained in some A_α. Select for each B_β a definite set $A_{\alpha(\beta)} \supset B_\beta$. Now $(\zeta = B_{\beta_0} \cdots B_{\beta_p} \in \Phi_b) \to ([\zeta] = \bigcap B_{\beta_i} \neq \emptyset) \to (\bigcap A_{\alpha(\beta_i)} \neq \emptyset) \to (A_{\alpha(\beta_0)} \cdots A_{\alpha(\beta_p)} \in \Phi_a)$. Therefore $B_\beta \to A_{\alpha(\beta)}$ defines a simplicial set-transformation and a simplicial chain-mapping t_a^b, $\pi_a^b : \Phi_b \to \Phi_a$. We call π_a^b a *projection by inclusion* or merely a *projection* $\Phi_b \to \Phi_a$. While these projections are not generally unique we have the following basic property:

(1.4) *Any two projections $\Phi_b \to \Phi_a$ are prismatically related in the strong sense* (IV, 16.2), *and hence by* (IV, 16.3) *they are prismatically chain-homotopic.*

Let π_a^b, $\pi_a'^b$ be two projections and t_a^b, $t_a'^b$ their simplicial carriers. Suppose $\zeta = B_{\beta_0} \cdots B_{\beta_p} \in \Phi_b$, and label temporarily the A_α so that $t_a^b B_{\beta_i} = A_{\alpha_i}$, $t_a'^b B_{\beta_i} = A_{\alpha_i'}$. Since $\bigcap B_{\beta_i} \neq \emptyset$ and both A_{α_i}, $A_{\alpha_i'} \supset B_{\beta_i}$, we also have $\bigcap(A_{\alpha_i} \cap A_{\alpha_i'}) \neq \emptyset$. Hence $A_{\alpha_0} \cdots A_{\alpha_p} A_{\alpha_0'} \cdots A_{\alpha_p'} \in \Phi_a$, and this proves (1.4).

Let $\mathfrak{D}_a^b$ be the homotopy operator for the chain-homotopy in (1.4). Referring to (IV, 16.3) we obtain:

(1.5) *$\mathfrak{D}_a^b \zeta$ is a chain of* $\text{Cl}(t_a^b \zeta)(t_a'^b \zeta)$.

From (IV, 15.2) and (1.4) we also have:

(1.6) *All the projections $\Phi_b \to \Phi_a$ are homologous.*

The collection of all the sets of $\mathfrak{A}$ containing any one of the sets $B_{\beta_0}, \cdots, B_{\beta_p}$ which are the vertices of $\zeta \in \Phi_b$ determine a simplex σ of Φ_a. Evidently $\zeta \to \text{Cl } \sigma$ defines *a closed carrier T_a^b for all the projections π_a^b*, and T_a^b is a generalized simplicial set-transformation. Since $t_a^b \zeta \subset T_a^b \zeta$, we have from (1.5):

$$\mathfrak{D}_a^b \zeta \subset T_a^b \zeta. \tag{1.7}$$

(1.8) Let E be any set and $\sigma \in \Phi_a$. If the kernel $[\sigma]$ meets E and $\sigma \succ \sigma'$ then $[\sigma'] \supset [\sigma]$ and hence $[\sigma']$ also meets E. Hence $\Phi_a(E) = \{\sigma \mid E \cap [\sigma] \neq \emptyset\}$ is a closed subcomplex of Φ_a. There is an analogous closed subcomplex $\Phi_b(E)$ of Φ_b.

(1.9) *All the operations T_a^b, t_a^b, π_a^b, $\mathfrak{D}_a^b$ map $\Phi_b(E)$ into $\Phi_a(E)$.*

Let $\zeta \in \Phi_b(E)$. Since $t_a^b \zeta$, $\pi_a^b \zeta$, $\mathfrak{D}_a^b \zeta \subset T_a^b \zeta$, it is only necessary to prove (1.9) for T_a^b. Since every vertex of $T_a^b \zeta$ is a set $\supset [\zeta]$, each simplex σ of $T_a^b \zeta$ has a kernel $[\sigma] \supset [\zeta]$ and so meeting E. Thus $\sigma \in \Phi_a(E)$ and hence $T_a^b \zeta \subset \Phi_a(E)$.

(1.10) Let σ^p be a simplex in any nerve, say Φ_a. It has then a dual σ_p and we define the *kernel* $[\sigma_p]$ of σ_p as identical with $[\sigma^p]$. If $C^p = g^i\sigma_i^p$, $C_p = g_i\sigma_p^i$ are a chain and a cochain of a nerve, say Φ_a, then we say that a set E is a *carrier* of C^p if $g^i \neq 0 \rightarrow [\sigma_i^p] \cap E \neq \emptyset$, or if $C^p = 0$, and that E is a *carrier* of C_p if $g_i \neq 0 \rightarrow [\sigma_p^i] \subset E$, or if $C_p = 0$. Notice that $\zeta^p \in \Phi_b \rightarrow [\pi_a^b\zeta^p] \supset [\zeta^p]$. Hence:

(1.11) *If C^p is a chain of Φ_b then a carrier of C^p is also one of $\pi_a^bC^p$.*

Consider now the dual $\pi_b^{*a}:\Phi_a^* \rightarrow \Phi_b^*$ of π_a^b. If $\sigma^p \in \Phi_a$, by (1.11) it can only occur in a chain $\pi_a^b\zeta^p$ such that $[\sigma^p] \supset [\zeta^p]$. Hence $\pi_b^{*a}\sigma_p$ consists solely of elements ζ_p such that the inclusion just written holds. Therefore

(1.12) *If C_p is a cochain of Φ_a then a carrier of C_p is also one of $\pi_b^{*a}C_p$.*

(1.13) *Extension to infinite aggregates.* When $\mathfrak{A}$, $\mathfrak{B}$ are infinite the only modifications required are that (1.6) need not hold as regards all the groups, and that $T_a^b\zeta$ is merely a closed subcomplex of Φ_a. Since a more general form of (1.6) is not required in the sequel we will not discuss it, and as for the modification in $T_a^b\zeta$ it is wholly immaterial. A complementary property solely required for infinite aggregates, will now be considered.

The situation remaining as before, we observe that since $\mathfrak{B}$ refines $\mathfrak{A}$ so does $\mathfrak{C} = \mathfrak{A} \vee \mathfrak{B}$. If $\Phi_c =$ nerve $\mathfrak{C}$ we have therefore $T_a^c, \cdots$ with their obvious meaning. A simplex of Φ_c is a join $\sigma\zeta$ where $\sigma = A_{\alpha_0} \cdots A_{\alpha_p} \in \Phi_a$, $\zeta = B_{\beta_0} \cdots B_{\beta_q} \in \Phi_b$, σ has no vertices in Φ_b and $\bigcap A_{\alpha_i} \cap \bigcap B_{\beta_j} \neq \emptyset$. Since $t_a^bB_\beta \supset B_\beta$, we also have $\bigcap A_{\alpha_i} \cap \bigcap(t_a^bB_{\beta_j}) \neq \emptyset$, and hence $\sigma(t_a^b\zeta) \in \Phi_a$. It follows that $\sigma\zeta \rightarrow \sigma(t_a^b\zeta)$ is a projection π_a^c; it is this particular projection which is meant henceforth by π_a^c. We shall also require the generalized set-transformation $T_c^c:\Phi_c \rightarrow \Phi_c$ defined by $\sigma\zeta \rightarrow \mathrm{Cl}\ \sigma\zeta(T_a^b\zeta)$. We prove:

(1.14) *π_a^c is a chain-deformation of Φ_c into its subcomplex Φ_a which is contiguous to the identity in T_c^c. If $\pi_a^c \mid \Phi_a = 1$ then π_a^c is evidently a chain-deformation retraction. This holds notably if Φ_a, Φ_b are disjoint, i.e., if the aggregates $\mathfrak{A}$, $\mathfrak{B}$ have no common sets.*

By (IV, 16.3) π_a^c is a chain-deformation over a simplicial complex Ψ which is the union of the closures of the simplexes such as $\eta = \sigma B_{\alpha_0} \cdots B_{\alpha_r}(t_a^bB_{\alpha_r}) \cdots (t_a^bB_{\alpha_q})$ corresponding to the simplex $\sigma\zeta \in \Phi_c$ considered above. Once more $[\sigma\zeta] \neq \emptyset$ and $t_a^bB_\beta \supset B_\beta$ imply $[\eta] \neq \emptyset$, and hence $\eta \in \Phi_c$. Thus $\Psi \subset \Phi_c$. Since the converse is obvious $\Psi = \Phi_c$, and so π_a^c has the asserted property.

If Φ_a, Φ_b are disjoint no B_β is an A_α, and then every simplex of Φ_a is a σ, hence $\pi_a^c \mid \Phi_a = 1$. Thus π_a^c is the asserted retraction.

2. Consider now a collection $\mathfrak{A} = \{\mathfrak{A}_\lambda\}$, where $\mathfrak{A}_\lambda = \{A_{\lambda\alpha}\}$ is a finite aggregate of sets. Suppose the collection directed by $\lambda > \mu \leftrightarrow (\mathfrak{A}_\lambda$ refines $\mathfrak{A}_\mu)$. We will then say that $\mathfrak{A}$ is a *directed* collection. Let $\Phi_\lambda =$ nerve $\mathfrak{A}_\lambda$, and for $\lambda > \mu$ let π_μ^λ be a projection by inclusion $\Phi_\lambda \rightarrow \Phi_\mu$. Consider now $\Phi = \{\Phi_\lambda ; \pi_\mu^\lambda\}$. Of the net properties Ni of (VI, 2.1), N3 is fulfilled by (1.6). If $\lambda > \mu > \nu$ and $\pi_\mu^\lambda A_{\lambda\alpha} = A_{\mu\beta}$, $\pi_\nu^\mu A_{\mu\beta} = A_{\nu\gamma}$, then $A_{\lambda\alpha} \subset A_{\mu\beta} \subset A_{\nu\gamma}$. Hence $A_{\lambda\alpha} \rightarrow A_{\nu\gamma}$ defines a π_ν^λ or $\pi_\nu^\mu\pi_\mu^\lambda$ is a π_ν^λ. Thus N2 of (VI, 2.1) holds. As for N1 it merely asserts the existence of the projections and so it holds also. Therefore

(2.1) *The nerves* $\{\Phi_\lambda\}$ *of a directed collection together with their projections by inclusion form a simplicial net* Φ.

(2.2) Once we have the net Φ we may bring to bear the theory of nets. If the results are to have topological character all that is necessary, given a topological space $\mathfrak{R}$, is to select topological directed collections. Two obvious collections are: $\mathfrak{U}$, $\mathfrak{F}$ the collections of the finite open and finite closed coverings of $\mathfrak{R}$. It is indeed clear that if $\mathfrak{U}_1$, $\mathfrak{U}_2$ are finite open coverings so is $\mathfrak{U}_1 \wedge \mathfrak{U}_2$ and since it is $\succ \mathfrak{U}_1, \mathfrak{U}_2$, $\mathfrak{U}$ is a directed collection. Similarly for $\mathfrak{F}$.

Generally speaking perfect symmetry between $\mathfrak{U}$ and $\mathfrak{F}$ is not to be expected, and this on two grounds: (a) they are not dual to one another since the dual of "covering" is "a collection of sets with an empty intersection;" (b) in most "convenient" spaces points are closed sets, and this causes considerable dissymetry throughout. We introduce therefore the

(2.3) CONVENTION. *Unless otherwise stated all the homology concepts to be considered shall refer to those obtained by means of the finite open coverings.*

Thus instead of "$\mathfrak{U}$-cycle, $\cdots$" we shall merely say "cycle, $\cdots$."

In point of fact while the $\mathfrak{U}$-theory has been selected as the basic theory, it will turn out (8), that when $\mathfrak{R}$ is normal, and this covers a very large territory, the $\mathfrak{U}$- and $\mathfrak{F}$-theories coincide.

3. Irreducible coverings. Frequently, especially when the space $\mathfrak{R}$ is normal, we shall find it convenient to replace arbitrary coverings by more special types. Two particularly convenient types will now be considered. A finite open covering $\mathfrak{U} = \{U_1, \cdots, U_r\}$ with nerve Φ is said to be

irreducible whenever no open refinement of $\mathfrak{U}$ has a nerve isomorphic with a proper subcomplex of Φ;

strongly irreducible whenever it is irreducible and $U_i \to \overline{U}_i$ defines a similitude $\mathfrak{U} \to \overline{\mathfrak{U}}$.

When $\mathfrak{U}$ is strongly irreducible $U_i \to \overline{U}_i$ defines an isomorphism $\Phi \to$ nerve $\overline{\mathfrak{U}}$.

(3.1) *Every finite open covering* $\mathfrak{U}$ *has a finite irreducible open refinement. Or equivalently: the irreducible coverings are cofinal in the family of all the open coverings.*

Suppose $\mathfrak{U} \prec \mathfrak{U}_1 \prec \mathfrak{U}_2 \prec \cdots$, where the $\mathfrak{U}_i$ are reducible and such that their nerves form a sequence of complexes $\Phi = \Phi_0, \Phi_1, \cdots$, where $\Phi_{i+1} \cong$ a proper subcomplex of Φ_i. Since Φ is finite the sequence must stop and if Φ_r is its last term $\mathfrak{U}_r$ is an irreducible refinement of $\mathfrak{U}$.

(3.2) *If* $\mathfrak{U}$ *is irreducible then every* U_i *contains a point which is on no other set of the covering.*

For, if U_i contains no such point $\{U_h \mid h \neq i\}$ is a refinement of $\mathfrak{U}$ whose nerve is a proper subcomplex of nerve $\mathfrak{U}$.

(3.3) *If* $\mathfrak{R}$ *is normal, every finite open covering* $\mathfrak{U}$ *has a strongly irreducible open refinement. Hence the strongly irreducible finite open coverings are cofinal in the family of all the finite open coverings.*

Let $\mathfrak{U}_1 = \{U_{1i}\}$ be a finite irreducible refinement of $\mathfrak{U}$ and shrink it to $\mathfrak{V} = \{V_i\}$ (I, 33.4). Since $\mathfrak{V} \succ \mathfrak{U}_1$, $\mathfrak{V}$ is irreducible and so $U_{1i} \to V_i$ defines a similitude $\mathfrak{U}_1 \to \mathfrak{V}$. Since $V_i \subset \bar{V}_i \subset U_i$, $V_i \to \bar{V}_i$ defines a similitude $\mathfrak{V} \to \bar{\mathfrak{V}}$. Hence $\mathfrak{V}$ is a strongly irreducible refinement of $\mathfrak{U}$.

(3.4) *Every compactum $\mathfrak{R}$ has strongly irreducible finite open coverings whose mesh is arbitrarily small.*

For $\mathfrak{R}$ has a finite open ϵ covering $\mathfrak{U}$, whatever ϵ, and so (3.4) is a consequence of (3.3).

4. **The net of the finite open coverings and its subnets.** We set then $\mathfrak{U} = \{\mathfrak{U}_\lambda\}$, $\Phi_\lambda =$ nerve $\mathfrak{U}_\lambda$, and define $\lambda > \mu$ by $\mathfrak{U}_\lambda > \mathfrak{U}_\mu$. The resulting projections by inclusion π_μ^λ make $\Phi = \{\Phi_\lambda ; \pi_\mu^\lambda\}$ a simplicial net. From (VI, 6.5, 8.3) follows then the

(4.1) DUALITY THEOREM FOR TOPOLOGICAL SPACES. *The cycles and cocycles of a topological space are dual categories with intersections.*

(4.2) *Augmented homology theory.* It is obtained by augmenting the net Φ (VI, 10). We say then that $\mathfrak{R}$ has been *augmented.* This operation is useful in certain questions. Thus wherever it is convenient to augment, say a finite simplicial complex K, it is equally convenient as regards the groups of the polyhedron $| K_e |$, K_e an Euclidean realization of K, to augment $| K_e |$ in the sense just considered.

(4.3) *Dissections and related subnets.* By a *dissection* of $\mathfrak{R}$ we shall refer to a decomposition (U, F) of $\mathfrak{R}$ into two disjoint sets U open and F closed: $\mathfrak{R} = U \cup F$, $U \cap F = \emptyset$. If we set $\Phi_{1\lambda} = \{\sigma \mid \sigma \in \Phi_\lambda, [\sigma] \cap F \neq \emptyset\}$, $\pi_{1\mu}^{\lambda} = \pi_\mu^\lambda \mid \Phi_{1\lambda}$, then $\Phi_{1\lambda}$ is a closed subcomplex of Φ_λ (1.8) and by (1.9; VI, 12.1, 12.2) $\Phi_1(F) = \{\Phi_{1\lambda} ; \pi_{1\mu}^\lambda\}$, is a closed subnet of Φ with intersections induced by those of Φ. Notice incidentally that Φ itself is merely $\Phi_1(\mathfrak{R})$.

Let $\Phi_{0\lambda} = \Phi_\lambda - \Phi_{1\lambda}$, $\pi_{0\mu}^\lambda =$ the projection $\Phi_{0\lambda} \to \Phi_{0\mu}$ induced by π_μ^λ. Then $\Phi_0(U) = \{\Phi_{0\lambda} ; \pi_{0\mu}^\lambda\} = \Phi - \Phi_1(F) =$ the open subnet of Φ complementary to $\Phi_1(F)$. In $\Phi_0(U)$ there are intersections induced by those of Φ. To sum up then:

(4.4) *To a dissection (U, F) of $\mathfrak{R}$ there corresponds a unique dissection $(\Phi_0(U), \Phi_1(F))$ of Φ, where in the subnets there are interesections induced by those of Φ.*

The definition of the subnets yields

$$F' \subset F \to \Phi_1(F') \subset \Phi_1(F), \tag{4.5}$$

$$U' \subset U \to \Phi_0(U') \subset \Phi_0(U). \tag{4.6}$$

It is also natural to refer to the cycles of $\Phi_1(F)$, or of Φ mod $\Phi_1(F)$ as the cycles of F or of $\mathfrak{R}$ mod F, and similarly with the cocycles and $\Phi_0(U)$.

Referring to (VI, 12.4) we have:

(4.7) *The cycles of $\mathfrak{R}$* mod *F and the cocycles of $\mathfrak{R} - F = U$ are dual categories with intersections.*

(4.8) *Cyclic and acyclic properties.* The definitions are the same as for complexes (III, 21) and refer to Φ and its groups. Thus $\mathfrak{R}$ is acyclic whenever all its homology groups are zero, $\cdots$.

5. General properties of the homology groups.

(5.1) *When* $\mathfrak{U}$ *is replaced by a cofinal family the homology theory is unchanged* (VI, 9.8).

(5.2) *If* $n = \dim \mathfrak{R}$ *is finite then all the groups for dimension greater than* n *are zero.*

For the refinements of the $\mathfrak{U}_\lambda$ whose order does not exceed n form a cofinal family giving rise to a net whose complexes are at most n-dimensional.

(5.3) *The universal theorem for fields holds. Hence the only Betti numbers that need to be considered are the Betti numbers over the rationals or* mod π, *a prime.*

(5.4) *The homology theory of the closed subnet* $\Phi_1(F)$ *of* Φ *associated with the closed set* F (4.3) *is the homology theory of* F *itself.*

This implies for instance that the class intersections and their rings may be obtained for F by means of $\Phi_1(F)$.

A noteworthy consequence of (5.4), indeed little more than an equivalent formulation is:

(5.5) *Let* $\mathfrak{R}$ *be imbedded topologically as a closed set in* $\mathfrak{S}$, *and let* $\{\mathfrak{W}_\mu\}$, $\mathfrak{W}_\mu = \{W_{\mu i}\}$ *be the finite open coverings of* $\mathfrak{S}$. *Then to obtain the homology theory of* $\mathfrak{R}$ *one may replace its finite open coverings by the coverings* $\{\mathfrak{R} \wedge \mathfrak{W}_\mu\}$, $\mathfrak{R} \wedge \mathfrak{W}_\mu = \{\mathfrak{R} \cap W_{\mu i}\}$.

Indeed this is clearly (5.4) with $\mathfrak{R}$ replaced by $\mathfrak{S}$, and F by $\mathfrak{R}$.

The proof of (5.4) will rest mainly upon the following property:

(5.6) *If in any net* $\Phi = \{\Phi_\lambda ; \pi_\mu^\lambda\}$ *certain* π_μ^λ, $\lambda > \mu$, *are isomorphisms and for each such pair one introduces new ordering relations and projections*: $\mu > \lambda$, $\pi_\lambda^\mu = (\pi_\mu^\lambda)^{-1}$, *together with all those that follow by transitivity* (N2 *of* VI, 2.1), *then the resulting net* Ψ *has the same homology theory as* Φ.

For convenience we denote by ω_μ^λ the projections in Ψ. Let $\gamma^p = \{\gamma_\lambda^p\} \in \mathfrak{Z}^p(\Phi, G)$. A projection ω_μ^λ is a finite product $\omega_{\lambda_r}^{\lambda_{r-1}} \cdots \omega_{\lambda_1}^{\lambda_0}$ $(\lambda_0 = \lambda, \lambda_r = \mu)$, where any factor is an isomorphism or else a projection of Φ. Thus $\omega_{\lambda_{i+1}}^{\lambda_i}\gamma_{\lambda_i}^p \sim \gamma_{\lambda_{i+1}}^p$, and hence $\omega_\mu^\lambda\gamma_\lambda^p \sim \gamma_\mu^p$. Therefore the γ_λ^p are likewise the coordinates of a $\bar{\gamma}^p \in \mathfrak{Z}^p(\Psi, G)$. The converse is obvious. It follows that $\gamma^p \leftrightarrow \bar{\gamma}^p$ defines a simultaneous isomorphism τ in the algebraic sense of $\mathfrak{Z}^p(\Phi, G)$ with $\mathfrak{Z}^p(\Psi, G)$. The subbase $\{V_\lambda\}$ of (VI, 3.3) for the first group is immediately seen to be one also for the second. Hence τ is an isomorphism. Since $\tau\mathfrak{F}^p(\Phi, G) = \mathfrak{F}^p(\Psi, G)$, τ induces a simultaneous isomorphism of the corresponding homology groups.

Similarly the representatives of a cocycle γ_p of Φ are shown to be those of a cocycle $\bar{\gamma}_p$ of Ψ, and $\gamma_p \leftrightarrow \bar{\gamma}_p$ defines an isomorphism τ^* of the corresponding groups $\mathfrak{Z}$, $\mathfrak{F}$ which induces one of the corresponding cohomology groups. Minor modifications of the same argument yield the proof that τ, τ^* together induce (in an obvious sense) an isomorphism between the intersection classes in Φ, Ψ and also between the corresponding intersection rings. This proves (5.6).

Proof of (5.4). We return to the notations of (4). Let $\mathfrak{U}_\lambda = \{U_{\lambda i}\}$. If $F \cap U_{\lambda i} \neq \emptyset$ we identify the vertex $F \cap U_{\lambda i}$ of nerve $F \wedge \mathfrak{U}_\lambda$ with the vertex $U_{\lambda i}$ of $\Phi_{1\lambda}$. There results an isomorphism nerve $F \wedge \mathfrak{U}_\lambda \leftrightarrow \Phi_{1\lambda}$. We will identify the two isomorphic complexes, so that henceforth $\Phi_{1\lambda} =$ nerve $F \wedge \mathfrak{U}_\lambda$. Under the circumstances $\pi_{1\mu}^{\lambda} = \pi_{\mu}^{\lambda} \mid \Phi_{1\lambda}$, $\lambda > \mu$, represents a projection by inclusion nerve $F \wedge \mathfrak{U}_{\lambda} \to$ nerve $F \wedge \mathfrak{U}_{\mu}$.

It may very well happen that $F \wedge \mathfrak{U}_{\lambda} > F \wedge \mathfrak{U}_{\mu}$ and yet $\mathfrak{U}_{\lambda} \not> \mathfrak{U}_{\mu}$. If so, no projection by inclusion nerve $F \wedge \mathfrak{U}_{\lambda} \to$ nerve $F \wedge \mathfrak{U}_{\mu}$ will occur in $\Phi_1(F)$. Taking advantage of (5.6) we will introduce such a projection without modifying the homology theory of the subnet.

Suppose first that $F \wedge \mathfrak{U}_{\lambda} = F \wedge \mathfrak{U}_{\mu}$, and let $\mathfrak{U}_{\nu} = \mathfrak{U}_{\lambda} \wedge \mathfrak{U}_{\mu}$, so that $\mathfrak{U}_{\nu} > \mathfrak{U}_{\lambda}, \mathfrak{U}_{\mu}$. We may choose as π_{λ}^{ν} a projection by inclusion $\Phi_{\nu} \to \Phi_{\lambda}$ which is the identity on the common vertices, and then $\pi_{1\lambda}^{\nu} = 1$. By (5.6) there may be introduced in $\Phi_1(F)$ the new ordering and projection: $\lambda > \nu$, $\pi_{1\nu}^{\lambda} = 1$, and those that follow by transitivity. Since there is a π_{μ}^{ν} with $\pi_{1\mu}^{\nu} = 1$, we have with the ordering prevailing in the new $\Phi_1(F)$: $\lambda > \mu$, $\pi_{1\mu}^{\lambda} = 1$. If this is done for all similar pairs the new net, still written for convenience $\Phi_1(F)$, will have the same homology theory as the initial $\Phi_1(F)$. The situation is now such that $\mathfrak{U}_{\lambda}$ may be replaced by the covering consisting of U and of the $U_{\lambda i}$ meeting F. This being done throughout, suppose then $F \wedge \mathfrak{U}_{\lambda} > F \wedge \mathfrak{U}_{\mu}$ and let $\mathfrak{U}_{\mu} = \{U_{\mu j}\}$. For each $U_{\lambda i}$ meeting F select a $U_{\mu j} \supset F \cap U_{\lambda i}$ and replace $U_{\lambda i}$ by $U_{\lambda i} \cap U_{\mu j}$. There results a $\mathfrak{U}_{\rho} > \mathfrak{U}_{\mu}$ and such that $F \wedge \mathfrak{U}_{\lambda} = F \wedge \mathfrak{U}_{\rho}$. Hence we already have the projection $\pi_{1\rho}^{\lambda} = 1$. Now there is a projection by inclusion $\pi_{1\mu}^{\rho}:\Phi_{1\rho} \to \Phi_{1\mu}$, and hence $\pi_{1\mu}^{\lambda} = \pi_{1\mu}^{\rho}\pi_{1\rho}^{\lambda}$ is likewise a projection by inclusion $\Phi_{1\lambda} \to \Phi_{1\mu}$. In other words we are now at liberty to assume that the projections in $\Phi_1(F)$ include a projection by inclusion $\pi_{1\mu}^{\lambda}$ whenever $F \wedge \mathfrak{U}_{\lambda} > F \wedge \mathfrak{U}_{\mu}$. In view of N3 of (VI, 2.1), as regards the homology theory of $\Phi_1(F)$ we may assume that the $\pi_{1\mu}^{\lambda}$ include *all* the projections by inclusion of the nerves of the $F \wedge \mathfrak{U}_{\lambda}$.

It is now clear that the ultimate $\Phi_1(F)$ is merely obtained from the net Ψ of the finite open coverings of F by repeating certain nerves and introducing corresponding isomorphisms in accordance with (5.6). Hence Ψ and the ultimate $\Phi_1(F)$, and consequently Ψ and the initial $\Phi_1(F)$, have the same homology theory and this is precisely (5.4).

(5.7) In view of (5.4) and for reasons of expediency we shall naturally say "cycles of F, of $\mathfrak{R}$ mod F," instead of "cycles of $\Phi_1(F)$, of Φ mod $\Phi_1(F)$," likewise "cocycles of U" instead of "cocycles of $\Phi_0(U)$." We notice explicitly:

(5.8) *The cycles of* $\mathfrak{R}$·mod F, *F closed in* $\mathfrak{R}$, *and the cocycles of* $\mathfrak{R} - F$ *are dual categories with intersections.*

Remark. One may be tempted to expect here an analogue of (5.4) for U and $\Phi_0(U)$. However, for the same reasons as in (2.2) complete duality is not to be looked for in the present instance.

(5.9) *Carriers.* Let now $\gamma^p = \{\gamma_{\lambda}^p\}$ be a cycle of $\mathfrak{R}$. We say that γ^p has for *carrier* the closed set F whenever there is a $\{\mu\}$ cofinal in $\{\lambda\}$ such that every γ_{μ}^p has the carrier F, or equivalently such that $\{\gamma_{\mu}^p\}$ are the coordinates of a cycle of $\Phi_1(F)$.

Passing to the cocycles, $\gamma_p = \{\gamma_p^{\lambda}\}$ has for *carrier* the open set U whenever one of its coordinates γ_p^{μ} has the carrier U or equivalently whenever γ_p is a cocycle of $\Phi_0(U)$.

Carriers for the relative cycles and cocycles are defined in the same way.

Hereafter a cycle [cocycle] will always be assumed given with a definite closed [open] carrier. If the carrier remains unspecified then it will mean that it is the space $\mathfrak{R}$ itself.

(5.10) A variety of new interpretations may be given to the projections and injections such as those of (VI, 13) corresponding to the relations between certain nets and subnets. It is chiefly a matter of adjusting the terminology. Thus if F is a closed set then there is an injection $\tau:\Phi_1(F) \to \Phi$ and a projection $\theta:\Phi \to \Phi_0(\mathfrak{R} - F)$. Under our interpretations they are operations on the groups of cycles of F and $\mathfrak{R}$ to those of $\mathfrak{R}$ and $\mathfrak{R}$ mod F. For this reason we will refer to τ as the *injection* $F \to \mathfrak{R}$ and to θ as the *projection* $\mathfrak{R} \to \mathfrak{R} - F$, or also as the *reduction* mod F of the cycles of $\mathfrak{R}$. Analogous interpretations hold for the dual operations τ^*, θ^* of (VI, 13), but we will not require special terms for them.

Suppose now $F' \subset F$, where both are closed sets in $\mathfrak{R}$. Referring to (VI, 12.5) $\Phi_1(F')$ is a closed subnet of $\Phi_1(F)$ and so there arises an injection $\tau:\Phi_1(F') \to \Phi_1(F)$, and a projection $\theta:\Phi_0(\mathfrak{R} - F') \to \Phi_0(\mathfrak{R} - F)$. They are referred to as: the injection $F' \to F$ for τ, and the projection $\mathfrak{R} - F' \to \mathfrak{R} - F$ or reduction mod F of the cycles mod F' for θ. Here again no specific terms are required for the dual operations τ^*, θ^*.

It is hardly necessary to observe that the theorem of (VI, 13.9) holds for the operations $\tau, \cdots$, as here understood, and with the obvious modifications.

Notice the obvious parallel between the injection $F \to \mathfrak{R}$ and the operation of *topological imbedding as a closed set*, likewise between the projection $\mathfrak{R} \to \mathfrak{R} - F$ and the operation of *neglecting the part of a set in F*. That this analogy is not at all accidental is clearly shown when $\mathfrak{R}$ and F are polyhedra (see VIII, 11, 12).

(5.11) *Let T be a mapping $\mathfrak{R} \to \mathfrak{S}$ such that $R = T\mathfrak{R}$ is closed in $\mathfrak{S}$. Then:*

(a) *T induces homomorphisms $\tau:\mathfrak{H}^p(\mathfrak{R}, G) \to \mathfrak{H}^p(\mathfrak{S}, G)$ and $\tau^*:\mathfrak{H}_p(\mathfrak{S}, H) \to \mathfrak{H}_p(\mathfrak{R}, H)$, ($H$ discrete);*

(b) *if Γ, Γ' denote the homology and cohomology classes in $\mathfrak{R}$, $\mathfrak{S}$ there subsist the relations of permanence* (VI, 9.4, 9.5, 9.6) *for the Kronecker index and the intersections;*

(c) *the intersection rings of $\mathfrak{S}$ are mapped homomorphically by τ^* into the corresponding rings of $\mathfrak{R}$;*

(d) *τ^* is uniquely determined by τ, and is called the "dual" of τ.*

Let the notations of (4) continue to hold for $\mathfrak{R}$; and let $\{\mathfrak{B}_\mu\}$ be the finite open coverings of $\mathfrak{S}$ and Ψ their net. Corresponding to R, or rather to the dissection $(\mathfrak{S} - R, R)$ of $\mathfrak{S}$ we define the closed subnet Ψ_1 of Ψ in accordance with (4.4). Since Ψ_1 is a subnet of Ψ we have an injection $\theta_1:\Psi_1 \to \Psi$ and a dual projection $\theta_1^*:\Psi^* \to \Psi_1^*$ (VI, 13). Since T is continuous $T^{-1}R \wedge \mathfrak{B}_\mu = \{T^{-1}(R \cap V_{\mu i})\}$, $\mathfrak{B}_\mu = \{V_{\mu i}\}$, is a finite open covering of $\mathfrak{R}$ whose net is a part $'\Phi$ of Φ and $\cong \Psi_1$. As a consequence we have the analogues $\theta_2:\Phi \to {}'\Phi$ and $\theta_2^*:{}'\Phi^* \to \Phi^*$ of τ, τ^* of (VI, 9.1). Let ξ be the isomorphism $'\Phi \to \Psi_1$, and ξ^* its "dual" the isomorphism $\Psi_1^* \to {}'\Phi^*$. If we set $\tau = \theta_1\xi\theta_2$, $\tau^* = \theta_2^*\xi^*\theta_1^*$ we conclude from (VI, 9, 13) that τ, τ^* have all the properties of (5.11).

(5.12) A more intuitive description of the transformations of the cycles may also be given in the following way. Let $\{\mathfrak{U}_\nu\}$ be the collection of all the finite open refinements of the coverings $\{T^{-1}R \wedge \mathfrak{B}_\mu\}$ of $\mathfrak{R}$, so that in particular $\{\nu\}$ is cofinal in $\{\lambda\}$. Suppose $\mathfrak{U}_\nu > T^{-1}R \wedge \mathfrak{B}_\mu$. If $\mathfrak{U}_\nu = \{U_{\nu j}\}$ there may be chosen for each $U_{\nu j}$ a set $V_{\mu i}$ such that $U_{\nu j} \subset T^{-1}(R \cap V_{\mu i})$ and $U_{\nu j} \to V_{\mu i}$ defines a simplicial projection $p^\nu_\mu : \Phi_\nu \to \Psi_{1\mu}$. If $T^{-1}R \wedge \mathfrak{B}_\mu = \mathfrak{U}_{\nu'}$ then $p^\nu_\mu = \xi\pi^\nu_{\nu'}$. From this follows readily that if $\gamma^p = \{\gamma^p_\lambda\}$ is a cycle of $\mathfrak{R}$ and Γ^p its class and if we choose for each μ a cycle $\delta^p_\mu = p^\nu_\mu \gamma^p_\nu$, then $\delta^p = \{\delta^p_\mu\}$ is a cycle of $\mathfrak{S}$ whose class is $\tau\Gamma^p$.

6. **Homology theory and compactness.**

(6.1) Let us return to the situation and notations of (I, 26, 27, 28). Then if $\{\mathfrak{U}_\lambda\}$ are the finite open coverings of $\mathfrak{R}$, $\{\Omega(\mathfrak{U}_\lambda)\}$ are finite open coverings of the compacted space $\mathfrak{S}$ with the following properties: (a) the nets of the nerves of the $\mathfrak{U}_\lambda$ and of the $\Omega(\mathfrak{U}_\lambda)$ are isomorphic (I, 27.8, 27.10) i.e., they differ only in the labels; (b) $\{\Omega(\mathfrak{U}_\lambda)\}$ is cofinal in the family of all the finite open coverings of $\mathfrak{S}$ (I, 27.11); (c) if T is the topological imbedding $\mathfrak{R} \to \mathfrak{S}$ and F a closed set in $\mathfrak{R}$, then the same properties hold regarding the $F \wedge \mathfrak{U}_\lambda$ and $(TF) \wedge \Omega(\mathfrak{U}_\lambda)$. We have therefore:

(6.2) *The topological space* $\mathfrak{R}$ *and the compacted space* $\mathfrak{S}$ *of* (I, 26.1) *have the same homology theory* (Čech [g], Wallman [a]).

(6.3) *Under the same conditions the dissections* $(\mathfrak{R} - F, F)$ *of* $\mathfrak{R}$ *and* $(\mathfrak{S} - TF, TF)$ *of* $\mathfrak{S}$ *likewise have the same homology theory.*

(6.4) In the light of the preceding results one may say that the Čech homology theory by finite open coverings is essentially a homology theory of compact spaces. This has certain noteworthy consequences. Thus take so simple a space as the real line L (an open one-cell) and let it be compacted to S. The space S turns out to be the space of all real single-valued functions. Now it was shown by Dowker [a] that $\mathfrak{H}^1(L) = \mathfrak{H}^1(S)$ (the integral homology group) is isomorphic with the additive group of all real single-valued functions reduced modulo the bounded functions. This group is certainly significant for the class of real continuous functions, but scarcely as a topological character of the line.

7. (7.1) Theorem. *Let* $\mathfrak{R}$ *be compact and* $\mathfrak{S}$ *a Hausdorff space. Then any mapping* $T:\mathfrak{R} \to \mathfrak{S}$ *induces a homomorphism* τ *of the homology groups of* $\mathfrak{R}$ *into the corresponding groups of* $\mathfrak{S}$, *and a dual homomorphism* τ^* *of the cohomology groups of* $\mathfrak{S}$ *into those of* $\mathfrak{R}$. *Moreover* τ, τ^* *are the same for all homotopic mappings* $\mathfrak{R} \to \mathfrak{S}$, *i.e., they depend merely upon the homotopy class of* T.

Since $\mathfrak{R}$ is compact and $\mathfrak{S}$ a Hausdorff space, $T\mathfrak{R}$ is closed in $\mathfrak{S}$ (I, 32.2), and so the existence of τ, τ^* is a consequence of (5.11).

Let now T, T' be homotopic mappings $\mathfrak{R} \to \mathfrak{S}$. By hypothesis there is a mapping t: $l \times \mathfrak{R} \to \mathfrak{S}$, $l: 0 \leqq u \leqq 1$, such that $t(0 \times x) = Tx$, $t(1 \times x) = T'x$. If γ^p is any cycle of $\mathfrak{R}$ then $x \to u \times x$ defines a topological mapping $\mathfrak{R} \to u \times \mathfrak{R}$ under which γ^p is mapped into a cycle which we denote by $u \times \gamma^p$ (cycle of both

$u \times \mathfrak{R}$ and $l \times \mathfrak{R}$; see (5.5)). In view of (5.11) we merely need to prove

(7.2) $$0 \times \gamma^p \sim 1 \times \gamma^p \text{ in } l \times \mathfrak{R}.$$

Let $\mathfrak{B} = \{l_1, \cdots, l_n\}$ denote a finite open covering of l defined as follows: l_1 is a set $0 \leqq u < \alpha$, l_n is a set $\beta < u \leqq 1$, and the other sets are intervals; only consecutive sets l_i, l_{i+1} meet. Let $\{\mathfrak{B}_\mu\}$ be the coverings of this type. Since there is a $\mathfrak{B}_\mu$ whose mesh is less than any assigned ϵ, $\{\mathfrak{B}_\mu\}$ is cofinal in the family of all the open coverings of l. Hence by (I, 24.2) if $\{\mathfrak{U}_\lambda\}$ are the finite open coverings of $\mathfrak{R}$, then $\{\mathfrak{B}_\mu \times \mathfrak{U}_\lambda\}$ is cofinal in the family of all the finite open coverings of $l \times \mathfrak{R}$. Let Φ_λ = nerve $\mathfrak{U}_\lambda$, $\Psi_{\lambda\mu}$ = nerve $\mathfrak{B}_\mu \times \mathfrak{U}_\lambda$. If $\mathfrak{U}_\lambda = \{U_{\lambda j}\}$, $\mathfrak{B}_\mu = \{l_i\}$, $i = 1, 2, \cdots, n$, and $\sigma_h = U_{\lambda j_0} \cdots U_{\lambda j_p} \in \Phi_\lambda$, then $\Psi_{\lambda\mu}$ contains the simplex $\sigma_{hi} = (l_i \times U_{\lambda j_0}) \cdots (l_i \times U_{\lambda j_p})$, and $\{\sigma_{hi}\}$ is a closed simplicial complex $\cong \Phi_\lambda$ and denoted by Φ_λ^i. In particular, $\Phi_\lambda^1 = \{\sigma_{h1} \mid \sigma_{h1} \in \Psi_{\lambda\mu}$; $[\sigma_{h1}] \cap 0 \times \mathfrak{R} \neq \emptyset\}$, and $\Phi_\lambda^n = \{\sigma_{hn} \mid \sigma_{hn} \in \Psi_{\lambda\mu}$; $[\sigma_{hn}] \cap 1 \times \mathfrak{R} \neq \emptyset\}$. In other words Φ_λ^1, Φ_λ^n are the analogues of $\Phi_{1\lambda}$ of (4), for $0 \times \mathfrak{R}$, $1 \times \mathfrak{R}$ and $\Psi_{\lambda\mu}$. If $\gamma^p = \{\gamma_\lambda^p\}$ then $0 \times \gamma^p$, $1 \times \gamma^p$ have coordinates $l_1 \times \gamma_\lambda^p$, $l_n \times \gamma_\lambda^p$ relative to $\Psi_{\lambda\mu}$, and (7.2) reduces to proving

(7.3) $$l_1 \times \gamma_\lambda^p \sim l_n \times \gamma_\lambda^p \text{ in } \Psi_{\lambda\mu}.$$

It is clear that $\sigma_{h1} \to \sigma_{hn}$ defines a chain-mapping $\theta:\Phi_\lambda^1 \to \Phi_\lambda^n$ such that $\theta(l_1 \times \gamma_\lambda^p) = l_n \times \gamma_\lambda^p$. Therefore (7.3) and hence (7.1) will be proved if we can show that

(7.4) *θ is a chain-deformation over $\Psi_{\lambda\mu}$.*

Now we find immediately that if σ_h is as before then $\bigcap(l_i \times U_{\lambda j_r}) \cap \bigcap(l_{i+1} \times U_{\lambda j_s}) \neq \emptyset$, and hence

$$(l_i \times U_{\lambda j_0}) \cdots (l_i \times U_{\lambda j_q})(l_{i+1} \times U_{\lambda j_q}) \cdots (l_{i+1} \times U_{\lambda j_p}) \in \Psi_{\lambda\mu}.$$

Consequently by (IV, 16.3): $\sigma_i \to \sigma_{i+1}$ defines a simplicial chain-deformation $\theta_i:\Phi_\lambda^i \to \Phi_\lambda^{i+1}$ over $\Psi_{\lambda\mu}$. Since $\theta = \theta_{n-1} \cdots \theta_2\theta_1$ (7.4) follows, and (7.1) is proved.

(7.5) *If $\mathfrak{R}$ is a compact Hausdorff space and A is a deformation-retract of $\mathfrak{R}$ then A and $\mathfrak{R}$ have the same homology groups.*

Let T be the retraction $\mathfrak{R} \to A$. By hypothesis this time there is a mapping $t:l \times \mathfrak{R} \to \mathfrak{R}$ such that $t(0 \times x) = x$; $t(1 \times x) = Tx \in A$; $t(u \times x) = x$ for $x \in A$. It is already known that T induces a homomorphism $\tau:\mathfrak{H}^p(\mathfrak{R}, G) \to \mathfrak{H}^p(A, G)$. Similarly the injection $A \to \mathfrak{R}$ induces a homomorphism $\eta:\mathfrak{H}^p(A, G) \to \mathfrak{H}^p(\mathfrak{R}, G)$ (5.10).

Let $\Gamma^p \in \mathfrak{H}^p(\mathfrak{R}, G)$ and $\gamma^p \in \Gamma^p$. Since T is a deformation $T\gamma^p \sim \gamma^p$ in $\mathfrak{R}$. Hence the elements of the class $\Gamma'^p = \tau\Gamma^p$ are also in Γ^p or $\Gamma'^p \subset \Gamma^p$. It follows that $\eta\Gamma'^p = \Gamma^p$ and so $\eta\tau = 1$.

Consider now $\Gamma'^p \in \mathfrak{H}^p(A, G)$. The elements of Γ'^p are members of the fixed class $\Gamma^p = \eta\Gamma'^p$. I say that $\tau\Gamma^p = \Gamma'^p$. For suppose $\tau\Gamma^p = \Gamma''^p \neq \Gamma'^p$, and let $\gamma'^p \in \Gamma'^p$, $\gamma''^p \in \Gamma''^p$. Then $\delta^p = \gamma'^p - \gamma''^p$ is a cycle of A which is $\not\sim 0$ in A and yet ~ 0 in $\mathfrak{R}$. We show that this is impossible. In the notations of the proof of (7.1) this will follow if we can prove that $0 \times \gamma^p \sim 0$ in $0 \times \mathfrak{R} \to$

$1 \times \gamma^p \sim 0$ in $1 \times \mathfrak{R}$ or finally that $l_1 \times \gamma_\lambda^p \sim 0$ in $\Phi_\lambda^1 \to l_n \times \gamma_\lambda^p \sim 0$ in Φ_λ^n, and this is an immediate consequence of the fact that θ is a chain-mapping. Thus $\tau\Gamma^p = \Gamma'^p = \tau\eta\Gamma'^p$, and finally $\tau\eta = 1$.

Since $\tau\eta = 1$, $\eta\tau = 1$, τ is an isomorphism and (7.5) is proved.

(7.6) Application. *Every parallelotope P is zero-cyclic. Hence P augmented is acyclic.*

For any point of P is a deformation-retract of P and so the homology groups of P are those of a point, which is what (7.6) asserts.

8. **Homology theory and normality.** What are the spaces, if any, whose $\mathfrak{U}$- and $\mathfrak{F}$-theories are the same? A sufficient condition, and scarcely more could be expected, is found in the important

(8.1) Theorem. *When $\mathfrak{R}$ is normal then*: (a) *its $\mathfrak{U}$- and $\mathfrak{F}$-homology theories (i.e., by finite open and closed coverings) are the same*; (b) *if F is closed in $\mathfrak{R}$ then the $\mathfrak{U}$- and $\mathfrak{F}$-theories for the cycles* mod F *and the cocycles of $\mathfrak{R} - F$ are likewise the same* (Čech [a]).

An immediate consequence is:

(8.2) *Theorem* (8.1) *holds when $\mathfrak{R}$ is metric, compact Hausdorff, or a compactum.*

For in all three cases $\mathfrak{R}$ is normal.

(8.3) Proof of (8.1a). Two finite collections of sets $\mathfrak{A} = \{A_1, \cdots, A_r\}$, $\mathfrak{B} = \{B_1, \cdots, B_r\}$ are said to be *congruent*, if, say $\mathfrak{A} > \mathfrak{B}$ and this so that $A_i \subset B_i$ and that $A_i \to B_i$ defines a similitude $\mathfrak{A} \to \mathfrak{B}$. This similitude induces an isomorphism τ: nerve $\mathfrak{A} \to$ nerve $\mathfrak{B}$.

(8.4) We will utilize now an elegant suggestion due to Lagerstrom and apply (5.6). Consider the following collections: (a) the finite open coverings $\{\mathfrak{U}_\lambda\}$; (b) the finite closed coverings $\{\mathfrak{F}_\mu\}$; (c) the union of the two $\{\mathfrak{U}_\lambda, \mathfrak{F}_\mu\}$. The first two are directed collections in the sense of (2). To show that the third is one also it is sufficient to show that $\mathfrak{U}_\lambda$, $\mathfrak{F}_\mu$ have a common refinement in the collection (c). Now by (I, 33.4) $\mathfrak{U}_\lambda$ may be shrunk to $\mathfrak{U}_{\lambda'}$ and so $\bar{\mathfrak{U}}_{\lambda'}$ refines $\mathfrak{U}_\lambda$. Hence $\bar{\mathfrak{U}}_{\lambda'} \wedge \mathfrak{F}_\mu$, which is a finite closed covering, is a common refinement of $\mathfrak{U}_\lambda$, $\mathfrak{F}_\mu$ such as required.

We may now introduce the nets Φ, Ψ, Ω of the collections (a,b,c) (see 2.1) and we will suppose that they have been modified after the manner of (5.6) as follows: In the notations of (8.3) if $\mathfrak{A}$, $\mathfrak{B}$ are in the same collection, say $\mathfrak{A} = \mathfrak{U}_\lambda$, $\mathfrak{B} = \mathfrak{U}_{\lambda'}$, we introduce in $\{\lambda\}$ the new ordering $\lambda' > \lambda$ and projection $\pi_\lambda^{\lambda'} = \tau^{-1}$.

We have already seen that there is a $\bar{\mathfrak{U}}_{\lambda'}$ refining $\mathfrak{U}_\lambda$. Hence Ψ is cofinal in Ω. By (I, 33.5) every $\mathfrak{F}_\mu$ is a congruent refinement of some $\mathfrak{U}_\lambda$, and in the ordering attached to Ω we have then $\lambda > \mu$. Thus Φ is likewise cofinal in Ω. It follows that Φ and Ψ have the same homology theories. Since by (5.6) these are the $\mathfrak{U}$- and $\mathfrak{F}$-theories, (8.1a) is proved.

(8.5) *Proof of* (8.1b). The same argument may be applied here provided that we strengthen the condition imposed on $\mathfrak{A}$, $\mathfrak{B}$ in (8.3) to: $F \cup \mathfrak{A}$ is a congruent refinement of $F \cup \mathfrak{B}$, and the related similitude sends F into F.

9. **Duality in the sense of Alexander.** Little more is required than to transfer the results of this type of duality for nets (VI, 15) to topological spaces. Let then $\mathfrak{R}$ be a topological space and F a closed set, and in the notations of (4) let $(\Phi_0(\mathfrak{R} - F), \Phi_1(F))$ be the dissection of Φ corresponding to the dissection $(\mathfrak{R} - F, F)$ of $\mathfrak{R}$. We may define the linking coefficients as in (VI, 15.3). The duality theorems for complexes (III, 38.3, 39) are thus valid for a dissection $(\mathfrak{R} - F, F)$, F closed. We state explicitly the main theorem, and leave the formulation of the rest to the reader.

(9.1) THEOREM. *If $\mathfrak{R}$ is $(p - 1, p)$-acyclic and (G, H) is a normal couple, then $\mathfrak{H}^{p-1}(F, G)$ and $\mathfrak{H}_p(\mathfrak{R} - F, H)$ are dually paired with the linking coefficient as the group multiplication.*

Since $\mathfrak{H}^{p-1}(F, G)$ and $\mathfrak{H}_{p-1}(F, H)$ are likewise dually paired (4.1), both $\mathfrak{H}_{p-1}(F, H)$ and $\mathfrak{H}_p(\mathfrak{R} - F, H)$ are isomorphic with the character-group or space of $\mathfrak{H}^p(\mathfrak{R}, F, G)$ (group of $\mathfrak{R}$ mod F over G) and hence also with one another. Therefore:

(9.2) THEOREM. *If $\mathfrak{R}$ is $(p - 1, p)$-acyclic, F closed in $\mathfrak{R}$, and H any discrete group, then $\mathfrak{H}_{p-1}(F, H) \cong \mathfrak{H}_p(\mathfrak{R} - F, H)$ (linear isomorphism when H is a field).*

This last formulation offers the advantage that it is expressed in terms of a *single* coefficient group. Similar formulations may of course be given for the analogues of the other theorems of (III, 38, 39). See notably Alexandroff [f].

Noteworthy consequence of (9.1):

(9.3) *When $\mathfrak{R}$ is acyclic the cohomology groups of $\mathfrak{R} - F$ are topological invariants of F alone, and in particular they do not depend upon the topological imbedding in $\mathfrak{R}$.*

(9.4) APPLICATION. An augmented parallelotope P has the groups of an augmented point and hence it is acyclic. Therefore (9.1) is applicable for all dimensions and (9.2, 9.3) also hold for P.

10. **Intersections and their carriers.** Up to the present, intersections of any sort have largely had an "algebraic" connotation. The consideration of the carriers will enable us to make the transfer to point set properties.

The general notations remaining those of (4), let in addition G, H be two groups commutatively paired to a third J, such as always occur in connection with intersections. The brunt of the argument will bear upon the intersections in a fixed Φ_λ.

(10.1) Let $\Phi_\lambda = \{\sigma_\lambda^p\}$ and let F, U be a closed and an open set (not necessarily complementary to one another). By (V, 4.1) $\sigma_\lambda^p \cdot \sigma_q^\lambda$ is a chain of Cl σ_λ^p and hence its vertices are among those of σ_λ^p. Moreover (V, 4.1) $\sigma_\lambda^p \cdot \sigma_q^\lambda \neq 0 \rightarrow \sigma_\lambda^q < \sigma_\lambda^p$.

Suppose that σ_λ^p, σ_q^λ have the carriers F, U so that

$$[\sigma_\lambda^p] \cap F \neq \emptyset, \qquad [\sigma_q^\lambda] \subset U.$$

Since the vertices of σ_λ^q are among those of σ_λ^p we have $[\sigma_q^\lambda] \supset [\sigma_\lambda^p]$, and therefore $[\sigma_\lambda^p] \subset U$. Similarly since the vertices of $\sigma_\lambda^p \cdot \sigma_q^\lambda$ are among those of σ_λ^p if σ_λ^{p-q} is a simplex of this intersection then $[\sigma_\lambda^{p-q}] \supset [\sigma_\lambda^p]$ and therefore $[\sigma_\lambda^{p-q}] \cap F \supset [\sigma_\lambda^p] \cap F \subset F \cap \bar{U}$. Hence $[\sigma_\lambda^{p-q}] \cap (F \cap \bar{U}) \neq \emptyset$, and so $\sigma_\lambda^p \cdot \sigma_q^\lambda$ has the carrier $F \cap \bar{U}$. It follows that if C_λ^p, C_q^λ have the carriers F, U then their intersection has the carrier $F \cap \bar{U}$.

(10.2) For the cochains it is most convenient to utilize the intersections of the Whitney type (V, 20). The vertices of Φ_λ are ranged in some fixed order and intersections determined in terms of that order. Referring to (V, 20.1) if $\sigma_p^\lambda \cdot \sigma_q^\lambda \neq 0$, it is a dual simplex whose vertices include all those and only those of σ_p^λ and σ_q^λ. Therefore

$$\sigma_p^\lambda \cdot \sigma_q^\lambda \neq 0 \rightarrow [\sigma_p^\lambda \cdot \sigma_q^\lambda] = [\sigma_p^\lambda] \cap [\sigma_q^\lambda].$$

Consequently if C_p^λ, C_q^λ have the carriers U, V, their intersection has the carrier $U \cap V$.

(10.3) Let now $\gamma^p = \{\gamma_\lambda^p\}$ be a cycle of $\mathfrak{R}$ over G with the closed carrier F, $\gamma_q = \{\gamma_q^\lambda\}$ a cocycle of $\mathfrak{R}$ over H with the open carrier V, and let Γ^p, Γ_q be their classes. We have then a class intersection $\Gamma^p \cdot \Gamma_q$ and our purpose is to select a suitable carrier for a cycle of that class.

By hypothesis there exists $\{\mu\}$ cofinal in $\{\lambda\}$ and $\nu \in \{\lambda\}$ such that every γ_μ^p has the carrier F and γ_p^ν has the carrier U. Since $\{\mu \mid \mu > \nu\}$ is also cofinal in $\{\lambda\}$, it may replace $\{\mu\}$ and so we may suppose every $\mu > \nu$. It is then a consequence of the definition of $\Gamma^p \cdot \Gamma_q$ that the cycles $\gamma_\mu^p \cdot \pi_\mu^{*\nu} \gamma_p^\nu$ are coordinates of a cycle $\gamma^{p-q} \in \Gamma^p \cdot \Gamma_q$.

Now by (1.12), $\pi_\mu^{*\nu} \gamma_q^\nu$ has the carrier V. Hence (10.1) $\gamma_\mu^p \cdot \pi_\mu^{*\nu} \gamma_q^\nu$ has the carrier $F \cap \bar{V}$. Since $\{\mu\}$ is cofinal in $\{\lambda\}$, $\gamma^p \cdot \gamma_q$ has the same carrier. Thus:

(10.4) *If γ^p, γ_q have the carriers F, V then in the intersection class $\Gamma^p \cdot \Gamma_q$ there is a cycle with the carrier $F \cap \bar{V}$. Any cycle of the class with the same carrier is called an intersection cycle of γ^p, γ_q, written $\gamma^p \cdot \gamma_q$.*

(10.5) Consider now two cocycles $\gamma_p = \{\gamma_p^\lambda\}$ and $\gamma_q = \{\gamma_q^\lambda\}$ over G and H, with the open carriers U and V and classes Γ_p and Γ_q. There exist representatives γ_p^μ, γ_q^ν with the carriers U, V. Choose any $\rho > \mu, \nu$. The intersection cocycle $(\pi_\rho^{*\mu} \gamma_p^\mu) \cdot (\pi_\rho^{*\nu} \gamma_q^\nu)$ taken as in (10.2) is a representative of a cocycle $\gamma_{p+q} \in \Gamma_p \cdot \Gamma_q$. By (1.12) $\pi_\rho^{*\mu} \gamma_p^\mu$ has the carrier U and $\pi_\rho^{*\nu} \gamma_q^\nu$ the carrier V and hence (10.2): $\pi_\rho^{*\mu} \gamma_p^\mu \cdot \pi_\rho^{*\nu} \gamma_q^\nu$ has the carrier $U \cap V$, or finally γ_{p+q} has the carrier $U \cap V$. Thus

(10.6) *If γ_p, γ_q have the carriers U, V then in the intersection class $\Gamma_p \cdot \Gamma_q$ there is a cocycle with the carrier $U \cap V$. Any cocycle of $\Gamma_p \cdot \Gamma_q$ with this property will be called an intersection cocycle of γ_p, γ_q, written $\gamma_p \cdot \gamma_q$.*

(10.7) The preceding results may be extended to relative cycles and cocycles.

Let F, F_1 be closed sets with $F_1 \subset F$, and V, V_1 open sets with $V_1 \subset V$ and let γ^p be a cycle mod F_1 over G with the carrier F and γ_q a cocycle mod V_1 over H with the carrier V. We will say that γ^p, γ_q are in *general position* if we have:

$$(F \cap \bar{V}_1) \cup (F_1 \cap \bar{V}) = \emptyset. \tag{10.8}$$

Similarly let $U \supset U_1$, $V \supset V_1$, where all the sets are open, and let γ_p be a cocycle mod U_1 over G with the carrier U and γ_q as before. Then γ_p, γ_q are said to be in general position if we have:

$$(U \cap V_1) \cup (U_1 \cap V) = \emptyset.$$

It is an elementary matter to verify that for relative elements in general position all the considerations of (VI, 8) and of (10.1, $\cdots$, 10.5) are valid without modification. The intersections $\gamma^p \cdot \gamma_q$ and $\gamma_p \cdot \gamma_q$ continue to be absolute cycles and cocycles.

(10.9) Application. (a) *If γ^p, γ_q have carriers F, V such that F and $\bar{V}$ are disjoint then $\Gamma^p \cdot \Gamma_q = 0$.* (b) *If γ_p, γ_q have disjoint carriers U, V then $\Gamma_p \cdot \Gamma_q = 0$. Similarly for the relative intersections.*

The "point set" value of this property is clear. Its proof is elementary and left to the reader.

§2. RELATIONS BETWEEN CONNECTEDNESS AND HOMOLOGY

11. (11.1) We have seen (III, 20) that in simplicial complexes there exists a close relationship between connectedness and the zero-cycles. The same situation is found to hold, following Čech [a], for every topological space $\mathfrak{R}$. To bring the two concepts together one must first associate zero-cycles with the points of $\mathfrak{R}$. Let the notations remain those of (4). Take any point $x \in \mathfrak{R}$ and let U_λ, U'_λ both contain x. Since the two sets meet they are joined by a σ^1 in Φ_λ, or $U_\lambda \sim U'_\lambda$ in Φ_λ. Suppose that we choose a $U_\lambda \ni x$ for each λ and let $\gamma^0 = \{U_\lambda\}$. If $\lambda > \mu$ then $\pi^\lambda_\mu U_\lambda$ is a U'_μ and so $\pi^\lambda_\mu U_\lambda \sim U_\mu$ in Φ_μ. Therefore γ^0 is a zero-cycle of $\mathfrak{R}$ associated with x. Similarly $\{gU_\lambda\}$, $g \in G$, is a zero-cycle of $\mathfrak{R}$ over G and it is denoted by $g\gamma^0$.

(11.2) *All the cycles such as γ^0 associated with x are homologous integral cycles.*

For if $\delta^0 = \{U'_\lambda\}$, $x \in U'_\lambda$, then $U_\lambda \sim U'_\lambda$ in Φ_λ and so $\gamma^0 \sim \delta^0$.

(11.3) *If $\mathfrak{R}$ is a T_1-space then any γ^0 associated with x has the closed carrier x* (obvious).

(11.4) For convenience we shall designate also by x any cycle γ^0 attached to x in the above sense, i.e., any collection $\{U_\lambda\}$, $x \in U_\lambda$. This is not strictly accurate since the cycle in question is not generally unique. However, x as a cycle will only be utilized in homology relations, and since its class is unique the deviation is immaterial. The open sets U_λ will also be referred to as *coordinates* of x. Thus, if x, x' have the coordinates $\{U_\lambda\}$, $\{U'_\lambda\}$ then $x \sim x' \leftrightarrow U_\lambda \sim U'_\lambda$ in Φ_λ for every λ.

In an obvious sense also gx will designate a cycle $\{gU_\lambda\}$ where U_λ are coordinates of x and the same remarks may be made for the cycle gx as for the cycle x.

(11.5) We have already defined the *components* (I, 20) and they will be denoted by $\{K_i\}$. Two other related concepts may be naturally introduced here:

(a) the *pseudo-components* of $\mathfrak{R}$, which are merely the components $\{\Phi^i\}$ of the net Φ of the finite open coverings (VI, 5);

(b) the *quasi-components* $\{Q_i\}$ which are defined as follows: the quasi-component Q_i of a point x is the set of all points $x' \sim x$. It is clear that Q_i is then also the quasi-component of every such x'. Each point x belongs to one and only one quasi-component.

(11.6) Let $\{\Phi_\lambda^i\}$, λ fixed, be the components of the complex Φ_λ. The union of the vertices of Φ_λ^i is an open set denoted by V_λ^i. Since the V_λ^i are disjoint open sets and form a covering of $\mathfrak{R}$, V_λ^i has for complement an open set, and so V_λ^i is also *closed*. Thus $\{V_\lambda^i\}$, λ fixed, is both a finite open covering and a finite closed covering.

Let $x \in \mathfrak{R}$ have the coordinates $\{U_\lambda\}$ and suppose that U_λ is a vertex of Φ_λ^i, or which is the same, that $U_\lambda \subset V_\lambda^i$. If U'_λ is another choice of a λ coordinate for x then $U'_\lambda \sim U_\lambda$ in Φ_λ. Hence U'_λ is also a vertex of Φ_λ^i (III, 20.5) and so $U'_\lambda \subset V_\lambda^i$. Thus Φ_λ^i, hence V_λ^i, depends solely upon x and not upon its coordinates. Notice that $x \in V_\lambda^i$, and so V_λ^i is merely a coordinate of x in the covering $\{V_\lambda^i\}$.

(11.7) Quasi-components were originally introduced by Hausdorff whose definition was, however, different. We will say that a closed set $\mathfrak{R}$ is *irreducible from x to x'* if there is no decomposition $\mathfrak{R} = F_1 \cup F_2$, $F_1 \cap F_2 = \emptyset$, F_i closed, with $x \in F_1$, $x' \in F_2$. According to Hausdorff the quasi-component S of x is the set of all the points x' such that $\mathfrak{R}$ is irreducible from x to x'.

It is not difficult to identify the two definitions. First S is contained in a single Q_i. For suppose $x, x' \in S$, $x \nsim x'$. Then for some $\lambda : x \in V_\lambda^i = F_1$, $x' \in \mathfrak{R} - V_\lambda^i = F_2$, where the F_i are closed and disjoint. Hence $\mathfrak{R}$ is not irreducible from x to x', contrary to assumption. Therefore $x \sim x'$ and so $x \in Q_i \rightarrow x' \in Q_i \rightarrow S \subset Q_i$, If $S \neq Q_i$, we may choose $x \in S$, $x' \in Q_i - S$, and there will be a pair of disjoint closed, hence also open, sets U_1, U_2 whose union is $\mathfrak{R}$ and such that $x \in U_1$, $x' \in U_2$. We have then in $\{U_1, U_2\}$ a finite open covering in the nerve of which the coordinates of x, x' are not homologous. This implies $x \nsim x'$; which contradicts $x, x' \in Q_i$. Thus $S = Q_i$ and Hausdorff's definition agrees with (11.5b).

The following example essentially due to Hausdorff will make clear the distinction between components and quasi-components. The set $\mathfrak{R}$ is planar and consists of the segments $l_n : 0 \leqq y \leqq 1$, $x = 1/n$, $n = 1, 2, \cdots$, together with a set of disjoint subsegments $\{L_m\}$ of $x = 0$, $0 \leqq y \leqq 1$. The components of $\mathfrak{R}$ are the $\{l_n, L_m\}$. On the other hand if the component V_λ^i of $\mathfrak{U}_\lambda$ meets L_m it will contain every l_n for n above a certain value, and hence also every L_m.

Therefore $\bigcup L_m$ is a quasi-component, and it is clear that it may contain arbitrarily many components. The other quasi-components are the l_n .

(11.8) Let x have the coordinates $\{U_\lambda\}$ and let $\Phi_\lambda^{i\lambda}$ be the component of Φ_λ with U_λ as a vertex. As shown in (11.6) $V_\lambda^{i\lambda}$, and hence $\Phi_\lambda^{i\lambda}$ depends only on x. Since $\pi_\mu^\lambda U_\lambda \sim U_\mu$, $\lambda > \mu$, $\pi_\mu^\lambda U_\lambda$ is a vertex of $\Phi_\mu^{i\mu}$. Thus $\pi_\mu^\lambda \Phi_\lambda^{i\lambda}$ is connected and meets $\Phi_\mu^{i\mu}$. Since the latter is a component of Φ_μ , we must have $\pi_\mu^\lambda \Phi_\lambda^{i\lambda} \subset \Phi_\mu^{i\mu}$. Therefore $\Phi^i = \{\Phi_\lambda^{i\lambda}\}$ is a pseudo-component (VI, 5) and it depends solely upon x. We will say that Φ^i *contains* x and denote by $[\Phi^i]$ the totality of the points x which Φ^i contains. If $\{\lambda\}$ is replaced by a cofinal subset $\{\mu\}$, Φ and Φ^i go over into, say, $'\Phi$ and a component $'\Phi^i$ of $'\Phi$. The set $[\Phi^i]$ is now defined as before and it is readily seen to be $[\Phi^i]$ itself. Thus this set is unchanged when $\{\mathfrak{U}_\lambda\}$ is replaced by a cofinal family.

(11.9) *If* $\Phi^i = \{\Phi_\lambda^{i\lambda}\}$ *is a pseudo-component and the* $V_\lambda^{i\lambda}$ *are as in* (11.6) *then*:

(a) $\{V_\lambda^{i\lambda}; \subset\}$ *is directed* (*the intersection of any two elements contains a third*);

(b) $\{V_\lambda^{i\lambda}\}$ *has the finite intersection property*;

(c) $[\Phi^i] = \bigcap V_\lambda^{i\lambda}$;

(d) *if* $[\Phi^i] \neq \emptyset$ *then it is a quasi-component, and the quasi-components are all of this form, i.e., they are the sets* $[\Phi^i] \neq \emptyset$.

If $\lambda > \mu$ then $\pi_\mu^\lambda \Phi_\lambda^{i\lambda} \subset \Phi_\mu^{i\mu}$. Therefore if U_λ is a vertex of $\Phi_\lambda^{i\lambda}$ then $\pi_\mu^\lambda U_\lambda = U_\mu$ is a vertex of $\Phi_\mu^{i\mu}$. By the definition of the π_μ^λ then $U_\lambda \subset U_\mu$ and hence $V_\lambda^{i\lambda} \subset V_\mu^{i\mu}$. It follows that if $\nu > \lambda, \mu$, then $V_\nu^{i\nu} \subset V_\lambda^{i\lambda} \cap V_\mu^{i\mu}$ which is (a). Property (b) is an immediate consequence of (a). A n. a. s. c. for $x \in [\Phi^i]$ is that a coordinate U_λ of x be a vertex of $\Phi_\lambda^{i\lambda}$ or that $U_\lambda \subset V_\lambda^{i\lambda}$, and since $x \in U_\lambda$, we also have $x \in V_\lambda^{i\lambda}$, and hence $[\Phi^i] \subset \bigcap V_\lambda^{i\lambda}$. Conversely, $x \in \bigcap V_\lambda^{i\lambda}$ implies that a λ coordinate U_λ of x is in $V_\lambda^{i\lambda}$ and hence that $x \in [\Phi^i]$, proving (c). Regarding (d) if $x, x' \in \mathfrak{R}$, and U_λ, U'_λ are their λ coordinates then $x, x' \in [\Phi^i] \rightarrow U_\lambda , U'_\lambda$ are vertices of $\Phi_\lambda^{i\lambda}$; hence $U_\lambda \sim U'_\lambda$ and so $x \sim x'$. Thus $[\Phi^i]$ is in a unique quasi-component Q_h . On the other hand, $x \in [\Phi^i]$, $x' \in [\Phi^j]$, $i \neq j$, $\rightarrow \Phi_\lambda^{i\lambda} \neq \Phi_\lambda^{j\lambda}$ for some $\lambda \rightarrow U_\lambda \not\sim U'_\lambda \rightarrow x \not\sim x'$, and so x, x' are not in the same quasi-component. Thus $[\Phi^i]$ is a quasi-component and $[\Phi^i]$, $[\Phi^j]$, $i \neq j$, are distinct quasi-components or null sets. Take now any quasi-component Q_i and let $x \in Q_i$. As shown in (11.8) x is also in some set $[\Phi^j]$ which is a quasi-component. Since x belongs to a unique quasi-component we have $Q_i = [\Phi^j]$. Therefore $\{Q_i\}$ consists merely of the sets $[\Phi^j] \neq \emptyset$, and (d) is proved.

(11.10) *If a component* K_i *meets a quasi-component* Q_j *then* $K_i \subset Q_j$. *Hence every quasi-component is a union of components.*

Let $\Psi_\lambda = \{\sigma \mid \sigma \in \Phi_\lambda , [\sigma] \cap K_i \neq \emptyset\}$. Since K_i is connected so is Ψ_λ (I, 18.1). If $x, x' \in K_i$ have U_λ , U'_λ as λ coordinates, then U_λ , U'_λ are vertices of Ψ_λ and so $U_\lambda \sim U'_\lambda$. Therefore x, x' are in the same quasi-component which is (11.10).

(11.11) *Components and quasi-components are closed sets.*

By (I, 19.3) $\bar{K}_i$ is connected and since $K_i \subset \bar{K}_i$ necessarily $K_i = \bar{K}_i$, or K_i is closed. Since V_λ^i is closed (11.6), it is a consequence of (11.9cd) that the quasi-components are likewise closed.

(11.12) *When* $\mathfrak{R}$ *is compact all the* $[\Phi^i] \neq \emptyset$, *and hence they are the quasi-components.*

For $\{V_\lambda^{i\lambda}\}$ is a collection of closed sets by (11.6), and has the finite intersection property (11.9b). Hence (11.9c): $[\Phi^i] \neq \emptyset$, and so (11.12) is a consequence of (11.9d).

(11.13) *Suppose* $\mathfrak{R}$ *is not compact and let it be compacted to* $\mathfrak{S}$ *in accordance with* (I, 26.1), *then let* $\mathfrak{R}$ *be identified with its topological image in* $\mathfrak{S}$. *If* $\{P_i\}$ *are the quasi-components of* $\mathfrak{S}$ *then* $P_i \to \Phi^i$ *is one-one and the quasi-components of* $\mathfrak{R}$ *are merely the intersections different from* $\emptyset$ *of the* P_i *with* $\mathfrak{R}$.

Referring to (6.1), the net of the finite open coverings of $\mathfrak{S}$ contains a cofinal net $\Psi \cong \Phi$. Hence (11.8) the components $\{\Psi^i\}$ are in one-one correspondence with $\{\Phi^i\}$ and also with $\{P_i\}$. Thus we may assume the notations so chosen that Ψ^i, Φ^i, P_i are associated in the correspondences. Furthermore if $\Phi^i = \{\Phi_\lambda^{i\lambda}\}$ and $\Phi_\lambda^{i\lambda}$ has the vertices $\{U_{\lambda m}\}$, then $\Psi^i = \{\Psi_\lambda^{i\lambda}\}$ where $\Psi_\lambda^{i\lambda}$ has the vertices $\{\Omega(U_{\lambda m})\}$ (notations of I, 27). From the properties of the operator Ω (I, 27, 28) we infer that $\Omega(U_{\lambda m}) \cap \mathfrak{R} = U_{\lambda m}$, then from $P_i = \bigcap_\lambda \bigcup_m \Omega(U_{\lambda m})$, $[\Phi^i] = \bigcap_\lambda \bigcup_m U_{\lambda m}$, finally that $[\Phi^i] = P_i \cap \mathfrak{R}$. This proves (11.13).

(11.14) Consider the following numbers: $r_\lambda = R^0(\Phi_\lambda) =$ the number of sets V_λ^i; $r = \sup r_\lambda$; $k =$ the number of components; $q =$ the number of quasi-components; the Betti number R^0 of $\mathfrak{R}$.

(11.15) *If one of the numbers* r, k, q, R^0 *is finite then*:

(a) *all four numbers are equal and have the common value* R^0;

(b) *every quasi-component is a component, and conversely*;

(c) *if* $\{K_i\}$, $i = 1, 2, \cdots, R^0$ *are the components then whatever* G:

$$\mathfrak{H}^p(\mathfrak{R}, G) \cong \mathbf{P}\mathfrak{H}^p(K_i, G);$$

this holds also for the homology groups by finite closed coverings;

(d) *if* $\{x_i\}$ *are points such that one and only one* $x_i \in K_i$, *then every zero-cycle* γ^0 *over* G *satisfies a relation*

$$\gamma^0 \sim g^i x_i, \qquad g^i \in G.$$

Moreover there is no relation

$$g^i x_i \sim 0, \qquad g^i \in G,$$

with coefficients not all zero.

We will first prove

$$r \leqq \inf \{k, q, R^0\}. \tag{11.16}$$

We recall that the number of pseudo-components is R^0 (VI, 5.3). From (11.9, 11.10) we have then $q \leqq k, R^0$. Choose a point x_i in V_λ^i. Since no two points x_i have their λ coordinates in the same V_λ^i we have $x_i \nsim x_j$ for $i \neq j$, and so $r_\lambda \leqq q$. Hence $r_\lambda \leqq \inf \{k, q, R^0\}$, from which (11.16) follows.

As a consequence of (11.16) if any one of the numbers k, q, R^0 is finite so is r. Suppose now r finite. We may then choose a μ such that $r_\mu = r$. We show that every V_μ^i is a component. Since the V_μ^i are disjoint it is only necessary to prove V_μ^i connected. Suppose this false. We have then $V_\mu^i = U_1 \cup U_2$; $U_1 \cap U_2 = \emptyset$; U_1, U_2 open in V_μ^i. Since V_μ^i is open in $\mathfrak{R}$, U_1 and U_2 are open in $\mathfrak{R}$ also. Hence $\{V_\mu^1, \cdots, V_\mu^{i-1}, U_1, U_2, V_\mu^{i+1}, \cdots, V_\mu^r\}$ is a finite open covering of $\mathfrak{R}$, say $\mathfrak{U}_\lambda$, such that Φ_λ consists of $r + 1$ vertices. Thus $r_\lambda = r + 1$, which is ruled out, and so the V_μ^i are components. As a consequence $r = k$ and since $r \leqq q \leqq k$, also $r = q$.

Keeping μ fixed as above, let $\mathfrak{U}_\lambda = \{U_{\lambda m}\}$. Then $\mathfrak{U}_\nu = \{U_{\lambda m} \cap V_\mu^i\} > \mathfrak{U}_\lambda$. Since V_μ^i is a component the covering $\{U_{\lambda m} \cap V_\mu^i\}$ of V_μ^i is connected. Since $V_\mu^i, V_\mu^j, i \neq j$, are disjoint the nerves of $\{U_{\lambda m} \cap V_\mu^i\}$ and $\{U_{\lambda m} \cap V_\mu^j\}$ are disjoint and so they are distinct components of Φ_ν. It follows that $\bigcup(U_{\lambda m} \cap V_\mu^i) = V_\mu^i$ is one of the sets V_ν^h, and clearly every V_ν^h is a V_μ^i. In other words, the two collections $\{V_\mu^i\}$, $\{V_\nu^i\}$ are identical and hence $r_\nu = r_\mu = r$. Since λ is any element of $\{\lambda\}$, $\{\nu\}$ is cofinal in $\{\lambda\}$. Thus for a $\{\nu\}$ cofinal in $\{\lambda\}$, Φ_ν will have exactly r components Φ_ν^i. Clearly also $\nu' > \nu \rightarrow \pi_\nu^{\nu'}\Phi_{\nu'}^i \subset \Phi_\nu^i$. Consequently the net $\{\Phi_\nu\}$ cofinal in Φ has exactly r components. Since the two nets have the same number of components we have by (VI, 5.3): $r = R^0$.

We have thus found that when r is finite $r = k = q = R^0$. In view of (11.16) this proves (11.15a). We have also shown incidentally that the V_μ^i are the components of $\mathfrak{R}$. Since each is contained in a quasi-component, and $k = q$, (11.15b) follows.

We have also found that for $\{\nu\}$ cofinal in $\{\lambda\}$ the net $\{\Phi_\nu\}$ consists of R^0 closed subnets $\{\Phi^i\}$, $\Phi^i = \{\Phi_\nu^i\}$, where the Φ_ν^i are disjoint. It follows that each Φ^i determines a pseudo-component and so (11.15c) is a consequence of (VI, 5.6). Referring now to the proof of (VI, 5.3) it is an elementary matter to show that the x_i chosen as in (11.15d) may be taken as the γ_i^0 loc. cit. and thus (11.15d) is a consequence of (VI, 5.4).

(11.17) The example in (11.7) raises the natural inquiry: when are all the quasi-components also components? We have just found that this is certainly the case when the number of components or quasi-components is finite. Another noteworthy circumstance under which the same property holds is described in:

(11.18) *When $\mathfrak{R}$ is compact normal, and so in particular when it is compact Hausdorff, every quasi-component is a component. Hence under these conditions the number of quasi-components is the same as the number of components.*

Let x, x' be points of the same quasi-component say Q_1. We will prove

$$x \sim x' \text{ in } Q_1. \tag{11.19}$$

On the strength of property (14.4), due to A. D. Wallace and proved independently of the present considerations, it is sufficient to show that if W is any neighborhood of the closed set Q_1 then

$$x \sim x' \text{ in } W. \tag{11.20}$$

Let $Q_1 = \bigcap V_\lambda^{i_\lambda}$. We show that for some μ: $V_\mu^{i_\mu} \subset W$. For in the contrary case every $V_\lambda^{i_\lambda} \cap (\mathfrak{R} - W) \neq \emptyset$. Hence by (11.9a) the collection of closed sets

$\{V_\lambda^{i_\lambda} \cap (\mathfrak{R} - W)\}$ has the finite intersection property. Since $\mathfrak{R}$ is compact this implies $(\bigcap V_\lambda^{i_\lambda}) \cap (\mathfrak{R} - W) = Q_1 \cap (\mathfrak{R} - W) \neq \emptyset$, a contradiction. Therefore some $V_\lambda^{i_\lambda} \subset W$, and since $x \sim x'$ in $V_\lambda^{i_\lambda}$, (11.20) follows and so does (11.19).

In view of (11.19) Q_1 is its own unique quasi-component and so its own unique component. Thus Q_1 is connected and so it is a component of $\mathfrak{R}$, which proves (11.18).

§3. GROUPS RELATED TO WEBS

12. (12.1) To facilitate comparison with (VI, 21, 22) we will denote throughout this section open sets and closed sets in webs of sets by A and B. If (A, B) is a dissection of $\mathfrak{R}$, then in the notations of (4): $A \subset A' \rightarrow \Phi_0(A) \subset \Phi_0(A')$, while $B \subset B' \rightarrow \Phi_1(B) \subset \Phi_1(B')$. Therefore if $\mathfrak{A} = \{A_\lambda\}, \mathfrak{B} = \{B_\lambda\}$ are complementary webs of sets then $\Phi_0(\mathfrak{A}) = \{\Phi_0(A_\lambda)\}$ and $\Phi_1(\mathfrak{B}) = \{\Phi_1(B_\lambda)\}$ are, respectively, an open and a closed web of nets, and furthermore they are readily shown to be complementary. If we define the groups of $\mathfrak{A}$, $\mathfrak{B}$ by reference to $\Phi_0(\mathfrak{A})$, $\Phi_1(\mathfrak{B})$, then the basic table (VI, 22.9) carries over to webs of sets.

We shall now make several applications of the preceding considerations.

13. **Locally compact spaces.**

(13.1) Let $\mathfrak{B} = \{B_\lambda\}$ consist of all the compact subsets of $\mathfrak{R}$. Then $\mathfrak{B}$ is clearly a web of sets. If $\{\lambda; >\}$ is ordered by the inclusions of the B_λ we are under Case III of the table (VI, 22.9) with the cycles the elements of an H-conet. Thus the groups of the cycles must be taken discrete and so we have IIIb. The cycles of $\mathfrak{B}$ are said to be *compact*. Any such cycle is defined by a single coordinate, and so it is merely a cycle γ^p of $\mathfrak{R}$ with a compact carrier B and $\gamma^p \sim 0$ signifies that $\gamma^p \sim 0$ in some compact $B' \supset B$. The web $\mathfrak{B}$ is direct.

(13.2) Let now $\mathfrak{A} = \{A_\lambda\}$ be the web of all the open sets with compact closures ordered by inclusion. This time again the web is direct and it corresponds then to Case Ia of the table with an H-net. Thus a cocycle of $\mathfrak{A}$ is a cocycle γ^p of $\mathfrak{R}$ whose carrier is an open set A with compact closure, and $\gamma_p \sim 0$ means that $\gamma_p \sim 0$ in a similar set $A' \supset A$. We call γ_p a *compact* cocycle. The groups of the related cycles are topologized, those of the compact cocycles discrete, and they form two dual categories with intersections.

(13.3) While it is true that the groups just considered may be defined for any space, they acquire significance only when $\mathfrak{R}$ is "liberally supplied" with compact subsets and this will generally hold only when $\mathfrak{R}$ is locally compact. In this regard there is a certain similarity between the present situation and the situation of (I, 29), in relation to the remark (I, 29.4).

14. **Neighborhood groups of a closed set.**

(14.1) Let F be a closed set and take $\mathfrak{A} = \{A_\lambda\}$ to consist of all the neighborhoods of F. If $B_\lambda = \mathfrak{R} - A_\lambda$, let $\{\lambda; >\}$ be ordered by the inclusions of the B_λ. We have then the inverse web $\mathfrak{A}$, and so Case IIa of the table: a weak H-conet (all groups discrete). A cycle is determined by a single coordinate; i.e., by a cycle γ^p of $\mathfrak{R}$ mod B, where B, F are disjoint, and $\gamma^p \sim 0$ means that $\gamma^p \sim 0$ mod B', for some $B' \supset B$. The cycles are said to be *through* F.

A cocycle of $\mathfrak{A}$ is a collection $\gamma_p = \{\gamma_p^\lambda\}$ where γ_p^λ is a cocycle of A_λ and $\lambda > \mu \to \gamma_p^\lambda \sim \gamma_p^\mu$ in A_μ. Moreover $\gamma_p \sim 0$ means that $\gamma_\lambda^p \sim 0$ in A_λ for every λ. The cocycles are said to be *around* F.

(14.2) Let now $\bar{\mathfrak{A}} = \{\bar{A}_\lambda\}$. Evidently $\bar{\mathfrak{A}}$ is an inverse closed web of sets and so we may introduce its cycles and cocycles. The description is the same as above with cycles and cocycles, A_λ and $\bar{A}_\lambda$ interchanged. The cycles are now said to be *around* F, and the cocycles to be *through* F. This time we have the weak H-net, IVb of the table, and the groups are still discrete.

(14.3) THEOREM. *If $\mathfrak{R}$ is normal then the homology groups of the cycles around F and the cohomology groups of the cocycles through F are the same as the corresponding homology and cohomology groups of F itself.*

The proof requires the following noteworthy result due to A. D. Wallace:

(14.4) *If $\mathfrak{R}$ is normal, F closed in $\mathfrak{R}$ and $\mathfrak{U}$ a finite open covering of $\mathfrak{R}$, then there exists a refinement $\mathfrak{B} = \{V_i\}$ of $\mathfrak{U}$ such that if $V_i \cap \cdots \cap V_j \cap A \neq \emptyset$ whatever the neighborhood A of F, then also $V_i \cap \cdots \cap V_j \cap F \neq \emptyset$. The same property holds with A replaced by $\bar{A}$.*

By a mere paraphrase of the proof of (3.1) we first show that $\mathfrak{U}$ has a finite open refinement $\mathfrak{W} = \{W_i\}$ such that $\{W_i \cap F\}$ is an irreducible covering of F. Let now $\mathfrak{W}$ be shrunk to $\mathfrak{B} = \{V_i\}$ (I, 33.4). Since $V_i \subset \bar{V}_i \subset W_i$ and $\{W_i \cap F\}$ is irreducible $V_i \cap \cdots \cap V_j \cap F \neq \emptyset \leftrightarrow W_i \cap \cdots \cap W_j \cap F \neq \emptyset \leftrightarrow \bar{V}_i \cap \cdots \cap \bar{V}_j \cap F \neq \emptyset$.

Let B be the union of the intersections $\bar{V}_i \cap \cdots \cap \bar{V}_j$ which do not meet F. Since their number is finite, B is closed and so $A = \mathfrak{R} - B$ is a neighborhood of F. We also have $\bar{V}_i \cap \cdots \cap \bar{V}_j \cap A \neq \emptyset \leftrightarrow \bar{V}_i \cap \cdots \cap \bar{V}_j \cap F \neq \emptyset \leftrightarrow V_i \cap \cdots \cap V_j \cap F \neq \emptyset$. This proves (14.4) for the open neighborhoods of F. Since $\mathfrak{R}$ is normal there is a neighborhood A' of F such that $\bar{A}' \subset A$ and clearly in the equivalences just written A may be replaced by $\bar{A}'$. This proves the statement regarding the closed neighborhoods.

PROOF OF (14.3). Let $\{\mathfrak{B}_\mu\}$ be the collection of all the coverings such as $\mathfrak{B}$ of (14.4). By (14.4) this collection is cofinal in the collection of all the finite open coverings and so it may replace it in all homology questions. Let then $\Psi = \{\Psi_\mu\}$ be the net of the nerves of the $\mathfrak{B}_\mu$ and introduce the closed web $\Psi_1(\bar{\mathfrak{A}}) = \{\Psi_1(\bar{A}_\lambda)\}$ (notations of 4, with Ψ in place of Φ). Let $\Psi_1(\bar{A}_\lambda) = \{\Psi_{1\mu}^\lambda\}$ and let σ be a simplex of $\Psi_{1\mu}^\lambda$. Since $\mathfrak{B}_\mu$ has the property of the $\mathfrak{B}$ of (14.4) if the kernel $[\sigma]$ of σ meets every $\bar{A}_\lambda$ it meets F, and conversely. Hence

$$\Psi_1(F) = \bigcap \Psi_1(\bar{A}_\lambda), \tag{14.5}$$

and so (14.3) is a consequence of (VI, 23.2).

15. **Convergency of the cycles of a normal space to a closed set** F. The following question arises in one form or another in the applications. There is given a class Γ^p of cycles of the space $\mathfrak{R}$ such that every closed neighborhood

$\bar{A}_\lambda$ of F contains an element $\gamma_\lambda^p \in \Gamma^p$, with the property that $\lambda > \mu$ (i.e., $\bar{A}_\lambda \subset \bar{A}_\mu) \rightarrow \gamma_\lambda^p \sim \gamma_\mu^p$ in $\bar{A}_\mu$. Does there exist a cycle $\gamma_0^p \in \Gamma^p$ in the set F itself? If so, $\{\gamma_\lambda^p\}$ is said to *converge* to γ_0^p. We have:

(15.1) *If* $\mathfrak{R}$ *is normal then*: (a) *a collection such as* $\{\gamma_\lambda^p\}$ *always converges to a cycle* γ_0^p *of* F; (b) *if the cycle* $\gamma_0^p \sim 0$ *in every closed neighborhood* $\bar{A}_\lambda$ *of* F *then* $\gamma_0^p \sim 0$ *in* F *also.*

Under the assumptions $\gamma^p = \{\gamma_\lambda^p\}$ is a cycle of the inverse web $\Phi_1(\mathfrak{A})$. Referring to (VI, 23.7) we may conclude from (14.5) that there is a cycle γ_0^p of $\Phi_1(F)$, that is to say of F, such that: (α) $\gamma_0^p \cup \{\gamma_\lambda^p\} = {}'\gamma^p$ is a cycle of $\Phi_1(F) \cup \Phi_1(\mathfrak{A})$; (β) $\gamma^p \rightarrow {}'\gamma^p$ defines an isomorphism between the corresponding homology groups of $\mathfrak{A}$ and $F \cup \mathfrak{A}$. It is a consequence of (α) that $\gamma_0^p \sim \gamma_\lambda^p$ in $\bar{A}_\lambda$ which is (15.1a). Regarding (15.1b) since $F \subset \bar{A}_\lambda$, $\gamma_0^p \cup \{\gamma_\lambda'^p\}$, ($\gamma_0^p$ a cycle of F, $\gamma_\lambda'^p$ a cycle of $\bar{A}_\lambda$, $\gamma_\lambda'^p = \gamma_0^p$) is a cycle ${}'\gamma^p$ of $\Phi_1(F) \cup \Phi_1(\mathfrak{A})$. Under the assumption of (15.1b): $\{\gamma_\lambda'^p\} \sim 0$ in $\Phi_1(\mathfrak{A})$, and hence by (β) ${}'\gamma^p \sim 0$ in $\Phi_1(F) \cup \Phi_1(\mathfrak{A})$, which implies $\gamma_0^p \sim 0$ in F.

Another case of convergency of the cycles is described in:

(15.2) *Suppose that the* $\{\bar{A}_\nu\}$ *coinitial in* $\{\bar{A}_\lambda\}$ *have their pth Betti numbers* mod π *finite. If the* γ_ν^p *are all members of the same homology class* Γ^p *over a field* G *of characteristic* π *then* $\{\gamma_\nu^p\}$ *converges to a cycle* γ_0^p *of* F *which is in* Γ^p. *Furthermore if* $\gamma_\nu^p \sim 0$ *in* $\bar{A}_\nu$ *then* $\gamma_0^p \sim 0$ *in* F.

Let $\mathfrak{H}^p(\bar{A}_\lambda, G) = H_\lambda$, and let η_μ^λ, $\lambda > \mu$, denote the injection $\bar{A}_\lambda \rightarrow \bar{A}_\mu$. Then $S = \{H_\nu; \eta_{\nu'}^\nu\}$ is an inverse system of vector spaces over G. Since dim H_ν depends solely upon the characteristic π of G, it is finite and so H_ν is linearly compact (II, 27.7). The elements (homology classes) of H_ν which contain a $\gamma^p \in \Gamma^p$ form a linear variety L_ν in H_ν. Since H_ν is finite-dimensional it is discrete (II, 25.6) and hence L_ν is closed in H_ν, and consequently it is linearly compact also (II, 27.3). Since manifestly $\nu > \nu' \rightarrow \eta_{\nu'}^\nu L_\nu \subset L_{\nu'}$, by (II, 27.6) there is a limit-element of S whose ν coordinate is in L_ν. This means that there is a collection $\{\gamma_\nu^p\}$, γ_ν^p a cycle of $\bar{A}_\nu$ in Γ^p, such that $\nu > \nu' \rightarrow \gamma_\nu^p \sim \gamma_{\nu'}^p$ in $\bar{A}_{\nu'}$. Since $\{\nu\}$ is cofinal in $\{\lambda\}$ there is a similar collection $\{\gamma_\nu^p\}$ relative to $\{\bar{A}_\lambda\}$, and (15.2) follows then from (15.1).

(15.3) Betti and Alexandroff numbers over a field G may be introduced both for the cycles around and for those through F. Since they are defined by means of nets they depend solely upon the characteristic π of G and so they are written $R_0^p(F, \pi)$, $\rho_0^p(F, \pi)$ for the cycles around F and $R_1^p(F, \pi)$, $\rho_1^p(F, \pi)$ for those through F.

When $\mathfrak{R}$ is normal the numbers $R_0^p(F, \pi)$ are those of F itself (15.1). Notice that the definition of (II, 35) for the numbers ρ_0^p may be put in the following form. Denote by $\rho_0^p(A, A', \pi)$ the maximum number of p-cycles of $\bar{A}'$ over π which are independent in $\bar{A}$, where $A \supset A'$. Then

$$\rho_0^p(F, \pi) = \sup_{A} \{\inf_{A'} \rho_0^p(A, A', \pi)\}. \tag{15.4}$$

Similarly of course for the numbers ρ_1^p.

From (II, 35.7) there comes

(15.5) $\rho_0^p(F, \pi) = R_0^p(F, \pi)$ *or* ∞.

(15.6) The neighborhood groups and related numbers R, ρ may be defined for any set A as those of its closure $\bar{A}$.

(15.7) *Groups at a point.* They are merely the neighborhood groups of x, i.e., those of its closure $\bar{x}$. In particular if $\mathfrak{R}$ is a normal T_1-space then x is closed, and so, (14.3), *the groups around x are those of x itself.* Hence $R_0^p(x, \pi) = 0$ for $p > 0$, $R_0^0(x, \pi) = 1$. This yields together with (15.5) the following noteworthy result due to Alexandroff [c]:

(15.8) *In a normal T_1-space $\rho_0^p(x, \pi) = 0, \infty$ for $p > 0$ and $\rho_0^0(x, \pi) = 1$.*

(15.9) It is hardly necessary to observe that all the groups that we have just defined are topological invariants. In particular the numbers $R_1^p(x, \pi)$, $\rho_1^p(x, \pi)$ are noteworthy topological characters of the points in relation to their neighborhoods.

(15.10) Supposing $\mathfrak{R}$ to be a T_1-space there may be defined the cyclic or acyclic properties of $\mathfrak{R}$ at x in the usual way in terms of the groups of the cycles through x. Thus $\mathfrak{R}$ is *n-cyclic* at x if all the n-dimensional homology groups through x are cyclic and the rest zero.

16. **Products of compact Hausdorff spaces.** We shall utilize webs to extend to such products the product formulas for Betti numbers of (IV, 6.9, 6.10; VI, 11.3). The method was suggested to the author for a product of compacta of finite dimension by Samelson.

(16.1) Any compact Hausdorff space may be identified topologically with a closed subset $\mathfrak{R}$ of a compact parallelotope $P = \mathbf{P}l_\lambda$, where l_λ is the segment $0 \leqq x_\lambda \leqq 1$. Let us denote by $U_{\lambda_1 \cdots \lambda_r}$ any set of the form $\mathbf{P}A_\lambda$, where $A_\lambda = l_\lambda$ for $\lambda \neq \lambda_1, \cdots, \lambda_r$ and A_{λ_i} is an open subset of l_{λ_i}. Let N be any neighborhood of $\mathfrak{R}$ in P. Since $\{U_{\lambda_1 \cdots \lambda_r}\}$ is a base, given any $x \in \mathfrak{R}$ there is a $U_{\lambda_1 \cdots \lambda_r}$ between x and N. Since $\mathfrak{R}$ is compact it is covered by a finite number of such sets. If $\{\lambda_1, \cdots, \lambda_s\}$ are all the indices present in them, then each is a $U_{\lambda_1 \cdots \lambda_s}$, and so we have a finite open covering $\{U^i_{\lambda_1 \cdots \lambda_s}\}$ of $\mathfrak{R}$ in P. Let $P_{\lambda_1 \cdots \lambda_s}$ denote the "face" of P consisting of all points for which $x_\lambda = 0$ for $\lambda \neq \lambda_1, \cdots, \lambda_s$, and let $\pi_{\lambda_1 \cdots \lambda_s}$ denote the projection $P \rightarrow P_{\lambda_1 \cdots \lambda_s}$ whereby each point x is sent into the point of $P_{\lambda_1 \cdots \lambda_s}$ with the same coordinates x_{λ_i}. Since $\pi_{\lambda_1 \cdots \lambda_s}$ is an open mapping (I, 12.1), $V^i_{\lambda_i \cdots \lambda_s} = \pi_{\lambda_1 \cdots \lambda_s} U^i_{\lambda_1 \cdots \lambda_s}$ is an open set of $P_{\lambda_1 \cdots \lambda_s}$ and evidently $U^i_{\lambda_1 \cdots \lambda_s} = \pi^{-1}_{\lambda_1 \cdots \lambda_s} V^i_{\lambda_1 \cdots \lambda_s}$. Hence if $\mathfrak{R}_{\lambda_1 \cdots \lambda_s} = \pi_{\lambda_1 \cdots \lambda_s} \mathfrak{R}$, then we have $V_{\lambda_1 \cdots \lambda_s} = \bigcup V^i_{\lambda_1 \cdots \lambda_s}$ as a neighborhood of $\mathfrak{R}_{\lambda_1 \cdots \lambda_s}$ in $P_{\lambda_1 \cdots \lambda_s}$ such that $U_{\lambda_1 \cdots \lambda_s} = \pi^{-1}_{\lambda_1 \cdots \lambda_s} V_{\lambda_1 \cdots \lambda_s}$ is a neighborhood of $\mathfrak{R}$ contained in N.

(16.2) We now borrow a few simple results from (VIII) proved independently of the present considerations. Since $P_{\lambda_1 \cdots \lambda_s}$ is a finite polyhedron $| \Pi |$ by (VIII, 4.1) applied to $| \Pi' |$ there is a barycentric derived $\Pi^{(q)}$ whose mesh is arbitrarily small. In particular, q may be so chosen that all the simplexes of Π^q whose closures meet $\mathfrak{R}_{\lambda_1 \cdots \lambda_s}$ have their closures in $V_{\lambda_1 \cdots \lambda_s}$. Their union is a so-called "polyhedral" neighborhood of $\mathfrak{R}_{\lambda_1 \cdots \lambda_s}$ in $P_{\lambda_1 \cdots \lambda_s}$ which we now call $V_{\lambda_1 \cdots \lambda_s}$. We still have $\mathfrak{R} \subset U_{\lambda_1 \cdots \lambda_s} \subset N$ and we note that $\bar{V}_{\lambda_1 \cdots \lambda_s}$ is now a finite polyhedron. We will also designate $U_{\lambda_1 \cdots \lambda_s}$ as a "polyhedral" neighborhood of $\mathfrak{R}$ in P.

(16.3) By the preceding argument the polyhedral neighborhoods of $\mathfrak{R}$ are cofinal in the family of all its neighborhoods. Therefore the inverse web $\mathfrak{A} = \{\bar{U}_{\lambda_1 \cdots \lambda_s}\}$ has the same homology groups as $\mathfrak{R}$ itself (14.3).

It is an elementary matter to prove that $\pi_{\lambda_1 \cdots \lambda_s} \bar{U}_{\lambda_1 \cdots \lambda_s} = \bar{V}_{\lambda_1 \cdots \lambda_s}$. Moreover if $x \in \bar{U}_{\lambda_1 \cdots \lambda_s}$ and $x' = \pi_{\lambda_1 \cdots \lambda_s} x$, the segment $\overline{xx'} \in \bar{U}_{\lambda_1 \cdots \lambda_s}$. Therefore $\bar{V}_{\lambda_1 \cdots \lambda_s}$ is a deformation-retract of $\bar{U}_{\lambda_1 \cdots \lambda_s}$ (I, 47.4) and so their homology groups are the same. Thus $\bar{U}_{\lambda_1 \cdots \lambda_s}$ has the homology groups of a finite polyhedron.

(16.4) Suppose now that we have a second compact Hausdorff $\mathfrak{R}'$ and let P', $\cdots$ have their obvious meaning. Then $\mathfrak{R} \times \mathfrak{R}'$ is topologically immersed in the parallelotope $P \times P'$ and $\mathfrak{B} = \{\bar{U}_{\lambda_1 \cdots \lambda_s} \times \bar{U}_{\lambda_1' \cdots \lambda_t'}\}$ is the analogue of $\mathfrak{A}$ for $\mathfrak{R} \times \mathfrak{R}'$. We may apply here the reasoning of (VI, 11) for net products and this will prove the extension of the product formulas. Explicitly:

(16.5) *The product formulas for Betti numbers and Poincaré polynomials of* (IV, 6.9, 6.10) *are valid for a product of compact Hausdorff spaces $\mathfrak{R} \times \mathfrak{R}'$ wherever they have a meaning, i.e., wherever the Betti numbers of the factors involved are finite. If any $R^p(\mathfrak{R}, \pi)$, $R^q(\mathfrak{R}, \pi)$, $p + q = s$, is infinite and the other is not zero then $R^s(\mathfrak{R} \times \mathfrak{R}', \pi)$ is infinite.*

§4. GROUPS RELATED TO THE UNION AND INTERSECTION OF TWO SETS

17. Given two sets A, B in $\mathfrak{R}$, there exist interesting isomorphisms between their homology groups and certain groups of $A \cup B$ and $A \cap B$. These relations are the generalization of similar relations for finite complexes due to W. A. Mayer [a] (see also Vietoris [b]) which we shall first derive (following [A–H, 287]).

Let then X be a finite complex and X_1, X_2 closed subcomplexes of X such that $X = X_1 \cup X_2$. It is clear that $X_{12} = X_1 \cap X_2$ is likewise a closed subcomplex of X.

We shall deal exclusively with chains, $\cdots$ over a division-closure group G. Since all the complexes are finite it will mean that "~ 0" $\leftrightarrow$ "bounding" (III, 17.2).

All the chains, $\cdots$ being over G the following notations shall be used:

$C, \gamma, \Gamma, \mathfrak{Z}, \mathfrak{F}, \mathfrak{H}$ chains, cycles, homology classes, groups of cycles, of bounding cycles, homology groups for X;

$C_i, \cdots, C_{12} \cdots$ the same for X_i, X_{12};

δ_i a cycle of X_i mod X_{12};

R^p, R_i^p, R_{12}^p the Betti numbers of X, X_i, X_{12} over G when G is a field.

In addition to the groups $\mathfrak{H}$ we shall also require the following:

$\mathfrak{L}^p$ = the subgroup of $\mathfrak{H}_{12}^p$ consisting of the classes Γ_{12}^p of the cycles γ_{12}^p which are ~ 0 in both X_1 and X_2;

$\mathfrak{D}^p$ = the subgroup of $\mathfrak{H}^p$ consisting of the classes of the cycles $\gamma_1^p - \gamma_2^p$;

$H^p = \mathfrak{H}_1^p \times \mathfrak{H}_2^p$; its elements are the pairs (Γ_1^p, Γ_2^p);

$\mathfrak{M}^p$ = the subgroup of H^p consisting of the pairs (Γ_1^p, Γ_2^p) such that there exists a $\gamma_{12}^p \in \Gamma_1^p$ and $\in \Gamma_2^p$.

18. We have at once

$$C^p = C_1^p - C_2^p. \tag{18.1}$$

Therefore

$$(C_1^p - C_2^p = 0) \rightarrow (C_1^p = C_2^p = C_{12}^p), \tag{α}$$

since in this case the elements of the two chains can only be in X_{12}.

In particular if $C^p = \gamma^p = C_1^p - C_2^p$, we have $\mathrm{F}C_1^p = \mathrm{F}C_2^p = \gamma_{12}^{p-1}$ and so C_i^p is a δ_i^p or

$$\gamma^p = \delta_1^p - \delta_2^p, \qquad \mathrm{F}\delta_i^p = \gamma_{12}^{p-1}. \tag{18.2}$$

Suppose $\gamma^p \sim \gamma'^p = \delta_1'^p - \delta_2'^p$, with $\mathrm{F}\delta_i'^p = \gamma_{12}'^{p-1}$. We have then

$$\gamma^p - \gamma'^p = \delta_1^p - \delta_1'^p - (\delta_2^p - \delta_2'^p) = \mathrm{F}C_1^{p+1} - \mathrm{F}C_2^{p+1},$$

and so by (α):

$$\delta_i^p - \delta_i'^p - \mathrm{F}C_i^{p+1} = C_{12}^p.$$

By taking boundaries we obtain then

$$\gamma_{12}^{p-1} - \gamma_{12}'^{p-1} = \mathrm{F}C_{12}^p \sim 0 \text{ in } X_{12}.$$

Thus although the decomposition (18.2) is not necessarily unique, when γ^p is in a specified class Γ^p then γ_{12}^{p-1} remains in a fixed class Γ_{12}^{p-1}. Since

$$\gamma_{12}^{p-1} \sim 0 \text{ in both } X_1 \text{ and } X_2, \tag{18.3}$$

we have $\Gamma_{12}^{p-1} \in \mathfrak{L}^{p-1}$. It is clear that $\Gamma^p \rightarrow \Gamma_{12}^{p-1}$ defines a homomorphism in the algebraic sense $\tau: \mathfrak{H}^p \rightarrow \mathfrak{L}^{p-1}$. We will show that τ is a homomorphism. Choose systematically C_1^p = (the part of γ^p in $X - X_2$) and still $\gamma_{12}^{p-1} = \mathrm{F}C_1^p$. If π is the projection $X \rightarrow X - X_2$ we have then $\gamma_{12}^{p-1} = \mathrm{F}\pi\gamma^p$. We have also shown that $\gamma^p \in \mathfrak{F}^p \rightarrow \gamma_{12}^{p-1} \in \mathfrak{F}_{12}^{p-1}$. Hence π induces a homomorphism: $\mathfrak{H}^p \rightarrow \mathfrak{H}_{12}^{p-1}$, and this homomorphism is in fact τ.

Now if $\gamma_{12}^{p-1} \in \Gamma_{12}^{p-1} \in \mathfrak{L}^{p-1}$, then $\gamma_{12}^{p-1} = \mathrm{F}\delta_i^p$, and so $\gamma^p = \delta_1^p - \delta_2^p$ is such that $\tau\Gamma^p = \Gamma_{12}^{p-1}$. Therefore τ is a mapping "onto": $\tau\mathfrak{H}^p = \mathfrak{L}^{p-1}$. To find the kernel of τ suppose $\Gamma_{12}^{p-1} = \tau\Gamma^p = 0$, or $\gamma_{12}^{p-1} \sim 0$ in X_{12}, where γ_{12}^{p-1} is as in (18.2). Then $\gamma_{12}^{p-1} = \mathrm{F}C_{12}^p$ and hence

$$\gamma^p = (\delta_1^p - C_{12}^p) - (\delta_2^p - C_{12}^p) = \gamma_1^p - \gamma_2^p. \tag{18.4}$$

In other words γ^p is then the difference of two absolute cycles in X_1, X_2. Conversely, when it has this form manifestly $\gamma_{12}^{p-1} \sim 0$. Therefore the kernel of τ is $\mathfrak{D}^p$. This yields our first basic isomorphism:

$$\mathfrak{H}^p/\mathfrak{D}^p \cong \mathfrak{L}^{p-1}. \tag{18.5^p}$$

Let η_i be the homomorphism $\mathfrak{H}_{12}^p \rightarrow \mathfrak{H}_i^p$ induced by the injection $X_{12} \rightarrow X_i$. Then $\Gamma_{12}^p \rightarrow (\eta_1\Gamma_{12}^p, \eta_2\Gamma_{12}^p)$ defines a homomorphism onto $\eta: \mathfrak{H}_{12}^p \rightarrow \mathfrak{M}^p$. Its kernel consists of the Γ_{12}^p such that $\eta_i \Gamma_{12}^p = 0$, $(i = 1, 2)$, i.e., of $\mathfrak{L}^p$. Hence our second basic isomorphism:

$$(18.6)^p \qquad \mathfrak{H}_{12}^p / \mathfrak{L}^p \cong \mathfrak{M}^p.$$

Consider now the cycles $\gamma_1^p - \gamma_2^p$ of X. If $\gamma_i^p \in \Gamma_i^p$ then $\gamma_1^p - \gamma_2^p$ is in a fixed class Γ^p of $\mathfrak{H}^p$ which is in $\mathfrak{D}^p$, and which depends solely upon Γ_1^p, Γ_2^p. Moreover clearly again $(\Gamma_1^p, \Gamma_2^p) \to \Gamma^p$ defines a homomorphism onto $\theta : H^p \to \mathfrak{D}^p$. The kernel of θ consists of those and only those pairs (Γ_1^p, Γ_2^p) such that $\gamma_1^p - \gamma_2^p = \mathrm{F}C^{p+1} = \mathrm{F}C_1^{p+1} - \mathrm{F}C_2^{p+1}$. By (α) this is equivalent to $\gamma_i^p - \mathrm{F}C_i^{p+1} = \gamma_{12}^p$, or to $\gamma_i^p \sim \gamma_{12}^p$. Hence the required kernel is $\mathfrak{M}^p$ and so we have our third basic isomorphism:

$$(18.7)^p \qquad H^p / \mathfrak{M}^p \cong \mathfrak{D}^p.$$

Suppose that G is a field and let l^p, m^p, h^p, d^p denote the dimensions of the groups $\mathfrak{L}^p$, $\mathfrak{M}^p$, H^p, $\mathfrak{D}^p$. We have then from $(18.5)^p$, $(18.6)^p$, $(18.7)^p$ and an obvious expression for h^p, the relations between these numbers and the Betti numbers:

$$(18.8)^p \qquad \begin{aligned} R^p - d^p &= l^{p-1}; \qquad & h^p &= R_1^p + R_2^p ; \\ R_{12}^p - l^p &= m^p; \qquad & d^p &= h^p - m^p. \end{aligned}$$

Hence by addition there comes the following noted formula due to W. Mayer:

$$(18.9)^p \qquad R^p = R_1^p + R_2^p - R_{12}^p + l^p + l^{p-1}.$$

(18.10) *Suppose now that* $X = X_1 \cup X_2$, *where the* X_i *are open in* X. Then $X_{12} = X_1 \cap X_2$ is likewise open in X. Hence $X^* = X_1^* \cup X_2^*$, $X_1^* \cap X_2^* = X_{12}^*$, where X_i^* and X_{12}^* are closed in X^*. Applying therefore the preceding results to X^*, we find that they hold under the conditions just stated for the cycles of X^*, i.e., for the cocycles of X. In other words, the *same results are still valid provided that the dimensions are those of cocycles.* The new relations are like $(18.5)^p, \cdots$ except that p must now be written as a subscript: $\mathfrak{H}_p$, $\mathfrak{L}_p, \cdots$. We designate them by $(18.5)_p, \cdots, (18.9)_p$.

19. The extension to nets offers little difficulty. Let $\Omega = \{\Omega_\lambda ; \pi_\mu^\lambda\}$ be a net and $\Phi, \Phi_1, \Phi_2, \Phi_{12}$ closed subnets of Ω with $\Phi = \{\Phi_\lambda\}, \cdots, \Phi_1$ and $\Phi_2 \subset \Phi$, $\Phi_\lambda = \Phi_{1\lambda} \cup \Phi_{2\lambda}$, $\Phi_{12\lambda} = \Phi_{1\lambda} \cap \Phi_{2\lambda}$. The previous designations for the cycles are applied to Φ, Φ_i, Φ_{12} and those for chains, $\cdots$ with the complementary index $\lambda : C_\lambda^p, \cdots$ to $\Phi_\lambda, \cdots$. In the first part of the extension Ω plays no real role but it reappears later and will help to make the combined situation more symmetrical.

The extension is made under the assumption that *G is compact or a field.*

The crux of the argument lies in the proof of the three basic isomorphisms.

Proof of $(18.5)^p$. Given $\gamma^p = \{\gamma_\lambda^p\} \subset \Phi$ we have from (18.2):

$$(19.1) \qquad \gamma_\lambda^p = \delta_{1\lambda}^p - \delta_{2\lambda}^p, \qquad \mathrm{F}\delta_{i\lambda}^p = \gamma_{12\lambda}^{p-1}.$$

Therefore by (VI, 14.1), and with F applied to the δ_i^{p-1} as there indicated: $\delta_i^p = \{\delta_{i\lambda}^p\}$ is a cycle of Φ_i mod Φ_{12}, $\gamma_{12}^{p-1} = \{\gamma_{12\lambda}^{p-1}\}$ a cycle of Φ_{12} and $\gamma_{12}^{p-1} =$

$F\delta_i^p$. If $\gamma'^p \sim \gamma^p$ determines similarly $\delta_i'^p$, $\gamma_{12}'^{p-1}$ we have $\delta_{i\lambda}^p \sim \delta_{i\lambda}'^p \bmod \Phi_{12\lambda}$ and so as in (18): $F\delta_{i\lambda}^p \sim F\delta_{i\lambda}'^p$ in $\Phi_{12\lambda}$, or $\gamma_{12\lambda}^{p-1} \sim \gamma_{12\lambda}'^{p-1}$ in $\Phi_{12\lambda}$, and finally $\gamma_{12}^{p-1} \sim \gamma_{12}'^{p-1}$. Thus again Γ^p determines a unique Γ_{12}^{p-1}. We define τ as before and show that $\tau\mathfrak{H}^p = \mathfrak{L}^{p-1}$. To find its kernel suppose $\gamma_{12}^{p-1} \in \Gamma_{12}^{p-1} = \tau\Gamma^p = 0$, or $\gamma_{12}^{p-1} \sim 0$ in Φ_{12}. By (18.4) then

$$\gamma_\lambda^p = \delta_{1\lambda}^p - C_{12\lambda}^p - (\delta_{2\lambda}^p - C_{12\lambda}^p) = \gamma_{1\lambda}^p - \gamma_{2\lambda}^p. \tag{19.2}$$

Let $\mathfrak{C}_{12\lambda}^p$, $\mathfrak{Z}_{1\lambda}^p$ be the groups of the chains of $\Phi_{12\lambda}$ and of the cycles of $\Phi_{1\lambda}$ over G, and set:

$$R_\lambda = (\delta_{1\lambda}^p - \mathfrak{C}_{12\lambda}^p) \cap \mathfrak{Z}_{1\lambda}^p. \tag{19.3}$$

The parenthesis is a coset of the group $\mathfrak{C}_{1\lambda}^p$ mod the closed subgroup $\mathfrak{C}_{12\lambda}^p$, and $\mathfrak{Z}_{1\lambda}^p$ is a closed subgroup of $\mathfrak{C}_{1\lambda}^p$. If G is compact [a field] the groups and coset are compact [linearly compact], and so is R_λ which is in any case different from $\emptyset$. If η_λ is the projection $\mathfrak{Z}_{1\lambda}^p \to \mathfrak{H}_{1\lambda}^p$ (homology group of $\Phi_{1\lambda}$ over G) then $S_\lambda = \eta_\lambda R_\lambda$ has the same compactness properties as R_λ. Clearly $\lambda > \mu \to \bar{\pi}_\mu^\lambda S_\lambda \subset S_\mu$, where $\bar{\pi}_\mu^\lambda$ is the projection $\mathfrak{H}_\lambda^p \to \mathfrak{H}_\mu^p$ induced by π_μ^λ. Therefore $\{S_\lambda; \bar{\pi}_\mu^\lambda\}$ is an inverse mapping system of non-void compact [linearly compact] spaces and so it has a limit-element Γ_1^p. If $\gamma_1^p = \{\gamma_{1\lambda}^p\} \in \Gamma_1^p$ then $\gamma_{1\lambda}^p = \delta_{1\lambda}^p - C_{12\lambda}^p$. Hence $\gamma_2^p = \{\gamma_{2\lambda}^p\}$, $\gamma_{2\lambda}^p = \delta_{2\lambda}^p - C_{12\lambda}^p$, is a cycle of Φ_2 such that (18.4) holds. The rest of the proof is completed as before.

PROOF OF $(18.6)^p$. Essentially as before.

PROOF OF $(18.7)^p$. θ is defined as before and the discovery of the kernel reduces again to finding what happens when $\gamma_1^p \sim \gamma_2^p$, or here $\gamma_{1\lambda}^p \sim \gamma_{2\lambda}^p$. As shown in (18) we have then

$$\gamma_{i\lambda}^p \sim \gamma_{12\lambda}^p \text{ in } \Phi_{i\lambda}. \tag{19.4}$$

If $\gamma_{12\lambda}'^p$ is any other choice for $\gamma_{12\lambda}^p$ in (19.4) then $\Gamma_{12\lambda}'^p - \Gamma_{12\lambda}^p \in \mathfrak{L}_\lambda^p$ which is compact [linearly compact] when G is compact [is a field] and so $T_\lambda = \{\Gamma_{12\lambda}'^p\} = \Gamma_{12\lambda}^p + \mathfrak{L}_\lambda^p$ is likewise compact [linearly compact]. Since again $\lambda > \mu \to \pi_\mu^\lambda T_\lambda \subset T_\mu$, we show as before that $\gamma_{12\lambda}^p$ may be chosen such that $\gamma_{12}^p = \{\gamma_{12\lambda}^p\}$ is a cycle of Φ_{12}. By (19.4):

$$\gamma_i^p \sim \gamma_{12}^p \text{ in } \Phi_i \tag{19.5}$$

and this enables us to conclude the argument as before.

Thus our isomorphisms are now proved.

When the Betti numbers involved are finite the other characters in $(18.8)^p$ are likewise finite and so Mayer's formula $(18.9)^p$ will hold in that case also.

20. Suppose now that Φ_1, Φ_2 are arbitrary closed subnets of Ω and set $\Psi = \Omega - \Phi_{12}$, $\Psi_i = \Omega - \Phi_i$. It is known that Ψ is a net, and it is an elementary matter to verify that Ψ_i is an open subnet of Ψ. This time $\Psi_{12\lambda} = \Psi_{1\lambda} \cap \Psi_{2\lambda}$ is open in Ψ_λ and $\Psi_{12} = \{\Psi_{12\lambda}\}$ is an open subnet of Ψ. In place of the cycles we must now take the cocycles over a discrete group G. Now the relations between the cocycles are always reducible to relations for one coordinate with possible invariance under a single projection $\pi_\lambda^{*\mu}$. On the strength of this $(18.5)_p$,

$\cdots$, $(18.9)_p$ for nets, defined as in (18.10), are proved like $(18.5)^p$, $\cdots$, in (18), and the details are left to the reader.

21. The applications to a topological space $\mathfrak{R}$ require little more than a transfer from nets to spaces.

(21.1) Let $F = F_1 \cup F_2$, $F_{12} = F_1 \cap F_2$ be closed sets in $\mathfrak{R}$ and let Ω be the net of the finite open coverings $\{\mathfrak{U}_\lambda\}$. Except for replacing Φ by Ω the notations are as in (4). Define

$$\Phi = \Omega_1(F);\ \Phi_i = \{\Phi_{i\lambda}\} = \Omega_1(F_i),\ \Phi_{12} = \{\Phi_{12\lambda}\} = \Omega_1(F_{12}).$$

The situation is then as in (19) and so $(18.5)^p$, $\cdots$, $(18.9)^p$ hold under the following conditions:

(a) G is compact or a field;

(b) $\mathfrak{H}^p$, R^p, $\mathfrak{H}_i^p$, R_i^p, $\mathfrak{H}_{12}^p$, R_{12}^p are the homology groups and Betti numbers of F, F_i, F_{12};

(c) the groups H^p, $\cdots$, $\mathfrak{D}^p$ are defined as in (17) with X, X_i, X_{12} replaced by F, F_i, F_{12}.

(21.2) Let now $U = \mathfrak{R} - F_{12}$, $U_i = \mathfrak{R} - F_i$, $U_{12} = U_1 \cap U_2 = \mathfrak{R} - (F_1 \cup F_2)$. Let Φ, Φ_i be defined as above and set $\Psi = \Omega - \Phi_{12}$, $\Psi_i = \Phi - \Phi_i$. The situation is that of (20). Ψ_{12} is introduced as in (20) and $(18.5)_p$ $\cdots$ $(18.9)_p$ are valid under the following conditions:

(a) G is discrete or a field;

(b) $\mathfrak{H}_p$, R_p are the cohomology groups and Betti numbers of $U = \mathfrak{R} - (F_1 \cap F_2)$;

(c) the groups H_p, $\cdots$, $\mathfrak{D}_p$ are defined as H^p $\cdots$ in (17) with X, X_i, X_{12} replaced by U, U_i, U_{12}, and cocycles in place of cycles.

22. Interesting results are obtained by specializing some of the sets.

(22.1) *$F_{12} = F_1 \cap F_2$ is $(p - 1, p)$-acyclic.* Then $\mathfrak{L}^{p-1} = 0$ as a subgroup of $\mathfrak{H}_{12}^{p-1}$ and $\mathfrak{L}^p = \mathfrak{M}^p = 0$ as subgroups of $\mathfrak{H}_{12}^p$. Consequently by $(18.7)^p$: $H^p \cong \mathfrak{D}^p$, and from $(18.5)^p$: $\mathfrak{H}^p \cong \mathfrak{D}^p$. Therefore $\mathfrak{H}^p = \mathfrak{H}_1^p \times \mathfrak{H}_2^p$, and $R^p = R_1^p + R_2^p$ when G is a field. In particular, when F_{12} is acyclic this holds for all p.

(22.2) *$\mathfrak{R}$ is $(p - 1, p)$-acyclic.* Let also (G, H) be a normal couple and let the cycles and cocycles be, respectively, over G, H. Suppose now $\mathfrak{H}_{12}^{p-1}(G) = 0$. By (9.1) then $\mathfrak{H}_p(H) = 0$ and so by $(18.5)_p$: $\mathfrak{L}_{p+1} = 0$. In other words, if a cocycle $\gamma_{\overline{p+1}}$ of U_{12} is ~ 0 in both U_1 and U_2 it is ~ 0 in U also.

Generally speaking, given F closed in $\mathfrak{R}$, we say with Alexander that a cocycle γ_q of $\mathfrak{R} - F$ *links* F when it is $\not\sim 0$ in $\mathfrak{R} - F$, and that it does not link F otherwise. The result just obtained may then be formulated as:

(22.3) THEOREM. *Let $\mathfrak{R}$ be $(p - 1, p)$-acyclic, (G, H) a normal couple, and the closed sets F_1, F_2 such that $\mathfrak{H}_{12}^{p-1}(G) = 0$. Then if γ_p is a cocycle of $\mathfrak{R}$ over H meeting neither F_1 nor F_2, and not linked with them, it is likewise not linked with F_{12}.*

This is in essence a theorem of the type of Phragmén-Brouwer. The strict theorem going by that name, and the generalizations due to Alexandroff [a, 178] will follow at once from (22.3), as soon as we gather more information about manifolds (VIII, 19.7).

Remark. With the topological applications in view, the results just obtained have been formulated for $\mathfrak{R}$ and closed sets. Analogous results may of course be stated for complexes or nets.

§5. THE VIETORIS HOMOLOGY THEORY FOR COMPACTA

23. Let $\mathfrak{R}$ be a compactum and let $\Omega = \{\sigma\}$ be the simplicial complex composed of all the simplexes whose vertices are points of $\mathfrak{R}$. We associate with σ the set $|\sigma|$ of its vertices, define diam σ = diam $|\sigma|$, and thus turn Ω into a metric complex (VI, 24.2) *the complete Vietoris complex* of $\mathfrak{R}$. Its V-theory, the prototype of all such theories, was introduced by Vietoris [a]. It is noteworthy for its convenience in many applications, but as we shall show, it is in fact reducible to the Čech theory.

Throughout this and the next section we shall understand by "simplex, complex, chain, cycle" the simplexes, finite subcomplexes, finite chains, finite cycles of Ω. By a *Vietoris cycle* or *cocycle* of $\mathfrak{R}$ is meant a *V-cycle, V-cocycle* of Ω (VI, 25). The "V-terminology" is in fact carried over bodily to Ω in the obvious way. There is no need to repeat the definitions and we shall merely consider the more special properties of Vietoris cycles. For reference we recall that such a cycle is a sequence $\{\gamma_n^p\}$ of finite cycles with a companion sequence of finite chains $\{C_n^{p+1}\}$ such that

$$\text{(23.1)} \qquad FC_n^{p+1} = \gamma_{n+1}^p - \gamma_n^p \text{ ; mesh } \gamma_n^p \text{, mesh } C_n^{p+1} \to 0.$$

Furthermore $\gamma^p \sim 0$ whenever there is another sequence of finite chains $\{C_n'^{p+1}\}$ such that

$$\text{(23.2)} \qquad FC_n'^{p+1} = \gamma_n^p, \qquad \text{mesh } C_n'^{p+1} \to 0.$$

The group properties are as in (VI, 25).

(23.3) *Notations.* The Vietoris groups of cycles will be designated by $\mathfrak{Z}_v^p$, $\mathfrak{F}_v^p$, $\mathfrak{H}_v^p$.

(23.4) It is convenient to define inclusion relations between the subsets of $\mathfrak{R}$ and the Vietoris elements. If K is a complex and A a closed set we define $K \subset A$ whenever $|K| \subset A$ (A contains all the vertices of K). Then $C^p \subset A \leftrightarrow |C^p| \subset A$, $\gamma^p = \{\gamma_n^p\} \subset A \rightarrow$ every $\gamma_n^p \subset A$.

(23.5) *Extension to local compacta.* Once Vietoris cycles and groups have been defined for compacta, one may introduce them for a local compactum $\mathfrak{R}$ by a mere paraphrase of (13). Generally speaking a Vietoris cycle γ^p of $\mathfrak{R}$ is merely a Vietoris cycle of a compactum $A \subset \mathfrak{R}$, and $\gamma^p \sim 0 \leftrightarrow \gamma^p \sim 0$ in some compactum $B \supset A$. The groups are defined as in (13) and they are discrete.

(23.6) *Topology of the Vietoris groups.* We have seen (VI, 25.1) that on general grounds the V-cycles of a metric complex are the elements of a weak

H-net, and so in accordance with our conventions they are only to be taken over discrete coefficient groups. This remains true, for instance, as regards the Vietoris groups of local compacta. We shall find, however, (28.1) that when $\mathfrak{R}$ is a compactum the Vietoris groups over division-closure groups may be identified with the Čech groups in the algebraic sense, and this will enable us to carry over to the Vietoris groups the topology of the Čech groups. This lies at the root of all the topological Vietoris groups encountered in the literature.

24. (24.1) *Let K be a subcomplex of Ω whose mesh is less than ϵ. If each vertex of K undergoes an ϵ displacement θ there is induced a simplicial chain-deformation of K over Ω to a complex L of mesh less than 3ϵ, and with homotopy operator $\mathfrak{D}$ such that the chains $\mathfrak{D}\sigma$ are likewise of mesh less than 3ϵ.*

If $\sigma = x_i \cdots x_j \in K$ then $\zeta = (\theta x_i) \cdots (\theta x_j)$ is a simplex and $L = \{\zeta\}$ a complex such that θ is a simplicial chain-mapping of K onto L. By (IV, 16.3) θ is a chain-deformation with simplicial operator $\mathfrak{D}$ such that $\mathfrak{D}\sigma$ and also ζ, are both in the closure of the simplex $x_i \cdots x_j (\theta x_i) \cdots (\theta x_j)$. Therefore the deformation is over Ω. Moreover since $d(x_i, x_j) < \epsilon$, $d(x_i, \theta x_i) < \epsilon$ and $d(\theta x_i, \theta x_j) \leqq d(x_i, x_j) + d(x_i, \theta x_i) + d(x_j, \theta x_j) < 3\epsilon$, our assertions as to mesh L and $\mathfrak{D}$ follow.

(24.2) *Let $\gamma^p = \{\gamma_n^p\}$ be a Vietoris cycle and let each vertex of γ_n^p undergo an η_n-displacement, where $\eta_n \to 0$. In accordance with* (24.1) *there results a certain chain-deformation θ_n of $|\gamma_n^p|$. Then $\{\theta_n\gamma_n^p\}$ is a Vietoris cycle $\sim \gamma^p$.*

If $\mathfrak{D}_n$ is the deformation operator for θ_n we have (IV, 14.4): $\mathrm{F}\mathfrak{D}_n\gamma_n^p = \theta_n\gamma_n^p - \gamma_n^p$ where mesh $\mathfrak{D}_n\gamma_n^p$, mesh $\theta_n\gamma_n^p < 3\eta_n$ and so approach 0. Hence (24.2) follows from (VI, 25.4).

(24.3) *The Vietoris homology groups are topological invariants.*

Let $\mathfrak{R}$, $\mathfrak{R}'$ be compacta and Ω, Ω' their complete Vietoris complexes. If there is a topological transformation $T:\mathfrak{R} \to \mathfrak{R}'$ then T induces a metric isomorphism of Ω with Ω' from which (24.3) follows (VI, 27.2).

(24.4) *If $\mathfrak{R}$, $\mathfrak{R}'$ are compacta than a mapping $T:\mathfrak{R} \to \mathfrak{R}'$ induces a homomorphism of the Vietoris homology groups of $\mathfrak{R}$ into the corresponding groups of $\mathfrak{R}'$.*

The notations being as above if $\sigma = x_i \cdots x_j \in \Omega$ then $T\sigma = (Tx_i) \cdots (Tx_j) \in \Omega'$ and $\sigma \to T\sigma$ defines a simplicial chain-mapping $\theta:\Omega \to \Omega'$. Since the spaces $\mathfrak{R}$, $\mathfrak{R}'$ are compacta diam $T\sigma \to 0$ uniformly with diam σ (I, 45.7). Hence θ is metric and so (24.3) is a consequence of (VI, 27.4).

25. Let A be a closed set in $\mathfrak{R}$ and let Ω_A be its complete Vietoris complex. The V-cycles of Ω around Ω_A (VI, 26.2) are known as *Vietoris cycles around A*. Their description is the same as for ordinary cycles in (23) except that the sequences $\{\gamma_n^p\}$, $\{C_n^{p+1}\}$, $\{C_n'^{p+1}\}$ are to have at most a finite number of terms not in any given $\mathfrak{S}(A, \epsilon)$. The corresponding homology groups are to be denoted by $\mathfrak{H}_A^p(\mathfrak{R}, G)$ and we prove:

$$\mathfrak{H}_A^p(\mathfrak{R}, G) \cong \mathfrak{H}_v^p(A, G). \tag{25.1}$$

Or explicitly: the corresponding Vietoris homology groups of A and of $\mathfrak{R}$ around A are isomorphic.

If we assign to each $x \in \mathfrak{R}$ a point $y \in A$ such that $d(x, y) = d(x, A)$ then $x \to y$ defines a simplicial chain-mapping $\theta:\Omega \to \Omega_A$ such that if $\sigma \in \Omega$ then diam $\theta\sigma \to 0$ uniformly with (diam $\sigma + d(\sigma, A)$). Coupling this with (24.2) we prove readily that if $\gamma^p = \{\gamma_n^p\}$ is a Vietoris cycle around A then $\delta^p = \{\theta\gamma_n^p\}$ is a Vietoris cycle of A whose class Δ^p depends solely upon the class Γ^p of γ^p. Since $\Gamma^p \to \Delta^p$ manifestly defines a homomorphism $\tau:\mathfrak{H}_A^p(\mathfrak{R}, G) \to \mathfrak{H}_v^p(A, G)$ there remains to show that τ is an isomorphism. Since every cycle δ^p is also a γ^p, and invariant under the operation θ, τ is onto. Referring to (IV, 16.3) θ is a chain-deformation over Ω whose operator $\mathfrak{D}$ is such that $\mathfrak{D}\sigma$ is a chain of Cl $\sigma(\theta\sigma)$. It follows that: (a) mesh $\mathfrak{D}\gamma_n^p \to 0$; (b) given any $\epsilon > 0$ there is at most a finite number of chains $\mathfrak{D}\gamma_n^p \not\subset \mathfrak{S}(A, \epsilon)$. Since $\mathrm{F}\mathfrak{D}\gamma_n^p = \theta\gamma_n^p - \gamma_n^p$, we have $\delta^p \sim \gamma^p$ as cycles around A. Now if $\delta^p \sim 0$ in A it is also ~ 0 as a cycle around A, and hence in that case $\gamma^p \sim 0$ also. This shows that τ is univalent. Therefore τ is an isomorphism and (25.1) follows.

§6. REDUCTION OF THE VIETORIS THEORY TO THE ČECH THEORY

26. Henceforth we assume the Vietoris groups topologized as indicated in (VI, 26.1). This being understood we may state the basic:

(26.1) THEOREM. *The Vietoris and Čech homology groups of a compactum $\mathfrak{R}$ over a discrete group G are isomorphic.*

A more complete result which includes (26.1) will in fact be proved but requires more detailed description.

(26.2) Consider first an arbitrary countable collection of finite open coverings $\{\mathfrak{U}_n\}$ such that: (a) $\mathfrak{U}_{n+1} > \mathfrak{U}_n$; (b) mesh $\mathfrak{U}_n \to 0$. If $\mathfrak{U}$ is any finite open covering then some mesh $\mathfrak{U}_n$ is less than Lebesgue number $\mathfrak{U}$, and so $\mathfrak{U}_n > \mathfrak{U}$. Thus $\{\mathfrak{U}_n\}$ is cofinal in the family of all the finite open coverings. Hence (5.1; VI, 18.2) if $\Phi_n =$ nerve $\mathfrak{U}_n$ and π_n^{n+1} is a projection by inclusion $\Phi_{n+1} \to \Phi_n$, then the homology theory of $\mathfrak{R}$ is the net theory of the spectrum $\Sigma = \{\Phi_n ; \pi_n^{n+1}\}$. Explicitly

(26.3) THEOREM. *The Čech homology theory of a compactum is the net theory of a sequential simplicial spectrum.*

It is of interest to observe here that (26.3) establishes the connection with the method of Alexandroff [a], and also of [L, VII, 323] for investigating the homology groups of compacta.

As a consequence of (26.3) it is sufficient for our purpose to consider the cycles $\delta^p = \{\delta_n^p\}$ of the spectrum Σ.

(26.4) Let Ω continue to denote the complete Vietoris complex of $\mathfrak{R}$. Let $\mathfrak{U}_n = \{U_{ni}\}$ and choose a point $x_{ni} \in U_{ni}$. Then $U_{ni} \to x_{ni}$ defines a simplicial chain-mapping $\tau_n:\Phi_n \to \Omega$. If $\delta^p = \{\delta_n^p\}$ is a cycle of Σ then $\gamma_n^p = \tau_n\delta_n^p$ is a finite cycle of Ω. Let $\mathfrak{H}_c$, $\mathfrak{H}_v$ denote the Čech and Vietoris groups and let Δ^p

be the class of δ^p. We will prove the following result which implies (26.1):

(26.5) *There is an infinite sequence* $\{n'\}$ *such that* (a) $\gamma^p = \{\gamma^p_{n'}\}$ *is a Vietoris cycle*; (b) *the class* Γ^p *of* γ^p *depends solely upon* Δ^p; (c) $\Delta^p \to \Gamma^p$ *defines an isomorphism* $\tau:\mathfrak{H}^p_c(\mathfrak{R}, G) \to \mathfrak{H}^p_v(\mathfrak{R}, G)$.

27. To avoid difficulties with the so-called "isolated points" of $\mathfrak{R}$ (points which are also open sets), the space is first imbedded topologically in a compactum $'\mathfrak{R}$ without isolated points (for instance, the Hilbert parallelotope P^ω (I, 46.3)) and $\mathfrak{R}$ is then identified with its image in $'\mathfrak{R}$. Referring to (5.1), as far as the Čech theory goes, and clearly then as regards (26.3), we may replace the finite open coverings of $\mathfrak{R}$ by the coverings whose sets are the intersections with $\mathfrak{R}$ of the sets of the finite open coverings of $'\mathfrak{R}$. We will therefore denote henceforth by $\{U\}$ the open sets of $'\mathfrak{R}$. Let $'\Psi$ be the nerve of the full collection of these open sets. The simplexes $\sigma \in {}'\Psi$ such that $[\sigma] \cap \mathfrak{R} \neq \emptyset$ form a closed subcomplex Ψ. In fact $U \to \mathfrak{R} \cap U$ defines a similitude $\Psi \to$ nerve $\{\mathfrak{R} \cap U\}$. We assign to $\sigma = U_i \cdots U_j \in \Psi$ the set $|\sigma| = U_i \cup \cdots \cup U_j$. If $\sigma' < \sigma$, the vertices of σ' are among those of σ', and hence $\overline{|\sigma'|} \subset \overline{|\sigma|}$. We now define diam σ = diam $|\sigma|$, and it turns Ψ into a metric complex (VI, 24.2). It is with this metric complex, still called Ψ, that we shall first be concerned. We notice at once the convenient property:

$$\text{(27.1)} \qquad \operatorname{diam} \sigma > 0,$$

which is the chief reason for imbedding $\mathfrak{R}$ in $'\mathfrak{R}$.

(27.2) *Let* $K = \{\sigma\}$ *be a simplicial complex and* τ_1, τ_2 *two simplicial chain-mappings* $K \to \Psi$. *Then* τ_1, τ_2 *are chain-homotopic and with a simplicial homotopy operator* $\mathfrak{D}$ *such that* $\|\mathfrak{D}\sigma\| \subset \|\tau_1\sigma\| \cup \|\tau_2\sigma\|$. *Hence if* K, τ_1, τ_2 *are metric and if* diam $(\|\tau_1\sigma\| \cup \|\tau_2\sigma\|) \to 0$ *uniformly with* diam σ, *then the chain-homotopy is likewise metric.*

As an immediate corollary we have:

(27.3) *Let* K, L *be closed subcomplexes of* Ψ *and* τ *a simplicial chain-mapping* $K \to L$ *such that* diam $(|\sigma| \cup \|\tau\sigma\|) \to 0$ *uniformly with* diam σ, $\sigma \in K$. *Then* τ *is a metric chain-deformation over* Ψ.

Proof of (27.2) Let $\{A_i\}$ be the vertices of K and let the open set $\tau_h A_i$, be denoted by U_{hi}. We will also set $U_i = U_{1i} \cup U_{2i}$. If $\sigma = A_{i_0} \cdots A_{i_p} \in K$ then $\bigcap_q U_{hi_q} \neq \emptyset$, and since $U_i \supset U_{hi}$, likewise $\bigcap U_{i_q} \neq \emptyset$. Therefore $A_i \to U_i$ defines a chain-mapping $\tau:K \to \Psi$. Since $U_i \supset U_{hi}$ we find at once from (IV, 16.3) that τ is chain-homotopic with τ_h, and this with a simplicial operator $\mathfrak{D}_h$ such that $\|\mathfrak{D}_h\sigma\| \subset \|\tau\sigma\| \cup \|\tau_h\sigma\| \subset \|\tau_1\sigma\| \cup \|\tau_2\sigma\|$. Hence (IV, proof of 15.1), τ_1 is chain-homotopic with τ_2 and with the homotopy operator $\mathfrak{D} = \mathfrak{D}_2 - \mathfrak{D}_1$ behaving as asserted.

(27.4) *Take a point* x *on each set* $\mathfrak{R} \cap U \neq \emptyset$. *Then*: (a) $U \to x$ *defines a metric simplicial chain-mapping* $\theta:\Psi \to \Omega$ (*the complete Vietoris complex*); (b) θ *induces an isomorphism of the group of V-cycles* $\mathfrak{H}^p_v(\Psi, G)$ *with the Vietoris group* $\mathfrak{H}^p_v(\mathfrak{R}, G)$.

It is clear that θ induces a homomorphism of the groups of the V-cycles of

Ψ into the corresponding Vietoris groups, and hence a homomorphism $\bar{\theta}:\mathfrak{H}_v^p(\Psi, G) \to \mathfrak{H}_v^p(\mathfrak{R}, G)$.

Consider now any finite subcomplex K of Ω, and let $\{x_i\}$ be its vertices and ϵ its mesh. Denote by U_i the union of the $\mathfrak{S}(x_j, \epsilon)$ such that x_j is a vertex of St x_i. Thus U_i is an open set of $'\mathfrak{R}$ meeting $\mathfrak{R}$, and hence a vertex of Ψ. If $x_i \cdots x_j \in K$ then $U_i \cap \cdots \cap U_j \neq \emptyset$ and so $x_i \to U_i$ defines a simplicial chain-mapping $\rho(K):K \to \Psi$. We notice that $x_i \in \rho(K)x_i$ and diam $\rho(K)x_i \leqq 4\epsilon$.

Let now $\gamma^p = \{\gamma_n^p\}, \{C_n^{p+1}\}$ be a Vietoris cycle with its companion sequence satisfying (23.1). If $\rho(| C_n^{p+1} |) = \rho_n$ then $\rho_n C_n^{p+1} = D_n'^{p+1}$ and $\rho_n \gamma_n^p = \delta_n^p$, $\rho_n \gamma_{n+1}^p = \delta_{n+1}'^p$ are such that

$$\mathrm{F}D_n'^{p+1} = \delta_{n+1}'^p - \delta_n^p ; \qquad \text{mesh } D_n'^{p+1}, \text{mesh } \delta_n^p \to 0.$$

Compare now δ_{n+1}^p with $\delta_{n+1}'^p$. If x is a vertex of γ_{n+1}^p then $\rho_n x$ and $\rho_{n+1} x$ meet. From this and the definition of the ρ_n we deduce that if $\sigma \in | \gamma_{n+1}^p |$ then diam $(|| \rho_n \sigma || \cup || \rho_{n+1} \sigma ||) < 4$ (mesh C_n^{p+1} + mesh $C_{n+1}^{p+1}) = \zeta_n \to 0$. Therefore (27.3) ρ_n, ρ_{n+1} as chain-mappings $| \gamma_{n+1}^p | \to \Psi$ are chain-homotopic with simplicial homotopy operator $\mathfrak{D}$ such that $\mathfrak{D}\sigma$ is of mesh less than ζ_n. Hence $D_n''^{p+1} = \mathfrak{D}\delta_{n+1}'^p$ is a chain of Ψ whose mesh is less than ζ_n and such that $\mathrm{F}D_n''^{p+1} = \delta_{n+1}^p - \delta_{n+1}'^p$.

Therefore if $D_n^{p+1} = D_n'^{p+1} + D_n''^{p+1}$, then

$$\mathrm{F}D_n^{p+1} = \delta_{n+1}^p - \delta_n^p ; \qquad \text{mesh } D_n^{p+1}, \text{mesh } \delta_n^r \to 0.$$

Consequently $\delta^p = \{\delta_n^p\}$ is a V-cycle of Ψ which we denote by $\rho\gamma^p$.

Suppose that $\gamma^p \sim 0$. We have then $\{C_n'^{p+1}\}$ such that (23.2) holds, and a slight modification of the preceding reasoning yields $\delta^p \sim 0$ in Ψ. Therefore $\gamma^p \to \delta^p$ induces a homomorphism $\bar{\rho}:\mathfrak{H}_v^p(\mathfrak{R}, G) \to \mathfrak{H}_v^p(\Psi, G)$.

If x is a vertex of γ_n^p then the point $\theta\rho_n x = \theta_n x$ is a point of $\mathfrak{R}$ such that sup $d(x, \theta_n x) \to 0$ with increasing n. Hence (23.5): $\theta\rho\gamma^p \sim \gamma^p$. Similarly if U is a vertex of δ_n^p then $\rho_n \theta U \cap U \neq \emptyset$ and so from (27.2): $\rho_n \theta \delta_n^p \sim_{\epsilon_n} \delta_n^p$, $\epsilon_n \to 0$. Hence $\rho\theta\delta^p \sim \delta^p$. From this follows $\bar{\rho}\bar{\theta} = 1$, $\bar{\theta}\bar{\rho} = 1$. We conclude then that $\bar{\theta}$ is an isomorphism and (27.4) is proved.

(27.5) Consider now a sequence $\{\mathfrak{V}_n\}$, $\mathfrak{V}_n = \{V_{ni}\}$, of finite open coverings of $'\mathfrak{R}$ such that:

(a) $\mathfrak{V}_1 = {}'\mathfrak{R}$;

(b) if $\epsilon_n =$ mesh $\mathfrak{V}_n$, $\eta_n =$ inf {diam V_{ni}, Lebesgue number $\mathfrak{V}_n$} then $\epsilon_{n+1} < \eta_n/2$.

As a consequence of (b) we also have

(c) $\mathfrak{V}_{n+1} > \mathfrak{V}_n$;

(d) $\{\epsilon_n\} \to 0$.

If $\{\mathfrak{W}_n\}$ is any sequence of finite open coverings such that mesh $\mathfrak{W}_n \to 0$, $\mathfrak{W}_1 = {}'\mathfrak{R}$, then a subsequence has properties (a), (b). It is clear that in (26.2) we may add the covering $\mathfrak{R}$ without disturbing the situation. Hence if we prove (26.5) with the $\mathfrak{U}_n$ of (26.2) replaced by $\mathfrak{U}_n = \{U_{ni}\}$, where the U_{ni} are the V_{nj} meeting $\mathfrak{R}$, then it will have been proved in its full generality.

(27.6) For convenience let Φ_n denote henceforth the set of the simplexes of nerve $\mathfrak{B}_n$ whose kernels meet $\mathfrak{R}$, and let π_n^{n+1} be a projection by inclusion $\Phi_{n+1} \to \Phi_n$. Thus the homology theory of $\mathfrak{R}$ is still the net theory of the spectrum $\Sigma = \{\Phi_n ; \pi_n^{n+1}\}$. The general projection $\Phi_n \to \Phi_m$, $n > m$, is given by $\pi_m^n = \pi_m^{m+1} \cdots \pi_n^{n+1}$. We also require the complexes:

X = the union of the simplexes of nerve $\bigcup \mathfrak{U}_n$ whose kernels meet $\mathfrak{R}$;

X_m = the similar complex for $\bigcup\{\mathfrak{U}_n \mid n \geqq m\}$.

If we apply (1.14) to the nerves of $\mathfrak{U}_m$ and $\bigcup\{\mathfrak{U}_n \mid n \geqq m\}$, we verify:

(27.7) *There exists a chain-deformation retraction* $\tau_m : X_m \to \Phi_m$ *with a simplicial deformation operator* $\mathfrak{D}_m$ *such that*: (a) $\mathfrak{D}_m\sigma \subset \mathrm{Cl}(\tau_m\sigma)$; (b) $\tau_m \mid \Phi_n = \pi_m^n$, $n > m$.

Regarding X we also have:

(27.8) *X is a metric chain-deformation retract of* Ψ.

Since the U_{ni} include $'\mathfrak{R} = U_{11}$, every open set U meeting $\mathfrak{R}$ is contained in some U_{ni}, and if $U_{ni} \supset U$ then $\eta_n \geqq \operatorname{diam} U$. It follows that at most a finite number of U_{ni} contain U. Among these select one πU for which n has its largest value or else $\pi U = U_{ni}$ itself when $U = U_{ni}$. Clearly $U \to \pi U$ defines a projection by inclusion $\Psi \to X$ which we still call π. Since $\pi \mid X = 1$, by (1.14) π is a chain-deformation retraction with simplicial operator $\mathfrak{D}$ such that $\mathfrak{D}\sigma \subset \mathrm{Cl}(\pi\sigma)$, and so mesh $\mathfrak{D}\sigma \leqq \operatorname{diam} \pi\sigma$. Now if $\operatorname{diam} \sigma < \epsilon_{n+1}$, every vertex of $\pi\sigma$ is in a set of $\mathfrak{U}_n$ and so $\operatorname{diam} \pi\sigma \leqq 2 \operatorname{mesh} \mathfrak{B}_n = 2\epsilon_n$, hence also mesh $\mathfrak{D}\sigma \leqq 2\epsilon_n$. It follows that both $\operatorname{diam} \pi\sigma$ and mesh $\mathfrak{D}\sigma \to 0$ uniformly with $\operatorname{diam} \sigma$. Hence π is a metric chain-deformation retraction, proving (27.8).

(27.9) A cycle of the spectrum Σ is a collection $\delta^p = \{\delta_n^p\}$, δ_n^p a cycle of Φ_n, such that:

(27.9a) $$\pi_n^{n+1}\delta_{n+1}^p \sim \delta_n^p \text{ in } \Phi_n .$$

The relation $\delta^p \sim 0$ is equivalent to

(27.9b) $$\delta_n^p \sim 0 \text{ in } \Phi_n .$$

In particular for cycles over a discrete G this means that Φ_n contains a chain D_n^{p+1} such that

(27.9c) $$\mathrm{F}D_n^{p+1} = \delta_n^p .$$

If Z^p, F^p are the net groups of the cycles and bounding cycles of Σ over G then F^p is closed in Z^p (VI, 3.5) and

(27.10) $$\mathfrak{H}_c^p(\mathfrak{R}, G) \cong Z^p/F^p.$$

(27.11) Since G is discrete the topology is defined as follows: the δ^p with an assigned δ_n^p make up an open set u_n in Z^p, and $\{u_n\}$ is a base for the group; F^p is closed in Z^p and $\mathfrak{H}_c^p(\mathfrak{R}, G)$ has the factor-group topology.

(27.12) Passing now to the V-cycles of Ψ, it is a consequence of (27.8) that as regards the homology theory we may replace Ψ by X. Consider then a V-cycle of X, $\gamma^p = \{\gamma_n^p\}$, with associated sequence $\{C_n^{p+1}\}$ and the relation (23.1). Since mesh $C_n^{p+1} \to 0$ it is in an X_{m_n} such that $m_n \to \infty$. Since $\{\gamma_n^p\}$ may be replaced by any infinite subsequence we may assume it such that $m_n \geqq n$

and so $C_n^{p+1} \subset X_n$. Applying τ_n (27.7) we obtain $\delta_n^p = \tau_n \gamma_n^p \subset \Phi_n$ and there will be a finite chain $'C_n^{p+1}$ in X_n such that $\mathrm{F}'C_n^{p+1} = \gamma_n^p - \delta_n^p$. From this and (23.1) it will follow that X_n contains a chain D_n^{p+1} such that $\mathrm{F}D_n^{p+1} = \delta_{n+1}^p - \delta_n^p$. Applying then τ_n we obtain (27.9a). If $\delta^p = \{\delta_n^p\} \sim 0$ (as a V-cycle) there exists a finite chain D_n^{p+1} of X such that $\mathrm{F}D_n^{p+1} = \delta_n^p$, and that mesh $D_n^{p+1} \to 0$. Hence there exists for each n an m_n such that $D_n^{p+1} \subset X_{m_n}$ and that $\{m_n\} \to \infty$. It follows that we may choose an $m > n$ such that $D_m^{p+1} \subset X_n$. Consequently $\tau_n \delta_n^p = \pi_n^m \delta_m^p = \mathrm{F}\tau_m D_m^{p+1}$, and so $\pi_n^m \delta_m^p \sim 0$ in Φ_n. From (27.9a) there follows then: $\pi_n^m \delta_m^p \sim \delta_n^p \sim 0$ in Φ_n. Thus if $\delta^p \sim 0$ as a V-cycle (27.9b) holds. The converse being manifestly true we find that whether δ^p is considered as a Čech cycle or a V-cycle the homology groups are Z^p/F^p in the algebraic sense. Referring however to (VI, 25.2) one verifies that $\{u_n\}$ of (27.11) is also a base for Z^p as a group of V-cycles. Hence the Čech groups of $\mathfrak{R}$ and V-groups of X, and consequently of Ψ, are isomorphic:

$$\mathfrak{H}_c^p(\mathfrak{R}, G) \cong \mathfrak{H}_v^p(\Psi, G). \tag{27.13}$$

Coupling this with (27.4), our basic theorem (26.1) follows.

All the material is also at hand for obtaining (26.5). First of all, in view of (24.2), the choice of the point $x_{ni} \in U_{ni}$ in (26.4) does not affect the class Γ^p of the Vietoris cycle γ^p there considered. And now if $\delta^p = \{\delta_n^p\}$ is the V-cycle of X, which is also a Čech cycle of $\mathfrak{R}$, considered in (27.11, 27.12), the isomorphism θ of (27.4) is precisely of the kind required by (26.5), as applied to the V-cycles of X of the particular form just considered. Since these cycles may also be viewed as Čech cycles of Σ without disturbing the homology groups, (26.5) follows.

28. Complementary remarks.

(28.1) Suppose that G is any division-closure group. If G_0 is the discrete group isomorphic with G in the algebraic sense, then $\mathfrak{H}_v^p(\mathfrak{R}, G_0)$ is isomorphic with $\mathfrak{H}_c^p(\mathfrak{R}, G_0)$ and hence (VI, 3.8) with $\mathfrak{H}_c^p(\mathfrak{R}, G)$ in the algebraic sense. It follows that we may assign to $\mathfrak{H}_v^p(\mathfrak{R}, G_0)$ the topology of $\mathfrak{H}_c^p(\mathfrak{R}, G)$. This is essentially the manner in which Pontrjagin [c] topologized the Vietoris groups, and for a compact G obtained compact "Vietoris" groups. It is important to observe that *this procedure succeeds only for division-closure groups*. For other groups the Vietoris definition yields $\mathfrak{H} = \mathfrak{Z}/\mathfrak{F}$, the Čech definition $\mathfrak{H} = \mathfrak{Z}/\bar{\mathfrak{F}}$, and the two may be distinct. Wherever the topology of the groups is to be taken into account it is advisable therefore to adopt the Čech definition.

(28.2) In order to show the full equivalence of the Vietoris and Čech homology groups for a compactum $\mathfrak{R}$, it is still necessary to show that they behave alike with respect to mappings. More precisely let T map $\mathfrak{R}$ into the compactum $\mathfrak{S}$, and let the analogues of $\Omega, \cdots$ for $\mathfrak{S}$ be designated by $\Omega', \cdots$. We have then $\tau: \mathfrak{H}_c^p(\mathfrak{R}, G) \to \mathfrak{H}_v^p(\mathfrak{R}, G)$ of (26.5) and the corresponding τ' for $\mathfrak{S}$. On the other hand T induces homomorphisms $\theta: \mathfrak{H}_c^p(\mathfrak{R}, G) \to \mathfrak{H}_c^p(\mathfrak{S}, G)$, $\theta_1: \mathfrak{H}_v^p(\mathfrak{R}, G) \to \mathfrak{H}_v^p(\mathfrak{S}, G)$. To prove the asserted equivalence is to prove the relation

$$\theta_1 \tau = \tau' \theta. \tag{28.3}$$

To establish this it is sufficient to show that if δ^p is the cycle of the spectrum Σ in (27.9), and γ^p a companion Vietoris cycle, and Δ^p, Γ^p their classes, then $\theta\Delta^p$, $\theta_1\Gamma^p$ are similarly related relatively to $\mathfrak{S}$. It is clear that as regards the question under consideration $\mathfrak{S}$ may be replaced by the compactum $T\mathfrak{R}$, and so we may assume T to be a mapping *onto*. Referring now to (5.12), and recollecting that $\{\mathfrak{U}_\nu\}$, $\{\mathfrak{B}_\mu\}$ there considered may be replaced by cofinal families, we find that in the argument there given θ is induced in substance by what, in our present notations, is a mapping $\eta : \delta^p \to \delta'^p$ of the cycles of Σ into those of Σ'. Moreover this mapping η goes together with a mapping of γ^p into a cycle $\epsilon\, \theta_1\Gamma^p$ such that $\theta_1\Gamma^p = \tau'\theta\Delta^p$. This is in outline the proof of (28.3).

(28.4) Returning to the situation of (27.6) let $\mathfrak{D}'_n = \mathfrak{D}_n \mid \Phi_{n+1} =$ the deformation operator for $\tau_n \mid \Phi_{n+1}$, and set $Y = \bigcup \mid \mathfrak{D}'_n\Phi_{n+1} \mid$. The complex Y has for Euclidean representation a *fundamental complex* K for $\mathfrak{R}$ in the sense of [L, 327]. In fact there is a metric isomorphism $Y \to K$. As shown loc. cit. K, hence equally well Y, may serve to define a homology theory for $\mathfrak{R}$, which by arguments similar to those used for the Vietoris theory, may likewise be reduced to the Čech theory. For rational coefficients the equivalence to the Vietoris theory was explicitly proved in [L, 330].

§7. HOMOLOGY THEORIES OF KUROSCH AND OF ALEXANDER-KOLMOGOROFF

29. A special type of covering will enable us to introduce, following Kurosch [a], a homology theory of the Čech type whose net is a spectrum. This theory is closely related to another of a type recently introduced by J. W. Alexander [e] and Kolmogoroff [a]. The common feature is the restoration of the chains and cochains as central elements. We shall find nevertheless that all these special homology theories are variants of the Čech theory.

30. (30.1) Definition. *An open set U of the topological space $\mathfrak{R}$ is said to be regular whenever* $U = \operatorname{Int} \bar{U}$. *A closed set A is said to be regular whenever* $A = \overline{\operatorname{Int} A}$.

(30.2) *The complement of a regular open [closed] set is a regular closed [open] set* (proof elementary).

(30.3) *If A, A' are regular closed sets so is $A \cup A'$. Hence if U, U' are regular open sets so is $U \cap U'$* (proof elementary).

(30.4) *If A $[U]$ is any closed [open] set then* $\operatorname{Int} A$ $[\bar{U}]$ *is a regular open [closed] set.*

If $V = \operatorname{Int} A$ then $\bar{V} \subset A$, and hence $\operatorname{Int} \bar{V} \subset \operatorname{Int} A = V$. Also $V \subset \bar{V} \to V \subset \operatorname{Int} \bar{V}$, and hence $V = \operatorname{Int} \bar{V}$; so V is regular. The treatment of U is essentially similar.

(30.5) *If U, U' are regular open sets then $U \subset U' \leftrightarrow \bar{U} \subset \bar{U}'$.*

The first inclusion manifestly implies the second. When the second holds we have ($\text{Int } \bar{U} = U$) $\subset$ ($\text{Int } \bar{U}' = U'$), proving the asserted equivalence.

(30.6) *If* $\mathfrak{A} = \{A_1, \cdots, A_p\}$ *is a finite closed covering and* $U_i = \text{Int } A_i$, $\mathfrak{U} = \{U_i\}$, *then* $\bar{\mathfrak{U}} > \mathfrak{A}$ *and* $\bar{\mathfrak{U}}$ *is a covering.*

It is only necessary to prove $\bar{\mathfrak{U}}$ a covering. In substance it is sufficient to show that if any set of $\mathfrak{A}$, say A_1, is replaced in $\mathfrak{A}$ by $\bar{U}_1$, then the collection is still a covering or that $W = \mathfrak{R} - (\bar{U}_1 \cup A_2 \cup \cdots \cup A_p) = \emptyset$. Now in the contrary case W is an open set not meeting A_i, $i > 1$, and so $W \subset A_1$, hence $W \subset U_1 = \text{Int } A_1$, which is ruled out. Therefore (30.6) is true.

31. **Gratings. The theory of Kurosch.**

(31.1) We understand by an *open grating*, or merely a *grating* of $\mathfrak{R}$, a finite collection of disjoint regular open sets $\mathfrak{B} = \{V_i\}$ such that $\bar{\mathfrak{B}}$ is a covering. We refer to $\bar{\mathfrak{B}}$ as a *closed* grating. It is on the gratings that the Kurosch theory is based. J. W. Alexander (to whom "grating" is due) utilized these coverings for a similar purpose.

(31.2) *If* $\mathfrak{B} = \{V_i\}$, $\mathfrak{B}' = \{V_j'\}$ *are two gratings, so is* $\mathfrak{B} \wedge \mathfrak{B}' = \{V_i \cap V_j'\}$.

The sets $\{V_i \cap V_j'\}$ are clearly disjoint, and they are regular (30.3); so we only need to prove that $\overline{\mathfrak{B} \wedge \mathfrak{B}'}$ is a covering. Now $\bar{\mathfrak{B}} \wedge \bar{\mathfrak{B}}' = \{\bar{V}_i \cap \bar{V}_j'\}$ is a covering, and hence by (30.6): $\{\overline{\text{Int }(\bar{V}_i \cap \bar{V}_j')}\}$ is likewise a covering. Since V_i, V_j' are regular, we have $\text{Int }(\bar{V}_i \cap \bar{V}_j') = \text{Int } \bar{V}_i \cap \text{Int } \bar{V}_j' = V_i \cap V_j'$. Therefore $\{\overline{V_i \cap V_j'}\} = \overline{\mathfrak{B} \wedge \mathfrak{B}'}$ is a covering, proving (31.2).

(31.3) *Every grating* $\mathfrak{B}$ *is of the form* $\mathfrak{B} = \mathfrak{B}_1 \wedge \cdots \wedge \mathfrak{B}_r$, *where the* $\mathfrak{B}_i$ *are binary gratings (i.e., composed of two sets).*

If $\mathfrak{B} = \{V_i\}$, $(i = 1, 2, \cdots, r)$, and $\mathfrak{B}_i = \{V_i, \text{Int }(\mathfrak{R} - V_i)\}$, then $\mathfrak{B}_i$ is a binary grating and $\mathfrak{B} = \mathfrak{B}_1 \wedge \cdots \wedge \mathfrak{B}_r$.

The property just proved was extensively utilized by Alexander in his "grating" theory.

(31.4) It is a consequence of (31.2) that any two gratings $\mathfrak{B}$, $\mathfrak{B}'$ have a common refinement which is a grating. Let then $\{\mathfrak{B}_\lambda\}$ be the collection of all the gratings and set $\Phi_\lambda = \text{nerve } \bar{\mathfrak{B}}_\lambda$. If $\{\lambda\}$ is ordered by $\lambda > \mu \leftrightarrow \mathfrak{B}_\lambda > \mathfrak{B}_\mu$, it becomes a directed set $\{\lambda; >\}$.

Suppose now $\lambda > \mu$, and let $\mathfrak{B}_\lambda = \{V_{\lambda i}\}$, $\mathfrak{B}_\mu = \{V_{\mu j}\}$. Every set $V_{\lambda i}$ is contained in a set $V_{\mu j}$ and $V_{\mu j}$ is *unique* since the sets of $\mathfrak{B}_\mu$ are disjoint. By (30.5) we also have: $V_{\lambda i} \subset V_{\mu j} \leftrightarrow \bar{V}_{\lambda i} \subset \bar{V}_{\mu j}$. It follows that $\bar{V}_{\lambda i} \to \bar{V}_{\mu j}$ defines a projection by inclusion $\pi_\mu^\lambda: \Phi_\lambda \to \Phi_\mu$. Thus $\Sigma = \{\Phi_\lambda; \pi_\mu^\lambda\}$ is a simplicial net with unique projections and so it is a simplicial spectrum, the *grating spectrum*. Since the Φ_λ and their projections are among those of the net Ψ of all the finite closed coverings, Σ is a partial net of Ψ.

Since Σ is a spectrum it may serve to define a net homology theory and a projective homology theory. We leave the latter aside for the present and prove:

(31.5) THEOREM. *The net homology theory of the grating spectrum is the same as the homology theory by finite closed coverings* ($\mathfrak{F}$-*theory*) (Kurosch [a]).

From (31.5) and (8.1) there follows then:

(31.6) *When $\mathfrak{R}$ is a normal T_1-space and hence when it is a metric space, a compact Hausdorff space or a compactum, then the finite open or closed coverings yield the same homology theory as the net theory of the grating spectrum.*

The proof consists in showing that Σ is cofinal in Ψ, or in the last analysis that every finite closed covering $\mathfrak{A} = \{A_1, \cdots, A_p\}$ has a refinement which is a closed grating. According to (30.6) $\mathfrak{A}$ has a refinement $\overline{\mathfrak{U}} = \{\overline{U}_1, \cdots, \overline{U}_q\}$, where the U_i are regular open sets; so we may start with $\overline{\mathfrak{U}}$. Let $V_i = U_i - \bigcup\{\overline{U}_j \mid j > i\}$, and define $V_0 = U_0 = \emptyset$. Then if $\mathfrak{W}_i = \{V_0, \cdots, V_i, U_{i+1}, \cdots, U_q\}$ we will show that every $\overline{\mathfrak{W}}_i$ is a covering. Since this is true for $\overline{\mathfrak{W}}_0$, we assume it for $\overline{\mathfrak{W}}_i$ and prove it for $\overline{\mathfrak{W}}_{i+1}$. If $\overline{\mathfrak{W}}_{i+1}$ is not a covering then $W = \mathfrak{R} - (\overline{V}_0 \cup \cdots \cup \overline{V}_{i+1} \cup \overline{U}_{i+2} \cup \cdots \cup \overline{U}_q)$ is not empty. Since $\overline{\mathfrak{W}}_i$ is a covering and W meets none of its sets except possibly $\overline{U}_{i+1}$, we must have $W \subset \overline{U}_{i+1}$, hence $W \subset \operatorname{Int} \overline{U}_{i+1} = U_{i+1}$, and finally from the expression of $V_{i+1}: W \subset V_{i+1}$ which is ruled out. Therefore all the $\overline{\mathfrak{W}}_i$ are coverings. In particular if $\overline{\mathfrak{W}}_q = \mathfrak{B} = \{V_1, \cdots, V_q\}$ then $\overline{\mathfrak{B}}$ is a covering and $\overline{\mathfrak{B}} > \mathfrak{A}$.

The sets V_i are clearly disjoint. Moreover from $V_i = U_i \cap (\mathfrak{R} - \bigcup\{\overline{U}_j \mid j > i\})$ and (30.2, 30.3, 30.4) we conclude that V_i is a regular open set. Therefore $\overline{\mathfrak{B}}$ is a closed grating refining $\mathfrak{A}$ and (31.5) follows.

32. Projective groups of the grating spectrum.

(32.1) We have just found that the net theory of the grating spectrum is equivalent to the $\mathfrak{F}$-theory. Its projective theory (VI, 16) will serve as an introduction to the Alexander-Kolmogoroff theory. Let $\mathbf{C}^p(\Sigma, G)$, $\mathbf{Z}^p(\Sigma, G)$ $\mathbf{F}^p(\Sigma, G)$ and $\mathbf{H}^p(\Sigma, G) = \mathbf{Z}^p(\Sigma, G)/\overline{\mathbf{F}}^p(\Sigma, G)$ be the *projective* groups of chains, cycles, bounding cycles, and homology groups of Σ over a given G. Let first $c^p = \{c_\lambda^p\} \in C^p$. If $\{\sigma_{\lambda i}^p\}$ are the p-simplexes of Φ_λ we have then

$$c_\lambda^p = g_\lambda^i \sigma_{\lambda i}^p, \qquad g_\lambda^i \in G, \tag{32.1a}$$

$$\lambda > \mu \rightarrow \pi_\mu^\lambda c_\lambda^p = c_\mu^p. \tag{32.1b}$$

(32.2) Let $\sigma_{\lambda i}^p = \overline{V}_{\lambda i_0} \cdots \overline{V}_{\lambda i_p}$ and suppose that $\sigma_{\lambda i}^p$ occurs also in Φ_μ, $\lambda > \mu$; that is to say, the sets $V_{\lambda i_h}$ are also sets of $\mathfrak{B}_\mu$. Then $\pi_\mu^\lambda \overline{V}_{\lambda i_h} = \overline{V}_{\lambda i_h}$, $\pi_\mu^\lambda \overline{V}_{\lambda j} \neq \overline{V}_{\lambda i_h}$ for $j \neq i_h$. Hence $\pi_\mu^\lambda \sigma_{\lambda i}^p = \sigma_{\lambda i}^p$, $\pi_\mu^\lambda \sigma_{\lambda j}^p \neq \sigma_{\lambda i}^p$ for $j \neq i$. By identifying the coefficients we find from (32.1b) that those of $\sigma_{\lambda i}^p$ in c_λ^p and c_μ^p are the same. Suppose now that $\sigma_{\lambda i}^p$ occurs in Φ_μ, where μ is arbitrary. Then if $\mathfrak{B}_\nu = \mathfrak{B}_\lambda \wedge \mathfrak{B}_\mu$, σ_λ^p occurs in Φ_ν also and $\nu > \lambda, \mu$. Therefore $\sigma_{\lambda i}^p$ has the same coefficient in c_λ^p, c_ν^p and c_μ^p, c_ν^p, hence the same in c_λ^p, c_μ^p. In other words, g_λ^i depends solely upon the simplex $\sigma_{\lambda i}^p$, and so it may be written as a function $\varphi^p(\sigma^p)$ of the oriented simplex σ^p alone (the former $\sigma_{\lambda i}^p$). It is thus as yet a function $\varphi^p(\overline{V}_{\lambda i_0}, \cdots, \overline{V}_{\lambda i_p})$ of the ordered set $\overline{V}_{\lambda i_0}, \cdots, \overline{V}_{\lambda i_p}$. To free it from this ordering we define $\varphi^p(\overline{V}_{\lambda i_0}, \cdots, \overline{V}_{\lambda i_p})$ as skew-symmetric in all its arguments.

(32.3) Let Ω_0 be the complete nerve of the regular closed sets and let us write now σ^p as $\sigma^p = \overline{V}_0 \cdots \overline{V}_p$. This simplex σ^p may be characterized by the property that while $\cap \bar{V}_i \neq \emptyset$ nevertheless the vertices $\bar{V}_i$ have disjoint interiors V_i. Indeed it is an arbitrary simplex of this nature. For if $V_{p+1} = \text{Int}(\mathfrak{R} - \cup V_i)$ then $\{\bar{V}_0, \cdots, \bar{V}_{p+1}\}$ is a closed grating whose nerve has σ^p as a simplex. Let this particular grating and its nerve be written $\mathfrak{B}(\sigma^p)$, $\Phi(\sigma^p)$.

Since the vertices of $\sigma^q < \sigma^p$ are among those of σ^p, they also have disjoint interiors. Therefore the simplexes of Ω_0 whose vertices have disjoint interiors make up a closed subcomplex Ω of Ω_0. To the projective chain c^p of the spectrum Σ there corresponds now a chain of Ω given by

$$'c^p = \sum \varphi^p(\sigma^p)\sigma^p \tag{32.4}$$

or more explicitly, and with summation over all the sets $\{\bar{V}_0, \cdots, \bar{V}_p\}$ with disjoint interiors:

$$'c^p = \frac{1}{(p+1)!} \sum \varphi^p(\bar{V}_0, \cdots, \bar{V}_p)\bar{V}_0 \cdots \bar{V}_p. \tag{32.5}$$

Evidently $c^p \to {}'c^p$ defines an isomorphism in the algebraic sense $\tau: \mathbf{C}^p(\Sigma, G) \to {}'\mathfrak{C}^p \subset \mathfrak{C}^p(\Omega, G)$. If U is an open set in G and $U(\sigma^p) = \{c^p \mid \varphi^p(\sigma^p) \in U\}$, then $\{U(\sigma^p)\}$ and $\{\tau U(\sigma^p)\}$ are subbases for the two chain-groups. Therefore τ is an isomorphism. For convenience we identify c^p with $'c^p = \tau c^p$, thus identifying $\mathbf{C}^p(\Sigma, G)$ with $'\mathfrak{C}^p$. The coordinate c_λ^p of c^p becomes now merely the part of the chain which is in Φ_λ.

(32.6) Since $c^p = \{c_\lambda^p\}$ is a projective chain so is $\text{F}c^p = \{\text{F}c_\lambda^p\}$. It is important however to bear in mind that $\text{F}c^p$ *is not the chain-boundary of* c^p *as an element of* $\mathfrak{C}^p(\Omega, G)$ *but only its boundary as a projective chain of the spectrum* Σ. We shall nevertheless characterize $\text{F}c^p$ as a chain-function of c^p.

Returning to the form (32.1a) for c_λ^p, we have:

$$\text{F}c_\lambda^p = h_\lambda^j \sigma_{\lambda j}^{p-1}; \qquad h_\lambda^j = g_\lambda^i[\sigma_{\lambda i}^p : \sigma_{\lambda j}^{p-1}].$$

Since the coefficient of $\sigma_{\lambda j}^{p-1}$ in $\text{F}c_\lambda^p$ is independent of λ, we may replace Φ_λ by any other nerve $\ni \sigma_{\lambda j}^{p-1}$.

Let the notations be so chosen that

$$\sigma^{p-1} = \bar{V}_0 \cdots \bar{V}_{p-1}, \qquad \mathfrak{B}_\lambda = \{V_0, \cdots, V_{p-1}, V_{\lambda p} \cdots V_{\lambda r}\}.$$

We have then $\mathfrak{B}(\sigma^{p-1}) = \{V_0, \cdots, V_{p-1}, W\}$, where $\overline{W} = \cup \bar{V}_{\lambda j}$ or $W = \text{Int}(\mathfrak{R} - \cup V_i)$ and so $\mathfrak{B}_\lambda > \mathfrak{B}(\sigma^{p-1})$. The projection $\pi: \Phi_\lambda \to \Phi(\sigma^{p-1})$ is defined by $\pi\bar{V}_i = \bar{V}_i$, $\pi\bar{V}_{\lambda j} = \overline{W}$. From this and $\pi\text{F}c_\lambda^p = \text{F}(\pi c_\lambda^p)$, we find that the coefficient of σ^{p-1} in $\text{F}c_\lambda^p$ is simply

$$\varphi^{p-1}(\bar{V}_0, \cdots, \bar{V}_{p-1}) = \varphi^p(\overline{W}, \bar{V}_0, \cdots, \bar{V}_p).$$

Thus $\text{F}c^p$ may be defined as the chain

$$\text{F}c^p = \frac{1}{p!} \sum \varphi^p(\overline{W}, \bar{V}_0, \cdots, \bar{V}_{p-1})\bar{V}_0 \cdots \bar{V}_{p-1}. \tag{32.7}$$

We may also define F as an operator on φ^p by

$$\text{(32.8)} \qquad \mathrm{F}\varphi^p = \varphi^p(\overline{\mathrm{Int}\,(\mathfrak{R} - \bigcup V_i)}, \bar{V}_0, \cdots, \bar{V}_{p-1}).$$

All these results will be utilized in a moment.

(32.9) There is no difficulty in describing the projective cochain-groups. The duals of the projections π_μ^λ are denoted as usual by $\pi_\lambda^{*\mu}$ and the duals of the $\sigma_{\lambda i}^p \in \Phi_\lambda$ by $\sigma_p^{\lambda i}$. The group of the p-cochains over a discrete H is the limit-group $\mathfrak{C}_p(\Sigma, H)$ of the direct system $\{\mathfrak{C}_p(\Phi_\lambda, H); \pi_\lambda^{*\mu}\}$. A p-cochain is thus a collection $c_p = \{c_p^\lambda\}$, where c_p^λ is a representative of c_p, and where if both c_p^μ, c_p^ν are representatives then for some $\lambda > \mu, \nu$: $\pi_\lambda^{*\mu} c_p^\mu = \pi_\lambda^{*\nu} c_p^\nu$. A p-cochain c_p is uniquely determined by any representative c_p^μ, and $c_p = 0 \leftrightarrow \pi_\lambda^{*\mu} c_p^\mu = 0$ for some $\lambda > \mu$. The boundary $\mathrm{F}c_p$ is then the cochain with the representative $c_{p+1}^\mu = \mathrm{F}c_p^\mu$. The groups $\mathfrak{Z}_p$, $\mathfrak{F}_p$, $\mathfrak{H}_p = \mathfrak{Z}_p/\mathfrak{F}_p$ are defined in the usual way, and they are the same as the net groups of Σ.

(32.10) The intersection of cochains is determined as for any net. Owing to the "simplicial" situation one may even apply Whitney's method of the "cup, cap" products (V, 20). All that is required is to order the set $\{\bar{V}\}$ of all the regular closed sets in a specified way and proceed as loc. cit.

33. Homology theory of Alexander and Kolmogoroff. Chiefly for purposes of extending the concepts of *differential* and *integral* to general topological spaces Alexander [d, e], and later Kolmogoroff [a], have developed a type of theory based directly upon chains and cochains. Since the only theory of this kind that we have encountered is the projective theory of the grating spectrum Σ, one will expect that the two are the same, and this surmise is in fact justified.

(33.1) Taking then the more general formulation of Kolmogoroff, let $\{A\}$ be the closed sets of $\mathfrak{R}$. By a *p-chain of $\mathfrak{R}$ over G* we will mean a skew-symmetric function $\varphi^p(A_0, \cdots, A_p)$ with values in G which vanishes whenever $\bigcap A_i = \emptyset$ or two of the arguments coincide, and is additive in all arguments in the sense that if A_i, A_i' have disjoint interiors then

$$\varphi^p(\cdots, A_i \cup A_i', \cdots) = \varphi^p(\cdots, A_i, \cdots) + \varphi^p(\cdots, A_i', \cdots).$$

The boundary of φ^p is the $(p - 1)$-chain

$$\varphi^{p-1}(A_0, \cdots, A_{p-1}) = \mathrm{F}\varphi^p = \varphi^p(\mathfrak{R}, A_0, \cdots, A_{p-1}).$$

Clearly $\mathrm{FF}c^p = 0$. The cycles are defined in the usual way and we have thus the usual groups of chains, cycles, bounding cycles, which we denote by $\mathfrak{C}_0^p(\mathfrak{R}, G)$, $\mathfrak{Z}_0^p(\mathfrak{R}, G)$, $\mathfrak{F}_0^p(\mathfrak{R}, G)$ with topologies to be defined in a moment.

(33.2) It is a consequence of the definitions that: (a) the A_i may be replaced by the regular closed sets $\overline{\mathrm{Int}\, A_i}$ without changing the values of φ^p; (b) the values of φ^p on all the collections $(A_0, \cdots, A_p)$ are known once we know those on collections $(\bar{V}_0, \cdots, \bar{V}_p)$, where the V_i are disjoint regular open sets. Let again Ω, Ω_0 be as in (32.3). We may consider φ^p as a function on the set of p-simplexes of Ω_0 to G. The values of φ^p on the p-simplexes of Ω to G define a function $'\varphi^p(\bar{V}_0, \cdots, \bar{V}_p)$ such as in (32.3), and hence a projective chain c^p of the grating spectrum Σ. It is clear that $\varphi^p \to c^p$ defines an isomorphism in the algebraic sense $\tau: \mathfrak{C}_0^p(\mathfrak{R}, G) \to \mathbf{C}^p(\Sigma, G)$. If $\{U\}$ are the open sets of $\mathbf{C}^p(\Sigma, G)$ we choose $\{\tau^{-1}U\}$ as those of $\mathfrak{C}_0^p(\mathfrak{R}, G)$ thus making τ an isomorphism. We then identify the elements of the groups corresponding under τ and hence the two groups themselves. This causes an identification of $\mathbf{F}^p(\Sigma, G)$ with $\mathfrak{F}_0^p(\mathfrak{R}, G)$, and hence of $\mathbf{Z}^p(\Sigma, G)$ with $\mathfrak{Z}_0^p(\mathfrak{R}, G)$.

Once our groups are topologized we define the homology group of the φ^p over G as: $\mathfrak{H}_0^p(\mathfrak{R}, G) = \mathfrak{Z}_0^p(\mathfrak{R}, G)/\bar{\mathfrak{F}}_0^p(\mathfrak{R}, G) = \mathbf{H}^p(\Sigma, G)$.

(33.3) We shall now define the cochains and related groups, and in this we follow essentially Alexander. Let $\psi_p(x_0, \cdots, x_p)$ be a skew-symmetric function of sets of $p + 1$ points of $\mathfrak{R}$ whose values are in a discrete group H, and with the following property: there exists a grating $\mathfrak{V} = \{V_i\}$ such that whenever $\{x_0, \cdots, x_p\}$, $\{x_0', \cdots, x_p'\}$ have the property that for each i there is a set V_i containing both x_i and x_i' then $\psi_p(x_0, \cdots, x_p) = \psi_p(x_0', \cdots, x_p')$. We will say that the grating $\mathfrak{V}$ *belongs* to ψ. It is clear that $\mathfrak{V}$ is by no means unique, for if $\mathfrak{V}_1 > \mathfrak{V}$ then $\mathfrak{V}_1$ likewise belongs to ψ.

Let $\{(\psi_p, \mathfrak{V})\}$ be the collection of the couples related in the above way. If $(\psi_p, \mathfrak{V})$, $(\psi_p', \mathfrak{V}')$ are two of the couples then $(\psi_p \pm \psi_p', \mathfrak{V} \wedge \mathfrak{V}')$ is one also. Hence $\Psi_p(H) = \{\psi_p\}$ is an additive group which is taken discrete.

(33.4) Returning to the notations of (32.1) let nerve $\overline{\mathfrak{V}}_\mu = \Phi_\mu = \{\sigma_{\mu i}^p\}$, and suppose that $\mathfrak{V}_\mu$ belongs to ψ_p. If $\sigma_{\mu i}^p = \bar{V}_{\mu i_0} \cdots \bar{V}_{\mu i_p}$ then $\psi_p(x_0, \cdots, x_p)$, $x_r \in V_{\mu i_r}$, depends solely upon $\sigma_{\mu i}^p$ and is written $\psi_p(\sigma_{\mu i}^p)$.

(33.5) Let $\mathfrak{V}_\mu$ belong to ψ_p. We shall say that ψ_p *vanishes* on $\mathfrak{V}_\mu$, or is *locally zero*, if every $\psi_p(\sigma_{\mu i}^p) = 0$. If ψ_p, ψ_p' vanish on $\mathfrak{V}_\mu$, $\mathfrak{V}_\nu$ then $\psi_p - \psi_p'$ vanishes on $\mathfrak{V}_\mu \wedge \mathfrak{V}_\nu$, and so the functions which are locally zero make up a subgroup $\Psi_p^0(H)$ of $\Psi_p(H)$. A coset φ_p of $\Psi_p(H)$ mod $\Psi_p^0(H)$ is known as a *p-cochain of* $\mathfrak{R}$ *over* H, and the group $\mathbf{C}_p(\mathfrak{R}, H) = \Psi_p(H)/\Psi_p^0(H)$ of the φ_p is known as the *group of the p-cochains of* $\mathfrak{R}$ *over* H.

(33.6) Consider again ψ_p and define as the *boundary* of ψ_p the function

$$\mathrm{F}\psi_p = \psi_{p+1}(x_0, \cdots, x_{p+1}) = \sum (-1)^q \psi_p(x_0, \cdots, x_{q-1}, x_{q+1}, \cdots, x_{p+1}).$$

Evidently $\mathrm{F}\psi_p \in \Psi_{p+1}$, and if $\mathfrak{V}_\mu$ belongs to ψ_p it belongs also to ψ_{p+1}. Furthermore if ψ_p vanishes on $\mathfrak{V}_\mu$ so does $\mathrm{F}\psi_p$. Hence if ψ_p is chosen in a fixed coset φ_p of $\Psi_p(H)$ mod $\Psi_p^0(H)$ then $\mathrm{F}\psi_p$ is in a fixed coset φ_{p+1} of $\Psi_{p+1}(H)$ mod $\Psi_{p+1}^0(H)$. We call φ_{p+1} the *boundary* of φ_p, written $\mathrm{F}\varphi_p$. Once more $\mathrm{FF} = 0$, and so the p-cocycles, bounding p-cocycles, their groups $\mathbf{Z}_p(\mathfrak{R}, H)$, $\mathbf{F}_p(\mathfrak{R}, H)$ and finally the cohomology groups $\mathbf{H}_p(\mathfrak{R}, H) = \mathbf{Z}_p(\mathfrak{R}, H)/\mathbf{F}_p(\mathfrak{R}, H)$, all chosen discrete, may be introduced in the customary way.

(33.7) We will now identify the preceding groups with the corresponding groups of the grating spectrum Σ. If $\mathfrak{V}_\mu$ belongs to ψ_p then ψ_p determines the cochain

$$c_p^\mu = \psi_p(\sigma_{\mu i}^p)\sigma_p^{\mu i} \text{ (summation on } i \text{ alone)},$$

of Φ_μ, and we notice that:

(a) *the vanishing of* ψ_p *on* $\mathfrak{V}_\mu$ *is equivalent to* $c_p^\mu = 0$.

We may also state:

(b) *if* $\mathfrak{B}_\mu$ *belongs to both* ψ_p, ψ_p' *and* ψ_p, ψ_p' *determine the same cochain* c_p^μ *then they are in the same coset* φ_p *of* $\Psi_p(H) \bmod \Psi_p^0(H)$.

For $\psi_p - \psi_p'$ vanishes then on $\mathfrak{B}_\mu$.

Suppose now $\lambda > \mu$. Then $\mathfrak{B}_\lambda$ belongs also to ψ_p and so the latter determines an analogue c_p^λ of c_p^μ. Let $\pi_\mu^\lambda \sigma_{\lambda j}^p = \sigma_{\mu i}^p$. The notations may be so chosen that $\sigma_{\lambda j}^p = \bar{V}_{\lambda j_0} \cdots \bar{V}_{\lambda j_p}$ and that $V_{\lambda j_r} \subset V_{\mu i_r}$, $r = 0, \cdots, p$. Under the circumstances evidently $\psi_p(\sigma_{\lambda j}^p) = \psi_p(\sigma_{\mu i}^p)$. From the definition of dual chain-mappings (IV, 10.1) there follows also that $\pi_\lambda^{*\mu} \sigma_p^{\mu i} = \sigma_p^{\lambda j} + \cdots$, while no $\pi_\lambda^{*\mu} \sigma_p^{\mu i'}$, $i' \neq i$, contains a term in $\sigma_p^{\lambda j}$. From this one infers that the coefficient of $\sigma_p^{\lambda j}$ in $\pi_\lambda^{*\mu} c_p^\mu$ is $\psi_p(\sigma_{\lambda j}^p) = \psi_p(\sigma_{\mu i}^p)$ and therefore the same as in c_p^μ. Hence:

(c) $c_p^\lambda = \pi_\lambda^{*\mu} c_p^\mu$, $\lambda > \mu$.

Suppose now ψ_p' in the same coset φ_p of $\Psi_p(H) \bmod \Psi_p^0(H)$ as ψ_p, and let $\mathfrak{B}_\nu$ belong to ψ_p'. By hypothesis $\psi_p - \psi_p'$ is locally zero, and so vanishes say on $\mathfrak{B}_\rho$. Let $\mathfrak{B}_\lambda = \mathfrak{B}_\mu \wedge \mathfrak{B}_\nu \wedge \mathfrak{B}_\rho$, hence $\lambda > \mu, \nu, \rho$. Thus $\mathfrak{B}_\lambda$ belongs to ψ_p, ψ_p' and $\psi_p - \psi_p'$ vanishes on $\mathfrak{B}_\lambda$. By (a, c) then $\pi_\lambda^{*\mu} c_p^\mu - \pi_\lambda^{*\nu} c_p^\nu = 0$. Therefore

(d) *the cochains* c_p^μ *determined by the elements of the same* φ_p *are representatives of the same cochain* c_p *of the grating spectrum* Σ *over* H.

Evidently $\varphi_p \to c_p$ defines a homomorphism $\tau: \mathbf{C}_p(\mathfrak{R}, H) \to \mathfrak{C}_p(\Sigma, H)$.

Consider now any $c_p \in \mathfrak{C}_p(\Sigma, H)$ and let it have the representative c_p^μ. Define a skew-symmetric function $\psi_p(x_0, \cdots, x_p)$ by the condition that if $x_i \in V_{\mu i_r}$, $(r = 0, \cdots, p)$, where $\sigma_{\mu i}^p = \bar{V}_{\mu i_0} \cdots \bar{V}_{\mu i_p} \in \Phi_\mu$, then $\psi_p(\sigma_{\mu i}^p)$ is equal to the coefficient of $\sigma_p^{\mu i}$ in c_p^μ. Let a second representative c_p^ν of c_p yield similarly ψ_p'. Then for some $\lambda > \mu, \nu$ we have $\pi_\lambda^{*\mu} c_p^\mu - \pi_\lambda^{*\nu} c_p^\nu = 0$, and hence $\psi_p - \psi_p'$ vanishes on $\mathfrak{B}_\lambda$ or $\psi_p - \psi_p' \in \Psi_p^0(H)$. Thus ψ_p, ψ_p' belong to the same coset φ_p of $\Psi(H) \bmod \Psi_p^0(H)$, or:

(e) *if* $c_p = \{c_p^\mu\}$ *is a cochain of* Σ *over* H, *then the functions* ψ_p *determined as above by the* c_p^μ, *are in a fixed cochain* φ_p *of* $\mathfrak{R}$ *over* H.

Here again $c_p \to \varphi_p$ defines a homomorphism $\theta: \mathfrak{C}_p(\Sigma, H) \to \mathbf{C}_p(\mathfrak{R}, H)$.

(f) $\tau\theta = 1$, $\theta\tau = 1$, *hence* τ *and* θ *are isomorphisms* (proof elementary).

(g) $\theta\mathrm{F} = \mathrm{F}\theta$, $\tau\mathrm{F} = \mathrm{F}\tau$.

The second relation is a consequence of the first and (f); so we only need to prove the first. If c_p has c_p^μ for representative and c_p^μ determines ψ_p as in the proof of (d), then we must show that $\mathrm{F}c_p$ determines $\mathrm{F}\psi_p \bmod \Psi_{p+1}^0(H)$. Now

$$\mathrm{F}c_p^\mu = \sum_{i,h} \psi_p(\sigma_{\mu i}^p)[\sigma_p^{\mu i} : \sigma_{p+1}^{\mu h}]\sigma_{p+1}^{\mu h},$$

and so, mod $\Psi_{p+1}^0(H)$, $\mathrm{F}c_p^\mu$ determines ψ_{p+1} such that

$$\psi_{p+1}(\sigma_{\mu h}^{p+1}) = \sum_i \psi_p(\sigma_{\mu i}^p)[\sigma_{\mu h}^{p+1} : \sigma_{\mu i}^p].$$

If we compare with $\mathrm{F}\psi_p$ we verify that it takes the value $\psi_{p+1}(\sigma_{\mu h}^{p+1})$ in $\sigma_{\mu h}^{p+1}$. By (b) these values determine $\psi_{p+1} \bmod \Psi_{p+1}^0(H)$, and so ψ_{p+1} is in the same cochain as $\mathrm{F}\psi_p$, which proves our assertion.

(h) It is a transparent consequence of (g) that τ induces an isomorphism of the groups $\mathfrak{Z}_p$, $\mathfrak{F}_p$, hence also of the groups $\mathfrak{H}_p$ of Σ, with the corresponding groups $\mathbf{Z}_p$, $\mathbf{F}_p$, $\mathbf{H}_p$ of $\mathfrak{R}$. This isomorphism is such that if c_p is identified with the φ_p which it determines then the corresponding groups are all identified.

34. Our combined results may now be stated as:

(34.1) THEOREM. *The homology theory of the Alexander-Kolmogoroff type is the projective theory of the grating spectrum. In particular* (VI, 17.1; 31.5):

(a) *the homology groups over a group G which is compact or a field are the same as those of the Čech theory by finite closed coverings;*

(b) *when $\mathfrak{R}$ is normal T_1, and hence when it is metric, compact Hausdorff or a compactum, the groups described under* (a) *are the same as the corresponding Čech groups* (8.1, 8.2).

(34.2) *Parallel with integration and the theory of Cartan differential forms.* J. W. Alexander has indicated an interesting parallel between homology and certain formal relations of the type occurring in the theory of differential operators due to E. Cartan. Under certain circumstances, namely when the space is an absolute differentiable manifold the parallel becomes an overlapping of the two situations (De Rham [a]).

Let us introduce the following new terminology:

p-chain C_p = p-fold domain;

p-cycle = closed p-fold domain;

bounding p-cycle = bounding closed p-fold domain;

p-cochain C_p = (p-function);

boundary of a $(p - 1)$-cochain $\mathrm{F}C_p$ = derived of C_p;

p-cocycle = (an exact p-function)

= (a p-function whose derived vanishes);

Kronecker index $\mathrm{KI}(C^p, C_p)$ = the integral of the function C_p over the p-fold domain C^p.

In this new terminology a number of well known theorems on chains and cochains in a complex become formally identical with certain theorems on integration of Cartan forms. Notably:

(a) *The integral of the derived of C_{p-1} over C^p is equal to $(-1)^p$ times the integral of C_{p-1} over* $\mathrm{F}C^p$. This is the same as (III, 29.1).

(b) *The integral of a derived p-function over a closed p-fold domain is zero.* This is the same as (III, 29.6) for $\gamma_p \sim 0$.

(c) *The integral of an exact p-function over the boundary of a $(p + 1)$-fold domain is zero.* This is the same as (III, 29.6) for $\gamma^p \sim 0$.

(d) *Stokes theorem is essentially the same as* (III, 29.1).

Finally the cup product cochain $\cup$ cochain (32.10) is essentially the same as E. Cartan's product of multi-dimensional differentials.

(34.3) REMARK. The work of De Rham has been shown by Hodge [H, IV] to be still valid when the "general" integrals are specialized to his "harmonic" integrals and this has had very important applications in algebraic geometry. In another direction also A. W. Tucker [e] has indicated the modifications required in the results of De Rham and Hodge when the differentiable manifold has a regular boundary.

CHAPTER VIII

TOPOLOGY OF POLYHEDRA AND RELATED QUESTIONS

After some further preliminary properties the results of (VII) will first be applied to the homology theories associated with polyhedra. From the topological standpoint a polyhedron may as well be replaced by a simplicial partition. Unless otherwise stated therefore *all polyhedral complexes under consideration will be simplicial*, i.e., they will be Euclidean complexes. In addition to the general type we shall also discuss geometric manifolds and their special intersection properties. Closely related topics taken up are: continuous and singular complexes, topological complexes, a rapid survey of differentiable manifolds (Whitney's results and the related work of Cairns and Whitehead). The chapter contains also a treatment of coincidences and fixed points for finite polyhedra and for a general class of spaces which have been named "quasi-complexes."

General references: Alexandroff-Hopf [A-H], Cairns [a, b], Hodge [H]; Lefschetz [L, L_1, b, g], Reidemeister [R, R_1], Seifert-Threlfall [S-T]; Veblen [V], Whitehead [b], Whitney [a, b].

§1. GEOMETRIC COMPLEMENTS

1. (1.1) *Notations.* If $K = \{\sigma\}$ is an Euclidean complex we shall write

$\sigma(x)$ = the simplex of K containing the point x;

$'\sigma$ = the centroid of σ;

$K^{(n)}$ = the nth barycentric derived of K.

These are meant to be "typical" designations. Thus if $L = \{\zeta\}$ then $'\zeta$ is the centroid of ζ, etc.

(1.2) As on previous occasions (III, 6.12, 6.14) $|K|$ is a subset of an Euclidean space $\mathfrak{E}^n$ or of the Hilbert parallelotope P^ω. We consider here also $\mathfrak{E}^n$ as a linear variety of a certain real vector space $\mathfrak{B}$ and P^ω as a convex subset of such a variety. Thus the points of $|K|$ are vectors of $\mathfrak{B}$ and they will be dealt with accordingly.

(1.3) DEFINITION. Mesh $K = \sup \{\text{diam } \sigma \mid \sigma \in K\}$.

(1.4) In (III, 6.12) the antecedent of K has been defined as the simplicial complex $\mathfrak{K} \cong K$, the isomorphism being the identity on the vertices of K. We now extend the term *antecedent* to cover any simplicial complex $\mathfrak{K}_1 \cong \mathfrak{K}$. We also say that K is an *Euclidean* realization of $\mathfrak{K}_1$.

2. (2.1) *If $\sigma = \sigma'\sigma'' = a_0 \cdots a_p$, $p > 0$, is an Euclidean simplex and $x \in \sigma$, there passes through x a unique segment $\overline{x'x''}$, where $x' \in \sigma'$ and $x'' \in \sigma''$. (We assume of course $\sigma', \sigma'' \neq \sigma$.)*

We may assume $\sigma' = a_0 \cdots a_q$, $q < p$. If x has the barycentric coordinates $\{x^0, \cdots, x^p\}$ the following numbers are uniquely defined:

$$t' = \sum x^i, \qquad i \leqq q; \qquad t'' = \sum x^{q+i};$$

$$x'^i = \frac{x^i}{t'}, \qquad i \leqq q; \quad x'^{q+i} = 0;$$

$$x''^i = 0, \qquad i \leqq q; \quad x''^{q+i} = \frac{x^{q+i}}{t''}.$$

Moreover $\{x'^i\}$, $\{x''^i\}$ are the barycentric coordinates of two points $x' \in \sigma'$, $x'' \in \sigma''$ such that $x = t'x' + t''x''$. Hence $x \in \overline{x'x''}$.

Suppose that there is a second segment $\overline{y'y''}$ containing x, where $x' \neq y' \in \sigma'$, $x'' \neq y'' \in \sigma''$. As a consequence $x = t'x' + t''x'' = s'y' + s''y''$. Expressing x', x'', y', y'' in terms of the vertices of σ we find a relation

$$u_0 a_0 + \cdots + u_q a_q + \cdots = 0$$

where $x' \neq y'$ implies that at least one of $u_0, \cdots, u_q \neq 0$. Since the vertices are independent this is excluded, and so $\overline{x'x''}$ is unique.

(2.2) *Let $\sigma^p = a_0 \cdots a_p$ be in a space $\mathfrak{R}$ which is either Euclidean or the Hilbert parallelotope P^ω. Then $d(x, y)$, $x \in \mathfrak{R}$, $y \in \sigma^p$, does not exceed the maximum distance ρ from x to the set of vertices.*

For every $a_i \in \overline{\mathfrak{S}(x, \rho)}$, and since the sphere and σ^p are both convex, we have $\bar{\sigma}^p \subset \overline{\mathfrak{S}(x, \rho)}$, proving (2.2).

(2.3) *The diameter of σ is the length of its longest edge.*

By (2.2) diam σ is the distance from a point of $\bar{\sigma}$ to a vertex and again by (2.2) this is at most equal to a certain $d(a_i, a_j)$.

(2.4) Mesh $(\mathrm{Cl}\, \sigma^p)' \leqq p/(p + 1)$ diam σ^p.

Any one-simplex of $(\mathrm{Cl}\, \sigma^p)'$ is of the form $'\sigma^q{}'\sigma^r$, $\sigma^q < \sigma^r < \sigma^p$, and we have to show that its length λ does not exceed the value in question. Let σ^s be the face of σ^r opposite σ^q. The segment $'\sigma^q{}'\sigma^s$ is the longest segment of $\bar{\sigma}^p$ carrying $'\sigma^q{}'\sigma^r$ and if μ is its length then by an elementary calculation:

$$\frac{\lambda}{\mu} = \frac{r - q}{r + 1} \leqq \frac{r}{r + 1} \leqq \frac{p}{p + 1}.$$

From this follows

$$\lambda \leqq \frac{p}{p + 1}\mu \leqq \frac{p}{p + 1} \operatorname{diam} \sigma^p.$$

(2.5) *Let $\{y_n\}$ be coordinates for the space of σ^p. Then on $\bar{\sigma}^p$ the barycentric coordinates may be expressed as linear functions of a set $\{y_{n_0}, \cdots, y_{n_p}\}$. Hence they are continuous in $y_{n_0}, \cdots, y_{n_p}$ on $\bar{\sigma}^p$.*

If $\{\alpha_{ni}\}$ are the coordinates of a_i and $\{y_n\}$ those of $x \in \bar{\sigma}^p$, the system in the unknowns x^i:

$$y_n = x^i \alpha_{ni}, \qquad \sum x^i = 1,$$

is compatible and has a unique solution. As is well known this implies (2.5).

3. A few of the simpler properties of an Euclidean complex $K = \{\sigma\}$ are:

(3.1) (a) *If K is finite $|K|$ is a compactum;*

(b) *if K is locally finite, $|K|$ is a local compactum;*

(c) *if L is a closed subcomplex, $|L|$ is a closed set and hence $|K - L|$ is an open set.*

(d) *if $\{K_i\}$ are the components of the Euclidean complex K (not necessarily finite) then $\{|K_i|\}$ are those of the polyhedron $|K|$.*

If K is finite then $|K|$ is the union of a finite set of compacta, the $\bar{\sigma}$, and hence it is a compactum. It is a consequence of (III, 6.12c) that if $x \in \sigma$ then $d(x, |K - \text{St } \sigma|) > 0$. Hence $x \in \text{Int } |\text{St } \sigma|$. If $\sigma' > \sigma$ then $\text{St } \sigma' \subset \text{St } \sigma$, and so $x \in \sigma' \to x \in \text{Int } |\text{St } \sigma'| \subset \text{Int } |\text{St } \sigma|$. Therefore $\text{Int } |\text{St } \sigma| = |\text{St } \sigma|$, and so $|\text{St } \sigma|$ is open. It follows that $|K - L|$ is open and hence $|L|$ is closed. Since $\text{St } \sigma$ is finite $\overline{|\text{St } \sigma|}$ is a compactum. Thus $x \in \sigma$ has a neighborhood whose closure is a compactum, and so $|K|$ is a local compactum.

Let $x, y \in |K_i|$ and A, A' vertices of $\sigma(x), \sigma(y)$. Since K_i is connected there is a sequence, $\sigma(x) = \sigma_0, \sigma_1, \cdots, \sigma_r = \sigma(y)$ of simplexes of K_i such that any two consecutive are incident. Hence $\{\bar{\sigma}_i\}$ is a finite collection of connected sets of which any two consecutive ones intersect. Their union is a connected subset of $|K_i|$ containing x and y. Thus the two points are in the same component of $|K|$.

Suppose now $x \in |K_i|, y \notin |K_i|$. Then by (c) $|K_i|$ and $|K - K_i|$ are disjoint closed sets whose union is $|K|$, containing, respectively, x and y and so the two points are in distinct components of $|K|$. Thus x, y are in the same component of $|K|$ when and only when they are in the same set $|K_i|$ and this implies (d).

(3.2) We have seen (1.2) that $|K|$ is in a space $\mathfrak{R}$ which is an $\mathfrak{E}^n$ or P^ω. We specify that diam σ, *in the sense* of (VI, 24.1) is to be its diameter as a subset of $\mathfrak{R}$ and it is clear that *this turns K into a metric complex* (VI, 24.2).

(3.3) DEFINITION. *Let $\{\sigma_1, \sigma_2, \cdots\}$ be the simplexes of K ranged in some order. Then K is said to be regular if* diam $\sigma_n \to 0$. *If A is a closed subset of $\mathfrak{R}$ (the space of* 3.2) *then K is said to be regular relatively to A, whenever* $\sup \{\text{diam } \sigma_n, d(A, \sigma_n)\} \to 0$. *The two types of regularity are evidently independent of the order in which the $\{\sigma_n\}$ have been ranged.*

(3.4) *Every countable locally finite simplicial complex $\mathfrak{K}$ has a regular Euclidean realization K in P^ω. If in addition $n =$* dim *$\mathfrak{K}$ is finite then $\mathfrak{K}$ has a regular realization in any parallelotope P^m, $m \geqq 2n + 1$.*

Let $\{\zeta\}$ be the simplexes of $\mathfrak{K}$ and $\{A_1, A_2, \cdots\}$ its vertices. If $\{z_1, z_2, \cdots\}$, $0 \leqq z_i \leqq 1/i$, are coordinates for P^ω let a_i be the point: $z_h = 0, h \neq i, z_i = 1/i$. If $\zeta^p = A_{i_0} \cdots A_{i_p} \in \mathfrak{K}$ then $a_{i_0}, \cdots, a_{i_p}$ are the vertices of an Euclidean simplex $\sigma^p \subset P^\omega$ and $K = \{\sigma^p\}$ behaves as required.

Suppose now $n = \dim K$ finite. Take first in P^m a sequence $\{b_i\}$ tending to a limit b. Choose now points $\{a_i\}$, one at a time, in P^m as follows. First we require that $d(a_i, b_i) < 2^{-i}$. Then suppose $a_1, \cdots, a_k$ so chosen that no subset of $2n + 2$ is in an $\mathfrak{E}^{2n}$ of the $\mathfrak{E}^m$ of P^m. The spaces $\mathfrak{E}^{2n}$ determined by all the subsets consisting of $2n + 1$ of the $a_1, \cdots, a_k$, are finite in number and a_{k+1} is chosen exterior to all these. Proceeding as before all but a finite number of the constructed σ are contained in any preassigned $\mathfrak{S}(b, \epsilon)$; we will therefore have once more an Euclidean realization in P^m which is clearly regular.

4. In many applications one requires subdivisions of arbitrarily small mesh. This may sometimes be accomplished by means of derivation. The possibilities are discussed below.

(4.1) *When K is either* (a) *finite-dimensional and with bounded mesh (hence in particular when K is finite), or else* (b) *regular, then mesh $K^{(p)} \to 0$.*

Consider first (a). If $\dim K = n$ and mesh $K = l$ then by (2.3, 2.4) we have mesh $K^{(p)} < (n/(n + 1))^p l \to 0$.

Consider now (b). Let $\{\sigma_i\}$ be the simplexes of K ranged in some order. Given any ϵ we may select r so high that diam $\sigma_{r+i} < \epsilon$. The union of the closures of the σ_j, $j \leqq r$, is a finite closed subcomplex K_r, and so by the result just proved we may choose p so high that mesh $K_r^{(p)} < \epsilon$. Hence mesh $K^{(p)} < \epsilon$.

(4.2) *Every Euclidean complex $K = \{\sigma\}$ has a simplicial partition K_1 of arbitrarily small mesh.*

If the situation is as in (4.1), and in particular if K is finite, we may merely choose for K_1 a suitable $K^{(p)}$. In the general case let $\mathfrak{E}_i^p$ be the space of σ_i^p and let it be referred to the coordinates $x_1, \cdots, x_p$. Take a fixed $\epsilon > 0$ and choose $\eta > 0$ such that the partition of $\mathfrak{E}_i^p$ by the subspaces $x_h/\eta = 0, \pm 1, \cdots$ is of mesh less than ϵ. These subspaces decompose Cl σ_i^p into the elements of a polyhedron whose mesh is less then ϵ. Since K is locally finite, any σ occurs thus in at most a finite number of Cl σ_1, and so it is decomposed into a finite set of convex polyhedral cells. Their totality is a partition Π of K whose mesh is less than ϵ and its derived Π' is a K_1 whose mesh is also less than ϵ.

(4.3) DEFINITION. *An Euclidean complex K whose antecedent is a circuit, a manifold, $\cdots$ will be called a geometric circuit, geometric manifold, $\cdots$.*

5. (5.1) *Let $K = \{\sigma\}$ be a finite Euclidean complex and A a closed proper subset of $|K|$. Then $|K| - A$, may be covered with an Euclidean complex $L = \{\zeta\}$ regular relative to A and such that each ζ is contained in a σ.*

Let $\{m_1, m_2, \cdots\}$ be a monotone increasing sequence of integers to be specified presently. Denote by P_n the open subcomplex of the barycentric derived $K^{(m_n)}$ consisting of the simplexes whose closures meet A and set $P_0 = K$. Define $Q_n = \text{Cl } P_n$; R_n as the set of simplexes of $K^{(m_{n+1})}$ in $|Q_n| - |Q_{n+1}|$. We select, as we may, $\{m_n\}$ such that $\eta_{n+1} = \text{mesh } K^{(m_{n+1})} < (1/3)\, d(A, |K| - |P_n|)$, with $\eta_1 < (1/3) \sup d(A, x)$, $x \in K$.

Let Π be the union of the simplexes of all the R_n and take any simplex σ_n of R_n. Owing to the condition on the meshes, $\bar{\sigma}_n$ cannot meet both $| \text{Cl } R_{n-1} |$ and $| \text{Cl } R_{n+1} |$. Moreover if it meets $| \text{Cl } R_{n+1} |$ then all its points, and hence also its faces are each in a simplex of Cl R_n. It follows that if $\sigma' < \sigma_n$ then σ' is in $| R_n |$ or $| R_{n+1} |$. In the former case it is a simplex of R_n, in the latter it is a union of simplexes of R_{n+1}. Thus $\bar{\sigma}_n - \sigma_n$ is the union of a finite set of faces of Π. Since $\bar{\sigma}_n$ can only meet consecutive sets $| \text{Cl } R_n |$, and the latter are finite, σ_n is the face of at most a finite number of simplexes of Π. Hence Π is locally finite. Since the R_n are finite and disjoint Π is countable. Since the simplexes of Π are disjoint and clearly sup $\{\text{diam } \sigma_n, d(A, \sigma_n)\} \to 0$, Π is a polyhedron, and its derived $L = \Pi'$ answers the question.

REMARK. The theorem may readily be extended to infinite complexes. The chief modification required in the proof will then be replacing the derived by suitably chosen partitions of K.

(5.2) *Let K be a finite Euclidean complex contained in a parallelotope P^n. Then P^n may be covered with a finite Euclidean complex which has a simplicial partition of K as a subcomplex.*

The different faces of P^n (in an obvious sense) make up a polyhedral complex Ω covering P^n (i.e., such that $| \Omega | = P^n$). Any given σ_i^p of K is a subset of an $\mathfrak{E}_i^p$ of the space $\mathfrak{E}^n$ of P^n and $\mathfrak{E}_i^p$ is the intersection of $n - p$ subspaces $\{\mathfrak{E}_{ih}^{n-1}\}$ $h = 1, 2, \cdots, n - p$ of $\mathfrak{E}^n$. The total set $\{\mathfrak{E}_{ih}^{n-1}\}$ for all i, h, causes a partition Π^n of Ω whose derived is related to K in the asserted way.

6. **Barycentric mappings.**

(6.1) Let $K = \{\sigma\}$, $L = \{\zeta\}$ be Euclidean complexes with respective vertices $\{a_i\}$, $\{b_i\}$. Suppose that there exists a simplicial set-transformation $t:K \to L$ and let τ be the induced simplicial chain-mapping. Let the b_i be so labelled (with possible repetitions) that $ta_i = b_i$. We introduce a point set-transformation $T:| K | \to | L |$ defined as follows: If $x = x^i a_i$ then $y = Tx = x^i b_i$. This implies in particular that: (a) if $b_i = \cdots = b_j$ while $b_h \neq b_i$ for $h \neq i, \cdots, j$ then the barycentric coordinate of y as to b_i is $x^i + \cdots + x^j$; (b) $x \in \sigma \to Tx \in t\sigma$. We prove:

(6.2) *T is a mapping $| K | \to | L |$, said to be barycentric.*

Let $\{u_i\}$, $\{v_i\}$ be coordinates of reference for the spaces of K, L, and let $y_0 = Tx_0$. Since K is locally finite, by (2.5) there is a finite set of the $\{u_i\}$, say $\{u_1, \cdots, u_r\}$ such that on $\overline{| \text{St } \sigma(x_0) |}$ the barycentric coordinates are continuous functions of $\{u_1, \cdots, u_r\}$ and hence of x. It follows that T is continuous on $| \text{St } \sigma(x_0) |$, and since $\{| \text{St } \sigma |\}$ is an open covering it is an elementary matter to prove T continuous on $| K |$.

Evidently $Ta_i = ta_i$ so that T induces t and hence also τ. Hereafter we drop all mention of t, and will call τ the chain-mapping *induced* by T.

(6.3) As an application suppose that $K \cong L$. This means that they have a common antecedent $\mathfrak{K}$. Let the notations be so chosen that a_i, b_i correspond under the isomorphism, or which is the same that they are the images of the same vertex of $\mathfrak{K}$. Then t is one-one, and so is T. Moreover under the circumstances T^{-1} corresponds to t^{-1} like T to t. Therefore T is topological. In fact it takes on $\bar{\sigma}^p$ the values of a nonsingular affine transformation of the space $\mathfrak{E}^p$ of σ. Thus:

(6.4) *If the Euclidean complexes K, L are isomorphic then $|K|$, $|L|$ are topologically equivalent. Furthermore there exists a topological mapping $|K| \to |L|$ which is barycentric.*

As a consequence:

(6.5) *All the Euclidean realizations $\{K\}$ of a given countable locally finite simplicial complex yield topologically equivalent polyhedra $\{|K|\}$, which may be mapped topologically into one another as indicated in* (6.4).

7. **Normal subcomplexes.** A closed subcomplex L of K is said to be *normal* in K whenever if a simplex of K has all its vertices in L then it is a simplex of L.

(7.1) *The derived L' of L is normal in K'.*

In the notations of (IV, 25) if $\zeta = {}'\sigma_i, \cdots, {}'\sigma_j, \sigma_i < \cdots < \sigma_j$, is a simplex of K' with its vertices ${}'\sigma_i, \cdots, {}'\sigma_j$ in L', then $\sigma_i, \cdots, \sigma_j \in L$, and so $\zeta \in L'$.

(7.2) *If L is normal in K so is $M = K - \text{St } L$.*

If $\sigma \in K - M$ exists with all its vertices in M then $\sigma \in \text{St } L - L$, and so σ must have vertices in L. Since $L \cap M = \emptyset$ this is a contradiction proving (7.2).

(7.3) *Under the same conditions as in* (7.2) *there passes through every point $x \in |\text{St } L - L|$ a unique segment $x'x''$ with $x' \in |L|$, $x'' \in |M|$. Moreover the transformation $\rho: |\text{St } L| \to |L|$ such that $\rho x = x'$, $\rho \,|\, |L| = 1$, is a deformation-retraction.*

We have $\sigma(x) = \sigma'\sigma''$, $\sigma' \in L$, $\sigma'' \in M$, and so the existence of the segment is a consequence of (2.1). It follows readily from the expression of x' that it is continuous in the barycentric coordinates of x and hence continuous in x itself, (2.5). Hence $x \to x'$ for $x \in |\text{St } L - L|$ and $x = x'$ for $x \in |L|$, defines a retraction $\rho: |\text{St } L| \to |L|$. By (I, 47.4) ρ is a deformation.

(7.4) Application. *Under the same conditions as in* (7.3) *the homology groups of the compact cycles of $|M|$ are isomorphic with the corresponding groups of $|K - L|$.*

In view of (7.2) we may interchange L, M in (7.3) and so there is a deformation retraction $\rho_1: |\text{St } M| = |K - L| \to |M|$. Hence (7.4) is a consequence of (3.1) and (VII, 7.5).

(7.5) *Let $|\Pi|$ be a finite Euclidean polyhedron in $\mathfrak{E}^n$. Then $|\Pi|$ has a neighborhood U in $\mathfrak{E}^n$ for which it is a deformation retract.*

Let P^n be a parallelotope in $\mathfrak{E}^n$ containing Π in such a way that $d(\Pi, \mathfrak{E}^n - P^n) = 2\alpha > 0$. By (4.1, 5.2, 7.2) a suitable simplicial partition L of Π is a normal subcomplex of an Euclidean complex K of mesh less than α covering P^n. As a consequence $U = |\text{St } L|$ (star in K) is in P^n and is a neighborhood of $|\Pi|$ in $\mathfrak{E}^n$. The existence of ρ is then a consequence of (7.3).

The retraction here considered is a special case of so-called "neighborhood retraction" in the sense of Borsuk. In point of fact it may be proved with Borsuk that $|\Pi|$ is a so-called "absolute neighborhood retract," i.e., whenever topologically imbedded in a compactum it is a retract of a suitable neighborhood of the compactum. For details regarding these questions see Lefschetz [L_2, III, IV].

8. Since the derived of a polyhedral complex is an Euclidean complex, the problem of the topological classification of polyhedra is equivalent to the same problem for simplicial polyhedra. We have shown, on the other hand (6.4), that if two Euclidean complexes K, K_1 are isomorphic then $|K|$, $|K_1|$ are topologically equivalent. Conversely, supposing $|K|$, $|K_1|$ topologically equivalent, what can be said regarding the isomorphism of K with K_1? Since $|K^{(m)}|$, $|K_1^{(n)}|$ are likewise topologically equivalent, $K \cong K_1$ would be no more reasonable than $K^{(m)} \cong K_1^n$ for some m, n. Or instead of the derived we may equally well compare any two partitions. Now if K, K_1 have isomorphic partitions Π, Π_1 they also have the isomorphic simplicial partitions Π', Π_1'. Thus we only need to consider simplicial partitions. Let K, K_1 be defined as *partition-equivalent* whenever they have isomorphic simplicial partitions. This relation is manifestly a true relation of equivalence. Evidently partition-equivalence implies topological equivalence of the polyhedra. The converse, one of Poincaré's well known unsolved problems, may be explicitly formulated as:

PROBLEM A. *Does the topological equivalence of the polyhedra* $|K|$, $|K_1|$ *imply the partition-equivalence of the complexes* K, K_1?

As a special case if $|K|$ is a closed n-cell we have the likewise unsolved

PROBLEM B. *If the polyhedron* $|K|$ *is a closed n-cell, is* K *partition-equivalent to a closed Euclidean n-simplex?*

For the dimension one the solutions of A, B are elementary. For the dimension two, solutions may be obtained but they lean heavily upon the Jordan-Schoenflies theorem regarding the subdivision of a two-sphere by a simply closed curve. For higher dimensions only partial extensions of this theorem are known (see Wilder [a, b]), and this is one source of difficulty. Another lies of course in our ignorance regarding the Poincaré group (see (23a)).

In connection with these problems it may be recalled that M. H. A. Newman [a] has taken as point of departure in his investigations on Euclidean complexes, their classification with respect to partition-equivalence.

§2. HOMOLOGY THEORY

9. Since we are unable to identify topological and partition-equivalence classes of complexes it is natural to investigate the topological invariants of the partition-equivalence classes. From (IV, 28.1) and (V, 16.2) we already infer that various homology and cohomology groups, class intersections and rings are partition-invariant. Our next object is to prove that they are also topologically invariant.

(9.1) Two types of cycles, $\cdots$ related to K will occur simultaneously in the sequel: (a) those of the space $|K|$, referred to as *geometrical*; (b) those of the complex K itself, referred to as *combinatorial.* Often the context will indicate the type: thus compact cycles are necessarily geometrical, while finite cycles are combinatorial.

(9.2) *Notations.* They remain those of (1.1) except that: (a) the simplexes of $K^{(n)}$ are written σ_{ni}, σ_{ni}^p; (b) if a_{ni}, γ_{ni}, $\cdots$ are elements of any sort related to $K^{(n)}$ the analogues for K will be written a_i, γ_i, $\cdots$.

(9.3) In connection with certain coverings there will occur on more than one occasion instead of their nerves, suitable Euclidean representations of the nerves. Usually they will be K or one of its derived. For convenience the term "nerve" will also be applied to these realizations. The meaning will be clear enough from the context to cause no confusion.

(9.4) *Method of proof.* It will always consist in identifying the combinatorial elements with similar elements in a suitable net or web topologically related to $|K|$.

10. Consider first a finite Euclidean complex and its combinatorial homology theory: homology and cohomology groups and class intersections (V, §2). Consider the finite open covering of $|K|$ by the stars of the vertices: $\mathfrak{B} = \{\text{St } a_i\}$. From $\text{St } a_i \cap \cdots \cap \text{St } a_j = \text{St } a_i \cdots a_j$, follows that $\text{St } a_i \cap \cdots \cap \text{St } a_j \neq 0 \leftrightarrow a_i \cdots a_j \in K$. Therefore nerve $\mathfrak{B} = K$ (9.3). Similarly if $\mathfrak{B}_n$ is the finite open covering by the stars of the vertices of $K^{(n)}$ then nerve $\mathfrak{B}_n = K^{(n)}$. Since by (4.1) mesh $\mathfrak{B}_n \to 0$, $\{\mathfrak{B}_n\}$ is cofinal in the family of all the finite open coverings, and so it may serve to define the geometrical groups and intersections in K.

Since $\text{St } '\sigma_n$ in $K^{(n+1)}$ is contained in σ_n it is also contained in the star of any vertex of σ_n. Therefore a mapping of $'\sigma_n$ into a vertex of σ_n defines a projection $\pi_n^{n+1}: K^{(n+1)} \to K^{(n)}$, and $\Sigma(K) = \{K^{(n)}, \pi_n^{n+1}\}$ is a simplicial spectrum whose net homology and cohomology groups and intersection theory are those of $|K|$. Now π_n^{n+1} is merely the operation τ of (IV, 23) for $K^{(n)}$ (a reciprocal of chain-derivation in $K^{(n)}$, IV, 26.2c). Therefore by (IV, 24.2), π_n^{n+1} induces an isomorphism on the homology groups, and similarly for its dual π_{n+1}^{*n} and the cohomology groups. It follows that the net groups of $\Sigma(K)$ are those of any $K^{(n)}$, and hence those of K itself (II, 13.4b, 14.6). Likewise also for the intersections. Therefore we have:

(10.1) THEOREM. *The combinatorial and geometrical homology theories of a finite Euclidean complex are the same. This implies the topological invariance of the combinatorial theory: topological invariance of the homology and cohomology groups, class intersections and indices, cohomology rings, Betti numbers, torsion coefficients.*

We may state in fact with more precision:

(10.2) *The mapping of each geometrical cycle or cocycle into the corresponding combinatorial cycle or cocycle induces an isomorphism of the corresponding homology groups, cohomology groups, intersection rings. Moreover if* Γ, $'\Gamma$ *denote corresponding combinatorial and geometrical classes then*

$$'(\Gamma^p \cdot \Gamma_q) = '\Gamma^p \cdot '\Gamma_q, \qquad '(\Gamma_p \cdot \Gamma_q) = '\Gamma_p \cdot '\Gamma_q, \qquad \text{KI}('\Gamma^p \cdot '\Gamma_p) = \text{KI}(\Gamma^p \cdot \Gamma_p).$$

An interesting complement is:

(10.3) *The properties of* (III, 20), *relating connectedness and homology, continue to hold for a finite Euclidean complex K if the term "component" refers to the components of the polyhedron $|K|$. In particular the number of components of $|K|$ is $R^0(K)$* (3.1d).

11. Suppose now that we are dealing with an arbitrary Euclidean K. Referring to (VI, §§5, 6) there are various associated combinatorial homology theories. That is to say, in each case there are two dual categories, with intersections since K is simplicial.

(11.1) Definition. *The combinatorial theory of two specific dual categories A, B of cycles and cocycles of K is said to be invariant whenever it is the same as the theory of two dual categories A', B' which have topological character for $|K|$. Thus the combinatorial theory of the cycles and cocycles of a finite K is invariant, since it has been shown to be the same as the theory of the geometrical cycles or cocycles of K.*

It is implicit in the definition that when the theory of A, B is invariant, their homology and cohomology groups, class intersections, cohomology rings, Betti numbers, are all topological invariants of $|K|$. As an application we may state the following theorem, which is proved essentially like (10.1):

(11.2) *Let K, K_1 be finite Euclidean complexes with the closed subcomplexes L, L_1. Let T be a topological mapping $|K| \to |K_1|$ such that $T|L| = |L_1|$. Then the combinatorial theory of the cycles* mod L *and cocycles of $K - L$ is invariant under T.*

12. **Complementary remarks.** It will be very convenient to associate with the cycles of K certain specific cycles of the spectrum $\Sigma(K)$.

(12.1) Let δ denote chain-derivation both in K and in all its derived. As applied then to $K^{(n)}$ it has π_n^{n+1} of (10) for a reciprocal. By (IV, 26.8, 23.1),

$$\pi_n^{n+1}\delta = 1, \qquad \delta\pi_n^{n+1} \sim 1. \tag{12.2}$$

We will compare more particularly $A = \mathfrak{C}^p(K^{(n)}, G)$ with $B = \delta\mathfrak{C}^p(K^{(n)}, G) \subset \mathfrak{C}^p(K^{(n+1)}, G)$. If we set $\bar{\pi} = \pi_n^{n+1} | B$ then we still have $\bar{\pi}\delta = 1$. Moreover since $\delta\bar{\pi}\delta = \delta(\bar{\pi}\delta) = \delta$, as operations between the two groups A, B we have in addition $\delta\bar{\pi} = 1$. Therefore δ is an isomorphism of A with B. Hence:

(12.3) *δ^m is a simultaneous isomorphism $\mathfrak{C}^p(K^{(n)}, G) \to \delta^m\mathfrak{C}^p(K^{(n)}, G)$.*

(12.4) Take now any cycle $\{\gamma_n^p\}$ of $\Sigma(K)$. We have $\pi_n^{n+1}\gamma_{n+1}^p \sim \gamma_n^p$, and hence by (12.2): $\gamma_{n+1}^p \sim \delta\gamma_n^p \sim \delta^n\gamma^p$. Therefore $'\gamma^p = \{\delta^n\gamma^p\} \sim \{\gamma_n^p\}$. Thus the geometric class $'\Gamma^p$ of $\{\gamma_n^p\}$ contains a representative among the cycles of $\Sigma(K)$ of the form $'\gamma^p = \{\delta^n\gamma^p\}$. We will say that $'\gamma^p$ is a geometric cycle *adherent* to γ^p in K. It is also convenient to refer to Γ^p, $'\Gamma^p$ as *adherent* to one another.

It is clear that $\gamma^p \to {}'\gamma^p$ defines an isomorphism in the algebraic sense θ of $\mathfrak{Z}^p(K, G)$ with a subgroup $'\mathfrak{Z}^p$ of $\mathfrak{Z}^p(\Sigma, G)$. If U_n is any open set of $\mathfrak{Z}^p(K^{(n)}, G)$

then $V_n = \{'\gamma^p \mid \delta^n\gamma^p \in U_n\}$ is open in $'\mathfrak{Z}^p$ and $\{V_n\}$ is a base for $'\mathfrak{Z}^p$. By (12.3), and since δ maps cycles into cycles, there is an open set U in $\mathfrak{Z}^p(K, G)$ such that $\delta^n\gamma^p \in U_n \rightarrow \gamma^p \in U$. Hence $V_n = \{'\gamma^p \mid \gamma^p \in U\}$. Thus if U is any open set of $\mathfrak{Z}^p(K, G)$ and $V = \{'\gamma^p \mid \gamma^p \in U\}$, then $\{V\}$ is a base for $'\mathfrak{Z}^p$. Therefore θ maps into one another the elements of two bases $\{U\}$, $\{V\}$ for $\mathfrak{Z}^p(K, G)$ and $'\mathfrak{Z}^p$ and so it is an isomorphism.

Let $'\mathfrak{F}^p = \mathfrak{F}^p(\Sigma, G) \cap {'\mathfrak{Z}^p}$ = the group of the bounding cycles of form $'\gamma^p$. Since $\mathfrak{F}^p(\Sigma, G)$ is closed in $\mathfrak{Z}^p(\Sigma, G)$, $'\mathfrak{F}^p$ is closed in $'\mathfrak{Z}^p$. If $'\gamma^p \sim 0$ then $\gamma^p \sim 0$, and so $\theta^{-1}{'\mathfrak{F}^p} \subset \bar{\mathfrak{F}}^p(K, G)$. On the other hand $(\gamma^p = \mathrm{F}C^{p+1}) \rightarrow (\delta^n\gamma^p = \delta^n\mathrm{F}C^{p+1} = \mathrm{F}\delta^nC^{p+1} \sim 0) \rightarrow {'\gamma^p} \sim 0$. Hence $\theta\mathfrak{F}^p(K, G) \subset {'\mathfrak{F}^p}$, and therefore $\theta\bar{\mathfrak{F}}^p(K, G) \subset \overline{\theta\mathfrak{F}^p(K, G)} \subset {'\mathfrak{F}^p}$, since the latter is closed in $'\mathfrak{Z}^p$. Therefore $\theta\bar{\mathfrak{F}}^p(K, G) = {'\mathfrak{F}^p}$, and so θ maps the cycles γ^p which are ~ 0 into the $'\gamma^p \sim 0$. It follows that θ induces an isomorphism $\bar{\theta}: \mathfrak{H}^p(K, G) \rightarrow \mathfrak{H}^p(\Sigma, G)$ and this isomorphism is readily recognized to be the one in (10.1).

To sum up, the preceding analysis yields:

(12.5) *The relation of adherence between the combinatorial cycles γ^p of K and the geometrical cycles of the spectrum $\Sigma(K)$ of the special form $'\gamma^p = \{\delta^n\gamma^p\}$ is an isomorphism of the groups $\mathfrak{Z}$, $\mathfrak{F}$ of K with the corresponding groups of the $'\gamma^p$. Similarly adherence between the combinatorial and geometrical classes is an isomorphism between the corresponding homology groups.*

(12.6) Let now L be a closed subcomplex of K. Referring to (IV, 24.3c), δ is likewise chain-derivation in L and its derived. Moreover $\pi_n^{n+1} \mid L^{(n+1)}$ is a reciprocal of δ as chain-derivation in $L^{(n)}$. It follows that $\pi_n^{n+1}L^{(n+1)} = L^{(n)}$, a result which may also be deduced from (IV, 26.2bc). The projection π_n^m, $m > n$, of $\Sigma(K)$ is given by $\pi_n^m = \pi_n^{n+1} \cdots \pi_{m-1}^m$, and so $\pi_n^m L^{(m)} = L^{(n)}$. Referring now to (VI, 12) if $\pi_{0n}^{n+1} = \pi_n^{n+1} \bmod L^{(n)}$ and $\pi_{1n}^{n+1} = \pi_n^{n+1} \mid L^{(n+1)}$, then $\Sigma_0 = \{K^{(n)} - L^{(n)}; \pi_{0n}^{n+1}\}$, $\Sigma_1 = \{L^{(n)}; \pi_{1n}^{n+1}\}$ are spectra such that (Σ_0, Σ_1) is a dissection of $\Sigma(K)$. If we replace, in everything that precedes, $\Sigma(K)$ by Σ_0, K by $K - L$, and the cycles of K by those of K mod L, we will extend (12.5) automatically to adherence for the cycles mod L. We point out explicitly that the operation δ remains the same. For by reference to (IV, 26.2b) we verify without difficulty that the chain-mapping $(K^{(n)} - L^{(n)}) \rightarrow (K^{(n+1)} - L^{(n+1)})$ induced by δ is merely $\delta \mid (K^{(n)} - L^{(n)})$. Thus the adherent cycles will still be γ^p and $'\gamma^p = \{\delta^n\gamma^p\}$.

(12.7) Let L_1 be a second closed subcomplex of K containing L and let ω, π be, respectively, the topological and combinatorial projections $|K - L| \rightarrow |K - L_1|$ and $(K - L) \rightarrow (K - L_1)$. If $'\gamma^p = \{\delta^n\gamma^p\}$ is a cycle mod L, then $\omega'\gamma^p$ is obtained by reducing mod $L_1^{(n)}$ the coordinate $\delta^n\gamma^p$, and this yields merely $\delta^n(\pi\gamma^p)$. Therefore $\omega\gamma^p = \{\delta^n(\pi\gamma^p)\}$. Or $\pi\gamma^p$ and $\omega'\gamma^p$ are still adherent. Thus the associated projections π, ω preserve adherence. This result will be useful later.

(12.8) *Let K, L be finite. Then a barycentric mapping $T: |K| \rightarrow |L|$ sends adherent elements into adherent elements; in other words T preserves adherence.*

For convenience we also designate by T the induced operations on the combinatorial and geometrical cycles and classes. If $\{a_i\}$, $\{b_i\}$ are the vertices of K, L evidently $\{T \mid \text{St } a_i \mid\} > \{\mid \text{St } b_i \mid\}$. If $'\gamma^p = \{\gamma^p, \delta\gamma^p, \cdots\}$ we find by reference to (VII, 5.12) that $T'\gamma^p$ is a cycle of $|L|$ whose coordinate relative to L as nerve of $\{\mid \text{St } b_i \mid\}$ is precisely $T\gamma^p$. Since the relations of adherence in L are determined by the coordinates in L, it follows that the classes of $T\gamma^p$, $T'\gamma^p$ are adherent. A similar argument is valid for the cocycles except that the homomorphisms are τ^* of (VII, 5.11) for the topological cocycles, and θ^*, the dual of the chain-mapping induced by T, for the combinatorial cocycles. Similarly for the intersections.

13. We will now consider a few invariance theorems for infinite complexes.

(13.1) *Let K be a general Euclidean complex. Then the combinatorial theory of the dual categories of the infinite cycles* [*cocycles*] *and finite cocycles* [*cycles*] *are topologically invariant.*

Corresponding to $|K|$ there may be introduced the direct web of sets $\mathfrak{A} = \{A_\lambda\}$ of the open sets with compact closures of (VII, 13.2). Since $\{\mid \text{St } \sigma(x) \mid \mid x \in \bar{A}_\lambda\}$ is a covering of the compact set $\bar{A}_\lambda$, there is a finite subcovering $\{\mid \text{St } \sigma_i \mid\}$ and the union of its simplexes is a finite subcomplex K_μ of K such that $|K_\mu| \supset \bar{A}_\lambda$. Hence if $\{K_\mu\}$ are the finite subcomplexes, $\{\mid K_\mu \mid\}$ is cofinal in $\mathfrak{A}$, and so both have the same homology theory. It is, however, an elementary consequence of (12.7) that the direct webs $\{K_\mu\}$ and $\{\mid K_\mu \mid\}$ have the same homology theory. Hence $\{K_\mu\}$ has the same as $\mathfrak{A}$, and so the homology theory of the first, like that of the second, has topological character. Hence the theory of the dual categories of the infinite cycles and finite cocycles, which is that of $\{K_\mu\}$ (VI, 20) has topological character. For infinite cocycles and finite cycles the treatment is the same except that the comparison is with the direct web $\mathfrak{B}$ of the compact subsets of $|K|$ (VII, 13.1).

We have proved incidentally the following result:

(13.2) *The homology* [*cohomology*] *groups of the compact cycles* [*cocycles*] *of $|K|$ are isomorphic with the corresponding combinatorial groups of the finite cycles* [*cocycles*] *of K.*

An argument essentially similar to that of (13.1) yields the following extension of (11.2):

(13.3) *The situation being as in* (11.2) *save that the complexes need not be finite, the combinatorial theory of the infinite cycles of K* mod *L and of the finite cocycles of $K - L$ is invariant under T.*

We may in fact sharpen the preceding result to:

(13.4) *Property* (13.3) *still holds if T is merely a mapping $|K| \to |K_1|$ such that*: (a) *T maps $|K - L|$ topologically onto $|K_1 - L_1|$*; (b) $T|L| \subset |L_1|$.

Suppose first K, K_1 finite. The dual categories under consideration are then those of the cycles of K mod L and of the cocycles of $K - L$, and the same in K_1. Consider the direct open web of sets $\mathfrak{A} = \{A_\lambda\}$ whose elements are the open subsets of $|K - L|$. Let $\Phi, \cdots$ and the other notations of (VII, 4.3) be applied to $|K|$. If we form the direct open web $\Phi_0(\mathfrak{A}) = \{\Phi_0(A_\lambda)\}$ then by (VI, 23.3) its homology theory is the same as that of the cycles of Φ

mod $\Phi_1(|L|)$ and cocycles of $\Phi_0(|K - L|)$, i.e., the same as that of the geometric cycles of K mod L and cocycles of $K - L$. Hence, by (11.2) the homology theory of the direct web $\Phi_0(\mathfrak{A})$ is then the same as the combinatorial theory of the cycles of K mod L and cocycles of $K - L$. Since $\Phi_0(\mathfrak{A})$ direct has topological character relative to $|K - L|$, it is clear that its theory is invariant under T. Therefore the combinatorial theory of the cycles of K mod L and cocycles of $K - L$ is likewise invariant under T.

Passing to the general case the proof is the same as that of (13.1), the A_λ being now merely subsets of $|K - L|$.

A result of somewhat different character required later is:

(13.5) *Let the conditions be the same as in* (13.4) *except that*: (a) L, L_1 *are normal in* K, K_1 ; (b) T *is merely a topological mapping of* $|K - L|$ *onto* $|K_1 - L_1|$. *Then* T *induces an isomorphism of the combinatorial homology groups of the finite cycles of* $M = K - \mathrm{St}\, L$ *with the corresponding groups of* $M_1 = K - \mathrm{St}\, L_1$.

For by (13.2) the groups in question are those of the compact cycles of $|M|$, $|M_1|$, and by (7.4) likewise those of the compact cycles of $|K - L|$, $|K_1 - L_1|$.

(13.6) *Property* (10.3), *relating connectedness and homology, holds for any Euclidean complex* (10.3; III, 20, 40.7).

14. **Groups at the points.** Since a polyhedron is metric the only groups requiring consideration are those through the points (VII, 15.7). We prove

(14.1) *The homology groups of a polyhedron* $|K|$, $K = \{\sigma\}$, *through a point* x *are the same as the corresponding combinatorial groups of* $\mathrm{St}\,\sigma(x)$, *and so in particular the same for all points of* σ.

Let σ' be any simplex of $\mathrm{St}\,\sigma(x)$. If σ' is replaced by the join $x(\mathfrak{B}\sigma')_a$ in K, there is obtained a special case of the simplicial partition of (IV, 29.5). Let S be the resulting partition operation. If $L = SK$, let $\mathrm{St}_1\, x$ denote the star of the vertex x in L. Since $S(\mathrm{St}\,\sigma(x)) = \mathrm{St}_1\, x$, by (IV, 24.3b), the groups of $\mathrm{St}\,\sigma(x)$ are those of $\mathrm{St}_1\, x$. To prove (14.1) it will thus be sufficient to prove that the groups through x are the same as the corresponding groups of $\mathrm{St}_1\, x$. Hence in the last analysis we merely have to show that:

(14.2) *The groups through a vertex* a *of* K *are the same as the corresponding groups of* $\mathrm{St}\, a$.

Let this time $\mathrm{St}_n\, a$ denote the star of a in $K^{(n)}$. Since $\mathrm{St}_n\, a$ is a subcomplex of the nth derived of the finite complex $\mathrm{Cl}\,\mathrm{St}\, a$, by (4.1) $\mathrm{diam}\,\mathrm{St}_n\, a \to 0$. As a consequence $'\mathfrak{A} = \{|\mathrm{St}_n\, a|\}$ is coinitial in the inverse web of sets $\mathfrak{A}$ (VII, 14.1) whose homology groups are those of the cycles through a. Thus $'\mathfrak{A}$ may serve to determine the groups of the cycles through a. We have then a conet and the operations of interest to us are the projections $\pi^n_{n+1}:\mathrm{St}_n\, a \to \mathrm{St}_{n+1}\, a$. To prove (14.2), and hence (14.1), it is sufficient (II, 14.6) to show that π^n_{n+1}, induces an isomorphism $\mathfrak{H}^p(\mathrm{St}_n\, a, G) \to \mathfrak{H}^{p+1}(\mathrm{St}_{n+1}\, a, G)$ (geometrical groups). Since this is proved in the same way for all n we merely need to establish:

(14.3) π^0_1 *induces an isomorphism of the geometrical homology groups of* $\mathrm{St}\, a$ *with the corresponding groups of* $\mathrm{St}_1\, a$.

Set $\mathrm{St}\, a = \{a\sigma_i \cup a\}$, $\mathrm{St}_1\, a = \{a\sigma_{1i} \cup a\}$, $B = \{\sigma_i\}$, $B_1 = \{\sigma_{1i}\}$, $C = K - \mathrm{St}\, a$, $C_1 = K' - \mathrm{St}_1\, a$. Let also π be the reduction mod C_1 of the chains of K'. By (12.7) if $'\gamma^p = \{\gamma^p, \delta\gamma^p, \cdots\}$, where γ^p is a cycle of K' mod C', then $\pi_1^0{}'\gamma^p = \{\pi\gamma^p, \delta(\pi\gamma^p), \cdots\}$. Since $\gamma^p \to {}'\gamma^p$ and $\pi\gamma^p \to \pi_1^0{}'\gamma^p$ induce isomorphisms of the corresponding homology groups (12.1, 12.2) the proof of (14.3) reduces to

(14.4) *$\pi\delta$ induces an isomorphism of the combinatorial homology groups of K mod C, or groups of* $\mathrm{St}\, a$, *with the corresponding groups of* $\mathrm{St}_1\, a$.

By definition the simplexes of B_1 are all the simplexes of the form $'(a\sigma_i) \cdots {}'(a\sigma_j)$, such that $a\sigma_i \prec \cdots \prec a\sigma_j$, or equivalently such that $\sigma_i \prec \cdots \prec \sigma_j$. To the simplex just written there corresponds thus the simplex $'\sigma_i \cdots {}'\sigma_j$ of B'. It follows that $'\sigma_i \to {}'(a\sigma_i)$ defines an isomorphism $\theta{:}B' \to B_1$. Hence $d = \theta\delta$ is a chain-mapping $B \to B_1$. We prove

$$\pi\delta a\sigma^p = ad\sigma^p. \tag{14.5}$$

We notice first that from the definition of δ (IV, 26.2b) and since θ is an isomorphism there comes:

$$d\sigma^0 = {}'(a\sigma^0); \qquad d\sigma^p = {}'(a\sigma^p)d\mathrm{F}\sigma^p, \qquad p > 0. \tag{14.6}$$

Since (14.5) is immediate for $p = 0$, we use induction on p. We have:

$$\delta(a\sigma^p) = {}'(a\sigma^p)\delta\mathrm{F}(a\sigma^p) = {}'(a\sigma^p)(\delta\sigma^p - \delta(a\mathrm{F}\sigma^p)),$$

and therefore

$$\begin{aligned}\pi\delta(a\sigma^p) &= -{}'(a\sigma^p)\pi\delta(a\mathrm{F}\sigma^p) = -{}'(a\sigma^p)ad(\mathrm{F}\sigma^p)\\ &= a(a\sigma^p)d(\mathrm{F}\sigma^p) = ad\sigma^p,\end{aligned}$$

which is (14.5).

If we lower all dimensions in $\mathrm{St}\, a$ and $\mathrm{St}_1\, a$ one unit, they become isomorphic with the augmented complexes B_a, B'_{1a}. If we identify each with their isomorphs, ad merely goes over into d. Since both θ, δ induce isomorphisms of the homology groups this holds also for d. It follows that ad induces isomorphisms of the homology groups of $\mathrm{St}\, a$ with the corresponding groups of $\mathrm{St}_1\, a$. Hence this holds also for $\pi\delta$ (14.5). This proves (14.4) and hence also (14.1).

15. **Invariance of certain dimensional numbers.** The classical results to be proved below will establish the identity of certain "combinatorial" dimensional numbers with corresponding dimensions in the sense of (I, 15.1). Since the latter are topologically invariant so will be the former. In each case the topological invariance was first proved by L. E. J. Brouwer (around 1910) and the identification with topological dimensions was made later by Menger and Urysohn.

Let the dimension in the sense of (I, 15.1) be called temporarily *topological.* Thus an Euclidean space $\mathfrak{E}^n$, an n-cell E^n, a parallelotope P^n have a *combinatorial* dimension namely n, and in addition a topological dimension. Similarly an Euclidean complex K, or an open subcomplex $K - L$, have a *combinatorial* dimension, namely as complexes, in the sense of (III, 1.1) and in addition there are the topological dimensions of $|K|$, $|K - L|$, which we call temporarily the topological dimension of K, $K - L$.

(15.1) THEOREM. *The combinatorial dimension of an open or closed Euclidean complex K is equal to its topological dimension and hence it is a topological invariant. As a consequence the topological dimension of an n-cell or n-parallelotope is precisely n.*

(15.2) THEOREM. *A region Ω^n of an $\mathfrak{E}^n$ cannot be mapped topologically on an Ω^m, $m < n$.*

(15.3) THEOREM. *No Ω^n can be represented in one-one bicontinuous manner by less than n parameters.*

(15.4) THEOREM. *The combinatorial and topological dimensions of an open or closed Euclidean complex and hence of an n-cell, an Euclidean space, a parallelotope are the same.*

(15.5) COROLLARY. *The (topological) dimension of the Hilbert parallelotope P^ω is infinity.*

Let K be an n-complex, x a point on a $\sigma^n \in K$, $R^p(x)$ the pth rational Betti number for the cycles through x. Here $\text{St}\,\sigma(x) = \sigma^n$, hence the groups of $\text{St}\,\sigma(x)$ are those of a single element. Thus $|K|$ is n-cyclic at x, and so $R^n(x) = 1$. On the other hand for every $x \in |K|$, $\dim \text{St}\,\sigma(x) \leqq n$, and so $R^p(x) = 0$ for $p > n$. Thus n is the largest index for which some $R^n(x) \neq 0$. Since the $R^n(x)$ are topologically invariant so is n. This proves (15.1).

Let now $x \in \Omega^n$. There is a σ^n between x and Ω^n and so the groups at x in Ω^n are the same as for σ^n. Therefore as above Ω^n is n-cyclic at all points. Since this property is topologically invariant Ω^n cannot be m-cyclic at all points and so (15.2) holds. As for (15.3) it is a direct corollary of (15.2).

Before proceeding, we recall the following results which we borrow from Menger's work: *Dimensionstheorie* (Springer, 1928) (see also Hurewicz-Wallman [H–W]):

(15.6) *Let $\mathfrak{R}$ be a separable metric space. Then:*

(a) $\dim \mathfrak{R}$ *as defined above is the same as the Menger-Urysohn dimension* (Menger, p. 157; [H–W, 66]; Menger's closed sets are readily replaced by open sets);

(b) $A \subset \mathfrak{R} \rightarrow \dim A \leqq \dim \mathfrak{R}$; (Menger, p. 81; [H–W, 26]);

(c) *If U is a neighborhood of $x \in \mathfrak{R}$ then the dimension of U at x is the same as the dimension of $\mathfrak{R}$ at the point* (immediate consequence of the Menger-Urysohn definition of the dimension).

PROOF OF (15.4, 15.5). Suppose first K Euclidean and finite. The stars of the vertices of $K^{(p)}$ make up a finite open covering $\mathfrak{U}_p$ of $|K|$ with $K^{(p)}$ for nerve. Since $\{\mathfrak{U}_p\}$ is cofinal in the family of all the finite open coverings of $|K|$ and $\dim K^{(p)} \leqq n$, necessarily $\dim |K| \leqq n$. Since $|K|$ is n-cyclic at the points of the σ^n, $\dim |K| \geqq n$, and so $\dim |K| = n$.

Suppose now K infinite. Since K is locally finite every point x has a neighborhood whose closure is a finite polyhedron of dimension less than or equal to n. Hence by (15.6bc) and the result just proved the dimension at x is at most n; so $\dim |K| \leqq n$, then as above $\dim |K| = n$.

Let now $K - L$ be an open Euclidean n-complex. We may suppose $K = \mathrm{Cl}\,(K - L)$, and so $\dim |K| \leqq n$, hence by (15.6b): $\dim |K - L| \leqq n$, and again as before $\dim |K - L| = n$.

Since σ^n, P^n, $\mathfrak{E}^n$ may be covered with an Euclidean n-complex their dimension is n. Since P^ω contains a P^n for every n its dimension exceeds every n. Thus (15.4), (15.5) are proved.

16. (16.1) *Let $K - L$, $K_1 - L_1$ be open Euclidean complexes and T a mapping $|K| \to |K_1|$ such that T is a topological mapping of $|K - L|$ onto $|K_1 - L_1|$ and that $T|L| \subset |L_1|$. If K is an n-circuit* mod L *then K_1 is an n-circuit* mod L_1, *and if the first is orientable or simple so is the second.*

We designate as before by γ, $'\gamma$ adherent combinatorial and geometrical cycles in the complexes. We also designate by σ, σ_1 the simplexes of $K - L$, $K_1 - L_1$.

We recall (III, 24) that $K - L$ is an open n-complex such that: (a) $\gamma^n = \sum \sigma_i^n$ is an n-cycle mod $(L, 2)$; (b) no proper closed subcomplex of $K - L$ contains such a cycle. By (15.1) then $\dim (K_1 - L_1) = n$ also, and by (13.4), K_1 contains an n-cycle mod $(L_1, 2)$, $'\gamma_1^n = T'\gamma^n$, where $\gamma_1^n = \sum g_i \sigma_{1i}^n$, $g_i = 0, 1$. Suppose that $'\gamma_1^n$ is in a proper closed subcomplex M_1 of $K_1 - L_1$. This will certainly be the case if in the expression of γ_1^n there is missing a simplex σ_1^n of $K_1 - L_1$, since γ_1^n will then be in $M_1 = K_1 - L_1 - \sigma_1^n$ which is a proper closed subcomplex of $K_1 - L_1$. Be it as it may, if M_1 exists as stated, there is a simplex $\sigma_1 \in K_1 - L_1$ such that $\mathrm{St}\, \sigma_1 \cap M_1 = \emptyset$. Consequently $'\gamma^n$ will be in a set which does not contain a certain open set U of $|K - L|$. It follows that the coordinate γ_p^n of $'\gamma^n$ in a certain derived $K^{(p)} - L^{(p)}$ will not contain any element in some star, and hence will lack some n-simplex of $K^{(p)} - L^{(p)}$. On the other hand if δ denotes chain-derivation, $\delta^p\gamma^n$ contains all the n-simplexes of $K^{(p)} - L^{(p)}$ and so $\delta^p\gamma^n \neq \gamma_p^n$, and hence $\delta^p\gamma^n \not\sim \gamma_p^n$ since $n = \dim (K^{(p)} - L^{(p)})$. Since derivation does not alter the homology groups, $K - L$ must contain a cycle mod $(L, 2)$, γ'^n, different from γ^n. In view of (a) this can only be if $\gamma'^n = \sum \sigma_i'^n$, where the $\sigma_i'^n$ do not include all the σ_i^n. Therefore $|\gamma'^n| \cap (K - L)$ violates (b), M_1 cannot exist and γ_1^n fulfills condition (a) in $K_1 - L_1$. Moreover no such cycle may be in a proper closed subcomplex of $K_1 - L_1$; so the latter satisfies (b) also. Therefore it is an n-circuit.

By (III, 24.2) $K - L$ is orientable when and only when K contains a combinatorial integral n-cycle mod L, and hence by (13.4) when and only when K contains a geometrical integral n-cycle mod L, $'\gamma^n$. When this takes place K_1 contains the geometrical integral n-cycle mod L_1, $T'\gamma^n$, and so $K_1 - L_1$ is also orientable.

For the invariance of the simple circuit we require the

(16.2) DEFINITION. *A point x of the n-complex $K - L$ is said to be regular whenever $|K - L|$ is n-cyclic at x; otherwise x is said to be singular.*

By (14.1) if σ has a regular point every point of σ is regular. Therefore the aggregate of the singular points is a union of simplexes, and its closure S is a closed subcomplex of $K - L$. We call S the *singular locus* of $K - L$. Since every point of a σ^n is regular, necessarily $\dim S \leqq n - 1$.

Now a n.a.s.c. for the n-circuit to be simple is that every St σ^{n-1} be n-cyclic, i.e., that $\dim S \leqq n - 2$. Since S and hence $\dim S$ are invariant under T, if $K - L$ is a simple circuit so is its transform $K_1 - L_1$.

§3. GEOMETRIC MANIFOLDS

17. For convenience we revert in this section to the notations for combinatorial manifolds (V, §4). The complexes are thus designated again by $X, \cdots$ and their elements by $x, \cdots$. However, $X, \cdots$ are now Euclidean complexes or their subcomplexes, and so the elements are Euclidean simplexes.

Let then Y be an Euclidean complex, with a closed subcomplex Z and set $X = Y - Z$. We first prove:

(17.1) THEOREM. *Let X_1, Y_1, Z_1 be analogous to X, Y, Z, and let T be a mapping $|Y| \to |Y_1|$ such that T is a topological transformation $|X| \to |X_1|$ and that $T|Z| \subset |Z_1|$. Then*: (a) *if X is an M^n, or an orientable M^n so is X_1*; (b) *if T maps topologically $|\mathrm{Cl}\, X| \to |\mathrm{Cl}\, X_1|$, and X is an M^n with regular boundary so is X_1.*

For simplicial complexes the manifold conditions are (V, 29, 34):

(17.2) every St x is n-cyclic;

(17.3) under suitable orientations of the elements $\sum x_i^n$ is an n-cycle mod Z.

Condition (17.2) makes X an M^n which may or may not be orientable, while (17.3) makes it orientable. The supplementary conditions for a regular boundary are:

(17.4) when $x \in \mathfrak{B}X$ then St x is n-cyclic mod Z;

(17.5) $\mathfrak{B}X$ is an absolute M^{n-1}.

Now owing to (14.1) condition (17.2) is equivalent to:

(17.6) $|X|$ *is n-cyclic at every point* (*every point is regular*).

Since (17.6) is topological so is (17.2). Hence if X is an M^n so is X_1. Suppose now that X is an orientable M^n, that is to say, that it satisfies (17.2, 17.3). By (V, 29.3, 29.9) condition (17.3) merely asserts in the presence of (17.2) that

(17.7) *every component of X is an orientable n-circuit, and* $\{(17.2), (17.3)\} \leftrightarrow \{(17.6), (17.7)\}$.

Since the second pair of conditions is topological (14.1, 16.1), so is the first pair. Therefore if X is an orientable M^n so is X_1.

Suppose finally X to have a regular boundary $B = \mathfrak{B}X$, with T topological on $|\text{Cl } X|$. Then necessarily $T | B | = | B_1 | = | \mathfrak{B}X_1 |$. By what has just been proved (17.5) is fulfilled by X_1. As for (17.4) owing to (14.1), it makes an assertion regarding the groups at the points of B relative to $| \text{Cl } X |$ which has obvious topological character. Therefore (17.4) holds for X_1 also, and hence it is an M^n with regular boundary.

18. **Duality theorems.** We first consider a noteworthy complement to the duality theorems for relative manifolds. Since derivation does not alter the manifold properties we may replace Y by Y' and hence assume the subcomplex Z normal in Y. Since $Y, Z, X_1 = Y - \text{St } Z$, and Y', Z', X' form each a triple such as K, L, M of (13.5) the homology groups of the compact cycles of $| X |$ are isomorphic with the corresponding groups of the finite cycles of X_1 or of X'. Now the former are the groups of the finite *absolute* cycles of X (i.e., the cycles γ^p such that the complex $| \gamma^p |$ is finite closed simplicial) with respect to bounding in a finite closed simplicial complex $\subset X$. Hence the homology groups of the finite cycles of X' and those of the finite cycles of X with respect to bounding in a finite closed simplicial complex $\subset X$ are isomorphic. Therefore:

(18.1) *In the duality theorems for relative manifolds* (V, 33.2cd) *the homology groups of the finite cycles of X' may be replaced either by those of the compact cycles of X, or else also whenever Z is normal in Y, by those of the finite absolute cycles of X with respect to bounding in a finite closed simplicial subcomplex of X.*

19. We shall now consider an extension of the duality theorems in a new direction. We suppose Y itself to be a *finite* absolute orientable M^n and take a closed subset Z of $| Y |$. By (5.1) $| Y | - Z$ may be covered with an Euclidean complex regular relative to Z.

(19.1) *The covering complex X of $| Y | - Z$ is an absolute orientable M^n.*

It is to be shown that (17.6, 17.7) hold. Since Y is an M^n, (17.6) holds in each point of X relative to Y, hence also relative to X, since $| X |$ is open in $| Y |$. Consider now any component X_1 of X. It is a consequence of (17.6) that X_1 is a simple n-circuit. Therefore we merely have to prove X_1 orientable. Now orientability for X_1, comes down to the following: given x_1^n, $x_{2r+1}^n \in X_1$ is it possible to orient all the elements so that if $x_1^n x_2^{n-1} x_3^n \cdots x_{2r+1}^n$ is any finite sequence of elements joining x_1^n, x_{2r+1}^n, in which any two consecutive are incident, then each x_{2h}^{n-1} is oppositely related (with incidence numbers $+1$ and -1) to x_{2h-1}^n and x_{2h+1}^n. Denote this property by (α). It is a consequence of the construction of the complex X that any sequence $x_1, \cdots, x_{2r+1}$ such as above consists of elements of a simplicial partition Y_1 of Y. Therefore if (α) fails in X_1, it fails also in Y_1. However, since Y is an orientable M^n so is Y_1. Consequently (α) holds in Y_1, hence also in X_1. Therefore X fulfills (17.7), and so it is an orientable M^n.

(19.2) *The conditions remaining the same, the geometrical pth cohomology groups of X are isomorphic with the corresponding $(n - p)$th homology groups for the* *compact cycles. Hence* (V, 32.3; VII, 4.7) *the cycles of* Y mod Z *and the compact cycles of* X *are quasi-dual categories.*

Consider as in (VII, 13.2) the direct web of sets $\mathfrak{A} = \{A_\lambda\}$ whose elements are the open subsets of $|X|$ with compact closures. Since the closure of X is in Y, $|\bar{X}|$ is compact and hence it is an $\bar{A}_\lambda$, say $|X| = A_{\lambda_0}$. Since every $\lambda < \lambda_0$, the groups of $\mathfrak{A}$ are those of $|X|$ itself. In particular, the geometrical cohomology groups of X are the same as those of its compact cocycles. By (13.2) the latter are also the combinatorial groups of the finite cocycles of X. Therefore the geometrical cohomology groups of X are isomorphic with the corresponding combinatorial groups of the finite cocycles of X. Since X is an absolute orientable M^n its combinatorial pth cohomology groups of finite cocycles are isomorphic with the corresponding $(n - p)$th combinatorial homology groups of the finite cycles (V, 32.1) and this proves (19.2).

(19.3) Suppose now Y to be $(p - 1, p)$-acyclic and let τ be the above isomorphism of the groups of the compact cycles of $|X|$ with those of the finite cocycles of X. Then if Γ^{n-p} is a homology class of the former and Δ^{p-1} a homology class of cycles of Z we define their class linking coefficient as $\text{Lk}(\Delta^{p-1}, \Gamma^{n-p}) = \text{Lk}(\Delta^{p-1}, \tau\Gamma^{n-p})$.

From (19.2) and (VII, 9.1; VI, 15.8; III, 39.3) we obtain the extension of Alexander's initial sphere duality theorem. Stated for convenience for the n-sphere, it holds in fact for any finite absolute orientable $(0, n)$-cyclic geometric M^n.

(19.4) THEOREM. *Let* X *be a topological n-sphere,* Z *a closed subset of* X, Γ^n *the basic n-class of* X mod Z, Γ^0 *the class of a point of* Z. *Then*:

(a) *for* $n = 1$, *or* $n > 1$ *and* $1 < p < n$, *the groups* $\mathfrak{H}^{p-1}(Z, G)$ *and* $\mathfrak{H}^{n-p}(X - Z, H)$, *for the absolute cycles of* Z *and compact cycles of* $X - Z$, *are dually paired with the class linking coefficient as the multiplication*;

(b) *for* $n > 1$ *one must replace* $\mathfrak{H}^0(Z, G)$, $\mathfrak{H}^n(X - Z, H)$ *by* $\mathfrak{H}^0(Z, G)/G\Gamma^0$, $\mathfrak{H}^n(X - Z, H)/H\Gamma^n$ (Alexander [a], Pontrjagin [c]).

From (19.4) we deduce the duality relation for the Betti numbers mod π:

$$R^{p-1}(Z, \pi) = R^{n-p}(X - Z, \pi) + \delta_1^p - \delta_n^p . \tag{19.5}$$

(19.6) APPLICATION. THE JORDAN-BROUWER THEOREM. *Let* Z *be a topological* $(n - 1)$*-sphere contained in the topological n-sphere,* X, $n > 1$. *Then* $X - Z$ *consists of two connected regions (open sets) whose common boundary is* Z.

We reproduce essentially Alexander's proof deduced from (19.5). Whatever the closed set Z we may cover $X - Z$ with a polyhedron $|K|$ such that K is regular relative to Z. The number of components of K and $|K|$ are the same, and so they are $R^0(K) = R^0(X - Z)$.

Suppose now Z to be an $(n - 1)$-sphere. By (19.5) we have then $R^0(X - Z) = R^{n-1}(Z) + 1 = 2$. Let U_1, U_2 be the two components. Clearly

$\mathfrak{B}U_i \subset Z$. Suppose $\mathfrak{B}U_1 \neq Z$ and let $a \in Z - \mathfrak{B}U_1$. We can find an $(n-1)$-cell E^{n-1} between a and $Z - \mathfrak{B}U_1$ such that $Z - E^{n-1}$ is a closed $(n-1)$-cell $\bar{E}_1^{n-1}$. By the above $X - \bar{E}_1^{n-1}$ has $R^0(X - \bar{E}_1^{n-1}) = R^{n-1}(\bar{E}_1^{n-1}) + 1 = 1$ component. Since $\bar{U}_1 \cap E^{n-1} = \emptyset$, one component must be U_1 and a must be in another component. Thus we have a contradiction proving (19.6).

For further information regarding the preceding questions, and notably the converse of (19.6), see Wilder [b].

Coupling now (19.2) with (VII, 22.3) we have Alexandroff's generalization of the so-called Phragmén-Brouwer theorem:

(19.7) *Let Y be a finite absolute orientable geometric (p-1, n-p)-acyclic M^n. Given two closed sets Z_1, Z_2 in $|Y|$ and a normal couple (G, H) suppose that $\mathfrak{H}^{p-1}(Z_1 \cup Z_2, G) = 0$. If γ^{n-p} is a compact cycle of $|Y|$ with a carrier which meets neither Z_1 nor Z_2 and γ^{n-p} is not linked with either of the two sets (in the sense of* VII, 22) *then γ^{n-p} is likewise not linked with $Z_1 \cap Z_2$* (Alexandroff [a, 178]).

(19.8) *Indicatrix.* Let $X = Y - Z$ be a connected orientable geometric M^n. Since X is connected it is an n-circuit and so it has a basic combinatorial n-cycle γ^n which we identify with the adherent geometrical cycle. For the same reason if E^n is an n-cell such that $\bar{E}^n$ is a closed n-cell $\subset X$ then there is a basic geometrical n-cycle δ^n of E^n, or cycle of $|Y|$ mod $(|Y| - E^n)$. Now the projection $\pi: |X| \to E^n$, or reduction mod $(|X| - E^n)$ of the cycles of $|X|$, yields a reduced cycle $\pi\gamma^n$. Since δ^n is the basic n-cycle of E^n we have $\pi\gamma^n = \alpha\delta^n$.

Since $\bar{E}^n$ is compact we show as in the proof of (13.1) that E^n is contained in a finite open subcomplex of Y and hence in a finite closed subcomplex Z. Since mesh $Z^{(p)} \to 0$ we may take p so high that $Z^{(p)}$ has an n-simplex $\subset E^n$. Since we may freely replace Y by any derived we may suppose that Y, and hence X has a simplex $x^n \subset E^n$. If we replace E^n by x^n and apply the projection $\pi': E^n \to x^n$, we will have $\pi'\pi\gamma^n = \pi'\alpha\delta^n = \alpha' x^n$, where α divides α'. Now $\pi'\pi$ is merely the projection $|X| \to x^n$. Since X is a simple circuit the coefficient of x^n in γ^n is $\beta = \pm 1$, and so $\pi'\pi\gamma^n = \beta x^n = \alpha' x^n$. Thus $\alpha' = \pm 1$, and since α divides α', likewise $\alpha = \pm 1$.

We conclude then that $\pi\gamma^n = \pm\delta^n$. Suppose δ^n given. Replacing γ^n if need be by $-\gamma^n$ we may so choose it that $\pi\gamma^n = +\delta^n$. Thus an assigned pair (E^n, δ^n) may serve to select one of the two basic n-cycles $\pm\gamma^n$ for M^n. The following terms are used for obvious reasons:

(E^n, δ^n) = an *oriented* n-cell;

(M^n, γ^n) = an *oriented* manifold;

an oriented n-cell utilized to determine γ^n as above, i.e., to orient M^n, is called an *indicatrix* of M^n.

20. Invariance of intersections.

(20.1) Let first X be a connected finite absolute orientable M^n and let γ^n, Γ^n be its basic cycle and class (V, 36). Since all the coefficients at the right in (V, 36.6) are ± 1, and X is an orientable n-circuit (V, 29.9), hence cyclic in the dimension n, every n-cycle of X is of the form $g\gamma^n$. Moreover γ^n is determined to within its sign in the sense that the only other cycle having the property just stated is $-\gamma^n$. If Γ^p is any other combinatorial class of X, by (V, 36.12) there is a unique combinatorial class Γ'_{n-p} such that

(20.2) $$\Gamma'_{n-p}\cdot\Gamma^n = \Gamma^p.$$

If Γ^q, Γ'_{n-q} is an analogous pair we have (V, 37.6):

(20.3) $$\Gamma^p\circ\Gamma^q = \Gamma'_{n-p}\cdot\Gamma'_{n-q}\cdot\Gamma^n.$$

Let each combinatorial class be identified with the adherent geometrical class. The combinatorial dot-intersections are then identified with the corresponding geometrical intersections (10.2). Suppose in particular that X_1 is a second finite absolute M^n such that $|X| = |X_1|$. Then Γ^n defines a class Γ_1^n for X_1 such that every other integral nth homology class of X_1 is a multiple of Γ_1^n. Hence (20.1) the basic class of X_1 is $\pm\Gamma_1^n$ and by (V, 36.6) we may suppose the n-simplexes of X_1 so oriented that it is $+\Gamma_1^n$. Thus as geometrical classes the two basic classes of X, X_1 will then coincide. It follows that the geometrical intersection class $\Gamma'_{n-p}\cdot\Gamma'_{n-q}\cdot\Gamma^n$ will be the same whether determined by means of X or X_1. Hence we have the following situation: Given two geometrical classes Γ^p, Γ^q if we identify them with their combinatorial images in any complex such as X there results a combinatorial intersection whose geometrical image is unique. This topological image is called the *geometrical intersection* of Γ^p, Γ^q and denoted by $\Gamma^p\circ\Gamma^q$.

(20.4) *The identification of the geometrical and combinatorial classes in* M^n *causes the identification of the intersections of the geometrical homology classes. Hence in particular* (V, 37.7) *holds also for the latter.*

From the topological invariance of the dot-intersections and (20.3) there comes also:

(20.5) *If* T *is a topological transformation* $|X| \to |X_1|$ *inducing an isomorphism* τ *of the homology groups such that* $\tau\Gamma^n = \Gamma_1^n$ *is the basic class of* X_1 *then*

$$\tau\Gamma^p\circ\tau\Gamma^q = \tau(\Gamma^p\circ\Gamma^q).$$

(20.6) Suppose now X not connected and let its components be $\{X_i\}$. Then $\Gamma^p = \sum\Gamma_i^p$, $\Gamma^q = \sum\Gamma_i^q$ where Γ_i^p, Γ_i^q are unique classes of X_i, and we define

$$\Gamma^p\circ\Gamma^q = \sum(\Gamma_i^p\circ\Gamma_i^q),$$

with the same conclusions as before.

(20.7) The modifications required for the other types of manifolds may be deduced from the above together with reference to (V, 38).

(20.8) Referring also to (V, 37.9) we observe that the right side has topological character. Hence (V, 37.9) with all classes taken geometrical defines also an index $\mathrm{KI}(\Gamma^p, \Gamma^{n-p})$ which has topological character in the same sense as in (20.5). That is to say:

(20.9) *If* τ *is as in* (20.5) *then*

$$\mathrm{KI}(\tau\Gamma^p, \tau\Gamma^{n-p}) = \mathrm{KI}(\Gamma^p, \Gamma^{n-p}).$$

§4. CONTINUOUS AND SINGULAR COMPLEXES

21. Let $K = \{\sigma\}$ be an Euclidean complex and $l = AB$ a segment parametrized as $0 \leqq u \leqq 1$. The product $\mathfrak{K} = l \times K$ is a polyhedral complex known as a *prism*. Let K be identified with $A \times K$ and set $K_1 = B \times K$. Consider the topological mapping (translation) $T: | K | \to | K_1 |$ defined by $T(0 \times x) = 1 \times x$. Evidently $\sigma \to T\sigma$, or which is the same $A \times \sigma \to B \times \sigma$ defines a chain-mapping $\tau: K \to K_1$ which is an isomorphism. By (IV, 5.6):

$$\text{(21.1)} \qquad \mathrm{F}(l \times \sigma) = B \times \sigma - A \times \sigma - l \times (\mathrm{F}\sigma).$$

Let C be any chain of K and let $\mathfrak{D}C = l \times C$. From (22.1) follows

$$\mathrm{F}\mathfrak{D}\sigma = \tau\sigma - \sigma - \mathfrak{D}\mathrm{F}\sigma,$$

and hence

$$\text{(21.2)} \qquad \mathrm{F}\mathfrak{D}C = \tau C - C - \mathfrak{D}\mathrm{F}C.$$

Therefore τ is a chain-deformation $K \to K_1$ in $\mathfrak{K}$ with the deformation operator $\mathfrak{D}$. There is nothing surprising in this, since the whole concept of chain-homotopy has been designed to carry over the above situation.

22. It is often convenient to replace $\mathfrak{K}$ by an Euclidean complex and its derived $\mathfrak{K}'$ may serve for the purpose. The property to be proved presently describes an alternate method which has the double advantage of requiring no new vertices and not modifying the bases K, K_1.

Let $\{A_i\}$ be the vertices of K ranged in some order and let $A_i^1 = TA_i$.

(22.1) *If* $\sigma^p = A_{i_0} \cdots A_{i_p} \in K$, $i_0 < \cdots < i_p$, *then* $\zeta_q^{p+1} = A_{i_0} \cdots A_{i_q} A_{i_q}^1 \cdots A_{i_p}^1$ *is contained in* $\mathfrak{K}$ *and* $\mathfrak{K}_1 = \bigcup \mathrm{Cl}\, \zeta_q^{p+1}$ *is a simplicial partition of* $\mathfrak{K}$.

Since $l \times \sigma$ is convex and all the vertices of ζ are in $\overline{l \times \sigma}$, we have $\zeta \subset \overline{l \times \sigma} \subset | \mathfrak{K} |$. To prove that $\mathfrak{K}_1$ is a partition we must show that:

(a) the simplexes of $\mathfrak{K}_1$ are disjoint;

(b) every point of $| \mathfrak{K} |$ is in a $| \mathrm{Cl}\, \zeta |$.

Let $x \in l \times \sigma^p$, $p > 0$. The segment $A_{i_0}x$ extended meets $| l \times \mathfrak{B}\sigma^p |$ in a point y. If x is common to two distinct simplexes of $\mathfrak{K}_1$ in $l \times \sigma^p$ then y has the same property relative to $| l \times \mathfrak{B}\sigma^p |$. This reduces the proof of (a) for the simplexes of $\mathfrak{K}_1$ in $l \times \sigma^p$ to the same for an $l \times \sigma^{p-1}$, and since it is trivial for $p = 0$, (a) is proved. Similarly (b) for x reduces to the same for y, i.e., for a point in an $l \times \sigma^{p-1}$ and since (b) is also trivial for points in an $l \times \sigma^0$, it is true for all points and (22.1) follows.

Let S denote the partition of (22.1). By (IV, 29.1) S is a subdivision. I say that the chain-mapping θ defined as follows is the chain-subdivision associated with S. Namely on K and $K_1 : \theta = 1$. For σ^p:

$$\text{(22.2)} \qquad \theta\mathfrak{D}\sigma = \sum (-1)^q A_{i_0} \cdots A_{i_q} A_{i_q}^1 \cdots A_{i_p}^1 .$$

At all events θ is readily verified to commute with F, and so it is a chain-mapping. It has also the required carrier S, and is the identity on the vertices as it should be. Since S does not raise dimensions this is sufficient to prove the asserted property of θ (IV, 18.1).

(22.3) Let us set now $\theta\mathfrak{D} = \mathfrak{D}_1$. Since $\theta = 1$ on K, K_1 we have in $\mathfrak{K}_1$ also

$$\text{(22.4)} \qquad F\mathfrak{D}_1 + \mathfrak{D}_1 F = \tau - 1.$$

Therefore τ is likewise a chain-deformation $K \to K_1$ in $\mathfrak{K}_1$ with associated operator $\mathfrak{D}_1$. Writing now $\mathfrak{K}$, $\mathfrak{D}$ for $\mathfrak{K}_1$, $\mathfrak{D}_1$ we may state

(22.5) *With the translation $l:|K| \to |K_1|$ there may be associated a chain-deformation $\tau:K \to K_1$ in an Euclidean complex $\mathfrak{K}$ with operator $\mathfrak{D}$ such that $\mathfrak{D}\sigma$, $\sigma \in K$, has all its vertices among those of σ and $\tau\sigma$ ($\tau\sigma$ is a simplex of K_1).*

It is hardly necessary to point out the parallel with (IV, 16.3). In point of fact it is out of (22.5) that there arose the general concept of chain-homotopy. See notably Lefschetz [e] and [L, 78].

The following homotopy properties will be required later. The notations are as in (1.1).

(22.6) *If t_1, t_2 are mappings $A \to |K|$ such that $\sigma(t_1x)$, $\sigma(t_2x)$ are always incident then t_1, t_2 are homotopic.*

For the condition of (I, 47.4) is manifestly satisfied here.

(22.7) Consider now K and $K^{(n)}$. Every σ_n is contained in a σ. Hence if every vertex of σ_n is sent into a vertex of the simplex σ carrying it, there is defined a simplicial chain-mapping $\tau:K^{(n)} \to K$ which is induced by a barycentric mapping $t:|K^{(n)}| \to |K|$. It may be shown as in (IV, 26.2c) for $n = 1$ that τ is a reciprocal of δ^n, where δ is chain-derivation in K. Since $t\sigma_n \prec \sigma_n$, by (22.6):

(22.8) *There is a barycentric deformation $|K^{(n)}|| \to |K|$ whose induced chain-mapping $\tau:K^{(n)} \to K$ is a reciprocal of δ^n.*

23. Continuous complexes.

(23.1) Let $\mathfrak{R}$ be a topological space. By a *continuous complex* $\mathfrak{K}$ *in* $\mathfrak{R}$ is meant a pair (K, t) where K is an Euclidean complex and t a mapping $|K| \to \mathfrak{R}$. The complex K is known as the *antecedent* of $\mathfrak{K}$. If $K = \{\sigma\}$ then the pairs $\zeta^p = (\sigma^p, t)$ are the *continuous p-cells* of $\mathfrak{K}$. By definition if the space is metric diam ζ = diam $t\sigma$ and mesh $\mathfrak{K}$ = sup diam ζ.

The continuous complex $\mathfrak{K}^{(n)} = (K^{(n)}, t)$ is the *nth derived* of $\mathfrak{K}$. More generally if K_1 is a simplicial partition of K then $\mathfrak{K}_1 = (K_1, t)$ is called a *subdivision* of $\mathfrak{K}$.

(23.2) All this may be extended in an obvious way to the mappings of polyhedra and we thus obtain *continuous polyhedra* in $\mathfrak{R}$, their derived and subdivisions. Suppose in particular that the continuous complexes $\mathfrak{K} = (K, t)$ and $\mathfrak{K}_1 = (K, t_1)$ are *homotopic* in $\mathfrak{R}$, that is to say, that t, t_1 are homotopic. There exists then a continuous polyhedron $\mathfrak{L} = (l \times K, T)$ in $\mathfrak{R}$, where l is the segment $0 \leqq u \leqq 1$, such that $T(0 \times x) = tx$, $T(1 \times x) = t_1x$, $x \in |K|$. The elements $(l \times \sigma, T)$ of $\mathfrak{L}$ are called *homotopy-cells*, or *deformation-cells* when $t = 1$.

(23.3) *If* $\mathfrak{K} = (K, t)$ *is a continuous complex in* $\mathfrak{R}$ *and* t_1 *is a mapping* $\mathfrak{R} \to \mathfrak{S}$, *then* $\mathfrak{K}_1 = (K, t_1 t)$ *is a continuous complex in* $\mathfrak{S}$ (obvious).

(23.4) Definition. *Suppose the continuous complex* $\mathfrak{L} = (L, t)$ *in* $|K|$. *If* t *is a barycentric mapping* $|L| \to |K|$ *we will say that* $\mathfrak{L}$ *is a continuous subcomplex of* K.

(23.5) *Let* $\mathfrak{L} = (L, t)$ *be a continuous complex in* K *and let* $\{a_i\}$, $\{b_j\}$ *be the vertices of* K, L. *If the "star condition":* $\{t \,|\, \text{St}\; b_j \,|\}$ *refines* $\{|\, \text{St}\; a_i \,|\}$, *holds then* $\mathfrak{L}$ *is homotopic to a continuous subcomplex* $\mathfrak{L}_1 = (L, t_1)$ *of* K, *where* t_1 *is such that if* $x \in |L|$ *then its path is in* $\sigma(tx)$.

By hypothesis corresponding to b_i there may be chosen a vertex of K, denoted by a_i, such that $|\, t\; \text{St}\; b_i \,| \subset |\, \text{St}\; a_i \,|$. The resulting repetitions among the a_i are immaterial. If $\zeta = b_j \cdots b_h \in L$ then $t\zeta \subset t \,|\, \text{St}\; b_j \cdots b_h \,| = t \,|\, \text{St}\; b_j \cap \cdots \cap \text{St}\; b_h \,| \subset |\, \text{St}\; a_j \cap \cdots \cap \text{St}\; a_h \,| = |\, \text{St}\; a_j \cdots a_h \,| \neq \emptyset$. Therefore $\sigma = a_j \cdots a_h$ is a simplex of K and so $b_j \to a_j$ defines a barycentric mapping $t_1 \colon |L| \to |K|$. Let $\mathfrak{L}_1 = (L, t_1)$. Since $t\zeta \subset |\, \text{St}\; (t_1\zeta) \,|$, (23.5) is a consequence of (22.6).

From (23.5) we deduce the following well known proposition which until recently provided the only method for proving the invariance of the homology groups of Euclidean complexes. (See [L, 86].)

(23.6) Theorem. *Every finite continuous complex* $\mathfrak{L}$ *on the polyhedron* $|K|$, K *finite, has a derived* $\mathfrak{L}^{(n)}$ *which is homotopic with a continuous subcomplex of* K *after the manner of* (23.5) (Alexander-Veblen).

For n may be chosen such that mesh $\{t \,|\, \text{St}\; b_j \,|\} <$ Lebesgue number $\{|\, \text{St}\; a_i \,|\}$, and then $\mathfrak{L}^{(n)}$ will satisfy the star condition.

For general Euclidean complexes we have:

(23.7) *Let* $K = \{\sigma\}$ *be an Euclidean complex and* $\mathfrak{L} = (L, t)$, $L = \{\zeta\}$, *a continuous complex in* $|K|$, *such that at most a finite number of* $t\zeta$ *meet a given* σ, *and conversely. Then* $\mathfrak{L}$ *has a subdivision which is homotopic with a continuous subcomplex of* K *after the manner of* (23.5).

If St b is the star of the vertex b in L then some finite closed subcomplex K_b of K is such that $(\text{Cl St}\; b, t) \subset |K_b|$, and so for some n the star condition will be fulfilled by $(\text{Cl St}\; b)^{(n)}$ relative to K_b. Every $\zeta \in L$ will be such that $t\zeta$ is in at most a finite number of $|K_b|$ and so ζ will undergo at most a finite number of derivations. Moreover $\zeta_1 < \zeta$ and $t\zeta \subset |K_b| \to t\zeta_1 \subset \overline{t\zeta} \subset |K_b|$. Hence ζ_1 will undergo at least as many derivations as ζ. Applying all the operations indicated for every K_b there will result a polyhedral complex L_1 which is a partition of L and such that every star of a vertex of L_1 is imaged by t in a star of a vertex of K. Therefore the derived L_1' is an Euclidean complex which is a subdivision of L such that the star condition holds relative to (L_1', t) and K, and the application of (23.5) to (L_1', t) yields (23.7).

The preceding results may be utilized for convenient approximations. For the sake of simplicity we will only consider finite complexes.

(23.8) *Let t be a mapping $|L| \to |K|$, K and L finite. Then*: (a) *there exists an n such that t is homotopic to a barycentric mapping $|L^{(n)}| \to |K|$*; (b) *given any $\epsilon > 0$ there exist m, n such that t is ϵ homotopic to a barycentric mapping* $|L^{(n)}| \to |K^{(m)}|$ (Alexander [b_1]).

Property (a) is merely another formulation for (23.6). Regarding (b) choose m so high that mesh $K^{(m)} < \epsilon$. By (23.6) we may choose n so high that $(L^{(n)}, t)$ is homotopic with a continuous complex $(L^{(n)}, t_0)$ after the manner of (23.5). The path of any $x \in |L|$ under the homotopy is a segment on a closed simplex of $K^{(m)}$ and hence its diameter is less than ϵ. Since otherwise t_0 conforms with (b) the latter is proved.

23a. **Homotopy groups.** While these important groups lie wholly outside our program, a few words concerning them will not be amiss.

(23a.1) *The Poincaré group.* Assume $\mathfrak{R}$ arcwise connected (any two points may be joined by an arc). We first introduce the *paths* in $\mathfrak{R}$ and a certain law of composition between them. A path is merely a continuous one-complex $\lambda = (l, t)$. The points $x = t(0)$ and $x' = t(1)$ are known as the *initial* and *terminal* points of λ, both as the *end points* of λ. We also say that λ *joins* x to x'. If $\lambda' = (l, t')$ is a second path with initial point x' and terminal point x'' then $\lambda'' = (l, t'')$ with t'' defined by

$$t''(u) = t(2u), \qquad 0 \leqq u \leqq 1/2;$$

$$t''(u) = t'(2u - 1), \qquad 1/2 \leqq u \leqq 1,$$

is likewise a path joining x to x'' and denoted by $\lambda'\lambda$. The path (l, t^*), where $t^*(u) = t(1 - u)$ whose end points are those of λ interchanged is denoted by λ^{-1} and called the *inverse* of λ.

Let $\mathrm{M} = \{\mu\}$ be the paths whose end points are both the fixed point x_0. If μ, μ' are in the set so are $\mu'\mu$ and μ^{-1} Hence our laws of composition and inversion may be applied unrestrictedly to $\{\mu\}$. Given $\mu = (l, t)$ in M we will write $\mu \sim 1$ if there is a homotopy of t with the mapping $t_0 : l \to x_0$ such that in all "intermediate" positions the path remains in M. Given μ, μ' in M we will write $\mu \sim \mu'$ if $\mu^{-1}\mu' \sim 1$. It is easy to see that $\sim$ is an equivalence relation and that the equivalence classes with the previously introduced laws of composition form a group. It is denoted by $\pi_1(\mathfrak{R})$ and known as the *Poincaré group* (also *fundamental group* or *group of paths*) of $\mathfrak{R}$ and is independent of x_0 (to within an isomorphism).

Not only is the Poincaré group generally noncommutative, but it is not too much to say that all the significant noncommutative groups ever discussed in topology have been $\pi_1(\mathfrak{R})$ or its derivates. It is not surprising therefore that ignorance regarding this group seems to account for the fact that many of the major problems of topology have so far eluded all attempts at solution.

(23a.2) *Covering manifold.* To simplify matters let $\mathfrak{R}$ be a topological n-manifold M^n (44.1). We define a new space $\mathfrak{M}^n = \{x\}$ as follows. Let $N =$

$\{\nu\}$ be the paths with x_0 as initial point. For ν, ν' in the set we will write $\nu \sim \nu'$ if ν, ν' are coterminal and $\nu^{-1}\nu' \sim 1$. This relation is readily seen to be an equivalence and the resulting equivalence classes are the points y of $\mathfrak{M}^n$. If $\nu \in y$, x is the terminal point of ν, and E is an n-cell containing x, then the classes of the paths $\lambda\nu$ where $\lambda = (l, t)$ has x as initial point and $tl \subset E$ make up a neighborhood $\mathfrak{E}$ in $\mathfrak{M}^n$ and $\{\mathfrak{E}\}$ is chosen as a base for $\mathfrak{M}^n$. It is not difficult to verify that $\mathfrak{M}^n$ is likewise a topological M^n, and is topologically independent of x_0. It is called the *universal covering manifold* of M^n.

If $\nu \in y$ and x is the terminal point of ν then the mapping $\rho: y \to x$ is easily verified to be "locally" topological.

(23a.3) *The homotopy groups of Hurewicz.* A noteworthy generalization of the Poincaré group has been given by Hurewicz [a]. Let $\mathfrak{R}$ be an arcwise connected compactum, and in addition *locally contractible.* (This property means that given any $\epsilon > 0$ there is a corresponding $\eta > 0$ such that $\mathfrak{S}(x, \eta)$, $x \in \mathfrak{R}$, is deformable to a point in $\mathfrak{S}(x, \epsilon)$.) Take now a fixed Euclidean $(n - 1)$-sphere S^{n-1}, $n > 1$, and let x_0, y_0 be fixed points of $\mathfrak{R}$, S^{n-1}. Consider all the mappings $t: S^{n-1} \to \mathfrak{R}$ such that $ty_0 = x_0$. If t, t' are two such mappings and we set $d(t, t') = \sup \{d(ty, t'y) \mid y \in S^{n-1}\}$, it is readily seen that $d(t, t')$ metrizes $\{t\}$. The resulting space has arcwise connected components and the Poincaré group of the component containing the identity mapping is known as the *nth homotopy group* of $\mathfrak{R}$, written $\pi_n(\mathfrak{R})$. The *first homotopy group* is $\pi_1(\mathfrak{R})$. We will merely recall the following basic properties of the homotopy groups, both due to Hurewicz.

(23a.4) *The groups* $\pi_n(\mathfrak{R})$, $n > 1$, *are abelian.*

(23a.5) *If* $n > 1$ *and the groups* $\pi_q(\mathfrak{R})$, $q < n$, *reduce to the identity, then* $\pi_n(\mathfrak{R})$ *is isomorphic with the homology group* $\mathfrak{H}_s^n(\mathfrak{R})$ *of the finite singular n-cycles* (25) *of* $\mathfrak{R}$.

It is also an elementary matter to prove:

(23a.6) $\mathfrak{H}_s^1(\mathfrak{R})$ *is isomorphic with the commutator group of the Poincaré group* $\pi_1(\mathfrak{R})$.

For further details regarding the homotopy groups and their applications the reader is referred to Hurewicz [a] and also to a recent comprehensive paper by Eilenberg [UM, pp. 57–99].

24. **Singular elements.** The singular elements to be introduced presently are very useful wherever homology and homotopy occur together.

By a *singular* p-cell in a metric space $\mathfrak{R}$ is meant a pair $E^p = (\sigma^p, t)$ where σ^p is an Euclidean simplex and t a mapping $\bar{\sigma}^p \to \mathfrak{R}$. The convention is made that if s is a barycentric mapping $\bar{\sigma}_1^p \to \bar{\sigma}^p$ then we still have $E^p = (\sigma_1^p, ts) = (s^{-1}\sigma^p, ts)$. It is a consequence of the preceding definition that if $\sigma^q \prec \sigma^p$ then $E^q = (\sigma^q, t)$ is likewise a singular q-cell; it is called a *q-face* of E^p. We shall also write correspondingly $E^q \prec E^p$. The simplex σ^p is said to be an *antecedent* of $E^p = (\sigma^p, t)$.

We have just assigned dimensions and incidences in $\Sigma = \{E\}$. To make it a complex there remains to assign suitable incidence numbers. First of all, we now assume in $E^p = (\sigma^p, t)$ that σ^p is *oriented* (with vertices taken in a definite order). We also agree to designate $(-\sigma^p, t)$ by $-E^p$. We may now define the finite singular chains over a group G as the linear forms

$$C^p = g^i E_i^p, \qquad g^i \text{ an integer}, \tag{24.1}$$

with the restriction that

$$g(-E^p) = (-g)E^p.$$

If we have

$$\mathrm{F}\sigma^p = \eta^i \sigma_i^{p-1}, \qquad E_i^{p-1} = (\sigma_i^{p-1}, t), \tag{24.2}$$

then we define $\mathrm{F}E^p$, the boundary of E^p, as the singular chain

$$\mathrm{F}E^p = \eta^i E_i^{p-1}, \tag{24.3}$$

and $\mathrm{F}C^p$, where C^p is (24.1), as:

$$\mathrm{F}C^p = g^i \mathrm{F}E_i^p. \tag{24.4}$$

Since $\mathrm{F}(-E^p) = -\mathrm{F}E^p$, $\mathrm{F}C^p$ thus defined is unique.

The incidence number $[E^p : E_i^{p-1}]$ is now defined as the coefficient of E_i^{p-1} in (24.3) (after terms are collected). To prove that Σ is a complex there remains to show that (III, 1, K4):

$$\textstyle\sum_{E'} [E:E'][E':E''] = 0. \tag{24.5}$$

Since E has at most a finite number of faces, (24.5) is equivalent to $\mathrm{FF}E^p = 0$, which under our definitions is an immediate consequence of the known relation $\mathrm{FF}\sigma^p = 0$. Therefore Σ is a closure-finite complex.

If $E = (\sigma, t)$ is a singular cell in $\mathfrak{R}$ then $t\sigma$ is a subset of $\mathfrak{R}$ which depends solely upon E but not upon its representation. We denote $t\sigma$ by $| E |$, and these sets have just the properties required to make Σ a metric complex as defined in (VI, 24.1). This metric complex is known as the *complete singular complex* of $\mathfrak{R}$. The subcomplexes, chains, cycles of Σ are referred to as *singular* complexes, chains, cycles, of $\mathfrak{R}$.

25. There are then two possible homology theories to be considered in relation to Σ. They correspond to:

(a) the finite singular chains and cycles;

(b) the V-cycles of Σ, referred to as VS-cycles.

The groups for the two types are to be chosen discrete throughout. In point of fact the VS-cycles are only interesting as auxiliaries and their groups are generally reducible (in the interesting cases) to those of the finite cycles.

(25.1) That the homology groups resulting for instance from (a) are not necessarily the Vietoris groups is shown by the following example. The space $\mathfrak{R}$ is a planar set which is the union of the following three sets: a segment $\lambda_1 : x = 0, 1 \leqq y \leqq 1$; the arc $\lambda_2 : y = \sin(1/x), 0 < x < 1$; an arc λ_3 joining

(1,0) to (0, -1) but otherwise not meeting $\lambda_1 \cup \lambda_2$. The Betti number for the rational Vietoris one-cycles is readily found to be $R^1 = 1$. On the other hand the fact that all the "closed" curves in the set are homotopic to points enables one to prove that for the finite rational singular one-cycles we have $R^1 = 0$. Thus the rational Betti numbers for the finite singular one-cycles and those for the Vietoris cycles are distinct.

(25.2) If A is a subset of $\mathfrak{R}$, and E, $\mathfrak{K}$, C are a singular cell, complex or chain, then: $E \subset A, \cdots$ signifies: $| E | \subset A, \cdots$. In particular when A is closed and $E \subset A$ then $E' \prec E \rightarrow | E' | \subset \overline{| E |} \subset A$. Hence $\Sigma_A = \{E \mid E \subset A\}$ is a closed subcomplex of Σ. The singular cycles, $\cdots$ mod A are those of Σ mod Σ_A.

(25.3) *Groups related to a point.* Since a point x is closed the E's in x form a closed subcomplex Σ_x. The VS-cycles around x are known as *singular cycles around* x. At the same time one may also define *singular cycles through* x, as follows: a singular p-cycle γ^p through x over a discrete G is a singular chain over G such that $x \notin \| FC^p \|$; the cycle ~ 0 whenever there are singular chains $C^{p+1}, D^p, x \notin \| D^p \|$, such that $FC^{p+1} = \gamma^p + D^p$. The groups $\mathfrak{Z}^p$, $\mathfrak{F}^p$, $\mathfrak{H}^p$ are then defined in the customary way.

26. We will now describe a certain number of results with proofs merely outlined or even omitted in the simpler cases.

(26.1) *If $\mathfrak{R}$, $\mathfrak{S}$ are metric spaces and T is a mapping $\mathfrak{R} \rightarrow \mathfrak{S}$, then T induces a homomorphism of the homology groups of the finite singular cycles of $\mathfrak{R}$ into the corresponding groups of $\mathfrak{S}$.*

(26.2) *The homology groups of the finite singular cycles are topologically invariant.*

(26.3) Let $\mathfrak{K} = (K, t)$ be a continuous complex in $\mathfrak{R}$, where $K = \{\sigma\}$. Then $\mathfrak{K}_1 = \{(\sigma, t)\}$ is a singular complex said to be *induced* by $\mathfrak{K}$.

(26.4) *Every singular complex $\mathfrak{K}_1$ is induced by some continuous complex $\mathfrak{K}$.* If $\mathfrak{K}_1 = \{E_i\}$ we may set $E_i = (\sigma_i, t_i)$ where the Cl σ_i are disjoint. Then if $K = \bigcup \mathrm{Cl}\, \sigma_i$, and t: $|K| \rightarrow \mathfrak{R}$ is defined by $t \mid \bar{\sigma}_i = t_i$, we have in $\mathfrak{K} = (K, t)$ a continuous complex inducing $\mathfrak{K}_1$.

(26.5) By taking operations on $\mathfrak{K}$ such that if say $\sigma_1, \cdots, \sigma_s$ are antecedents of $E \in \mathfrak{K}_1$ in $\mathfrak{K}$ then the effect of the operations is independent of the permutations of $\sigma_1, \cdots, \sigma_s$, there may be introduced corresponding operations in $\mathfrak{K}_1$. Thus we may define chain-derivation d in $\mathfrak{K}_1$, as well as other types of subdivisions, likewise singular chain-homotopy, etc.

Let $\mathfrak{K} = \{E_i^p\}$ be a singular complex, $\mathfrak{K}^{(n)} = \{E_{nj}^q\}$ its nth derived, δ chain-derivation in all simplicial complexes, d singular chain-derivation in $\mathfrak{K}$.

(26.6) *d^n is a singular chain-deformation $\mathfrak{K} \rightarrow \mathfrak{K}^n$ with operator $\mathfrak{D}$ such that* $\| \mathfrak{D} E \| \subset \overline{\| E \|}$.

Since $(26.6)_n$ is obtained by repetition from $(26.6)_1$, we only need to prove the latter and so suppose $n = 1$. Let l be the segment $0 \leqq u \leqq 1$ and suppose $E = (\sigma, t)$. Consider the mapping $T: l \times (\bar{\sigma}) \rightarrow \mathfrak{R}$ such that $T(l \times x) = tx$, $x \in \bar{\sigma}$. Apply to $l \times (\mathrm{Cl}\, \sigma)$ a simplicial partition which differs only from a barycentric subdivision in that no new vertices are introduced in $0 \times \text{Cl } \sigma$, and let θ be the induced chain-subdivision. We have $\theta(0 \times \sigma) = 0 \times \sigma$, $\theta(1 \times \sigma) = \delta(1 \times \sigma)$. Hence the singular images of $\theta(0 \times \sigma)$ and $\theta(1 \times \sigma)$ under T are, respectively, E and dE. We have now the relation (22.1) in $l \times \text{Cl } \sigma$. Applying θ to both sides, taking the singular images under T, and denoting in particular by $\mathfrak{D}E$ the singular image of $\theta(l \times \sigma)$, we obtain $\mathfrak{D}\text{F} + \text{F}\mathfrak{D} = d - 1$. It is readily seen that $\mathfrak{D}E$ depends solely upon E, and not upon the particular representation (σ, t) chosen for E. Thus d is a singular chain-deformation. Since $T(l \times \sigma) = \| \mathfrak{D}E \| \subset T\bar{\sigma}$, (26.6) is proved.

(26.7) *The notations remaining the same suppose $\mathfrak{K}$ finite and contained in the polyhedron $| K |$, $K = \{\sigma\}$. Then*

(a) *for n above a certain value there is a singular chain-deformation $\eta: \mathfrak{K}^{(n)} \rightarrow K$ with operator $\mathfrak{D}$ such that $\| E_{ni} \| \subset \bar{\sigma} \rightarrow \| \mathfrak{D}E_{ni} \| \subset \bar{\sigma}$;*

(b) *there is likewise a singular chain-deformation $\theta: \mathfrak{K} \rightarrow K$ with the property that if L is a closed subcomplex of both K and $\mathfrak{K}$ then $\theta \mid L = 1$;*

(c) *we also notice explicitly that mesh $\mathfrak{K}^{(n)} \rightarrow 0$.*

Let (K_1, t), $K_1 = \{\zeta\}$, be an antecedent of $\mathfrak{K}$. Since K_1 is finite $t \mid K_1 \mid = \mid \mathfrak{K} \mid$ is compact. The open covering of this compact set by the intersections with the stars of the vertices of K has a finite subcovering. Hence $\mathfrak{K}$ is contained in a finite set of the stars and so in a finite closed subcomplex of K. Thus we may assume K finite. Under the circumstances by (23.6) there is an n such that (23.5) may be applied to $(K_1^{(n)}, t)$. Thus we have a mapping $T: \mid l \times K_1^{(n)} \mid \rightarrow \mid K \mid$ corresponding to the homotopy of (23.5) as applied here. If we take now a simplicial partition of $l \times K_1^{(n)}$ similar to the derived but with no new vertices in $0 \times K_1^{(n)}$ or $1 \times K_1^{(n)}$, and pass to the singular images, the mappings of the bases $0 \times K_1^{(n)}, 1 \times K_1^{(n)}$ of $l \times K_1^{(n)}$ induce the chain-mappings $1, \eta$, and the image of $(l \times \zeta'_{ni}; T)$ subdivided is a singular chain $\mathfrak{D}E_{ni}$. If $K_1^{(n)} = \{\zeta_{ni}\}$ and $\zeta'_{ni}, \zeta''_{ni}$ are two antecedents of E_{ni} we find readily that $(l \times \zeta'_{ni}, T) = (l \times \zeta''_{ni}, T)$, where the parentheses represent the singular chains obtained from the subdivision of the cells. Thus $\mathfrak{D}E$ is shown as in the proof of (26.6) to yield a suitable $\mathfrak{D}$ making η a singular chain-deformation. If $\| E_{ni} \| \subset \bar{\sigma}$, then by (23.5): $T(l \times \zeta'_{ni}) \subset \bar{\sigma}$, and since $\| \mathfrak{D}E_{ni} \| = T(l \times \zeta'_{ni})$, the rest of (26.7a) follows.

If we set $\theta = \eta d^n$ then from (26.6, 26.7a) we deduce that θ is a singular chain-deformation $\mathfrak{K} \rightarrow K$. If L is as in (26.7b) then $\eta \mid L^{(n)} = \tau$, a reciprocal of δ^n, and $d^n = \delta^n$. Hence $\theta \mid L = \tau\delta^n = 1$. This completes the proof of (26.7b).

(26.8) We will now apply (26.6, 26.7) to the reduction of the finite singular cycles of $\mid K \mid$ to those of K itself. If γ^p is such a cycle its class will be written Γ^p. If $\bar{\gamma}^p$ is a cycle of any finite Euclidean complex then we agree to identify its combinatorial class and geometrical class, with a common designation $\Delta(\bar{\gamma}^p)$.

If δ denotes chain-derivation in K and τ is a reciprocal of δ then $\Delta = \Delta\delta$, and since $\tau\delta \sim 1$, likewise $\Delta\tau = \Delta\tau\delta = \Delta$.

Taking then γ^p in $|K|$ by (26.7b), there is a singular chain-deformation $\theta_n\colon |\gamma^p| \to K^{(n)}$.

(26.8a) $\Delta(\theta_n\gamma^p)$ *depends solely upon* Γ^p.

If $\mathfrak{D}_n$ is the homotopy operator for θ_n then

$$\mathrm{F}\mathfrak{D}_n\gamma^p = \theta_n\gamma^p - \gamma^p \sim 0.$$

If γ'^p is any other cycle of Γ^p there will be corresponding $\mathfrak{D}'_n$, θ'_n and

$$\mathrm{F}\mathfrak{D}'_n\gamma'^p = \theta'_n\gamma'^p - \gamma'^p \sim 0.$$

Hence $\theta'_n\gamma'^p \sim \theta_n\gamma^p$ in the singular sense. Therefore there is a singular chain C^{p+1} such that $\mathrm{F}C^{p+1} = \theta'_n\gamma'^p - \theta_n\gamma^p$. If we treat C^{p+1} like the cycles and notice that the chain-operations involved reduce to the identity on the chains of $K^{(n)}$ (26.7b) we find that C^{p+1} is chain-homotopic with a chain in $K^{(n)}$ whose boundary is the same. Hence we may assume $C^{p+1} \subset K^{(n)}$ and so $\theta'_n\gamma'^p_n \sim \theta_n\gamma^p_n$ in $K^{(n)}$. This proves (26.8a).

Notice incidentally that we may assume $\gamma'^p = \gamma^p$ and θ'_n merely any singular chain-homotopy whatever $|\gamma^p| \to K^{(n)}$, and the preceding argument shows that $\Delta(\theta_n\gamma^p)$ is independent of the particular chain-homotopy $\theta_n\colon |\gamma^p| \to K^{(n)}$.

(26.9) $\Delta(\theta_n\gamma^p)$ *is independent of* n.

Let this time τ denote the reciprocal of chain-derivation δ in $K^{(n)}$. Since τ is a singular chain-deformation $K^{(n+1)} \to K^{(n)}$, (22.8), $\tau\theta_{n+1}$ is a chain-homotopy $|\gamma^p| \to K^{(n)}$, and so $\Delta(\tau\theta_{n+1}\gamma^p) = \Delta(\theta_n\gamma^p)$. Since $\Delta\tau = \Delta$ we have $\Delta(\theta_{n+1}\gamma^p) = \Delta(\theta_n\gamma^p)$ which implies (26.9).

Thus $\Delta(\theta_n\gamma^p)$ depends solely upon Γ^p and so we denote it by $\Delta(\Gamma^p)$.

(26.10) $\Gamma^p \to \Delta(\Gamma^p)$ *defines an isomorphism of the corresponding groups.*

At all events there is defined a homomorphism ω of the groups. If γ^p is a cycle of K then the chain-mapping identity is a singular chain-homotopy sending γ^p into itself. Hence ω is of "onto" type. If $\gamma^p \sim 0$ in K then it is also ~ 0 in the singular sense, and so ω is univalent. Since we are dealing with homology groups of finite cycles, which are all discrete, ω is an isomorphism. This proves (26.10). More explicitly:

(26.11) *The homology groups of the finite singular cycles of an Euclidean complex K are isomorphic with the corresponding combinatorial groups. Hence in particular the latter are topologically invariant.*

This result implies (10.1) and in particular the topological invariance of the Betti numbers and torsion coefficients of a finite Euclidean complex. It was essentially along these lines, i.e., by means of the singular cycles, that the invariance of the Betti numbers was first proved by Alexander [a], and that of the Betti numbers and torsion coefficients later also by Veblen [V]. (See also [L, 87; L₁, X].)

An interesting complement required later is:

(26.12) *Assuming for simplicity K finite, the operation $\Gamma^p \to \Delta(\Gamma^p)$ has topological character with respect to a topological mapping $t\colon |K| \to |K_1|$, where K_1 is also finite.*

Consider the continuous complex (K, t) and let n be chosen in accordance with (23.6). We have then the following two associated operations:

(a) If $\{a_i\}$, $\{b_j\}$ are the vertices of $K^{(n)}$, K_1, and if $K^{(n)}$, K_1 are considered as the nerves of the coverings $\{ \mid \text{St } a_i \mid \}$, $\{ \mid \text{St } b_j \mid \}$, then the barycentric mapping $\mid K^{(n)} \mid \rightarrow \mid K_1 \mid$ of (23.5) (implicit in 23.6) induces a chain-mapping $\theta: K^{(n)} \rightarrow K_1$, which is of the same type as the p_μ^ν of (VII, 5.12) corresponding to the mapping t. Since the combinatorial classes of $K^{(n)}$, K_1 are identified with the corresponding geometrical classes, and since t is topological, θ induces a mapping of $\Delta(\Gamma^p)$ into its image class $t\Delta(\Gamma^p)$ in K_1.

(b) The second operation referred to is the homotopy $T: (K^{(n)}, t) \rightarrow \mid K_1 \mid$ of (23.6).

Take as a representative of Γ^p a cycle γ^p of $K^{(n)}$. Then we verify immediately that the singular chain-deformation $(K^{(n)}, t) \rightarrow K_1$ induced by T sends $t\gamma^p$ into $\theta\gamma^p$. Thus if $t\Gamma^p$ denotes the class of $t\gamma^p$ then $\Delta(t\Gamma^p)$ is the class of $\theta\gamma^p$ in K_1. Since this is the same as $t\Delta(\Gamma^p)$, (26.12) is proved.

(26.13) *Relative cycles.* All the preceding results hold with minor modifications for the cycles of K mod L, L a closed subcomplex of K. Of course in (26.12) t must be such that $t \mid L \mid = \mid L_1 \mid$, L_1 a closed subcomplex of K_1.

(26.14) *Cycles through the points.* Essentially the same arguments are valid for these cycles. As in (14) one may reduce the treatment to the case of a vertex a. Care will merely have to be taken not to send into a any singular vertex different from a. This yields a new proof of (14.1) and in particular implies also:

(26.15) *The homology groups of the singular cycles through the points of* $\mid K \mid$ *are isomorphic with the corresponding geometrical groups.*

27. **Intersections of singular chains in a manifold.** The questions which we shall now consider are the last dealing with intersections in the present work. It is interesting to observe that they were the first investigated by the author in his initial paper on intersections [a]. This is in keeping with the development of topology since that time: the algebraic properties are now completely to the fore, and the more special topological properties are dealt with afterwards.

(27.1) Let again X be the manifold of (20). Given a finite singular cycle γ^p in $\mid X \mid$ consider the associated closed set $P = \| \gamma^p \|$ and let $U(\epsilon) = \mathfrak{S}(P, \epsilon)$. By (4.1) corresponding to a given $\epsilon > 0$ there is an s such that if Y_s is the union of the closures of the simplexes of $X^{(s)}$ which meet P then $\mid Y_s \mid \subset U(\epsilon/4)$. If we treat now γ^p as in (26.8) we obtain a cycle δ^p of Y_s such that $\gamma^p \sim \delta^p$ in Y_s and hence also in $\overline{U(\epsilon)}$. Moreover (26.12) the class Δ^p of δ^p as a cycle of Y_s and hence of $\overline{U(\epsilon)}$ depends solely upon γ^p. Since derivation does not alter the manifold properties we may apply to δ^p in $X^{(s)}$ the operation analogous to $\zeta\eta$ of (V, 36). This will give rise to a cocycle δ'_{n-p} such that if γ^n is the basic cycle of X or rather its image in $X^{(s)}$ then, by virtue of (V, 36.13):

$$\gamma^p \sim \delta^p \sim \delta'_{n-p} \cdot \gamma^n \text{ in } \overline{U(\epsilon)}. \tag{27.2}$$

Finally it is a consequence of (V, 36.12) and the topological character of the intersection classes in a finite complex, that the class of δ'_{n-p}, say in $U(2\epsilon)$, also has topological character.

Thus we have assigned to γ^p a definite cocycle δ'_{n-p} over the same group G as γ^p, whose class in $U(2\epsilon)$ depends solely upon γ^p.

(27.3) Let now G, H be commutatively paired to J, and let γ^p, γ^q be finite singular cycles over G, H. If $Q = \| \gamma^q \|$ and $V(\epsilon) = \mathfrak{S}(Q, \epsilon)$, we find δ'_{n-p}, δ'_{n-q} carried by $U(2\epsilon)$, $V(2\epsilon)$. We will denote the classes of $\gamma^p, \delta'_{n-p}, \cdots$, by Γ^p, $\Delta'_{n-p}, \cdots$.

Since Γ^n is integral there is a class intersection $\Gamma_t \cdot \Gamma^n$ of Γ^n with a cohomology class Γ_t over any group whatever. Hence $\Delta'_{n-p} \cdot \Delta'_{n-q} \cdot \Gamma^n = \Gamma_\epsilon^{p+q-n}$ is a class of cycles over J. By (VII, 10.4, 10.6) it has a representative γ_ϵ^{p+q-n} in $\overline{W}(2\epsilon)$, where $W(2\epsilon) = U(2\epsilon) \cap V(2\epsilon)$, whose class in $\overline{W(2\epsilon)}$ is unique. Hence if $\epsilon' < \epsilon$ we have

$$\gamma_{\epsilon'}^{p+q-n} \sim \gamma_\epsilon^{p+q-n} \text{ in } \overline{W(2\epsilon)}.$$

If $R = P \cap Q$ then $\{W(\epsilon)\}$ is coterminal in the web of all the neighborhoods of R. Hence by (VII, 15.1) $\{\gamma_\epsilon^{p+q-n}\}$ converges to a cycle γ^{p+q-n} in R, such that

$$\gamma^{p+q-n} \sim \gamma_\epsilon^{p+q-n} \text{ in } \overline{W(2\epsilon)}$$

for every ϵ, and its class Γ^{p+q-n} (as a cycle of R) is unique. Any cycle $\gamma^{p+q-n} \in \Gamma^{p+q-n}$ is called an *intersection cycle* of the singular cycles γ^p, γ^q and is denoted by $\gamma^p \circ \gamma^q$. The class Γ^{p+q-n} is likewise denoted by $\Gamma^p \circ \Gamma^q$ and called the *intersection class* of the singular classes Γ^p, Γ^q. It is an elementary matter to prove:

(27.4) $\circ$ *has all the properties of* (V, 37.7).

(27.5) *Kronecker index.* If $q = n - p$ then $\gamma^p \circ \gamma^{n-p}$ is a zero-dimensional cycle, and so it has an index. We naturally define

$$\mathrm{KI}(\gamma^p, \gamma^{n-p}) = \mathrm{KI}(\gamma^p \circ \gamma^{n-p}).$$

(27.6) *If γ^p, γ^q are disjoint (i.e., $P \cap Q = \emptyset$) then $\gamma^p \circ \gamma^q \sim 0$.*

(27.7) Suppose now that C^p, C^q are finite singular chains over G, H in X and let

$$P = \| C^p \|, \qquad Q = \| C^q \|, \qquad R = P \cap Q,$$

$$P_1 = \| \mathrm{F}C^p \|, \qquad Q_1 = \| \mathrm{F}C^q \|.$$

We will say that C^p, C^q are in *general position* if

$$P \cap Q_1 \cup P_1 \cap Q = \emptyset.$$

Using (VII, 10.7) it is now a simple matter to extend the preceding results and obtain an *intersection cycle* $C^p \circ C^q$ *in* R and its *intersection class*. Moreover if $q = n - p$ we will have an index

$$\mathrm{KI}(C^p, C^{n-p}) = \mathrm{KI}(C^p \circ C^{n-p}).$$

(27.8) *The intersections of singular chains and cycles in a finite geometric absolute orientable* $M^n = X$, *as well as the Kronecker index which have just been introduced, are topologically invariant with respect to any topological mapping* $t:| X | \rightarrow | X_1 |$, *where* X_1 *is a similar* M^n, *such that if* γ^n *is the basic n-cycle of* X *then* $t\gamma^n$ *is the basic n-cycle of* X_1.

For under the circumstances all the elements used in defining the intersections have topological character.

(27.9) *Extension to any orientable* M^n. In the first place the finiteness restriction is manifestly unimportant. However if M^n and the chains are allowed to be infinite, various cases may have to be distinguished. We merely observe that if $R = P \cap Q$ is compact then so is the intersection cycle $\gamma^p \circ \gamma^q$.

Suppose now that $X = Y - Z$, i.e., that Y is an orientable M^n mod Z. The notations being essentially as before the set $| X |$ is covered with an Euclidean complex X_1 regular with respect to Z after the manner of (5.1) duly extended. We show then as in (19.1) that X_1 is an absolute (generally infinite) orientable M^n and intersections are now defined by reference to that manifold.

(27.10) *Application.* Let the Euclidean space $\mathfrak{E}^n$ be referred to $\{x_1, \cdots, x_n\}$ and let X_i be the one-complex arising from $-\infty < x_i < +\infty$ and its subdivision by $x_i = 0, \pm 1, \cdots$. Then $\mathfrak{E}^n = | X |$, $X = X_1 \times \cdots \times X^n$. Let $q < p$, $r = p - q$ and consider the subspaces

$$\mathfrak{E}^p : x_{p+i} = 0, \quad i > 0; \qquad \mathfrak{E}^{n-q} : x_i = 0, \quad i \leqq q;$$

$$\mathfrak{E}^r = \mathfrak{E}^p \cap \mathfrak{E}^{n-q} : x_i = 0, \quad i \leqq q \text{ or } > p.$$

We suppose them oriented by $\{x_1, \cdots, x_p\}$, $\{x_{q+1}, \cdots, x_n\}$, $\{x_{q+1}, \cdots, x_p\}$. If A_i is the point $x_i = 0$ in the ith line and L_i its basic one-cycle (geometrical identified with adherent combinatorial cycles) then the classes of the basic cycles $\mathfrak{E}_0^p$, $\mathfrak{E}_0^{n-q}$, $\mathfrak{E}_0^r$ are precisely those of (V, 38.7). Thus $\mathfrak{E}_0^p \cap \mathfrak{E}_0^{n-q}$ is a cycle of $\mathfrak{E}^r$ in the class of $\mathfrak{E}_0^r$, or by (V, 38.7):

$$\mathfrak{E}_0^p \circ \mathfrak{E}_0^{n-q} = \mathfrak{E}_0^r. \tag{27.11}$$

If $p + q = n$ then $\mathfrak{E}_0^p \circ \mathfrak{E}_0^{n-p}$ is the origin of coordinates, taken once according to (V, 38.8), and so

$$\text{KI}(\mathfrak{E}_0^p, \mathfrak{E}_0^{n-p}) = 1. \tag{27.12}$$

These are precisely the intersection rules of Lefschetz [a] *for convex intersections.*

It is to be noted throughout that the preceding results imply a reorientation of the dual X^* or equivalently of the reciprocal $\bar{X}$ as described in (V, 38.3).

§5. COINCIDENCES AND FIXED POINTS

28. In this and the next section we propose to deal with extensions to mappings of the results of (V, 24) for coincidences and fixed elements of chain-mappings.

(28.1) If $\mathfrak{R}$ is a topological space and T a mapping $\mathfrak{R} \rightarrow \mathfrak{R}$ then a *fixed point* of T has its obvious meaning: it is a point x such that $Tx = x$. Similarly if $\mathfrak{R}$, $\mathfrak{S}$ are two topological spaces and T_1 and T_2 are mappings $\mathfrak{R} \rightarrow \mathfrak{S}$ and $\mathfrak{S} \rightarrow \mathfrak{R}$, then a *coincidence* of T_1, T_2 is a pair (x, y), $x \in \mathfrak{R}$, $y \in \mathfrak{S}$, such that $y = T_1x$, $x = T_2y$. This is the general situation, with T_1, T_2 running in opposite directions: from $\mathfrak{R}$ to $\mathfrak{S}$ and $\mathfrak{S}$ to $\mathfrak{R}$. However, for finite geometric manifolds we shall also consider coincidences of two mappings $T_1, T_2 : \mathfrak{R} \to \mathfrak{S}$, both going in the same direction, and they are defined as any pair (x, y), $x \in \mathfrak{R}$, $y \in \mathfrak{S}$, such that $y = T_1x = T_2x$.

(28.2) All our results will apply solely to spaces which have

PROPERTY A. The space $\mathfrak{R}$ is compact, all its rational Betti numbers R^p are finite and all but a finite number of the R^p are zero.

(28.3) Suppose that T is a mapping $\mathfrak{R} \to \mathfrak{R}$, where $\mathfrak{R}$ possesses Property A. Since the numbers R^p are finite there is a finite so-called *rational homology base* or maximal set of rational cycles $\{\gamma_i^p\}$, $i = 1, 2, \cdots, R^p$, independent with respect to homology. The mapping T induces a homomorphism on the homology groups (VII, 7.1) and hence a transformation on the elements $\{\gamma_i^p\}$ given by homologies with rational coefficients:

$$(28.4) \qquad T\gamma_i^p \sim \lambda_i^j(p)\gamma_j^p, \qquad \lambda^p = \| \lambda_i^j(p) \|.$$

(For simplicity we also denote by T the simultaneous homomorphisms of the homology groups which T induces, and similarly for the other mappings.)

Since all but a finite number of the R^p are zero, the expression

$$(28.5) \qquad \psi(T) = \sum (-1)^p \text{ trace } \lambda^p$$

has a meaning. Similarly if $\mathfrak{S}$ is a second space with Property A and rational homology bases $\{\delta_i^p\}$, and if T_1, T_2 are mappings $\mathfrak{R} \to \mathfrak{S}$, $\mathfrak{S} \to \mathfrak{R}$ then

$$(28.6) \qquad \begin{gathered} T_1\gamma_i^p \sim \mu_{1i}^j(p)\delta_j^p, \qquad T_2\delta_i^p \sim \mu_{2i}^j(p)\gamma_j^p, \\ \mu_h^p = \| \mu_{hi}^j(p) \|, \end{gathered}$$

and the expression

$$(28.7) \qquad \varphi(T_1, T_2) = \psi(T_2 T_1) = \sum (-1)^p \text{ trace } \mu_1^p \mu_2^p$$

has meaning. As is well known μ_1^p, μ_2^p may be interchanged at the right without changing the traces.

If $\{\gamma_i'^p\}$ is a second maximal independent set analogous to $\{\gamma_i^p\}$ we have

$$(28.8) \qquad \gamma_i'^p = a_i^j \gamma_j^p, \qquad a = \| a_i^j \|,$$

where a is a rational nonsingular square matrix and λ^p will be replaced by $a\lambda^p a^{-1}$ whose trace is the same. Similarly for trace $\mu_1^p \mu_2^p$. Coupling this with (VII, 7.1) we have:

(28.9) *The numbers: trace λ^p, trace $\mu_1^p \mu_2^p$, $\psi(T)$, $\varphi(T_1, T_2)$ are independent of the particular homology bases in terms of which they have been calculated. Furthermore they depend merely upon the homotopy classes of the mappings.*

29. **Coincidences and fixed points for finite Euclidean complexes.** We will now suppose that we have finite Euclidean complexes K, L and a mapping $T: |K| \to |K|$ or a pair of mappings $T_1, T_2 : |K| \to |L|, |L| \to |K|$.

It is clear that $|K|$, $|L|$ have Property A, and so (28.9) is applicable to them. We will now prove:

(29.1) FIXED POINT THEOREM. *If* $\psi(T) \neq 0$ *then* T *has a fixed point.*

(29.2) COROLLARY. *If* $\varphi(T_1, T_2) \neq 0$ *then* T_1, T_2 *have a coincidence.*

We shall assume that T has no fixed point and show that this leads to a contradiction, and similarly for the coincidences.

(29.3) If T has no fixed point, $d(x, Tx) > 0$ for every $x \in |K|$. Since $|K|$ is a compactum $\epsilon = \inf d(x, Tx) > 0$. Choose an n such that mesh $K^{(n)} < \epsilon/3$ (4.1). By (23.8) there is an $m > n$ such that: (a) a suitable T' homotopic to T maps $|K^{(m)}|$ barycentrically into $|K^{(n)}|$; (b) $T'x \in \overline{\sigma_n(Tx)}$, and so $d(Tx, T'x) < \epsilon/3$. As a consequence $\bar{\sigma}_n$ and $\overline{T'\sigma_n}$ are always disjoint. Therefore if θ is the chain-mapping $K^{(m)} \to K^{(n)}$ induced by T', then $\sigma_m \subset \bar{\sigma}_n$ implies that $\theta\sigma_m$ has no element in $\bar{\sigma}_n$.

On the other hand if δ denotes chain-derivation in K, then δ^{m-n} applied to $K^{(n)}$ is a chain-mapping such that $\| \delta^{m-n}\sigma_n \| \subset \bar{\sigma}_n$. Therefore θ, δ^{m-n} are chain-mappings $K^{(m)} \to K^{(n)}$, $K^{(n)} \to K^{(m)}$ without coincidences, and so $\varphi(\theta, \delta^{m-n}) = 0$, where φ is defined as in (V, 24).

Since T, T' are homotopic they induce the same homomorphisms on the homology groups. Since δ induces an isomorphism of the homology groups of K with the corresponding groups of $K^{(n)}$ we may as well identify the classes Γ^p, $\delta^n\Gamma^p$ of K, $K^{(n)}$. Hence if Γ_i^p is the class of γ_i^p we will have from (28.4):

$$\theta\Gamma_i^p = \lambda_i^j(p)\Gamma_j^p .$$

To δ^{m-n} there will correspond similar relations with every $\lambda^p = 1$. Therefore

$$\varphi(\theta, \delta^{m-n}) = \psi(T), \tag{29.4}$$

and so here $\varphi(\theta, \delta^{m-n}) \neq 0$, a contradiction proving (29.1).

(29.5) Passing now to (29.2), T_2T_1 is a mapping $|K| \to |K|$ whose fixed points are in one-one correspondence with the coincidences of T_1, T_2. This together with $\psi(T_2T_1) = \varphi(T_1, T_2)$ and (29.1) yields (29.2).

(29.6) Since θ, δ^{m-n} are chain-mappings of finite complexes into one another, $\varphi(\theta, \delta^{m-n})$ is given by a relation (V, 22.4), in which at the right all the numbers are integers. Hence $\varphi(\theta, \delta^{m-n})$ is an integer, and therefore this holds also for $\psi(T)$, consequently also by (28.7) for $\varphi(T_1, T_2)$. Or:

(29.7) *For mappings of finite polyhedra into one another both* $\psi(T)$ *and* $\varphi(T_1, T_2)$ *are integers.*

(29.8) We have just been considering coincidences for mappings T_1, T_2 going in opposite directions (from $|K|$ to $|L|$ and $|L|$ to $|K|$). A similar result may be derived for finite absolute orientable manifolds when the mappings proceed in the same direction. For this purpose we utilize the result of (V, 40). Let the notations be those loc. cit., except that X, Y are geometric manifolds and let T_1, T_2 be two mappings $|X| \to |Y|$. Replacing if need be both X and Y by suitable derived and proceeding as in (29.3) there will be obtained barycentric mappings T_1', $T_2' : |X| \to |Y|$ homotopic with T_1, T_2 and such that if the latter have no coincidences the same holds regarding T_1', T_2'. In fact we may then so choose X, Y that throughout

$$(29.9) \qquad (\text{Cl St } T_1'x) \cap (\text{Cl St } T_2'x) = \emptyset.$$

As a consequence if τ, θ are the chain-mappings $X \to Y$ induced by T_1', T_2' then no $y \in Y$ will be found in both τx and θx. Coupling this with (V, 36.3) we find that τ, $\bar{\theta}$ (the latter as in V, 40) have no coincidences. On the other hand the coincidences of τ, $\bar{\theta}$ are in one-one correspondence with those of τ, $\bar{\theta}^*$, and so the latter will have none. Therefore with χ, ω as in (V, 40) we must have:

$$(29.10) \qquad 0 = \varphi(\tau, \bar{\theta}^*) = \chi(\tau, \bar{\theta}) = \omega(\tau, \theta).$$

We will set again $\omega(\tau, \theta) = \omega(T_1, T_2)$ and it is an elementary matter to prove the analogues of (28.9):

(29.11) *$\omega(T_1, T_2)$ depends merely upon the homotopy classes of T_1, T_2 and it is an integer.*

We have then from (29.10):

(29.12) *Let X, Y be finite absolute orientable geometric n-manifolds and T_1, T_2 two mappings $|X| \to |Y|$. If $\omega(T_1, T_2) \neq 0$ then T_1, T_2 have at least one coincidence.*

Thus there has been obtained a coincidence theorem for two mappings of absolute orientable geometric manifolds into one another even when the two mappings proceed *in the same direction.*

30. Applications. We will prove some of Brouwer's classical degree and fixed point theorems (Brouwer [a, b, d]). We first extend the concept of Brouwer degree (V, 25) in the following way: If K, L are both cyclic in the dimension n and acyclic in the dimensions greater then n, and have the basic integral classes Γ^n, Δ^n then (28.4) for $p = n$ yields

$$(30.1) \qquad T\Gamma^n = c\Delta^n$$

and c is the *degree* of the mapping $T: |K| \to |L|$. From (28.9) follows

(30.2) *The degree depends merely upon the homotopy class of T, and if T is a mapping $|K| \to |K|$ it is even independent of the basic n-cycle chosen on K.*

Let now T be a mapping of an n-parallelotope P^n (closed n-cell) on itself. Since P^n is zero-cyclic, a point x is a rational homology base for the dimension zero, and evidently $\sigma x \sim x$. Therefore $\psi(T) = 1$, and so:

(30.3) *Every mapping of a finite-dimensional parallelotope (closed cell) into itself has a fixed point.*

Let now $|K|$ be a topological n-sphere S^n. If we denote also by S^n the basic n-cycle of the sphere then (28.1) for a mapping $T: S^n \to S^n$ becomes

$$\sigma\Gamma^0 = \Gamma^0, \qquad \sigma S^n = cS^n,$$

where c is the degree. Therefore

$$(30.4) \qquad \psi(T) = 1 + (-1)^n c.$$

From (30.4) follows the

(30.5) THEOREM. *The following mappings of a topological sphere S^n into itself have a fixed point*: (a) *sense-preserving* [*reversing*] *mappings* $S^{2n} \to S^{2n}$ $[S^{2n+1} \to S^{2n+1}]$; (b) *mappings of degree different from* $(-1)^{n+1}$.

§6. QUASI-COMPLEXES AND THE FIXED POINT THEOREM

31. **Quasi-complexes.** The motivation for introducing this new and last class of spaces is the search for a sufficiently general type for which our basic fixed point theorem (29.1) is still valid. It has been known for some time (Lefschetz [g]) that it holds for so-called LC^* compacta and their generalization: HLC^* compact spaces. These spaces may be briefly described as follows:

(a) *LC^* spaces.* Let $K = \{\sigma\}$ be a finite Euclidean complex and L a closed subcomplex containing all the vertices of K. An LC^* space is a compactum $\mathfrak{R}$ characterized by the following property: given any $\epsilon > 0$ there is an $\eta > 0$ such that if there is a pair (K, L) and a mapping $t_0\colon | L | \to \mathfrak{R}$ with mesh $\{t_0(L \cap \mathrm{Cl}\, \sigma)\} < \eta$, then t_0 has an extension t to $| K |$ such that the continuous complex $\mathfrak{K} = (K, t)$ is of mesh less then ϵ. The pair (L_0, t_0) is known as a *partial realization of K to mesh less than η*. It may be said that LC^* spaces have been identified (Lefschetz [d]) with the so-called *absolute neighborhood retracts* of Borsuk (compacta which are neighborhood retracts of every compactum in which they are topologically imbedded). They include notably all finite Euclidean complexes as well as the Hilbert parallelotope.

(b) *HLC^* spaces.* These are analogues of the LC^* type which may be described in outline as follows. First K, L are now simplicial. Next $\mathfrak{R}$ is merely compact and the partial realizations and extensions are chain-mappings into nerves. For a fuller description the reader is referred to a complete treatment of a class of spaces which includes HLC^* spaces in a forthcoming paper by E. G. Begle [a]. We merely state that the HLC^* class includes the LC^* class and in addition, for instance, *all parallelotopes.*

The characteristic property of all these spaces, is the presence of an operation resembling indefinite chain-derivation. That is to say, if $\Phi = \{\Phi_\lambda ; \pi_\mu^\lambda\}$ is the net of the finite open coverings (VIII, 4) then there exist chain-mappings $\omega_\lambda^\mu\colon \Phi_\mu \to \Phi_\lambda$, $\lambda > \mu$, for a sufficiently large class of (λ, μ). Since indefinite chain-derivation has been at the root of the proof of the fixed point theorem it is natural to search for our generalization in that direction. As we shall see this surmise is completely justified: quasi-complexes, the class of spaces to which it leads, satisfy the fixed point theorem. Moreover the class includes HLC^* spaces, hence also LC^* spaces (= absolute neighborhood retracts), finite Euclidean complexes and all parallelotopes.

32. (32.1) Let then $\mathfrak{R}$ be compact and as in (VII, 4) let $\{\mathfrak{U}_\lambda\}$ be its finite open coverings and $\Phi = \{\Phi_\lambda ; \pi_\mu^\lambda\}$ the net of their nerves. We will be dealing throughout the remainder of the section with a certain cofinal family M of $\Lambda = \{\lambda; >\}$ and the elements of M will be designated by a, b, f, g. The indices f, g, will be used particularly to designate a certain dependence upon other indices. Corresponding to $a, \cdots$, the coverings, nerves, $\cdots$ are of course written $\mathfrak{U}_a, \cdots, \Phi_a, \cdots$.

We describe two properties B, C of $\mathfrak{R}$.

PROPERTY B. There is a family M (cofinal in Λ) such that for every a there is an $f(a) > a$ with one or more chain-mappings $\omega_f^a:\Phi_a \to \Phi_f$, called *antiprojections*, and such that:

(a) $\omega_f^a \pi_a^f \sim 1$;

(b) if $g > f > a$ and ω_f^a, ω_g^f are antiprojections so is $\omega_g^f \omega_f^a$;

(c) if ω_f^a, $\bar{\omega}_f^a$ are antiprojections then $\omega_f^a \sim \bar{\omega}_f^a$.

The antiprojections ω_f^a behave exactly like the duals of the projections, and give to the collections of the homology groups $\{\mathfrak{H}^p(\Phi_a, G)\}$ the character of direct systems. In the terminology of (VI, 2) $\{\Phi_a\}$ is thus both a net and a conet. The role of (a) is to make projections and antiprojections cancel out along the homology classes. These remarks already hint at a theorem such as (33.1) below.

PROPERTY C. First of all Property B holds with M and the antiprojections ω_f^a as in its statement. All the indices $a, \cdots$ being understood in M, we have in addition: for every a there is an index $g > a$, and for every b an index $h(a, g, b) > b, g$ such that ω_h^g exists (in accordance with Property A), and that if $\Phi_g = \{\sigma_g\}$ then $\{[\sigma_g] \cup [\omega_h^g \sigma_g]\} > \mathfrak{U}_a$, where if C^p is a chain of Φ_λ, $[C^p]$ denotes the union of the kernels of the simplexes of C^p. Thus here the union of the kernels of σ_g and of the simplexes of $\omega_h^g \sigma_g$ must be contained in a set of $\mathfrak{U}_a$.

Roughly speaking, for a compactum Property C asserts that the nerve of a finite open covering $\mathfrak{U}_g$ of sufficiently small mesh may be ϵ chain-mapped (in an obvious sense) into the nerve of an $\mathfrak{U}_h$ of arbitrarily small mesh.

(32.2) DEFINITION. *A quasi-complex is a compact Hausdorff space which possesses Property* C *and hence also Property* B.

33. (33.1) *If a compact space possesses Property* B *then its homology groups are isomorphic with subgroups of the homology groups of a certain finite simplicial complex* (*in fact with those of a nerve* Φ_a). *Hence*:

(a) *the groups above a certain n vanish*;

(b) *the space possesses Property* A *of* (28.2).

(33.2) *Noteworthy special case. Proposition* (33.1) *holds for a quasi-complex.*

We assume then that $\mathfrak{R}$ is compact and satisfies Property B, with M, ω as in the statement of Property B. Take a fixed $a \in$ M and let $\gamma^p = \{\gamma_\lambda^p\}$ be any cycle of $\mathfrak{R}$ over G. The mapping $\gamma^p \to \gamma_a^p$ defines a simultaneous homomorphism π_a of the groups of cycles of $\mathfrak{R}$ into the same for Φ_a, and this induces in turn a simultaneous homomorphism $\Pi_a : \mathfrak{H}^p(\mathfrak{R}, G) \to \mathfrak{H}^p(\Phi_a, G)$. To prove (33.1) it is sufficient to show that Π_a is a simultaneous isomorphism with subgroups of $\mathfrak{H}^p(\Phi_a, G)$. We must first prove Π_a univalent or equivalently:

$$\text{(33.3)} \qquad \gamma_a^p \sim 0 \to \gamma^p \sim 0.$$

(33.4) It is an immediate consequence of Property B that there is a family $\{f\}$ cofinal in M and hence in Λ, such that ω_f^a exists for every f of the family. Now: $\pi_a^f \gamma_f^p \sim \gamma_a^p \sim 0$, and so from (a) of Property B: $\omega_f^a \pi_a^f \gamma_f^p \sim \gamma_f^p \sim 0$. Since all the coordinates of γ^p for $\{f\}$ cofinal in Λ are ~ 0 we have $\gamma^p \sim 0$, which is (33.3).

To prove (33.1) there remains to show that

(33.5) Π_a *is open.*

For this purpose it is more convenient to pass to the homology classes Γ^p, $\Gamma_a^p, \cdots$ of $\gamma^p, \gamma_a^p, \cdots$. The homomorphisms in the homology groups induced by π_a^f, ω_f^a will be denoted by Π_a^f, Ω_f^a, so that we have as a consequence of (a) of Property B:

$$\text{(33.6)} \qquad \Omega_f^a \Pi_a^f = 1.$$

Let a and $\{f\}$ be related as before. If $b \in$ M and U_b is an open set of $\mathfrak{H}^p(\Phi_b, G)$ then $V_b = \{\Gamma^p \mid \Gamma_b^p \in U_b\}$ is open in $\mathfrak{H}^p(\mathfrak{R}, G)$, and since $\{f\}$ is cofinal in Λ, $\{V_f\}$ is a base for $\mathfrak{H}^p(\mathfrak{R}, G)$. To prove (33.5) we merely need to show that $\Pi_a V_f$ is open. Since Ω_f^a is continuous $(\Omega_f^a)^{-1} U_f = U_a$ is open. Now $\Gamma^p \in V_f \leftrightarrow \Gamma_f^p \in U_f \leftrightarrow (\Omega_f^a \Pi_a^f \Gamma_f^p = \Omega_f^a \Gamma_a^p) \in U_f \leftrightarrow \Gamma_a^p \in U_a$. Therefore $\Pi_a V_f = U_a \cap \Pi_a \mathfrak{H}^p(\mathfrak{R}, G) =$ an open set. This proves (33.5) and hence also (33.1).

34. Coincidences and fixed points for quasi-complexes.

(34.1) THEOREM. *The basic fixed point theorem* (29.1) *and its corollary* (29.2) *hold also for quasi-complexes.*

The derivation of (29.2) is as in (29.5); so we only need to prove (29.1). The proof is by a reduction to the analogous property for chain-mappings of certain nerves and requires a careful selection of coverings.

(34.2) If $\mathfrak{A} = \{A_\alpha\}$ is any aggregate of sets then the star of A_α, written $\mathfrak{St}\, A_\alpha$ (not to be confused with star in a complex), is the union of A_α and all the A_β which meet it. If $\mathfrak{R}$ is a topological space and $\mathfrak{U}$, $\mathfrak{B}$ open coverings then $\mathfrak{B} = \{V_\alpha\}$ is said to be a *star-refinement* of $\mathfrak{U}$ whenever $\{\mathfrak{St}\, V_\alpha\} > \mathfrak{U}$ (every star of $\mathfrak{B}$ is contained in a set of $\mathfrak{U}$). Star-refinements have been repeatedly utilized in questions related to the metrization problem (see notably J. Tukey [T]). However, we merely require here:

(34.3) *If $\mathfrak{R}$ is normal, every finite open covering $\mathfrak{U}$ has a finite star-refinement $\mathfrak{B}$.*

Let $\mathfrak{U} = \{U_1, \cdots, U_r\}$ and let it be shrunk to $\mathfrak{U}' = \{U_i'\}$, $\overline{U}_i' \subset U_i$. Then $\mathfrak{W}_i = \{U_i, \mathfrak{R} - \overline{U}_i'\}$ is a binary open covering and $\mathfrak{B} = \mathfrak{U}' \wedge \mathfrak{W}_1 \wedge \cdots \wedge \mathfrak{W}_r$ is a finite open covering. Let $\mathfrak{B} = \{V_h\}$ and suppose $V_h \cap V_j \neq \emptyset$. The set V_h is contained in a set U_i' of $\mathfrak{U}'$, and V_j in one of the sets of $\mathfrak{W}_i$, i.e., $V_j \subset U_i$, or $V_j \subset \mathfrak{R} - \overline{U}'_i$. The second inclusion is ruled out since V_j meets the subset V_h of U'_i. Therefore the first holds. Thus V_h and all the sets of $\mathfrak{V}$ meeting it are in U_i and so $\mathfrak{V}$ satisfies (34.3).

(34.4) We will assume now that $\mathfrak{R}$ is a quasi-complex with a mapping $T:\mathfrak{R} \to \mathfrak{R}$ and set $R = T\mathfrak{R}$. The notations are as in the statement of Property C and the various elements there considered are selected in the following way. We choose any $\mathfrak{U}_a$, $a \in \mathrm{M}$, then take $\mathfrak{U}_g = \{U_{gi}\}$ as stated in Property C. Next we choose $\mathfrak{U}_b > \mathfrak{U}_a \wedge \{T^{-1}(R \cap U_{gi})\}$ and finally $\mathfrak{U}_h$ is selected as in Property C.

Suppose $\mathfrak{U}_b = \{U_{bi}\}$. We may find for each U_{bi} two sets of $\mathfrak{U}_a$, $\mathfrak{U}_g$ which we denote for convenience by U_{ai}, U_{gi}, such that $U_{bi} \subset U_{ai} \cap T^{-1}(R \cap U_{gi})$. Since $\bigcap U_{bi_r} \neq \emptyset \rightarrow \bigcap U_{gi_r} \neq \emptyset$, $U_{bi} \to U_{gi}$ defines a simplicial chain-mapping $\overset{b}{\tau}_g:\Phi_b \to \Phi_g$. Hence $\overset{h}{p}_g = \overset{b}{\tau}_g \overset{h}{\pi}_b$ is a chain-mapping $\Phi_h \to \Phi_g$. (It is the analogue of $\overset{\nu}{p}_\mu$ of (VII, 5.12) for our present mapping T.) Thus finally $\theta = \overset{h}{p}_g \overset{g}{\omega}_h$ is a chain-mapping $\Phi_g \to \Phi_g$.

If $\gamma^p = \{\gamma^p_\lambda\}$ is any cycle of $\mathfrak{R}$, then

$$\theta\gamma^p_g = \overset{h}{p}_g \overset{g}{\omega}_h \gamma^p_g \sim \overset{h}{p}_g \overset{g}{\omega}_h \overset{h}{\pi}_g \gamma^p_h \sim \overset{h}{p}_g \gamma^p_h .$$

Therefore (VII, 5.12):

(34.5) *$\theta\gamma^p_g$ is the g coordinate of the cycle $T\gamma^p$.*

It is convenient to specialize a, g still further in accordance with:

(34.6) *g may be so chosen that all the rational cycles $\overset{g}{\pi}_a \gamma^p_g$ are essential.*

Corresponding to a there exists by (VI, 3.12) an index $\lambda > a$ such that the rational cycles $\overset{\lambda}{\pi}_a \gamma^p_\lambda$ are all essential. This remains true if λ is replaced by any $\lambda' > \lambda$. On the other hand in Property C if $a' > a$, $a' \in \mathrm{M}$, then any $g(a')$ is a suitable $g(a)$. Choosing then $a' > \lambda$, and $g = g(a')$, (34.5) will be satisfied.

35. Since the Betti numbers of $\mathfrak{R}$ are finite (33.1) it possesses a finite maximal independent set of rational cycles $\{\gamma^p_i\}$, $i = 1, 2, \cdots, R^p$, $\gamma^p_i = \{\gamma^p_{i\lambda}\}$. The cycles $\{\gamma^p_{ig}\}$, g fixed, are independent for Φ_g. However, to have a maximal independent set for Φ_g it may be necessary to add a new set $\{\delta^p_{ig}\}$.

We have seen (34.5) that $\theta\gamma^p_{ig}$ is the g coordinate of $T\gamma^p_i$. Therefore if

$$T\gamma^p_i \sim \lambda^j_i(p)\gamma^p_j , \tag{35.1}$$

then by the argument in the proof of (33.1) with a in place of g we find:

$$\theta\gamma^p_{ig} \sim \lambda^j_i(p)\gamma^p_{jg} . \tag{35.2}$$

On the other hand

$$\theta\delta^p_{ig} \sim \rho^j_i(p)\gamma^p_{jg} + \zeta^j_i(p)\delta^p_{jg} . \tag{35.3}$$

Now we have:

$$\overset{g}{\omega}_h \delta^p_{ig} \sim \overset{g}{\omega}_h \overset{a}{\omega}_g \overset{g}{\pi}_a \delta^p_{ig} . \tag{35.4}$$

By (34.6) and (II, 27.12) there exists a cycle $\delta^p = \{\delta^p_\lambda\}$ of $\mathfrak{R}$ such that $\dot{\delta}^p_a \sim \overset{g}{\pi}_a \delta^p_{ig}$ and we will have $\delta^p \sim c^j\gamma^p_j$ in $\mathfrak{R}$. Replacing, as we may, δ^p_{ig} by $\delta^p_{ig} - c^j\gamma^p_{jg}$ the situation will be unchanged except that now $\delta^p \sim 0$ and hence $\pi_a^g \delta_{ig}^p \sim 0$. With this new choice of the δ_{ig}^p we find then from (35.4): $\omega_h^g \delta_{ig}^p \sim 0$, and hence (35.3) will be replaced (in view of $\theta = p_g^h \, \omega_{\;h}^{\,g}$) by

$$\theta \delta_{ig}^p \sim 0. \tag{35.5}$$

From this follows

$$\psi(\theta) = \sum (-1)^p \operatorname{trace} \lambda^p = \psi(T). \tag{35.6}$$

Since θ is a chain-mapping of a finite complex into itself, $\psi(\theta)$ is an integer (V, 22.7), and hence the same holds for $\psi(T)$, and consequently also for $\varphi(T_1, T_2) = \psi(T_2 T_1)$. Therefore as in (29.7):

(35.7) *For mappings of quasi-complexes into one another both* $\psi(T)$ *and* $\varphi(T_1, T_2)$ *are integers.*

36. We will assume now that T has no fixed point and so dispose of the situation that θ has no fixed element. In view of (35.6) this will prove (34.1).

(36.1) *There is a finite open covering* $\mathfrak{V}$ *of* $\mathfrak{R}$ *such that no star of* $\mathfrak{V}$ *meets its transform.*

If $x_1 = Tx$ then $x \neq x_1$ and so the two points have disjoint neighborhoods U, U_1. Hence $U \cap T^{-1}U_1 = U'$ is a neighborhood of x which does not meet TU'. Since $\mathfrak{R}$ is compact the covering $\mathfrak{U}' = \{U'\}$ has a finite subcovering $\{U_i'\}$. Since $\mathfrak{R}$ is compact Hausdorff it is normal (I, 33.6), and so $\{U_i'\}$ has a finite star-refinement $\mathfrak{V}$, and $\mathfrak{V}$ satisfies (36.1).

It is clear that if $\mathfrak{V}' > \mathfrak{V}$ then $\mathfrak{V}'$ still has property (36.1). Since M is cofinal in Λ, we may choose $a \in \mathrm{M}$ such that $\mathfrak{U}_a$ of (34.4) satisfies (36.1). The situation being then as before we prove:

(36.2) θ *has no fixed elements.*

Let $\Phi_g = \{\sigma_g\}$. Since the coverings of (34.4) have been selected in accordance with Property C, some $U_{ai} \supset [\sigma_g] \cup [\omega_h^g \sigma_g]$. Hence if σ_h is a simplex of the chain $\omega_h^g \sigma_g$ we have $[\sigma_h] \subset U_{ai}$. From this follows that each vertex of $\pi_b^h \sigma_h$ is contained in a set of $\mathfrak{U}_a$ meeting U_{ai}, and so $[\pi_b^h \sigma_h] \subset \mathfrak{St}\, U_{ai}$. Hence $[\tau_g^b \pi_b^h \sigma_h] = [p_g^h \sigma_h] \subset T\, \mathfrak{St}\, U_{ai} \subset \mathfrak{R} - U_{ai}$, and so $[\theta \sigma_g] \subset \mathfrak{R} - U_{ai}$. Consequently $[\theta \sigma_g] \cap [\sigma_g] = \emptyset$, which proves (36.2).

As already observed this completes the proof of the fixed point theorem (34.1).

(36.3) APPLICATION. A topological space $\mathfrak{R}$ is said to have the *fixed point property* whenever every mapping $\mathfrak{R} \to \mathfrak{R}$ has a fixed point.

Suppose that $\mathfrak{R}$ is a zero-cyclic quasi-complex. Since $\mathfrak{R}$ is zero-cyclic it is connected, and so we prove readily that if $x \in \mathfrak{R}$ then $Tx \sim x$. From this follows $\psi(T) = 1$, and so $\mathfrak{R}$ has the fixed point property. Since any parallelotope may be deformed into a point it falls under this category. Thus

(36.4) *A zero-cyclic quasi-complex, and in particular an arbitrary parallelotope, has the fixed point property.*

There are noteworthy applications to analysis to which we propose to return elsewhere [L_2, IV].

§7. TOPOLOGICAL COMPLEXES

37. In the applications to analysis or geometry complexes do not always occur in the convenient aspect of polyhedra or Euclidean realizations. Thus a circumference with two subdivision points gives rise to a complex which is not polyhedral and other examples could be multiplied. Topological complexes are intended to bridge the gap. Historically they are also of interest since they were the complexes introduced by Poincaré [b] and dealt with at length by Veblen [V]. As we shall see (40.3) topologically speaking we remain within the class of Euclidean complexes and this will justify if need be our having devoted our major efforts to that class.

38. (38.1) It will be convenient to break up the description of our complexes into two parts. We will first consider an Euclidean complex $\mathfrak{K} = \{\sigma\}$ with which there is to be associated a new complex $K = \{E\}$ with a derived $K' \cong \mathfrak{K}$. Thus the passage from $\mathfrak{K}$ to K is the inverse of derivation. The complex K is characterized by these properties:

I. E_i^p is a p-cell covered by an open subcomplex of $\mathfrak{K}$ which is a join $A(S^{p-1})_a = AS^{p-1} \cup A$ of a vertex A of $\mathfrak{K}$ and an augmented $(p-1)$-sphere which is a finite closed subcomplex of $\mathfrak{K}$. Thus $\bar{E}_i^p = A_a(S^{p-1})_a$ is a closed p-cell and it is covered by a closed subcomplex of $\mathfrak{K}$.

II. The cells E are disjoint and their union is $|\mathfrak{K}|$.

III. $\bar{E}$ is the union of the cells of Cl E.

IV. The description of the incidence numbers is less immediate. As above let $E_i^p = |A(S^{p-1})_a|$. Since $\bar{E}_i^p$ is a closed p-cell, by (16.1) it is a simple p-circuit mod S^{p-1}. For convenience let E_i^p also designate a basic p-cycle for that circuit. Since the E_h^q are disjoint their q-simplexes may be written uniquely $\{\sigma_{hj}^q\}$, $\sigma_{hj}^q \subset E_h^q$. In particular

$$E_i^p = \sum \alpha_{ij}\sigma_{ij}^p,$$

where $\alpha_{ij} = \pm 1$ since E_i^p is a simple circuit.

It is clear that each $\sigma^{p-1} \subset S^{p-1}$ is the face of one and only one σ_{ij}^p, and hence:

$$\mathrm{F}E_i^p = \sum_{j,k} \epsilon_{ijk}\,\sigma_{jk}^{p-1},$$

where $\epsilon_{ijk} = \pm 1$ if $E_j^{p-1} \prec E_i^p$, and is equal to 0 otherwise. If we reduce $\mathrm{F}E_i^p$ mod $(S^{p-1} - E_j^{p-1})$ there arises a cycle of E_j^{p-1} which is a multiple of its basic cycle. Thus

$$\sum_k \epsilon_{ijk}\,\sigma_{jk}^{p-1} = \eta_i^j(p-1)E_j^{p-1}, \qquad \text{(no summation on } i, j), \tag{38.2}$$

and therefore

$$\mathrm{F}E_i^p = \eta_i^j(p-1)E_j^{p-1}. \tag{38.3}$$

We set:

$$[E_i^p : E_j^{p-1}] = \eta_i^j(p-1)$$

and determine all the other incidence numbers in K by means of (III, 1.1, K23).

(38.4) *K is a simple complex which has $\mathfrak{K}$ as a subdivision $\cong K'$.*

(a) *K is a complex.* Referring to (III, 1.1, 8.3) we merely have to verify that if the boundary operator in K is defined by (38.2) then FF $= 0$. Now we have

$$\mathrm{FF}E_i^p = \zeta^j E_j^{p-2}.$$

Since the complexes E_j^{p-2} are disjoint the σ_{j1}^{p-2} are distinct. The coefficient of σ_{j1}^{p-2} in E_j^{p-2} is a number $\lambda \neq 0$, and its coefficient in $\mathrm{FF}E_i^p$ expressed as a cycle of $\mathfrak{K}$ is $\lambda\zeta^j = 0$, since FF $= 0$ holds in $\mathfrak{K}$. Hence $\zeta^j = 0$ and (a) follows.

(b) *K is augmentable.* Since E_i^1 is a one-chain in a simplicial complex we have $\mathrm{KI}(\mathrm{F}E_i^1) = 0 = \sum \eta_i^j(0)$, and therefore if $\gamma_0 = \sum E_0^j$ then $\mathrm{F}\gamma_0 = \sum_j \eta_i^j(0)E_1^i = 0$. Thus K is augmentable and with γ_0 as fundamental zero-cocycle.

(c) *$(\mathrm{Cl}\ E)_a$ is acyclic.* For $|\ \mathrm{Cl}\ E\ | = \bar{E}$ is a closed cell, so (c) holds for the geometrical groups, hence also for the combinatorial groups (10.1).

(d) *K is closure-finite* (obvious).

It is now a consequence of (abcd) that K is simple (III, 47.1). If the vertex of $\mathfrak{K}$ in E is written $'E$ then the simplexes of $\mathfrak{K}$ are uniquely represented as $'E_i \cdots 'E_j$, $E_i < \cdots < E_j$ and so $\mathfrak{K} \cong K'$. Since K is simple, $\mathfrak{K}$ is a subdivision (IV, 27.1) and (38.4) is proved.

$$(38.5) \qquad [E_i^p : E_j^{p-1}] = \begin{cases} \pm 1 \ \textit{if } E_j^{p-1} < E_i^p, \\ 0 \ \textit{otherwise.} \end{cases}$$

Since σ^{p-1} is a relative $(p-1)$-circuit so is E_j^{p-1}, and hence the coefficient λ of σ_{j1}^{p-1} in E_j^{p-1} is different from 0. Since the E_j^{p-1} are disjoint the σ_{j1}^{p-1} are distinct. Thus the coefficient of σ_{j1}^{p-1} in the chain of $\mathfrak{K}$ at the right is $\lambda\eta_i^j(p-1)$. Hence $\lambda\eta_i^j(p-1) = \epsilon_{ij1}$. From the known values of ϵ_{ij1} we have then $\lambda\eta_i^j(p-1) = 0 = \eta_i^j(p-1)$ if $E_j^{p-1} \nless E_i^p$, and $\lambda\eta_i^j(p-1) = \pm 1 = \pm\eta_i^j(p-1)$ if $E_j^{p-1} < E_i^p$, and this proves (38.5).

(38.6) *The complex K corresponding to a given collection $\{E\}$ is unique.*

At all events it is unique except possibly for the incidence numbers. In their determination the only allowable modification is a change in the signs of certain basic cycles E_i^p and this is equivalent to reorienting K by means of an $\alpha(E)$ taking the value -1 in those E_i^p. Under our conventions this does not modify K.

39. The ground is now prepared for topological complexes. Here again the only difficulty will be caused by the incidence numbers.

(39.1) Definition. *A topological complex is a countable locally finite complex $K = \{E\}$ with the following properties:*

I. *The elements of K are disjoint cells and their union is a topological space, written $|\ K\ |$ and said to be polyhedral.*

II. *E^p is a p-cell, $\bar{E}^p$ is a closed p-cell which is the union of the cells of* Cl E^p.

III. *If $\varphi(E)$ is the union of the cells of $K -$* St *E then $\overline{\varphi(E)} \cap E = \emptyset$.*

The specification of the incidence numbers will be given presently.

The analogy of these properties with (III, 6.1abc) for polyhedral complexes is evident.

(39.2) DEFINITION. *A topological complex* $K = \{E\}$ *is said to be simplicial if there exists an Euclidean complex* $\mathfrak{K} = \{\sigma\}$ *and a topological mapping* $t: | \mathfrak{K} | \to | K |$ *such that* $t\sigma$ *is a cell* E *of* K.

(39.3) DEFINITION. *If* $K = \{E\}$, $K_1 = \{E_1\}$ *are topological complexes then* K_1 *is said to be a partition of* K *whenever every* E *is the union of a finite set of* E_1*'s and every* E_1 *is contained in some* E.

(39.4) Returning to K of (39.1) let its elements ranged in some order be denoted by $E_1, E_2, \cdots$, and select a point $A_i \in E_i$. In the Hilbert parallelotope P^ω referred to $\{u_1, u_2, \cdots\}$, $0 \leqq u_i \leqq 1/i$, let B_i be the point $u_i = 1/i$, $u_j = 0$ for $i \neq j$. Corresponding to every set $E_i < \cdots < E_j$ introduce the Euclidean simplex $\sigma = B_i \cdots B_j \subset P^\omega$. Evidently $\mathfrak{K} = \{\sigma\}$ is an Euclidean complex $\cong K'$. We will now define a topological mapping $t: | K | \to | \mathfrak{K} |$ as follows. Let K^q be the q-section of K. We choose $tA_i = B_i$ and this defines $t \,||\, K^0 \,|$. Suppose $t \,||\, K^{p-1} \,|$ known and let dim $E_i = p$. Take a parallelotope P^p and let Σ^{p-1} be its boundary sphere. Introduce a topological mapping $s: \bar{E}^p \to P^p$ and suppose $sA_i = C_i$. Then ts^{-1} is a topological mapping $\Sigma^{p-1} \to \mathfrak{K}$ such that $ts^{-1}\Sigma^{p-1} = S^{p-1}$ is a subcomplex of $\mathfrak{K}$, and that $ts^{-1}C_i = B_i$. This mapping is extended to a topological mapping $t_i: P^p \to \mathfrak{K}$ as follows: if $R \in \Sigma^{p-1}$ then $\overline{C_i R}$ is mapped barycentrically on $\overline{B_i(ts^{-1}R)}$. If we choose $t \,|\, E_i = t_i s$ and operate likewise for all E^p we obtain $t \,||\, K^p \,|$, then proceeding in this fashion t itself.

(39.5) We will take advantage of t to identify $| \bigcup E |$ with $| \mathfrak{K} |$ so that points corresponding under t coincide. Under the circumstances $\{E\}$ becomes a collection related to $\mathfrak{K}$ as in (38.1). We now specify:

IV. The incidence numbers in K are in accordance with (38.1, IV).

40. Several interesting conclusions may quickly be drawn from the definition.

(40.1) *The topological complex* K *is uniquely determined by the collection of cells* $\{E\}$ *and properties* (39.1, I, II, III).

$$\textit{In particular } [E_i^p : E_j^{p-1}] = \begin{cases} \pm 1 \textit{ if } E_j^{p-1} < E_i^p, \\ 0 \textit{ otherwise.} \end{cases}$$

For the decomposition of $\mathfrak{K}$ by $\{E\}$ is independent of the construction and the resulting incidence numbers obtained in accordance with (38.1, IV) determine K uniquely (38.6). That the incidence numbers behave as stated is a consequence of (38.5).

The identification of the topological complex with the complex K of (38) together with (38.4) yield:

(40.2) *A topological complex is simple.*

(40.3) *A topological complex K has a simplicial partition which is a subdivision, and at the same time a realization of the derived K' such that the vertex of K' corresponding to E_i^p is imaged into a preassigned point of the cell.*

The distinction between combinatorial and geometrical groups (9.1) may of course be made for topological complexes. From (40.3) together with (IV, 24.1) and the results of (10, $\cdots$, 13) we deduce:

(40.4) *The invariance properties* (10.1, 11.2, 13.1, 13.2, 13.3, 17.1) *hold for topological complexes.*

Further results of this nature could be obtained but the topological identification of Euclidean and topological complexes implicit in (40.3) makes their derivation superfluous.

(40.5) *A polyhedral complex is topological.*

Let $\Pi = \{E\}$ be a polyhedral complex, Π' its derived, $'E$ the new vertex of Π' in E, D and δ set and chain-derivation in Π. If

$$\mathrm{F}E_i^p = \eta_i^j(p - 1)E_j^{p-1},$$

where the $\eta_i^j(p - 1)$ are the incidence numbers in Π, then (IV, 29.6, 29.1):

$$\delta\mathrm{F}E_i^p = \mathrm{F}(\delta E_i^p) = \eta_i^j(p - 1)\delta E_j^{p-1}. \tag{40.6}$$

Since DE_i^p coincides with E_i^p, $\{DE_i^p\}$ and Π' are related like K and $\mathfrak{K}$ in (38). The cycles denoted there by E_i^p are the cycles δE_i^p. The comparison of (40.6) with (38.3) shows that the incidence numbers of the polyhedral complex Π conform with (38.1, IV). This proves (40.5).

§8. DIFFERENTIABLE COMPLEXES AND MANIFOLDS

41. The parametric representation of topological complexes and manifolds brings to the fore a wealth of questions and problems, many as yet unsolved, which are of great interest in both topology and differential geometry in the large. We will do little more than touch upon these questions here, and describe some of the more important recent contributions, together with certain complements regarding homology and intersections. This will necessitate, however, a number of preliminary definitions most of which are standard (see notably: Veblen-Whitehead: *The foundations of differential geometry*, Cambridge Tract 29, (1932); Cairns [a, b], Whitehead [b], Whitney [a, b]).

42. **Systems of class C^r.**

(42.1) Let $\{u_1, \cdots, u_m\}$ be real variables and L_i the real line $-\infty < u_i < +\infty$. The product $L_1 \times \cdots \times L_m$ is called the *space* of the parameters u_i, and is denoted by U^m. It is merely an Euclidean space with a specified coordinate system. The point $(u_1, \cdots, u_m)$ is denoted by u.

(42.2) If V^n is the space of $\{v_1, \cdots, v_n\}$ then the space corresponding to $\{u_1, \cdots, u_m, v_1, \cdots, v_n\}$ is called the product of U^m, V^n, written $U^m \times V^n$.

(42.3) A finite set of real functions $S = \{f_i(u_1, \cdots, u_n)\}$ is said to be: (a) *of class C^q* in a region Ω of U^m whenever the f_i and all their partial derivatives of order at most q are defined and continuous in Ω; (b) *regular* if it is of class C^1 and the Jacobian matrix $\| \partial f_i / \partial u_j \|$ is of maximal rank at all points of Ω. We also say that S is *differentiable* in Ω if it is of class C^1 in Ω, and that it is *analytical* or *of class* C^ω in Ω, if the f_i are analytical in Ω and (b) holds. Throughout the sequel one may freely interchange "differentiable" with "class C^1," and "analytical" with "class C^ω."

If $A \subset U^n$ then S is [regular] of class C^q in $\bar{A}$ if it is [regular] of class C^q in some neighborhood of A.

(42.4) Let V^n be the space of $(v_1, \cdots, v_n)$ and let t be a transformation $\Omega \to V^n$, Ω a region of U^m, given by relations

$$v_i = f_i(u_1, \cdots, u_m).$$

We will say that t is [regular] of class C^r or is a [regular] C^r-*mapping* if $\{f_i\}$ is [regular] of class C^r in Ω.

43. **Parametric cells.**

(43.1) A parametric n-cell is an n-cell E^n together with a topological mapping $t{:}\bar{E}^n \to U^n$. The coordinates of U^n are called the *parameters* of E^n. A convenient designation for the parametric cell is (E^n, t, U^n).

(43.2) Consider two parametric cells (E^m, t, U^m) and (E'^n, t', U'^n) and let $A = E^m \cap E'^n$. We say that A is a [regular] r-C^r-*intersection* of the two parametric cells whenever the following holds. For each $x \in A$ there is an (E^r, s, V^r) such that: (a) E^r is a neighborhood of x in A; (b) on $t^{-1}E^r$ and $t'^{-1}E^r$ the u_i and u'_i satisfy systems

$$u_i = f_i(v_1, \cdots, v_r), \qquad u'_i = \varphi_i(v_1, \cdots, v_r)$$

such that $\{f_i\}$, $\{\varphi_i\}$ are [regular] of class C^r in $t^{-1}E^r$, $t'^{-1}E^r$. Two noteworthy special cases are:

(a) $E'^n \subset E^m$. Here $r = n$ and one may take for (E^r, s, V) the parametric cell (E'^n, t', U'^n) itself. Thus the u coordinates of the points of E'^n form then a system of functions of class C^r of their u' coordinates in $t'E'^n$. We have then a [regular] C^r-*imbedding* of the second parametric cell in the first.

(b) $n = m$ and A is an open set in both E^n and E'^n. Then the [regular] C^r-intersection becomes a [regular] C^r-*overlap*.

(43.3) The product of two parametric cells (E^m, t, U^m), (E'^n, t', U'^n) is by definition the parametric cell $(E^m \times E'^n, t \times t', U^m \times U'^n)$, i.e., $E^m \times E'^n$ parametrized by $(u_1, \cdots, u_m, u'_1, \cdots, u'_n)$, the mapping $t \times t'$ being defined by $(t \times t')(x, y) = (tx, t'y)$.

44. C^r**-manifolds.**

(44.1) A *topological n-manifold* is a metric space μ with a countable locally finite open covering consisting of n-cells.

(44.2) An *n-manifold* M^n of class C^r or C^r-*n-manifold* is a topological n-manifold μ with a countable locally finite open covering by parametric n-cells, $\{E_i^n\}$ (supposed to exist) such that if E_i^n, E_j^n intersect then they have a regular C^r-overlap. The collection $\{E_i^n\}$ is called the *basic covering* of M^n and μ is written $| M^n |$.

Let μ give rise to two C^r-manifolds M^n, M'^n by means of $\{E_i^n\}$, $\{E_i'^n\}$. If the two basic coverings together may serve as a basic covering for a C^r-n-manifold associated with μ then M^n and M'^n are considered as identical C^r-manifolds. This allows for a certain latitude in the choice of the basic covering.

(44.2a) EXAMPLE. U^n is a parametric n-cell and so it is an analytical M^n with itself as basic covering consisting of a single cell. On the other hand the cells defined by the inequalities $u_i > -1/2$, $u_i < 1/2$, make up a basic covering consisting of 2^n elements.

(44.3) Let M^n, M'^n be C^r-manifolds with the basic coverings $\{E_i^m\}$, $\{E_j^n\}$. Then $\{E_i^m \times E_j^n\}$ is a basic covering for the topological manifold $| M^m | \times | M'^n |$ which turns it into a C^r-$(m + n)$-manifold known as the product of M^m, M'^n and written $M^m \times M'^n$. Furthermore this product is independent of the particular choice of basic coverings for M^m, M'^n.

EXAMPLE. If M^n is a C^r-manifold and L the real line $-\infty < u < +\infty$ turned into an analytical manifold then $L \times M^n$ is a C^r-$(n + 1)$-manifold.

(44.4) The notations remaining those of (44.3), we say that M^m, M'^n have a [regular] C^r-*intersection* if the intersections of the elements of the basic covering of M^m with those of the basic covering of M'^n are all [regular] C^r. If $M^m \subset M'^n$ we also say that M^m *is [regularly]* C^r*-imbedded in* M'^n, or merely M^m *is [regularly]* C^r *in* M'^n. Here again the situation is independent of the basic coverings.

(44.5) Let the C^r-manifolds discussed so far be termed *absolute*. A *relative* C^r-n-manifold, or C^r*-n-manifold with regular boundary* M^{n-1} is a region Ω of $| M^n |$, where M^n is an absolute C^r-manifold, whose boundary $\mathfrak{B}\Omega$ is a regular C^r-$(n - 1)$-manifold in M^n.

(44.6) EXAMPLE. Let again L be the real line of (44.3) turned into a C^ω-manifold and let λ be the interval $0 < u < 1$. If M^n is a C^r-manifold then $\lambda \times | M^n |$ is a region of $L \times M^n$ whose boundary $0 \times M^n \cup 1 \times M^n$ is a C^r-n-manifold regularly C^r-imbedded in $L \times M^n$. Thus we have a manifold with regular boundary which will be written $\lambda \times M^n$.

(44.7) Let M^m, M'^n be as in (44.3). A mapping T:$| M^m | \to | M'^n |$ is said to be a [regular] C^r*-mapping* $M^m \to M'^n$, whenever if x is in the parametric cell (E^m, t, U^m) of M^m and Tx in the parametric cell (E'^n, t', U'^n) then the u' coordinates of Tx make up a system of functions of the u coordinates of x which is [regular] of class C^r at tx in U^m. If $| M^m | \subset | M'^n |$, the resulting imbedding is readily seen to be a [regular] C^r-imbedding in the sense of (44.4).

(44.8) Let λ be an interval of $-\infty < u < \infty$ containing 0, 1. Two C^r-mappings T_1, T_2 of M^m into M'^n are said to be [regular] C^r*-homotopic* if there exists a [regular] C^r-mapping T_0 of $\lambda \times M^m$ which agrees with T_1, T_2 on $0 \times M^m$, $1 \times M^m$.

(44.9) Let M^m be a topological manifold in $\mathfrak{E}^n$. The subspace $\mathfrak{E}^r$ through $x_0 \in M^m$ is said to be *transversal* to M^m at x_0, if $\mathfrak{E}^r$ and the tangent $\mathfrak{E}^m$ to $| M^m |$ at x_0 intersect at no other point than x_0. The manifold is said to be in *normal position* in $\mathfrak{E}^n$ if for each $x \in | M^m |$ there is a transversal $\mathfrak{E}^{n-m}$ to M^m at x with a system of coordinates which vary continuously with x.

45. C^r-complexes and other structures.

(45.1) A C^r-complex is a topological complex $K = \{E_i^p\}$ such that: (a) E_i^p is a subset of a C^r-p-manifold denoted by (E_i^p); (b) if $E' < E$ then (E') is regularly C^r-imbedded in (E). We add the following convention: if the ( ) are replaced by C^r-manifolds ( )$'$, which are identical with them in the common parts, then the C^r-complex remains the same.

(45.2) An *analytical spread* is a subset Σ of U^n with the property that every point $x \in \Sigma$ has a neighborhood N in U^n such that the points of $N \cap \Sigma$ are the solutions of a system

$$f_s(u_1, \cdots, u_n) = 0,$$

where the f_s are analytical in N.

(45.3) EXAMPLES. Euclidean and real or complex projective spaces, algebraic loci in such spaces are all analytical spreads.

(45.4) Let M^n be a C^r-manifold and K a C^r-n-complex such that: (a) every (E_i^n) is M^n itself; (b) the set $| M^n | = | K |$. Then M^n is said to be *C^r-covered by K.*

46. **Imbedding and covering theorems.** Only the statements are given, the reader being referred to the original papers for the proofs.

(46.1) *Every differentiable M^m may be differentially and topologically mapped onto an analytical M^m in some Euclidean space $\mathfrak{E}^{2m+1}$. More precisely if M^m is of class C^r ($r \neq \omega$) then the mapping may be chosen of the same class* (Whitney [a, b]).

(46.2) *If M^m is a manifold of class C^0 in normal position in $\mathfrak{E}^n$ then it may be indefinitely approximated by a topologically equivalent analytical M'^m in $\mathfrak{E}^n$* (Whitney [a]).

(46.3) *Let t be a mapping $M^m \to M^n$ where t and the manifolds are C^r, and let η be a continuous positive function on M^m. Then:*

(a) *t is C^r-homotopic to a mapping T such that $d(tx, Tx) < \eta(x)$, $x \in M^n$;*

(b) *if $n \geqq 2m$ the mapping T may be chosen regular and if $n > 2m$ it may be chosen topological* (Whitney [a]).

(46.4) *Every differentiable manifold may be covered with a differentiable simplicial complex. More precisely, if the manifold is of class C^r, $r > 0$, then the complex may be chosen of the same class* (Cairns [a, b], Whitehead [b]).

(46.5) *Every bounded analytical spread may be covered with an analytical complex* (Brown-Koopman [a], Lefschetz-Whitehead [a]).

It may be noted that (46.5) leads rapidly to the following property which is also implicit in (46.4):

(46.6) *Every analytical manifold may be covered with a simplicial analytical complex.*

From (46.1, 46.5) we deduce the following property implicit in (46.4):

(46.7) *Every differentiable manifold is a polyhedral space.*

A problem as yet unsolved for dimensions greater than 2 is:

(46.8) *Are topological manifolds polyhedral spaces? That is to say, can they be covered with topological complexes?*

For $n = 1$ the answer is affirmative and elementary and for $n = 2$, at least when M^n is compact, it is likewise affirmative, a result due to Radon. Nothing is known for $n > 2$.

47. **Differentiable manifolds: Homology and intersection theory.**

(47.1) If M^n is differentiable then it may be covered with a topological complex K. Since the latter has a simplicial derived we may already assume K simplicial. We apply directly to K whatever properties will be required which hold for an Euclidean complex under a topological transformation of the underlying topological space (polyhedron).

(47.2) Now if $x \in | M^n |$, x has a neighborhood which is an n-cell E^n. Hence E^n is n-cyclic in x and therefore the same holds for $| M^n |$. By (17.6) K is thus an absolute combinatorial n-manifold. Let us assume this manifold *orientable*. Then by (17.1) every topological simplicial complex covering M^n is an absolute orientable M^n. We say then that M^n itself is *orientable*.

(47.3) Suppose M^n *orientable* and *connected*. It has then a basic n-cycle M_0^n which is specified by any indicatrix. Consider in particular a parametric n-cell (E^n, t, U^n). Let A_0 be the point $tx_0 = (u_{01}, \cdots, u_{0n})$, $x_0 \in E^n$, and let $A_i \in U^n$ be a point $(u_{0j} + \epsilon\delta_i^j)$ where $\epsilon > 0$ is chosen so small that the Euclidean simplex $\sigma^n = A_0 \cdots A_n \subset tE^n$. Replacing if need be one of the u_i by $-u_i$ we may assume the coordinates such that $t^{-1}\sigma^n$ is an indicatrix of M^n.

(47.4) The situation remaining the same let M^p, M^{n-q}, $p \geqq q$, be orientable connected differentiable manifolds regularly and differentiably imbedded in M^n and suppose that they have a regular differentiable intersection which is an orientable connected differentiable M^r, $r = p - q$. The related basic cycles are written M_0^p, M_0^{n-q} and M_0^r. By (27.3) we will have an intersection cycle $M_0^p \cap M_0^{n-q}$ and

$$M_0^p \cap M_0^{n-q} = \alpha M_0^r. \tag{47.5}$$

We assume now explicitly that for some $x_0 \in | M^r |$ there may be chosen a parametric cell (E^n, t, U^n) such that $| M^p |$, $| M^{n-q} |$, $| M^r |$ intersect it in cells E^p, E^{n-q}, E^r mapped by t into cells $E_0^n, \cdots$, where $E_0^n \subset U^n$ and the others are in the spaces:

$$\mathfrak{E}^p : u_i = 0, \quad i > p; \qquad \mathfrak{E}^{n-q} : u_i = 0, \quad i \leqq q;$$

$$\mathfrak{E}^r = \mathfrak{E}^p \cap \mathfrak{E}^{n-q} : u_i = 0, \quad i \leqq q \text{ or } > p.$$

Furthermore the situation may be so disposed that E_0^n, E_0^p, $\cdots$ are indicatrices for U^n, $\mathfrak{E}^p$, $\cdots$. Let $\epsilon_n E^n, \cdots$ be indicatrices for $M^n, \cdots$, where $\epsilon_n, \cdots = \pm 1$. If $\mathfrak{E}_0^n, \cdots$ are the basic cycles of $\mathfrak{E}^n, \cdots$, then (27.11) holds. If we apply the projection $\mathfrak{E}^n \to E_0^n$ (VII, 5.10) we obtain

$$E_0^p \circ E_0^{n-q} = \alpha E_0^r,$$

where $E_0^p, \cdots$ stand also for the basic cycles of the cells. Similarly the projection $|M^n| \to E^n$ followed by t yields:

$$\epsilon_n(\epsilon_p E_0^p) \circ (\epsilon_{n-q} E_0^{n-q}) = \epsilon_r E_0^r,$$

and hence $\alpha = \epsilon_p \epsilon_{n-q} \epsilon_r \epsilon_n$, or finally:

$$M_0^p \circ M_0^{n-q} = \epsilon_p \epsilon_{n-q} \epsilon_r \epsilon_n M_0^n. \tag{47.6}$$

Once more we recognize here the rule for determining intersections given in Lefschetz [a], and in closely related form in [L, IV].

If $q = p$ then $r = 0$, $\epsilon_0 = 1$, and M^0 is a single point. Its index is

$$\mathrm{KI}(M_0^p, M_0^{n-p}) = \epsilon_p \epsilon_{n-p} \epsilon_n. \tag{47.7}$$

(47.8) If $|M^r|$ is not connected each part is determined by the above rule and the intersection is the sum of the individual cycles thus obtained. Similarly if M^0 is a finite point set $\{A_1, \cdots, A_s\}$ each gives rise to an index λ_i determined by our rule and

$$\mathrm{KI}(M_0^p, M_0^{n-p}) = \sum \lambda_i. \tag{47.9}$$

(47.10) *Noteworthy special case: the manifolds are complex algebraic varieties.* Then orientations may be assigned in advance to all manifolds such that the index (47.7) is always $+1$, and hence (47.9) always greater than 0. For details see [L, 379].

(47.11) We have assumed that the intersection $M^p \cap M^{n-q}$ is an M^r. Using Whitney's theorem (46.3) this restriction may be removed by a suitable method of approximation (Whitney [a]). This was carried out in [L, 383] for algebraic varieties. One may state more precisely this: If M^p, M^{n-q} are C^r-imbedded (r finite and greater than 0) in M^n, then an arbitrarily small deformation of M^p may be chosen such that afterwards $M^p \cap M^{n-q}$ is an M^{p-q} C^r-imbedded in M^n. If $M^n \subset \mathfrak{E}^m$ then this holds also for $k = \omega$.

(47.12) We may conclude with the historical remark that the structures with which Poincaré initiated his epoch-making investigations in topology were essentially differentiable manifolds (see his paper [a]). It was only in order to straighten out difficulties with torsion which were pointed out to him by Heegaard that Poincaré in [b] invented the first complexes. In a certain sense the work of Whitney and Cairns may be said to have placed for the first time these first contributions of Poincaré on a solid basis.

§9. GROUP MANIFOLDS

48. A number of properties of the homology groups of group manifolds have been obtained by various authors. These investigations culminated recently in a highly interesting theorem due to H. Hopf [c] (see 49.1) which we propose to discuss in the present section. It is noteworthy that Hopf's methods belong chiefly to the domain of chain-multiplication. Further references notably to the related work of E. Cartan and Pontrjagin will be found in Hopf's paper.

(48.1) CONVENTIONS. Unless otherwise stated in the present section all cycles, $\cdots$ are rational, and all manifolds are finite connected absolute orientable geometric.

(48.2) DEFINITIONS. *Returning for a moment to standard group terminology, let* $G = \{g\}$ *be a multiplicative group. Then* G *is said to be a topological group, or a group space, whenever it is assigned a Hausdorff topology making* gg_1^{-1} *continuous. If* G *has further properties such as being a manifold,* $\cdots$ *, we describe it as a "group manifold,"* $\cdots$ *. Thus a closed connected Lie group is an analytical group manifold* M *such that the group operations are analytical functions on* M^2.

(48.3) Γ-*complexes.* Hopf himself has shown that the properties under consideration are those of a large class of manifolds, the Γ-*manifolds*, described below. Using essentially the same methods but with cocycles replacing more or less cycles, we shall show that they are properties of a large class of complexes called by analogy Γ-*complexes.*

Let $X = \{x\}$ be a finite connected simple complex (special case: X is simplicial) and let its classes (rational unless otherwise stated) be written Γ^p, Γ_p, or alternately Δ^p, Δ_p. The intersections existing in X are denoted as usual by a dot-product. The rational cohomology ring will be written $R(X)$. We say that X is a Γ-complex whenever:

(a) *For some* n *the Betti number* $R^n(X) = 1$. Choosing a fixed rational homology class $\Gamma^n \neq 0$ every other is of the form $t\Gamma^n$, t rational.

(b) $\Gamma_p \cdot \Gamma^n = \Gamma^{n-p} = 0 \leftrightarrow \Gamma_p = 0$.

(c) *There exists a multiplication* μ *of* X, X *to* X, *i.e., a chain-mapping* $X^2 \to X$ *such that if* Γ^0 *is the class of a vertex then both* $\mu\Gamma^0 \times \Gamma^n$ *and* $\mu\Gamma^n \times \Gamma^0$ *are different from* 0.

Notice that

$$\mu\Gamma^0 \times \Gamma^n = c_l\Gamma^n, \qquad \mu\Gamma^n \times \Gamma^0 = c_r\Gamma^n,$$

and so (c) is equivalent to the relation

(c′) $c_lc_r \neq 0$.

It has been observed by Hurewicz that the proof of Hopf's theorem (49.1) merely requires the following property: *There exists a multiplication* μ *of* X, X *to* X *such that* $\Gamma^p \to \mu\ (\Gamma^0 \times \Gamma^p)$ *and* $\Gamma^p \to \mu\ (\Gamma^p \times \Gamma^0)$ *define isomorphisms of the rational homology groups of* X *with themselves.* However (abc) appear to be better adapted to the applications.

(48.4) Γ-*manifolds.* A Γ-manifold is a connected M^n such that there exists a mapping t: $(M^n)^2 \to M^n$, written also as a product xy (x, $y \in M^n$), whose values are in M^n, and with the following property: if x [y] is fixed then xy defines a mapping t_l [t_r]:$M^n \to M^n$ with a degree c_l [c_r], and we must have $c_lc_r \neq 0$.

(48.5) *A* Γ-*manifold is a* Γ-*complex.*

That (48.3ab) hold is a consequence of (V, 29.9, 36.14). Regarding (48.3c), or its equivalent (48.3c′), consider first a mapping $t:|K| \to |L|$ where K, L are finite geometric complexes. By (23.8) there is a barycentric mapping $t_1:|K^{(r)}| \to |L|$ homotopic to t. If t_1 induces the chain-mapping $\theta:K^{(r)} \to L$ then θ induces the same simultaneous homomorphism $\bar{\theta}$ in the homology groups as t_1, and hence as t. If δ denotes chain-derivation in K then $\theta\delta^r$ is a chain-mapping $K \to L$ inducing likewise $\bar{\theta}$. The same considerations hold also if K, L are polyhedral complexes, since their derived are still simplicial. Applying them to the mapping: $(M^n)^2 \to M^n$ induced by xy, we find that there is a chain-multiplication μ of M^n, M^n to M^n such that if Γ^0 is the class of a vertex then, by the definition of the degrees: $\mu\Gamma^0 \times \Gamma^n = c_l\Gamma^n$, $\mu\Gamma^n \times \Gamma^0 = c_r\Gamma^n$, and so under our assumption (48.3c′), holds. This proves (48.5).

(48.6) *A group manifold is a* Γ-*manifold and hence also a* Γ-*complex.*

For if xy is the group operation then for $x = 1$ we have $t_l = 1$ and hence $c_l = 1$, and similarly $c_r = 1$.

(48.7) CONCLUSION. *The homology properties of* Γ-*complexes are also homology properties of* Γ-*manifolds and group manifolds.*

(48.8) Returning to the Γ-complex X we will designate by Γ_0 the class of its fundamental cocycle (sum of the duals of the vertices) and we note that by (V, 4.13, 8.9):

$$\Gamma_0 \cdot \Gamma_p = \Gamma_p \cdot \Gamma_0 = \Gamma_p . \tag{48.9}$$

Thus Γ_0 is the unit of the ring $R(X)$.

49. We are now ready for our main argument and prove:

(49.1) THEOREM OF HOPF. *The rational homology group and ring of a* Γ-*complex, hence those of a* Γ-*manifold or group manifold, are isomorphic with those of a certain finite product* S *of odd-dimensional spheres.*

The proof will rest upon the following proposition:

(49.2) *There may be chosen* l *cohomology classes* $\{\Delta^i\}$ *in* $R(X)$ *and* $\{C^i\}$ *in a suitable* $R(S)$ *such that* $\dim \Delta^i = \dim C^i$, *and that* $\Delta^i \to C^i$ *define isomorphisms of* $R(X)$, $R(S)$.

To complete the picture we will also prove with Hopf:

(49.3) *Every finite product of odd-dimensional spheres is a* Γ-*manifold.*

(49.4) REMARK. Actually Hopf's treatment was given only for Γ-manifolds. However, by making use of the co-theory, his proof extends to Γ-complexes.

50. (50.1) Let $\mu^*:X^* \to X^{*2}$ be the dual of μ. As regards both μ, μ^* we designate also by μ, μ^* the induced homomorphism in the homology and cohomology groups. If we choose bases for the rational cohomology classes then every cycle of X^{*2}, i.e., cocycle of X^2, is a linear rational combination of products of elements of the bases (IV, 6.7). Therefore with $p > 0$:

$$\mu^* \Gamma_p = \Gamma_0 \times \lambda\Gamma_p + (\rho\Gamma_p) \times \Gamma_0 + \sum_{q,r\neq 0} \Gamma_q^i \times \Gamma_r^j . \tag{50.2}$$

According to (V, 14.1c) the operations $\lambda:\Gamma_p \to \lambda\Gamma_p$, $\rho:\Gamma_p \to \rho\Gamma_p$ are homomorphisms $R(X) \to R(X)$. We first prove:

(50.3) *λ and ρ are isomorphisms.*

It will be sufficient to consider λ. Since $R(X)$ is a vector space, we merely need to show that λ is univalent. Suppose then $\lambda\Gamma_p = 0$. By (V, 17.6):

$$\begin{aligned}\mu^*\Gamma_p\cdot\Gamma^0 \times \Gamma^n &= (\Gamma_0 \times \lambda\Gamma_p)\cdot(\Gamma^0 \times \Gamma^n)\\ &= (\Gamma_0\cdot\Gamma^0) \times (\lambda\Gamma_p\cdot\Gamma^n)\\ &= \Gamma^0 \times (\lambda\Gamma_p\cdot\Gamma^n).\end{aligned}$$

Applying now (V, 14.1a) and with an obvious permutation of terms in accordance with (V, 8.8a) we have if $\lambda\Gamma^p = 0$:

$$\begin{aligned}\mu(\mu^*\Gamma_p\cdot\Gamma^0 \times \Gamma^n) &= \Gamma_p\cdot c_l\Gamma^n = c_l(\Gamma_p\cdot\Gamma^n)\\ &= \mu(\Gamma^0 \times (\lambda\Gamma_p\cdot\Gamma^n)) = 0.\end{aligned}$$

Since $c_l \neq 0$ by (48.3b): $\Gamma_p = 0$. Thus λ is univalent and (50.3) follows.

(50.4) *Generators for $R(X)$.* If $\{\Gamma^i\}$ is any subset of $R(X)$, including Γ_0, then the operations of the ring applied between the elements of the set give rise to a subring $R\{\Gamma^i\}$ of $R(X)$. If $R\{\Gamma^i\} = R(X)$, then the Γ^i other than Γ_0 are called *generators* of $R(X)$. A set of generators is said to be *irreducible* if no proper subset is a set of generators for the ring.

We say that a class Γ_p, $p > 0$, is *maximal* if we cannot write $\Gamma_p = \sum \Gamma_q^h\cdot\Gamma_r^j$, where $0 < q, r < p$.

An irreducible system of generators may be constructed as follows: choose for each $p > 0$ a maximal set of maximal Γ_p linearly independent modulo the non-maximal elements. The total set $\{\Delta^i\}$, $i = 1, 2, \cdots, l$, thus obtained, is clearly an irreducible set of generators consisting of actual cohomology classes, and not merely of sums of classes of mixed dimensions. We prove:

(50.5) $$\Delta^1 \cdots \Delta^l \neq 0.$$

Since every $\Delta^i \neq 0$ and the order of the Δ^i is immaterial the proof reduces to:

(a) $\Delta^2 \cdots \Delta^k \neq 0 \to \Delta^1 \cdots \Delta^k \neq 0$, where we also assume $\dim \Delta^{i+1} \leqq \dim \Delta^i$, and where $\{\Delta^1, \cdots, \Delta^k\}$ are merely any k generators of the initial set.

Let $\Delta'^i = \lambda\Delta^i$, and denote by $\mathfrak{A}$ the *ideal* of $R(X)$ based on $\{\Delta'^2, \cdots, \Delta'^l\}$, or set of elements $\{\sum_{i>1} \Gamma^i\cdot\Delta'^i\}$. Corresponding to $\mathfrak{A}$ the elements $\{\Gamma_p \times \Delta'^i\}$, $i > 1$, generate an ideal $\mathfrak{A}^*$ of $R(X^2)$. Now $\mu^*\Delta^i = \Gamma_0 \times \Delta'^i + \rho\Gamma^i \times \Gamma_0 + \sum \Gamma_q \times \Gamma_r$, $1 \leqq i \leqq k$, where $r < \dim \Delta^i$. Since λ is an isomorphism it preserves dimensions and furthermore $\{\Delta'^j\}$, $j = 1, 2, \cdots, l$ is likewise a set of generators for $R(X)$. On dimensional grounds then Γ_r is a linear rational combination of intersections of the Δ'^j, $j > 1$, i.e., $\Gamma_r \in \mathfrak{A}$ and hence $\Gamma_q \times \Gamma_r \in \mathfrak{A}^*$. Thus we have

$$\left.\begin{aligned}\mu^*\Delta^1 &\equiv \Gamma_0 \times \Delta'^1 + \rho\Delta^1 \times \Gamma_0\\ \mu^*\Delta^i &\equiv \rho\Delta^i \times \Gamma_0, k \geqq i > 1\end{aligned}\right\} \bmod \mathfrak{A}^*.$$

From this follows by repeated application of (V, 17.6):

$$\mu^*(\Delta^1 \cdots \Delta^k) \equiv \rho(\Delta^1 \cdots \Delta^k) \times \Gamma_0 \pm \rho(\Delta^2 \cdots \Delta^k) \times \Delta'^1 \bmod \mathfrak{A}^*.$$

Hence $\Delta^1 \cdots \Delta^k = 0 \to \rho(\Delta^2 \cdots \Delta^k) \times \Delta'^1 \equiv 0 \bmod \mathfrak{A}^*$. Since $\Delta^2 \cdots \Delta^k \neq 0$ and ρ is an isomorphism we also have $\rho(\Delta^2 \cdots \Delta^k) \neq 0$, and so by a transparent application of (II, 37.5): $\Delta'^1 \in \mathfrak{A}$. Since λ is an isomorphism this implies that Δ^1 is in the ideal based on $\{\Delta^2, \cdots, \Delta^l\}$, contrary to assumption. This proves (a), and hence (50.5).

(50.6) *If p is even Γ_p is not maximal.*

Let $\mathfrak{B}$ be the ideal of $R(X)$ generated by the elements $\Gamma_r, r \neq 0, p$. Thus $\mathfrak{B}$ consists of all the elements of dimension different from 0, p and of the non-maximal Γ_p. Denote also by $\mathfrak{B}^*$ the ideal of $R(X^2)$ based on the elements $\Gamma_q \times \Gamma_r$, $\Gamma_r \in \mathfrak{B}$. We have:

$$\mu^*\Gamma_p \equiv \Gamma_0 \times \lambda\Gamma_p + \rho\Gamma_p \times \Gamma_0 \bmod \mathfrak{B}^*.$$

Supposing now p even, we find by multiplying m times and recollecting (V, 8.8a, 17.6):

$$\mu^*(\Gamma_p)^m \equiv \rho(\Gamma_p)^m \times \Gamma_0 + m\rho(\Gamma_p)^{m-1} \times \lambda\Gamma_p \bmod \mathfrak{B}^*, \tag{50.7}$$

where the powers refer to repeated dot-products. Since X is finite-dimensional there exists an m such that $(\Gamma_p)^{m-1} \neq 0$, $(\Gamma_p)^m = 0$ and we shall have:

$$\rho(\Gamma_p)^{m-1} \times \lambda\Gamma_p \equiv 0 \bmod \mathfrak{B}^*.$$

Since ρ is an isomorphism $\rho(\Gamma_p)^{m-1} \neq 0$, and so by (II, 37.5): $\lambda\Gamma_p \in \mathfrak{B}$, proving (50.6).

(50.8) *Every generator Δ^i is of odd dimension.*

For Δ^i is maximal and so by (50.6) its dimension is odd.

$$\Delta^i \cdot \Delta^j = -\Delta^j \cdot \Delta^i, \qquad \Delta^i \cdot \Delta^i = 0 \qquad \text{(V, 8.8a)}. \tag{50.9}$$

51. (51.1) By (50.8) $p_i = \dim \Delta^i$ is odd. Let S^{p_i} denote a p_i-sphere (boundary of a $(p_i + 1)$-simplex), and let A^i be the fundamental zero-cohomology class, and B^i a p_i-cohomology class different from 0 for S^{p_i}. Referring also to (V, 8.9) or (V, 20.5) there comes

$$A^i \cdot A^i = A^i, \qquad A^i \cdot B^i = B^i \cdot A^i = B^i, \qquad B^i \cdot B^i = 0, \tag{51.2}$$

(the last on dimensional grounds). Set now $S = \mathbf{P}S^{p_i}$ and define

$$C^i = A^1 \times \cdots \times A^{i-1} \times B^i \times A^{i+1} \times \cdots \times A^l,$$

$$C^{ij\cdots k} = D^1 \times \cdots \times D^l,$$

where $D^h = A^h$ for $h \neq i, j, \cdots, k$ and $D^h = B^h$ otherwise.

By (IV, 6.7) the $C^{ij\cdots k}$, $i + \cdots + k = m > 0$, form a base for the rational group $\mathfrak{H}_m(S)$, while $\mathfrak{H}_0(S)$ consists merely of the multiples of the cohomology class $C = \mathbf{P}A^i$ which is the Γ_0 of S. It follows that the rational ring $R(S)$

is generated by the intersections of the classes C^i, $C^{ij\cdots k}$. It is easily seen that the C^i are maximal and that $C^{ij\cdots k} = \pm C^i \cdots C^k$. Hence $\{C^i\}$ is an irreducible system of generators. It is also an elementary matter to verify that $C^i \leftrightarrow \Delta^i$ defines an isomorphism $R(S) \leftrightarrow R(X)$ and consequently also $\mathfrak{H}_m(S) \leftrightarrow \mathfrak{H}_m(X)$. This completes the proof of (49.2) and hence also of (49.1).

(51.3) PROOF OF (49.3). An outline of the argument will suffice. It breaks up into two parts:

(a) *An odd-dimensional sphere* S^m *is a* Γ*-manifold.*

(b) *The product of two* Γ*-manifolds is a* Γ*-manifold.*

Clearly (a, b) together prove (49.3).

Take first (a). If $x, y \in S^m$ we denote by xy the reflection of x in the diameter through y. In the notations of (48) t_r is topological, and so $c_r = \pm 1$. A fairly simple argument whose detail is omitted (see Hopf [c, 30]) shows then that $c_l = \pm 2$. This is sufficient to prove (a).

Take now two Γ-manifolds M, M', and let t_l, t_l', $\cdots$ analogous to t_l, $\cdots$ of (48) have their obvious meaning. If (x, x'), $(y, y') \in M \times M'$ we define multiplication in the latter by $(x, x') \times (y, y') = (xy, x'y')$. By an argument such as the one leading to (48.5) it is shown that the corresponding chain-multiplication $X^2 \times Y^2 \rightarrow X \times Y$ is $\mu \times \mu'$ (notation of IV, 21) and the analogues of c_l, c_r for $\mu \times \mu'$ are $c_lc_{l'} \neq 0$, $c_rc_{r'} \neq 0$. Since $M \times M'$ is a connected geometric manifold (b) follows and (49.3) is proved.

52. Many interesting properties may now be deduced from (49.1). We reproduce a few, all taken from Hopf [c], where references to earlier related results, notably those due to E. Cartan, are also given. In the statements X may be a Γ-complex, a Γ-manifold or a group manifold.

(52.1) *The elements* Γ^0, Δ^i, $\Delta^{i_1}\cdot\Delta^{i_2}$, $(i_1 < i_2)$, $\cdots$ *form a base for the vector space* $R(X)$. *Hence* $2^l = \dim R(X)$, *and so* l *depends solely upon* X.

(52.2) *No even-dimensional class* Γ_{2p} *is maximal. In other words a* Γ_{2p} *is a sum of intersections of classes of dimension less than* $2p$.

(52.3) *The Poincaré polynomial of* X (III, 15.3) *is given by*:

$$P(t, X) = (1 + t^{p_1}) \cdots (1 + t^{p_l}),$$

where the p_i *are all odd.*

(52.4) *The Euler characteristic* $\chi(X) = P(-1, X) = 0$.

(52.5) *The sum of the Betti numbers of* X *is a power of two, namely* 2^l.

$$R^p(X) = R^{n-p}(X). \tag{52.6}$$

In other words, the duality theorem of Poincaré for the Betti numbers of a manifold (V, 33.1a) holds for X. Needless to say this is far from making X a manifold. It must be remembered that n is the integer occurring in (48.3ab), and need not be the dimension of X which may well exceed n. However, if X is a Γ-manifold, hence also when it is a group manifold, then n is the dimension of the manifold.

(52.7) *The dimensions p_i of the generators satisfy the relation*

$$p_1 + \cdots + p_l = n.$$

For S, hence X, has an n-cycle and none of higher dimension.

(52.8) Various relations for the Betti numbers R^s follow from the expression of $P(t, X)$. Thus:

$$R^1 \leqq n, \qquad R^2 = \binom{R^1}{2}, \qquad R^s \geqq \binom{R^1}{s}.$$

53. **Complements.**

(53.1) *The Pontrjagin ring.* To the multiplication μ of (48) associated with the Γ-complex X there corresponds a rational homology ring $P(X, \mu)$ in the sense of (V, 1.10). This ring was introduced for the first time for group manifolds by Pontrjagin, and for this reason Hopf designates it as the Pontrjagin ring. When X is a group manifold the ring is *associative*, but it need not be so otherwise. It has been proved by Samelson [a] that whenever the ring $P(X, \mu)$ is associative and Γ^0 (the class of a vertex) is a unit of this ring, then the isomorphism of $R(X)$ with $R(S)$ (S is the sphere-product) may be so chosen as to yield an isomorphism of $P(X, \mu)$ with the analogue $P(S, \bar{\mu})$ for S, where $\bar{\mu}$ is the multiplication in S which corresponds in the obvious way to μ in X.

(53.2) *Γ-spaces.* We will designate by that term a compact connected Hausdorff space $\mathfrak{R}$ such that: (a) it has properties (48.3abc), where μ refers merely to a simultaneous homomorphism of the rational homology groups of $\mathfrak{R}^2$ into these of $\mathfrak{R}$; (b) the rational cohomology ring of $\mathfrak{R}$ has a finite number of generators. It may be seen by reference to (VII, 16) that all the machinery is at hand for the extension to such spaces of the results centering around Hopf's theorem proved for Γ-complexes.

An arbitrarily large supply of examples of Γ-spaces which are not group or Γ-manifolds may be based upon the following property:

(53.3) *If K is a finite Euclidean Γ-complex and $\mathfrak{R}$ has $|K|$ for deformation retract then $\mathfrak{R}$ is a Γ-space.*

For $\mathfrak{R}$ has the same homology properties as $|K|$, and hence as K (10.1; VII, 7.1, 7.5).

As an explicit example take an odd-dimensional sphere S^{2p+1} and on the sphere a countable dense set $\{x_n\}$. Extend the radius to x_n by a segment of length 2^{-n}. The resulting space $\mathfrak{R}$ has S^{2p+1} for deformation retract and so it is a Γ-space, which is manifestly not a manifold.

§10. NOMENCLATURE OF COMPLEXES AND MANIFOLDS

54. **Complexes.** The complexes in the sense utilized in the present work (III, 1.1) have generally been called "abstract complexes." They may be finite or infinite and in the latter case the most important subtypes are: star-finite, closure-finite and locally finite complexes. The following special types have been introduced:

A. *Simplicial complexes* (III, 5). The simplexes have usually been designated by σ, ζ. If K is a simplicial complex, L a closed subcomplex then $K - L$ is known as an open simplicial complex, and by contrast K, L as closed simplicial complexes.

B. *Polyhedral complexes.* The collection $\Pi = \{E\}$ of the faces of a polyhedron turned into a complex (III, 6). The polyhedron as a space is denoted by $|\Pi|$.

C. *Euclidean complexes.* Polyhedral complexes whose elements are Euclidean simplexes in an Euclidean space or in the Hilbert parallelotope (III, 6.9).

D. *Topological complexes*: Collections $K = \{E\}$ whose elements are cells with incidences and dimensions similar to those in a polyhedral complex (VIII, §7). The union of the cells is a metric space denoted by $|K|$ and called a polyhedral space.

E. *Differentiable complexes, C^r-complexes.* Topological complexes whose cells are subjected to certain differentiability conditions (VIII, 45).

F. *Simple complexes.* Closure-finite complexes whose elements satisfy certain algebraic conditions (III, 47.1). The class includes all the types A, $\cdots$, E.

G. *Γ-complexes.* Simple complexes, with certain special properties described in (48.3).

55. **Manifolds.** All the manifolds to be described are supposed to be n-dimensional.

A. *Combinatorial manifolds.* The types investigated in (V, §4). They may be finite or infinite, absolute or relative, orientable or non-orientable, simplicial or merely simple complexes.

B. *Geometric manifolds.* Euclidean realizations of the preceding simplicial types.

C. *Manifolds in the sense of Brouwer.* Euclidean complexes such that the star of each vertex is isomorphic with a set of simplexes in an Euclidean $\mathfrak{E}^n$ having a common vertex P and making up a neighborhood of P in $\mathfrak{E}^n$.

D. *Manifolds in the sense of Newman.* Euclidean complexes such that if a is a vertex and St $a = aB$, then B is partition-equivalent to an $(n - 1)$-sphere.

E. *Manifolds in the sense of Poincaré* [b] *and Veblen* [V]. Topological complexes such that every point has for neighborhood an n-cell.

F. *Topological manifolds.* An M^n of this type is a separable metric space with a countable locally finite open covering consisting of n-cells. (See Flexner [a, b].) Noteworthy special cases: C^r-manifolds, differentiable manifolds, analytical manifolds (44.1), Γ-manifolds, group manifolds (48.2, 48.3).

G. *Generalized manifolds.* Locally compact spaces discussed by Čech [b], Lefschetz [c], Wilder [a] and others and characterized by certain properties of so-called "local connectedness" or "local connectedness in the sense of homology" and also by the property: each point is n-cyclic. They have been investigated at length in a forthcoming paper by E. Begle [a] to which the reader is referred for all details.

H. *Pseudo-manifolds.* This term has been applied by Brouwer and other authors to what we have called a simple geometric n-circuit.

I. *Manifolds of grade p.* Simplicial n-complexes, investigated by Čech, and which behave like an M^n only as regards the two consecutive dimensions $p - 1, p$.

APPENDIX A

ON HOMOLOGY GROUPS OF INFINITE COMPLEXES AND COMPACTA

BY

SAMUEL EILENBERG AND SAUNDERS MACLANE

The results of (III, 18) on universal coefficient groups for finite complexes suggest the consideration of similar problems for the homology theory of infinite complexes. To what extent are the homology groups constructed from a general group G of coefficients determined by the homology (or cohomology) groups with specially chosen "universal" coefficient groups? Results of this nature have already been found by Čech [d] and Steenrod [a]. The present summary describes results on this problem recently obtained by the authors. They are based on a complete analysis of the homology groups under consideration, utilizing an important concept not previously occurring in topology: the group of group extensions of one given group by another group. This analysis yields a formula (5.1) expressing the homology group of the infinite cycles over G in terms of the finite integral cohomology groups. This formula also applies to the Čech homology groups of a compactum. Modified formulas can also be found for other varieties of homology theories. For further details the reader is referred to a forthcoming article by the authors [a].

1. In the present appendix all groups will be of the type which we will call *generalized topological groups*. A group of this sort is a group $G = \{g\}$ such that G is a topological space and that $g - g'$ is continuous. In other words the restriction of (II, 1.1) that G obeys the T_0 separation axiom is no longer imposed. Without this added freedom we could not, as we must, consider topologies on factor-groups modulo non-closed subgroups.

We consider a star-finite complex X and a *topological* coefficient group G. As usual $\mathfrak{J}$ and $\mathfrak{P}$ will denote the respective groups of the integers and of the reals mod 1. We define and topologize the groups $\mathfrak{C}^q(X, G)$, $\mathfrak{Z}^q(X, G)$, $\mathfrak{F}^q(X, G)$ of chains, cycles and bounding cycles as usual and consider the following homology groups, which are generalized topological groups:

$$H^q(X, G) = \mathfrak{Z}^q(X, G)/\mathfrak{F}^q(X, G), \tag{1.1}$$

$$\mathfrak{H}^q(X, G) = \mathfrak{Z}^q(X, G)/\bar{\mathfrak{F}}^q(X, G). \tag{1.2}$$

The second of these groups is obviously topological. Clearly the second one can be obtained from the first, since

$$\mathfrak{H}^q \cong H^q/\bar{O} \tag{1.3}$$

where $\bar{O}$ is that subgroup of H^q which is the closure of the null-subgroup.

We shall also consider the discrete groups $\mathfrak{C}_q(X, \mathfrak{J})$, $\mathfrak{Z}_q(X, \mathfrak{J})$ and $\mathfrak{F}_q(X, \mathfrak{J})$ of the *finite* integral cochains, cocycles and bounding cocycles and the (discrete) cohomology group

$$(1.4) \qquad \mathfrak{H}_q(X, \mathfrak{J}) = \mathfrak{Z}_q(X, \mathfrak{J})/\mathfrak{F}_q(X, \mathfrak{J}).$$

We shall denote by $\mathfrak{T}_q(X, \mathfrak{J})$ the subgroup of $\mathfrak{H}_q$ consisting of all the elements of finite order.

The boundary operator F maps $\mathfrak{C}_q$ onto $\mathfrak{F}_{q+1}$, hence $\mathfrak{C}_q/\mathfrak{Z}_q = \mathfrak{F}_{q+1}$. Since $\mathfrak{F}_{q+1}$ as a subgroup of a free group $\mathfrak{C}_{q+1}$ is free, we deduce easily that

(1.5) $\mathfrak{Z}_q(X, \mathfrak{J})$ *is a direct factor of* $\mathfrak{C}_q(X, \mathfrak{J})$.

2. Given $c^q \in \mathfrak{C}^q(X, G)$ and $d_q \in \mathfrak{C}_q(X, \mathfrak{J})$ the Kronecker index $\mathrm{KI}(c^q, d_q)$ is an element of G, and it establishes a multiplication of $\mathfrak{C}^q$ and $\mathfrak{C}_q$ to G. Using (III, 29.1) we readily show that

$$\mathfrak{Z}^q(X, G) = \text{annihilator of } \mathfrak{F}_q(X, \mathfrak{J}).$$

Further we define

$$\mathfrak{A}^q(X, G) = \text{annihilator of } \mathfrak{Z}_q(X, \mathfrak{J}),$$

and prove that

$$(2.1) \qquad \mathfrak{F}^q(X, G) \subset \mathfrak{A}^q(X, G) = \bar{\mathfrak{A}}^q(X, G) \subset \mathfrak{Z}^q(X, G).$$

Given $c^q \in \mathfrak{C}^q(X, G)$ we define $\varphi(d_q) = \mathrm{KI}(c^q, d_q)$ for $d_q \in \mathfrak{C}_q(X, \mathfrak{J})$ and obtain a homomorphism $\varphi: \mathfrak{C}_q \to G$. This establishes an isomorphism $\mathfrak{C}^q(X, G) \cong \mathrm{Hom}\,\{\mathfrak{C}_q(X, \mathfrak{J}), G\}$, where $\mathrm{Hom}\,\{H, G\}$ stands for the group of all homomorphisms $H \to G$, the topology being defined as for the group of characters (II, 18.1).

From this we deduce the following results:

(2.2) $\mathfrak{A}^q(X, G)$ *is a direct factor of* $\mathfrak{Z}^q(X, G)$,

$$(2.3) \qquad \mathfrak{Z}^q(X, G)/\mathfrak{A}^q(X, G) \cong \mathrm{Hom}\,\{\mathfrak{H}_q(X, \mathfrak{J}), G\}.$$

3. We now proceed with an analysis of the group $\mathfrak{A}^q(X, G)$. Let $z^q \in \mathfrak{A}^q(X, G)$ and $w_{q+1} \in \mathfrak{F}_{q+1}(X, \mathfrak{J})$. Choose $d_q \in \mathfrak{C}_q(X, \mathfrak{J})$, so that $\mathrm{F}d_q = w_{q+1}$. Define $\theta(w_{q+1}) = \mathrm{KI}(z^q, d_q)$. Clearly if $\mathrm{F}d_q' = w_{q+1}$ then $d_q - d_q'$ is a cocycle and $\mathrm{KI}(z^q, d_q - d_q') = 0$, for z^q is one of the annihilators of all cocycles. Hence θ is a uniquely defined homomorphism $\mathfrak{F}_{q+1}(X, \mathfrak{J}) \to G$. We also verify that the correspondence $z^q \to \theta$ defines an isomorphism $\mathfrak{A}^q(X, G) \cong \mathrm{Hom}\,\{\mathfrak{F}_{q+1}(X, \mathfrak{J}), G\}$. Furthermore, the subgroup $\mathfrak{F}^q(X, G)$ of $\mathfrak{A}^q(X, G)$ is mapped onto a subgroup $\mathrm{Hom}\,\{\mathfrak{Z}_{q+1}(X, \mathfrak{J}) \mid \mathfrak{F}_{q+1}(X, \mathfrak{J}), G\}$, consisting of those homomorphisms of $\mathfrak{F}_{q+1}(X, \mathfrak{J})$ which can be extended to homomorphisms of $\mathfrak{Z}_{q+1}(X, \mathfrak{J})$ into G. Consequently

$$(3.1) \qquad \mathfrak{A}^q(X, G)/\mathfrak{F}^q(X, G) \cong \mathrm{Hom}\,\{\mathfrak{F}_{q+1}(X, \mathfrak{J}), G\}/\mathrm{Hom}\,\{\mathfrak{Z}_{q+1}(X, \mathfrak{J}) \mid \mathfrak{F}_{q+1}(X, \mathfrak{J}), G\}.$$

It can be proved directly that the factor-group on the right in (3.1) depends only upon $\mathfrak{H}_{q+1} = \mathfrak{Z}_{q+1}/\mathfrak{F}_{q+1}$ and G; however, the meaning of this group becomes much clearer if we relate it to the concept of a group extension.

4. **Group extensions.** Let G and H be given abelian groups. An abelian group, E, is said to be an *extension of* G *by* H if E has G as a subgroup and $H = E/G$ is the corresponding factor-group. Two extensions E and E' will be regarded as equivalent if there is an isomorphism (in the algebraic sense) of E with E' in which both the elements of G and the cosets of H are left fixed. For fixed G and H, the equivalence classes of extensions form an additive group, written Ext $\{G, H\}$, whose zero is the class of the direct product $G \times H$, regarded as an extension of G by H. To represent any extension E, choose an element e_h in each coset h of E mod G. Then E is the union of its cosets $e_h + G$, and the composition is given by an addition table for the elements e,

$$e_{h+h'} = e_h + e_{h'} + \gamma(h, h'), \tag{4.1}$$

where γ is a function on H^2 to G. To within an equivalence E is determined by any symmetric $\gamma(h, h')$, such that the addition law (4.1) defined by it is associative, i.e., such that

$$\gamma(h + h', h'') + \gamma(h, h') = \gamma(h, h' + h'') + \gamma(h', h'').$$

It is clear that $\Gamma = \{\gamma\}$ is an additive group. Since we may replace e_h by $e_h + \varphi(h)$, where φ is any function on H to G, E is unchanged when γ is augmented by a function $\delta(h, h') = \varphi(h) + \varphi(h') - \varphi(h + h')$. Clearly $\Delta = \{\delta\}$ is a subgroup of Γ, and one may readily prove: Ext $\{G, H\} \cong \Gamma/\Delta$.

When H is discrete and G topological Ext $\{G, H\}$ may be regarded as a generalized topological group.

Suppose now $H = Z/F$ where Z is a free group and F is a subgroup of Z. We shall denote again by $\mathrm{Hom}\{Z \mid F, G\}$ the subgroup of Hom $\{F, G\}$ which consists of the homomorphisms $F \to G$ which may be extended to homomorphisms $Z \to G$. There is then an isomorphism

$$\mathrm{Ext}\,\{G, H\} \cong \mathrm{Hom}\,\{F, G\} \,/\, \mathrm{Hom}\,\{Z \mid F, G\}. \tag{4.2}$$

Formula (3.1) therefore takes the following form:

$$\mathfrak{A}^q(X, G)/\mathfrak{F}^q(X, G) \cong \mathrm{Ext}\,\{G, \mathfrak{H}_{q+1}(X, \mathfrak{J})\}. \tag{4.3}$$

5. We are now in a position to formulate our main results:

(5.1) THEOREM. *The group $H^q(X, G)$ is determined by the groups G, $\mathfrak{H}_q(X, \mathfrak{J})$ and $\mathfrak{H}_{q+1}(X, \mathfrak{J})$. Explicitly*

$$H^q(X, G) \cong \mathrm{Hom}\,\{\mathfrak{H}_q(X, \mathfrak{J}), G\} \times \mathrm{Ext}\,\{G, \mathfrak{H}_{q+1}(X, \mathfrak{J})\},$$

where the groups involved are considered as generalized topological groups.

This is a direct consequence of (2.2), (2.3) and (4.3). We compute $\mathfrak{H}^q$ by means of (1.3). Since the group Hom $\{\mathfrak{H}_q, G\}$ is topological, all we need is to replace Ext $\{G, \mathfrak{H}_{q+1}\}$ by Ext $\{G, \mathfrak{H}_{q+1}\}/\bar{O}$. However we can prove that this last group $\cong$ Ext $\{G, \mathfrak{T}_{q+1}\}/\bar{O}$. Hence

(5.2) THEOREM. *The group $\mathfrak{H}^q(K, G)$ is determined by the groups G, $\mathfrak{H}_q(X, \mathfrak{J})$ and $\mathfrak{T}_{q+1}(X, \mathfrak{J})$. Explicitly*

$$\mathfrak{H}^q(X, G) \cong \text{Hom } \{\mathfrak{H}_q(X, \mathfrak{J}), G\} \times (\text{Ext } \{G, \mathfrak{T}_{q+1}(X, \mathfrak{J})\}/\bar{O}).$$

6. The preceding results may be further specialized if additional information about the group G is available.

If the group G is infinitely divisible (i.e., if $mg' = g$ has a solution $g' \in G$ for $g \in G$, m an integer) then Hom $\{F, G\}$ = Hom $\{Z \mid F, G\}$ and Ext $\{G, H\} = 0$ for every group H. Hence

(6.1) *If G is infinitely divisible then*

$$\mathfrak{F}^q(X, G) = \mathfrak{A}^q(X, G),$$
$$H^q(X, G) = \mathfrak{H}^q(X, G) \cong \text{Hom } \{\mathfrak{H}_q(X, \mathfrak{J}), G\}.$$

If G is discrete and has no elements of finite order then the closure of Hom $\{Z \mid F, G\}$ is the group Hom $\{Z_0 \mid F, G\}$, where Z_0 consists of all elements of Z of finite order modulo F (all "torsion cycles"). Hence

(6.2) *If G is discrete and has no elements of finite order then*

$$\mathfrak{H}^q(X, G) \cong \text{Hom } \{\mathfrak{H}_q(X, \mathfrak{J}), G\} \times \text{Ext } \{G, \mathfrak{T}_{q+1} (X, \mathfrak{J})\}.$$

If $G = \mathfrak{J}$ then Ext $\{G, \mathfrak{T}_{q+1}\} \cong$ Char $\mathfrak{T}_{q+1}$, the character-group of $\mathfrak{T}_{q+1}$, and we obtain

(6.3) *In the case of integral coefficients we have*

$$\mathfrak{H}^q(X, \mathfrak{J}) \cong \text{Hom } \{\mathfrak{H}_q(X, \mathfrak{J}), \mathfrak{J}\} \times \text{Char } \mathfrak{T}_{q+1}(X, \mathfrak{J}).$$

An extremely important case arises when G is compact. The group Ext $\{G, \mathfrak{H}_{q+1}\}$ = Ext $\{G, \mathfrak{T}_{q+1}\}$ is then compact and is the group of characters of the discrete group Hom $\{G, \mathfrak{T}_{q+1}\}$. Hence

(6.4) *If G is compact then $\mathfrak{F}^q(X, G) = \bar{\mathfrak{F}}^q(X, G)$ and*

$$H^q(X, G) \cong \mathfrak{H}^q(X, G) \cong \text{Hom } \{\mathfrak{H}_q(X, \mathfrak{J}), G\} \times \text{Char Hom } \{G, \mathfrak{T}_{q+1}(X, \mathfrak{J})\}.$$

7. We shall use (6.4) to compute the cohomology groups $\mathfrak{H}_q(X, G)$ (finite cocycles) for a discrete group G of coefficients. Since the groups $\mathfrak{H}_q(X, G)$ and $\mathfrak{H}^q(X, \text{Char } G)$ are isomorphic with one another's character groups (III, 41.2) we obtain

(7.1) THEOREM. *The cohomology group $\mathfrak{H}_q(X, G)$ of finite cocycles is determined by the groups G, $\mathfrak{H}_q(X, \mathfrak{J})$ and $\mathfrak{T}_{q+1}(X, \mathfrak{J})$. Explicitly, if G is discrete,*

$$\mathfrak{H}_q(X, G) \cong \text{Char Hom } \{\mathfrak{H}_q(X, \mathfrak{J}), \text{Char } G\} \times \text{Hom } \{\text{Char } G, \mathfrak{T}_{q+1}(X, \mathfrak{J})\}.$$

The first factor on the right is isomorphic with the subgroup of $\mathfrak{H}_q(X, G)$ determined by the "pure" cocycles. (A pure cocycle is a finite linear combination, with coefficients in G, of integral cocycles.)

We also have the isomorphism

$$\text{Hom}\,\{\text{Char}\, G, \mathfrak{T}_{q+1}\} \cong \text{Hom}\,\{\text{Char}\, \mathfrak{T}_{q+1}, G\}.$$

Since the group Char $\mathfrak{T}_{q+1}$ is compact, and G is discrete, while, as usual, all homomorphisms are continuous, it follows that in the group at the right above, G enters only through its elements of finite order. Hence:

(7.2) *If G has no elements of finite order then every cocycle over G is pure and*

$$\mathfrak{H}_q(X, G) \cong \text{Char Hom}\,\{\mathfrak{H}_q(X, \mathfrak{J}), \text{Char}\, G\}.$$

8. Given a compactum $\mathfrak{R}$ we shall consider the Čech homology and cohomology groups $\mathfrak{H}^q(\mathfrak{R}, G)$ (topologized, with G a division-closure group) and $\mathfrak{H}_q(\mathfrak{R}, G)$, (discrete). In order to apply our theory developed for star-finite complexes to $\mathfrak{R}$, we shall consider a fundamental complex X for $\mathfrak{R}$ in the sense of [L, 327].

The groups of $\mathfrak{R}$ and X are related by the following isomorphisms:

$$(8.1) \qquad \mathfrak{H}^q(\mathfrak{R}, G) \cong \mathfrak{H}^{q+1}(X, G), \qquad \mathfrak{H}_q(\mathfrak{R}, G) \cong \mathfrak{H}_{q+1}(X, G).$$

Hence, using (5.1), (5.2) and (7.1) we obtain the following explicit relations:

$$(8.2) \quad \mathfrak{H}^q(\mathfrak{R}, G) \cong \text{Hom}\,\{\mathfrak{H}_q(\mathfrak{R}, \mathfrak{J}), G\} \times (\text{Ext}\,\{G, \mathfrak{T}_{q+1}(\mathfrak{R}, \mathfrak{J})\}/\bar{O}),$$

$$(8.3) \quad \mathfrak{H}_q(\mathfrak{R}, G) \cong \text{Char Hom}\,\{\mathfrak{H}_q(\mathfrak{R}, \mathfrak{J}), \text{Char}\, G\} \times \text{Hom}\,\{\text{Char}\, G, \mathfrak{T}_{q+1}(\mathfrak{R}, \mathfrak{J})\}.$$

9. **Universal coefficient groups.** The preceding results acquire more significance and force when they are viewed as universal coefficient-group theorems, in the manner of (III, 18).

(9.1) UNIVERSAL COEFFICIENT THEOREM FOR STAR-FINITE COMPLEXES. *In a star-finite complex the cohomology groups of finite integral cocycles determine all the groups obtained using finite cocycles or infinite cycles with arbitrary coefficients.*

Since the group $\mathfrak{H}_q(X, \mathfrak{J})$ is the character-group of $\mathfrak{H}^q(X, \mathfrak{P}) = H^q(X, \mathfrak{P})$ we see that the groups $\mathfrak{H}^q(X, \mathfrak{P})$ are also universal for X.

Replacing the complex X by its dual we obtain the

(9.2) UNIVERSAL COEFFICIENT THEOREM FOR CLOSURE-FINITE COMPLEXES. *In a closure-finite complex the homology groups of finite integral cycles determine all the groups obtained using finite cycles or infinite cocycles with arbitrary coefficients.*

Again in (9.2) we could use the groups $\mathfrak{H}_q(X, \mathfrak{P})$ instead of $\mathfrak{H}^q(X, \mathfrak{J})$.

(9.3) UNIVERSAL COEFFICIENT THEOREM FOR COMPACTA. *In a compactum the integral Čech cohomology groups determine the Čech homology and cohomology groups with coefficients in a division-closure group.*

Again instead of the groups $\mathfrak{H}_q(\mathfrak{R}, \mathfrak{J})$ we could use the groups $\mathfrak{H}^q(\mathfrak{R}, \mathfrak{P})$.

10. We will conclude with a few remarks regarding the extension to the relative theories. If Y is a closed subcomplex of X, then $X - Y$ is an open subcomplex, and in particular it is itself a complex, and is star- or closure-finite with X. As we know the theory of its cycles is the same as that of the cycles of X mod Y (III, 23). Therefore (9.1) and (9.2) hold with the cycles and cocycles of X replaced, respectively, by cycles of X mod Y and cocycles of $X - Y$.

Now let $\mathfrak{R}$ be a compactum and F a closed subset of $\mathfrak{R}$. If X is a fundamental complex for $\mathfrak{R}$ then there is a closed subcomplex Y such that the theory of X mod Y is essentially the theory of $\mathfrak{R}$ mod F. From this and the above remarks we infer that (9.3) holds with the homology and cohomology groups of $\mathfrak{R}$ replaced, respectively, by the homology groups of $\mathfrak{R}$ mod F and cohomology groups of $\mathfrak{R} - F$.

APPENDIX B

FIXED POINTS OF PERIODIC TRANSFORMATIONS

BY

P. A. SMITH

Let T be a periodic transformation of a space $\mathfrak{R}$ into itself. We shall give a brief account of a homology theory in $\mathfrak{R}$ related in a special way to T and leading to topological invariants of the pair $(\mathfrak{R}, T)$. With the aid of these special homologies we establish theorems concerning the structure of the totality L of fixed points of T. Roughly speaking, the structure of L cannot be more complicated, from the point of view of homology theory, than that of the space $\mathfrak{R}$ under transformation. As a special instance of this phenomenon we shall show in detail that if $\mathfrak{R}$ (compact, finite-dimensional) possesses the homology groups of an n-sphere (over suitable coefficient-groups), L possesses those of an r-sphere, $r \leqq n$. Further applications of the theory will lead rapidly to a general theorem concerning the existence of fixed points and to theorems concerning the mapping of one periodic transformation upon another.

Although the present account deals only with the situation in the large, an easy modification can be made to yield *local* invariants of $(\mathfrak{R}, T)$ by means of which the local structure of L can be studied. For details in this direction see P. A. Smith, [b, c].

General references: Brouwer [f], Eilenberg [b], Kerékjártó [a], Richardson and Smith [a], Smith [a–g].

PRELIMINARY DEFINTIONS. *A space will mean a Hausdorff space. $\mathfrak{R}$ will consistently denote a space, T a homeomorphic transformation of $\mathfrak{R}$ into itself. T will always be periodic—that is, some power of T will be the identity. The smallest power which gives the identity is the period of T, to be denoted by p. The identity itself will not be considered as being periodic. If A is a subset of $\mathfrak{R}$, the sets $A, TA, \cdots, T^{p-1}A$ will be called the T-images of A. We write $\sigma A = \bigcup T^i A$. Evidently σA is invariant (identical with its T-images). We shall use the term* fixed *only in connection with individual points. The totality of fixed points will be denoted by L. T will be called primitive if each point of $\mathfrak{R} - L$ has p distinct T-images. T is automatically primitive if its period p is a prime.*

1. **Simplicial transformations.** We shall say that $(\mathfrak{R}, T)$ is *simplicial* if $\mathfrak{R}$ is a closed finite Euclidean simplicial complex whose (geometric) simplexes are permuted among themselves by T. The totality of invariant simplexes will be denoted by $\mathfrak{R}_I$.

A simplicial $(\mathfrak{R}, T)$ will be called *primitive* if each simplex in $\mathfrak{R} - \mathfrak{R}_I$ has p

distinct (hence mutually exclusive) T-images. Evidently a simplicial $(\mathfrak{R}, T)$ is automatically primitive if p is prime.

A simplicial $(\mathfrak{R}, T)$ will be called *regular* if the subcomplex $\mathfrak{R}_I$ is closed.

(1.1) *If* $(\mathfrak{R}, T)$ *is regular,* $|\mathfrak{R}_I|$ *is identical with the fixed-point set* L.

PROOF. In any case regularity implies that the vertices of $\mathfrak{R}_I$ are in L. Suppose L contains $|\mathfrak{R}_I^h|$ where $\mathfrak{R}_I^h$ is the maximal closed h-dimensional subcomplex of $\mathfrak{R}_I$. If E is a simplex of $\mathfrak{R}_I^{h+1}$ and if $J = |\mathfrak{B}E|$, then $J \subset L$. Suppose E is not pointwise invariant under T. Then if we put $E \cup J$ into homeomorphic correspondence with one-half of an $(h + 1)$-sphere S, T induces there a periodic transformation T_1 leaving fixed the points of the boundary h-sphere. By defining T_1 as the identity on the other half of S, we obtain a periodic transformation operating in S and admitting a hemisphere of fixed points. But we shall see later (11.1) that the fixed-point set of any periodic transformation operating in S must be nowhere dense. From this contradiction we conclude that $E \subset L$, hence eventually $|\mathfrak{R}_I| \subset L$. On the other hand $L \subset |\mathfrak{R}_I|$ since simplexes which contain points of L are necessarily invariant. Hence $L = |\mathfrak{R}_I|$.

While the preceding result will not be used in what follows, it does suggest that a desirable combinatorial situation for the study of the fixed-point set occurs when a given $(\mathfrak{R}, T)$ can be reduced to a regular $(\mathfrak{R}, T)$ by a suitable simplicial subdivision. This can be done if, for example, $\mathfrak{R}$ and T are defined by analytic functions and $\mathfrak{R}$ is compact. For general $(\mathfrak{R}, T)$, however, there is no known way of introducing a subdivision into $\mathfrak{R}$ — even if $\mathfrak{R}$ is a complex to start with—which will render $(\mathfrak{R}, T)$ regular, or even simplicial.

2. Special systems and coverings. It is fairly obvious that cycles and homologies in special relation to a given period T can most readily be defined when $(\mathfrak{R}, T)$ is simplicial. The power which lies in the assumption of periodicity is due chiefly to the fact that a general $(\mathfrak{R}, T)$ can, so to speak, be approximated by a simplicial one. Our immediate purpose is to show (3.1, 3.2) how this can be done in a manner especially suitable to the study of fixed points.

A *system* will mean a finite collection of point sets in $\mathfrak{R}$. The component sets of a system $\mathfrak{U}$ are the $\mathfrak{U}$-*vertices*. A system whose vertices are permuted among themselves by T will be called a T-*system*. The T^q-images of the vertices of a system $\mathfrak{U}$ form a system $T^q\mathfrak{U}$. The vertices of $\mathfrak{U}, T\mathfrak{U}, \cdots, T^{p-1}\mathfrak{U}$ taken together form a system denoted by $\sigma\mathfrak{U}$. Evidently $\sigma\mathfrak{U}$ is a T-system.

Let $\mathfrak{U}$ be a T-system. We write $\mathfrak{U} = \mathfrak{U}' \cup \mathfrak{U}''$ where $\mathfrak{U}'$ consists of the invariant $\mathfrak{U}$-vertices, $\mathfrak{U}''$ of the remaining. Evidently $\mathfrak{U}', \mathfrak{U}''$ are T-systems. We shall call $\mathfrak{U}$ *primitive* if each $\mathfrak{U}''$-vertex has p mutually exclusive T-images.

Let $\mathfrak{U}_\lambda, \mathfrak{U}_\mu$ be T-systems with $\mathfrak{U}_\lambda > \mathfrak{U}_\mu$. A projection $\pi: \mathfrak{U}_\lambda \to \mathfrak{U}_\mu$ will be called a T-projection if $\pi T = T\pi$. Evidently a T-projection $\mathfrak{U}_\lambda \to \mathfrak{U}_\mu$ carries $\mathfrak{U}'_\lambda$-vertices into $\mathfrak{U}'_\mu$-vertices.

(2.1) *Let* $\mathfrak{U}_\lambda, \mathfrak{U}_\mu$ *be* T-*systems with* $\mathfrak{U}_\lambda > \mathfrak{U}_\mu$. *If* $\mathfrak{U}_\mu$ *is primitive, there exists a* T-*projection* $\mathfrak{U}_\lambda \to \mathfrak{U}_\mu$.

PROOF. We may write $\mathfrak{U}_\lambda = \mathfrak{U}_\lambda^1 \cup \mathfrak{U}_\lambda^2$ where $\mathfrak{U}_\lambda^2$ consists of all $\mathfrak{U}_\lambda$-vertices which are contained in $\mathfrak{U}''_\mu$-vertices, $\mathfrak{U}_\lambda^1$ of the remaining $\mathfrak{U}_\lambda$-vertices. Evidently

$\mathfrak{U}_\lambda^1$, $\mathfrak{U}_\lambda^2$ are T-systems, refinements of $\mathfrak{U}_\mu'$ and $\mathfrak{U}_\mu''$, respectively. Moreover since each $\mathfrak{U}_\mu''$-vertex has p mutually exclusive images, the same is true of the $\mathfrak{U}_\lambda^2$-vertices, and hence $\mathfrak{U}_\lambda^2$ can be represented, without repetitions, as consisting of the T-images of a suitably chosen subsystem of its vertices, say $\mathfrak{U}_{\lambda 1}^2$, $\cdots$, $\mathfrak{U}_{\lambda s}^2$. Let π_2 be a projection of this subsystem into $\mathfrak{U}_\mu''$ and let π_2 be extended over $\mathfrak{U}_\lambda^2$ by the formula $\pi_2 T^q U_{\lambda i}^2 = T^q \pi_2 U_{\lambda i}^2$ $(q = 1, \cdots, p - 1)$. In this manner π_2 becomes a T-projection $\mathfrak{U}_\lambda^2 \to \mathfrak{U}_\mu''$. The $\mathfrak{U}_\mu'$-vertices being invariant, it is even simpler to define a T-projection $\pi_1 : \mathfrak{U}_\lambda^1 \to \mathfrak{U}_\mu'$. Taken together, π_1 and π_2 define a T-projection $\mathfrak{U}_\lambda \to \mathfrak{U}_\mu$.

We shall be particularly concerned with systems that are finite open coverings of $\mathfrak{R}$ (we shall simply call them coverings).

(2.2) *Every covering* $\mathfrak{U}$ *is refined by a* T*-covering.*

In fact the intersection of $\mathfrak{U}$, $T\mathfrak{U}$, $\cdots$, $T^{p-1}\mathfrak{U}$ is evidently a T-covering, refinement of $\mathfrak{U}$.

Definition. *A* T*-system* $\mathfrak{U}$ *will be said to satisfy condition* L_a *if* $\mathfrak{U}'$ *consists precisely of those* $\mathfrak{U}$*-vertices which meet* L; $\mathfrak{U}$ *satisfies* L_b *if all non-empty intersections of* $\mathfrak{U}'$*-vertices meet* L. *A* T*-covering which satisfies* L_a *and* L_b *will be called special.*

Let $\mathfrak{U}$ be a T-system, X its nerve. T induces in X a simplicial transformation T_X which is the identity or else is of period q, q a divisor of p. We denote by X_I the totality of X-simplexes which are invariant under T_X. In addition, denote by X_L the totality of X-simplexes which meet L. X_L is a closed subcomplex of X but in general X_I is not.

(2.3) *If the covering* $\mathfrak{U}$ *is primitive and special, then* $X_I = X_L$ *and* (X, T_X) *is primitive, regular.*

Proof. A non-invariant X-simplex E has at least one non-invariant vertex say U; the T-images of U (regarded as a $\mathfrak{U}$-vertex) are mutually exclusive sets. A relation $T_X^q E = E$ would imply T^q (kernel E) $=$ kernel E, which in turn would imply $T^q U \cap U \neq \emptyset$. We conclude that the T-images of $| E |$ are distinct, hence (X, T_X) is primitive. The vertices of an X_I-simplex E_I are permuted among themselves by T_X and since as $\mathfrak{U}$-vertices they have a non-empty intersection, each must meet L by the primitivity of $\mathfrak{U}$. Then condition L_a implies that E_I is vertex-wise invariant and (X, T_X) is therefore regular. Moreover, condition L_b implies that E_I meets L, hence $X_I \subset X_L$. On the other hand the vertices of an X_L-simplex E_L, since they meet L, are invariant by L_a. Hence E_L is invariant and $X_L \subset X_I$. Hence $X_I = X_L$.

3. **The existence of special coverings.** The next two propositions are of fundamental importance in what follows. Let us recall that $\dim \mathfrak{R} \leqq m$ if every covering is refined by a covering the dimension of whose nerve does not exceed m.

(3.1) *If* $\mathfrak{R}$ *is compact,* T *primitive, every covering is refined by a special primitive covering.*

(3.2) *If* $\mathfrak{R}$ *is compact,* T *primitive and if* $\dim \mathfrak{R} \leqq m$, *every covering* $\mathfrak{U}_\lambda$ *is refined by a special primitive covering* $\mathfrak{U}_\mu$ *such that* $\dim (X_\mu - X_{\mu I}) \leqq k$, $k = pm + p - 1$ ($X_\mu =$ *nerve* $\mathfrak{U}_\mu$).

We need give only the proof of (3.2) since it contains essentially the proof of (3.1). It will be convenient to say that a T-covering, $\mathfrak{U}$ satisfies L_c if, among the $\mathfrak{U}$-vertices which meet L, each contains a point of L not contained in any other. In addition let $\mathfrak{U}^*$ denote the covering $\{\text{St } U_i\}$ ($U_i \in \mathfrak{U}$).

We now establish the following propositions.

(A) Every covering $\mathfrak{U}_\lambda$ is refined by a primitive T-covering satisfying L_a, L_c.

(B) Every primitive T-covering, $\mathfrak{U}_\lambda$ satisfying L_a, L_c is refined by a covering $\mathfrak{U}_\mu$ of the same sort and such that $\dim X_\mu \leqq k$.

(C) For every primitive T-covering $\mathfrak{U}_\lambda$ satisfying L_a, L_c and with $\dim X_\lambda \leqq k$, there exists a special primitive covering $\mathfrak{U}_\mu$ such that $\mathfrak{U}_\mu > \mathfrak{U}_\lambda^*$, $\dim (X_\mu - X_{\mu I}) \leqq k$.

(3.2) follows from (A), (B), (C), the theorem that for compact $\mathfrak{R}$ there exists for given covering $\mathfrak{U}_\lambda$ a covering $\mathfrak{U}_\mu$ such that $\mathfrak{U}_\mu^* > \mathfrak{U}_\lambda$, and the obvious fact that $\mathfrak{U}_\lambda > \mathfrak{U}_\mu$ implies $\mathfrak{U}_\lambda^* > \mathfrak{U}_\mu^*$.

Proof of (A). Evidently an arbitrary point x of L possesses an invariant neighborhood $O(x)$ which is contained in some $\mathfrak{U}_\lambda$-vertex; in fact such a neighborhood is the intersection of the T-images of $U(x)$ where $U(x)$ is any $\mathfrak{U}_\lambda$-vertex containing x. Since $\mathfrak{R}$ is compact and L closed, there exists a finite set of neighborhoods $O(x)$, say $O_1, \cdots, O_s$ such that $L \subset \bigcup_i O_i$. Since T is primitive an arbitrary point y in the closed set $\mathfrak{R} - \bigcup_i O_i$ possesses a neighborhood $R(y)$ with p mutually exclusive T-images; evidently none of these T-images meets L. Using the separation properties of compact Hausdorff spaces and the continuity of T, we can choose $R(y)$ so as to have the further property that the system of T-images of $R(y)$ is a refinement of $\mathfrak{U}_\lambda$. Let $R_1, \cdots, R_t$ be a finite set of sets $R(y)$ such that $R - \bigcup_i O_i \subset \bigcup_j R_j$. The sets O_i together with the T-images of the sets R_j form a T-covering $\mathfrak{U}_\nu$, refinement of $\mathfrak{U}_\lambda$. Evidently $\mathfrak{U}_\nu$ is primitive and satisfies L_a.

Now suppose that $\mathfrak{U}_\mu$ is a covering which is a modification of $\mathfrak{U}_\nu$ obtained by replacing each O_i by an invariant open set Q_i with $Q_i \subset O_i$, $Q_i \cap L \neq \emptyset$. Then $\mathfrak{U}_\mu$, like $\mathfrak{U}_\nu$, is a refinement of $\mathfrak{U}_\lambda$, is primitive and satisfies L_a. We shall show that this modification can be carried out in such a way that the resulting $\mathfrak{U}_\mu$ also satisfies L_c. Choose distinct points $a_1, \cdots, a_s$ with $a_i \in O_i \cap L$. Then choose mutually exclusive invariant neighborhoods $A_1, \cdots, A_s$ of $a_1, \cdots, a_s$ such that for each i, $\bar{A}_i$ is contained in the intersection of those O's which contain a_i. Now consider the invariant open sets

$$Q_i = O_i - \bigcup \{\bar{A}_j \mid j \neq i\}.$$

It is easy to see that Q_i contains a_j if and only if $i = j$. The obvious relation $\bigcup_i Q_i \subset \bigcup_i O_i$ holds with the inclusion sign reversed. For, a point $x \in O_i$ either is not contained in any A_j $(j \neq i)$ in which case $x \in Q_i$, or else it is contained say in A_l $(l \neq i)$ in which case $x \in Q_l$. In either case $x \in \bigcup_i Q_i$. Thus we have shown that $\bigcup_i Q_i = \bigcup_i O_i$ and hence that the system $\mathfrak{U}_\mu$ obtained from $\mathfrak{U}_\nu$ by replacing O_i by Q_i $(i = 1, \cdots, s)$ is a covering, refinement of $\mathfrak{U}_\lambda$. Each vertex Q_i meets L since $a_i \in Q_i$. Hence the passage from $\mathfrak{U}_\nu$ to $\mathfrak{U}_\mu$ is of the type described above and we conclude that $\mathfrak{U}_\mu$ is primitive and satisfies L_a. Moreover, it is as easy to see that Q_i contains a_j if and only if $i = j$. Hence $\mathfrak{U}_\mu$ satisfies L_c.

Proof of (B). The hypothesis $\dim \mathfrak{R} \leqq m$ implies the existence of a covering $\mathfrak{U}_{\lambda'} > \mathfrak{U}_\lambda$, with $\dim X_{\lambda'} \leqq m$. Let $\mathfrak{U}_\nu = \sigma \mathfrak{U}_{\lambda'}$. Then $\dim X_\nu \leqq k$. Moreover $T\mathfrak{U}_{\lambda'} > T\mathfrak{U}_\lambda = \mathfrak{U}_\lambda$ implies $\mathfrak{U}_\nu = \sigma \mathfrak{U}_{\lambda'} > \mathfrak{U}_\lambda$. Write $\mathfrak{U}_\lambda = \mathfrak{U}_\lambda' \cup \mathfrak{U}_\lambda''$ where as always $\mathfrak{U}_\lambda'$ consists of the invariant $\mathfrak{U}_\lambda$-vertices. Write $\mathfrak{U}_\nu = \mathfrak{U}_\nu^1 \cup \mathfrak{U}_\nu^2$ where $\mathfrak{U}_\nu^1$ consists of all $\mathfrak{U}_\nu$-vertices which are subsets of $\mathfrak{U}_\lambda'$-vertices, $\mathfrak{U}_\nu^2$ the remaining. Evidently $\mathfrak{U}_\nu^1$, $\mathfrak{U}_\nu^2$ are T-systems and $\mathfrak{U}_\nu^1 > \mathfrak{U}_\lambda'$, $\mathfrak{U}_\nu^2 > \mathfrak{U}_\lambda''$. By (2.1) there exists a T-projection $\pi : \mathfrak{U}_\nu^1 \to \mathfrak{U}_\lambda'$. Write $\mathfrak{U}_\lambda' = \{U_{\lambda i}'\}$. Let O_i be the union of the vertices which constitute $\pi^{-1} U_{\lambda i}'$. Let $\mathfrak{U}_\nu^{11} = \{O_i\}$. Evidently $\mathfrak{U}_\nu^{11} > \mathfrak{U}_\lambda'$. The $\mathfrak{U}_\nu^{11}$-vertices are open invariant sets and their union is identical with the union of the $\mathfrak{U}_\nu'$-vertices. Hence $\mathfrak{U}_\nu^{11}$ and $\mathfrak{U}_\nu^2$ together form a T-covering $\mathfrak{U}_\mu$ such that $\mathfrak{U}_\mu > \mathfrak{U}_\lambda$, $\mathfrak{U}_\mu' > \mathfrak{U}_\nu^{11}$, $\mathfrak{U}_\mu'' = \mathfrak{U}_\nu^2 > \mathfrak{U}_\lambda''$. The primitivity of $\mathfrak{U}_\lambda$ and the relation $\mathfrak{U}_\mu'' > \mathfrak{U}_\lambda''$ imply that $\mathfrak{U}_\mu$ is primitive. We assert that $\mathfrak{U}_\mu$ satisfies L_a. Evidently a $\mathfrak{U}_\mu$-vertex which meets L must be an O_i, hence a $\mathfrak{U}_\mu'$-vertex. Conversely, every $\mathfrak{U}_\lambda'$-vertex meets L. For since $U_{\lambda i}'$ meets L (condition L_a for $\mathfrak{U}_\lambda$) $U_{\lambda i}'$ contains a point a_i of L not contained in $U_{\lambda j}'$, $j \neq i$ (condition L_c for $\mathfrak{U}_\lambda$). A $\mathfrak{U}_\nu$-vertex which contains a_i—there is at least one, say $U_{\nu i}$—cannot be a subset of any $\mathfrak{U}_\lambda''$-vertex (since those do not meet L) nor of any $\mathfrak{U}_\lambda'$-vertex other than $U_{\lambda j}'$ (otherwise some $U_{\lambda j}'$ $(j \neq i)$ would contain a_i). Hence $U_{\nu i} \subset U_{\lambda i}'$ and $a_i \in \bigcup (\pi^{-1} U_{\lambda i}') = O_i$, proving our assertion.—Now $\mathfrak{U}_\mu$ was formed from $\mathfrak{U}_\nu$ by applications of the operation of replacing a number of vertices by their union. Since this operation does not raise the dimension of the nerve, we have $\dim X_\mu \leqq \dim X_\nu \leqq k$. Finally, we showed in the proof of (A) how to pass from a primitive covering satisfying L_a to a primitive covering satisfying L_a, L_c by replacing the vertices of the first covering by suitable subsets of themselves. This operation does not raise the dimension of the nerve and hence if it is applied to $\mathfrak{U}_\mu$ it yields the required refinement of $\mathfrak{U}_\lambda$.

Proof of (C). Say $\mathfrak{U}_\lambda' = \{U_{\lambda i}'\}$. For each i, choose a point a_i contained in $L \cap U_{\lambda i}'$ but not in $U_{\lambda j}'$, $j \neq i$ (condition L_c). Choose invariant open sets $A_1, \cdots, A_s$ such that $a_i \in A_i \subset U_{\lambda i}'$, $A_i \cap U_{\lambda j}' = 0$ for $i \neq j$, and such that no A_i meets any $\mathfrak{U}_\lambda''$-vertex (recall that the $\mathfrak{U}_\lambda''$-vertices do not meet L). Evidently $A_i \cap A_j = \emptyset$ when $i \neq j$. For each i choose a set of A's by the following rule: A_j is in the ith set if and only if $U_{\lambda i}' \cap U_{\lambda j}' \neq \emptyset$. Let B_i be the union of the A's in the ith set and let

$$O_i = U_{\lambda i}' \cup B_i. \tag{3.3}$$

The sets O_i together with the $\mathfrak{U}_\lambda''$-vertices form a covering $\mathfrak{U}_\mu$ and since the O_i are invariant,

$$\mathfrak{U}_\mu' = \{O_i\}, \qquad \mathfrak{U}_\mu'' = \mathfrak{U}_\lambda''.$$

Hence $\mathfrak{U}_\mu$ is primitive and it is easy to see that $\mathfrak{U}_\mu > \mathfrak{U}_\lambda^*$. Each O_i meets L because $U'_{\lambda i}$ does; hence $\mathfrak{U}_\mu$ satisfies L_a. We assert that $\mathfrak{U}_\mu$ satisfies L_b. Suppose $J = O_q \cap O_r \cap \cdots \cap O_t \neq \emptyset$. If $U'_{\lambda q} \cap U'_{\lambda r} \cap \cdots \cap U'_{\lambda t} \neq \emptyset$, we have from (3.3) and the definition of the B_i,

$$A_q \cup A_r \cup \cdots \cup A_t \subset O_q \cap O_r \cap \cdots \cap O_t = J$$

so that J contains $a_q, \cdots, a_t$, hence meets L. If $U'_{\lambda q} \cap \cdots \cap U'_{\lambda r} = \emptyset$, it follows from (3.3) that J is the intersection of sets of B's and $\mathfrak{U}'_\lambda$-vertices, each set containing at least one B. Since each B is the union of mutually exclusive sets A_i, subsets of the corresponding sets $U'_{\lambda i}$, it follows that J is a union of A's, hence meets L and our assertion is proved.—We show finally that dim $(X_\mu - X_{\mu I}) \leqq k$. The existence of a non-invariant X_μ-simplex E implies a relation of the form

$$\text{(3.4)} \qquad \text{kernel } E = (U'_{\lambda i_0} \cup B_{i_0}) \cap \cdots \cap (U_{\lambda i_h} \cup B_{i_h}) \cap S \neq \emptyset$$

where S is an intersection of l $\mathfrak{U}''_\lambda$-vertices, $l \geqq 1$. The B's do not meet any $\mathfrak{U}''_\lambda$-vertex because the A's do not and hence (3.4) implies that $U'_{\lambda i_0} \cap \cdots \cap U'_{\lambda i_h} \cap S \neq \emptyset$. Hence $1 + h + l \leqq \dim X_\lambda \leqq k$. Hence $\dim E \leqq k$. This concludes the proof of (C).

4. **ρ-chains in a complex.** We assume throughout this section that $(\mathfrak{R}, T)$ is simplicial.

Let G be an abelian coefficient-group for chains and homologies in $\mathfrak{R}$. T induces in an obvious manner a chain-mapping which we shall also denote by T. We may thus regard T as an additive operator acting on chains over G and permutable with the boundary operator F. With regard to a 0-dimensional chain C^0, we recall that $\mathrm{KI}(TC^0) = \mathrm{KI}(C^0)$, a relation which holds equally well if T is a simplicial mapping of $\mathfrak{R}$ into some other complex. It will be convenient in this appendix to take $\mathfrak{R}$ augmented, so that *C^0 is a cycle if and only if* $\mathrm{KI}(C^0) = 0$. Thus if E is a vertex, $E - TE$ is a cycle over arbitrary G, $E + TE + \cdots + T^{p-1}$ a cycle over $\mathfrak{J}_p$ (the group of residues mod p), $+E$ is never a cycle. Boundaries of 1-chains are cycles and, in a connected complex, all 0-cycles bound.

The operators

$$\sigma = 1 + T + \cdots + T^{p-1}, \qquad \delta = 1 - T$$

(where 1 is the identity operator) bear useful reciprocal relations to each other and play an important part in what follows. We shall also denote these operators by ρ and $\bar{\rho}$ agreeing that ρ may stand for σ, $\bar{\rho}$ for δ or *vice versa*, but that the meaning of ρ and $\bar{\rho}$ shall remain fixed in any given discussion. Note that in any case $\rho\bar{\rho} = \bar{\rho}\rho = 1 - T^p = 0$ (the annihilator). ρ and $\bar{\rho}$ are of course permutable with T and **F**.

DEFINITION. *Chains which are annulled by* ρ *will be called* ρ*-chains. Chains which are annulled* mod $\mathfrak{R}_I$ *by* ρ *will be called* ρI*-chains.*

We shall assume in the remainder of this section that the simplicial $(\mathfrak{R}, T)$ *is primitive.*

(4.1) *A necessary and sufficient condition that a chain* C *be a* ρI*-chain is that there exist a chain* A *such that* $C = \bar{\rho}A$ mod $\mathfrak{R}_I$.

PROOF. The sufficiency is implied by the relation $\rho\bar{\rho} = 0$. To prove the necessity suppose first that C is a δI-chain. In any case since $(\mathfrak{R}, T)$ is primitive, the oriented simplexes corresponding to the non-invariant simplexes of $\mathfrak{R}$ may be represented without repetition by

$$\pm E_i^h, \pm TE_i^h, \cdots, \pm T^{p-1}E_i^h, \qquad i = 1, \cdots, \alpha_h.$$

Hence we may write $C = \sum \sum g_{ai}T^aE_i$ where the g's are elements of G and a ranges over the residues mod p. The relation $\delta C = 0$ mod $\mathfrak{R}_I$ implies

$$\sum \sum g_{ai}(T^a - T^{a+1})E_i = \sum \sum (g_{ai} - g_{a-1,i})T^aE_i = 0.$$

Hence $g_{1i} = g_{2i} = \cdots = g_{pi}$ so that $C = \sigma A$ mod $\mathfrak{R}_I$ where $A = \sum_i g_{1i}E_i$. Suppose now that C is a σI-chain. Since

$$\sigma C = \sum \sum g_{ai}(\sigma E_i) \text{ mod } \mathfrak{R}_I,$$

the relation $\sigma C = 0$ mod $\mathfrak{R}_I$ implies that

$$\sum_a g_{ai} = 0 \tag{4.2}$$

for each i. We wish to show the existence of a chain $A = \sum \sum x_{ai}T^aE_i$ such that $\sum \sum g_{ai}T^aE_i = \delta(\sum \sum x_{ai}T^aE_i)$. This last expression equals $\sum \sum (x_{ai} - x_{a+1,i})T^aE_i$ and therefore it is sufficient to show for each value of i the existence of a solution in the variables $x_{1i}, \cdots, x_{pi}$ of the system

$$x_{1i} - x_{pi} = g_{1i}, \quad x_{2i} - x_{1i} = g_{2i}, \cdots, x_{pi} - x_{p-1,i} = g_{pi}.$$

In view of (4.2) such a solution is given explicitly by

$$x_{ai} = g_{1i} + g_{2i} + \cdots + g_{ai}, \qquad a = 1, \cdots, p.$$

We shall use a subscript I with the symbol of a chain to show that the chain is in $\mathfrak{R}_I$.

(4.3) *If* $G = \mathfrak{J}_p$, *all chains in* $\mathfrak{R}_I$ *are* ρ*-chains.* For, $\delta C_I = C_I - C_I = 0$ and $\sigma C_I = pC_I = 0$.

(4.4) *If* $G = \mathfrak{J}_p$ *and* C *is a chain,* $\rho C \subset \mathfrak{R} - \mathfrak{R}_I$. For in any case we can write $C = A + A_I$ where $A \subset \mathfrak{R} - \mathfrak{R}_I$. Then using (4.3), $\rho C = \rho A \subset \mathfrak{R} - \mathfrak{R}_I$.

(4.5) *If* $G = \mathfrak{J}_p$, *a necessary and sufficient condition that* C *be a* ρ*-chain is that it be expressible in the form* $\bar{\rho}A + A_I$.

PROOF. Sufficiency follows from (4.3) and the relation $\rho\bar{\rho} = 0$. To prove necessity write $C = B + B_I$ where $B \subset \mathfrak{R} - \mathfrak{R}_I$. Since C and B_I are ρ-chains

(4.3) so is B. In fact B is evidently a ρI-chain, hence expressible in the form $\bar{\rho}D + D_I$ (4.1). Hence C has the stated form.

(4.6) *If* $G = \mathfrak{Z}_p$, *a necessary and sufficient condition that* C *be a* ρ*-chain in* $\mathfrak{R} - \mathfrak{R}_I$ *is that* C *be of the form* $\bar{\rho}A$ (4.4, 4.5).

5. **Special homologies in a complex. Definition.** Assume that $(\mathfrak{R}, T)$ is simplicial. A ρ*-cycle* is simply a ρ-chain which is a cycle. If a ρ-cycle γ is the boundary of a ρ-chain, write $\gamma \sim_\rho 0$. If, modulo $\mathfrak{R}_I$, a ρI-cycle γ is the boundary of a ρI-chain, write $\gamma \sim_\rho 0 \bmod \mathfrak{R}_I$. These homologies, which we shall refer to as ρ and ρI-homologies have the same algebraic properties as ordinary homologies. Note that a chain which is identically zero may be regarded as a ρ-cycle and as such, it is ρ-homologous to zero. A similar remark applies modulo $\mathfrak{R}_I$.

We assume during the remainder of this section that $(\mathfrak{R}, T)$ is simplicial, primitive, regular. Regularity implies that the boundaries of chains in $\mathfrak{R}_I$ are in $\mathfrak{R}_I$.

(5.1) *If* $G = \mathfrak{Z}_p$ *and if* $\bar{\rho}C^h + C_I^h$ *is a cycle (hence a* ρ*-cycle by* 4.5*), then* $\bar{\rho}C$ *and* C_I *are cycles (hence* ρ*-cycles). If* $\bar{\rho}C + C_I \sim_\rho 0$*, then* $\bar{\rho}C \sim_\rho 0$ *and* $C_I \sim_\rho 0$.

PROOF. We have

$$0 = \mathrm{F}(\bar{\rho}C + C_I) = \mathrm{F}\bar{\rho}C + \mathrm{F}C_I. \tag{5.1a}$$

Since $\mathrm{F}\bar{\rho}C = \bar{\rho}\mathrm{F}C \subset \mathfrak{R} - \mathfrak{R}_I$ (4.4) and $\mathrm{F}C_I \subset \mathfrak{R}_I$, (5.1a) implies $\mathrm{F}\bar{\rho}C = 0$, $\mathrm{F}C_I = 0$.—Suppose now that $\bar{\rho}C + C_I \sim_\rho 0$. This implies a relation $\mathrm{F}(\bar{\rho}A + A_I) = \bar{\rho}C + C_I$ (using 4.5). This can be written $\bar{\rho}(\mathrm{F}A - C) = C_I - \mathrm{F}A_I$. The left side of this last equation is in $\mathfrak{R} - \mathfrak{R}_I$, the right in $\mathfrak{R}_I$. Hence both chains vanish. The result then follows if we recall that all chains in $\mathfrak{R}_I$ are ρ-chains. Incidentally, we have also proved

(5.2) *Let* $G = \mathfrak{Z}_p$ *and let* γ *be a cycle in* $\mathfrak{R}_I$, γ' *a* ρ*-cycle in* $\mathfrak{R} - \mathfrak{R}_I$. *Then* $\gamma \sim_\rho 0$ *if and only if* $\gamma \sim 0$ *in* $\mathfrak{R}_I$; $\gamma' \sim_\rho 0$ *if and only if* $\gamma' \sim_\rho 0$ *in* $\mathfrak{R} - \mathfrak{R}_I$.

(5.3) *Let* $\gamma^h = \bar{\rho}C^h$, $\gamma^{h-1} = \rho C^{h-1}$ *be* ρI*- and* $\bar{\rho}I$*-cycles such that* $\gamma^{h-1} = FC^h \bmod \mathfrak{R}_I$. *If* $\gamma^h \sim_\rho 0 \bmod \mathfrak{R}_I$, *then* $\gamma^{h-1} \sim_{\bar{\rho}} 0 \bmod \mathfrak{R}_I$.

The proof is essentially the same as that of

(5.4) *Let* $G = \mathfrak{Z}_p$ *and let* $\gamma^h = \bar{\rho}C^h + C_I^h$ *and* $\gamma^{h-1} = \rho C^{h-1} + C_I^{h-1}$ *be* ρ*- and* $\bar{\rho}$*-cycles such that* $\gamma^{h-1} = \mathrm{F}C^h$. *If* $\gamma^h \sim_\rho 0$, *then* $\gamma^{h-1} \sim_{\bar{\rho}} 0$.

PROOF. The relation $\gamma^h \sim_\rho 0$ implies by (5.1) and (5.2) that $\rho C^{h-1} \sim_\rho 0$ in $\mathfrak{R} - \mathfrak{R}_I$. By (4.6) this implies a relation of the form $\mathrm{F}\bar{\rho}B = \bar{\rho}C^h$. Let $A = C^h - \mathrm{F}B$. Then $\bar{\rho}A = \bar{\rho}C^h - \mathrm{F}\bar{\rho}B = 0$ so that A is a $\bar{\rho}$-chain. Also $\mathrm{F}A = \mathrm{F}C^h = \gamma^{h-1}$. Hence $\gamma^{h-1} \sim_{\bar{\rho}} 0$.

(5.5) *Let* $G = \mathfrak{Z}_p$ *and assume that* $\mathfrak{R}_I = 0$. *Let* C^0 *be a 0-dimensional chain. Recall that* $\bar{\rho}C^0$ *is a cycle (see* 4*), obviously a* ρ*-cycle. If* $\bar{\rho}C^0 \sim_\rho 0$, *then* $\mathrm{KI}(C^0) = 0$.

PROOF. The relation $\bar{\rho}C \sim_\rho 0$ implies a relation $\mathrm{F}\bar{\rho}A = \bar{\rho}C$ (using 4.5 and the hypothesis $\mathfrak{R}_I = 0$). Write

$$\mathrm{F}A = C^0 + B. \tag{5.5a}$$

Then $\bar{\rho}B = 0$ and hence we may write $B = \rho D$ (4.5). This implies, since $G = \mathfrak{Z}_p$, that $\mathrm{KI}(B) = 0$. Since also $\mathrm{KI}(\mathrm{F}A) = 0$, (5.5a) implies $\mathrm{KI}(C^0) = 0$.

(5.6) *Let* $G = \mathfrak{J}_p$, *and let* C *be a cycle. If* $C \sim 0$, *then* $\rho C \sim_{\bar\rho} 0$. For, $FA = C$ implies $F\rho A = \rho C$ and ρA is a $\bar\rho$-chain since $\bar\rho\rho = 0$.

6. **ρ-homology groups in a complex.** We assume in this section that $(\mathfrak{R}, T)$ is simplicial.

We shall denote the additive groups of ρ- and ρI-homology classes of $(\mathfrak{R}, T)$ over G by

$$\mathfrak{H}^h_\rho(\mathfrak{R}, T; G), \qquad \mathfrak{H}^h_{\rho I}(\mathfrak{R}, T; G).$$

Let Γ^h_ρ be a ρ-homology class of dimension h. If one ρ-cycle in Γ^h_ρ is homologous to 0 in the *ordinary* sense, so is every ρ-cycle in Γ^h_ρ. The totality of classes Γ^h_ρ whose ρ-cycles are ~ 0 is a subgroup of $\mathfrak{H}^h_\rho$ which we denote by $\mathfrak{B}^h(\mathfrak{R}, T; G)$. The corresponding subgroup of $\mathfrak{H}^h_{\rho I}$ is $\mathfrak{B}^h_{\rho I}(\mathfrak{R}, T; G)$.

Let γ^h be a ρ-cycle, element of Γ^h_ρ. By (4.5) we may write

$$\gamma^h = \bar\rho C + C_I. \tag{6.1}$$

Suppose now that Γ^h_ρ has the property that for at least one γ^h in Γ^h_ρ, there exists a representation (6.1) in which C *is a cycle.* The totality of classes Γ^h_ρ with this property is evidently a subgroup of $\mathfrak{H}^h_\rho$; we denote it by $\mathfrak{K}^h_\rho(\mathfrak{R}, T; G)$. The corresponding subgroup $\mathfrak{K}^h_{\rho I}$ of $\mathfrak{H}^h_{\rho I}$ consists of the ρI-homology classes which contain ρI-cycles of the form $\bar\rho C$ when C is a cycle mod $\mathfrak{R}_I$.

Let γ^h be an ordinary cycle, element of the ordinary homology class Γ^h. Since $\bar\rho\rho = 0$, $\rho\gamma_h$ is a $\bar\rho$-cycle. Suppose Γ^h has the property that for at least one γ^h in Γ^h, $\rho\gamma^h \sim_{\bar\rho} 0$. The totality of classes Γ^h with this property evidently is a subgroup of $\mathfrak{H}^h$; we denote it by $\mathfrak{H}^h_{(\rho)}(\mathfrak{R}, T; G)$. The corresponding subgroup of $\mathfrak{H}^h(\mathfrak{R}, \mathfrak{R}_I, G)$ is $\mathfrak{H}^h_{(\rho I)}(\mathfrak{R}, T; G)$.

We assume during the remainder of this section that $(\mathfrak{R}, T)$ *is simplicial, primitive, regular and that* $G = \mathfrak{J}_p$.

We shall define certain homomorphic mappings α and β of the groups $\mathfrak{H}^h_\rho(\mathfrak{R}, T, \mathfrak{J}_p)$.

α. By (4.5), a ρ-cycle γ^h in Γ^h_ρ has a representation $\gamma^h = \bar\rho C^h + C^h_I$. Let $\gamma^{h-1} = FC^h$. γ^{h-1} is a $\bar\rho$-cycle; for since $\bar\rho C^h$ is a cycle (5.1), $\bar\rho\gamma^{h-1} = F\bar\rho C^h = 0$. We assert that the class $\Gamma^{h-1}_{\bar\rho}$ containing γ^{h-1} is independent of the choice of γ^h in Γ^h_ρ and of C^h in the representation $\bar\rho C^h + C^h_I$ for γ^h. For suppose $\gamma^h \sim_\rho \gamma'^h$, $\gamma'^h = \bar\rho C'^h + C'_I$, $\gamma'^{h-1} = FC'^h$. It follows from (5.4) that $\gamma'^{h-1} - \gamma^{h-1} \sim_{\bar\rho} 0$ proving our assertion. Thus the correspondence $\alpha: \Gamma^h_\rho \to \Gamma^{h-1}_{\bar\rho}$ is a homomorphic mapping of $\mathfrak{H}^h_\rho$ into a subgroup $\mathfrak{H}^{h-1}_{\bar\rho}$. Since $\gamma^{h-1} \sim 0$, the image of $\mathfrak{H}^h_\rho$ under α is a subgroup of $\mathfrak{B}^{h-1}_{\bar\rho}$. *We assert that* α *covers* $\mathfrak{B}^{h-1}_{\bar\rho}$; that is,

$$\alpha\mathfrak{H}^h_\rho = \mathfrak{B}^{h-1}_{\bar\rho}. \tag{6.2}$$

For, let $\Gamma^{h-1}_{\bar\rho}$ be an element of $\mathfrak{B}^{h-1}_{\bar\rho}$ and let γ^{h-1} be a $\bar\rho$-cycle in $\Gamma^{h-1}_{\bar\rho}$. We have $\gamma^{h-1} \sim 0$, say $FC^h = \gamma^{h-1}$. Since $F\bar\rho C^h = \bar\rho\gamma^{h-1} = 0$, $\gamma^h = \bar\rho C^h$ is a cycle. In fact γ^h is a ρ-cycle and we have $\alpha\Gamma^h_\rho = \Gamma^{h-1}_{\bar\rho}$ where Γ^h_ρ is the ρ-homology class containing γ^h.

The kernel of α *(as acting on* $\mathfrak{H}_\rho^h$*) is* $\mathfrak{K}_\rho^h$. For, $\alpha\Gamma_\rho^h = 0$ implies relations

$$\gamma^h = \bar{\rho}C + C_I, \qquad FC \sim_\rho 0, \qquad \gamma^h \in \Gamma_\rho^h.$$

The second of these implies a relation $F(\rho A + A_I) = FC$. Then $B = C - \rho A - A_I$ is a cycle. By (4.5) $\bar{\rho}B = \bar{\rho}C$ so that $\gamma^h = \bar{\rho}B + C_I$. Hence $\Gamma_\rho^h \in \mathfrak{K}_\rho^h$. Conversely, every element of $\mathfrak{K}_\rho^h$ is evidently carried by α into the zero of $\mathfrak{B}_{\bar{\rho}}^{h-1}$.

REMARK. Strictly speaking, the symbol α should bear indices ρ, h corresponding to the ρ, h of the group on which it acts. In general, however, these indices need not be written.

β. The elements of a ρ-homology class Γ_ρ^h are contained in a uniquely determined ordinary homology class Γ^h and the correspondence $\beta: \Gamma_\rho^h \to \Gamma^h$ is evidently a homomorphism of $\mathfrak{H}_\rho^h$ into $\mathfrak{H}^h$. The kernel of β is $\mathfrak{B}_\rho^h$. We assert that $\beta\mathfrak{H}^h = \mathfrak{H}_{(\rho)}^h$. For consider an element Γ_ρ^h of $\mathfrak{H}_\rho^h$. A ρ-cycle γ^h, member of Γ_ρ^h satisfies $\rho\gamma^h = 0$, hence (trivially) $\rho\gamma^h \sim_{\bar{\rho}} 0$. Hence $\Gamma_\rho^h \in \mathfrak{H}_{(\rho)}^h$. Conversely, let $\Gamma_{(\rho)}^h$ be an element of $\mathfrak{H}_{(\rho)}^h$. To show that $\Gamma_{(\rho)}^h$ has a pre-image under β, it is sufficient to show that $\Gamma_{(\rho)}^h$ contains a ρ-cycle. In any case $\Gamma_{(\rho)}^h$ contains a cycle γ^h such that $\rho\gamma^h \sim_{\bar{\rho}} 0$. This implies a relation $F(\rho C + C_I) = \rho\gamma^h$. The relations $\rho\gamma^h \subset \mathfrak{R} - \mathfrak{R}_I$, $F\rho C \subset \mathfrak{R} - \mathfrak{R}_I$ (4.4) and $FC_I \subset \mathfrak{R}_I$ imply $F\rho C = \rho\gamma^h$. Let $\gamma'^h = \gamma^h - F(\rho C + C_I)$. Then $\rho\gamma'^h = 0$, $\gamma'^h = \gamma^h - FC$, $\gamma'^h \sim \gamma^h$, so that γ'^h is the desired ρ-cycle in $\Gamma_{(\rho)}^h$.

(6.3) THEOREM. *For a simplicial regular primitive* $(\mathfrak{R}, T)$,

(6.3a) $$\mathfrak{H}_\rho^h(\mathfrak{R}, T; \mathfrak{J}_p)/\mathfrak{K}_\rho^h(\mathfrak{R}, T; \mathfrak{J}_p) = \mathfrak{B}_{\bar{\rho}}^{h-1}(\mathfrak{R}, T; \mathfrak{J}_p),$$

(6.3b) $$\mathfrak{H}_\rho^h(\mathfrak{R}, T; \mathfrak{J}_p)/\mathfrak{B}_\rho^h(\mathfrak{R}, T; \mathfrak{J}_p) = \mathfrak{H}_{(\rho)}^h(\mathfrak{R}, T; \mathfrak{J}_p).$$

These formulas hold for arbitrary G *if* ρ *is everywhere replaced by* ρI.

PROOF. (6.3a) and (6.3b) follow from the properties of α, β established above. The proof of these formulas for the ρI-homology groups depends on the properties of the corresponding homomorphisms α_I, α_I of $\mathfrak{H}_{\rho I}^h$. We shall not discuss α_I, β_I further than to remark that for their definition no restriction need be put on G.

7. A decomposition. We assume in this section that $(\mathfrak{R}, T)$ is simplicial, primitive, regular and that $G = \mathfrak{J}_p$.

Let Γ_ρ^h be an element of $\mathfrak{H}_\rho^h$ with the property that $\gamma^h \subset \mathfrak{R} - \mathfrak{R}_I$ for at least one element γ^h in Γ_ρ^h. The totality of classes Γ_ρ^h with this property is a subgroup of $\mathfrak{H}_\rho^h$ which we denote by $\mathfrak{O}_\rho^h$. By (4.6), each element Γ_ρ^h of $\mathfrak{O}_\rho^h$ contains a ρ-cycle of the form ρC.

Let Γ_ρ^h be an element of $\mathfrak{H}_\rho^h$ with the property that $\gamma^h \subset \mathfrak{R}_I$ for at least one γ^h in Γ_ρ^h. The totality of classes Γ_ρ^h with this property is a subgroup of $\mathfrak{H}_\rho^h$ which we denote by $\mathfrak{H}_{0\rho}^h$. From (5.2), $\mathfrak{H}_{0\rho}^h$ *may be regarded as being identical with* $\mathfrak{H}^h(\mathfrak{R}_I, \mathfrak{J}_p)$.

As an immediate consequence of (5.1), the groups $\mathfrak{O}_\rho^h$, $\mathfrak{H}_\rho^h$ furnish a decomposition

(7.1) $$\mathfrak{H}_\rho^h = \mathfrak{O}_\rho^h \times \mathfrak{H}_{0\rho}^h .$$

Consider the mappings α. A glance at the definition of α reveals that $\alpha\mathfrak{H}_{0\rho}^h = 0$. Hence

(7.2) $$\alpha\mathfrak{O}_\rho^h = \alpha\mathfrak{H}_\rho^h = \mathfrak{B}_{\bar\rho}^{h-1}.$$

Let ζ be the projection $\mathfrak{O}_\rho^h \times \mathfrak{H}_{0\rho}^h \to \mathfrak{H}_{0\rho}^h$, and let $\kappa = \zeta\alpha$ (α followed by ζ). Then

(7.3) $$\kappa\mathfrak{H}_\rho^h = \kappa\mathfrak{O}_\rho^h \subset \mathfrak{H}_{0\bar\rho}^{h-1}.$$

REMARK. It will simplify matters in the simplicial case if from now on we make no distinction between a cycle and its homology class (ρ- or ordinary). Thus we may regard $\mathfrak{H}$ as composed of cycles, $\mathfrak{H}_\rho$ of ρ-cycles, $\mathfrak{R}_\rho$ of ρ-cycles of the form $\bar\rho C + C_I$ where C is a cycle, $\mathfrak{O}_\rho$ of ρ-cycles of the form ρC, $\mathfrak{H}_{(\rho)}^h$ of cycles γ such that $\rho\gamma \sim_\rho 0$. Conversely, every h-dimensional ρ-cycle may be regarded as an element of $\mathfrak{H}_\rho^h$ and so on.

(7.4) $$\mathfrak{H}_{0\rho}^h \cap \mathfrak{B}_\rho^h \subset \kappa\mathfrak{H}_\rho^{h+1}.$$

PROOF. An element γ in $\mathfrak{H}_{0\rho}^h \cap \mathfrak{B}_\rho^h$ is a cycle in $\mathfrak{N}_I$ which is ~ 0 (in $\mathfrak{N}$) say $\mathrm{F}C = \gamma$. Then $\mathrm{F}\bar\rho C = \bar\rho\gamma = 0$. Hence $\bar\rho C$ is a ρ-cycle, hence an element of $\mathfrak{H}_\rho^{h+1}$ and from the definition of α, ζ, κ,

$$\kappa(\bar\rho C) = \zeta\alpha(\bar\rho C) = \zeta\gamma = \gamma.$$

8. **Projections of ρ-homology classes.** Let $\mathfrak{U}$ be a T-system, X its nerve, T_X the transformation induced in X by T. Let $\rho_X = 1 - T_X$ or $1 + T_X + \cdots + T_X^{p-1}$ according as $\rho = \delta$ or σ. We may think of ρ_X as the operator induced in X by ρ. Now let $\mathfrak{B}$ be a second T-system, Y its nerve. Suppose that $\mathfrak{U} > \mathfrak{B}$ and that π is a T-projection $\mathfrak{U} \to \mathfrak{B}$. The chain-mapping induced by π will also be denoted by π. The relation $\pi T = T\pi$ implies that

(8.1) $$\pi\rho_X = \rho_Y\pi.$$

An important consequence of (8.1) is that π carries ρ_X-chains and homologies into ρ_Y-chains and homologies.

In general, induced operators such as ρ_X, ρ_Y may simply be denoted by ρ, since it will always be clear in the context what the meaning of ρ is to be. We may, accordingly, describe (8.1) simply by saying that ρ permutes with π.

(8.2) *Let $\mathfrak{U}$, $\mathfrak{B}$ be T-systems with nerves X, Y and such that $\mathfrak{B}$ is primitive and $\mathfrak{U} > \mathfrak{B}$. Let π_1, π_2 be T-projections $\mathfrak{U} \to \mathfrak{B}$. If γ is a ρ-cycle in X, $\pi_1\gamma \sim_\rho \pi_2\gamma$ in Y.*

PROOF. We may suppose that the passage from π_1 to π_2 can be effected by redefining π_1 over the T-images of a single $\mathfrak{U}$-vertex. For it will be seen that in any case the passage from π_1 to π_2 can be obtained by a finite number of such steps. Suppose then that π_1 differs from π_2 only as concerns the T-images of U.

Assume first that U is contained in a non-invariant $\mathfrak{V}$-vertex. Then since $\mathfrak{V}$ is primitive, the images $U_q = T^q U$ are mutually exclusive. Let $\pi_i U_q = V_q^i$ ($i = 1, 2$). We define an additive operator $\mathfrak{D}$ over X-chains as follows. An oriented simplex E either has just one vertex among the images of U (regarded now as vertices of X), or has none. In the latter case define $\mathfrak{D}E = 0$. In the former, suppose $E = (U_h S)$ where S is a simplex with no vertex U_i. Then define $\mathfrak{D}E = (V_h^1 V_h^2 S')$ where $S' = \pi_1 S = \pi_2 S$. Let the definition of $\mathfrak{D}$ be extended additively to all X-chains. It is a straightforward verification to show that the formulas

$$\text{(8.3)} \qquad \mathrm{F}\mathfrak{D} = \pi_2 - \pi_1 - \mathfrak{D}\mathrm{F}, \qquad \mathfrak{D}T = T\mathfrak{D}$$

hold for individual simplexes, hence for chains. From the first of these formulas it follows that $\pi_2\gamma - \pi_1\gamma = \mathrm{F}\mathfrak{D}\gamma$ and from the second that $\rho\mathfrak{D}\gamma = \mathfrak{D}\rho\gamma = 0$. Hence $\pi_2\gamma \sim_\rho \pi_1\gamma$.—We must still dispose of the case in which U is not a subset of any non-invariant $\mathfrak{V}$-simplex. In this case $\pi_1 U$ and $\pi_2 U$ are invariant so that $\pi_i U_q = V_i$ say ($i = 1, 2$). Suppose the X-simplex E has just one vertex among the images of U, say $E = (U_h S)$. Then take $\mathfrak{D}E = (V_1 V_2 S)$. In all other cases take $\mathfrak{D}E = 0$. The formulas (8.3) again hold, but in their verification it is necessary to examine the case in which the vertices of E include more than one image of U, say for example $E = (U_h U_k S)$. Here $\mathfrak{D}E = 0$ by definition, hence $\mathrm{F}\mathfrak{D}E = 0$. Also $\pi_i E$ equals $(V_i V_i S)$ which is degenerate, hence zero, and $\mathfrak{D}\mathrm{F}E$ vanishes by cancellation and definition.

9. **ρ-homology groups in a compact space.** We assume in this section that $\mathfrak{R}$ is compact, T primitive.

Let $\Sigma = \{\mathfrak{U}_\lambda\}$ be the totality of primitive special coverings of $\mathfrak{R}$ and let T_λ be the transformation induced by T in X_λ = nerve $\mathfrak{U}_\lambda$. Each (X_λ, T_λ) is primitive, regular (2.3). Primitivity implies that each T_λ is of period p. Notice that by (1.1) $X_{\lambda I}$ is the fixed-point set of X_λ (not to be confused with the fixed-point set L of $\mathfrak{R}$).

By (3.1) Σ is a directed set relative to ordering by refinement, and is cofinal with the totality of all finite open coverings of $\mathfrak{R}$. Hence Σ is adequate for carrying the ordinary homology theory of $\mathfrak{R}$. We shall show now that Σ carries a ρ-homology theory for $(\mathfrak{R}, T)$.

Let $\mathfrak{U}_\lambda$, $\mathfrak{U}_\mu$ be coverings in Σ with $\mathfrak{U}_\lambda > \mathfrak{U}_\mu$. Since $\mathfrak{U}_\mu$ is primitive there exists by (2.1) a T-projection $\pi_\mu^\lambda : \mathfrak{U}_\lambda \to \mathfrak{U}_\mu$. Since π_μ^λ is permutable with ρ and F, thus carrying ρ-cycles into ρ-cycles and preserving ρ-homologies, π_μ^λ induces a mapping

$$\text{(9.1)} \qquad \bar{\pi}_\mu^\lambda : \mathfrak{H}_\rho^h(X_\lambda, T_\lambda; G) \to \mathfrak{H}_\rho^h(X_\mu, T_\mu; G)$$

and it is a consequence of (8.2) that $\bar{\pi}_\mu^\lambda$ is independent of the particular choice of the T-projection π_μ^λ. Thus the groups $\mathfrak{H}_\rho^h(X_\lambda, T_\lambda; G)$ and associated mappings $\bar{\pi}_\mu^\lambda$ form an inverse system invariantly related to $(\mathfrak{R}, T)$. Let

$$\mathfrak{H}_\rho^h(\mathfrak{R}, T; G) = \lim \{\mathfrak{H}_\rho^h(X_\lambda, T_\lambda; G); \bar{\pi}_\mu^\lambda\}.$$

The elements of $\mathfrak{H}_\rho^h$ may be regarded as ρ-homology classes of ρ-cycles of $(\mathfrak{R}, T)$, a ρ-cycle γ^h being a collection $\{\gamma_\lambda^h\}$ where γ_λ^h is a (ρ, X_λ)-cycle (ρ-cycle in X_λ) and where $\mathfrak{U}_\lambda$, $\mathfrak{U}_\mu \in \Sigma$, $\mathfrak{U}_\lambda > \mathfrak{U}_\mu$ imply

$$\pi_\mu^\lambda \gamma_\lambda^h \sim_\rho \gamma_\mu^h \tag{9.2}$$

$\gamma^h \sim_\rho 0$ if and only if $\gamma_\lambda^h \sim_\rho 0$ for each λ. As in the simplicial case, frequently we shall not distinguish between a ρ-cycle and its ρ-homology class. Let A be a subset of $\mathfrak{R}$ and let $X_{\lambda A}$ be the subcomplex of X_λ consisting of the X_λ-simplexes whose kernels meet A. If $\gamma_\lambda^h \subset X_{\lambda A}$ for each λ and if the relations (9.2) can be expressed in the form $\mathrm{F}C = \pi_\mu^\lambda \gamma_\lambda^h - \gamma_\mu^h$, where C is a ρ-chain in $X_{\lambda A}$, we call γ^h a *ρ-cycle of A*.

It is clear that the totality Σ is a topologically definite entity uniquely determined by $\mathfrak{R}$ and T. It follows that the groups $\mathfrak{H}_\rho^h(\mathfrak{R}, T; G)$ are topological invariants of $(\mathfrak{R}, T)$.

Concerning the subgroups of $\mathfrak{H}^h$ and $\mathfrak{H}_\rho^h$, it can immediately be verified that the homomorphism (9.1) carries $\mathfrak{B}_h^\rho(X_\lambda, T_\lambda; G)$ into a subgroup of $\mathfrak{B}_\rho^h(X_\lambda, T_\lambda; G)$ and similarly for $\mathfrak{K}_\rho^h(X_\lambda, T_\lambda; G)$. Thus $\mathfrak{H}_\rho^h(\mathfrak{R}, T; G)$ admits subgroups

$$\mathfrak{B}_\rho^h(\mathfrak{R}, T; G) = \lim \{\mathfrak{B}_\rho^h(X_\lambda, T_\lambda; G); \bar{\pi}_\mu^\lambda\},$$

$$\mathfrak{K}_\rho^h(\mathfrak{R}, T; G) = \lim \{\mathfrak{K}_\rho^h(X_\lambda, T_\lambda; G); \bar{\pi}_\mu^\lambda\}.$$

In the same way, the homomorphism $\mathfrak{H}^h(X_\lambda, G) \to \mathfrak{H}^h(X_\mu, G)$ induced by π_μ^λ carries $\mathfrak{H}_{(\rho)}^h(X_\lambda, T_\lambda; G)$ into a subgroup of $\mathfrak{H}_{(\rho)}^h(X_\lambda, T_\lambda; G)$. Thus $\mathfrak{H}^h(\mathfrak{R}, G)$ admits a subgroup

$$\mathfrak{H}_{(\rho)}^h(\mathfrak{R}, T; G) = \lim \{\mathfrak{H}_{(\rho)}^h(X_\lambda, T_\lambda; G); \bar{\pi}_\mu^\lambda\}.$$

Concerning the corresponding groups of *relative* cycles,—the relation $\pi_\mu^\lambda T = T\pi_\mu^\lambda$ implies that π_μ^λ carries invariant X_λ-simplexes into invariant X_μ-simplexes, hence $\pi_\mu^\lambda X_{\lambda I} \subset X_{\mu I}$. Thus $(\rho I, X_\lambda)$-chains are carried by π_μ^λ into $(\rho I, X_\mu)$-chains and ρI-homologies are preserved. This leads to inverse systems of groups $\mathfrak{H}_{\rho I}^h(X_\lambda, T_\lambda; G)$ etc., based on an easy modification of (8.2) for relative cycles. The relations $X_{\lambda I} = X_{\lambda L}$ of (2.3) imply that elements in the resulting limit-groups are ρ-cycles modulo L and these limit-groups are therefore properly denoted by $\mathfrak{H}_{\rho L}^h(\mathfrak{R}, L; G)$, etc.

(9.3) Remark. A topology in the coefficient group G will of course lead to a topology in $\mathfrak{H}_\rho^h$, $\mathfrak{H}_{\rho L}^h$, etc. In what follows, however, groups are to be considered as being discrete.

(9.4) Remark. Like the groups $\mathfrak{H}_\rho^h$, the groups $\mathfrak{B}_\rho^h$, $\mathfrak{K}_\rho^h$, $\mathfrak{H}_{(\rho)}^h$ and the corresponding groups $\mathfrak{H}_{\rho L}^h$ etc. are topological invariants of $(\mathfrak{R}, T)$.

(9.5) Remark. Suppose that $\dim \mathfrak{R} \leqq m$. Then it is a consequence of (3.2) that Σ can be replaced in the preceding discussion by $\Sigma_k = \{\mathfrak{U}_\lambda\}$, the totality of primitive special coverings $\mathfrak{U}_\lambda$ such that $\dim (X_\lambda - X_{\lambda I}) \leqq k$.

Let α_λ, β_λ denote the mappings α, β (defined in 6) for (X_λ, T_λ). That α_λ, β_λ actually exist is guaranteed by the regularity and primitivity of (X_λ, T_λ).

It is a consequence of the definition of α, β and the fact that T-projections permute with ρ, F that for $\mathfrak{U}_\lambda$, $\mathfrak{U}_\mu$ in Σ with $\mathfrak{U}_\lambda > \mathfrak{U}_\mu$ and T-projection π_μ^λ

$$(9.6) \qquad \bar{\pi}_\mu^\lambda \alpha_\lambda = \alpha_\mu \bar{\pi}_\mu^\lambda, \qquad \bar{\pi}_\mu^\lambda \beta_\lambda = \beta_\mu \bar{\pi}_\mu^\lambda .$$

If $\gamma^h = \{\gamma_\lambda^h\}$ is a ρ-cycle, element of $\mathfrak{H}_\rho^h(\mathfrak{R}, T; \mathfrak{J}_p)$, (9.6) implies that $\{\alpha_h \gamma_\lambda^h\}$ is a $\bar{\rho}$-cycle; denote it by $\alpha\gamma^h$. Evidently α is a homomorphic mapping

$$\mathfrak{H}_\rho^h(\mathfrak{R}, T; \mathfrak{J}_p) \rightarrow \mathfrak{B}_{\bar{\rho}}^{h-1}(\mathfrak{R}, T; \mathfrak{J}_p).$$

Using (9.1) and the fact that the groups $\mathfrak{H}_h^\rho(\mathfrak{R}, T; \mathfrak{J}_p)$ are finite (and hence compact) it follows from the general theory of inverse systems that α actually covers $\mathfrak{B}_{\bar{\rho}}^{h-1}$ (see II, 13.6). Moreover, the fact that kernel $\alpha_\lambda = \mathfrak{K}_\rho^h(X_\lambda, T_\lambda; \mathfrak{J}_p)$ for every $\mathfrak{U}_\lambda$ in Σ implies that the kernel of α is $\mathfrak{K}_\rho^h(\mathfrak{R}, T; \mathfrak{J}_p)$. Similar remarks lead to a homomorphism $\beta: \mathfrak{H}_\rho^h(\mathfrak{R}, T; \mathfrak{J}_p) \rightarrow \mathfrak{H}_{(\rho)}^h(\mathfrak{R}, T; \mathfrak{J}_p)$ with kernel $\mathfrak{B}_\rho^h(\mathfrak{R}, T; \mathfrak{J}_p)$ and therefore the formulas (6.3a, 6.3b) *hold for every compact* $\mathfrak{R}$ *and primitive* T.

In connection with the corresponding relations between the groups $\mathfrak{H}_{\rho L}^h$ etc., we merely remark that they hold if G is any compact group or any field.

10. **Homologies in** L. We assume in this section that $\mathfrak{R}$ is compact, T primitive and that $G = \mathfrak{J}_p$.

We shall now see the importance of the relations $X_{\lambda I} = X_{\lambda L}$ (2.3) in connection with questions about homological properties of L. Let π_μ^λ be a T-projection ($\mathfrak{U}_\lambda$, $\mathfrak{U}_\mu \in \Sigma$). We have already remarked that $\pi_\mu^\lambda X_{\lambda I} \subset X_{\mu I}$. Consequently if we recall that the elements of $\mathfrak{H}_{0\rho}^h(X_\lambda, T_\lambda; \mathfrak{J}_p)$ are cycles in $X_{\lambda I}$,—that $\mathfrak{H}_{0\rho}^h(X_\lambda, T_\lambda; \mathfrak{J}_p)$ is in fact identical with $\mathfrak{H}^h(X_{\lambda I}, \mathfrak{J}_p)$, we see that the induced mapping (9.1) carries $\mathfrak{H}_{0\rho}^h(X_\lambda, T_\lambda; \mathfrak{J}_p)$ into a subgroup of $\mathfrak{H}_{0\rho}^h(X_\mu, T_\mu; \mathfrak{J}_p)$. Let

$$\mathfrak{H}_{0\rho}^h(\mathfrak{R}, T; \mathfrak{J}_p) = \lim \{\mathfrak{H}_{0\rho}^h(X_\lambda, T_\lambda; \mathfrak{J}_p); \bar{\pi}_\mu^\lambda\}.$$

The relations $X_{\lambda I} = X_{\lambda L}$ now imply that $\mathfrak{H}_{0\rho}^h$ is the group of ordinary homology classes of L (see VII, 5.4). That is,

$$(10.1) \qquad \mathfrak{H}_{0\rho}^h(\mathfrak{R}, T; \mathfrak{J}_p) \cong \mathfrak{H}^h(L, \mathfrak{J}_p).$$

Moreover, since X_λ-chains of the form ρA are carried by π_μ^λ into X_μ-chains of the same form, it follows from (4.6) and (5.2) that $\bar{\pi}_\mu^\lambda \mathfrak{D}^h(X_\lambda, T_\lambda; \mathfrak{J}_p) \subset \mathfrak{D}_\mu^h(X_\lambda, T_\lambda; \mathfrak{J}_p)$. Let

$$\mathfrak{D}_\rho^h(\mathfrak{R}, T; \mathfrak{J}_p) = \lim \{\mathfrak{D}_\rho^h(X_\lambda, T_\lambda; \mathfrak{J}_p); \bar{\pi}_\mu^\lambda\}.$$

Evidently if γ^h is a ρ-cycle, element of $\mathfrak{D}_\rho^h(\mathfrak{R}, T; \mathfrak{J}_p)$, γ_λ^h may be taken as a ρ-cycle in $X_\lambda - X_{\lambda I}$. If $X_{\lambda J}$ is the totality of X_λ-simplexes whose kernels meet $\mathfrak{R} - L$, we have $X_\lambda - X_{\lambda I} = X_\lambda - X_{\lambda L} \subset X_{\lambda J}$ and consequently (5.2) and the relations $\gamma_\lambda^h \subset X_\lambda - X_{\lambda I}$ imply that γ^h is a ρ-cycle of $\mathfrak{R} - L$ (or at least is ρ-homologous to such a cycle).

The decomposition (7.1) which holds in each (X_λ, T_λ), implies the decomposition

$$(10.2) \qquad \mathfrak{H}_\rho^h = \mathfrak{D}_\rho^h \times \mathfrak{H}_{0\rho}^h \quad \text{for} \quad (\mathfrak{R}, T; \mathfrak{J}_p);$$

and the relations (7.2) and (9.1) imply

(10.3) $$\alpha\mathfrak{D}_\rho^h = \alpha\mathfrak{H}_\rho^h = \mathfrak{B}_{\bar\rho}^{h-1} \quad \text{for} \quad (\mathfrak{R}, T; \mathfrak{J}_p).$$

Hence if we denote by ζ the mapping $\mathfrak{D}_\rho^h \times \mathfrak{H}_{0\rho}^h \to \mathfrak{H}_{0\rho}^h$ by projection, and by κ the mapping $\zeta\alpha$, we have

(10.4) $$\kappa\mathfrak{H}_\rho^h = \kappa\mathfrak{D}_\rho^h \subset \mathfrak{H}_{0\bar\rho}^{h-1} \quad \text{for} \quad (\mathfrak{R}, T; \mathfrak{J}_p).$$

Denoting by κ_λ the mapping of $\mathfrak{H}_\rho^h(X_\lambda, T_\lambda; \mathfrak{J}_p)$ induced by κ, it is obvious that κ_λ is precisely the mapping κ defined in 7 for simplicial $(\mathfrak{R}, T)$ with $\mathfrak{R} = X_\lambda$, $T = T_\lambda$. Moreover (9.6) implies

(10.5) $$\bar\pi_\mu^\lambda\kappa_\lambda = \kappa_\mu\bar\pi_\mu^\lambda, \qquad \lambda > \mu,\ \pi_\mu^\lambda \text{ a } T\text{-projection.}$$

(10.6) *If* $\mathfrak{D}_\rho^{h-1} = 0$, *the kernel of* κ, *as applied to* $\mathfrak{H}_\rho^h$, *is* $\mathfrak{K}_\rho^h$.

Proof. The hypothesis $\mathfrak{D}_\rho^{h-1} = 0$ implies that $\alpha\mathfrak{H}_\rho^h \subset \mathfrak{H}_{0\bar\rho}^{h-1}$. Hence $\zeta\alpha\mathfrak{H}_\rho^h = \alpha\mathfrak{H}_\rho^h$ so that κ has the same kernel as α, namely $\mathfrak{K}_\rho^h$.

(10.7) *If* $\mathfrak{D}_\rho^{h-1} = 0$ *and* $\mathfrak{H}_{(\rho)}^h = \mathfrak{H}^h$, *then* κ *transforms* $\mathfrak{D}_\rho^h$ *isomorphically.*

Proof. By (10.6) the kernel of κ, acting on $\mathfrak{D}_\rho^h$, is $\mathfrak{D}_\rho^h \cap \mathfrak{K}_\rho^h$. Now from (4.6, 5.2) and the definition of $\mathfrak{D}_\rho^h$, $\mathfrak{K}_\rho^h$, it is easy to see that an element in $\mathfrak{K}_\rho^h \cap \mathfrak{D}_\rho^h$ is of the form $\gamma^h = \{\bar\rho C_\lambda^h\}$, where $C^h = \{C_\lambda^h\}$ is an ordinary cycle. Thus C^h is an element of $\mathfrak{H}^h$, consequently an element of $\mathfrak{H}_{(\rho)}^h$. This implies that $\bar\rho C^h \sim_\rho 0$ (see definition of $\mathfrak{H}_{(\rho)}^h$), hence that γ^h is the zero of $\mathfrak{D}_\rho^h$. Thus the kernel of κ vanishes.

(10.8) *If* $\mathfrak{H}^{h-1} = 0$, *then* κ, *acting on* $\mathfrak{H}_\rho^h$, *covers* $\mathfrak{H}_{0\bar\rho}^{h-1}$—*that is,* $\kappa\mathfrak{H}_\rho^h = \mathfrak{H}_{0\bar\rho}^{h-1}$.

Proof. We have only to show that a given cycle γ^{h-1} in $\mathfrak{H}_{0\bar\rho}^{h-1}$ has a pre-image in $\mathfrak{H}_\rho^h$. We may write $\gamma^{h-1} = \{\gamma_\lambda^{h-1}\}$ where $\gamma_\lambda^{h-1} \subset X_{\lambda I}$. Since ρ-homologies imply ordinary homologies, γ^{h-1} may be regarded as an element of $\mathfrak{H}^{h-1}$, hence $\gamma^{h-1} \sim 0$. Hence each γ_λ^{h-1} is an element of $\mathfrak{H}_{0\bar\rho}^{h-1}(X_\lambda, T_\lambda; \mathfrak{J}_p) \cap \mathfrak{B}_{\bar\rho}^{h-1}(X_\lambda, T_\lambda; \mathfrak{J}_p)$ and so has a pre-image in $\mathfrak{H}_\rho^h(X_\lambda, T_\lambda; \mathfrak{J}_p)$ under κ_λ (7.2). Let Δ_λ denote the totality of these pre-images of γ_λ^{h-1}. If π_μ^λ is a T-projection, the relations (10.5) imply that $\pi_\mu^\lambda\Delta_\lambda \subset \Delta_\mu$. The sets Δ_λ being finite (hence compact), it follows that the limit of the inverse system $\{\Delta_\lambda; \pi_\mu^\lambda\}$ is not empty (I, 39.6). The limit elements are evidently pre-images of γ^{h-1} under κ.

(10.9) *Let* A, B *be invariant sets with* $A \subset B$. *Let* $\mathfrak{H}_\rho^h[B]$ *be the subgroup of* $\mathfrak{H}_\rho^h$ *consisting of all* ρ*-homology classes* Γ_ρ^h *such that* Γ_ρ^h *contains at least one* ρ*-cycle of* B. *If every (ordinary) cycle of* A *is homologous to zero in* B, *then* $\mathfrak{H}_{\bar\rho}^{h-1}[A] \subset \alpha\mathfrak{H}_\rho^h[B]$.

The proof is essentially the same as that of the preceding proposition; it is based on an obvious modification of (7.2).

(10.10) *Assume that* $\mathfrak{R}$ *is finite-dimensional,* p *a prime. Assume further that* $\mathfrak{H}^h(\mathfrak{R}, \mathfrak{J}_p) = 0$ *for* $h > n$ *while* $\mathfrak{H}^n(\mathfrak{R}, \mathfrak{J}_p)$ *is cyclic of order* p. *Then*

$$\mathfrak{H}_{(\rho)}^n = \mathfrak{H}_\rho^n \cong \mathfrak{H}^n, \quad \mathfrak{B}_\rho^n = 0; \quad \mathfrak{H}_\rho^h = 0,\ h > n, \qquad (\text{for } (\mathfrak{R}, T; \mathfrak{J}_p)).$$

Proof. Suppose $\dim \mathfrak{R} \leqq m$; evidently $n \leqq m$. According to (9.5) the definition of the ρ-homology groups can be based on Σ_k, $k = pm + p - 1$, in place of Σ. Recall that the λ-coordinate of a cycle γ^h of $\mathfrak{O}_\rho^h$ is an h-dimensional (ρ, X_λ)-cycle in $X_\lambda - X_{\lambda I}$. Hence if $h > k$ and $\mathfrak{U}_\lambda$ is in Σ_k, $\gamma_\lambda^h = 0$. We conclude that $\mathfrak{O}_\rho^h = 0$, $h > k$. Now let l be an integer larger than k such that $l - k$ is even. We have $\mathfrak{O}_\rho^l = 0$. The relation $\mathfrak{H}^{l-1} = 0$ implies that $\mathfrak{H}_{\bar\rho}^{l-1} = \mathfrak{B}_{\bar\rho}^{l-1} = \alpha\mathfrak{O}_\rho^l = \alpha 0 = 0$, $\mathfrak{O}_{\bar\rho}^{l-1} = 0$. From this in turn we infer that $\mathfrak{O}_\rho^{l-2} = \mathfrak{H}_\rho^{l-2} = 0$, finally that $\mathfrak{O}_{\bar\rho}^{n+1} = \mathfrak{H}_{\bar\rho}^{n+1} = 0$ and hence

$$\mathfrak{B}_\rho^n = \alpha\mathfrak{H}_{\bar\rho}^{n+1} = 0. \tag{10.11}$$

From (6.3b) we conclude that $\mathfrak{H}_\rho^n \cong \mathfrak{H}_{(\rho)}^n$. It remains only to show that $\mathfrak{H}_{(\rho)}^n = \mathfrak{H}^n$.

Let γ^n be an n-cycle $\not\sim 0$. Then $T\gamma \sim x\gamma$ (x a nonzero integer), $T^2\gamma \sim x^2\gamma, \cdots, \gamma = T^p\gamma \sim x^p\gamma$. Hence $x^p = 1 \bmod p$ and since p is prime, $x = 1 \bmod p$. Thus for each λ, $T\gamma_\lambda \sim \gamma_\lambda$, hence $\bar\rho\gamma_\lambda \sim 0$. We assert that $\{\bar\rho\gamma_\lambda\}$ is a ρ-cycle. We have only to show that if π_μ^λ is a T-projection ($\mathfrak{U}_\lambda > \mathfrak{U}_\mu$), $\pi_\mu^\lambda\bar\rho\gamma_\lambda \sim_\rho \bar\rho\gamma_\mu$. We have $\pi_\mu^\lambda\gamma_\lambda - \gamma_\mu \sim 0$; hence by (5.6) $\bar\rho(\pi_\mu^\lambda\gamma_\lambda - \gamma_\mu) \sim_\rho 0$ as required. Now $\bar\rho\gamma \sim 0$ implies that $\bar\rho\gamma$ is an element of the subgroup $\mathfrak{B}_\rho^n$ of $\mathfrak{H}_\rho^n$. Hence (10.11) implies $\bar\rho\gamma \sim_\rho 0$ so that $\bar\rho\gamma$ is a nonzero element of $\mathfrak{H}_{(\rho)}^n$. Thus $\mathfrak{H}_{(\rho)}^n \neq 0$ and being a subgroup of a cyclic group $\mathfrak{H}^n$ of prime order, $\mathfrak{H}^n$ must be identical with $\mathfrak{H}_{(\rho)}^n$.

11. **The property** A_n. We shall say that $\mathfrak{R}$ possesses the property A_n over G if for any non-empty set A in $\mathfrak{R}$, all n-cycles of $\mathfrak{R} - A$ (over G) are homologous to zero. Example: an n-sphere (G arbitrary).

(11.1) THEOREM. *Let q be a prime factor of p. Suppose that $\mathfrak{R}$ is compact and finite-dimensional and that $\mathfrak{H}^h(\mathfrak{R}, \mathfrak{J}_q) = 0$ for $h > n$ while $\mathfrak{H}^n(\mathfrak{R}, \mathfrak{J}_q)$ is cyclic of order q. Suppose further that $\mathfrak{R}$ possesses property A_n over $\mathfrak{J}_q$. Then L is nowhere dense in $\mathfrak{R}$.*

PROOF. Let $s = p/q$. Then T^s is of prime period q, hence primitive. Since points fixed under T are fixed under T^s, it will be sufficient to show that the fixed-point set L_s of T^s is nowhere dense. By (10.10), $\mathfrak{H}_\rho^n(\mathfrak{R}, T^s; \mathfrak{J}_q)$ is cyclic of order q, and $\mathfrak{B}_\rho^n(\mathfrak{R}, T^s; \mathfrak{J}_q) = 0$. Let Γ_ρ^n be an arbitrary ρ-homology class, element of $\mathfrak{H}_\rho^n(\mathfrak{R}, T^s, \mathfrak{J}_q)$. From (10.1) and the properties of $\mathfrak{O}_\rho$, $\mathfrak{H}_\rho$, Γ_ρ^n contains a ρ-cycle γ which has a representation $\gamma = \gamma_1 + \gamma_2$ where γ_1 is a ρ-cycle of $\mathfrak{R} - L_s$ and γ_2 is an ordinary cycle of L_s. Since $\mathfrak{R} - L_s$ is open and non-empty, property A_n and the fact that γ_2 is a cycle of $\mathfrak{R} - (\mathfrak{R} - L_s)$ imply $\gamma_2 \sim 0$. Suppose now that L_s contains a non-empty open subset L_s'. Then γ_1 is a cycle of $\mathfrak{R} - L_s'$ and hence $\gamma_1 \sim 0$. Thus $\gamma \sim 0$ and hence $\Gamma_\rho^n \in \mathfrak{B}_\rho^n(\mathfrak{R}, T^s; \mathfrak{J}_q)$, hence it is the zero of $\mathfrak{H}_\rho^n(\mathfrak{R}, T^s; \mathfrak{J}_q)$. Since Γ_ρ^n was arbitrary this implies that $\mathfrak{H}_\rho^n(\mathfrak{R}, T^s; \mathfrak{J}_q) = 0$, which is impossible.

For locally Euclidean $\mathfrak{R}$, the preceding theorem follows from a theorem of Newman [c]. Concerning a generalization of Newman's theorem see P. A. Smith [e].

12. **Homological spheres.** A compact finite-dimensional space will be called a *homological n-sphere* over G if, when augmented, it is n-cyclic over G

(III, 21). We shall regard the empty set as a homological (-1)-sphere over every G.

(12.1) THEOREM. *Let T be a transformation operating in $\mathfrak{R}$, of period $p = q^a$ with $a \geqq 1$, q a prime. If $\mathfrak{R}$ is a homological n-sphere over $\mathfrak{J}_q$, then the fixed-point set L is a homological r-sphere over $\mathfrak{J}_q$, $-1 \leqq r \leqq n$.*

PROOF. We shall first establish the theorem for the case $a = 1$, $p = q$. In any case we may assume that $L \neq \emptyset$.

Let $\rho_0, \rho_1, \cdots$ stand alternately for δ, σ beginning with $\rho_0 = \delta$ (σ would do just as well). Let r be the dimensional index of the first vanishing group in the sequence

$$(12.2) \qquad \mathfrak{O}^n_{\rho_n}, \mathfrak{O}^{n-1}_{\rho_{n-1}}, \cdots \qquad (\mathfrak{R}, T; \mathfrak{J}_p).$$

We assert that $\mathfrak{O}^0_{\rho_0} = 0$ so that the definition of r has meaning. Consider an element of $\mathfrak{O}^0_{\rho_0}$. It contains a ρ_0-cycle of the form $\gamma^0 = \{\bar{\rho}_0 A^0_\lambda\}$. Since $\mathfrak{H}^0 = 0$, X_λ is connected (VII, 11.15). Hence if E is an X_λ-vertex, we have $E \sim 0$ mod $X_{\lambda I}$. It follows that $A^0_\lambda \sim 0$ mod $X_{\lambda I}$, say $\mathrm{F}B^1_\lambda = A^0_\lambda + A^0_{\lambda I}$. Then $\mathrm{F}\bar{\rho}_0 B^1_\lambda = \bar{\rho}_0 A^0_\lambda = \gamma^0_\lambda$ and so $\gamma^0_\lambda \sim 0$ for each λ proving our assertion.

We shall show that $\mathfrak{H}^r_{0\rho_r}$ is cyclic of order p while $\mathfrak{H}^i_{0\rho_i} = 0$ for $i \neq r$. This will imply, since the groups $\mathfrak{H}^0_{0\rho_h}$ are identical with the ordinary homology groups of L (see 10.1) that L is a homological r-sphere.

Note first that

$$(12.3) \qquad \mathfrak{B}^i_{\rho_i} = \mathfrak{H}^i_{\rho_i}, \qquad i \neq n,$$

since $\mathfrak{H}^i = 0$ for $i \neq n$. By (10.10), $\mathfrak{H}^n_{\rho_n}$ is cyclic of order p. Assume for the moment that $r < n$. Then $\mathfrak{O}^n_{\rho_n} \neq 0$ and hence $\mathfrak{O}^n_{\rho_n} = \mathfrak{H}^n_{\rho_n}$. We have $\alpha\mathfrak{O}^n_{\rho_n} = \mathfrak{B}^{n-1}_{\rho_{n-1}}$ (10.3) and consequently $\mathfrak{B}^{n-1}_{\rho_{n-1}}$ is cyclic, possibly zero. If $r < n - 1$, we have $0 \neq \mathfrak{O}^{n-1}_{\rho_{n-1}} \subset \mathfrak{H}^{n-1}_{\rho_{n-1}}$ so that (12.3) implies $\mathfrak{B}^{n-1}_{\rho_{n-1}} \neq 0$. Hence $\mathfrak{B}^{n-1}_{\rho_{n-1}}$ is of order p and hence so is $\mathfrak{O}^{n-1}_{\rho_{n-1}}$. On repeating this argument we conclude that the groups (12.2) with dimensional index exceeding r are cyclic of order p. Incidentally, we have shown that

$$(12.4) \qquad \mathfrak{H}^i_{\rho_i} = \mathfrak{O}^i_{\rho_i}, \qquad i = n, n-1, \cdots, r+1.$$

Next we assert that the remaining groups of (12.2) vanish. In any case $\mathfrak{O}^r_{\rho_r} = 0$ by definition of r. Then using (12.3), $\mathfrak{O}^{r-1}_{\rho_{r-1}} \subset \mathfrak{H}^{r-1}_{\rho_{r-1}} = \mathfrak{B}^{r-1}_{\rho_{r-1}} = \alpha\mathfrak{O}^r_{\rho_r} = 0$. Replacing r by $r - 1$ and so on, our assertion is established.

By (10.8) we have

$$(12.5) \qquad \mathfrak{H}^i_{0\rho_i} = \kappa\mathfrak{O}^{i+1}_{0\rho_{i+1}} = \kappa 0 = 0, \qquad i < r.$$

Using the fact that $\mathfrak{H}^i_{\rho_i} = 0$ for $i > n$ (10.10) we see that (12.5) holds equally well for $i > n$. Moreover $\mathfrak{H}^i_{0\rho_i}$ vanishes for $i = r + 1, \cdots, n$. This follows from (12.4) and the decomposition $\mathfrak{H}_\rho = \mathfrak{O}_\rho \times \mathfrak{H}_{0\rho}$ (10.2). We have now shown that $\mathfrak{H}^i_{\rho_i} = 0$ when $i \neq r$.

It remains to show that $\mathfrak{H}^r_{\rho r}$ is cyclic of order p. If $r = n$, this is implied by the relation $\mathfrak{O}^n_{\rho n} = 0$, the fact $\mathfrak{H}^n_{\rho n}$ is cyclic of order p, and the decomposition $\mathfrak{H}_\rho = \mathfrak{O}_\rho \times \mathfrak{H}_{0\rho}$. If $r < n$ it is implied by the fact that $\mathfrak{O}^{r+1}_{\rho r+1}$ is cyclic of order p and that κ carries $\mathfrak{O}^{r+1}_{\rho r+1}$ isomorphically onto $\mathfrak{H}^r_{0\rho r}$ (10.7, 10.8). This concludes the proof for $a = 1$.

Now assume that $a > 1$ and that the theorem has been proved for $p = q^b$, $b < a$. The transformation T^s, $s = q^{a-1}$, is of prime period q. Hence its fixed-point set L_s is a homological r-sphere, $r \leqq n$. Now T transforms L_s into itself and the transformation T' induced in L_s is either the identity or it is of period q^c where $c < a$. In the first case, $L = L_s$ and the theorem is established. In the second case the fixed-point set L' of T' is homological r'-sphere, $r' \leqq r$. But it will be seen on a moment's reflection that $L' = L$ and the theorem is therefore established.

Denoting by α_h the homomorphism α acting on the ρ- and $\bar{\rho}$-homology groups of dimension h, let $\alpha_{rn} = \alpha_r\alpha_{r+1} \cdots \alpha_n$. Let $\rho_0, \rho_1, \cdots$ stand alternately for δ, σ as in the proof of (12.1).

(12.6) Let $\mathfrak{R}$ be a homological n-sphere over $\mathfrak{J}_p$, p a prime, so that L is a homological r-sphere over $\mathfrak{J}_p$, $-1 \leqq r \leqq n$ (12.1). Then

$$\text{(12.6a)} \qquad \alpha_{s+1n}\mathfrak{H}^n_{\rho n}(\mathfrak{R}, T; \mathfrak{J}_p) = \mathfrak{H}^s_{\rho s}(\mathfrak{R}, T; \mathfrak{J}_p), \qquad s = r, \cdots, n,$$

$$\text{(12.6b)} \qquad \zeta\alpha_{r+1n}\mathfrak{H}^n_{\rho n}(\mathfrak{R}, T; \mathfrak{J}_p) = \mathfrak{H}^r(L, \mathfrak{J}_p).$$

PROOF. It follows from (6.3a) and (12.3) that $\alpha_i\mathfrak{H}^i_{\rho i} = \mathfrak{H}^{i-1}_{\rho i-1}$ $(i = r + 1, \cdots, n)$; this implies (12.6a).—It was shown in the proof of (12.1) that $\kappa\mathfrak{O}^{r+1}_{\rho r+1} = \mathfrak{H}^r(L, \mathfrak{J}_p)$. By (12.4) $\mathfrak{O}$ can here be replaced by $\mathfrak{H}^{r+1}_{\rho r+1}$. Since $\kappa = \zeta\alpha_{r+1}$ we have $\zeta\alpha_{r+1}\mathfrak{H}^{r+1}_{\rho r+1} = \mathfrak{H}^r(L, \mathfrak{J}_p)$ which with (12.6a) gives (12.6b).—We have referred here to the proof of (12.1) which was carried out on the assumption that $r \geqq 0$. It can readily be verified however that (12.6ab) hold equally well when $r = -1$.

13. The existence of fixed points. We shall say that $\mathfrak{R}$ is *acyclic* over G if for every compact set A in $\mathfrak{R}$ there exists a compact set B with $A \subset B$ such that cycles of A over G are ~ 0 in B. (Example: Euclidean n-space, G arbitrary.)

(13.1) THEOREM. *Let $\mathfrak{R}$ be a finite-dimensional locally compact space and assume that $\mathfrak{R}$ is acyclic over $\mathfrak{J}_q$, q a prime. Every transformation of period $p = q^a$ operating in $\mathfrak{R}$ admits at least one fixed point.*

PROOF. Assume first that $a = 1$, $p = q$. Suppose that $\dim \mathfrak{R} \leqq m$ and let $k = pm + p - 1$. Let B_0 be a non-empty compact set in $\mathfrak{R}$. Then $A_0 = \sigma B_0$ is invariant, compact, non-empty. We may choose a compact B_1 containing A_0 and such that cycles of A_0 are ~ 0 in B_1. Evidently B_1 can be replaced by the compact *invariant* set $A_1 = \sigma B_1$. Proceed in this way to obtain compact invariant sets $A_0, \cdots, A_{k+1}$ such that

$$\emptyset \neq A_0 \subset A_1 \subset \cdots \subset A_{k+1}$$

and such that cycles of A_i are ~ 0 in A_{i+1}. Let $S = \bar{A}_{k+1}$. It is easy to see that in the topology of S, regarded as a subspace of $\mathfrak{R}$, it is still true that cycles of A_i are ~ 0 in $\dot{A}_{i+1}$. T induces a transformation of period p in S and since dim $S \leqq m$, the ρ-homology groups for (S, T) can be based on the system Σ_k of coverings of S (9.5) where $k = pm + p - 1$.

Now assume that $L = 0$. The relations $X_{\lambda I} = X_{\lambda L}$ (2.3) then imply that $X_{\lambda I} = 0$, dim $X_\lambda \leqq k$ for each $\mathfrak{U}_\lambda$ in Σ_k. Hence $\mathfrak{H}_\rho^{k+1}(S, T; \mathfrak{J}_p) = 0$. Let α_h denote the mapping α as applied to cycles of dimension h. Repeated application of (10.9) gives

$$\mathfrak{H}_\rho^0[A_0] = \alpha_1\alpha_2 \cdots \alpha_{k+1}\mathfrak{H}_\nu^{k+1}[A_k] \qquad \text{(for } (S, T; \mathfrak{J}_p))$$

where $\nu = \rho$ or $\bar{\rho}$ according as k is odd or even. The relation $\mathfrak{H}_\nu^{k+1} = 0$ then implies

$$\mathfrak{H}_\rho^0(A_0) = 0 \qquad \text{(for } (S, T; \mathfrak{J}_p)). \tag{13.2}$$

Now choose a definite $\mathfrak{U}_\mu$ in Σ_k. By the theory of inverse systems of groups, there exists a $\mathfrak{U}_\lambda$ in Σ_k such that the image in X_μ under a T-projection π_μ^λ of an arbitrary 0-dimensional (ρ, X_λ)-cycle in $X_{\lambda A_0}$ (the subcomplex of X_λ-simplexes meeting A_0) will be the μ-coordinate of a ρ-cycle of A_0 (VI, 3.12), hence by (13.2) will be $\sim_\rho 0$. But now let C_λ^0 be an X_λ-chain consisting of a single X_λ-vertex in A_0, with nonzero coefficient. Then $\mathrm{KI}(C_\lambda^0) \neq 0$. But $\mathrm{KI}(\bar{\rho}C_\lambda^0) = 0$ since $G = \mathfrak{J}_p$. This implies that $\bar{\rho}C_\lambda^0$ is a (ρ, X_λ)-cycle in $X_{\lambda A_0}$ and hence by what was said above, $\pi_\mu^\lambda(\bar{\rho}C_\lambda^0) \sim_\rho 0$. Hence by (5.5) $\mathrm{KI}(\pi_\mu^\lambda C_\lambda^0) = 0$. But the value of $\mathrm{KI}(C_\lambda^0)$ is unaltered by projection, so $\mathrm{KI}(C_\lambda^0) = 0$ which is impossible. This contradiction completes the proof for the case $a = 1$.

We need not insist on the details for the case $a > 1$. The proof there rests on (i) the theorem that for $a = 1$, the acyclic property for $\mathfrak{R}$ implies the acyclic property for L and (ii) an induction on a. The proof of (i) involves the mappings α and κ and closely resembles parts of the proof of (12.1). Moreover the induction in that proof indicates how (ii) is to be carried out (see Smith [f]).

14. **The index.** Consider an $(\mathfrak{R}, T)$ with $\mathfrak{R}$ compact, $L = \emptyset$ and p a prime. We associate with each 0-dimensional ρ-cycle γ^0 a unique element of $\mathfrak{J}_p$ in the following way. Since by (2.3) $X_{\lambda I} = 0$ for each $\mathfrak{U}_\lambda$ in Σ, we may write $\gamma_\lambda^0 = \bar{\rho}A_\lambda$. Let $\eta\gamma_\lambda^0 = \mathrm{KI}(A_\lambda)$. It follows from (5.5) that $\eta\gamma_\lambda^0$ is independent of the choice of A_λ, in the expression $\bar{\rho}A_\lambda$ for γ_λ^0. Moreover $\eta\gamma_\lambda^0$ is independent of λ. To prove this it is sufficient to show that $\eta\gamma_\lambda^0 = \eta\gamma_\mu^0$ whenever $\lambda > \mu$. We have

$$\eta\pi_\mu^\lambda\gamma_\lambda = \eta\pi_\mu^\lambda\bar{\rho}A_\lambda = \eta\bar{\rho}(\pi_\mu^\lambda A_\lambda) = \mathrm{KI}(\pi_\mu^\lambda A_\lambda) = \mathrm{KI}(A_\lambda) = \eta\gamma_\lambda .$$

By (5.5) the relation $\pi_\mu^\lambda\gamma_\lambda \sim_\rho \gamma_\mu$ implies $\eta\pi_\mu^\lambda\gamma_\lambda = \eta\gamma_\mu$ which proves our assertion. Another application of (5.5) shows that η is independent of the choice of the ρ-cycles γ_λ^0 in their respective ρ-homology classes. Hence η is a function of the elements of $\mathfrak{H}_\rho^0$, with values which are residues mod p. Regarded as a mapping $\mathfrak{H}_\rho^0 \to \mathfrak{J}_p$, η is evidently a homomorphism.

(14.1) *If* $\mathfrak{H}^0(\mathfrak{R}, \mathfrak{J}_p) = 0$, η *is an isomorphism.*

PROOF. It is sufficient to show that $\eta\gamma^0 = 0$ implies $\gamma^0 \sim_p 0$. The first of these relations implies relations of the form

$$\gamma_\lambda^0 = \bar{\rho}A_\lambda \,; \qquad \mathrm{KI}(A_\lambda) = 0,$$

the second of which implies that A_λ is a cycle. The hypothesis $\mathfrak{H}^0 = 0$ implies that X_λ is connected and therefore $A_\lambda \sim 0$. Hence $\bar{\rho}A_\lambda \sim_p 0$ (5.6), so that $\gamma^0 \sim_p 0$.

Now let $\mathfrak{J}$ denote the additive group of integers. On reducing the integers modulo p, there is induced in an obvious way a homomorphic mapping ξ_h: $\mathfrak{H}^h(\mathfrak{R}, \mathfrak{J}) \to \mathfrak{H}^h(\mathfrak{R}, \mathfrak{J}_p)$.

We shall say that a homological n-sphere $\mathfrak{R}$ over $\mathfrak{J}_p$ is *integral* if $\mathfrak{H}^n(\mathfrak{R}, \mathfrak{J})$ is infinite cyclic and ξ_n is an isomorphism of $\mathfrak{H}^n(\mathfrak{R}, \mathfrak{J})$ onto $\mathfrak{H}^n(\mathfrak{R}, \mathfrak{J}_p)$. We shall agree to consider the empty set as an integral homological sphere over $\mathfrak{J}_p$ of dimension -1.

We shall assume in the remainder of this section that $\mathfrak{R}$ is an integral homological n-sphere over $\mathfrak{J}_p$, p a prime. Theorem (12.1) asserts that L is a homological r-sphere over $\mathfrak{J}_p$, $-1 \leqq r \leqq n$. It is not known, in the case $r \geqq 0$ whether or not L is necessarily integral. We shall say that $(\mathfrak{R}, T)$ *is integral* if $r = -1$ or else if $r \geqq 0$ and L is integral. It can be shown that $(\mathfrak{R}, T)$ is integral if, for example, $\mathfrak{R}$ is locally Euclidean and T locally analytic.

Let γ^n be a fundamental n-cycle of $\mathfrak{R}$, that is, an n-cycle over $\mathfrak{J}$ which is a generator of $\mathfrak{H}^n(\mathfrak{R}, \mathfrak{J})$. The only other fundamental cycle of $\mathfrak{R}$ is $-\gamma^n$; the choice of a definite γ^n amounts to the choice of a definite orientation $\epsilon_\mathfrak{R}$ of $\mathfrak{R}$. Suppose now that $(\mathfrak{R}, T)$ is integral and assume for the moment that $r \geqq 0$. Then L admits an orientation ϵ_L defined by a fundamental cycle γ^r. By (10.10) and (6.3) β maps $\mathfrak{H}^n_\rho(\mathfrak{R}, T; \mathfrak{J}_p)$ isomorphically onto $\mathfrak{H}^n(\mathfrak{R}, \mathfrak{J}_p)$. Hence, writing $\chi = \beta^{-1}$, we have from (12.6b)

$$\zeta\alpha_{r+1n}\chi\xi_n\mathfrak{H}^n(\mathfrak{R}, \mathfrak{J}) = \mathfrak{H}^r(L, \mathfrak{J}_p).$$

Hence

$$\zeta\alpha_{r+1n}\chi\xi_n(\gamma^n) \sim g\xi_r(\gamma^r)$$

where $g \in \mathfrak{J}_p$. The fact that γ^n is a generator of $\mathfrak{H}^n(\mathfrak{R}, \mathfrak{J})$ implies $g \neq 0$. Evidently g depends only on $(\mathfrak{R}, T)$, $\epsilon_\mathfrak{R}$, ϵ_L and is replaced by its negative when one of the orientations is reversed. We shall call g the *index* of $(\mathfrak{R}, T)$ with respect to the given orientations and write $g = \mathrm{ind}\,(\mathfrak{R}, T, \epsilon_\mathfrak{R}, \epsilon_L)$. If $r = -1$, we define an index by the formula

$$\mathrm{ind}\,(\mathfrak{R}, T, \epsilon_\mathfrak{R}) = \eta\alpha_{1n}\chi\xi_n(\gamma^n). \tag{14.2}$$

Here again the index is an element of $\mathfrak{J}_p$ and differs from 0 by virtue of (14.1), (12.6a).

15. **Mappings of** $(\mathfrak{R}, T)$. Let T, T' be transformations of period p operating

in spaces $\mathfrak{R}$, $\mathfrak{R}'$ and let s be a (single-valued continuous) mapping $\mathfrak{R} \to \mathfrak{R}'$ such that

$$sT = T's. \tag{15.1}$$

We may call s a mapping of $(\mathfrak{R}, T)$ into $(\mathfrak{R}', T')$ and write $s: (\mathfrak{R}, T) \to (\mathfrak{R}', T')$. Evidently (15.1) implies $sL \subset L'$.

Suppose that $\mathfrak{R}$, $\mathfrak{R}'$ are compact, T and T' primitive. Then a mapping $s:(\mathfrak{R}, T) \to (\mathfrak{R}', T')$ induces homomorphic mappings S of the ρ-homology groups of $(\mathfrak{R}, T)$ into those of $(\mathfrak{R}', T')$ in exactly the same way as it induces mappings of the ordinary homology groups (see VII, 5.11). Briefly let ρ' stand for $1 - T'$ or $1 + T'^2 + \cdots + T'^{p-1}$ according as ρ stands for δ or σ and let $\{\mathfrak{U}_\lambda\}$, $\{\mathfrak{U}'_{\lambda'}\}$ be the primitive special coverings of $\mathfrak{R}$, $\mathfrak{R}'$. As a result of (15.1), $s\mathfrak{U}_\lambda$ is a T'-system (not necessarily a covering). Denote nerve $s\mathfrak{U}_\lambda$ by $sX_\lambda$and the transformation which T' induces in sX_λ by sT_λ. Let $\rho_{s\lambda}$ denote the operator ρ defined with the aid of sT_λ. As a result of (15.1), the chain-mapping which s induces carries ρ-chains in X_λ into $\rho_{s\lambda}$-chains in sX_λ and hence s induces in an obvious way a homomorphic mapping

$$\bar{s}:\mathfrak{H}^h_\rho(X_\lambda, T_\lambda; G) \to \mathfrak{H}^h_\rho(sX_\lambda, sT_\lambda; G).$$

Suppose that there exists a $\mathfrak{U}'_{\lambda'}$ such that $s\mathfrak{U}_\lambda > \mathfrak{U}'_{\lambda'}$. Since $\mathfrak{U}'_{\lambda'}$ is primitive, there exists by (2.1) a T'-projection $\pi':s\mathfrak{U}_\lambda \to \mathfrak{U}'_{\lambda'}$. On the basis of (8.1), π' induces a homomorphic mapping

$$\bar{\pi}':\mathfrak{H}^h_\rho(sX_\lambda, sT_\lambda, G) \to \mathfrak{H}^h_\rho(X'_{\lambda'}, T'_{\lambda'}, G)$$

so that

$$\bar{\pi}'\bar{s}\mathfrak{H}^h(X_\lambda, T_\lambda; G) \subset \mathfrak{H}^h_\rho(X'_{\lambda'}, T_{\lambda'}; G).$$

Now by the continuity of s there can be associated to every λ' a $\lambda = \varphi(\lambda')$ such that $s\mathfrak{U}_\lambda > \mathfrak{U}'_{\lambda'}$. The set of coverings $\{\mathfrak{U}_{\varphi(\lambda')}\}$ is cofinal with $\{\mathfrak{U}_\lambda\}$ and hence a ρ-cycle γ is uniquely determined by coordinates $\gamma_{\varphi(\lambda')}$. The elements $\bar{\pi}'\bar{s}\gamma_{\varphi(\lambda')}$ can be shown to be the coordinates of a ρ-cycle in $\mathfrak{R}'$, call it $S\gamma$. The correspondence $\gamma \to S\gamma$ can be shown to be independent of φ and in fact it will be seen that S is precisely the induced homomorphism $\mathfrak{H}^h_\rho(\mathfrak{R}, T, G) \to \mathfrak{H}^h_\rho(\mathfrak{R}', T', G)$ which was to be defined.

From now on we need not disinguish, in our notation, between a mapping $s:(\mathfrak{R}, T) \to (\mathfrak{R}', T')$ and the homomorphisms it induces. Let α', β', $\cdots$ denote the homomorphisms α, β, $\cdots$ operating on the groups of $(\mathfrak{R}', T')$. Then

$$s\alpha = \alpha's, \qquad s\xi = \xi's, \qquad s\zeta = \zeta's, \tag{15.2}$$

$$s\beta^{-1} = \beta'^{-1}s \qquad \text{if } \beta,\ \beta' \text{ are isomorphisms.} \tag{15.3}$$

The verification of these relations is perfectly straightforward. With regard to α for example, (15.2) follows from the fact that α is defined by means of correspondences $\rho A + B \to \mathrm{F}A$ (A, B chains) and the fact that all chain-mappings which enter into the definition of the induced s are permutable with ρ and F.

With regard to ζ (defined in 10) we use the relation $sL \subset L'$. Let us observe also that when $L = L' = \emptyset$,

$$\eta' s = \eta. \tag{15.4}$$

This is simply an expression of the fact that Kronecker indices of 0-chains are invariant under chain-mappings.

16. Throughout this section let s be a mapping $(\mathfrak{R}, T) \to (\mathfrak{R}', T')$ where $\mathfrak{R}, \mathfrak{R}'$ *are integral homological n-spheres over* $\mathfrak{J}_p$. Let L, L' be the corresponding fixed-point sets; they are homological r- and r'-spheres over $\mathfrak{J}_p$ (12.1). Let $\epsilon_{\mathfrak{R}}, \epsilon_{\mathfrak{R}'}$ be definitely chosen orientations of $\mathfrak{R}, \mathfrak{R}'$ defined by fundamental cycles γ^n, γ'^n over $\mathfrak{J}$. The uniquely determined integer x such that $s\gamma^n \sim x\gamma'^n$ is the *degree* of s relative to the given orientations: $x = \deg(s, \epsilon_{\mathfrak{R}}, \epsilon_{\mathfrak{R}'})$. Let $\alpha', \beta', \cdots$ denote the homomorphisms $\alpha, \beta \cdots$ acting on the groups of $(\mathfrak{R}', T')$.

(16.1) Theorem. *If* $\deg(s, \epsilon_{\mathfrak{R}}, \epsilon_{\mathfrak{R}'}) \not\equiv 0 \bmod p$, *then* $r = r'$.

Proof. If $r' = -1$, then $r = -1$ since $sL \subset L'$. Assume therefore that $r' \geqq 0$. We have $s\gamma^n \sim x\gamma'^n$ where $x = \deg(s, \epsilon_{\mathfrak{R}}, \epsilon_{\mathfrak{R}'})$. Since $x \not\equiv 0 \bmod p$, $\xi'_n(s\gamma^n) \not\sim 0$ in $\mathfrak{R}'$. Hence (12.6b),

$$\zeta'\alpha'_{r'+1\,n}\chi'\xi'_n(s\gamma^n) \not\sim 0 \text{ in } L'. \tag{16.2}$$

But on applying (15.2, 15.3) the left side of (16.2) becomes $s[\zeta\alpha_{r'+1n}\chi\xi_n(\gamma^n)]$. The expression in brackets is an r'-cycle over $\mathfrak{J}_p$ in L. Hence if $r \neq r'$, that cycle is ~ 0 in L so that its image under s is ~ 0 in L' contradicting (16.2). Hence $r = r'$.

(16.3) Theorem. *If* $r' = -1$ (*that is, if* $L = L' = 0$) *then* $\operatorname{ind}(\mathfrak{R}, T, \epsilon_{\mathfrak{R}}) = \operatorname{ind}(\mathfrak{R}', T', \epsilon_{\mathfrak{R}'}) \cdot \deg(s, \epsilon_{\mathfrak{R}}, \epsilon_{\mathfrak{R}'})$.

Proof. We have $s\gamma^n \sim x\gamma'^n$, $x = \deg(s, \epsilon_{\mathfrak{R}}, \epsilon_{\mathfrak{R}'})$. Hence (14.2),

$$\eta'\alpha'_{1n}\chi'\xi'_n(s\gamma^n) = x\eta'\alpha'_{1n}\xi'_n(\gamma'^n) = \deg(s, \epsilon_{\mathfrak{R}}, \epsilon_{\mathfrak{R}'}) \cdot \operatorname{ind}(\mathfrak{R}', T', \epsilon_{\mathfrak{R}'}). \tag{16.4}$$

On applying (15.2, 15.3, 15.4) the first expression (16.4) becomes $\eta\alpha_{1n}\chi\xi_n(\gamma^n)$ which equals $\operatorname{ind}(\mathfrak{R}, T, \epsilon_{\mathfrak{R}})$.

Consider the special case in which $\mathfrak{R} = \mathfrak{R}'$, $T = T'$, $\epsilon_{\mathfrak{R}} = \epsilon_{\mathfrak{R}'}$. Then $\operatorname{ind}(\mathfrak{R}, T, \epsilon_{\mathfrak{R}}) = \operatorname{ind}(\mathfrak{R}', T', \epsilon_{\mathfrak{R}'}) \not\equiv 0$ and hence

$$\deg s \equiv 1 \bmod p, \tag{16.5}$$

a relation obtained by Eilenberg [b] for simplicial $(\mathfrak{R}, T)$ and simplicial s. If we specialize still further and take for $\mathfrak{R}$ a Euclidean n-sphere and for T the reflection of $\mathfrak{R}$ across its center, then $p = 2$ and we conclude from (16.5) that an "antipode-preserving" mapping of an n-sphere into itself is necessarily of odd degree, a well known theorem of Borsuk [b].

(16.6) THEOREM. *Suppose that* $r' \geqq 0$ *and that* $\deg(s, \epsilon_{\mathfrak{R}}, \epsilon_{\mathfrak{R}'}) \neq 0$ *so that* $r = r'$ (16.1). *Suppose further that* $(\mathfrak{R}, T)$, $(\mathfrak{R}', T')$ *are integral so that* L, L' *admit orientations* ϵ_L, $\epsilon_{L'}$. *Then*

$$(16.7) \quad \operatorname{ind}(\mathfrak{R}, T, \epsilon_{\mathfrak{R}}, \epsilon_L) \deg(s_L, \epsilon_L, \epsilon_{L'}) = \operatorname{ind}(\mathfrak{R}', T', \epsilon_{\mathfrak{R}'}, \epsilon_{L'}) \deg(s, \epsilon_{\mathfrak{R}}, \epsilon_{\mathfrak{R}'})$$

where s_L *is the mapping* $L \rightarrow L'$ *induced by* s.

PROOF. Let γ^r, γ'^r be fundamental cycles in L, L' corresponding to $\epsilon_{\mathfrak{R}}$, $\epsilon_{\mathfrak{R}'}$. The relation $s\gamma^n \sim x\gamma'^n$ ($x = \deg(s, \epsilon_{\mathfrak{R}}, \epsilon_{\mathfrak{R}'})$) implies by (14.2)

$$(16.8) \qquad \zeta'\alpha'_{r+1,n}\chi'\xi'_n(s\gamma^n) \sim xg'\gamma'^n$$

where $g' = \operatorname{ind}(\mathfrak{R}', T', \epsilon_{\mathfrak{R}'}, \epsilon_{L'})$. Applying (15.2, 3, 4), the left side of (16.8) becomes $s[\zeta\alpha_{r+1n}\chi\xi_n(\gamma^n)]$. Now $[\ \ ] \sim g\gamma^r$ where $g = \operatorname{ind}(\mathfrak{R}, T, \epsilon_{\mathfrak{R}}, \epsilon_T)$. Hence $s[\ \ ] \sim x_Lg\gamma'^r$ where $x_L = \deg(s_L, \epsilon_L, \epsilon_{L'})$. Comparing this with (16.8) we conclude, since $x \neq 0$, $g' \neq 0$, that $x_Lg = xg'$ which is precisely (16.7).

Consider the special case in which $\mathfrak{R} = \mathfrak{R}'$, $T = T'$ so that $L = L'$. Take $\epsilon_{\mathfrak{R}} = \epsilon_{\mathfrak{R}'}$ and $\epsilon_L = \epsilon_{L'}$. Then (16.7) yields the relation

$$(16.8) \qquad \deg s = \deg s_L \bmod p.$$

Suppose $\mathfrak{R}$ is an n-sphere and T the reflection of $\mathfrak{R}$ across a great r-sphere, $r < n$. Then $p = 2$ and we conclude that if s is a mapping of $\mathfrak{R}$ into itself which preserves T-corresponding point-pairs, then $\deg s$ and $\deg s_L$ are either both even or both odd.

17. **Some further results.** Suppose $\mathfrak{R}$ is a homological n-sphere over $\mathfrak{J}_p$, p a prime, so that by (12.1) L is a homological r-sphere, $r \leqq n$. Simple examples show that r may equal n. However if it is assumed that $\mathfrak{R}$ satisfies certain *local* homological conditions, we have always $r < n$. Moreover in this case r can equal $n - 1$ only when $p = 2$, and if $r = 0$, L consists of exactly two points. If $\mathfrak{R}$ is a 3-sphere, L is empty or consists of two points or is a simple closed curve or is homeomorphic to a 2-sphere. (See P. A. Smith, [b, c].) The corresponding theorem for 2-spheres was first established by Brouwer [f] and independently by Kerékjártó [a].)

As a consequence of (13.1) a transformation of period q^a (q a prime) operating in Euclidean n-space, admits fixed points. However if $n = 3$, *every* periodic T admits fixed points; if $n = 4$, every locally analytic periodic T admits fixed points (P. A. Smith [f]).

18. **Problems**. The foregoing account seems to indicate that as concerns the properties of fixed-point sets, the method of ρ-homologies applies to best advantage when one has to do with a transformation of period p^a where p is a prime, and when one is concerned with homological properties over $\mathfrak{J}_p$. Whether the results obtained hold for other periods and coefficient groups is an open question. It would be interesting for example to know whether (12.1) holds if the period of T is a prime and the coefficient-group is the group of rational numbers; a partial answer is found in Smith [b].

(18.1) Special problems. Does every periodic transformation in Euclidean 4-space, every periodic locally analytic transformation in Euclidean n-space admit fixed points (cf. 17)? Assuming $\mathfrak{R}$ to be an n-sphere, p to be prime so that L is a homological r-sphere over $\mathfrak{J}_p$ (12.1), is it true that $n - r$ must be even or odd according as T preserves or reverses orientation, as would be the case if T were analytic?

BIBLIOGRAPHY

The following bibliography is rather complete as regards the books published from 1930 onwards. The list of papers has been reduced to the strict minimum required in the present work.

LIST OF BOOKS

ALEXANDROFF, P., and HOPF, H.

[A–H] *Topologie*, Berlin, Springer, 1935, (Die Grundlehren der mathematischen Wissenschaften, bd. 45).

BOURBAKI, N.

[B] *Éléments de Mathématiques.* I. *Théorie des Ensembles*: III. *Topologie Générale*, Paris, Hermann, 1939–1940, (Actualités Scientifiques, nos. 846, 858).

HODGE, W. V. D.

[H] *The Theory and Application of Harmonic Integrals*, Cambridge University Press, 1941.

HUREWICZ, W., and WALLMAN, H.

[H–W] *Dimension Theory*, Princeton University Press, 1941, (Princeton Mathematical Series, no. 4).

KURATOWSKI, K.

[K] *Topologie*, Warsaw, 1933.

LEFSCHETZ, S.

[L] *Topology*, New York, 1930, (American Mathematical Society Colloquium Publications, vol. 12).

[L_1] *Topology*, Notes by N. Steenrod and H. Wallman, Princeton, 1935.

[L_2] *Topics in Topology*, Princeton University Press, 1942, (Annals of Mathematics Studies, no. 10).

MENGER, KARL

[Me] *Kurventheorie*, Berlin, Teubner, 1932.

MOORE, R. L.

[Mo] *Foundations of Point Set Theory*, New York, 1932, (American Mathematical Society Colloquium Publications, vol. 13).

MORSE, MARSTON

[Mor] *The Calculus of Variations in the Large*, New York, 1934, (American Mathematical Society Colloquium Publications, vol. 18).

NEWMAN, M. H. A.

[N] *Topology of Plane Sets of Points*, Cambridge University Press, 1939.

PONTRJAGIN, L.

[P] *Topological Groups*, Princeton University Press, 1939, (Princeton Mathematical Series, no. 2).

REIDEMEISTER, K.

[R] *Einführung in die combinatorische Topologie*, Braunschweig, Vieweg, 1932.

[R_1] *Topologie der Polyeder*, Leipzig, Akademische Verlagsgesellschaft, 1938.

SIERPINSKI, W.

[S] *Introduction to General Topology*, University of Toronto Press, 1934.

SEIFERT, H., and THRELFALL, W.

[S–T] *Lehrbuch der Topologie*, Leipzig, Teubner, 1934.

TUKEY, J. W.

[T] *Convergence and Uniformity in Topology*, Princeton University Press, 1940, (Annals of Mathematics Studies, no. 2).

VEBLEN, O.

[V] *Analysis Situs*, 2d edition, New York, 1931, (American Mathematical Society Colloquium Publications, vol. 5, part 2).

WEIL, A.

[W] *L'Intégration dans les Groupes Topologiques et ses Applications*, Paris, Hermann, 1941, (Actualités Scientifiques, no. 869).

WHYBURN, G. T.

[Wh] *Analytic Topology*, New York, 1942, (American Mathematical Society Colloquium Publications, vol. 28).

WILDER, R. L., and AYRES, W. L., editors.

[UM] *Lectures in Topology*, The University of Michigan Conference of 1940, University of Michigan Press, 1941.

LIST OF PAPERS

ALEXANDER, J. W.

[a] *A proof of the invariance of certain constants of analysis situs*, Transactions of the American Mathematical Society, vol. 16 (1915), pp. 148–154.

[b] *A proof and extension of the Jordan-Brouwer separation theorem*, Transactions of the American Mathematical Society, vol. 23 (1922), pp. 333–349.

[b_1] *Combinatorial analysis situs.* I, Transactions of the American Mathematical Society, vol. 28 (1926), pp. 301–329.

[c] *On the chains of a complex and their duals*, Proceedings of the National Academy of Sciences, vol. 21 (1935), pp. 509–511.

[d] *On the connectivity ring of an abstract space*, Annals of Mathematics, (2), vol. 37 (1936), pp. 698–708.

[e] *A theory of connectivity in terms of gratings*, Annals of Mathematics, (2), vol. 39 (1938), pp. 883–912.

ALEXANDER, J. W., and ZIPPIN, LEO

[a] *Discrete Abelian groups and their character groups*, Annals of Mathematics, (2), vol. 36 (1935), pp. 71–85.

ALEXANDROFF, P. S.

[a] *Untersuchungen über Gestalt und Lage abgeschlossener Mengen beliebiger Dimension*, Annals of Mathematics, (2), vol. 30 (1928), pp. 101–187.

[b] *Dimensionstheorie. Ein Beitrag zur Geometrie der abgeschlossenen Mengen*, Mathematische Annalen, vol. 106 (1932), pp. 161–238.

[c] *On local properties of closed sets*, Annals of Mathematics, (2), vol. 36 (1935), pp. 1–35.

[d] *Diskrete Räume*, Matematicheskii Sbornik (Recueil Mathématique), (n. s.), vol. 2 (1937), pp. 501–518.

[e] *Zur Homologie-Theorie der Kompakten*, Compositio Mathematica, vol. 4 (1937), pp. 256–270.

[f] *General combinatorial topology*, Transactions of the American Mathematical Society, vol. 49 (1941), pp. 41–105.

ALEXANDROFF, P. S., and PONTRJAGIN, L.

[a] *Les variétés à n dimensions généralisées*, Comptes Rendus de l'Académie des Sciences, Paris, vol. 202 (1936), pp. 1327–1329.

ALEXANDROFF, P. S., and URYSOHN, P.

[a] *Mémoire sur les espaces topologiques compactes*, Verhandelingen, section I, Akademie van Wetenschappen, Amsterdam, vol. 14 (1929), pp. 1–96.

BEGLE, E.

[a] *Locally connected spaces and generalized manifolds*, American Journal of Mathematics, vol. 64 (1942).

BORSUK, KAROL

[a] *Sur les rétractes*, Fundamenta Mathematicae, vol. 17 (1931), pp. 152–170.

BROUWER, L. E. J.

[a] *On continuous vector distributions on surfaces*, Proceedings, Akademie van Wetenschappen, Amsterdam, vol. 11 (1909), pp. 850–858; vol. 12 (1910), pp. 716–734; vol. 13 (1910), pp. 171–186.

[b] *On continuous one-to-one transformations of surfaces into themselves*, Proceedings, Akademie van Wetenschappen, Amsterdam, vol. 11 (1909), pp. 788–798; vol. 12 (1910), pp. 286–297; vol. 13 (1911), pp. 767–777; vol. 14 (1911), pp. 300–310; vol. 15 (1913), pp. 352–360; vol. 22 (1920), pp. 811–814; vol. 23 (1921), pp. 232–234.

[c] *Beweis der Invarianz der Dimensionenzahl*, Mathematische Annalen, vol. 70 (1911), pp. 161–165.

[d] *Über Abbildungen von Mannigfaltigkeiten*, Mathematische Annalen, vol. 71 (1912), pp. 97–115.

[e] *Beweis der Invarianz des n-dimensionalen Gebiets*, Mathematische Annalen, vol. 71 (1912), pp. 305–313.

[f] *Über die periodischen Transformationen der Kugel*, Mathematische Annalen, vol. 80 (1919), pp. 39–41.

BROWN, A. B., and KOOPMAN, B. C.

[a] *On the covering of analytic loci by complexes*, Transactions of the American Mathematical Society, vol. 34 (1932), pp. 231–251.

CAIRNS, S. S.

[a] *On the triangulation of regular loci*, Annals of Mathematics, (2), vol. 35 (1934), pp. 579–587.

[b] *Triangulation of the manifold of class one*, Bulletin of the American Mathematical Society, vol. 41 (1935), pp. 549–552.

ČECH, EDUARD

[a] *Théorie générale de l'homologie dans un espace quelconque*, Fundamenta Mathematicae, vol. 19 (1932), pp. 149–183.

[b] *Théorie générale des variétés et de leurs théorèmes de dualité*, Annals of Mathematics, (2), vol. 34 (1933), pp. 621–730.

[c] *Sur les nombres de Betti locaux*, Annals of Mathematics, (2), vol. 35 (1934), pp. 678–701.

[d] *Les groupes de Betti d'un complexe infini*, Fundamenta Mathematicae, vol. 25 (1935), pp. 33–44.

[e] *Les théorèmes de dualité en topologie*, Časopis pro Pěstování Matematiky a Fysiky, vol. 64 (1935), pp. 17–25.

[f] *Multiplications on a complex*, Annals of Mathematics, (2), vol. 37 (1936), pp. 681–697.

[g] *On bicompact spaces*, Annals of Mathematics, (2), vol. 38 (1937), pp. 823–844.

CHEVALLEY, CLAUDE

[a] *Sur la définition des groupes de Betti des ensembles fermés*, Comptes Rendus de l'Académie des Sciences, Paris, vol. 200 (1935), pp. 1005–1007.

DOWKER, C. H.

[a] *Hopf's theorem for non-compact spaces*, Proceedings of the National Academy of Sciences, vol. 23 (1937), pp. 293–294.

EILENBERG, SAMUEL

[a] *Cohomology and continuous mappings*, Annals of Mathematics, (2), vol. 41 (1940), pp. 231–251.

[b] *On a theorem of P. A. Smith concerning fixed points for periodic transformations*, Duke Mathematical Journal, vol. 6 (1940), pp. 428–437.

EILENBERG, SAMUEL, and MACLANE SAUNDERS

[a] *Group extensions and homology*, Annals of Mathematics, (2), vol. 44 (1943).

FLEXNER, W. W.

[a] *On topological manifolds*, Annals of Mathematics, (2), vol. 32 (1931), pp. 393–406; *The Poincaré duality theorem for topological manifolds*, ibid., pp. 539–548.

[b] *Character group of a relative homology group*, Annals of Mathematics, (2), vol. 41 (1940), pp. 207–214.

[c] *Simplicial intersection chains for an abstract complex*, Bulletin of the American Mathematical Society, vol. 46 (1940), pp. 523–524.

FREUDENTHAL, HANS

[a] *Alexanderscher und Gordonscher Ring und ihre Isomorphie*, Annals of Mathematics, (2), vol. 38 (1937), pp. 647–655.

[b] *Entwicklungen von Räumen und ihren Gruppen*, Compositio Mathematica, vol. 4 (1937), pp. 145–234.

[c] *Bettische Gruppe mod.* 1 *und Hopfsche Gruppe*, Compositio Mathematica, vol. 4 (1937), pp. 235–238.

[d] *Die Triangulation der differenzierbaren Mannigfaltigkeiten*, Proceedings, Akademie van Wetenschappen, Amsterdam, vol. 42 (1939), pp. 880–901.

GELFAND, L., and RAIKOV, D.

[a] *On the theory of characters of commutative topological groups*, Comptes Rendus de l'Académie des Sciences de l'URSS (Doklady), (n.s.), vol. 28 (1940), pp. 195–198.

GORDON, I.

[a] *On intersection invariants of a complex and its complementary spaces*, Annals of Mathematics, (2), vol. 37 (1936), pp. 519–525.

HOPF, H.

[a] *Zur Algebra der Abbildungen von Mannigfaltigkeiten*, Journal für die reine und angewandte Mathematik, vol. 163 (1930), pp. 71–88.

[b] *Beiträge zur Klassifizierung der Flächenabbildungen*, Journal für die reine und angewandte Mathematik, vol. 165 (1931), pp. 225–236.

[c] *Über die Topologie der Gruppenmannigfaltigkeiten und ihre Verallgemeinerungen*, Annals of Mathematics (2), vol. 42 (1941), pp. 22–52.

HUREWICZ, W.

[a] *Beiträge zur Topologie der Deformationen.* I. *Höherdimensionale Homotopiegruppen*, Proceedings, Akademie van Wetenschappen, Amsterdam, vol. 38 (1935), pp. 112–119; II. *Homotopie- und Homologiegruppen*, vol. 38 (1935), pp. 521–528; III. *Klassen und Homologietypen von Abbildungen*, vol. 39 (1936), pp. 117–126; IV. *Asphärische Räume*, vol. 39 (1936), pp. 215–224.

JOHNSON, L. W.

[a] *A linear algebraic theory of complexes*, Princeton thesis, 1941.

KERÉKJÁRTÓ, B.

[a] *Über die periodischen Transformationen der Kreisscheibe und Kugelfläche*, Mathematische Annalen, vol. 80 (1919), pp. 36–38.

KOLMOGOROFF, ANDRÉ

[a] *Les groupes de Betti des espaces localement bicompactes*, Comptes Rendus de l'Académie des Sciences, Paris, vol. 202 (1936), pp. 1144–1147; *Propriétés des groupes de Betti des espaces localement bicompactes*, ibid., pp. 1325–1327; *Les groupes de Betti des espaces métriques*, ibid., pp. 1558–1560; *Cycles relatifs. Théorème de dualité de M. Alexander*, ibid., pp. 1641–1643.

[b] *Über die Dualität im Aufbau der kombinatorischen Topologie*, Matematicheskii Sbornik (Recueil Mathématique), (n.s.), vol. 1 (1936), pp. 97–102.

[c] *Homologiering des Komplexes und des lokal-bicompakten Räumes*, Matematicheskii Sbornik (Recueil Mathématique), (n.s.), vol. 1 (1936), pp. 701–705.

KOOPMAN, B. C., and BROWN, A. B.

See Brown and Koopman [a].

KÜNNETH, D.

[a] *Über die Bettischen Zahlen einer Produktmannigfaltigkeit*, Mathematische Annalen, vol. 90 (1923), pp. 65–85; *Über die Torsionszahlen von Produktmannigfaltigkeiten*, Mathematische Annalen, vol. 91 (1924), pp. 125–134.

KUROSCH, ALEXANDER
[a] *Kombinatorischer Aufbau der bikompakten topologischen Räume*, Compositio Mathematica, vol. 2 (1935), pp. 471–476.
LEFSCHETZ, S.
[a] *Intersections and transformations of complexes and manifolds*, Transactions of the American Mathematical Society, vol. 28 (1926), pp. 1–49.
[b] *On singular chains and cycles*, Bulletin of the American Mathematical Society, vol. 39 (1933), pp. 124–129.
[c] *On generalized manifolds*, American Journal of Mathematics, vol. 55 (1933), pp. 469–504.
[d] *On locally connected and related sets*. I. Annals of Mathematics, (2), vol. 35 (1934), pp. 118–129.
[e] *Chain-deformations in topology*, Duke Mathematical Journal, vol. 1 (1935), pp. 1–18.
[f] *On chains of topological spaces*, Annals of Mathematics, (2), vol. 39 (1938), pp. 383–396.
[g] *On the fixed point formula*, Annals of Mathematics, (2), vol. 38 (1937), pp. 819–822.
[h] *The role of algebra in topology*, Bulletin of the American Mathematical Society, vol. 43 (1937), pp. 345–359.
LEFSCHETZ, S., and WHITEHEAD, J. H. C.
[a] *On analytical complexes*, Transactions of the American Mathematical Society, vol. 35 (1933), pp. 510–517.
LEVIN, MADELIN
[a] *An extension of the Lefschetz intersection theory*, Revista de Ciencias, vol. 39 (1937), pp. 93–118.
MACLANE, SAUNDERS, and EILENBERG, SAMUEL
See Eilenberg and MacLane [a].
MAYER, W.
[a] *Über abstrakte Topologie*, Monatshefte für Mathematik und Physik, vol. 36 (1929), pp. 1–42; 219–258.
[b] *Topologische Gruppensysteme*, Monatshefte für Mathematik und Physik, vol. 47 (1938), pp. 40–86.
[c] *Charktersysteme und Dualitätstheoreme*, Journal of Mathematics and Physics, Massachusetts Institute of Technology, vol. 18 (1939), pp. 1–27.
NEWMAN, M. H. A.
[a] *On the foundations of combinatory analysis situs*, Proceedings, Akademie van Wetenschappen, Amsterdam, vol. 29 (1926), pp. 611–626; vol. 29 (1926), pp. 627–641; vol. 30 (1927), pp. 670–673.
[b] *Intersection complexes. I. Combinatory theory*, Proceedings of the Cambridge Philosophical Society, vol. 27 (1931), pp. 491–501.
[c] *A theorem on periodic transformations of spaces*, Quarterly Journal of Mathematics, Oxford Series, vol. 2 (1931), pp. 1–8.
POINCARÉ, H.
[a] *Analysis situs*, Journal de l'École Polytechnique, Paris, (2), vol. 1 (1895), pp. 1–123.
[b] *Complément à l'analysis situs*, Rendiconti, Circolo Matematico, Palermo, vol. 13 (1899), pp. 285–343.
PONTRJAGIN, L.
[a] *Über den algebraischen Inhalt topologischer Dualitätssätze*, Mathematische Annalen, vol. 105 (1931), pp. 165-205.
[b] *The theory of topological commutative groups*, Annals of Mathematics, (2), vol. 35 (1934), pp. 361-388.
[c] *The general topological theorem of duality for closed sets*, Annals of Mathematics, (2), vol. 35 (1934), pp. 904-914.
[d] *Products in complexes*, Matematicheskii Sbornik (Recueil Mathématique), (n.s.), vol. 9 (1941), pp. 321-330.

PONTRJAGIN, L., and ALEXANDROFF, P. S.
See Alexandroff and Pontrjagin [a].
RAIKOV, D., and GELFAND, L.
See Gelfand and Raikov [a].
DE RHAM, GEORGES
[a] *Sur l'analysis situs des variétés à n dimensions*, Journal de Mathématiques Pures et Appliquées, (9), vol. 10 (1931), pp. 115–200.
[b] *Relations entre la topologie et la théorie des intégrales multiples*, Enseignement Mathématique, vol. 35 (1936), pp. 213–228.
[c] *Sur la théorie des intersections et les intégrales multiples*, Commentarii Mathematici Helvetici, vol. 4 (1932), pp. 151–157.
RICHARDSON, M., and SMITH, P. A.
[a] *Periodic transformations of complexes*, Annals of Mathematics, (2), vol. 39 (1938), pp. 611–633.
SAMELSON, HANS
[a] *Beiträge zur Topologie der Gruppen-Mannigfaltigkeiten*, Annals of Mathematics, (2), vol. 42 (1941), pp. 1091–1137.
SMITH, P. A.
[a] *A theorem on fixed points for periodic transformations*, Annals of Mathematics, (2), vol. 35 (1934), pp. 572–578.
[b] *Transformations of finite period*, Annals of Mathematics, (2), vol. 39 (1938), pp. 127–164.
[c] *The topology of transformation groups*, Bulletin of the American Mathematical Society, vol. 44 (1938), pp. 497–514.
[d] *Transformations of finite period* II, Annals of Mathematics, (2), vol. 40 (1939), pp. 690–711.
[e] *Transformations of finite period* III, *Newman's theorem*, Annals of Mathematics, (2), vol. 42 (1941), pp. 446–458.
[f] *Fixed-point theorems for periodic transformations*, American Journal of Mathematics, vol. 63 (1941), pp. 1–8.
[g] *Periodic and nearly periodic transformations*, Lectures on Topology, University of Michigan Press, 1941, pp. 159–190.
SMITH, P. A., and RICHARDSON, M.
See Richardson and Smith [a].
STEENROD, N. E.
[a] *Universal homology groups*, American Journal of Mathematics, vol. 58 (1936), pp. 661–701.
[b] *Regular cycles of compact metric spaces*, Annals of Mathematics, (2), vol. 41 (1940), pp. 833–851.
TUCKER, A. W.
[a] *An abstract approach to manifolds*, Annals of Mathematics, (2), vol. 34 (1933), pp. 191–243.
[b] *Cell spaces*, Annals of Mathematics, (2), vol. 37 (1936), pp. 92–100.
[c] *On chain-mappings carried by cell-mappings*, Proceedings of the National Academy of Sciences, vol. 25 (1939), pp. 371–374.
[d] *The algebraic structure of complexes*, Proceedings of the National Academy of Sciences, vol. 25 (1939), pp. 643–647.
[e] *A relative theory of tensors and their integrals*, Annals of Mathematics, (2), vol. 44 (1943).
TYCHONOFF, A.
[a] *Über einen Funktionenraum*, Mathematische Annalen, vol. 111 (1935), pp. 762–766.
URYSOHN, P., and ALEXANDROFF, P. S.
See Alexandroff and Urysohn [a].
VAN KAMPEN, E. R.
[a] *Locally bicompact Albelian groups and their character groups*, Annals of Mathematics, (2), vol. 36 (1935), pp. 448–463.

VAUGHN, H. E., JR.

[a] *On local Betti numbers*, Duke Mathematical Journal, vol. 2 (1936), pp. 117–137.

VIETORIS, L.

[a] *Über den höheren Zusammenhang kompakter Räume und eine Klasse von zusammenhangstreuen Abbidungen*, Mathematische Annalen, vol. 97 (1927), pp. 454–472.

[b] *Über die Homologiegruppen der Vereinigung zweier Komplexe*, Monatshefte für Mathematik und Physik, vol. 37 (1930), pp. 159–162.

WALLACE, A. D.

[a] *Separation spaces*, Annals of Mathematics, (2), vol. 42 (1941), pp. 687–697.

WALLMAN, H.

[a] *Lattices and topological spaces*, Annals of Mathematics, (2), vol. 39 (1938), pp. 112–126.

WHITEHEAD, J. H. C.

[a] *On duality and intersection chains in combinatorial analysis situs*, Annals of Mathematics, (2), vol. 33 (1932), pp. 521–524.

[b] *On C^1-complexes*, Annals of Mathematics, (2), vol. 41 (1940), pp. 809–824.

WHITEHEAD, J. H. C., and LEFSCHETZ, S.

See Lefschetz and Whitehead [a].

WHITNEY, HASSLER

[a] *Differentiable manifolds*, Annals of Mathematics, (2), vol. 37 (1936), pp. 645–680.

[b] *The imbedding of manifolds in families of analytic manifolds*, Annals of Mathematics, (2), vol. 37 (1936), pp. 865–878.

[c] *On matrices of integers and combinatorial topology*, Duke Mathematical Journal, vol. 3 (1937), pp. 35–45.

[d] *On products in a complex*, Annals of Mathematics, (2), vol. 39 (1938), pp. 397–432.

[e] *Tensor products of Abelian groups*, Duke Mathematical Journal, vol. 4 (1938), pp. 495–528.

WILDER, R. L.

[a] *Generalized closed manifolds in n-space*, Annals of Mathematics, (2), vol. 35 (1934), pp. 876–903.

[b] *The sphere in topology*, American Mathematical Society Semicentennial Publications, vol. 2 (1938), pp. 136–184.

ZIPPIN, LEO, and ALEXANDER, J. W.

See Alexander and Zippin [a].

ADDED IN PROOF

KLINE, M.

[a] *Note on homology theory for locally bicompact spaces*, Fundamenta Mathematicae, vol. 32 (1939), pp. 64–68.

MAYER, W.

[d] *A new homology theory*, Annals of Mathematics, (2), vol. 43 (1942), pp. 370–380.

[e] *A new homology theory*. II, Annals of Mathematics, (2), vol. 43 (1942).

INDEX OF SPECIAL SYMBOLS AND NOTATIONS

INDEX

(This index does not cover the Appendices)